US Military Recipes
Volume 1 Armed Forces Recipe Service
Great for Cooking for Large Groups

Army TM 10-412
Navy NAVSUP Pub 7
USAF AFJMAN 34-606
USMC MCO P10110 42B

Contents
Basic Information
Conversion Charts
Equipment Guidelines
Ingredients
Food Safety

Sections
B- Appetizers
C- Beverages
D- Breads and Sweet Doughs
E - Cereals and Pasta Products
G - Cakes and Frostings
H- Cookies
I- Pastry and Pies
J - Puddings and Other Desserts
K - Dessert Sauces and Toppings
L- Meat, Fish, and Poultry

edited by
Brian Greul

All branches of the US Military use this standardized set of recipes. This is the 2003 edition. The full collection is 1691 pages. This book is the first 1094 pages of the full 1600 page collection. For reasons related to the maximum size of a book, the collection has been split into two books. The front 77 pages are repeated in each volume because the contain the instructions that are common to all recipes. This allows the books to be used independently. The editor recommends that you use flags to mark your favorite recipes. The recipes are fully scalable up or down and the instructions make the calculations as easy as one of the many pie recipes.

An 8.5x11 3 hole punched loose leaf copy may be purchased for your 3 ring binder. Email books@ocotillopress.com for current information.

Edited 2021 Ocotillo Press
ISBN 978-1-954285-28-6

No rights reserved. This content of this book is in the public domain as it is a work of the US Government. It is reproduced by the publisher as a convenience to enthusiasts and others who may wish to own a quality copy of it. It has been adjusted to accomodate the printing and binding process.

Printed in the United States of America for domestic orders and overseas for foreign orders.

Ocotillo Press
Houston, TX 77017
Books@OcotilloPress.com

Disclaimer: The user of this book is responsible for following safe and lawful practices at all times. The publisher assumes no responsibility for the use of the content of this book. The publisher has made an effort to ensure that the text is complete and properly typeset, however omissions, errors, and other issues may exist that the publisher is unaware of.

Armed Forces Recipe Service

UNITED STATES ARMY
TM 10-412

UNITED STATES NAVY
NAVSUP Publication 7

UNITED STATES AIR FORCE
AFJMAN 34-606 Volume I and Volume II

UNITED STATES MARINE CORPS
MCO P10110.42B

Stock No. 0530-LP-188-7302

DEPARTMENTS OF THE ARMY, THE NAVY AND THE AIR FORCE
Washington, DC

ARMED FORCES RECIPE SERVICE

The Armed Forces Recipe Service has been revised and updated and is issued for the purpose of standardizing and improving food prepared and served in military food service operations.

This recipe service is available for use by all Military Services and was coordinated and developed by technical representatives from each of the following:

Army Center of Excellence, Subsistence, Fort Lee, VA 23801-6020
Navy Supply Systems Command, Mechanicsburg, PA 17055-0791
Headquarters, Air Force Services Agency, San Antonio, TX 78216-4138
Headquarters, U. S. Marine Corps, Washington, DC 20380-1775
U. S. Army Soldier and Biological Chemical Command, Natick Soldier Center, Natick, MA 01760-5018

DEPARTMENTS OF THE ARMY, THE NAVY,
AND THE AIR FORCE
Washington, DC
June 2003

USA, TM-10-412
NAVSUP Publication 7
AFJMAN 34-606
MCO 10110.42B

ARMED FORCES RECIPE SERVICE

This 2003 revision replaces the 1993 (original), 1997 (Change 1) and 1999 (Change 2) Armed Forces Recipe Service cards. This printing includes all new recipe development and revisions to the 1993, 1997 and 1999 sets of Armed Forces Recipe Service cards. This update contains a new recipe format which reflects the nutritional analysis per serving, located at the top of each recipe card.

New recipes have been incorporated. Some of the recipes are designated with a number 800 and above which represent recipes that include speed scratch items, convenience prepared foods and additional new recipes. Some recipes have been deleted while other recipes that were printed on cards as variations are now individual recipes. Sources of recipes include the U.S. Army Solider and Biological Command, commercial quantity food cookbooks and food product manufacturers.

Replace current recipe cards sets with this 2003 Update. Replace the Index of Recipes, NAVSUP Publication P-7, dated July 1999 with Index of Recipes, dated 2003. An index of recipes is issued to assist food service personnel to easily locate recipes by category to ensure a varied menu. This card should be retained and inserted in the front of the publication.

BY ORDER OF THE SECRETARIES OF THE ARMY, THE NAVY, AND THE AIR FORCE:

JOHN P. JUMPER
General, USAF
Chief of Staff

J. D. MCCARTHY
Rear Admiral, SC, United States Navy
Commander, Naval Supply Systems Command

ERIC K. SHINSEKI
General, United States Army
Chief of Staff

R. L. KELLY
Lieutenant General, U. S. Marine Corps
Deputy Chief of Staff for Installations & Logistics

Distribution:
    Navy: NAVSUP 51 Distribution

A. GENERAL INFORMATION No. 0

INDEX

Card No.
Basic Information
A-19   Handling Frozen Foods, Guidelines for
A-4   Measuring Equivalents, Table of
A-2   Terms Used in Food Preparation, Definitions of
A-35   Use of Convenience Prepared Foods, Guidelines for

Conversion Charts
A-5   Can Sizes, Table of Weights and Measures for
A-9   Container Yields, Canned Fruits, Guidelines for
A-13   Fruit Bars, Guidelines for
A-16   Measure Conversion, Guidelines for
A-27   Metric Conversion, Guidelines for
A-1   Information for Standardized Recipes
A-15   Weight Conversion

Equipment, Guidelines for
A-33   Combi-Ovens
A-23   Convection Ovens
A-34   Skittles
A-21   Steam Cookers
A-25   Steam Table, Baking and Roasting Pans, Capacities for
A-24   Tilting Fry Pans

Ingredients
A-20   Antibrowning Agent, Use of

Card No.
A-28   Dehydrated Cheese, Use of
A-11   Dehydrated Green Peppers, Onions, and Parsley Use of
A-8   Egg Equivalents, Table of
A-30   Herbs, Guide to Cooking with Popular
A-10   Milk, Nonfat, Dry, Reconstitution Chart

Safety
A-32   "HACCP" (Hazard Analysis Critical Control Point) Guidelines

GUIDELINE CARDS

SECTION C - BEVERAGES
C-G-1 Brewing Coffee, Guidelines for
C-G-4 Coffee Urn Capacities, Guidelines for

SECTION D - BREADS AND SWEET DOUGHS
D-G-1 Recipe Conversion
D-G-2 Preparation of Yeast Doughs, Guidelines for
D-G-3 Retarded Sweet Dough Methods
D-G-4 Good Quality Bread Products and Rolls, Characteristics of
D-G-5 Poor Quality Bread Products and Rolls, Characteristics of

A. GENERAL INFORMATION No. 0

INDEX

Card No.
GUIDELINE CARDS - CONTINUED

SECTION G - CAKES
G-G-2 Batter Cakes, Characteristics of Good Quality/Bad Quality
G-G-5 Cutting Cakes, Guidelines for
G-G-7 High Altitude Baking
G-G-4 Scaling Cake Batter, Guidelines for
G-G-1 Successful Cake Baking, Guidelines for

SECTION G - FROSTINGS
G-G-6 Prepared Frostings and Frosting Cakes, Guidelines for

SECTION H - COOKIES
H-G-1 General Information Regarding Cookies

SECTION I - PASTRY AND PIES
I-G-1 Making One Crust Pies
I-G-2 Making Two Crust Pies

SECTION L - POULTRY
L-G-5 Timetable for Roasting Turkeys

SECTION M - SALADS
M-G-1 Trays or Salad Bars, Guidelines for

SECTION O - DRESSINGS, GRAVIES AND SAUCES
O-G-1 Sauces and Gravies, Guidelines for Preparing

Card No.

SECTION Q - VEGETABLES
Q-G-4 Potato Bar, Guidelines for
Q-G-5 Dehydrated Vegetables, Guidelines for
Q-G-6 Steam Cooking Vegetables, Guidelines for

A. GENERAL INFORMATION No. 1 (1)

INFORMATION FOR STANDARDIZED RECIPES

Standardized recipes are a necessity for a well-run food service operation. All of the recipes have been developed, tested and standardized for product quality, consistency and yield. Recipes are the most effective management tool for guiding the requisitioning of supplies and controlling breakouts and inventory. The U. S. Dietary Guidelines were among the many considerations in both the selection and development of the recipes included in the file. Many of the recipes have been modified to reduce fat, salt and calories. For new and experienced cooks, consistent use of standardized recipes is essential for quality and economy. The **Armed Forces Recipe Service** contains over 1600 tested recipes yielding 100 portions printed on cards.

Yield - The quantity of cooked product a recipe produces. The yield for each recipe in the Armed Forces Recipe is generally given as 100 portions and in some recipes in count or volume, e.g., 2 pans, 8 loaves, 6-1/2 gallons. Portion size is key to determining the quantity of food to be prepared. Many recipes also specify the weight per portion. For example, 3/4 cup (6-1/2 ounces) Beef Stroganoff.

Ingredients Column – Ingredients are listed in the order used. The specific form or variety of each ingredient is indicated. For example:

Flour, wheat, general purpose	Eggs, whole	Sugar, granulated
Flour, wheat, bread	Egg whites	Sugar, brown

A. GENERAL INFORMATION No. 1 (1)

Measure, Weights, and Issue Columns – Measures and Weights indicate the Edible Portion (E.P.) quantity of the ingredient required to prepare the recipe for 100 portions. The issue column represents the As Purchased (A.P.) quantity required if this amount is different from the E.P. quantity.

Method Column - Describes how the ingredients are to be combined and cooked. For example, the method will describe the order in which to sift dry ingredients, to thicken a sauce, or to fold in beaten egg whites. The method contains directions for the most efficient order of work, eliminating unnecessary tools and equipment and unnecessary steps in preparation.

A. GENERAL INFORMATION No. 1(2)

INFORMATION FOR STANDARDIZED RECIPES
RECIPE CONVERSION

Since few dining facilities serve exactly 100 persons, and, in some instances, the acceptable size portion may be smaller or larger, it is often necessary to reduce or increase a recipe. You may adjust the recipe to yield the number of portions needed, or to use the amount of ingredients available, or to produce a specific number of smaller portions. When increasing or decreasing a recipe, the division or multiplication of pounds and ounces is simplified when decimals are used.

1. To convert the quantities to decimals, use this table:

Weight in Ounces	Decimal of Pound	Weight in Ounces	Decimal of Pound
1	.06	9	.56
2	.13	10	.63
3	.19	11	.69
4 (1/4 lb)	.25	12 (3/4 lb)	.75
5	.31	13	.81
6	.38	14	.88
7	.44	15	.94
8 (1/2 lb)	.50	16 (1 lb)	1.00

For example: 1 lb 4 oz is converted to 1.25 lb; 2 lb 10 oz is converted to 2.63 lb.

A. GENERAL INFORMATION No. 1(2)

2. To adjust the recipe to yield a specific number of portions:

> First -- Obtain a working factor by dividing the number of portions needed by 100. For example:
>
> 348 (portions needed) ÷ 100 = 3.48 (Working Factor)
>
> Then -- Multiply the quantity of each ingredient by the working factor. For example:
>
> 1.25 lb (recipe) X 3.48 (Working Factor) = 4.35 lb (quantity needed).
>
> The part of the pound is converted to ounces by multiplying the decimal by 16. For example:
>
> .35 lb X 16 ounces = 5.60 ounces
>
> After the part of the pound has been converted to ounces, use the following scale to "round off":
>
> | | | | | | |
> |---|---|---|---|---|---|
> | .00 to .12 | = | 0 | .63 to .87 | = | 3/4 ounce |
> | .13 to .37 | = | 1/4 ounce | .88 to .99 | = | 1 ounce |
> | .38 to .62 | = | 1/2 ounce | | | |
>
> Thus 5.60 ounces will be "rounded off" to 5 1/2 ounces, and 4 lb 5 1/2 ounces will be the quantity needed (equal to 4.35 lb).

A. GENERAL INFORMATION No. 1(3)

INFORMATION FOR STANDARDIZED RECIPES RECIPE CONVERSION

3. To adjust the recipe for volume:

   First -- Obtain a working factor by dividing the number of portions needed by 100 as shown in Step 2 of A.1, Recipe Conversion.

   $$333/100 = 3.33$$

   Then -- Multiply the quantity of each ingredient by the working factor. You will round off to the nearest 1/4 teaspoon. For example, the recipe calls for 6 gallons of water per 100 portions. Portions to prepare are 333.

   $$333/100 = 3.33 \text{ Working Factor (W/F)}$$

1. W/F x No. of gallons	= gallon	3.33 W/F x 6	= 19.98 GL
2. Decimal (of gal) x 4	= quart (QT)	.98 GL x 4	= 3.92 QT
3. Decimal (of quart) x 2	= pint (PT)	.92 QT x 2	= 1.84 PT
4. Decimal (of pint) x 2	= cup (C)	.84 PT x 2	= 1.68 C
5. Decimal (of tbsp) x 16	= tablespoon (TBSP)	.68 C x 16	= 10.88 TBSP
6. Decimal (of tbsp) x 3	= teaspoon (TSP)	.88 TBSP x 3	= 2.64 TSP
7. Round off decimal portion (see paragraph 2)		.64 TSP	= 3/4 TSP

A. GENERAL INFORMATION No. 1(3)

The amount of water needed for 333 portions is: 19 GL, 3 QT, 1 PT, 1 C, 10 TBSP and 2 3/4 TSP.

NOTE: 4 QT = 1 GL   2 C = 1PT   3 TSP = 1 TBSP
        2 PT = 1 QT   16 TBSP = 1C

4. To adjust the recipe on the basis of a quantity of an ingredient to be used:

   First -- Obtain a Working Factor by dividing the pounds you have to use by the pounds required to yield 100 portions.

   For example:

   102 lb ÷ 30 (lb per 100 servings) = 3.40 (Working Factor)

   Then -- Multiply the quantity of each ingredient in the recipe by the Working Factor.

5. To adjust the recipe to yield a specific number of portions of a specific size:

   First -- Divide the desired portion size by the standard portion of the recipe.
        3 oz (desired size) ÷ 4 oz (standard portion) = .75
        348 (servings needed) x .75 = 261
        261 ÷ 100 = 2.61 (Working Factor)

   Then -- Multiply the quantity of each ingredient in the recipe by the Working Factor.

A. GENERAL INFORMATION No. 2 (1)

DEFINITION OF TERMS USED IN FOOD PREPARATION

Bake	To cook by dry heat in an oven, either covered or uncovered.
Barbecue	To roast or cook slowly, basting with a highly seasoned sauce.
Baste	To moisten food with liquid or melted fat during cooking to prevent drying of the surface and to add flavor.
Batch Preparation	A predetermined quantity or number of servings of food that is to be prepared at selected time intervals in progressive cookery for a given meal period to ensure fresh, high quality cooked food to customers.
Beat	To make a mixture smooth by using a fast regular circular and lifting motion which incorporates air into a product.
Blanch	To partially cook in deep fat, boiling water or steam.
Blend	To mix two or more ingredients thoroughly.
Boil	To cook in liquid at boiling point (212° F.) in which bubbles rise and break at the surface.
Braise	To brown in small amount of fat, then to cook slowly in small amount of liquid below the boiling point in a covered utensil.
Bread	To cover with crumbs or other suitable dry coating ingredient; or to dredge in a mixture of flour seasonings, and/or condiments, dip in a mixture of milk and slightly beaten eggs and then dredge in crumbs.
Broil	To cook by direct exposure to heat.
Brown	To produce a brown color on the surface of food by subjecting it to heat.

A. GENERAL INFORMATION No. 2 (1)

Chop	To cut food into irregular small pieces.
Cream	To mix until smooth, so that the resulting mixture is softened and thoroughly blended.
Crimp	To pinch together in order to seal.
Cube	To cut any food into square-shaped pieces.
Dice	To cut into small cubes or pieces.
Dock	To punch a number of vertical impressions in a dough with a smooth round stick about the size of a pencil to allow for expansion and permit gases to escape during baking.
Dredge	To coat with crumbs, flour, sugar or corn meal.
Fermentation	The process by which yeast acts on the sugar and starches in the dough to produce carbon dioxide gas and alcohol, resulting in expansion of the dough. During this period, the dough doubles in bulk.
Flake	To break lightly into small pieces.
Fold	To blend two or more ingredients together with a cutting and folding motion.
Fry	To cook in hot fat.
Garnish	To decorate with small pieces of colorful food.

Glaze	A glossy coat given to foods, as by covering with a sauce or by adding a sugary syrup, icing, etc.
Gluten	A tough elastic protein that gives dough its strength and ability to retain gas.
Grate	To rub food on a grater and thus break it into tiny pieces.
Grill	To cook, uncovered, on a griddle, removing grease as it accumulates. No liquid is added.
Knead	To work dough by folding and pressing firmly with palms of hands, turning between foldings.
Marinade	A preparation containing spices, condiments, vegetables, and aromatic herbs, and a liquid (acid or oil or combination of these) in which a food is placed for a period of time to enhance its flavor or to increase its tenderness.
Marinate	To allow to stand in a marinade to add flavor or tenderness.
Mince	To cut or chop into very small pieces.
Panbroil	To cook uncovered in a hot frying pan, pouring off fat as it accumulates.
Pare	To cut away outer covering.
Peel	To remove the outer layer of skin of a vegetable or fruit, etc.

Progressive Cookery	The continuous preparation of food in successive steps during the entire serving period (i.e., continuous preparation of vegetables, cook-to-order hamburgers, steaks, fried eggs, pancakes). This procedure ensures fresh, high quality cooked food to customers on a continuous basis. See Batch Preparation.
Proof	To allow shaped and panned yeast products like bread and rolls to double in size under controlled atmospheric conditions.
Reconstitute	To restore to liquid state by adding water. Also to reheat frozen prepared foods.
Rehydrate	To soak, cook, or use other procedures with dehydrated foods to restore water lost during drying.
Roast	To cook by dry heat; usually uncovered, in an oven.
Roux	Roux is a French word for a mixture of flour and fat, cooked to eliminate the raw, uncooked taste of flour.
Sauté	To brown or cook in small amount of fat.
Scald	To heat a liquid over hot water or direct heat to a temperature just below the boiling point.
Scale	To measure a portion of food by weighing.
Scant	Not quite up to stated measure.
Score	To make shallow cuts across top of a food item.
Seasoned Flour or Crumbs	A mixture of flour or crumbs with seasonings.

A. GENERAL INFORMATION No. 2 (3)

Shred	To cut or tear into thin strips or pieces using a knife or a shredder attachment.
Sift	To put dry ingredients through a sieve.
Simmer	To cook gently in a liquid just below the boiling point (190° F. - 210° F.); bubbles will form slowly and break at the surface.
Slurry	A lump-free mixture made by whipping cornstarch or flour into cold water or other liquids
Steam	To cook over or surrounded by steam.
Stew	To simmer in enough liquid to cover solid foods.
Stir	To mix two or more ingredients with a circular motion.
Temper	To remove from freezer and place under refrigeration for a period of time sufficient to facilitate separation and handling of frozen product. Internal temperature of the food should be approximately 26° F. to 28° F.
Thaw	To remove from freezer and place under refrigeration approximately 18-48 hours. Internal temperature should be above 30° F.
Toss	To mix ingredients lightly.
Wash	The liquid brushed on the surface of unbaked pies or turnovers to give a golden brown color to the crust or on the surface of proofed breads and rolls before baking and on baked bread and rolls to give a shine to the crust.
Whip	To beat rapidly with wire whip to increase volume by incorporating air.

A. GENERAL INFORMATION No. 4

TABLE OF MEASURING EQUIVALENTS

TSP	TBSP	FLUID OUNCES	CUPS	SCOOPS	LADLES	FLUID MEASURE
3	1	1/2		1-No. 40		
	1-1/2	3/4		1-No. 30	Size 0	
	2	1		1-No. 24		
	2-2/3	1-1/3		1-No. 20		
	3	1-1/2		1-No. 16	Size 1	
	4	2	1/4	1-No. 12		
	5-1/3	2-2/3	1/3	1-No. 10		
	6	3	3/8	1-No. 8	Size 2	
	8	4	1/2	1-No. 6		
	10-2/3	5-1/3	2/3			
	12	6	3/4			
	14	7	7/8			
	16	8	1		Size 3	1/2 pt
	18	9	1-1/8			
		12	1-1/2		Size 4	3/4 pt
		16	2			1 pt
		24	3			1-1/2 pt
		32	4			1 qt
		64	8			2 qt
		128	16			1 gal

NOTE: 1. Use ladles to serve individual portions of liquid or semi-liquid foods.
2. Scoop number indicates the number of portions per quart.

A. GENERAL INFORMATION No. 5

TABLE OF WEIGHTS AND MEASURES FOR CAN SIZES

CAN SIZE	AVERAGE NET WEIGHT OR FLUID MEASURE PER CAN (SEE NOTE)	AVERAGE CUPS PER CAN	APPROX. CANS PER CASE	NO. CANS EQUIV. NO. 10 CN
No. 10	6 lb 8 oz	12-1/2	6	1
No. 3 cyl	3 lb 2 oz (46 fl oz)	5-3/4	12	2
No. 3 (vacuum)	1 lb 7 oz	2-3/4	24	4-1/2
No. 2-1/2	1 lb 12 oz	3-1/2	24	4
No. 2	1 lb 4 oz	2-1/3	24	5
No. 303	1 lb	2	24	7
No. 300	14 oz	1-3/4	24	7
No. 2 (vacuum)	12 oz	1-1/2	24	8
No. 1 picnic	11 oz	1-1/4	48	10

NOTE: The net weight on can or jar labels differs among foods due to different densities of foods. For example: A No. 10 cn contains 6 lb 3 oz sauerkraut or 7 lb 5 oz cranberry sauce.

A. GENERAL INFORMATION No. 8(1)

TABLE OF EGG EQUIVALENTS

FRESH WHOLE EGGS (SHELLED)			DEHYDRATED EGG MIX		
Medium Size	Weight	Volume	Weight	Volume (Approx.)	Water to be Added
1 egg	1.6 oz	3 tbsp	1/2 oz	2 tbsp	2-1/2 tbsp
2 eggs	3.2 oz	6 tbsp	1 oz	1/4 cup	5 tbsp
10 eggs*	1 lb	1-7/8 cups	5 oz	1-1/4 cups	1-1/2 cups
12 eggs	1 lb 3.2 oz	2-1/4 cups	6 oz	1-1/2 cups	scant - 2 cups
20 eggs	2 lb	3-3/4 cups	10 oz	2-1/2 cups	3 cups
40 eggs	4 lb	7-1/2 cups	20 oz	1-1/4 qt (1-No. 3 cyl can)	1-1/2 qt

* 10 large eggs = 1 lb 2 oz

NOTES:
1. Frozen Whole Eggs and Frozen Egg Whites may be used in equivalent weights to shelled fresh whole eggs.
2. Dehydrated Egg Mix may be used in most recipes requiring whole eggs as shown in the table above. DO NOT USE RECONSTITUTED EGGS IN UNCOOKED SALAD DRESSINGS OR OTHER RECIPES WHICH DO NOT REQUIRE COOKING. RECONSTITUTED DEHYDRATED EGG MIX SHOULD BE USED WITHIN ONE HOUR UNLESS REFRIGERATED. DO NOT HOLD OVERNIGHT. For greater accuracy, weigh dehydrated egg mix.

3. *Reconstitution Methods for Dehydrated Egg Mix*

   a. Method 1. Place dehydrated egg mix in bowl; stir with a wire whip; add 1/2 of the water; whip until a smooth paste is formed; add remaining water; whip until mixture is blended.

   b. Method 2. Add dehydrated egg mix to water; stir to moisten; let stand 5 minutes; whip until smooth.

<u>For Baked Products</u>

   a. Method 1. Reconstitute dehydrated egg mix; substitute for eggs in recipe.

   b. Method 2. Sift dehydrated egg mix with dry ingredients; add water in step in Method column where whole eggs are incorporated.

<u>For Batter Dips</u>. Dehydrated egg mix may be reconstituted and used.

A. GENERAL INFORMATION No. 9(1)

GUIDELINES FOR CONTAINER YIELDS FOR CANNED FRUITS

TYPE OF FRUIT	PORTION SIZE (Approximate)	CAN SIZE	NO. OF CANS FOR 100 PORTIONS
Applesauce	1/2 cup	No. 303 cn	25
		No. 10 cn	4
Applesauce, Instant	1/2 cup	No. 2-1/2 cn	4
Apricots, halved	3 to 5 halves	No. 2-1/2 cn	16
		No. 10 cn	4
Blueberries	1/2 cup	No. 10 cn	4
Cherries, sweet, dark or light, pitted or unpitted	1/2 cup	No. 303 cn	25
		No. 10 cn	4
Cranberry Sauce, strained	1/4 cup	No. 303 cn or 300 cn	13
Cranberry Sauce, whole	1/4 cup	No. 10 cn	2
Figs, Kadota	3 to 4 figs	No. 303 cn	25
Fruit Cocktail	1/2 cup	No. 2-1/2 cn	16
		No. 10 cn	4
Fruit Mix, chunks	1/2 cup	No. 303 cn	25
		No. 10 cn	4
Grapefruit	1/2 cup	No. 303 cn	25
		No. 3 cyl cn	8

A. GENERAL INFORMATION No. 9(1)

TYPE OF FRUIT	PORTION SIZE (Approximate)	CAN SIZE	NO. OF CANS FOR 100 PORTIONS
Peaches, halves	2 halves	No. 2-1/2 cn	16
		No. 10 cn	4
Peaches, quarters or slices	1/2 cup	No. 2-1/2 cn	16
		No. 10 cn	4
Pears, halves	2 halves	No. 2-1/2 cn	16
		No. 10 cn	4
Pears, quarters or slices	1/2 cup	No. 2-1/2 cn	16
		No. 10 cn	4
Pineapple, chunks or tidbits	1/2 cup	No. 2 cn	20
		No. 10 cn	4
Pineapple slices	1 large or 2 small slices	No. 2 cn	20
		No. 10 cn	4
Plums, whole	2 to 3 plums	No. 2-1/2 cn	16
		No. 10 cn	4
Prunes, whole, unpitted	3 prunes	No. 10 cn	1-1/2

A. GENERAL INFORMATION No. 10(1)

NONFAT DRY MILK
RECONSTITUTION CHART FOR COOKING

Nonfat Dry Milk + (Conventional)	Water =	Fluid Skim Milk
1-2/3 tbsp	1/2 cup	1/2 cup
3 tbsp	1 cup	1 cup
1-2/3 oz (6 tbsp)	1-7/8 cups	2 cups
3-1/4 oz (3/4 cup)	3-3/4 cups	1 qt
5 oz (1-1/8 cups)	5-3/4 cups	1-1/2 qt
6-1/2 oz (1-1/2 cups)	7-1/2 cups	2 qt
8 oz (1-7/8 cups)	9-1/2 cups	2-1/2 qt
10 oz (2-1/4 cups)	11-1/2 cups	3 qt
11-1/4 oz (2-2/3 cups)	3-1/3 qt	3-1/2 qt
13 oz (3 cups)	3-3/4 qt	1 gal
1 lb 10 oz (1-1/2 qt)	1-7/8 gal	2 gal
2 lb 7 oz (2-1/4 qt)	2-7/8 gal	3 gal
4 lb 2 oz (3-3/4 qt)	4-3/4 gal	5 gal
5 lb 2 oz (4-3/4 qt)	6 gal	6-1/4 gal

A. GENERAL INFORMATION No. 10(1)

NOTE:
1. Recipes in this file use conventional nonfat dry milk.
2. Instant nonfat dry milk may be substituted on a pound for pound basis for the nonfat dry milk specified in any recipe. It should be weighed because the measures for instant nonfat dry milk are different from measures for nonfat dry milk (conventional). Nonfat dry milk, instant settles. If instant milk must be measured, follow directions on the container.
3. For best results, nonfat dry milk should be weighed instead of measured. Measures vary from one manufacturer to another. However, as a general rule, 1 ounce of nonfat dry milk will measure 3-2/3 tablespoons, and 4-1/2 ounces of nonfat dry milk will measure 1 cup.
4. Dry milk must be reconstituted in clean containers using clean utensils and must be treated like fresh milk after it is reconstituted. It must be refrigerated and protected from contamination.
5. Dry milk reconstitutes more easily in warm water. It should be stirred into the water with a circular motion using a whip or slotted spoon. It may also be reconstituted in a mixer if a large quantity is being prepared. However, it should be mixed at low speed to prevent excessive foaming.
6. If nonfat dry milk is to be used for a beverage, it should be weighed using 1 lb dry milk and 3-3/4 qt water per gallon. Chill thoroughly before serving. For 100 portions (8 oz), use 6 lb 4 oz nonfat dry milk and 23-1/2 qt water.

A. GENERAL INFORMATION No. 11(1)

GUIDELINES FOR USE OF DEHYDRATED ONIONS, GREEN PEPPERS, AND PARSLEY

ONIONS

Dehydrated, chopped and dehydrated compressed, chopped onions may be used in any recipe which specifies "onions, fresh, chopped or sliced."

REHYDRATION GUIDE:	Dehydrated Onions +	Water = (70-90° F.)	Rehydrated Onions **OR**	Fresh Onion Equivalent*
Dehydrated chopped onions	2 oz (9-2/3 tbsp)	1-1/2 cups	8 oz (1-1/4 cups)	1 lb (3 cups) (1 lb 1-3/4 oz A.P.)
	3-1/3 oz (1 cup)	2-1/2 cups	13 oz (2 cups)	1 lb 10 oz (4-3/4 cups) (1 lb 13 oz A.P.)
	1 lb (4-7/8 cups)	3 qt	4 lb (2-1/2 qt)	8 lb (1-1/2 gal) (8 lb 14 oz A.P.)
	2 lb 8 oz (3 qt-1 No. 10 cn)	7-1/2 qt	10 lb (6-1/4 qt)	20 lb (3-3/4 gal) (22 lb 3 oz A.P.)
Dehydrated, compressed chopped onions	1-3/4 oz	1-1/2 cups	8 oz (1-1/8 cups)	1 lb (3 cups) (1 lb 1-3/4 oz A.P.)
	2-1/3 oz	2 cups	10-1/2 oz (1-3/8 cups)	1 lb 5 oz (1 qt) (1 lb 7 oz A.P.)
	1 lb	3 qt	4 lb 8 oz (2-1/2 qt)	9 lb (6-3/4 qt) (10 lb A.P.)
	1 lb 3 oz (1 No. 2-1/2 cn)	3-1/2 qt	5 lb 5-1/2 oz (3 qt)	10 lb 11 oz (2 gal) (11 lb 14 oz A.P.)

* Volume is for chopped onions.

FOR RECIPES WITH SMALL AMOUNTS OF LIQUID: Cover dehydrated onions with 70° F. to 90 F. water. Stir dehydrated compressed onions occasionally to break apart. Let dehydrated onions stand 30 minutes; compressed dehydrated onions 1 hour or more. Drain. Note: Weight of rehydrated onions will be less than weight of dry onions but appearance and flavor will be similar.

FOR SOUPS, STEWS, SAUCES OR RECIPES WITH A LOT OF LIQUID: Add dehydrated chopped or dehydrated compressed onions directly.

A. GENERAL INFORMATION No. 11(1)

GREEN PEPPERS

Dehydrated green peppers may be used in any recipe which specifies "peppers, sweet, diced or chopped."

REHYDRATION GUIDE:	Dehydrated Peppers +	Cold Water = (35-55° F.)	Rehydrated Peppers OR	Sweet Peppers Equivalent*
	1 oz (2/3 cup)	2 cups	6-1/2 oz (1-1/3 cups)	6-1/2 oz (1-1/4 cups) (8 oz A.P.)
	1 lb (2-1/2 qt)	2 gal	6 lb 8 oz (5-1/2 qt)	6 lb 8 oz (1-1/4 gal)(7 lb 15 oz A.P.)

* Volume is for chopped peppers

FOR SALADS OR UNCOOKED DISHES: Cover with cold water. Refrigerate 1 hour or overnight. Drain.
FOR RECIPES WITH SMALL AMOUNTS OF LIQUID: Cover with cold water. Let stand 30 minutes. Drain.
FOR SOUPS, STEWS, SAUCES OR RECIPES WITH A LOT OF LIQUID: Add dehydrated peppers directly.

PARSLEY

Dehydrated parsley may be used in any recipe which specifies "chopped, fresh parsley."

REHYDRATION GUIDE:	Dehydrated Parsley +	Cold Water = (30-35° F.)	Rehydrated Parsley OR	Fresh Parsley Equivalent*
	1 oz (1-2/3 cup)	3-1/3 cups	8 oz (1-3/4 cups)	9 oz (4-1/4 cups) (9-1/2 oz A.P.)

* Volume is for chopped parsley

FOR SALADS OR UNCOOKED DISHES: Cover with ice cold water. Let stand 3 to 5 minutes. Drain.
FOR SOUPS, STEWS, SAUCES OR RECIPES WITH A LOT OF LIQUID: Add dehydrated parsley directly.

A. GENERAL INFORMATION No. 13(1)

GUIDELINES FOR FRUIT BARS

Fruit bars provide important sources of nutrients such as Vitamins A and C, and fiber. All fruits are low in fat and calories and none contain cholesterol. They may be set up for service at breakfast, lunch, dinner and brunch meals. A variety of fresh, canned and frozen fruits may be used.

Preparation: Wash all fresh fruits except bananas. Drain well. Refrigerate until ready to serve. Keep bananas in a cool, dry place until ready to serve.

ITEM	PORTION SIZE	100 PORTIONS	
		A.P. WEIGHT OR CONTAINER	E.P.
Apples, canned, drained	1/4 cup (1-1/2 oz)	13 lb 8 oz (2-No. 10 cn)	12 lb
Apples, fresh, eating	1 apple (6 oz)	37 lb 8 oz	
Applesauce, canned	1/4 cup (2 oz)	14 lb 10 oz (2-1/6-No. 10 cn)	
Apricots, canned, halves, drained	3 halves (1-1/2 oz)	20 lb 4 oz (3-No. 10 cn)	11 lb 10 oz
Apricots, fresh	2 apricots (2-1/2 oz)	16 lb 11 oz	
Bananas, fresh, peeled, sliced	1/2 cup (2-1/2 oz)	28 lb	18 lb 3 oz
Bananas, fresh	1 banana (6 oz)	40 lb	

A. GENERAL INFORMATION No. 13(1)

ITEM	PORTION SIZE	100 PORTIONS	
		A.P. WEIGHT OR CONTAINER	E.P.
Blueberries, canned drained	1/2 cup (4-1/2 oz)	52 lb 10 oz (8-1/4-No. 10 cn)	28 lb 6 oz
Cantaloupe, fresh, quartered, unpared	1/4 small cantaloupe (3 oz)	21 lb 14 oz	
Cantaloupe, fresh, pared, 1 inch pieces	1/2 cup (2-1/2 oz)	35 lb	17 lb 14 oz
Casaba melons, fresh, unpared, sliced	1/10 melon (4 oz)	31 lb 4 oz	
Casaba melons, fresh, pared 1 inch pieces	1/2 cup (2-1/2 oz)	29 lb 11 oz	17 lb 12 oz
Cherries, canned, sweet, drained	1/2 cup (3-1/2 oz)	38 lb 13 oz (5-3/4 No.-10 cn)	23 lb 14 oz
Cherries, fresh, sweet	1/2 cup (2-1/2 oz)	17 lb 10 oz	

GUIDELINES FOR FRUIT BARS - CONTINUED

ITEM	PORTION SIZE	100 PORTIONS	
		A.P. WEIGHT OR CONTAINER	E.P.
Coconut, prepared, sweetened, flakes	1 tbsp	1 lb 5 oz	
Fruit cocktail, canned, drained	1/2 cup (4 oz)	42 lb 3 oz (6-1/4-No. 10 cn)	27 lb 12 oz
Fruits, chunks, mixed, canned, drained	1/2 cup (3 oz)	39 lb 2 oz (5-3/4-No. 10 cn)	26 lb 3 oz
Grapefruit, canned, drained	1/2 cup (4 oz)	46 lb 14 oz (15-No. 3 cyl cn or 47-No. 303 cn)	25 lb 10 oz
Grapefruit, fresh, halved	1/2 grapefruit (8-3/4 oz)	54 lb 11 oz	
Grapefruit, fresh, segments	1/2 cup (4 oz)	48 lb	25 lb
Grapes, fresh	1/2 cup (2-1/2 oz)	16 lb 11 oz	
Honeyball melons, fresh, unpared, sliced	1/10 melon (3 oz)	40 lb 15 oz	
Honeyball melons, fresh, pared, 1 inch pieces	1/2 cup (2-1/2 oz)	37 lb 14 oz	17 lb 7 oz
Honeydew melons, fresh, unpared, sliced	1/10 melon (3 oz)	40 lb 15 oz	

		100 PORTIONS	
ITEM	PORTION SIZE	A.P. WEIGHT OR CONTAINER	E.P.
Honeydew melons, fresh, pared, 1 inch pieces	1/2 cup (2-1/2 oz)	37 lb 14 oz	17 lb 7 oz
Kiwifruit, fresh, pared, sliced	2 slices (1/2 oz)	5 lb 14 oz	5 lb 1 oz
Mangoes, fresh, pared, diced	1/2 cup (3 oz)	27 lb 12 oz	19 lb 3 oz
Mangoes, fresh, pared, sliced	4 slices (2 oz)	18 lb 9 oz	12 lb 12 oz
Nectarines, fresh	1 nectarine (4-1/2 oz)	28 lb 2 oz	
Oranges, fresh, peeled, sliced	3 slices (2 oz)	20 lb 9 oz	14 lb 9 oz
Oranges, fresh	1 orange (6 oz)	37 lb 8 oz	
Oranges, Mandarin, canned, drained	1/4 cup (1-1/2 oz)	20 lb 4 oz (3 No. 10 cn)	10 lb 15 oz
Papaya, fresh, pared, seeded, cubed	1/2 cup (2-1/2 oz)	24 lb	15 lb 11 oz

A. GENERAL INFORMATION No. 13(3)

GUIDELINES FOR FRUIT BARS - CONTINUED

ITEM	PORTION SIZE	100 PORTIONS	
		A.P. WEIGHT OR CONTAINER	E.P.
Papaya, fresh, pared, sliced	3 slices (2 oz)	22 lb 8 oz	14 lb 11 oz
Peaches, canned, halves, drained	2 halves (4 oz)	45 lb 9 oz (6-3/4-No 10 cn)	27 lb 7 oz
Peaches, canned, quarters/slices, drained	1/2 cup (4 oz)	43 lb 14 oz (6-1/2-No. 10 cn)	27 lb
Peaches, fresh	1 peach (4 oz)	25 lb	
Peaches, frozen	1/2 cup (4 oz)	27 lb 13 oz (4-1/4-No. 10 cn)	
Pears, canned, halves, drained	2 halves (3-1/2 oz)	41 lb 7 oz (6-1/4-No. 10 cn)	25 lb
Pears, canned, quarters/slices, drained	1/2 cup (3-1/2 oz)	36 lb 7 oz (5-1/2-No. 10 cn)	22 lb 8 oz
Pears, fresh	1 pear (5-1/2 oz)	36 lb	
Persian melons, fresh, unpared, sliced	1/10 melon (3 oz)	45 lb 13 oz	
Persian melons, fresh, pared, diced	1/2 cup (2-1/2 oz)	41 lb 4 oz	17 lb 5 oz
Pineapple, canned, chunks/tidbits, drained	1/2 cup (3-1/2 oz)	37 lb 2 oz (5-1/2-No. 10 cn)	22 lb 10 oz

Pineapple, canned, slices, drained	2 slices (2 oz)	25 lb 5 oz (3-3/4 No. 10 cn)	14 lb 7 oz
Pineapple, fresh, pared, cored, 1 inch pieces	1/2 cup (2-1/2 oz)	33 lb 4 oz	17 lb 5 oz
Plums, canned, drained	3 plums (2-1/2 oz)	32 lb 1 oz (4-3/4-No. 10 cn)	17 lb 13 oz
Plums, fresh	1 plum (2-1/2 oz)	15 lb 10 oz	
Prunes, whole, canned, drained	3 prunes (1-1/2 oz)	10 lb 1 oz (1-2/5-No. 10 cn)	9 lb 10 oz
Raisins	1 tbsp	2 lb 4 oz (1/2-No. 10 cn)	
Raspberries, frozen	1/2 cup (4 oz)	27 lb 13 oz (4-1/4-No. 10 cn)	
Strawberries, fresh, sliced	1/2 cup (2-1/2 oz)	18 lb 4 oz	17 lb 3 oz
Strawberries, fresh, whole	1/2 cup (2-1/2 oz)	16 lb 9 oz	15 lb 10 oz
Strawberries, frozen, sliced	1/2 cup (4 oz)	27 lb 13 oz (4-1/4-No. 10 cn)	
Tangelos, fresh	1 tangelo (6 oz)	37 lb 8 oz	
Tangerines, fresh	1 tangerine (3-1/2 oz)	22 lb 15 oz	
Watermelons, fresh, unpared, wedge (1 inch by 4 inches)	1 wedge (4 oz)	51 lb	
Watermelons, fresh, pared, 1 inch pieces	1/2 cup (2-1/2 oz)	34 lb	17 lb 11 oz

A. GENERAL INFORMATION No. 15(1)

CONVERSION OF QUANTITIES IN RECIPES
Weight Conversion Chart

The following chart for weights permit easy adjustment of recipes to yield the number of portions actually needed. Since recipes are based on 100 portions, find the amount as specified in the recipe under the column headed 100 portions, and then use the amount shown in the column with the heading for the number of portions to be prepared, i.e., if a recipe for 100 uses 1 pound of flour, find 1 pound under the column headed 100 portions and then look in the column under 125 portion and you will see that your should use 1 pound 4 ounces to prepare 125 portions of the item.

Oz = ounce   Lb = pound

10 Portions	25 Portions	50 Portions	75 Portions	100 Portions	125 Portions	150 Portions	175 Portions	250 Portions	275 Portions	300 Portions
1/10 oz	1/4 oz	1/2 oz	3/4 oz	1 oz	1 1/4 oz	1 1/2 oz	1 3/4 oz	2 1/2 oz	2 3/4 oz	3 oz
1/5 oz	1/2 oz	1 oz	1 1/2 oz	2 oz	2 1/2 oz	3 oz	3 1/2 oz	5 oz	5 1/2 oz	6 oz
3/10 oz	3/4 oz	1 1/2 oz	2 1/4 oz	3 oz	3 3/4 oz	4 1/2 oz	5 1/4 oz	7 1/2 oz	8 1/4 oz	9 oz
2/5 oz	1 oz	2 oz	3 oz	4 oz	5 oz	6 oz	7 oz	10 oz	11 oz	12 oz
1/2 oz	1 1/4 oz	2 1/2 oz	3 3/4 oz	5 oz	6 1/4 oz	7 1/2 oz	8 3/4 oz	12 1/2 oz	13 3/4 oz	15 oz
3/5 oz	1 1/2 oz	3 oz	4 1/2 oz	6 oz	7-1/2 oz	9 oz	10 1/2 oz	15 oz	1 lb	1 lb 2 oz
7/10 oz	1 3/4 oz	3 1/2 oz	5 1/4 oz	7 oz	8 3/4 oz	10 1/2 oz	12 1/4 oz	1 lb 2 oz	1 lb 4 oz	1 lb 5 oz

A. GENERAL INFORMATION No. 15(1)

10 Portions	25 Portions	50 Portions	75 Portions	100 Portions	125 Portions	150 Portions	175 Portions	250 Portions	275 Portions	300 Portions
4/5 oz	2 oz	4 oz	6 oz	8 oz	10 oz	12 oz	14 oz	1 lb 4 oz	1 lb 6 oz	1 lb 8 oz
7/8 oz	2 1/4 oz	4 1/2 oz	6 3/4 oz	9 oz	11 1/4 oz	13 1/2 oz	15 3/4 oz	1 lb 6 oz	1 lb 8 oz	1 lb 11 oz
1 oz	2 1/2 oz	5 oz	7 1/2 oz	10 oz	12 1/2 oz	15 oz	1 lb 2 oz	1 lb 10 oz	1 lb 12 oz	1 lb 14 oz
1 1/8 oz	2 3/4 oz	5 1/2 oz	8 1/4 oz	11 oz	13 3/4 oz	1 lb	1 lb 4 oz	1 lb 12 oz	1 lb 14 oz	2 lb 2 oz
1 1/4 oz	3 oz	6 oz	9 oz	12 oz	15 oz	1 lb 2 oz	1 lb 5 oz	1 lb 14 oz	2 lb 2 oz	2 lb 4 oz
1 1/3 oz	3 1/4 oz	6 1/2 oz	9 3/4 oz	13 oz	1 lb	1 lb 4 oz	1 lb 6 oz	2 lb	2 lb 4 oz	2 lb 8 oz
1 3/8 oz	3 1/2 oz	7 oz	10 1/2 oz	14 oz	1 lb 2 oz	1 lb 5 oz	1 lb 8 oz	2 lb 4 oz	2 lb 6 oz	2 lb 10 oz
1 1/2 oz	3 3/4 oz	7 1/2 oz	11 oz	15 oz	1 lb 2 oz	1 lb 6 oz	1 lb 10 oz	2 lb 5 oz	2 lb 10 oz	2 lb 14 oz
1 5/8 oz	4 oz	8 oz	12 oz	1 lb	1 lb 4 oz	1 lb 8 oz	1 lb 12 oz	2 lb 8 oz	2 lb 12 oz	3 lb
2 oz	5 oz	10 oz	15 oz	1 lb 4 oz	1 lb 10 oz	1 lb 14 oz	2 lb 4 oz	3 lb 2 oz	3 lb 8 oz	3 lb 12 oz
2 2/5 oz	6 oz	12 oz	1 lb 2 oz	1 lb 8 oz	1 lb 14 oz	2 lb 4 oz	2 lb 10 oz	3 lb 12 oz	4 lb 2 oz	4 lb 8 oz
2 4/5 oz	7 oz	14 oz	1 lb 5 oz	1 lb 12 oz	2 lb 4 oz	2 lb 10 oz	3 lb 2 oz	4 lb 6 oz	4 lb 14 oz	5 lb 4 oz
3 1/5 oz	8 oz	1 lb	1 lb 8 oz	2 lb	2 lb 8 oz	3 lb	3 lb 8 oz	5 lb	5 lb 8 oz	6 lb
3 3/5 oz	9 oz	1 lb 2 oz	1 lb 11 oz	2 lb 4 oz	2 lb 14 oz	3 lb 6 oz	4 lb	5 lb 10 oz	6 lb 4 oz	6 lb 12 oz
4 oz	10 oz	1 lb 4 oz	1 lb 14 oz	2 lb 8 oz	3 lb 2 oz	3 lb 12 oz	4 lb 6 oz	6 lb 4 oz	6 lb 14 oz	7 lb 8 oz

A. GENERAL INFORMATION No. 15(2)

CONVERSION OF QUANTITIES IN RECIPES
Weight Conversion Chart

10 Portions	25 Portions	50 Portions	75 Portions	100 Portions	125 Portions	150 Portions	175 Portions	250 Portions	275 Portions	300 Portions
4 2/5 oz	11 oz	1 lb 6 oz	2 lb 2 oz	2 lb 12 oz	3 lb 8 oz	4 lb 2 oz	4 lb 14 oz	6 lb 14 oz	7 lb 10 oz	8 lb 4 oz
4 4/5 oz	12 oz	1 lb 8 oz	2 lb 4 oz	3 lb	3 lb 12 oz	4 lb 8 oz	5 lb 4 oz	7 lb 8 oz	8 lb 4 oz	9 lb
5 1/5 oz	13 oz	1 lb 10 oz	2 lb 8 oz	3 lb 4 oz	4 lb 2 oz	4 lb 14 oz	5 lb 11 oz	8 lb 2 oz	9 lb	9 lb 12 oz
5 3/5 oz	14 oz	1 lb 12 oz	2 lb 10 oz	3 lb 8 oz	4 lb 6 oz	5 lb 4 oz	6 lb 2 oz	8 lb 12 oz	9 lb 10 oz	10 lb 8 oz
6 oz	15 oz	1 lb 14 oz	2 lb 14 oz	3 lb 12 oz	4 lb 11 oz	5 lb 10 oz	6 lb 10 oz	9 lb 6 oz	10 lb 5 oz	11 lb 4 oz
6 2/5 oz	1 lb	2 lb	3 lb	4 lb	5 lb	6 lb	7 lb	10 lb	11 lb	12
8 oz	1 lb 4 oz	2 lb 8 oz	3 lb 12 oz	5 lb	6 lb 4 oz	7 lb 8 oz	8 lb 12 oz	12 lb 8 oz	13 lb 12 oz	15 lb
9 3/5 oz	1 lb 8 oz	3 lb	4 lb 8 oz	6 lb	7 lb 8 oz	9 lb	10 lb 8 oz	15 lb	16 lb 8 oz	18 lb
11 1/5 oz	1 lb 12 oz	3 lb 8 oz	5 lb 4 oz	7 lb	8 lb 12 oz	10 lb 8 oz	12 lb 4 oz	17 lb 8 oz	19 lb 4 oz	21 lb
12 4/5 oz	2 lb	4 lb	6 lb	8 lb	10 lb	12 lb	14 lb	20 lb	22 lb	24 lb
1 lb	2 lb 8 oz	5 lb	7 lb 8 oz	10 lb	12 lb 8 oz	15 lb	17 lb 8 oz	25 lb	27 lb 8 oz	30 lb
1 lb 4 oz	3 lb	6 lb	9 lb	12 lb	15 lb	18 lb	21 lb	30 lb	33 lb	36 lb
1 lb 8 oz	3 lb 12 oz	7 lb 8 oz	11 lb 4 oz	15 lb	18 lb 12 oz	22 lb 8 oz	26 lb 4 oz	37 lb 8 oz	41 lb 4 oz	45 lb
2 lb	5 lb	10 lb	15 lb	20 lb	25 lb	30 lb	35 lb	50 lb	55 lb	60 lb
3 lb	7 lb 8 oz	15 lb	22 lb 8 oz	30 lb	37 lb 8 oz	45 lb	52 lb 8 oz	75 lb	82 lb 8 oz	90 lb

A. GENERAL INFORMATION No. 16(1)

CONVERSION OF QUANTITIES IN RECIPES
Measure Conversion Chart

The following chart for measures permits easy adjustments of recipes to yield the number of portions actually needed. Since recipes are based on 100 portions, find the amount as specified in the recipe under column headed 100 portions and then use the amount shown in the column with the heading for the number of portions to be prepared, i.e., if a recipe for 100 uses 3 cups of flour, find 3 cups under the column headed 100 portions and then look in the column under 125 portions and you will see that you should use 3 ¾ cups to prepare 125 portions of the item.

tsp – teaspoon   tbsp – tablespoon   qt – quart   gal - gallon

10 Portions	25 Portions	50 Portions	75 Portions	100 Portions	125 Portions	150 Portions	175 Portions	250 Portions	275 Portions	300 Portions
……	¼ tsp	½ tsp	¾ tsp	1 tsp	1 ¼ tsp	1 ½ tsp	1 ¾ tsp	2 ½ tsp	2 ¾ tsp	1 tbsp
……	½ tsp	1 tsp	1 ½ tsp	2 tsp	2 ½ tsp	1 tbsp	3 ½ tsp	1 2/3 tbsp	1 7/8 tbsp	2 tbsp
¼ tsp	¾ tsp	1 ½ tsp	2 tsp	1 tbsp	3 ¾ tsp	1 1/3 tbsp	1 2/3 tbsp	2 1/3 tbsp	2 2/3 tbsp	3 tbsp
½ tsp	1 ½ tsp	1 tbsp	1 2/3 tbsp	2 tbsp	2 2/3 tbsp	3 tbsp	3 2/3 tbsp	5 tbsp	5 2/3 tbsp	6 tbsp
¾ tsp	2 ¼ tsp	1 2/3 tbsp	2 1/3 tbsp	3 tbsp	¼ cup	4 2/3 tbsp	5 tbsp	7 2/3 tbsp	½ cup	9 tbsp
1 tsp	1 tbsp	2 tbsp	3 tbsp	¼ cup	5 tbsp	6 tbsp	7 tbsp	10 tbsp	11 tbsp	¾ cup
1 ½ tsp	3 ¾ tsp	2 2/3 tbsp	4 tbsp	5 tbsp	6 tbsp	7 2/3 tbsp	9 tbsp	12 2/3 tbsp	14 tbsp	1 cup
1 ¾ tsp	4 ½ tsp	3 tbsp	4 2/3 tbsp	6 tbsp	7 2/3 tbsp	½ cup	10 2/3 tbsp	15 tbsp	1 cup	1 cup + 2 tbsp
2 tsp	5 ¼ tsp	3 2/3 tbsp	5 tbsp	7 tbsp	9 tbsp	10 2/3 tbsp	¾ cup	1 cup + 1 2/3 tbsp	1 cup + 3 tbsp	1 1/3 cups

A. GENERAL INFORMATION No. 16(1)

10 Portions	25 Portions	50 Portions	75 Portions	100 Portions	125 Portions	150 Portions	175 Portions	250 Portions	275 Portions	300 Portions
2 ¼ tsp	2 tbsp	4 tbsp	6 tbsp	½ cup	10 tbsp	¾ cup	14 tbsp	1 ¼ cups	1 cup + 6 tbsp	1 ½ cups
2 ½ tsp	2 tbsp	4 2/3 tbsp	7 tbsp	9 tbsp	11 tbsp	13 2/3 tbsp	1 cup	1 cup + 6 tbsp	1 ½ cups	1 ¾ cups
1 tbsp	2 2/3 tbsp	5 tbsp	7 2/3 tbsp	10 tbsp	¾ cup	1 cup	1 cup + 2 tbsp	1 ½ cups	1 ¾ cups	2 cups
3 ¼ tsp	3 tbsp	5 2/3 tbsp	8 tbsp	11 tbsp	14 tbsp	1 cup	1 cup + 3 tbsp	1 ¾ cups	2 cups	2 1/8 cups
3 ½ tsp	3 tbsp	6 tbsp	9 tbsp	¾ cup	1 cup	1 cup + 2 tbsp	1 ¼ cups	2 cups	2 cups + 2 tbsp	2 ¼ cups
3 ¾ tsp	3 tbsp	6 2/3 tbsp	10 tbsp	13 tbsp	1 cup	1 ¼ cups	1 ½ cups	2 cups	2 ¼ cups	2 ½ cups
1 1/3 tbsp	3 2/3 tbsp	7 tbsp	10 2/3 tbsp	14 tbsp	1 cup + 2 tbsp	1 1/3 cups	1 ½ cups	2 cups + 3 tbsp	2 1/3 cups	2 ½ cups
4 ½ tsp	3 ¾ tbsp	7 2/3 tbsp	11 tbsp	15 tbsp	1 ¼ cups	1 ½ cups	1 ¾ cups	2 1/3 cups	2 ¾ cups	2 7/8 cups
4 ¾ tsp	¼ cup	½ cup	¾ cup	1 cup	1 ¼ cups	1 ½ cups	1 ¾ cups	2 ½ cups	2 ¾ cups	3 cups
2 tbsp	5 tbsp	10 tbsp	1 cup	1 ¼ cups	1 ½ cups	2 cups	2 ¼ cups	3 cups	3 ½ cups	3 ¾ cups
7 tsp	6 tbsp	¾ cup	1 cup + 2 tbsp	1 ½ cups	2 cups	2 ¼ cups	2 ¾ cups	3 ¾ cups	1 qt	4 ½ cups

A. GENERAL INFORMATION No. 16(2)

CONVERSION OF QUANTITIES IN RECIPES
Measure Conversion Chart

10 Portions	25 Portions	50 Portions	75 Portions	100 Portions	125 Portions	150 Portions	175 Portions	250 Portions	275 Portions	300 Portions
8 ¼ tsp	7 tbsp	14 tbsp	1 1/3 cups	1 ¾ cups	2 ¼ cups	2 ¾ cups	3 cups	4 ½ cups	4 ¾ cups	5 ¼ cups
9 ½ tsp	½ cup	1 cup	1 ½ cups	2 cups	2 ½ cups	3 cups	3 ½ cups	5 cups	5 ½ cups	1 ½ qt
10 ¾ tsp	½ cup + 1 tbsp	1 cup + 2 tbsp	1 ¾ cups	2 ¼ cups	2 ¾ cups	3 ½ cups	1 qt	5 ¾ cups	1 ½ qt	6 ¾ cups
¼ cup	10 tbsp	1 ¼ cups	2 cups	2 ½ cups	3 cups + 2 tbsp	3 ¾ cups	4 ½ cups	6 ¼ cups	1 ¾ qt	7 ½ cups
4 ¾ tbsp	¾ cup	1 ½ cups	2 ¼ cups	3 cups	3 ¾ cups	4 ½ cups	5 ¼ cups	7 ½ cups	8 ¼ cups	2 ¼ qt
5 2/3 tbsp	14 tbsp	1 ¾ cups	2 ½ cups	3 ½ cups	4 ½ cups	1 ¼ qt	1 ½ qt	2 ¼ qt	9 ¾ cups	10 ½ cups
6 ¼ tbsp	1 cup	2 cups	3 cups	1 qt	1 ¼ qt	1 ½ qt	1 ¾ qt	2 ½ qt	2 ¾ qt	3 qt
½ cup	1 ¼ cups	2 ½ cups	3 ¾ cups	1 ¼ qt	6 ¼ cups	7 ½ cups	8 ¾ cups	12 ½ cups	3 ½ qt	3 ¾ qt
9 ¾ tbsp	1 ½ cups	3 cups	4 ½ cups	1 ½ qt	7 ½ cups	2 ¼ qt	10 ½ cups	3 ¾ qt	1 gal	4 ½ qt
11 tbsp	1 ¾ cups	3 ½ cups	5 ¼ cups	7 cups	8 ¾ cups	10 ½ cups	3 qt	1 gal + 1 ½ cups	1 gal + 3 ¼ cups	5 ¼ qt
12 ¾ tbsp	2 cups	1 qt	1 ½ qt	2 qt	2 ¼ qt	3 qt	3 ½ qt	1 ¼ gal	5 ½ qt	1 ½ gal
1 ¼ cups	3 cups	1 ½ qt	2 ¼ qt	3 qt	3 ¾ qt	4 ½ qt	5 ¼ qt	7 ½ qt	2 gal	2 ¼ gal
1 ½ cups	1 qt	2 qt	3 qt	1 gal	1 ¼ gal	1 ½ gal	1 ¾ gal	2 ½ gal	2 ¾ gal	3 gal
3 cups	2 qt	1 gal	1 ½ gal	2 gal	2 ¼ gal	3 gal	3 ½ gal	5 gal	5 ½ gal	6 gal
4 ½ cups	3 qt	1 ½ gal	2 ¼ gal	3 gal	3 ¾ gal	4 ½ gal	5 ¼ gal	7 ¼ gal	8 gal	9 gal
1 ½ qt	1 gal	2 gal	3 gal	4 gal	5 gal	6 gal	7 gal	10 gal	11 gal	12 gal
7 ½ cups	1 ¼ gal	2 ½ gal	3 ¾ gal	5 gal	6 ¼ gal	7 ½ gal	8 ¾ gal	12 ½ gal	13 ¾ gal	15 gal

GUIDELINES FOR HANDLING FROZEN FOODS

Proper storage and thawing procedures for frozen foods are essential for keeping foods safe and palatable. Some foods, such as vegetables, do not need to be thawed before cooking. Many recipes require meat to be only partially thawed or tempered, to facilitate separation before cooking; this prevents excessive moisture loss. Unless otherwise indicated, preparation methods and cooking times are for thawed meat, fish and poultry.

Frozen foods should be stored at or below 0° F. and thawed at 36° F. DO NOT refreeze foods that have been thawed; cook and serve as soon as possible to promote maximum quality and safety.

FROZEN FRUITS: Thaw unopened under refrigeration (36° F. to 38° F.) or covered with cold water.

FROZEN FRUIT JUICES AND CONCENTRATES: These do not require thawing.

FROZEN VEGETABLES: These do not require thawing before cooking. For faster cooking, Brussels sprouts, broccoli, asparagus, cauliflower, and leafy greens may be partially thawed under refrigeration.

FROZEN MEATS: Improper thawing of meat encourages bacterial growth and also results in unnecessary loss of meat juices, poor quality and loss of yield and nutrients. To thaw meat, remove from shipping container, but leave inside wrappings (usually polyethylene bags) on meat. Thaw under refrigeration (36° F. to 38° F.) until almost completely thawed. Spread out large cuts, such as roasts, to allow air to circulate. The length of the thawing period will vary accordingly to the size of meat cut, the temperature and degree of air circulation in the chill space, and the quantity of meat being thawed in a given space. Boneless meats generally require 26 to 48 hours to thaw at 36° F. to 38° F.

A. GENERAL INFORMATION No. 19(1)

Meat may be cooked frozen or tempered except for a few cuts which require complete thawing (i.e., bulk ground beef, bulk beef patty mix, braising Swiss steak, bulk pork sausage and diced beef for stewing.)

Roasts, when cooked from the frozen state, will require one-third to one-half more cooking time than thawed roasts. The addition of seasonings, if required, must be delayed until the outside is somewhat thawed and the surface is sufficiently moist to retain the seasonings. The insertion of meat thermometers must also be delayed until roasts are partially thawed. Grill steaks, pork chops and liver should be tempered before cooking to ensure a moist, palatable product. (Temper - To remove from freezer and place under refrigeration for a period of time sufficient to facilitate separation and handling of frozen product. Internal temperature of the food should be approximately 26° F. to 28° F.). Pork sausage patties and pork and beef sausage links should be cooked frozen.

FROZEN SEAFOOD: Fish fillets and steaks may be cooked frozen or thawed. Any fish that is to be breaded or batter dipped should be thawed. Clams, crabmeat, oysters, scallops and shrimp should be kept wrapped while thawing. Fish and shellfish should be thawed under refrigeration (36° F. to 38° F.) and require 12 hours to thaw.

Frozen, whole lobster, king crab legs, spiny lobster tail, breaded fish portions or nuggets, batter-dipped fish portions, or breaded oysters and shrimp SHOULD NOT be thawed before cooking.

FROZEN POULTRY: Poultry must be thawed under refrigeration (36° F. to 38° F.). Proper thawing of poultry reduces bacterial growth, maintains quality and retains nutrients through less drip loss.

GUIDELINES FOR HANDLING FROZEN FOODS

RAW CHICKEN: Remove whole chickens from shipping containers and thaw in individual wrappers (plastic bags). To thaw parts or quarters, remove intermediate containers from shipping containers; remove overwrapping from intermediate containers and open intermediate containers to expose inner wrapping. Length of thawing period under refrigeration (36° F. to 38° F.) will vary according to size of chicken and refrigeration conditions.
Approximate Thawing Times: Chicken, whole - 37 hours; Chicken, quarters - 52 hours; Chicken, cut-up - 52 hours
PRECOOKED BREADED CHICKEN, NUGGETS OR FILLETS: DO NOT THAW before cooking.
PRECOOKED UNBREADED CHICKEN FILLETS: Temper. DO NOT THAW before cooking.
PREPARED FROZEN CHILIES RELLENOS, BURRITOS, PIZZAS, ENCHILADAS, LASAGNA, TAMALES, MANICOTTI, CANNELLONI: DO NOT THAW before cooking.
TURKEY: Remove turkeys from shipping containers. Thaw in individual wrappers under refrigeration (36° F. to 38° F.)
Approximate Thawing Times: Turkey, whole (16 lbs or less - 2 days; Turkey, whole (over 16 lbs) - 3 to 4 days; Turkey, boneless - 12 to 16 hours; Turkey, ground – thaw; Turkey sausage patties and links - cook frozen
FROZEN EGGS: Thaw under refrigeration (36 F. to 38 F.) or covered with cold water. Thirty pound cans require at least 2 days to thaw, 10 lb cans or cartons require at least 1 day.
FROZEN PIZZA BLEND CHEESE: If pizza blend cheese is received and stored as a frozen product, it should be thawed under refrigeration (36° F. to 38° F.) to ensure retention of its characteristic flavor, texture, and appearance. Thawing at room temperature will encourage bacterial growth (inherent in the product) resulting in an undesirable flavor and swelling of the container.

A. GENERAL INFORMATION No. 20

GUIDELINES FOR USE OF ANTIBROWNING AGENT
(NON-SULFATING AGENTS)

The purpose of an antibrowning agent is to prevent browning and maintain color and crispness in fresh potatoes and fruits.

DIRECTIONS FOR USE

1. Dissolve 1-3/4 oz (3 tbsp) antibrowning agent per gallon of cold water in a clean stainless steel, glass or plastic container. DO NOT use galvanized metal containers.

2. Dip fresh white potatoes (peeled, whole, quarters, French fry cut, slices) or fruits (apples, avocados, bananas, peaches, pears) peeled, sliced and free from bruises in the antibrowning solution. Soak for 3 minutes.

3. Drain and refrigerate product until ready to use.

NOTE: 1. Keep antibrowning agent stored in its original container. Make the solution fresh daily. A plastic measuring spoon should be kept with the antibrowning agent for easy measuring.

2. Antibrowning agent is not required for lettuce, cauliflower, green peppers, cabbage, celery or pineapple.

A. GENERAL INFORMATION No. 21

GUIDELINES FOR USE OF STEAM COOKERS

Use of steamers in quantity food preparation can save cooking time, labor, help maintain appearance of food, and preserve nutrients normally lost by other cooking methods. Steamers are ideal for batch preparation. Foods may be steamed and served in the same pan, if steam table pans are used for preparation.

Steamers are either 5 lb pressure or 15 lb pressure (high speed) type. When food is steamed at 5 lb pressure, the internal temperature of the steamer is 225° F. to 228° F. At 15 lb pressure, the temperature is 245° F. to 250° F.

Most canned, fresh or frozen vegetables, in addition to other foods such as rice, pasta, poultry, meats, fish, and shellfish, can be cooked in steamers.

Foods may be steamed in perforated or solid pans. Perforated pans are usually used, particularly for vegetables, unless the cooking liquid is retained or manufacturer's directions specify solid type pans. Pans are normally filled no more than 2/3 full to allow steam to circulate for even cooking.

Cooking times will vary depending on the type steamer, food, and temperature and quantity of the product. For best results follow the manufacturer's cooking times and directions. Cooking time should be scheduled to include bringing food up to cooking temperature, as well as steaming time. <u>Timing begins</u> when the pressure gauge registers 3 lb on the 5 lb steamer and 9 lb on the 15 lb steamer. <u>Be sure to use timer, if available, to prevent overcooking.</u>

After cooking is completed, the steam should be exhausted slowly for safety and to preserve skins of vegetables such as peas. Leave steamer doors ajar for cooling and to preserve door gaskets.

A. GENERAL INFORMATION No. 23(1)

GUIDELINES FOR CONVECTION OVENS

A convection oven has a blower fan which circulates hot air throughout the oven, eliminating cold spots and promoting rapid cooking. Overall, cooking temperatures and times are shorter than in conventional ovens. The size, thickness, type of food, and amount loaded into the oven at one time will influence the cooking time.

TEMPERATURE SETTINGS: Follow the recommended temperature guide provided in the manufacturer's operating manual. If not available, follow the guidelines furnished on this card or check specific recipe for convection oven information. Note: At this time, not all AFRS oven recipes contain convection information. If food is cooked around the edges, but the center is still raw or not thoroughly cooked, or if there is much color variation, reduce the heat by 15° F. to 25° F. and return food to the oven. If necessary, continue to reduce the heat on successive loads until the desired results are achieved. Record most successful temperature on the recipe card for future reference.

TIME SETTING: Follow the recommended times provided in the manufacturer's operating manual. Should the manual not be available, follow the guidelines furnished on this card or check the specific recipe for convection oven information. Check progress halfway through the cooking cycle since time will vary with the quantity of food loaded, the temperature, and the type of pan used. NOTE: meat thermometers for roasting and visual examination of baked products are the most accurate methods of determining cooking times, both in convection ovens and in conventional ovens. Record most successful cooking time on the recipe card for future reference.

VENT DAMPER CONTROL SETTING: The vent damper control is located on or near the control panel. The damper should be kept closed for most foods of low moisture content such as roasts. If open during roasting, meats will be dry with excessive shrinkage.

A. GENERAL INFORMATION No. 23(1)

The damper should be kept open when baking high moisture content foods (cakes, muffins, yeast bread, etc.). Leaving the damper closed throughout a baking cycle will produce cakes which are too moist and will not rise. A "cloud" or water droplets on the window indicate excessive moisture which should be vented out of the oven through the open damper.

FAN SPEED SETTINGS: SEE GENERAL NOTES BELOW.

INTERIOR OVEN LIGHTS: Turn on lights only when loading, unloading, or checking product. Continual burning of lights will result in short bulb life.

TIMER: The oven timer will ring only as a reminder; it has no control over the functioning of the oven. To ensure proper operation, wind the timer to the maximum setting, then turn back to the desired setting for the product.

GENERAL OPERATION:
1. Select and make the proper rack arrangement for the product to be cooked.
2. Turn or push the main power switch "ON" (gas oven - turn burner valve "ON"). Set thermostat to the recommended temperature. The thermostat signal light will light. Adjust fan speed on two-speed blower, if available (see General Notes below).
3. PREHEAT oven until thermostat signal light goes out indicating that the oven has reached the desired temperature. The oven should preheat to 350° F. within 10 to 15 minutes. (Note: To conserve energy, DO NOT turn on the oven until absolutely necessary - about 15 minutes before actual cooking is to start.)
4. OPEN oven doors and load the oven quickly to prevent excessive loss of heat. Load the oven from the top, centering the pans on the rack toward the front of the oven. Place partial loads in the center of the oven. Allow 1 to 2 inches between pans and along oven sides to permit good air circulation. <u>Remember - overloading is the major cause of non-uniform baking and roasting</u>.

A. GENERAL INFORMATION No. 23(2)

GUIDELINES FOR CONVECTION OVENS

5. Close oven doors and set the timer for the desired cooking time. Check the baking/roasting progress periodically until product is ready.

CLEANING AND MAINTENANCE: Refer to the manufacturer's operating manual for cleaning and maintenance instructions.

GENERAL NOTES: Most convection ovens are equipped with an electric interlock which energizes/de-energizes both the heating elements and the fan motor when the doors are closed/open. Therefore, the heating elements and fan will not operate independently and will only operate with the doors closed.

(Only one known company manufactures an oven in which the fan can be controlled independently.) Some convection ovens are equipped with single-speed fan motors while others are equipped with two-speed fan motors. This information is particularly important to note when baking cakes, muffins or meringue pies, or similar products, and when oven-frying bacon. High speed air circulation may cause damage to the food (e.g., cakes slope to one side of the pan) or blow melted fat throughout the oven. Read the manufacturer's manuals and determine exactly what features you have and then, for the above products, proceed as follows.

<u>Two-Speed Interlocked Fan Motor</u>: Set fan speed to "low."

<u>Single-Speed Interlocked Fan Motor</u>: Preheat oven 50° F. higher than the recommended cooking temperature. Load oven quickly, close doors, and reduce thermostat to recommended cooking temperature. (This action will allow the product to "set up" before the fan/heating elements come on again.)

<u>Single-Speed Independent Fan Motor</u>:

1. Preheat oven 25° F. above temperature specified in recipe.
2. Turn fan "OFF."
3. Reduce heat 25° F.
4. Load oven quickly and close doors.
5. Turn fan "ON" after 7 to 10 minutes and keep "ON" for remaining cooking time.

EXCEPTION: Leave fan "OFF" for bacon to prevent fat from blowing throughout the oven. READ AND UNDERSTAND THE MANUFACTURER'S MANUALS. THEY WILL MAKE YOUR JOB EASIER.

Note: Equipment is becoming more and more complex as the "state-of-the-art" progresses. It is absolutely essential that proper operating manuals be read and understood by everyone who either uses or maintains food service equipment. If you do not have the proper manuals available, proceed with extreme caution so as not to damage or misuse this equipment. Local food service equipment dealers, and/or your service's food service office should be contacted for assistance.

A. GENERAL INFORMATION No. 23(3)

GUIDELINES FOR CONVECTION OVENS

FOOD	PAN SIZE (INCHES)	RECOMMENDED NO. OF SHELVES FOR ONE LOAD	RECOMMENDED TEMPERATURE (° F.)	TIME
BREADS				
Breads, yeast	10-1/2 by 5 by 3-1/2	3	375	30 min
Coffee cakes	18 by 26	4	325	15 min
Muffins	12-cup muffin pan	4	350	30 min
Rolls, yeast	18 by 26	4	350	10 to 15 min
Sweet rolls	18 by 26	4	325	15 min
CAKES				
Angel food	16 by 4-1/2 by 4-1/8	3	300	25 to 30 min
Layer	8 or 9	4	300	25 to 35 min
Loaf	16 by 4-1/2 by 4-1/8	3	325	65 min
Sheet	18 by 26	4	300 to 325	25 to 35 min
DESSERTS				
Brownies	18 by 26	4	325	25 to 30 min
Cookies, bar	18 by 26	5	325	15 min
Cookies, drop	18 by 26	5	325	12 min
Cookies, sliced	18 by 26	5	350	8 to 10 min
Pies, fruit	9	4	375	25 min

A. GENERAL INFORMATION No. 23(3)

FOOD	PAN SIZE (INCHES)	RECOMMENDED NO. OF SHELVES FOR ONE LOAD	RECOMMENDED TEMPERATURE (° F.)	TIME
MEATS				
Bacon, oven fried	18 by 26	5	325	15 to 20 min
Chicken, quarters or pieces	18 by 26	5	350	30 min
Fish, baked or oven fried	18 by 26	4	325	15 to 20 min
Meatloaf	18 by 26	3	300	1 hr 15 min
Roasts, boneless,				
Beef	18 by 26	3	325	1 hr 45 min
Pork	18 by 26	3	325	1-1/2 hr to 2 hrs
Steak, grill (strip loin, ribeye roll, top sirloin butt)	18 by 26	7	400	See Recipe No. L00700
Turkey, boneless	18 by 26	3	325	3-1/2 to 4 hrs
MISCELLANEOUS				
Pizza	18 by 26	4	450	15 min
Potatoes, baked	18 by 26	5	400	35 to 40 min

A. GENERAL INFORMATION No. 24

GUIDELINES FOR USE OF TILTING FRY PANS

The tilting fry pan is a versatile piece of equipment. Although usually described as an oversized skillet because of its large flat cooking surface, this piece of equipment can perform almost any type of cooking except deep fat frying. The tilting fry pan can be used for braising, grilling, sautéing, pan frying, simmering, steaming, boiling, warming, and holding. The ability to tilt the pan allows for easy removal of food to the serving pans without heavy lifting. It can be used for successive cooking functions without having to move the food from one piece of equipment to another. The temperature dial is adjustable over a range of 200 F. to 400 F.

GENERAL OPERATION:

1. Turn or push main power switch to "on" position. The red light will signal that power is on.
2. Set thermostat to desired temperature. Yellow light will signal when heating unit has reached temperature. It will cycle on and off to maintain the temperature.
3. Preheat approximately 12 minutes before using as a griddle or fry pan.
4. To use as a steamer use 1 to 2 inches water with a rack for holding food above the water. Leave cover closed while steaming.
5. To use as a griddle, follow directions and temperature as shown on the recipe card.
6. For sautéing or pan frying, temperature should be between 300 F. and 365 F.
7. For simmering, temperature should be 200 F.

CLEANING AND MAINTENANCE: Refer to the manufacturer's operating manual for instructions.

A. GENERAL INFORMATION No. 25

GUIDELINES FOR CAPACITIES OF STEAM TABLE AND BAKING AND ROASTING PANS

PANS	DEPTH (Inches)	USABLE CAPACITY (Quarts)	USABLE CAPACITY (1/2 Cup Portions)
STEAM TABLE: 12 by 20 inch (full size)	2-1/2 4 6 8	7 13 18-1/2 27	56 104 148 216
12 by 10 inch (1/2 size)	2-1/2 4 6 8	3-1/2 6-1/2 9 12	28 52 72 96
6 by 12 inch (1/3 size)	2-1/2 4 6	2-1/2 4 6	20 16 24
6 by 10 inch (1/4 size)	2-1/2 4 6	1-2/3 2-2/3 4	13 21 32
BAKING AND ROASTING: 18 by 24 inch	4-1/2	24	192
16 by 16 inch	4	8	64

NOTE: Usable capacity: Pans are filled to about 1/2 inch from the brim. If pans are to be used for carrying liquids (i.e., soups, gravies), the capacity should be reduced to half full.

A. GENERAL INFORMATION No. 27(1)

METRIC CONVERSION

The metric system is an international language of measurement. Its symbols are based on the International System of Units (SI). Of these, food service preparation will be primarily involved with the following metric base units:

Weight (mass)	gram (g)
	kilogram (kg)
Volume	milliliter (mL)
	liter (L)
Length	centimeter (cm)
	meter (m)
Temperature	degree Celsius (°C.)

While the U. S. metric system is voluntary and the food service industry in the United States has not converted to metric system, except for a few soft conversions (e. g., labeling), military food service dining facilities/general messes outside CONUS may experience the metric system in food and equipment support provided by the host country. The information furnished in this guideline card is primarily for these food service personnel.

A. GENERAL INFORMATION No. 27(1)

CONVERSION OF U. S. CUSTOMARY TO METRIC UNITS

	U. S. Customary	**Metric**
Weight (or Mass)	1 ounce (oz) =	28.35 grams (g)
	1 pound (lb) =	453.6 grams (g) or .4536 kilograms
	2.2 pound (lb) =	1 kilogram (kg) or 1000 grams (g)
Volume	1 tsp =	4.93 milliliters (mL)
	1 tbsp =	14.79 milliliters (mL)
	1 cup =	236.59 milliliters (mL) or .237 liters (L)
	1 pint =	.473 liters (L)
	1 quart =	.946 liters (L)
	1 gallon =	3.785 liters (L)
	1.06 quarts =	1 liter (L) or 1000 milliliters (mL)
Length	1 inch =	2.54 centimeters (cm)
	1 foot =	.3048 meters (m)
	1 yard =	30.48 centimeters (cm) or .9144 meters (m)
	1.1 yards =	1 meter (m) or 100 centimeters (cm)

GUIDELINES FOR METRIC CONVERSION - CONTINUED
Temperature Conversions

°F.	°C.	°F	°C.
0	-18	212	100
26	-3	225	107
28	-2	228	109
30	-1	245	118
32	0	250	121
36	2	275	135
38	3	300	149
40	4	325	163
70	21	350	177
90	32	360	182
140	60	365	185
160	71	375	191
170	77	400	204
175	79	425	218
180	82	450	232
185	85	500	260
		550	288

A. GENERAL INFORMATION No. 28(1)

GUIDELINES FOR CHEESES
USE OF DEHYDRATED CHEESES

Two types of dehydrated cheeses are used - dehydrated American cheese and dehydrated cottage cheese.

   a. <u>Cheese, Cottage, Dehydrated</u>

      (1) USE - Dehydrated cottage cheese may be substituted in any recipe using fresh cottage cheese.

      (2) PREPARATION - Measure 8-1/2 cups water (70° F.) into a shallow serving pan. Pour 1-No. 10 cn (1 lb 1 oz) canned dehydrated cottage cheese evenly over the water. Stir gently to wet all particles of cheese. Let stand 5 minutes, then stir gently. If more water is needed, sprinkle 1/2 to 1 cup water over cheese. Chill rehydrated cheese thoroughly before serving (3 to 4 hours).

      (3) SUBSTITUTION - Rehydration ratio - 1 pound dehydrated cottage cheese to 4 pounds (2 qt) water.

<u>Dehydrated Cheese</u>	<u>Water Added</u> =	<u>Rehydrated Cheese</u>	OR	<u>Fresh Cheese Equivalent</u>
1-No. 10 cn (1 lb 1 oz (2-3/4 qt))	8-1/2 cups	5 lb oz (3 qt)		6 lb (3qt)
2-No. 10 cn (2 lb 2 oz (5-1/2 qt))	4-1/4 qt	10 lb 2 oz (6-1/4 qt)		12 lb (1-1/2 gal)

A. GENERAL INFORMATION No. 28(1)

b. Cheese, American, Processed, Dehydrated

(1) USE - Dehydrated American processed cheese may be substituted in any recipe using processed American cheese. Rehydrate cheese before adding to any recipe to eliminate any un-rehydrated cheese in the end product. To store dehydrated cheese after being opened, place unused portion in a tightly covered container to prevent absorption of moisture. Refrigerate if possible.

(2) PREPARATION - Add water to cheese and mix until blended. For a moist semi-solid cheese, such as for an appetizer or omelet, use 1 lb (1 qt) dehydrated cheese and 1 cup water. For a semi-fluid cheese for sauces (better volume substitute), use 1 pound (1 qt) dehydrated cheese and 2 cups water.

(3) SUBSTITUTION:

Dehydrated Cheese +	WARM Water Added =	Rehydrated Cheese	OR	Fresh Cheese Equivalent
Semi-solid 6 oz (1-1/2 cups)	3/8 cup	1-1/8 cups		1 lb
3 lb (3 qt) 1-No. 10 cn	3 cups	2-1/4 qt		8 lb
Fluid 6 oz (1-1/2 cups)	3/4 cup	1-1/2 cups		1 lb
3 lb (3 qt) 1-No. 10 cn	1-1/2 qt	3 qt		8 lb

GUIDELINES FOR USING HERBS

The following information is provided as a guide in developing familiarity and creativity with using herbs. Start with a small amount, taste, then add more if necessary.

Herb	Appetizers Salad	Breads/Eggs Sauces/Cheese	Vegetables Pasta	Meat Poultry	Fish Shellfish
Basil	Green, Potato & Tomato Salads, Salad Dressing, Stewed Fruit	Breads, Fondue & Egg Dishes, Dips, Marinades, Sauces	Mushrooms, Tomatoes, Squash, Pasta, Bland Vegetables	Broiled, Roast Meat & Poultry Pies, Stews, Stuffing	Baked, Broiled & Poached Fish, Shellfish
Bay Leaf	Seafood Cocktail, Seafood Salad, Tomato Aspic, Stewed Fruit	Egg Dishes, Gravies, Marinades, Sauces	Dried Bean Dishes, Beets, Carrots, Onions, Potatoes, Rice, Squash	Corned Beef, Tongue Meat & Poultry Stews	Poached Fish, Shellfish Fish Stews

Guide to Cooking with Popular Herbs (continued)

Herb	Appetizers Salad	Breads/Eggs Sauces/Cheese	Vegetables Pasta	Meat Poultry	Fish Shellfish
Chives	Mixed Vegetables, Green, Potato & Tomato Salads, Salad Dressings	Egg & Cheese Dishes, Cream Cheese, Cottage Cheese, Gravies, Sauces	Hot Vegetables, Potatoes	Broiled Poultry, Rissoles, Poultry & Meat Pies, Stews, Casseroles	Baked Fish, Fish Casseroles, Fish Stews, Shellfish
Dill	Seafood Cocktail, Green, Potato & Tomato Salads, Salad Dressings	Breads, Egg & Cheese Dishes, Cream Cheese, Fish and Meat Sauces	Beans, Beets, Cabbage, Carrots, Cauliflower, Peas, Squash, Tomatoes	Beef, Veal Roasts, Lamb, Steaks, Chips, Stews, Roast & Creamed Poultry	Baked, Broiled, Poached & Stuffed Fish, Shellfish
Garlic	All Salads, Salad Dressings	Fondue Poultry Sauces, Fish and Meat Marinades	Beans, Eggplant, Potatoes, Rice, Tomatoes	Roast Meats, Meat & Poultry Pies, Hamburgers, Stews & Casseroles	Broiled Fish, Shellfish, Fish Stews, Casseroles
Marjoram	Seafood Cocktail, Green, Poultry & Seafood Salads	Breads, Cheese Spreads, Egg & Cheese Dishes, Gravies, Sauces	Carrots, Eggplant, Peas, Onions, Potatoes, Dried Bean Dishes, Spinach	Roast Meats & Poultry Meat & Poultry Pies, Stews & Casseroles	Baked, Broiled & Stuffed Fish, Shellfish

Guide to Cooking with Popular Herbs (continued)

Herb	Appetizers Salad	Breads/Eggs Sauces/Cheese	Vegetables Pasta	Meat Poultry	Fish Shellfish
Mustard	Fresh Green Salads, Prepared Meat, Macaroni & Potato Salads, Salad Dressing	Biscuits, Egg & Cheese Dishes, Sauces	Baked Beans, Cabbage, Eggplant, Squash, Dried Beans, Mushrooms, Pasta	Chops, Steaks, Ham, Pork, Poultry Cold Meats	Shellfish
Oregano	Green, Poultry & Seafood Salads	Breads, Egg & Cheese Dishes, Meat, Poultry & Vegetable Sauces	Artichokes, Cabbage, Eggplant, Squash, Dried Beans, Mushrooms, Pasta	Broiled, Roast Meats, Meat & Poultry Pies, Stews, Casseroles	Baked, Broiled & Poached Fish, Shellfish
Parsley	Green, Potato, Seafood & Vegetable Salads	Biscuits, Breads, Egg & Cheese Dishes, Gravies, Sauces	Asparagus, Beets, Eggplant, Squash, Dried Beans, Mushrooms, Pasta	Meat Loaf, Meat & Poultry Pies, Stews and Casseroles, Stuffing	Fish Stews, Stuffed Fish
Rosemary	Fruit Cocktail, Fruit & Green Salads	Biscuits, Egg Dishes, Herb Butter, Cream Cheese, Marinades, Sauces	Beans, Broccoli, Peas, Cauliflower, Mushrooms, Baked Potatoes, Parsnips	Roast Meat, Poultry & Meat Loaf, Meat & Poultry Pies, Stews & Casseroles, Stuffing	Stuffed Fish, Shellfish

A. GENERAL INFORMATION No. 30(2)

Guide to Cooking with Popular Herbs (continued)

Herb	Appetizers / Salad	Breads/Eggs / Sauces/Cheese	Vegetables / Pasta	Meat / Poultry	Fish / Shellfish
Sage		Breads, Fondue, Egg & Cheese Dishes, Spreads, Gravies, Sauces	Beans, Beets, Onions, Peas, Spinach, Squash, Tomatoes	Roast Meat, Poultry, Meat Loaf, Stews, Stuffing	Baked, Poached, & Stuffed Fish
Tarragon	Seafood Cocktail, Avocado Salads (all), Salad Dressings	Cheese Spreads, Marinades, Sauces, Egg Dishes	Asparagus, Beans, Beets, Carrots, Mushrooms, Peas, Squash, Spinach	Steaks, Poultry, Roast Meats, Casseroles & Stews	Baked, Broiled & Poached Fish, Shellfish
Thyme	Seafood Cocktail, Green, Poultry, Seafood & Vegetable Salads	Biscuits, Breads Egg & Cheese Dishes, Sauces, Spreads	Beets, Carrots, Mushrooms, Onions, Peas, Eggplant, Spinach, Potatoes	Roast Meat, Poultry & Meat Loaf, Meat & Poultry Pies, Stews & Casseroles	Baked, Broiled & Stuffed Fish, Shellfish, Fish Stews

HAZARD ANALYSIS CRITICAL CONTROL POINT
(HACCP)

HACCP System: A food safety system that identifies hazards and develops control points throughout the receiving, storage, preparation, service and holding of food. This system is designed to prevent foodborne illness.

- **Critical Control Point (CCP):** A point in a specific food service process where loss of control may result in an unacceptable health risk. Implementing a control measure at this point may eliminate or prevent the food safety hazard.
- **Critical Limits:** Elements such as time and temperature that must be adhered to in order to keep food safe. The Temperature Danger Zone is defined by the Food and Drug Administration's Food Code as 41° F. to 140° F.
- **Foodborne Illness:** An illness transmitted to humans through food. Any food may cause a foodborne illness, however *potentially hazardous foods* are responsible for most foodborne illnesses. Symptoms may include abdominal pain/cramps, nausea and vomiting.
- **Potentially Hazardous Food:** A food that is used as an ingredient in recipes or served alone that is capable of supporting the growth of organisms responsible for foodborne illness. Typical foods include high protein foods such as meat, fish, poultry, eggs and dairy products.

COOKING TEMPERATURES *These temperatures represent the minimum required temperature. The time represents the minimum amount of time the temperature must be maintained.*	
Eggs, Raw shell eggs	155° F. for 15 seconds
Eggs, Egg products, pasteurized	145° F. for 15 seconds
Poultry	165° F. for 15 seconds
Pork	145° F. for 15 seconds
Whole Beef Roasts and Corned Beef Roasts	145° F. for 3 minutes
Fish	145° F. for 15 seconds
Stuffed meat, fish, poultry or pasta, OR stuffings containing meat, fish or poultry	165° F. for 15 seconds
Meat or fish that has been reduced in size by methods such as chopping (i.e., beef cubes), grinding (i.e., ground beef, sausage), restructuring (i.e., formed roast beef, gyro meat), or a mixture of two or more meats (i.e., sausage made from two or more meats)	155° F. for 15 seconds
CCP: SERVING AND HOLDING (hot foods)	140° F.
COOLING *FDA recommends a cooled product temperature of 41° F. In order to achieve a cooled internal product temperature of 34-38° F., the temperature of the refrigerator must be lower than 41° F.*	Cooling from 140° F. to 70° F. should take no longer than 2 hours. Cooling from 70° F. to 41° F. should take no longer than 4 hours.

A. GENERAL INFORMATION No. 32 (1)

GUIDELINES FOR COMBI-OVENS

A combi-oven is a versatile piece of equipment that combines three modes of cooking in one oven: steam, circulated hot air or a combination of both. The combi mode is used to re-heat foods and to roast, bake and "oven fry." The steam mode is ideal for rapid cooking of vegetables and shellfish. The hot air mode operates as a normal convection oven for baking cookies, cakes and pastries. The combi mode decreases overall cooking times, reduces product shrinkage and eliminates flavor transfer when multiple items are cooked simultaneously.

OVEN MODES

COMBI MODE: Use to roast and braise meats, bake poultry and fish and reheat prepared foods. The combination of steam and hot air will improve yield and reduce overall cooking times. To **OVEN FRY,** use food items that are labeled "ovenable" by the manufacturer. Refer to cooking guidelines for oven frying individual items. Place items on perforated sheet pan in a single layer. DO NOT place excess amount of product on pan. A solid sheet pan may be placed under perforated pan to catch excess oils and eliminate smoke.

HOT AIR MODE: Use to bake cakes, cookies and breads and to roast and bake meats and poultry. The hot air mode circulates air in the same manner as a convection oven.

GUIDELINES FOR COMBI-OVENS (continued)

STEAMING MODE: Use to steam fresh, frozen or canned vegetables and shellfish. Use of the Combi-oven to steam foods can save time, labor, and help maintain appearance, and preserve nutrients normally lost by other cooking methods. The oven is ideal for steaming more than one type of vegetable at the same time without flavor transfer. Foods may be steamed in perforated or solid pans. Perforated pans are generally used, particularly for vegetables, unless the cooking liquid is retained or manufacturer's directions specify solid pans. Pans are normally filled no more than 2/3 full to allow steam to circulate for even cooking.

Steam temperature is preset at 212° F. The cooking time will vary depending on the type of food and the number of pans in the oven. The cooking time should include the time it requires to heat food up to cooking temperature, as well as steaming.

TEMPERATURE SETTING: At this time the AFRS recipes do not contain combi-oven information. Refer to the attached cooking guidelines for individual items or begin by using the recommended convection oven temperature noted on individual recipes. If food is cooked around the edges, but the center is still raw or not thoroughly cooked, or if there is too much color variation (some is normal), turn pan or reduce the heat by 10° F. to 15° F. and return food to the oven and continue cooking until done.

TIME SETTING: Follow the recommended convection cooking times on recipe cards. Check progress halfway through the cooking cycle since times will vary in the Combi mode with the quantity of food being cooked, the temperature, and the type of pan used.

GUIDELINES FOR COMBI-OVENS (continued)

MEAT PROBE: The meat probe measures a product core temperature during the cooking process.

FAN SPEED SETTING: See general operations notes below.

GENERAL OPERATION NOTES:

1. **OVEN RACKS:** Position oven racks for the number of pans and product to be cooked.
2. **WATER SUPPLY:** Verify water supply is on.
3. **SELECT COOKING MODE AND TEMPERATURE:** Turn oven on; SELECT the cooking mode. To cook in the combi or hot air mode, set thermostat to desired temperature. To cook in the steam mode, set thermostat to 200° F. The thermostat light will come on indicating oven temperature is below set point.
4. **PREHEAT:** Heat oven until thermostat light goes out indicating that the oven has reached the set temperature. The oven should preheat to 350° F. within 10 to 15 minutes.
5. **FAN SPEED:** If two-speed fan is available, adjust the fan to recommended speed noted on individual recipe card. NOTE: The Combi-oven is equipped with electric interlock, which energizes/de-energizes both the heating element and fan motor when the doors are closed and open. Therefore, the heating elements and fan will not operate with the doors open, only when closed.

GUIDELINES FOR COMBI-OVENS (continued)

6. **MEAT PROBE:** Insert the meat probe in the thickest section of the product. NOTE: The tip of the probe should not be placed near bone or fat. This will result in inaccurate temperature readings. Turn the meat probe switch on and set the desired core temperature by using the up or down arrows. Press the set button to store the set point temperature. Set the timer to the STAY ON position. When the selected core temperature is reached the buzzer will sound and the oven automatically turns off.
7. **CLEANING AND MAINTENANCE:** Refer to the manufacturer's operating manual for cleaning and maintenance instructions. NOTE: Wipe out all spills as soon as they occur for ease of cleaning.

COMBI-OVEN COOKING GUIDELINES

Food	Cook Mode	Recommended Temperature	Time
MEATS			
Steak	Hot Air	400	See Recipe No. L 007 00
Bacon, oven fried	Hot Air	325	25-30 minutes
Roasts, boneless			
Beef	Combi	325	1 hr 45 minutes
Pork	Combi	325	2 to 2-1/2 hours
Spareribs	Combi	350	1 to 1-1/2 hours
Meatloaf	Combi	300	1 hour

GUIDELINES FOR COMBI-OVENS (continued)
COMBI-OVEN COOKING GUIDELINES

Food	Cook Mode	Recommended Temperature	Time
POULTRY			
Turkey, boneless	Combi	325	2 to 2-1/2 hours
Chicken, pieces (with bone)	Combi	350	20-30 minutes
FISH			
Fish, baked	Combi	325	10-20 minutes
Shrimp, raw, frozen	Steam	Preset	3-5 minutes
MISCELLANEOUS			
Casserole type dishes			
Macaroni & cheese	Combi	325	15-20 minutes
Lasagna	Combi	300	40-50 minutes
BREADS			
Breads, yeast	Hot Air	375	30 minutes
Coffee cakes	Hot Air	325	15 minutes
Muffins	Hot Air	350	30 minutes
Rolls Yeast	Hot Air	350	10-15 minutes
Sweet rolls	Hot Air	325	15 minutes

GUIDELINES FOR COMBI-OVENS (continued)

COMBI-OVEN COOKING GUIDELINES

Food	Cook Mode	Recommended Temperature	Time
EGGS			
Hard Cooked Eggs	Steam	Preset	12 minutes
CAKES			
Angel Food	Hot Air	300	30-35 minutes
Layer	Hot Air	300	25-35 minutes
Loaf	Hot Air	325	65-75 minutes
Sheet	Hot Air	300-325	25-35 minutes
DESSERTS			
Brownies	Hot Air	325	25-30 minutes
Cookies	Hot air	325	12-15 minutes
Pies, Fruit	Hot air	375	25 minutes
VEGETABLES			
Frozen	Steam	Preset	12-15 minutes
Canned	Steam	Preset	10-12 minutes
Fresh*	Steam	Preset	*See individual recipe cards

GUIDELINES FOR COMBI-OVENS (continued)

COMBI-OVEN COOKING GUIDELINES

Food	Cook Mode	Recommended Temperature	Time
EGGS			
Hard Cooked Eggs	Steam	Preset	12 minutes
CAKES			
Angel Food	Hot Air	300	30-35 minutes
Layer	Hot Air	300	25-35 minutes
Loaf	Hot Air	325	65-75 minutes
Sheet	Hot Air	300-325	25-35 minutes
DESSERTS			
Brownies	Hot Air	325	25-30 minutes
Cookies	Hot air	325	12-15 minutes
Pies, Fruit	Hot air	375	25 minutes
VEGETABLES			
Frozen	Steam	Preset	12-15 minutes
Canned	Steam	Preset	10-12 minutes
Fresh*	Steam	Preset	*See individual recipe cards

GUIDELINES FOR COMBI-OVENS (continued)

COMBI-OVEN COOKING GUIDELINES

Food	Cook Mode	Recommended Temperature	Time
OVEN FRYING			
French Fries	Combi	400	7-9 minutes
Fish Portions	Combi	400	10-12 minutes
Shrimp, Battered	Combi	400	7-8 minutes
Chicken Pieces	Combi	400	20 minutes
Chicken Nuggets	Combi	400	8-14 minutes
Onion Rings	Combi	400	6-8 minutes
Jalapeno Popper	Combi	400	9-12 minutes
Egg rolls	Combi	400	12-18 minutes

GUIDELINES FOR COMBI-OVENS (continued)

COMBI-OVEN COOKING GUIDELINES

Food	Cook Mode	Recommended Temperature	Time
OVEN FRYING			
French Fries	Combi	400	7-9 minutes
Fish Portions	Combi	400	10-12 minutes
Shrimp, Battered	Combi	400	7-8 minutes
Chicken Pieces	Combi	400	20 minutes
Chicken Nuggets	Combi	400	8-14 minutes
Onion Rings	Combi	400	6-8 minutes
Jalapeno Popper	Combi	400	9-12 minutes
Egg rolls	Combi	400	12-18 minutes

GUIDELINES FOR SKITTLE

A skittle is a multipurpose piece of equipment that can be used as a pressureless steamer, braising pan or griddle. The griddle mode is ideal for cooking steaks, sandwiches, eggs, pancakes, breakfast meats and potatoes. The steam mode may be used to cook vegetables, seafood, rice and pasta. The braising mode is used for slow moist-heat cooking of meats, poultry and vegetables.

TO OPERATE AS A STEAMER:

1. Add 5 gallons (2"- 3") of water to the skittle using the spray hose.
2. Position steaming racks for the number of pans and product to be cooked.
3. Close the lid and the steam vent.
4. Set the thermostat at 350° Fahrenheit and allow 6-8 minutes to preheat. The skittle is ready when the heater power light goes out.
5. When the skittle is preheated, raise the lid to the top of the steamer racks and place food pans in the racks and close the lid. **(NOTE: To retain maximum steam, do not raise the lid beyond steamer racks. The lid should be kept in a horizontal position)**
6. If steam escapes from the closed lid, open the rear vent until excess is released.

The skittle is ideal for steaming more than one type of vegetable at the same time without flavor transfer. Foods may be steamed in perforated or solid pans. Perforated pans are normally used, particularly for vegetables, unless the cooking liquid is retained or manufacturer's directions specify solid pans. Pans should not be filled more than 2/3 to the top to allow steam to circulate for even cooking.

A. GENERAL INFORMATION No. 34 (1)

Cooking times will vary depending on the type of food and the number of pans used. The cooking time should include the time it requires to heat food up to cook temperature, as well as steaming. Be sure to record the most successful steaming times on individual recipe cards for future reference.

TO OPERATE AS A BRAISING PAN:
    1. Set the thermostat at 375° Fahrenheit and allow 6-8 minutes to preheat. The skittle is ready when the heater power light goes out. Brown food according to individual AFRS recipe card instructions.
    2. Lower temperature to 325° Fahrenheit and add cooking liquid. Lower hood and cook according to individual recipe card instructions.
    3. To remove liquid, tilt the pan 10° using the tilt handle and drain the liquid through the drain valve into a food pan.

The Skittle may be used for braising pot roast, Swiss steaks, spareribs, stews and for preparing gravy, soups and sauces. Cooking times will vary according to individual foods and amount prepared.

TO OPERATE AS A GRIDDLE:
    1. Set the thermostat to 350° Fahrenheit and allow 6-8 minutes to preheat. The griddle is ready when the heater power light goes out.
    2. Raise the lid and cook foods according to individual AFRS guideline cards.
    3. To drain any accumulated grease, place a #10 can into the can holder attached to the drain valve. Tilt the pan 10° using the tilt handle and allow grease to drain into the can. The griddle can be used to cook hamburgers, steak, sandwiches, eggs, pancakes, breakfast meats and potatoes. Heat is distributed evenly over the entire pan surface ensuring food products cook uniformly.

GUIDELINES FOR SKITTLE (continued)

GENERAL OPERATION NOTES:

1. **STEAMING MODE:** The recommended thermostat temperature for steaming is 350° Fahrenheit. Higher temperatures may be used but water will evaporate quickly and cooking time will not be decreased.

2. **WATER SUPPLY:** The easiest way to fill the skittle with water is with the attached flexible spray hose.

3. **SELECT COOKING TEMPERATURE**: SELECT desired cooking temperature according to cook mode or individual recipe cards. The thermostat light will come on indicating oven temperature is below set point.

4. **PREHEAT:** Heat Skittle until thermostat light goes out indicating that the unit has reached the set temperature. The Skittle should preheat to 350° F. within 6 to 8 minutes. (Note: Lower the lid for faster preheating.)

5. **CLEANING AND MAINTENANCE:** Remove food waste. Fill the pan with warm water using the spray hose. Add mild detergent and scrub with a nylon scrub pad if necessary. Tilt the pan 10° using the tilt handle and allow water to drain into container placed directly under the drain valve. Rinse with clean water and drain again. Refer to the manufacturer's operating manual for cleaning and maintenance instructions.

A. GENERAL INFORMATION No. 35

GUIDELINES FOR USE OF CONVENIENCE PREPARED FOODS

Convenience prepared foods reduce labor since they only require heating. Specific cooking instructions should be located on each advanced foods package. Items to be considered when using convenience prepared foods are cooking times, nutrient content and serving size. Cooking times, nutrient content and serving size will vary among manufacturers for identical food items, therefore, in order to maintain the quality of these convenience prepared foods, instructions must be read and followed every time a convenience prepared food is utilized.

B. APPETIZERS No. 0

INDEX

Card No.	
B 001 00	Cranberry and Orange Juice Cocktail
B 001 01	Cranberry and Apple Juice Cocktail
B 002 00	Chinese Egg Rolls (Baked)
B 002 01	Chinese Egg Rolls (Fried)
B 002 02	Philippine Style Egg Rolls (Baked)
B 002 03	Philippine Style Egg Rolls (Fried)
B 003 00	Tomato Juice Cocktail
B 003 01	Vegetable Juice Cocktail
B 003 02	Spicy Tomato Juice Cocktail
B 004 00	Shrimp Cocktail
B 004 01	Spiced Shrimp
B 005 00	Pizza Treats

APPETIZERS No.B 001 00

CRANBERRY AND ORANGE JUICE COCKTAIL

Yield 100   **Portion** 1/2 Cup

Calories	Carbohydrates	Protein	Fat	Cholesterol	Sodium	Calcium
69 cal	17 g	0 g	0 g	0 mg	3 mg	9 mg

Ingredient	Weight	Measure	Issue
CRANBERRY JUICE COCKTAIL	14-7/8 lbs	1 gal 2-2/3 qts	
JUICE,ORANGE,FROZEN,CONCENTRATE,3/1,THAWED	4-1/8 lbs	1 qts 2-5/8 cup	
WATER,COLD	9-7/8 lbs	1 gal 3/4 qts	

Method

1 Combine juices and water; stir until blended. Cover and refrigerate at 41 F. or lower.

APPETIZERS No.B 001 01

CRANBERRY AND APPLE JUICE COCKTAIL

Yield 100   Portion   1/2 Cup

Calories	Carbohydrates	Protein	Fat	Cholesterol	Sodium	Calcium
70 cal	17 g	0 g	0 g	0 mg	7 mg	7 mg

Ingredient | Weight | Measure | Issue

CRANBERRY JUICE COCKTAIL — 14-7/8 lbs — 1 gal 2-2/3 qts
JUICE,APPLE,FROZEN,CONCENTRATE,3/1,THAWED — 4-1/8 lbs — 1 qts 2-5/8 cup
WATER,COLD — 9-7/8 lbs — 1 gal 3/4 qts

Method

1 Combine juices add water; stir until blended. Cover and refrigerate at 41 F. or lower.

APPETIZERS No.B 002 00

CHINESE EGG ROLLS (BAKED)

Yield 100 **Portion** 1 Egg Roll

Calories	Carbohydrates	Protein	Fat	Cholesterol	Sodium	Calcium
140 cal	13 g	10 g	5 g	50 mg	247 mg	22 mg

Ingredient Weight Measure Issue

EGG ROLLS,CHINESE,FROZEN 18-3/4 lbs

Method

1. Place 50 egg rolls on each sheet pan.
2. Using a convection oven, bake at 350 F. for 20 to 25 minutes or until brown on high fan, closed vent. CCP: Internal temperature must reach 145 F. or higher for 15 seconds. Hold for service at 140 F. or higher.

APPETIZERS No.B 002 01

CHINESE EGG ROLLS (FRIED)

Yield 100   **Portion** 1 Egg Roll

Calories	Carbohydrates	Protein	Fat	Cholesterol	Sodium	Calcium
180 cal	13 g	10 g	10 g	50 mg	247 mg	22 mg

Ingredient   Weight   Measure   Issue

EGG ROLLS,CHINESE,FROZEN   18-3/4 lbs

Method

1. Fry egg rolls in deep fat at 350 F. for 7 minutes or until golden brown and heated through. DO NOT OVERCOOK. Egg rolls will rise to the surface when cooked. CCP: Internal temperature must reach 145 F. or higher for 15 seconds.
2. Drain well in basket or on absorbent paper. CCP: Hold for service at 140 F. or higher.

APPETIZERS No.B 002 02

PHILIPPINE STYLE EGG ROLLS (BAKED)

Yield 100 **Portion** 1 Egg Roll

Calories	Carbohydrates	Protein	Fat	Cholesterol	Sodium	Calcium
93 cal	8 g	7 g	4 g	33 mg	165 mg	15 mg

Ingredient	**Weight**	**Measure**	**Issue**
EGG ROLLS,PHILIPPINE STYLE,FROZEN	12-1/2 lbs		

Method

1. Place 50 egg rolls per sheet pan.
2. Using a convection oven, bake at 350 F. for 10 to 15 minutes or until heated through on high fan, closed vent. CCP: Internal temperature must reach 145 F. or higher for 15 seconds. Hold for service at 140 F. or higher.

APPETIZERS No.B 002 03

PHILIPPINE STYLE EGG ROLLS (FRIED)

Yield 100   **Portion** 1 Egg Roll

Calories	Carbohydrates	Protein	Fat	Cholesterol	Sodium	Calcium
133 cal	8 g	7 g	8 g	33 mg	165 mg	15 mg

Ingredient   Weight   Measure   Issue

EGG ROLLS,PHILIPPINE STYLE,FROZEN   12-1/2 lbs

Method

1. Fry Philippine egg rolls in deep fat at 350 F. for 4 to 5 minutes, or until golden brown and heated through. DO NOT OVERCOOK.
2. Drain well in basket or on absorbent paper. CCP: Hold for service at 140 F. or higher.

APPETIZERS No.B 003 00

TOMATO JUICE COCKTAIL

Yield 100   **Portion** 1/2 Cup

Calories	Carbohydrates	Protein	Fat	Cholesterol	Sodium	Calcium
22 cal	6 g	1 g	0 g	0 mg	457 mg	12 mg

Ingredient   **Weight**   **Measure**   **Issue**

JUICE,TOMATO,CANNED   27-7/8 lbs   3 gal 1 qts
JUICE,LEMON   6-1/2 oz   3/4 cup

Method

1. Combine tomato juice and lemon juice; cover; refrigerate at 41 F. or lower for several hours or overnight.
2. Stir well before serving.

APPETIZERS No.B 003 01

VEGETABLE JUICE COCKTAIL

Yield 100   Portion   1/2 Cup

Calories	Carbohydrates	Protein	Fat	Cholesterol	Sodium	Calcium
24 cal	6 g	1 g	0 g	0 mg	340 mg	14 mg

Ingredient   **Weight**   **Measure**   **Issue**
JUICE,VEGETABLE,CANNED   27-3/4 lbs   3 gal 1 qts
JUICE,LEMON   6-1/2 oz   3/4 cup

Method

1 Combine canned vegetable juice and lemon juice; cover; refrigerate at 41 F. or lower for several hours or overnight.
2 Stir well before serving.

APPETIZERS No.B 003 02

SPICY TOMATO JUICE COCKTAIL

Yield 100   **Portion** 1/2 Cup

Calories	Carbohydrates	Protein	Fat	Cholesterol	Sodium	Calcium
22 cal	6 g	1 g	0 g	0 mg	467 mg	12 mg

Ingredient   Weight   Measure   Issue

JUICE,TOMATO,CANNED   27-7/8 lbs   3 gal 1 qts
SAUCE,TABASCO   6 oz   3/4 cup
JUICE,LEMON   6-1/2 oz   3/4 cup

Method

1 Combine tomato juice, hot sauce and lemon juice; cover; refrigerate at 41 F. or lower for several hours or overnight.
2 Stir well before serving.

APPETIZERS No. B 004 00

SHRIMP COCKTAIL

Yield 100   Portion   4 Shrimp

Calories	Carbohydrates	Protein	Fat	Cholesterol	Sodium	Calcium
83 cal	12 g	10 g	1 g	84 mg	480 mg	43 mg

Ingredient | Weight | Measure | Issue

Ingredient	Weight	Measure	Issue
SHRIMP,FROZEN,RAW,PEELED,DEVEINED	12 lbs		
WATER,BOILING	6-1/4 lbs	3 qts	
SEAFOOD COCKTAIL SAUCE		3 qts 1-3/8 cup	
LETTUCE,ICEBERG,FRESH	4 lbs		4-1/3 lbs
LEMONS,FRESH	5-1/8 lbs	13 each	

Method

1. Place shrimp in boiling water and cover. Return to a boil; uncover; reduce heat; simmer 2 to 3 minutes. CCP: Internal temperature must reach 145 F. or higher for 15 seconds. DO NOT OVERCOOK. Drain immediately.
2. Place shrimp in single layer on pans. CCP: Refrigerate at 41 F. or lower for use in Step 5.
3. Prepare 1 recipe Seafood Cocktail Sauce, Recipe No. O 011 00. Cover; refrigerate for use in Step 6.
4. Line individual serving dishes with lettuce.
5. Arrange 4 shrimp on lettuce in each dish.
6. Place 2 tablespoons of sauce in each souffle cup. Serve shrimp with 1 lemon wedge. Cut 8 wedges per lemon. CCP: Hold for service at 41 F. or lower.

Notes

1. In Step 3, prepared seafood cocktail sauce may be used.

APPETIZERS No. B 004 01

SPICED SHRIMP

Yield 100   **Portion** 4 Shrimp

Calories	Carbohydrates	Protein	Fat	Cholesterol	Sodium	Calcium
60 cal	5 g	10 g	1 g	84 mg	100 mg	56 mg

Ingredient	Weight	Measure	Issue
SHRIMP,FROZEN,RAW,PEELED,DEVEINED	12 lbs		
WATER,BOILING	2-1/8 lbs	1 qts	
VINEGAR,DISTILLED	4-1/8 lbs	2 qts	
PEPPER,RED,GROUND	1-1/8 oz	1/4 cup 2-1/3 tbsp	
MUSTARD,DRY	2-3/8 oz	1/4 cup 2-1/3 tbsp	
CELERY SEED	7/8 oz	1/4 cup 1/3 tbsp	
PAPRIKA,GROUND	1/2 oz	2 tbsp	
GINGER,GROUND	1/4 oz	1 tbsp	
MACE,GROUND	1/4 oz	1 tbsp	
CINNAMON,GROUND	1/4 oz	1 tbsp	
CLOVES,GROUND	1/8 oz	1/3 tsp	
BAY LEAF,WHOLE,DRIED	3/8 oz	12 lf	
LETTUCE,FRESH,LEAF,RED	4 lbs	2 gal 1/8 qts	6-1/4 lbs
LEMONS,FRESH	5-1/8 lbs	13 each	

Method

1. Place shrimp in boiling water, add vinegar and spices, cover; return to a boil. Uncover; reduce heat; simmer 2 to 3 minutes. CCP: Internal temperature must reach 145 F. or higher for 15 seconds. DO NOT OVERCOOK. Drain immediately.
2. Place shrimp in single layer on pans. CCP: Refrigerate at 41 F. or lower for use in Step 5.
3. Line individual serving dishes with lettuce.
4. Arrange 4 shrimp on lettuce in each dish. CCP: Hold for service at 41 F. or lower.
5. Serve shrimp with 1 lemon wedge. Cut 8 wedges per lemon.

Notes

1. In Step 3, prepared seafood cocktail sauce may be used.

APPETIZERS No.B 005 00

PIZZA TREATS

Yield 100   Portion 1 Slice

Calories	Carbohydrates	Protein	Fat	Cholesterol	Sodium	Calcium
147 cal	17 g	8 g	5 g	10 mg	322 mg	158 mg

Ingredient	Weight	Measure	Issue
CHEESE,PIZZA BLEND,SHREDDED	4 lbs	1 gal	
TOMATO PASTE,CANNED	1 lbs	1-3/4 cup	
OIL,SALAD	3-7/8 oz	1/2 cup	
OLIVES,RIPE,PITTED,SLICED,DRAINED	7-1/8 oz	1-1/2 cup	
ONIONS,FRESH,CHOPPED	1 lbs	2-5/8 cup	1 lbs
PEPPERS,GREEN,FRESH,CHOPPED	11-7/8 oz	2-1/4 cup	14-3/8 oz
BREAD,FRENCH,SLICED 1/2 INCH	6-1/4 lbs	100 sl	

Method

1. Combine cheese, tomato paste, salad oil, olives, onions and peppers. Blend well.
2. Spread 3 tablespoons of mixture on each slice of bread.
3. Place on ungreased pans. Using a convection oven, bake at 350 F. 5 minutes or until cheese is melted on low fan, open vent.

C-G BEVERAGES No. 1

GUIDELINES FOR BREWING COFFEE

1. Measure or weigh quantities of water and coffee carefully. Prepare only in amounts necessary to maintain continuous service. Urn coffee held 1 hour or longer and automatic coffee maker coffee held 30 minutes or longer deteriorates in flavor and loses its aroma.
2. Use the proportion of 3/4 pound of coffee to 2-3/4 gallons of water for a standard strength brew. 1 lb 14 oz of coffee and 6-3/4 gallons of freshly drawn boiling water will yield approximately 100 (8 ounce) servings.
3. Ingredients for a good coffee brew are fresh coffee and fresh boiling water. Water that has been boiled a long time will have a flat taste which will affect the brew.
4. For an ideal brew, boiling water should pass through coffee within 4 to 6 minutes.
5. Keep equipment clean. Clean immediately after each use to prevent rancidity.
6. Urns and urn baskets should be washed with hot water and special urn cleaner or baking soda. (DO NOT use soap or detergent powder.) Rinse with clear water. When not in use, leave 1 or 2 gallons of clear water in urn. Drain before making coffee.
7. When using new urn bags: A new urn bag should be thoroughly rinsed in hot water before using. After using, urn bags should be thoroughly rinsed in clear, hot water; keep submerged in cold water until next use.
8. Faucets and glass gauges should be cleaned often with gauge brushes, not water, and urn cleaner or baking soda. Rinse with clear water. Caps on faucets and gauges are removable to permit cleaning.
9. NOTE: For a stronger brewed cup of coffee, use the proportion of 2 lb 8 oz coffee to 6-3/4 gal water.

C-G. BEVERAGES No. 4

GUIDELINES FOR COFFEE URN CAPACITIES

Urn Capacity (Gallons)	Coffee, roasted, ground		Number of Cups	
	Weights	Measures	5 oz	8 oz
1	4-3/4 oz	1-3/8 cups	25	16
1-1/2	7-1/4 oz	2-1/4 cups	38	24
2	9-3/4 oz	3 cups	51	32
3	13-1/2 oz	4-1/8 cups	76	48
4	1 lb 2 oz	5-1/2 cups	102	64
6	1 lb 11 oz	8-1/3 cups	153	96
8	2-1/4 lb	2-3/4 qt	204	128

NOTE: 1 lb 14 oz (2 1/4 qt) roasted and ground coffee and 6 3/4 water will yield 100-8 ounce portions or 6 1/4 gal coffee.

C. BEVERAGES No. 0

INDEX

Card No. .... Card No.

C 001 00	Hot Cocoa	C 010 00	Orangeade
C 001 01	Hot Whipped Cocoa		
C 002 00	Coffee (Instant)		
C 003 00	Coffee (Automatic Coffee Maker)		
C 004 00	Hot Tea		
C 005 00	Coffee (Automatic Urn)		
C 005 01	Coffee (Manual Urn)		
C 006 00	Fruit Punch		
C 006 01	Lime Lemon Punch		
C 006 02	Cherry-Ade		
C 006 03	Grape-Ade		
C 006 04	Lemon-Ade		
C 006 05	Orange-Ade		
C 006 06	Strawberry-Ade		
C 007 00	Orange and Pineapple Juice Cocktail		
C 007 01	Grapefruit and Pineapple Juice Cocktail		
C 008 00	Lemonade		
C 008 01	Limeade		
C 009 00	Iced Tea (Instant)		
C 009 01	Iced Tea (Instant For Dispenser)		
C 009 02	Iced Tea (Instant w/Lemon and Sugar for Dispenser)		

BEVERAGES No.C 001 00

HOT COCOA

Yield 100　　　　　　　　　　　　　　　　　**Portion**　1 Cup

Calories	Carbohydrates	Protein	Fat	Cholesterol	Sodium	Calcium
112 cal	24 g	4 g	0 g	2 mg	112 mg	137 mg

Ingredient	Weight	Measure	Issue
COCOA	12-1/8 oz	1 qts	
SALT	1/3 oz	1/4 tsp	
SUGAR,GRANULATED	3-1/2 lbs	2 qts	
WATER,COLD	3-1/8 lbs	1 qts 2 cup	
MILK,NONFAT,DRY	2-1/4 lbs	3 qts 3 cup	
WATER,WARM	43-7/8 lbs	5 gal 1 qts	
EXTRACT,VANILLA	7/8 oz	2 tbsp	

Method

1 Combine cocoa, salt, and sugar.
2 Add water; mix. Heat to boiling point; reduce heat and simmer 5 minutes.
3 Reconstitute milk; add to cocoa syrup, stirring constantly. Add vanilla (optional); mix until well blended.
4 Heat to just below boiling. DO NOT BOIL.
5 Serve hot.

Notes

1 Cocoa may be served with miniature marshmallows.

BEVERAGES No.C 001 01

HOT WHIPPED COCOA

Yield 100   **Portion** 3/4 Cup

Calories	Carbohydrates	Protein	Fat	Cholesterol	Sodium	Calcium
62 cal	15 g	5 g	4 g	0 mg	6 mg	35 mg

Ingredient	**Weight**	**Measure**	**Issue**
COCOA	6 lbs	1 gal 3-7/8 qts	

Method

1. Place Cocoa Beverage Powder in dispenser container. Follow manufacturer's directions for preparation and dispensing of cocoa.
2. Serve hot.

Notes

1. Cocoa may be served with miniature marshmallows. 8 ounce marshmallows will yield 4 to 5 marshmallows per serving of cocoa.

BEVERAGES No.C 002 00

COFFEE (INSTANT)

Yield 100   Portion 1 Cup

Calories	Carbohydrates	Protein	Fat	Cholesterol	Sodium	Calcium
5 cal	1 g	0 g	0 g	0 mg	8 mg	8 mg

Ingredient   Weight   Measure   Issue
COFFEE,INSTANT,FREEZE DRIED   8 oz   2-5/8 cup
WATER,BOILING   52-1/4 lbs   6 gal 1 qts

Method
1 Add coffee to water. Stir until dissolved.
2 Keep hot. DO NOT BOIL.

Notes
1 Omit Steps 1 and 2 if using an instant coffee dispenser. Place 8 ounces of freeze-dried instant coffee in dispenser jar. Follow dispenser manufacturer's directions for preparation and dispensing of coffee.

BEVERAGES No.C 003 00

COFFEE (AUTOMATIC COFFEE MAKER)

Yield 100   **Portion** 8 Ounces

Calories	Carbohydrates	Protein	Fat	Cholesterol	Sodium	Calcium
11 cal	2 g	1 g	0 g	0 mg	2 mg	6 mg

Ingredient	**Weight**	**Measure**	**Issue**
COFFEE,ROASTED,GROUND	1 lbs	2 qts 2-1/2 cup	

Method

1. Place filter paper in brewing funnel.
2. Spread coffee evenly in filter.
3. Slide funnel into brewer; place empty pot on heating element.
4. Press switch to start automatic brewing cycle.
5. Let water drip through completely; discard grounds.

Notes

1. Serve coffee within 30 minutes.
2. Check water temperature. The water filtered through the grounds must be 200 F. to ensure that the coffee from the brewing chamber will be at least 190 F.
3. For 1 pot: Use 2-1/2 ounces or 3/4 cup roasted, ground coffee. One pot makes 11 5-ounce portions or 7 8-ounce portions.
4. Coffee Maker Production Rates: 2 to 3 minutes to reach water temperature. 4 minutes average brewing time. 1 pot in average of 7 minutes. 8 pots per hour.
5. For 5-ounce portions: In Step 1, use 1-1/2 pound or 1-7/8 quarts roasted, ground coffee to make 10 pots.
6. For stronger brew, use 2-13/16 pounds or 3-1/2 quarts roasted, ground coffee for 8-ounce portion; for 5-ounce portion, use 2 pounds or 2-1/2 quarts roasted, ground coffee.

BEVERAGES No.C 004 00

HOT TEA

Yield 100   Portion 1 Cup

Calories	Carbohydrates	Protein	Fat	Cholesterol	Sodium	Calcium
0 cal	0 g	0 g	0 g	0 mg	7 mg	5 mg

Ingredient   Weight   Measure   Issue

TEA,BLACK,LOOSE   8 oz   1-1/4 cup
WATER,BOILING   54-1/3 lbs   6 gal 2 qts

Method

1 Place tea in a cloth bag large enough to hold three times the amount.
2 Tie top of bag with cord long enough to facilitate removal; tie cord to handle of urn or kettle.
3 Place tea bag in urn or kettle.
4 Boil water. Pour water over tea bag. Cover. Allow to steep 3 to 5 minutes. Do not agitate or stir.
5 Remove tea bag.
6 Cover; keep hot, but do not boil.

Notes

1 If loose tea, not enclosed in a cloth bag, is placed in the urn or kettle, strain tea after it has steeped 5 minutes.
2 Tea must never be boiled as this produces a bitter flavor.
3 Schedule preparation so not more than 15 minutes will elapse between preparation and service; hold tea at temperatures 175 F. to 185 F.
4 For 5-ounce portions, use 1-3/4 cups tea, loose and 4 gallons of water.
5 100 8-ounce individual tea bags may be used. Place on serving line for self-service.

BEVERAGES No.C 005 00

COFFEE (AUTOMATIC URN)

Yield 100 **Portion** 1 Cup

Calories	Carbohydrates	Protein	Fat	Cholesterol	Sodium	Calcium
9 cal	2 g	0 g	0 g	0 mg	1 mg	5 mg

Ingredient	**Weight**	**Measure**	**Issue**
COFFEE,ROASTED,GROUND	13-3/4 oz	2 qts 1 cup	

Method

1. Make sure water level in urn liner does not exceed 2 inches from top or is lower than the center of glass water gauge.
2. Push HEAT SELECTOR switch to BREW position.
3. Rinse urn liner by placing spray arm over top of urn. Push START button. Push STOP button after 30 seconds and drain liner.
4. Set timer for desired amount of water, 3 quarts of water for every minute; weigh coffee and spread evenly in filter paper. See Guidelines for Coffee Urn Capacities.
5. Place wire basket containing filter paper and coffee in top of urn. Cover and position spray arm through hole in cover.
6. When BREW TEMPERATURE light is on, press START button.
7. Five minutes after brewing is completed, turn heat selector to HOLD position. Discard grounds and filter paper; rinse wire basket.
8. When empty, rinse out urn.

Notes

1. Always thoroughly drain leftover coffee from urn; do not make fresh coffee on top of old.
2. Never operate the urn without water. Damage to the heating elements and/or the thermostat control may result.
3. For a 5-ounce portion, use 5-1/2 cups roasted, ground coffee per 100 portions in Step 4.
4. Cleaning after each batch of coffee should be a regular routine. Coffee urns should have a special cleaning twice a week. See the operating manual for cleaning instructions.

BEVERAGES No.C 005 01

COFFEE (MANUAL URN)

Yield 100   **Portion** 1 Cup

Calories	Carbohydrates	Protein	Fat	Cholesterol	Sodium	Calcium
9 cal	2 g	0 g	0 g	0 mg	1 mg	5 mg

Ingredient   **Weight**   **Measure**   **Issue**

COFFEE,ROASTED,GROUND   13-3/4 oz   2 qts 1 cup

Method

1. Fill boiler with water to desired level. See Guidelines for Coffee Urn Capacities. Turn on heat.
2. Spread ground coffee evenly in urn bag or filter paper in wire basket; set in top of urn. Close urn cover.
3. When boiler water reaches a vigorous boil, open blow-over valve and spray water over coffee for 3 to 4 minutes. Close blow-over valve. Remove and discard grounds.
4. If urn has no agitation system, re-pour about 1/3 of the coffee directly back into boiler. Rinse urn bag and store in cold water.
5. Gradually replenish water no more than 1 gallon at a time whenever gauge shows less than half full.

Notes

1. 1-7/8 pound or 2-1/4 quarts roasted and ground coffee and 6-3/4 gallon water will yield 100 8-ounce portions or 6-1/4 gallon coffee.

BEVERAGES No.C 006 00

FRUIT PUNCH

Yield 100　　　　　　　　　　　　　　　　　Portion　1-1/4 Cups

Calories	Carbohydrates	Protein	Fat	Cholesterol	Sodium	Calcium
117 cal	30 g	0 g	0 g	0 mg	10 mg	15 mg

Ingredient	Weight	Measure	Issue
SUGAR,GRANULATED	4-1/4 lbs	2 qts 1-5/8 cup	
WATER	12-1/2 lbs	1 gal 2 qts	
JUICE,GRAPEFRUIT,CONCENTRATE,FROZEN	3-2/3 lbs	1 qts 2 cup	
JUICE,LEMON	1-1/8 lbs	2 cup	
JUICE,PINEAPPLE,CANNED,UNSWEETENED	6-5/8 lbs	3 qts	
WATER,COLD	33-1/2 lbs	4 gal	
ICE CUBES	9-5/8 lbs	3 gal	

Method

1　Dissolve sugar in water. Cool.
2　Add juices and water to sugar solution. Mix thoroughly. Cover and refrigerate.
3　Add ice just before serving.

Notes

1　In Step 2, 1-1/2 gallons of canned grapefruit juice may be used. Reduce water to 2-3/4 gallons per 100 servings.
2　In Step 2, 2 quarts of fresh lemon juice may be used. Reduce water to 3-1/2 gallon per 100 servings.

BEVERAGES No.C 006 01

LIME LEMON PUNCH

Yield 100   Portion 1-1/4 Cups

Calories	Carbohydrates	Protein	Fat	Cholesterol	Sodium	Calcium
130 cal	34 g	0 g	0 g	0 mg	14 mg	10 mg

Ingredient	Weight	Measure	Issue
SUGAR,GRANULATED	7 lbs	1 gal	
WATER	12-1/2 lbs	1 gal 2 qts	
JUICE,LEMON	1-1/8 lbs	2 cup	
JUICE,LIME	5-7/8 lbs	2 qts 3-3/4 cup	
WATER	39-3/4 lbs	4 gal 3 qts	
FOOD COLOR,GREEN	1/2 oz	1 tbsp	
ICE CUBES	9-5/8 lbs	3 gal	

Method

1 Dissolve sugar in water. Cool.
2 Add juices, food coloring, and water to sugar solution. Mix thoroughly. Cover and refrigerate.
3 Add ice just before serving.

Notes

1 In Step 2, 2 quarts of fresh lemon juice may be used. Reduce water to 3-1/2 gallon per 100 servings.

BEVERAGES No.C 007 00

ORANGE AND PINEAPPLE JUICE COCKTAIL

Yield 100   **Portion** 1/2 Cup

Calories	Carbohydrates	Protein	Fat	Cholesterol	Sodium	Calcium
66 cal	16 g	1 g	0 g	0 mg	3 mg	17 mg

Ingredient	**Weight**	**Measure**	**Issue**
JUICE,ORANGE	15-3/8 lbs	1 gal 3 qts	
JUICE,PINEAPPLE,CANNED,UNSWEETENED	14-1/3 lbs	1 gal 2-1/2 qts	
ICE CUBES	4 lbs	1 gal 1 qts	

Method

1. Combine orange and pineapple juices; stir.
2. Add ice just before serving.

BEVERAGES No.C 007 01

GRAPEFRUIT AND PINEAPPLE JUICE COCKTAIL

Yield 100 **Portion** 1/2 Cup

Calories	Carbohydrates	Protein	Fat	Cholesterol	Sodium	Calcium
63 cal	15 g	1 g	0 g	0 mg	3 mg	18 mg

Ingredient	Weight	Measure	Issue
JUICE,GRAPEFRUIT,CONCENTRATE,FROZEN	4-1/8 lbs	1 qts 2-5/8 cup	
JUICE,PINEAPPLE,CANNED,UNSWEETENED	14-1/3 lbs	1 gal 2-1/2 qts	
WATER	12-1/2 lbs	1 gal 2 qts	
ICE CUBES	4 lbs	1 gal 1 qts	

Method

1. Combine grapefruit and pineapple juices with water; stir.
2. Cover and refrigerate.
3. Add ice just before serving.

BEVERAGES No.C 008 00

LEMONADE

Yield 100   Portion   1-1/4 Cups

Calories	Carbohydrates	Protein	Fat	Cholesterol	Sodium	Calcium
126 cal	33 g	0 g	0 g	0 mg	11 mg	7 mg

Ingredient   Weight   Measure   Issue

SUGAR,GRANULATED   7 lbs   1 gal
WATER   12-1/2 lbs   1 gal 2 qts
JUICE,LEMON   2-1/8 lbs   1 qts
WATER,COLD   37-5/8 lbs   4 gal 2 qts
ICE CUBES   9-5/8 lbs   3 gal

Method

1 Dissolve sugar in water. Cool.
2 Add juice and water to sugar solution. Mix thoroughly. Cover and refrigerate.
3 Add ice just before serving.

BEVERAGES No.C 008 01

LIMEADE

Yield 100   Portion   1-1/4 Cups

Calories	Carbohydrates	Protein	Fat	Cholesterol	Sodium	Calcium
131 cal	34 g	0 g	0 g	0 mg	14 mg	10 mg

Ingredient

Ingredient	Weight	Measure	Issue
SUGAR,GRANULATED	7 lbs	1 gal	
WATER	12-1/2 lbs	1 gal 2 qts	
JUICE,LIME	7-1/2 lbs	3 qts 3 cup	
WATER,COLD	37-5/8 lbs	4 gal 2 qts	
ICE CUBES	9-5/8 lbs	3 gal	

Method

1. Dissolve sugar in water. Cool.
2. Add juice and water to sugar solution. Mix thoroughly. Cover and refrigerate.
3. Add ice just before serving.

BEVERAGES No.C 009 00

ICED TEA (INSTANT)

Yield 100   **Portion** 1 Cup

Calories	Carbohydrates	Protein	Fat	Cholesterol	Sodium	Calcium
6 cal	1 g	0 g	0 g	0 mg	14 mg	8 mg

Ingredient   Weight   Measure   Issue

Ingredient	Weight	Measure	Issue
TEA MIX,INSTANT,UNSWEETENED	8-3/4 oz	1 qts 3-3/8 cup	
WATER,COLD	66-7/8 lbs	8 gal	
ICE CUBES	9-5/8 lbs	3 gal	

Method

1 Add tea to water; stir until dissolved.
2 Serve over crushed or cubed ice.

Notes

1 For each 8-ounce glass, use about 5 ounces of strong tea. Fill glass with crushed ice. Serve 2 8-ounce glasses per portion.

BEVERAGES No.C 009 01

ICED TEA (INSTANT FOR DISPENSER)

Yield 100 **Portion** 1-1/4 Cups

Calories	Carbohydrates	Protein	Fat	Cholesterol	Sodium	Calcium
5 cal	1 g	0 g	0 g	0 mg	4 mg	2 mg

Ingredient	**Weight**	**Measure**	**Issue**
TEA MIX,INSTANT,UNSWEETENED	6-3/4 oz	1 qts 1-5/8 cup	
ICE CUBES	9-5/8 lbs	3 gal	

Method

1. Place instant tea, on dispenser. Follow manufacturer's directions for preparation, dispensing of tea, and cleaning of dispenser.
2. Serve over crushed or cubed ice.

Notes

1. For each 8-ounce glass, use about 5 ounces of strong tea. Fill glass with crushed ice. Serve 2 8-ounce glasses per portion.

BEVERAGES No.C 009 02

ICED TEA (INSTANT W/LEMON AND SUGAR FOR DISPENSER)

Yield 100   **Portion** 1-1/4 Cups

Calories	Carbohydrates	Protein	Fat	Cholesterol	Sodium	Calcium
175 cal	44 g	0 g	0 g	0 mg	4 mg	2 mg

Ingredient	**Weight**	**Measure**	**Issue**
TEA MIX, INSTANT, W/LEMON AND SUGAR	10 lbs		
ICE CUBES	9-5/8 lbs	3 gal	

Method

1 Place instant tea mix with lemon and sugar on dispenser. Follow directions for preparation and dispensing of tea.
2 Serve over crushed or cubed ice.

Notes

1 For each 8-ounce glass, use about 5 ounces of strong tea. Fill glass with crushed ice. Serve 2 8-ounce glasses per portion.

BEVERAGES No.C 010 00

ORANGEADE

Yield 100 **Portion** 1-1/4 Cups

Calories	Carbohydrates	Protein	Fat	Cholesterol	Sodium	Calcium
137 cal	34 g	1 g	0 g	0 mg	6 mg	15 mg

Ingredient	Weight	Measure	Issue
SUGAR,GRANULATED	4 lbs	2 qts 1 cup	
WATER	12-1/2 lbs	1 gal 2 qts	
JUICE,ORANGE	35-1/8 lbs	4 gal	
ICE CUBES	9-5/8 lbs	3 gal	

Method

1 Dissolve sugar in water. Cool.
2 Add juice to sugar solution. Mix thoroughly. Cover and refrigerate.
3 Add ice just before serving.

Notes

1 In Step 1, use 5 pounds or 2-3/4 quarts of granulated sugar and 2 gallons of hot water for 100 servings.

D-G. BREADS AND SWEET DOUGHS No. 1

RECIPE CONVERSION

Most bread and sweet dough recipes have an additional column on the left side of each recipe card for TRUE PERCENTAGES. These are based on the total weight of all the ingredients, the sum of which is 100 percent. True percentages are used in adjusting a recipe to yield a specific number of servings to produce a specific number of smaller or larger servings, or to use the amount of ingredients available. To adjust a recipe to yield a specific number of servings, use this method (using Sweet Dough (Recipe D-36) as an example):

A. TRUE PERCENTAGE METHOD

$\underline{\text{Step 1}}$ - Obtain a working factor by dividing the number of servings needed by 100.

For example: 438 servings needed ÷ 100 = 4.38 working factor. See Recipe Conversion No. A-1.

$\underline{\text{Step 2}}$ - Multiply the working factor by the total weight of the recipe to obtain the pounds desired. (Note: the total weight of the recipe is listed at the bottom of the weight column on each recipe card.)

For example: 4.38 (working factor) x 12.958 (weight of recipe) = 56.76 (lbs desired).

$\underline{\text{Step 3}}$ - Multiply 56.76 (lbs desired) by the percent of each ingredient in the recipe.

Ingredient	Percent			Weight	Converted
Yeast	2.37%	X	56.76 =	1.34 lb	= 1 lb 5 1/2 oz
Water	18.92%	X	56.76 =	10.74 lb	= 10 lb 12 oz
Sugar	8.99%	X	56.76 =	5.10 lb	= 5 lb 1 1/2 oz
Salt	.95%	X	56.76 =	.54 lb	= 8 3/4 oz
Shortening	7.57%	X	56.76 =	4.30 lb	= 4 lb 5 oz
Eggs	9.46%	X	56.76 =	5.37 lb	= 5 lb 6 oz
Flour	50.16%	X	56.76 =	28.47 lb	= 28 lb 7 1/2 oz
Milk	1.58%	X	56.76 =	.90 lb	= 14 1/2 oz
TOTAL	100.00%			56.76 lb	

GUIDELINES FOR PREPARATION OF YEAST DOUGHS

1. The water temperature in which the yeast is dissolved is important. If temperatures above 110° F. are used, the yeast will be killed. If under 105° F., the yeast's growth or development will be retarded.
2. The amount of water required may vary from that specified in the recipe due to variable amounts of moisture in the flour.
3. Full mixing or dough development produces better volume and lighter yeast products.
4. Lightly grease the bowl in which the dough is allowed to rise. Heavy greasing may cause streaks in the bread.
5. Yeast dough is ready to be punched when it is light and doubled in size. To test, press the dough lightly with a finger tip. If the impression remains and the dough recedes slightly, it is ready to be punched.
6. Punching should be just enough to expel gases.
7. The dough for rolls is usually softer than that for bread.

RETARDED SWEET DOUGH METHODS

Retarded sweet dough is yeast dough that is refrigerated for a period of time prior to baking. Refrigeration temperatures retard fermentation of the dough. The quality of the end product not changed. Retarded sweet dough may be held in refrigeration below 40° F. as long as 24 hours. Retarded sweet dough may be prepared using Sweet Dough (Recipe No. D 036 00). Two methods of preparation are:

Method 1

1. Sprinkle yeast over water. DO NOT USE TEMPERATURES ABOVE 110° F. Mix well. Let stand 5 minutes; stir. Set aside for use in Step 3.
2. Place water, eggs, sugar, milk, and salt in mixer bowl. Using dough hook, mix at low speed just until blended.
3. Add flour and yeast solution. Mix at low speed 1 minute or until all flour mixture is incorporated into liquid.
4. Add shortening; mix at low speed 1 minute. Continue mixing at medium speed 10 minutes or until dough is smooth and elastic. Dough temperature should be between 78° F. to 82° F.
5. FERMENT: Set in warm place (80°F.) about 50 to 55 minutes.
6. PUNCH: Divide dough into desired working-size pieces. Shape each piece into a smooth rectangular piece. Let rest 15 minutes.
7. MAKE UP: As desired.
8. Cover; refrigerate immediately.
9. When ready to use, remove from refrigeration; PROOF until pieces are double in bulk.
10. BAKE: Using a convection oven, bake 15 minutes at 325 F. on high fan, open vent.
11. FINISH: As desired.

NOTE: Made up pieces prepared by this method can be stored safely for about 60 hours at 32° F.

Method 2
1. Sprinkle yeast over water. DO NOT USE TEMPERATURES ABOVE 110° F. Mix well. Let stand 5 minutes; stir. Set aside for use in Step 3.
2. Place water, eggs, sugar, milk, and salt in mixer bowl. Using dough hook, mix at low speed just until blended.
3. Add flour and yeast solution. Mix at low speed 1 minute or until all flour mixture is incorporated into liquid.
4. FERMENT: Set in warm place (80° F.) about 50 to 55 minutes.
5. PUNCH: Divide dough into 3 pieces, about 4 lb 5 oz each; shape ea piece into a smooth rectangle. Let rest 15 minutes.
6. Flatten each piece; brush lightly with melted shortening or salad oil. Place on greased sheet pans; cover and refrigerate.
7. When ready to use, remove dough from refrigeration; make up as desired. IT IS NOT NECESSARY TO BRING DOUGH TO ROOM TEMPERATE BEFORE MAKE UP.
8. PROOF: Until pieces are double in bulk.
9. BAKE: Using a convection oven, bake 15 minutes at 325 F. on high fan, open vent.
10. FINISH: As desired.

CHARACTERISTICS OF GOOD QUALITY BREAD PRODUCTS AND ROLLS

CHARACTERISTIC	BISCUITS	MUFFINS	YEAST BREADS AND ROLLS
Color	Uniform golden brown top and bottom. Inside creamy white. Free from yellow or brown spots.	Uniform golden brown outside. Inside creamy white or slightly yellow but free from streaks.	Even rich brown color, creamy white inside and free from streaks.
Shape and size	Uniform in shape and size, with straight sides and a smooth level top. The volume is at least twice the size of the unbaked product.	Uniform shape and size. Well-rounded pebbled top, free from peaks or cracks.	Well proportioned, symmetrical with a well-rounded top.
Crust	Tender and moderately smooth. Free from excess flour.	Tender, with a thin, slightly rough or pebbled shiny appearance.	Crisp-tender with an even thickness over entire surface. Free from cracks and bulges.
Texture	Slightly moist, tender and flaky crumb, with a medium fine grain.	Moist, tender and light crumb, with medium fine, evenly distributed air spaces.	Soft, springy texture, tender and slightly moist with fine grain, thin walled cells.
Flavor	Pleasing, well-blended flavor with no bitterness.	Pleasing, well-blended flavor with no bitterness or other off-flavors.	Wheaty, sweet nut-like flavor. No off-flavors.

CHARACTERISTIC		BISCUITS	MUFFINS	YEAST BREADS AND ROLLS
Tough		Not enough shortening or leavening. Too much liquid. Dough too cold or oven not hot enough. Overmixing.	Not enough shortening or sugar. Overmixing.	Not enough shortening. Insufficient proofing time. Overbaking.
Heavy		Wrong proportion of ingredients. Improper mixing. Oven not hot enough or dough too stiff.	Not enough baking powder or shortening. Overmixing.	Underproofing or overmixing.
Flavor	Poor	Wrong proportion of ingredients or improper mixing.	Wrong proportion of ingredients or improper mixing.	Wrong proportion of ingredients. Fermentation time too long.
Grain	Coarse or uneven	Too much leavening, not enough liquid, or improper mixing. "Not flaky" due to not enough shortening or improper mixing of shortening and flour.	Insufficient beating of eggs. Too much or not enough leavening, Overmixing. Tunnels due to not enough liquid or shortening or overmixing.	Improper make-up, excessive water or under-or overmixing.
Texture	Too dry	Dough too stiff. Overbaking. Oven not hot enough. Not enough sugar or shortening.	Batter too stiff. Overbaking. Too much leavening. Not enough sugar and/or shortening.	Overproofing. Not enough water or improper mixing time
	Too crumbly	Too much leavening, sugar or shortening. Not enough liquid.	Not enough liquid. Too much baking powder. Oven not hot enough.	Not enough water, improper mixing time

CHARACTERISTICS OF POOR QUALITY BREAD PRODUCTS AND ROLLS

CHARACTERISTIC		BISCUITS	MUFFINS	YEAST BREADS AND ROLLS
Crusts	Tough or hard	Too much flour. Overmixing. Oven too hot. Overbaking.	Too much flour or not enough sugar or shortening. Overmixing.	Not enough shortening. Overbaking. Insufficient fermentation. Too much rolling in flour.
	Irregular	Rough or blisters due to too much liquid, incorrect kneading or rolling.	Peaks due to mixture being too stiff, overmixing or oven too hot.	Blisters due to improper make-up. Too much rolling in flour.
	Too smooth		Too much liquid or overmixing.	
Inside Appearance	Color streaks or spots	Too much leavening. Ingredients not well mixed.	Eggs and milk not well blended.	"Crusting" during fermentation of dough. Undermixing. Too much dusting flour during make-up.
Outside Appearance	Shape irregular	Too much liquid. Dough not rolled to uniform thickness. Improper cutting of dough. Uneven oven heat.	Too much flour. Not enough liquid. Overmixing. Too much batter in pan. Oven too hot.	Improper shaping. Too much dough for bread pan. Insufficient proofing time.
Color	Too dark	Oven too hot. Overbaking. Dough too stiff. Oven not hot enough, insufficient sugar.	Too much sugar. Oven too hot. Overbaking.	Too much sugar or milk. Insufficient fermentation time. Oven too hot.
	Too pale	Dough too stiff. Oven not hot enough, insufficient sugar.	Overmixing. Oven not hot enough. Underbaking.	Not enough sugar or milk. Dough too warm during mixing and excessive fermentation. Oven not hot enough.

D. BREADS AND SWEET DOUGHS No. 0

INDEX

Card No. .... Card No.

Card No.	Title		Card No.	Title
D 001 00	Baking Powder Biscuits		D 014 03	Jalapeno Corn Bread
D 001 01	Baking Powder Biscuits (Biscuit Mix)		D 015 00	Corn Bread (Corn Bread Mix)
D 001 02	Cheese Biscuits		D 015 01	Corn Muffins (Corn Bread Mix)
D 001 03	Drop Biscuits		D 015 02	Hush Puppies (Corn Bread Mix)
D 002 00	Irish Soda Bread		D 015 03	Jalapeno Corn Bread (Corn Bread Mix)
D 003 00	Submarine Rolls (Hoagie, Torpedo)		D 016 00	Croutons
D 003 01	Submarine Rolls (Roll Mix)		D 016 01	Garlic Croutons
D 004 00	French Bread		D 016 02	Parmesan Croutons
D 005 00	Raisin Bread		D 017 00	Egg Wash
D 007 00	Toasted Garlic Bread		D 017 01	Egg White Wash
D 007 01	Toasted Parmesan Bread		D 018 00	Cake Doughnuts (Homemade)
D 007 02	Texas Toast		D 018 01	Sugar Coated Doughnuts
D 008 00	White Bread		D 018 02	Cake Doughnuts (Doughnut Mix)
D 009 00	White Bread (Short-Time Formula)		D 018 03	Chocolate Doughnuts
D 011 00	Pumpkin Bread		D 018 04	Cinnamon Sugar Doughnuts
D 012 00	Crumb Cake Snickerdoodle		D 018 05	Glazed Nut Doughnuts
D 012 01	Crumb Cake Snickerdoodle (Cake Mix, Yellow)	(1)	D 018 06	Glazed Coconut Doughnuts
			D 018 07	Glazed Doughnuts
D 013 00	Bagels		D 019 00	Raised Doughnuts
D 014 00	Corn Bread		D 019 01	Beignets (New Orleans Doughnuts)
D 014 01	Corn Muffins		D 019 02	Raised Doughnuts (Sweet Dough Mix)
D 014 02	Hush Puppies		D 019 03	Longjohns

D. BREADS AND SWEET DOUGHS No. 0

D 019 04	Crullers		D 028 00	Bran Muffins
D 020 00	Dumplings		D 028 01	Raisin Bran Muffins
D 021 00	English Muffins		D 028 02	Blueberry Bran Muffins
D 021 01	Cinnamon Raisin English Muffins		D 028 03	Banana Bran Muffins
D 022 00	French Toast		D 028 04	Apricot Bran Muffins
D 022 01	French Toast (Thick Slice)		D 028 05	Cranberry Bran Muffins
D 022 02	English Muffin French Toast		D 029 00	Muffins
D 022 03	French Toast (Frozen Eggs and Egg Whites)		D 029 01	Blueberry Muffins
D 023 00	French Toast Puff		D 029 02	Raisin Muffins
D 024 00	Apple Fritters		D 029 03	Banana Muffins
D 025 00	Pancakes		D 029 04	Apple Muffins
D 025 01	Buttermilk Pancakes (Dry Buttermilk)		D 029 05	Cinnamon Crumb Top Muffins
D 025 02	Blueberry Pancakes		D 029 06	Cranberry Muffins
D 025 04	Buttermilk Pancakes (Pancake Mix)		D 029 07	Date Muffins
D 025 05	Pancakes (Pancake Mix)		D 029 08	Nut Muffins
D 025 06	Waffles, Frozen (Brown And Serve)		D 029 09	Oatmeal Raisin Muffins
D 025 07	Waffles (Pancake Mix)		D 030 00	Banana Bread
D 025 08	Waffles		D 032 00	Hard Rolls
D 025 09	Whole Wheat Pancakes		D 033 00	Hot Rolls
D 025 10	Pancakes (Frozen Eggs and Egg Whites)	(1)	D 033 01	Hot Rolls (Brown and Serve)
D 025 11	Pancakes (Egg Substitute)		D 033 02	Hot Rolls (Roll Mix)
D 026 00	Hot Cross Buns		D 033 03	Oatmeal Rolls
D 027 00	Kolaches		D 033 04	Cloverleaf or Twin Rolls
D 027 01	Kolaches (Sweet Dough Mix)		D 033 05	Frankfurter Rolls

D. BREADS AND SWEET DOUGHS No. 0

INDEX

Card No.			Card No.	
D 033 06	Hamburger Rolls		D 036 13	Snails
D 033 07	Pan, Cluster, or Pull Apart Rolls		D 036 14	Bowknots, Figure 8's, and S Shapes
D 033 08	Parker House Rolls		D 036 15	Cinnamon Twists
D 033 09	Poppy Seed Rolls		D 036 16	Butterhorns
D 033 10	Sesame Seed Rolls		D 036 17	Crescents
D 034 00	Hot Rolls (Short-Time Formula)		D 037 00	Quick Coffee Cake (Biscuit Mix)
D 034 01	Brown And Serve Rolls (Short-Time Formula)		D 037 01	Quick Apple Coffee Cake (Biscuit Mix)
D 034 03	Whole Wheat Rolls (Short-Time Formula)		D 037 02	Quick French Coffee Cake (Biscuit Mix)
D 035 00	Onion Rolls		D 037 03	Quick Cherry Coffee Cake (Biscuit Mix)
D 035 01	Onion Rolls (Roll Mix)		D 037 04	Quick Orange-Coconut Coffee Cake (Biscuit Mix)
D 036 00	Sweet Dough		D 037 05	Quick Coffee Cake
D 036 01	Sweet Dough (Sweet Dough Mix)		D 038 00	Tempura Batter
D 036 02	Glazed Rolls		D 039 00	Danish Diamonds (Danish Pastry Dough)
D 036 03	Pecan Rolls		D 039 01	Bear Claws (Danish Pastry Dough)
D 036 04	Cinnamon Rolls		D 039 02	Fruit Turnovers (Frozen Puff Pastry Dough)
D 036 05	Cinnamon Nut Rolls		D 039 03	Fruit Puffs (Frozen Puff Pastry Dough)
D 036 06	Cinnamon Raisin Rolls	(2)	D 040 00	Cornstarch Wash
D 036 07	Butterfly Rolls		D 041 00	Cherry Filling (Cornstarch)
D 036 08	Sugar Rolls		D 041 01	Cherry Filling (Pie Filling, Prepared)
D 036 09	Streusel Coffee Cake		D 041 03	Apple Filling (Pie Filling, Prepared)
D 036 10	Small Coffee Cake		D 041 04	Blueberry Filling (Pie Filling, Prepared)
D 036 11	Twist Coffee Cake		D 041 05	Raspberry Filling (Prepared Bakery)
D 036 12	Bear Claws			

D. BREADS AND SWEET DOUGHS No. 0

D 042 00	Cinnamon Sugar Filling		D 057 00	Apple Coffee Cake
D 042 01	Cinnamon Sugar Nut Filling		D 058 00	Oven Baked French Toast
D 042 02	Cinnamon Sugar Raisin Filling		D 059 00	Whole Wheat Rolls
D 043 00	Nut Filling		D 060 00	Oats and Fruit Breakfast Squares
D 044 00	Oat Bran Raisin Muffins		D 502 00	Pumpkin Patch Muffins
D 045 00	Syrup Glaze		D 503 00	Date Nut Bread
D 046 00	Vanilla Glaze		D 507 00	Applesauce Cinnamon Crumb Top Muffin
D 046 01	Almond Glaze		D 508 00	Applesauce Blueberry Muffins
D 046 02	Rum Glaze		D 509 00	Cran-Apple Muffins
D 047 01	Pineapple Filling (Cornstarch)		D 800 00	Bread Loaves (Frozen Dough)
D 048 00	Orange-Coconut Topping		D 801 00	Dill Rolls (Frozen Dough)
D 049 00	Streusel Topping		D 802 00	Garlic Herb Rolls (Frozen Dough)
D 049 01	Pecan Topping		D 803 00	Hush Puppies, Frozen
D 050 00	Maple Syrup		D 804 00	Dinner Rolls (Frozen Dough)
D 051 00	Frying Batter		D 805 00	Potato Rolls (Frozen Dough)
D 052 00	Oatmeal Bread		D 806 00	Whole Wheat Rolls (Frozen Dough)
D 053 00	Applesauce Muffins		D 807 00	Oat Rolls (Frozen Dough)
D 053 01	Applesauce Raisin Muffins		D 808 00	Sesame or Caraway Rolls (Frozen Dough)
D 053 02	Applesauce Orange Muffins	(2)	D 809 00	Bran Muffins (White Cake Mix)
D 054 00	Pineapple Carrot Muffins		D 810 00	Banana Nut Muffins (White Cake Mix)
D 055 00	Whole Wheat Bread		D 811 00	Honey Cinnamon Muffins (White Cake Mix)
			D 812 00	Blueberry Muffins (White Cake Mix)
D 056 00	Whole Wheat Bread (Whole Wheat Flour Short Time Formula)		D 813 00	French Toast, Frozen
			D 814 00	Pancakes, Buttermilk, Frozen

D. BREADS AND SWEET DOUGHS No. 0

INDEX

Card No.　　　　　　　　　　　　　　　　　　　　　　　　　Card No.

D 815 00　　Muffins, Frozen, Batter
D 816 00　　Muffins, Frozen

BREADS AND SWEET DOUGHS No.D 001 00

BAKING POWDER BISCUITS

Yield 100   **Portion** 1 Biscuit

Calories	Carbohydrates	Protein	Fat	Cholesterol	Sodium	Calcium
148 cal	24 g	4 g	4 g	0 mg	345 mg	115 mg

Ingredient	**Weight**	**Measure**	**Issue**
FLOUR,WHEAT,GENERAL PURPOSE	6-5/8 lbs	1 gal 2 qts	
MILK,NONFAT,DRY	3-5/8 oz	1-1/2 cup	
BAKING POWDER	5-7/8 oz	3/4 cup	
SALT	1-1/2 oz	2-1/3 tbsp	
SHORTENING	12 oz	1-5/8 cup	
WATER	3-7/8 lbs	1 qts 3-1/2 cup	
COOKING SPRAY,NONSTICK	2 oz	1/4 cup 1/3 tbsp	

Method

1 Sift together flour, milk, baking powder, and salt into mixer bowl.
2 Blend shortening at low speed into dry ingredients until mixture resembles coarse cornmeal.
3 Add water; mix at low speed only enough to form soft dough.
4 Place dough on lightly floured board. Knead lightly 1 minute or until dough is smooth.
5 Roll or pat out to a uniform thickness of 1/2-inch.
6 Lightly spray each pan with non-stick cooking spray. Cut with 2-1/2 inch floured biscuit cutter. Place 50 biscuits on each pan.
7 Using a convection oven, bake at 350 F. for 15 minutes or until lightly browned on low fan, open vent.

Notes

1 For browner tops: In Step 1, add 1/2 cup granulated sugar per 100 portions to dry ingredients.

BREADS AND SWEET DOUGHS No.D 001 01

BAKING POWDER BISCUITS (BISCUIT MIX)

Yield 100   Portion   1 Biscuit

Calories	Carbohydrates	Protein	Fat	Cholesterol	Sodium	Calcium
153 cal	23 g	3 g	6 g	1 mg	456 mg	64 mg

Ingredient	Weight	Measure	Issue
BISCUIT MIX	7-7/8 lbs	1 gal 3-1/2 qts	

Method

1 Prepare biscuit mix according to instructions on container. Using a convection oven, bake at 350 F. 15 minutes or until lightly browned on low fan, open vent.

BREADS AND SWEET DOUGHS No.D 001 02

CHEESE BISCUITS

Yield 100   **Portion** 1 Biscuit

Calories	Carbohydrates	Protein	Fat	Cholesterol	Sodium	Calcium
166 cal	24 g	5 g	6 g	5 mg	373 mg	147 mg

Ingredient	**Weight**	**Measure**	**Issue**
FLOUR,WHEAT,GENERAL PURPOSE	6-5/8 lbs	1 gal 2 qts	
MILK,NONFAT,DRY	3-5/8 oz	1-1/2 cup	
BAKING POWDER	5-7/8 oz	3/4 cup	
SALT	1-1/2 oz	2-1/3 tbsp	
CHEESE,CHEDDAR,GRATED	1 lbs	1 qts	
SHORTENING	12 oz	1-5/8 cup	
WATER	3-7/8 lbs	1 qts 3-1/2 cup	
COOKING SPRAY,NONSTICK	2 oz	1/4 cup 1/3 tbsp	

Method

1 Sift together flour, milk, baking powder, and salt into mixer bowl. Add grated cheddar cheese to sifted dry ingredients.
2 Blend shortening at low speed into dry ingredients until mixture resembles coarse cornmeal.
3 Add water; mix at low speed only enough to form soft dough.
4 Place dough on lightly floured board. Knead lightly, 1 minute or until dough is smooth.
5 Roll or pat out to a uniform thickness of 1/2 inch.
6 Lightly spray each pan with non-stick cooking spray. Cut with 2-1/2 inch floured biscuit cutter. Place 50 biscuits on each pan.
7 Using a convection oven, bake at 350 F. for 15 minutes or until lightly browned on low fan, open vent.

Notes

1 For browner tops: In Step 1, add 1/2 cup of granulated sugar per 100 portions to dry ingredients.

BREADS AND SWEET DOUGHS No.D 001 03

DROP BISCUITS

Yield 100	**Portion** 1 Biscuit

Calories	Carbohydrates	Protein	Fat	Cholesterol	Sodium	Calcium
148 cal	24 g	4 g	4 g	0 mg	345 mg	115 mg

Ingredient	Weight	Measure	Issue
FLOUR,WHEAT,GENERAL PURPOSE	6-5/8 lbs	1 gal 2 qts	
MILK,NONFAT,DRY	3-5/8 oz	1-1/2 cup	
BAKING POWDER	5-7/8 oz	3/4 cup	
SALT	1-1/2 oz	2-1/3 tbsp	
SHORTENING	12 oz	1-5/8 cup	
WATER	4-7/8 lbs	2 qts 1-3/8 cup	
COOKING SPRAY,NONSTICK	2 oz	1/4 cup 1/3 tbsp	

Method

1 Sift together flour, milk, baking powder, and salt into mixer bowl.
2 Blend shortening at low speed into dry ingredients until mixture resembles coarse cornmeal.
3 Add water; mix at low speed only enough to form a soft dough.
4 Lightly spray each pan with non-stick cooking spray. Drop biscuit dough by heaping tablespoon, 1 inch apart, on sprayed sheet pans in rows 6 by 9.
5 Using a convection oven, bake at 350 F. for 15 minutes or until lightly browned on low fan, open vent.

Notes

1 For browner tops: In Step 1, add 1/2 cup granulated sugar per 100 portions to dry ingredients.

BREADS AND SWEET DOUGHS No.D 002 00

IRISH SODA BREAD

Yield 100   Portion   2 Slices

Calories	Carbohydrates	Protein	Fat	Cholesterol	Sodium	Calcium
335 cal	59 g	6 g	9 g	44 mg	456 mg	70 mg

Ingredient	Weight	Measure	Issue
MILK,NONFAT,DRY	4-1/4 oz	1-3/4 cup	
WATER	4-2/3 lbs	2 qts 1 cup	
VINEGAR,DISTILLED	5-5/8 oz	1/2 cup 2-2/3 tbsp	
FLOUR,WHEAT,GENERAL PURPOSE	8-7/8 lbs	2 gal	
SUGAR,GRANULATED	3 lbs	1 qts 2-3/4 cup	
BAKING SODA	1-1/3 oz	2-2/3 tbsp	
BAKING POWDER	1-3/4 oz	1/4 cup	
SALT	1-7/8 oz	3 tbsp	
RAISINS	3-7/8 lbs	3 qts	
CARAWAY SEED	2-1/2 oz	1/2 cup 2-2/3 tbsp	
BUTTER	2 lbs	1 qts	
EGGS,WHOLE,FROZEN,BEATEN,ROOM TEMPERATURE	1-1/4 lbs	2-1/4 cup	
COOKING SPRAY,NONSTICK	2 oz	1/4 cup 1/3 tbsp	

Method

1. Reconstitute milk; add vinegar. Let stand 15 minutes. Set aside for use in Step 4.
2. Place flour, sugar, baking soda, baking powder, salt, raisins, and caraway seeds in mixer bowl. Mix at low speed just enough to blend.
3. Using pastry knife attachment, cut butter or margarine into dry ingredients until it resembles coarse meal.
4. Stir eggs into milk. Add egg-milk mixture to dry ingredients; blend until just mixed, about 45 seconds. DO NOT OVERMIX.
5. Lightly spray each pan with non-stick cooking spray. Place 3 pounds or 1-1/2 quarts batter in each sprayed loaf pan.
6. Bake 55 to 60 minutes at 375 F. or until done.
7. Cool thoroughly before slicing.
8. Cut 25, 1/2 inch thick slices per loaf.

BREADS AND SWEET DOUGHS No.D 003 00

SUBMARINE ROLLS (HOAGIE, TORPEDO)

Yield 100   Portion 1 Roll

Calories	Carbohydrates	Protein	Fat	Cholesterol	Sodium	Calcium
389 cal	73 g	12 g	5 g	0 mg	423 mg	17 mg

Ingredient	Weight	Measure	Issue
YEAST,ACTIVE,DRY	6-3/4 oz	1 cup	
WATER,WARM	2-1/8 lbs	1 qts	
WATER,COLD	8-7/8 lbs	1 gal 1/4 qts	
SUGAR,GRANULATED	8-7/8 oz	1-1/4 cup	
SALT	3-3/4 oz	1/4 cup 2-1/3 tbsp	
FLOUR,WHEAT,BREAD	21-1/8 lbs	4 gal 1-1/2 qts	
SHORTENING,SOFTENED	9 oz	1-1/4 cup	
COOKING SPRAY,NONSTICK	2 oz	1/4 cup 1/3 tbsp	

Method

1. Sprinkle yeast over water. DO NOT USE TEMPERATURES ABOVE 110 F. Mix well. Let stand 5 minutes; stir. Set aside for use in Step 3.
2. Place water, sugar, salt, and flour in mixer bowl.
3. Mix at low speed 1 minute or until all flour is incorporated into liquid, using dough hook. Add yeast solution; mix at low speed 1 minute.
4. Add shortening; mix at medium speed 10 minutes or until dough is smooth and elastic. Dough temperature should be between 78 F. and 82 F.
5. FERMENT: Cover. Set in warm place, 80 F. for 1-1/2 hours or until double in bulk.
6. PUNCH: Fold sides into center. Turn dough over. Divide dough into approximately 3-pound pieces. Let rest about 10 minutes.
7. MAKEUP: Divide each ball into 10 4-1/2-ounce pieces; flatten. Roll up like jelly roll into 1-1/4x8-inch rolls. Lightly spray pans with non-stick cooking spray. Place 15 rolls about 2 inches apart on each sprayed pan.
8. Prepare 1/2 Recipe Cornstarch Wash, Recipe No. D 040 00. Brush on top and sides of each roll.
9. PROOF: At 90 F. until double in size, about 40 minutes.
   Using a convection oven, bake at 350 F. for 12 to 15 minutes or until lightly browned on high fan, open vent. Immediately brush with Cornstarch Wash. Cool on wire racks.

Notes

1. Rolls may be prepared using semi-automatic bakery equipment (roll divider and rounding machine, bread molder-dough sheeter machine and bun slicer). Follow Step 1. In Step 2, combine1 1/3 oz (3 tbsp) bakery emulsifier with flour and milk. Follow Steps 3 through 6. In Step 7, divide dough into 5 lb 6 oz pieces. Place in roll divider and rounding machine. Divide into 36 balls. Press 2 balls together to form 4-1/2 oz balls. Let rest 5 to 10 minutes. Feed balls, one at a time into bread molder-dough sheeter machine, with a 9-inch pressure plate. Follow Steps 8 through 10. Slice rolls partially through using bun slicer.

BREADS AND SWEET DOUGHS No.D 003 01

SUBMARINE ROLLS (ROLL MIX)

Yield 100   Portion 1 Roll

Calories	Carbohydrates	Protein	Fat	Cholesterol	Sodium	Calcium
340 cal	61 g	10 g	7 g	0 mg	532 mg	56 mg

Ingredient | Weight | Measure | Issue

Ingredient	Weight	Measure	Issue
YEAST,ACTIVE,DRY	8-1/2 oz	1-1/4 cup	
WATER,WARM	2-1/3 lbs	1 qts 1/2 cup	
ROLL,MIX	19-1/8 lbs		
WATER,COLD	8-1/3 lbs	1 gal	
COOKING SPRAY,NONSTICK	2 oz	1/4 cup 1/3 tbsp	

Method

1. Sprinkle yeast over water. Do not use temperatures above 110 F. Mix well. Let stand 5 minutes; stir. Prepare roll mix according to directions on package.
2. PUNCH: Fold sides into center. Turn dough over. Divide dough into approximately 3 pound pieces. Let rest about 10 minutes.
3. Lightly spray each pan with non-stick cooking spray. MAKEUP: Divide each ball into 10 4-1/2 ounce pieces; flatten. Roll up like jelly roll into 1-1/4x8 inch rolls. Place 15 rolls about 2 inches apart on each sprayed pan.
4. Prepare 1/2 recipe Cornstarch Wash, Recipe No. D 040 00. Brush on top and sides of each roll.
5. PROOF: At 90 F. until double in bulk, about 40 minutes.
6. Using a convection oven, bake 12 to 15 minutes at 350 F. or until lightly browned on high fan, open vent. Immediately brush with Cornstarch Wash. Cool on wire racks.

Notes

1. Rolls may be prepared using semi-automatic bakery equipment. Follow Step 1. In Step 1, add bakery emulsifier to roll mix. Follow Step 2. In Step 3 divide dough into 5lb 6 oz pieces. Place in roll divider and rounding machine. Divide into 36 balls. Press 2 balls together to form 4-1/2 oz balls. Let rest 5 to 10 minutes. Feed balls one at a time into bread molder-dough sheeter machine, with a 9-inch pressure plate. Follow Steps 4 through 5. Slice rolls partially through using bun slicer.

BREADS AND SWEET DOUGHS No.D 004 00

FRENCH BREAD

Yield 100   **Portion** 2 Slices

Calories	Carbohydrates	Protein	Fat	Cholesterol	Sodium	Calcium
189 cal	37 g	6 g	2 g	0 mg	328 mg	8 mg

Ingredient	Weight	Measure	Issue
YEAST,ACTIVE,DRY	2 oz	1/4 cup 1 tbsp	
WATER,WARM	12-1/2 oz	1-1/2 cup	
WATER,COLD	4-5/8 lbs	2 qts 3/4 cup	
SUGAR,GRANULATED	2-2/3 oz	1/4 cup 2-1/3 tbsp	
SALT	3 oz	1/4 cup 1 tbsp	
FLOUR,WHEAT,BREAD	10-7/8 lbs	2 gal 1 qts	
SHORTENING	2-3/4 oz	1/4 cup 2-1/3 tbsp	

Method

1. Sprinkle yeast over water. DO NOT USE TEMPERATURES ABOVE 110 F. Mix well. Let stand 5 minutes; stir. Set aside for use in Step 3.
2. Place water, sugar, salt, and flour in mixer bowl.
3. Using dough hook, mix at low speed 1 minute or until all flour mixture is incorporated into liquid; add yeast solution; mix at medium speed 5 minutes.
4. Add shortening; continue mixing at medium speed 3 minutes. Dough temperature should be between 78 F. and 82 F.
5. FERMENT: Cover and set in warm place, 80 F. for 2-1/4 hours or until double in bulk.
6. PUNCH: Fold sides into center and turn completely over. Let rest 15 minutes.
7. MAKE-UP: Scale into 12-19 ounce pieces; shape each piece into a smooth ball; let rest 10 minutes. Form each piece into a rope, 1-1/4 inches in diameter and 18 inches long. Place 3 loaves on each cornmeal dusted pan. Use 1/8 cup cornmeal per pan.
8. PROOF: At 90 F. to 100 F. for 50 to 60 minutes or until double in bulk.
9. Brush top of each loaf with Cornstarch Wash, Recipe No. D 040 00 or Egg White Wash, Recipe No. D 017 01. Cut 6 diagonal slashes, 1/4-inch deep, on top of each loaf.
   BAKE: 30 minutes at 425 F. or until done.
   When cool, cut 17 one-inch thick slices per loaf.

BREADS AND SWEET DOUGHS No.D 005 00

RAISIN BREAD

Yield 100   **Portion** 2 Slices

Calories	Carbohydrates	Protein	Fat	Cholesterol	Sodium	Calcium
201 cal	40 g	6 g	2 g	0 mg	264 mg	26 mg

Ingredient	**Weight**	**Measure**	**Issue**
YEAST,ACTIVE,DRY	2-7/8 oz	1/4 cup 3 tbsp	
WATER,WARM	1-1/8 lbs	2-1/4 cup	
WATER,COLD	3-1/8 lbs	1 qts 2 cup	
SUGAR,GRANULATED	5-1/4 oz	3/4 cup	
SALT	2-1/3 oz	1/4 cup	
MILK,NONFAT,DRY	3-1/4 oz	1-3/8 cup	
CINNAMON,GROUND	1/2 oz	2 tbsp	
FLAVORING,LEMON	1/2 oz	1 tbsp	
FLOUR,WHEAT,BREAD	8-1/8 lbs	1 gal 2-3/4 qts	
SHORTENING	6-1/3 oz	3/4 cup 2 tbsp	
RAISINS	2-7/8 lbs	2 qts 1 cup	

Method

1. Sprinkle yeast over water. DO NOT USE TEMPERATURES ABOVE 110 F. Mix well. Let stand five minutes; stir. Set aside for use in Step 4.
2. Place water, sugar, salt, milk, cinnamon, and lemon flavoring in mixer bowl. Using dough hook, mix at low speed just enough to blend.
3. Add flour. Mix at low speed 1 minute or until all flour is incorporated into liquid.
4. Add yeast solution; mix at low speed 1 minute.
5. Add shortening; mix at low speed 1 minute. Continue mixing at medium speed 10 to 15 minutes or until dough is smooth and elastic. Dough temperature should be between 78 F. and 82 F.
6. Soak raisins in 3 quarts lukewarm water 15 minutes. Drain. Mix at low speed 1 minute.
7. FERMENT: Cover and set in a warm place, 80 F. for 2 hours or until double in bulk.
8. PUNCH: Fold sides into center and turn dough completely over. Let rest 20 minutes.
9. MAKE UP: Scale into approximately 8-2 pound pieces; shape each piece into a smooth ball; let rest 10 minutes. Mold each piece into a loaf; place each loaf into lightly greased bread pan.
    PROOF: At 90 F. to 100 F. for 50 to 60 minutes or until double in bulk.
    BAKE: If convection oven is used, bake at 325 F. for 30 minutes or until done on high fan, closed vent.
    Prepare 1/4 recipe Syrup Glaze, Recipe No. D 045 00 (optional). Brush top of each loaf with hot Syrup Glaze.
    When cool, slice 25 slices (about 1/2 inch thick) per loaf.

Notes

1. In Step 9, when using 9x4-1/2x2-3/4 bread pans, scale into 10-25-ounce pieces.

BREADS AND SWEET DOUGHS No.D 007 00

TOASTED GARLIC BREAD

Yield 100   **Portion** 2 Slices

Calories	Carbohydrates	Protein	Fat	Cholesterol	Sodium	Calcium
259 cal	31 g	5 g	13 g	0 mg	487 mg	48 mg

Ingredient	Weight	Measure	Issue
MARGARINE,SOFTENED	3 lbs	1 qts 2 cup	
GARLIC POWDER	1/2 oz	1 tbsp	
BREAD,FRENCH	13 lbs		

Method

1. Place butter or margarine in mixer bowl. Whip at medium speed until creamy. Add garlic powder; blend thoroughly.
2. Slice each loaf in half lengthwise. Spread each half loaf with about 2 ounces or 1/4 cup of garlic-butter mixture. Cut each half loaf into 8 slices. Place 5 half loaves on each sheet pan.
3. Using a convection oven, bake at 350 F. for 10 to 12 minutes or until lightly browned on high fan, open vent.
4. Serve hot.

Notes

1. In Step 2, 100 hard rolls may be split and used for 100 portions.

BREADS AND SWEET DOUGHS No.D 007 01

TOASTED PARMESAN BREAD

Yield 100   **Portion** 2 Slices

Calories	Carbohydrates	Protein	Fat	Cholesterol	Sodium	Calcium
277 cal	31 g	7 g	14 g	3 mg	561 mg	103 mg

Ingredient	**Weight**	**Measure**	**Issue**
MARGARINE,SOFTENED	3 lbs	1 qts 2 cup	
CHEESE,PARMESAN,GRATED	14-1/8 oz	1 qts	
BREAD,FRENCH	13 lbs		

Method

1 Place butter or margarine in mixer bowl. Whip at medium speed until creamy. Add grated Parmesan cheese; mix thoroughly.
2 Slice each loaf in half lengthwise. Spread each half loaf with about 2 ounces or 1/4 cup cheese-butter mixture. Cut each half loaf into 8 slices. Place 5 half loaves on each sheet pan.
3 Using a convection oven, bake at 350 F. for 10 to 12 minutes or until lightly browned on high fan, open vent.
4 Serve hot.

Notes

1 In Step 2, 100 hard rolls may be split and used.

BREADS AND SWEET DOUGHS No.D 007 02

TEXAS TOAST

Yield 100   **Portion** 2 Slices

Calories	Carbohydrates	Protein	Fat	Cholesterol	Sodium	Calcium
202 cal	31 g	5 g	6 g	0 mg	359 mg	44 mg

Ingredient   **Weight**   **Measure**   **Issue**

BREAD, FRENCH   13 lbs

Method

1. Use unsliced French Bread. Diagonally cut each loaf into 8 even slices.
2. Using a convection oven, bake at 350 F. for 10 to 12 minutes or until lightly browned on high fan, open vent.
3. Serve hot.

Notes

1. Toast may be grilled. Place on lightly greased 400 F. griddle Grill 2 to 3 minutes until lightly browned.

BREADS AND SWEET DOUGHS No.D 008 00

WHITE BREAD

Yield 100 **Portion** 2 Slices

Calories	Carbohydrates	Protein	Fat	Cholesterol	Sodium	Calcium
181 cal	33 g	6 g	2 g	0 mg	334 mg	22 mg

Ingredient	Weight	Measure	Issue
YEAST,ACTIVE,DRY	1-2/3 oz	1/4 cup 1/3 tbsp	
WATER,WARM	12-1/2 oz	1-1/2 cup	
WATER,COLD	4-1/8 lbs	2 qts	
SUGAR,GRANULATED	5-1/4 oz	3/4 cup	
SALT	3 oz	1/4 cup 1 tbsp	
MILK,NONFAT,DRY	4-1/4 oz	1-3/4 cup	
FLOUR,WHEAT,BREAD	9-1/3 lbs	1 gal 3-3/4 qts	
SHORTENING	6-1/3 oz	3/4 cup 2 tbsp	

Method

1. Sprinkle yeast over water. DO NOT USE TEMPERATURES ABOVE 110 F. Mix well. Let stand 5 minutes; stir. Set aside for use in Step 4.
2. Place water, sugar, salt, and milk in mixer bowl. Mix at low speed just enough to blend.
3. Add flour. Using dough hook, mix at low speed 1 minute or until all flour is incorporated into liquid.
4. Add yeast solution; mix at low speed 1 minute.
5. Add shortening; mix at low speed 1 minute. Continue mixing at medium speed 10 to 15 minutes or until dough is smooth and elastic. Dough temperature should be between 78 F. to 82 F.
6. FERMENT: Cover and set in warm place, 80 F. for 2 hours or until double in bulk.
7. PUNCH: Fold sides into center and turn dough completely over. Let rest 30 minutes.
8. MAKE UP: Scale into approximately 8 1-3/4 pound pieces; shape each piece into a smooth ball; let rest 12 to 15 minutes. Mold each piece into an oblong loaf; place each loaf seam-side down into lightly greased pan.
9. PROOF: At 90 F. to 100 F. about 1 hour or until double in bulk.
   BAKE: Using a convection oven, bake at 375 F. until done, on low fan with open vent.
   When cool, slice 25 slices, about 1/2 inch thick, per loaf.

Notes

1. In Step 8, when using 9 x 4-1/4 x 2-3/4 bread pans, scale into 12-18 ounce pieces.
2. For Semi-Automated Equipment: Follow Steps 1 through 7. In Step 8, scale into 8-27-ounce pieces; shape each piece into a smooth ball; let rest 12 to 15 minutes. Using a 10-inch pressure plate, feed balls one at a time into bread molding machine. Pan seam-side down into lightly greased bread pans. Follow Steps 9 through 11.

BREADS AND SWEET DOUGHS No.D 009 00

WHITE BREAD (SHORT-TIME FORMULA)

Yield 100   Portion 2 Slices

Calories	Carbohydrates	Protein	Fat	Cholesterol	Sodium	Calcium
184 cal	34 g	6 g	2 g	0 mg	217 mg	20 mg

Ingredient

Ingredient	Weight	Measure	Issue
YEAST,ACTIVE,DRY	3 oz	1/4 cup 3-1/3 tbsp	
WATER,WARM	1 lbs	2 cup	
SUGAR,GRANULATED	3/4 oz	1 tbsp	
WATER	4-1/8 lbs	2 qts	
MILK,NONFAT,DRY	3-5/8 oz	1-1/2 cup	
SUGAR,GRANULATED	3-1/2 oz	1/2 cup	
FLOUR,WHEAT,BREAD	7-1/4 lbs	1 gal 2 qts	
SHORTENING,SOFTENED	5-7/8 oz	3/4 cup 1 tbsp	
FLOUR,WHEAT,BREAD	2-3/8 lbs	2 qts	
SALT	1-7/8 oz	3 tbsp	

Method

1. Sprinkle yeast over water. DO NOT USE TEMPERATURES ABOVE 110 F. Mix well. Let stand 5 minutes. Add sugar; stir until dissolved. Let stand 10 minutes; stir. Set aside for use in Step 3.
2. Place water in mixer bowl. Add milk and sugar. Using a dough hook, mix at low speed about 1 minute until blended.
3. Add flour; mix at low speed about 2 minutes or until flour is incorporated; add shortening and yeast solution. Mix at low speed about 2 minutes until smooth.
4. Mix at medium speed 10 minutes.
5. Let rise in mixer bowl 20 minutes.
6. Sift together flour and salt; add to mixture in mixer bowl. Mix at low speed 2 minutes or until flour in incorporated. Mix at medium speed 10 minutes or until smooth and elastic.
7. FERMENT: Cover. Set in warm place (80 F.) 25 to 30 minutes or until double in bulk.
8. MAKE UP: Scale into 8-28 ounce pieces. Roll scaled dough to pan size; place 1 loaf into each lightly greased bread pan.
9. PROOF: At 90 F. for 25 to 30 minutes or until double in bulk.
   BAKE: Using a convection oven, bake at 400 F. for 3 to 5 minutes on high fan, open vent. Reduce oven temperature to 325 F. and bake 15 to 18 minutes or until done.
   When cool, slice 25 slices, about 1/2-inch thick, per loaf.

Notes

1. In Step 8, when using 9 x 4-1/2 x 2-3/4 bread pans, scale into 10-22 ounce pieces.

BREADS AND SWEET DOUGHS No.D 011 00

PUMPKIN BREAD

Yield 100   Portion 1 Slice

Calories	Carbohydrates	Protein	Fat	Cholesterol	Sodium	Calcium
272 cal	40 g	4 g	12 g	30 mg	302 mg	23 mg

Ingredient	Weight	Measure	Issue
EGGS,WHOLE,FROZEN	1-1/2 lbs	2-7/8 cup	
FLOUR,WHEAT,GENERAL PURPOSE	3-1/3 lbs	3 qts	
SALT	1-1/4 oz	2 tbsp	
BAKING POWDER	1/3 oz	1/3 tsp	
BAKING SODA	1-1/3 oz	2-2/3 tbsp	
CINNAMON,GROUND	1/3 oz	1 tbsp	
ALLSPICE,GROUND	1/4 oz	1 tbsp	
NUTMEG,GROUND	1/3 oz	1 tbsp	
CLOVES,GROUND	1/4 oz	1 tbsp	
SUGAR,GRANULATED	5-1/4 lbs	3 qts	
OIL,SALAD	1-7/8 lbs	1 qts	
PUMPKIN,CANNED,SOLID PACK	3-3/4 lbs	1 qts 3 cup	
WATER	1 lbs	2 cup	
NUTS,UNSALTED,CHOPPED,COARSELY	10-1/3 oz	2 cup	
RAISINS	10-1/4 oz	2 cup	
COOKING SPRAY,NONSTICK	2 oz	1/4 cup 1/3 tbsp	

Method

1 Beat eggs in mixer bowl at medium speed 3 minutes or until lemon colored.
2 Blend flour, salt, baking powder, baking soda, cinnamon, allspice, nutmeg, and cloves together in separate bowl.
3 Add flour mixture, sugar, salad oil, pumpkin, water, nuts, and raisins to beaten eggs.
4 Beat at low speed about 1/2 minute. Beat 1 minute or until well blended. DO NOT OVER BEAT.
5 Lightly spray each pan with non-stick cooking spray. Pour about 7-1/2 cups of batter into each sprayed pan.
6 Using a convection oven, bake at 325 F. about 70 minutes or until done on low fan, open vent. Let cool in pans 5 to 10 minutes before removing from pans.
7 Cool thoroughly; wrap in waxed paper; store overnight before slicing.
8 Cut 25 slices per loaf.

BREADS AND SWEET DOUGHS No.D 012 00

CRUMB CAKE SNICKERDOODLE

Yield 100   Portion   1 Piece

Calories	Carbohydrates	Protein	Fat	Cholesterol	Sodium	Calcium
303 cal	46 g	5 g	12 g	36 mg	212 mg	81 mg

Ingredient	Weight	Measure	Issue
SHORTENING	1-1/8 lbs	2-1/2 cup	
SUGAR,GRANULATED	3 lbs	1 qts 2-3/4 cup	
EGGS,WHOLE,FROZEN	1-3/4 lbs	3-1/4 cup	
FLOUR,WHEAT,GENERAL PURPOSE	5 lbs	1 gal 1/2 qts	
BAKING POWDER	2-3/4 oz	1/4 cup 2 tbsp	
MILK,NONFAT,DRY	1-3/4 oz	3/4 cup	
NUTMEG,GROUND	1/4 oz	1 tbsp	
SALT	3/4 oz	1 tbsp	
WATER	2 lbs	3-3/4 cup	
EXTRACT,VANILLA	3/4 oz	1 tbsp	
RAISINS	1-3/4 lbs	1 qts 1-1/2 cup	
NUTS,UNSALTED,CHOPPED,COARSELY	1-1/4 lbs	1 qts	
COOKING SPRAY,NONSTICK	2 oz	1/4 cup 1/3 tbsp	
SUGAR,BROWN,PACKED	1-1/3 lbs	1 qts 1/4 cup	
MARGARINE	8 oz	1 cup	
CINNAMON,GROUND	1 oz	1/4 cup 1/3 tbsp	
YELLOW CAKE (CRUMBS)		3 cup	

Method

1 Place shortening and sugar in mixer bowl; cream at medium speed until light and fluffy.
2 Add eggs; beat at medium speed 2 minutes or until light and fluffy.
3 Sift together flour, baking powder, milk, nutmeg, and salt.
4 Add vanilla to water; add alternately with dry ingredients to mixture. Mix 1-1/2 minutes at low speed.
5 Fold raisins and nuts into batter.
6 Lightly spray each pan with non-stick cooking spray. Pour about 3-1/2 quarts of batter into each lightly sprayed pan.
7 Mix brown sugar, butter or margarine, cinnamon, and cake crumbs until mixture resembles cornmeal. Sprinkle about 1 quart of mixture over batter in each pan.
8 Using a convection oven, bake at 300 F. for 20 to 25 minutes or until done on low fan, open vent.
9 Cool; cut 6 by 9. If desired, top with Vanilla Glaze, Recipe No. D 046 00.

BREADS AND SWEET DOUGHS No.D 012 01

CRUMB CAKE SNICKERDOODLE (CK MIX, YELLOW)

Yield 100   Portion 1 Piece

Calories	Carbohydrates	Protein	Fat	Cholesterol	Sodium	Calcium
257 cal	36 g	3 g	12 g	11 mg	290 mg	25 mg

Ingredient   **Weight**   **Measure**   **Issue**
CAKE MIX,YELLOW   10 lbs
NUTMEG,GROUND   1/4 oz   1 tbsp

Method

1 Prepare mix according to instructions on container.
2 Add nutmeg.
3 Using a convection oven, bake at 300 F. for 25 minutes or until done on low fan, open vent.

BREADS AND SWEET DOUGHS No.D 013 00

BAGELS

Yield 100   Portion 1 Bagel

Calories	Carbohydrates	Protein	Fat	Cholesterol	Sodium	Calcium
242 cal	48 g	8 g	2 g	0 mg	375 mg	11 mg

Ingredient	Weight	Measure	Issue
YEAST,ACTIVE,DRY	3-3/8 oz	1/2 cup	
WATER,WARM	5-3/4 lbs	2 qts 3 cup	
SUGAR,GRANULATED	7 oz	1 cup	
SALT	3-3/8 oz	1/4 cup 1-2/3 tbsp	
FLOUR,WHEAT,BREAD	13-7/8 lbs	2 gal 3-1/2 qts	
COOKING SPRAY,NONSTICK	2 oz	1/4 cup 1/3 tbsp	

Method

1. Sprinkle yeast over water in mixer bowl. DO NOT USE TEMPERATURES ABOVE 110 F. Mix well. Let stand 5 minutes; stir.
2. Using a wire whip, add sugar and salt to yeast solution; stir until ingredients are dissolved.
3. Using a dough hook, add flour; mix at low speed 1 minute or until all flour is incorporated into liquid. Continue mixing at medium speed 13 to 15 minutes until dough is smooth and elastic. (Dough will be very stiff). Dough temperature should be 78 F. to 82 F.
4. Cover; let rest 15 minutes.
5. Place dough on unfloured work surface; divide dough into 3 ounce pieces; knead briefly; shape into balls by rolling in circular motion on work surface.
6. Place balls, in rows 4 by 6, on 4 ungreased sheet pans.
7. FERMENT: Cover. Set in warm place (80 F.) about 15 to 20 minutes or until dough increases slightly in bulk.
8. MAKE UP: Shape bagels like a doughnut; flatten to 2-1/2-inch circles, 3/4-inch thick. Pinch center of each bagel with thumb and forefinger and pull gently to make a 1-inch diameter hole and a total 3-1/2-inch diameter, keeping uniform shape. Place on 4 ungreased sheet pans in rows 4 by 6 per pan.
9. PROOF: At 90 F. until bagels begin to rise, about 20 to 30 minutes.
   Lightly spray 5 sheet pans with non-stick cooking spray. Sprinkle each pan with 1/2 cup cornmeal.
   Add water to steam-jacketed kettle or stock pot; bring to a boil; reduce heat to a simmer. Add 1/2 cup granulated sugar to water. Stir until dissolved. Gently drop bagels, one at a time, into water. Cook 30 seconds; turn; cook 30 seconds. Remove bagels with slotted spoon; drain. Place on sheet pans in rows 4 by 5.
   BAKE: 30 to 35 minutes or until golden brown and crisp in 400 F. oven. Remove from pans; cool on wire racks.

Notes

1. In Step 1, a 60-quart mixer should be used for 100 portions as dough is very stiff. If using 20 to 30 quart mixers, prepare no more than 50 portions at a time.
2. In Steps 7 and 9, bagels should not double in bulk.
3. In Step 12, if convection oven is used, bake at 350 F. for 15 to 20 minutes on high fan, open vent.

BREADS AND SWEET DOUGHS No.D 014 00

CORN BREAD

Yield 100   Portion 1 Piece

Calories	Carbohydrates	Protein	Fat	Cholesterol	Sodium	Calcium
212 cal	30 g	5 g	8 g	30 mg	359 mg	127 mg

Ingredient	Weight	Measure	Issue
FLOUR,WHEAT,GENERAL PURPOSE	3-7/8 lbs	3 qts 2 cup	
CORN MEAL	3-2/3 lbs	3 qts	
MILK,NONFAT,DRY	6 oz	2-1/2 cup	
SUGAR,GRANULATED	7 oz	1 cup	
BAKING POWDER	5-7/8 oz	3/4 cup	
SALT	1-1/2 oz	2-1/3 tbsp	
EGGS,WHOLE,FROZEN	1-1/2 lbs	2-7/8 cup	
WATER	7-7/8 lbs	3 qts 3 cup	
OIL,SALAD	1-1/2 lbs	3 cup	
COOKING SPRAY,NONSTICK	2 oz	1/4 cup 1/3 tbsp	

Method

1. Blend flour, cornmeal, milk, sugar, baking powder, and salt in mixer bowl.
2. Combine eggs and water; add to ingredients in mixer bowl. Blend at low speed about 1 minute. Scrape down bowl.
3. Add oil; mix at medium speed until blended.
4. Lightly spray each pan with non-stick cooking spray. Pour 1 gallon of batter into each pan.
5. Using a convection oven, bake at 375 F. for 20 minutes or until done on low fan, open vent.
6. Cool; cut into 6 by 9.

Notes

1. In step 1, omit sugar if southern-style cornbread is desired.

BREADS AND SWEET DOUGHS No.D 014 01

CORN MUFFINS

Yield 100 Portion 1 Muffin

Calories	Carbohydrates	Protein	Fat	Cholesterol	Sodium	Calcium
160 cal	22 g	4 g	6 g	24 mg	252 mg	95 mg

Ingredient	Weight	Measure	Issue
FLOUR,WHEAT,GENERAL PURPOSE	2-7/8 lbs	2 qts 2-1/2 cup	
CORN MEAL	2-3/4 lbs	2 qts 1 cup	
MILK,NONFAT,DRY	4-1/2 oz	1-7/8 cup	
SUGAR,GRANULATED	5-1/4 oz	3/4 cup	
BAKING POWDER	4-3/8 oz	1/2 cup 1 tbsp	
SALT	1 oz	1 tbsp	
EGGS,WHOLE,FROZEN	1-1/4 lbs	2-1/4 cup	
WATER	6 lbs	2 qts 3-1/2 cup	
OIL,SALAD	1-1/8 lbs	2-1/4 cup	
COOKING SPRAY,NONSTICK	2 oz	1/4 cup 1/3 tbsp	

Method

1. Blend flour, cornmeal, milk, sugar, baking powder, and salt in mixer bowl.
2. Combine eggs and water; add to ingredients in mixer bowl. Blend at low speed about 1 minute. Scrape down bowl.
3. Add oil; mix at medium speed until blended.
4. Lightly spray 9-12 cup muffin pans with non-stick cooking spray. Fill each cup 2/3 full.
5. Bake for 15 to 20 minutes at 425 F. or at 375 F. in a convection oven for 15 minutes or until done on low fan, open vent.

BREADS AND SWEET DOUGHS No.D 014 02

HUSH PUPPIES

Yield 100   **Portion** 3 Each

Calories	Carbohydrates	Protein	Fat	Cholesterol	Sodium	Calcium
200 cal	28 g	5 g	7 g	30 mg	359 mg	129 mg

Ingredient	Weight	Measure	Issue
FLOUR,WHEAT,GENERAL PURPOSE	3-7/8 lbs	3 qts 2 cup	
CORN MEAL	3-2/3 lbs	3 qts	
MILK,NONFAT,DRY	6 oz	2-1/2 cup	
BAKING POWDER	5-7/8 oz	3/4 cup	
SALT	1-1/2 oz	2-1/3 tbsp	
EGGS,WHOLE,FROZEN	1-1/2 lbs	2-7/8 cup	
WATER	5-3/4 lbs	2 qts 3 cup	
ONIONS,FRESH,CHOPPED	2-1/8 lbs	1 qts 2 cup	2-1/3 lbs
PEPPER,BLACK,GROUND	1/3 oz	1 tbsp	
SHORTENING,VEGETABLE,MELTED	1-1/3 lbs	3 cup	

Method

1. Blend flour, cornmeal, milk, baking powder, and salt in mixer bowl.
2. Combine eggs, water, onions, and pepper; add to ingredients in mixer bowl. Blend at low speed for minute. Scrape down bowl.
3. Add shortening; mix at medium speed until blended.
4. Drop batter by rounded tablespoon into deep fat at around 360 F.; fry about 3 minutes. Drain on absorbent paper.

BREADS AND SWEET DOUGHS No.D 014 03

JALAPENO CORN BREAD

Yield 100   Portion 1 Piece

Calories	Carbohydrates	Protein	Fat	Cholesterol	Sodium	Calcium
224 cal	30 g	5 g	9 g	33 mg	391 mg	143 mg

Ingredient	**Weight**	**Measure**	**Issue**
FLOUR,WHEAT,GENERAL PURPOSE	3-7/8 lbs	3 qts 2 cup	
CORN MEAL	3-2/3 lbs	3 qts	
MILK,NONFAT,DRY	6 oz	2-1/2 cup	
SUGAR,GRANULATED	7 oz	1 cup	
BAKING POWDER	5-7/8 oz	3/4 cup	
SALT	1-1/2 oz	2-1/3 tbsp	
EGGS,WHOLE,FROZEN	1-1/2 lbs	2-7/8 cup	
WATER	7-7/8 lbs	3 qts 3 cup	
OIL,SALAD	1-1/2 lbs	3 cup	
CORN,CANNED,WHOLE KERNEL,DRAINED	11-5/8 oz	2 cup	
CHEESE,CHEDDAR,GRATED	8 oz	2 cup	
PEPPERS,JALAPENOS,CANNED,DRAINED,CHOPPED	2-3/8 oz	1/2 cup	
ONIONS,FRESH,GRATED	1-3/8 oz	1/4 cup 1/3 tbsp	1-5/8 oz
COOKING SPRAY,NONSTICK	2 oz	1/4 cup 1/3 tbsp	

Method

1 Blend flour, cornmeal, milk, sugar, baking powder, and salt in mixer bowl.
2 Combine eggs and water; add to ingredients in mixer bowl. Blend at low speed about 1 minute. Scrape down bowl.
3 Add shortening, drained corn, cheese, drained jalapeno peppers, and onions to mixture. Blend only until ingredients are distributed throughout mixture.
4 Lightly spray each pan with non-stick cooking spray. Pour 4-3/4 quarts batter into each pan.
5 Bake for 30 minutes at 425 F. or at 375 F. in a convection oven 20 minutes or until done on low fan, open vent.
6 Cool, cut 6 by 9.

BREADS AND SWEET DOUGHS No.D 015 00

CORN BREAD (CORN BREAD MIX)

Yield 100   **Portion** 1 Piece

Calories	Carbohydrates	Protein	Fat	Cholesterol	Sodium	Calcium
218 cal	36 g	4 g	7 g	1 mg	567 mg	29 mg

Ingredient	**Weight**	**Measure**	**Issue**
CORN BREAD MIX	11-1/4 lbs	2 gal 1/3 qts	
COOKING SPRAY, NONSTICK	2 oz	1/4 cup 1/3 tbsp	

Method

1. Prepare mix according to instructions on container.
2. Lightly spray each pan with non-stick cooking spray. Pour 1 gallon of batter into each pan.
3. Bake 20 to 25 minutes at 425 F. or if a convection oven is used, bake at 375 F. for 20 minutes or until done on low fan, open vent or until done.
4. Cool; cut 6 by 9.

Notes

1. Cornbread Mix is a slightly sweetened product. In Step 1, 2-1/8 cup of granulated sugar may be added to mix if a sweeter product is desired.

BREADS AND SWEET DOUGHS No.D 015 01

CORN MUFFINS (CORN BREAD MIX)

Yield 100   Portion   1 Muffin

Calories	Carbohydrates	Protein	Fat	Cholesterol	Sodium	Calcium
189 cal	32 g	3 g	5 g	1 mg	454 mg	23 mg

Ingredient	**Weight**	**Measure**	**Issue**
CORN BREAD MIX	9 lbs	1 gal 2-2/3 qts	
SUGAR,GRANULATED	12-1/3 oz	1-3/4 cup	
COOKING SPRAY,NONSTICK	2 oz	1/4 cup 1/3 tbsp	

Method

1 Prepare Cornbread Mix and combine with granulated sugar.
2 Lightly spray 9-12 cup muffin pans with non-stick cooking spray. Fill each cup 2/3 full.
3 Bake 15 to 20 minutes at 425 F. or in a 375 F. convection oven for 15 minutes or until done on low fan, open vent.

BREADS AND SWEET DOUGHS No.D 015 02

HUSH PUPPIES (CORN BREAD MIX)

Yield 100   Portion   3 Pieces

Calories	Carbohydrates	Protein	Fat	Cholesterol	Sodium	Calcium
257 cal	36 g	4 g	11 g	1 mg	567 mg	31 mg

Ingredient   Weight   Measure   Issue

CORN BREAD MIX   11-1/4 lbs   2 gal 1/3 qts
ONIONS,FRESH,CHOPPED   2-1/8 lbs   1 qts 2 cup   2-1/3 lbs
PEPPER,BLACK,GROUND   1/3 oz   1 tbsp

Method

1 Prepare mix according to instructions on container. Add finely chopped onions and black or white pepper.
2 Drop batter by rounded tablespoon into deep fat, at around 360 F.; fry about 3 minutes. Drain on absorbent paper.

BREADS AND SWEET DOUGHS No.D 015 03

JALAPENO CORN BREAD (CORN BREAD MIX)

Yield 100   Portion 1 Piece

Calories	Carbohydrates	Protein	Fat	Cholesterol	Sodium	Calcium
230 cal	36 g	4 g	8 g	3 mg	599 mg	46 mg

Ingredient	Weight	Measure	Issue
CORN BREAD MIX	11-1/4 lbs	2 gal 1/3 qts	
CORN,CANNED,WHOLE KERNEL,DRAINED	11-5/8 oz	2 cup	
CHEESE,CHEDDAR,GRATED	8 oz	2 cup	
PEPPERS,JALAPENOS,CANNED,DRAINED,CHOPPED	2-3/8 oz	1/2 cup	
ONIONS,FRESH,GRATED	2-7/8 oz	1/2 cup	3-1/8 oz
COOKING SPRAY,NONSTICK	2 oz	1/4 cup 1/3 tbsp	

Method

1. Prepare mix according to instructions on container. Add drained whole kernel corn, grated Cheddar or American cheese, jalapeno peppers, and onions. Blend only until ingredients are distributed.
2. Lightly spray each pan with non-stick cooking spray. Pour 4-3/4 quarts of batter into each pan.
3. Bake 30 minutes at 425 F. or in a 375 F. convection oven for 20 minutes or until done on low fan, open vent.
4. Cool; cut 6 by 9.

BREADS AND SWEET DOUGHS No. D 016 00

CROUTONS

Yield 100   **Portion** 8 Croutons

Calories	Carbohydrates	Protein	Fat	Cholesterol	Sodium	Calcium
24 cal	4 g	1 g	0 g	0 mg	49 mg	10 mg

Ingredient	**Weight**	**Measure**	**Issue**
BREAD,WHITE,STALE,SLICED	2 lbs	1 gal 2-1/2 qts	

Method

1. Trim crusts from bread; cut bread into 1/2-inch cubes.
2. Place bread cubes on sheet pans. Brown lightly in 325 F. oven, about 20 to 25 minutes or in 375 F. convection oven, about 6 minutes on high fan, open vent.

Notes

1. In Step 1, 2 lbs bread will yield about 1 gallons lightly browned croutons.

BREADS AND SWEET DOUGHS No.D 016 01

GARLIC CROUTONS

Yield 100 **Portion** 1/4 Cup

Calories	Carbohydrates	Protein	Fat	Cholesterol	Sodium	Calcium
49 cal	4 g	1 g	3 g	8 mg	77 mg	11 mg

Ingredient	Weight	Measure	Issue
BREAD,WHITE,STALE,SLICED	2 lbs	1 gal 2-1/2 qts	
BUTTER,MELTED	12 oz	1-1/2 cup	
GARLIC CLOVES,FRESH,MINCED	1/8 oz	1/4 tsp	

Method

1. Trim crusts from bread; cut bread into 1/2-inch cubes.
2. Place bread cubes on sheet pans. Brown lightly in 325 F. oven, about 20 to 25 minutes or in 375 F. convection oven for about 6 minutes on high fan, open vent.
3. Melt butter or margarine; blend in minced garlic. Pour mixture evenly over lightly browned croutons in steam table pans; toss lightly.

Notes

1. In Step 1, 2 lbs bread will yield about 1 gallon lightly browned croutons.

BREADS AND SWEET DOUGHS No.D 016 02

PARMESAN CROUTONS

Yield 100   **Portion**   8 Croutons

Calories	Carbohydrates	Protein	Fat	Cholesterol	Sodium	Calcium
55 cal	4 g	1 g	4 g	9 mg	105 mg	31 mg

Ingredient	Weight	Measure	Issue
BREAD,WHITE,STALE,SLICED	2 lbs	1 gal 2-1/2 qts	
BUTTER,MELTED	12 oz	1-1/2 cup	
CHEESE,PARMESAN,GRATED	5-1/4 oz	1-1/2 cup	

Method

1. Trim crusts from bread; cut bread into 1/2-inch cubes.
2. Place bread cubes on sheet pans. Brown lightly in 325 F. oven, 20 to 25 minutes or in 375 F. convection oven, 6 minutes on high fan, open vent.
3. Melt butter or margarine; blend in grated Parmesan cheese. Pour mixture over lightly browned croutons in steam table pans; toss lightly.

Notes

1. In Step 1, 2 lbs bread will yield about 1 gallon lightly browned croutons.

BREADS AND SWEET DOUGHS No.D 017 00

EGG WASH

Yield 100   **Portion**   3 Cups

Calories	Carbohydrates	Protein	Fat	Cholesterol	Sodium	Calcium
428 cal	16 g	36 g	23 g	989 mg	457 mg	458 mg

Ingredient   Weight   Measure   Issue

Ingredient	Weight	Measure	Issue
EGGS,WHOLE,FROZEN	8 oz	3/4 cup 3 tbsp	
MILK,NONFAT,DRY	7/8 oz	1/4 cup 2-1/3 tbsp	
WATER	1 lbs	2 cup	

Method

1. Combine eggs, milk, and water; mix well. CCP: Refrigerate at 41 F. or lower until ready to use.
2. Brush over shaped dough before or after proofing.

Notes

1. In Step 1, 2-1/2 ounces canned dehydrated egg mix combined with 3/4 cup warm water may be used for whole eggs.

BREADS AND SWEET DOUGHS No.D 017 01

EGG WHITE WASH

Yield 100　　　　　　　　　　　　　　　　　**Portion** 3 Cups

Calories	Carbohydrates	Protein	Fat	Cholesterol	Sodium	Calcium
81 cal	2 g	17 g	0 g	0 mg	280 mg	20 mg

Ingredient	Weight	Measure	Issue
EGG WHITES	5-2/3 oz	1/2 cup 2-2/3 tbsp	
WATER	1-1/8 lbs	2-1/4 cup	

Method

1. Beat egg whites and water together. CCP: Refrigerate at 41 F. or lower until ready for use.
2. Brush over shaped dough before or after proofing.

BREADS AND SWEET DOUGHS No.D 018 00

CAKE DOUGHNUTS (HOMEMADE)

Yield 100   **Portion** 1 Doughnut

Calories	Carbohydrates	Protein	Fat	Cholesterol	Sodium	Calcium
186 cal	26 g	3 g	7 g	24 mg	197 mg	78 mg

Ingredient	Weight	Measure	Issue
FLOUR,WHEAT,GENERAL PURPOSE	5-1/2 lbs	1 gal 1 qts	
BAKING POWDER	3-7/8 oz	1/2 cup	
MILK,NONFAT,DRY	1-5/8 oz	1/2 cup 2-2/3 tbsp	
SALT	5/8 oz	1 tbsp	
NUTMEG,GROUND	1/4 oz	1 tbsp	
SHORTENING	7-1/4 oz	1 cup	
SUGAR,GRANULATED	1-1/2 lbs	3-3/8 cup	
EGGS,WHOLE,FROZEN	1-1/4 lbs	2-1/4 cup	
WATER	2 lbs	3-3/4 cup	
EXTRACT,VANILLA	1/2 oz	1 tbsp	

Method

1. Sift together flour, baking powder, milk, salt, and nutmeg. Set aside for use in Step 5.
2. Place shortening and sugar in mixer bowl; cream at medium speed until light and fluffy.
3. Add eggs; beat at medium speed until light and fluffy.
4. Combine water and vanilla. Add to creamed mixture.
5. Add dry ingredients to creamed mixture alternately with liquids; add about 1/3 flour mixture each time. Blend at low speed after each addition. DO NOT OVERMIX. Let dough rest 10 minutes.
6. Roll dough 3/8-inch thick on well-floured board; cut with doughnut cutter.
7. Fry 1 minute on each side or until golden brown. Drain on absorbent paper.

Notes

1. In Step 5, dough may be chilled 1 hour for ease in handling.
2. Omit Steps 6 and 7 if dough machine is used.

BREADS AND SWEET DOUGHS No.D 018 01

SUGAR COATED DOUGHNUTS

Yield 100   Portion 1 Doughnut

Calories	Carbohydrates	Protein	Fat	Cholesterol	Sodium	Calcium
203 cal	31 g	3 g	7 g	24 mg	197 mg	78 mg

Ingredient	Weight	Measure	Issue
FLOUR,WHEAT,GENERAL PURPOSE	5-1/2 lbs	1 gal 1 qts	
BAKING POWDER	3-7/8 oz	1/2 cup	
MILK,NONFAT,DRY	1-5/8 oz	1/2 cup 2-2/3 tbsp	
SALT	5/8 oz	1 tbsp	
NUTMEG,GROUND	1/4 oz	1 tbsp	
SHORTENING	7-1/4 oz	1 cup	
SUGAR,GRANULATED	1-1/2 lbs	3-3/8 cup	
EGGS,WHOLE,FROZEN	1-1/4 lbs	2-1/4 cup	
WATER	2 lbs	3-3/4 cup	
EXTRACT,VANILLA	1/2 oz	1 tbsp	
SUGAR,GRANULATED	1 lbs	2-1/4 cup	

Method

1. Sift together flour, baking powder, milk, salt, and nutmeg. Set aside for use in Step 5.
2. Place shortening and sugar in mixer bowl; cream at medium speed until light and fluffy.
3. Add eggs; beat at medium speed until light and fluffy.
4. Combine water and vanilla. Add to creamed mixture.
5. Add dry ingredients to creamed mixture alternately with liquids; add about 1/3 flour mixture each time. Blend at low speed after each addition. DO NOT OVERMIX. Let dough rest 10 minutes.
6. Roll dough 3/8 inch thick on well-floured board; cut with doughnut cutter.
7. Fry 1 minute on each side or until golden brown. Drain on absorbent paper. While doughnuts are warm, roll in granulated sugar or in sifted powdered sugar.

Notes

1. In Step 5, dough may be chilled 1 hour for ease in handling.
2. Omit Steps 6 and 7 if dough machine is used.

BREADS AND SWEET DOUGHS No.D 018 02

CAKE DOUGHNUTS (DOUGHNUT MIX)

Yield 100	Portion 1 Doughnut

Calories	Carbohydrates	Protein	Fat	Cholesterol	Sodium	Calcium
199 cal	31 g	4 g	7 g	0 mg	323 mg	36 mg

Ingredient	Weight	Measure	Issue
DOUGHNUT MIX,CANNED	9 lbs	1 gal 4 qts	

Method

1 Use canned Doughnut Mix. Prepare according to instructions on container.

BREADS AND SWEET DOUGHS No.D 018 03

CHOCOLATE DOUGHNUTS

Yield 100 Portion 1 Doughnut

Calories	Carbohydrates	Protein	Fat	Cholesterol	Sodium	Calcium
190 cal	27 g	4 g	8 g	24 mg	198 mg	80 mg

Ingredient	Weight	Measure	Issue
FLOUR,WHEAT,GENERAL PURPOSE	5-1/2 lbs	1 gal 1 qts	
COCOA	6-1/8 oz	2 cup	
BAKING POWDER	3-7/8 oz	1/2 cup	
MILK,NONFAT,DRY	1-5/8 oz	1/2 cup 2-2/3 tbsp	
SALT	5/8 oz	1 tbsp	
NUTMEG,GROUND	1/4 oz	1 tbsp	
SHORTENING	7-1/4 oz	1 cup	
SUGAR,GRANULATED	1-1/2 lbs	3-3/8 cup	
EGGS,WHOLE,FROZEN	1-1/4 lbs	2-1/4 cup	
WATER	2 lbs	3-3/4 cup	
EXTRACT,VANILLA	1/2 oz	1 tbsp	

Method

1. Sift together flour, cocoa, baking powder, milk, salt, and nutmeg. Set aside for use in Step 5.
2. Place shortening and sugar in mixer bowl; cream at medium speed until light and fluffy.
3. Add eggs; beat at medium speed until light and fluffy.
4. Combine water and vanilla. Add to creamed mixture.
5. Add dry ingredients to creamed mixture alternately with liquids; add about 1/3 flour mixture each time. Blend at low speed after each addition. DO NOT OVERMIX. Let dough rest 10 minutes.
6. Roll dough 3/8-inch thick on well-floured board; cut with doughnut cutter.
7. Fry 1 minute on each side or until golden brown. Drain on absorbent paper towels. Glaze or coat if desired.

Notes

1. In Step 5, dough may be chilled 1 hour for ease in handling.
2. Omit Steps 6 and 7 if dough machine is used.

BREADS AND SWEET DOUGHS No.D 018 04

CINNAMON SUGAR DOUGHNUTS

Yield 100　　　　　　　　　　　　　　　　**Portion** 1 Doughnut

Calories	Carbohydrates	Protein	Fat	Cholesterol	Sodium	Calcium
154 cal	28 g	3 g	3 g	24 mg	198 mg	81 mg

Ingredient	Weight	Measure	Issue
FLOUR,WHEAT,GENERAL PURPOSE	5-1/2 lbs	1 gal 1 qts	
BAKING POWDER	3-7/8 oz	1/2 cup	
MILK,NONFAT,DRY	1-5/8 oz	1/2 cup 2-2/3 tbsp	
SALT	5/8 oz	1 tbsp	
NUTMEG,GROUND	1/4 oz	1 tbsp	
SHORTENING	7-1/4 oz	1 cup	
SUGAR,GRANULATED	1-1/2 lbs	3-3/8 cup	
EGGS,WHOLE,FROZEN	1-1/4 lbs	2-1/4 cup	
WATER	2 lbs	3-3/4 cup	
EXTRACT,VANILLA	1/2 oz	1 tbsp	
CINNAMON SUGAR FILLING		2 cup	

Method

1. Sift together flour, baking powder, milk, salt, and nutmeg. Set aside for use in Step 5.
2. Place shortening and sugar in mixer bowl; cream at medium speed until light and fluffy.
3. Add eggs; beat at medium speed until light and fluffy.
4. Combine water and vanilla. Add to creamed mixture.
5. Add dry ingredients to creamed mixture alternately with liquids; add about 1/3 flour mixture each time. Blend at low speed after each addition. DO NOT OVERMIX. Let dough rest 10 minutes.
6. Roll dough 3/8 inch thick on well-floured board; cut with doughnut cutter.
7. Fry 1 minute on each side or until golden brown. Drain on absorbent paper towels. While doughnuts are still warm, roll in Cinnamon Sugar Filling, Recipe No. D 042 00.

Notes

1. In Step 5, dough may be chilled 1 hour for ease in handling.
2. Omit Steps 6 and 7 if dough machine is used.

BREADS AND SWEET DOUGHS No.D 018 05

GLAZED NUT DOUGHNUTS

Yield 100	Portion 1 Doughnut

Calories	Carbohydrates	Protein	Fat	Cholesterol	Sodium	Calcium
298 cal	47 g	5 g	11 g	26 mg	208 mg	82 mg

Ingredient	Weight	Measure	Issue
FLOUR,WHEAT,GENERAL PURPOSE	5-1/2 lbs	1 gal 1 qts	
BAKING POWDER	3-7/8 oz	1/2 cup	
MILK,NONFAT,DRY	1-5/8 oz	1/2 cup 2-2/3 tbsp	
SALT	5/8 oz	1 tbsp	
NUTMEG,GROUND	1/4 oz	1 tbsp	
SHORTENING	7-1/4 oz	1 cup	
SUGAR,GRANULATED	1-1/2 lbs	3-3/8 cup	
EGGS,WHOLE,FROZEN	1-1/4 lbs	2-1/4 cup	
WATER	2 lbs	3-3/4 cup	
EXTRACT,VANILLA	1/2 oz	1 tbsp	
VANILLA GLAZE		2-3/4 cup	
NUTS,UNSALTED,CHOPPED,COARSELY	1 lbs	3-1/8 cup	

Method

1. Sift together flour, baking powder, milk, salt, and nutmeg. Set aside for use in Step 5.
2. Place shortening and sugar in mixer bowl; cream at medium speed until light and fluffy.
3. Add eggs; beat at medium speed until light and fluffy.
4. Combine water and vanilla. Add to creamed mixture.
5. Add dry ingredients to creamed mixture alternately with liquids; add about 1/3 flour mixture each time. Blend at low speed after each addition. DO NOT OVERMIX. Let dough rest 10 minutes.
6. Roll dough 3/8-inch thick on well-floured board; cut with doughnut cutter.
7. Fry 1 minute on each side or until golden brown. Drain on absorbent paper.
8. Prepare Vanilla Glaze, Recipe No. D 046 00. Keep glaze warm; dip 1 side of doughnut into glaze, then into chopped, unsalted nuts. Place on racks to drain.

Notes

1. In Step 5, dough may be chilled 1 hour for ease in handling.
2. Omit Steps 6 and 7 if dough machine is used.

BREADS AND SWEET DOUGHS No.D 018 06

GLAZED COCONUT DOUGHNUTS

Yield 100   **Portion** 1 Doughnut

Calories	Carbohydrates	Protein	Fat	Cholesterol	Sodium	Calcium
300 cal	49 g	4 g	10 g	26 mg	222 mg	79 mg

Ingredient	**Weight**	**Measure**	**Issue**
FLOUR,WHEAT,GENERAL PURPOSE	5-1/2 lbs	1 gal 1 qts	
BAKING POWDER	3-7/8 oz	1/2 cup	
MILK,NONFAT,DRY	1-5/8 oz	1/2 cup 2-2/3 tbsp	
SALT	5/8 oz	1 tbsp	
NUTMEG,GROUND	1/4 oz	1 tbsp	
SHORTENING	7-1/4 oz	1 cup	
SUGAR,GRANULATED	1-1/2 lbs	3-3/8 cup	
EGGS,WHOLE,FROZEN	1-1/4 lbs	2-1/4 cup	
WATER	2 lbs	3-3/4 cup	
EXTRACT,VANILLA	1/2 oz	1 tbsp	
VANILLA GLAZE		2-3/4 cup	
COCONUT,PREPARED,SWEETENED FLAKES	1-1/4 lbs	1 qts 2 cup	

Method

1. Sift together flour, baking powder, milk, salt, and nutmeg. Set aside for use in Step 5.
2. Place shortening and sugar in mixer bowl; cream at medium speed until light and fluffy.
3. Add eggs; beat at medium speed until light and fluffy.
4. Combine water and vanilla. Add to creamed mixture.
5. Add dry ingredients to creamed mixture alternately with liquids; add about 1/3 flour mixture each time. Blend at low speed after each addition. DO NOT OVERMIX. Let dough rest 10 minutes.
6. Roll dough 3/8-inch thick on well-floured board; cut with doughnut cutter.
7. Fry 1 minute on each side or until golden brown. Drain on absorbent paper.
8. Prepare Vanilla Glaze, Recipe No D 046 00. Keep glaze warm; dip 1 side of doughnut into glaze, then into prepared, sweetened flaked coconut. Place on racks to drain.

Notes

1. In Step 5, dough may be chilled 1 hour for ease in handling.
2. Omit Steps 6 and 7 if dough machine is used.

BREADS AND SWEET DOUGHS No.D 018 07

GLAZED DOUGHNUTS

Yield 100 **Portion** 1 Doughnut

Calories	Carbohydrates	Protein	Fat	Cholesterol	Sodium	Calcium
217 cal	34 g	3 g	8 g	25 mg	201 mg	78 mg

Ingredient	**Weight**	**Measure**	**Issue**
FLOUR,WHEAT,GENERAL PURPOSE	5-1/2 lbs	1 gal 1 qts	
BAKING POWDER	3-7/8 oz	1/2 cup	
MILK,NONFAT,DRY	1-5/8 oz	1/2 cup 2-2/3 tbsp	
SALT	5/8 oz	1 tbsp	
NUTMEG,GROUND	1/4 oz	1 tbsp	
SHORTENING	7-1/4 oz	1 cup	
SUGAR,GRANULATED	1-1/2 lbs	3-3/8 cup	
EGGS,WHOLE,FROZEN	1-1/4 lbs	2-1/4 cup	
WATER	2 lbs	3-3/4 cup	
EXTRACT,VANILLA	1/2 oz	1 tbsp	
VANILLA GLAZE		2-3/4 cup	

Method

1. Sift together flour, baking powder, milk, salt, and nutmeg. Set aside for use in Step 5.
2. Place shortening and sugar in mixer bowl; cream at medium speed until light and fluffy.
3. Add eggs; beat at medium speed until light and fluffy.
4. Combine water and vanilla. Add to creamed mixture.
5. Add dry ingredients to creamed mixture alternately with liquids; add about 1/3 flour mixture each time. Blend at low speed after each addition. DO NOT OVERMIX. Let dough rest 10 minutes.
6. Roll dough 3/8-inch thick on well-floured board; cut with doughnut cutter.
7. Fry 1 minute on each side or until golden brown. Drain on absorbent paper.
8. Prepare Vanilla Glaze, Recipe No. D 046 00. Keep glaze warm; dip doughnuts to cover. Place on racks to drain.

Notes

1. In Step 5, dough may be chilled 1 hour for ease in handling.
2. Omit Steps 6 and 7 if dough machine is used.

BREADS AND SWEET DOUGHS No.D 019 00

RAISED DOUGHNUTS

Yield 100   Portion   1 Doughnut

Calories	Carbohydrates	Protein	Fat	Cholesterol	Sodium	Calcium
191 cal	26 g	4 g	8 g	11 mg	170 mg	13 mg

Ingredient	Weight	Measure	Issue
YEAST,ACTIVE,DRY	3-3/4 oz	1/2 cup 1 tbsp	
WATER,WARM	1-5/8 lbs	3 cup	
SUGAR,GRANULATED	1 lbs	2-1/4 cup	
SALT	1-1/2 oz	2-1/3 tbsp	
SHORTENING	9 oz	1-1/4 cup	
EGGS,WHOLE,FROZEN	8-5/8 oz	1 cup	
WATER,COLD	1-1/4 lbs	2-3/8 cup	
EXTRACT,VANILLA	1-3/8 oz	3 tbsp	
FLOUR,WHEAT,BREAD	3-7/8 lbs	3 qts 1 cup	
FLOUR,WHEAT,GENERAL PURPOSE	2-1/4 lbs	2 qts	
MILK,NONFAT,DRY	1-3/4 oz	3/4 cup	
NUTMEG,GROUND	1/4 oz	1 tbsp	

Method

1. Sprinkle yeast over water. DO NOT USE TEMPERATURES ABOVE 110 F. Mix well. Let stand for 5 minutes; stir. Set aside for use in Step 3.
2. Cream sugar, salt, and shortening in mixer bowl at medium speed.
3. Add eggs, yeast solution, water, and vanilla; mix at low speed until blended.
4. Sift together flours, milk, and nutmeg; add to mixture. Using dough hook, mix at low speed 1 minute or until all flour mixture is incorporated into liquid. Continue mixing at medium speed 10 minutes or until dough is smooth and elastic. Dough temperature should be between 78 F. to 82 F.
5. FERMENT: Cover. Set in warm place (80 F.) for 1-1/2 hours or until double in bulk.
6. PUNCH: Divide into 3 pieces (3 lb 8 oz); shape each piece into a smooth ball; let rest 10 to 20 minutes.
7. MAKE-UP: Roll each piece to 1/2-inch thickness. Cut with floured 3 inch doughnut cutter.
8. PROOF: Place on floured sheet pan; let rise 30 minutes or until light.
9. FRY: Until golden brown on underside. Turn; fry on other side. Drain on absorbent paper.
10. When cool, roll in granulated sugar or sifted powdered sugar or in Cinnamon Sugar Filling, Recipe No. D 042 00, or dip in Vanilla Glaze, Almond Glaze, or Rum Glaze, Recipe Nos. D 046 00, D 046 01, D 046 02. Place glazed doughnuts on racks to drain.

BREADS AND SWEET DOUGHS No.D 019 01

BEIGNETS (NEW ORLEANS DOUGHNUTS)

Yield 100   Portion 2 Each

Calories	Carbohydrates	Protein	Fat	Cholesterol	Sodium	Calcium
289 cal	38 g	5 g	13 g	13 mg	219 mg	17 mg

Ingredient	Weight	Measure	Issue
YEAST,ACTIVE,DRY	5-1/8 oz	3/4 cup	
WATER,WARM	2 lbs	3-3/4 cup	
SUGAR,GRANULATED	1-1/4 lbs	2-3/4 cup	
SALT	1-7/8 oz	3 tbsp	
SHORTENING	10-7/8 oz	1-1/2 cup	
EGGS,WHOLE,FROZEN	10-3/4 oz	1-1/4 cup	
WATER,COLD	1-1/2 lbs	2-3/4 cup	
EXTRACT,VANILLA	1-7/8 oz	1/4 cup 1/3 tbsp	
FLOUR,WHEAT,BREAD	4-7/8 lbs	1 gal	
FLOUR,WHEAT,GENERAL PURPOSE	2-3/4 lbs	2 qts 2 cup	
MILK,NONFAT,DRY	2-3/8 oz	1 cup	
SUGAR,POWDERED,SIFTED	1-1/3 lbs	1 qts 1 cup	

Method

1. Sprinkle yeast over water. DO NOT USE TEMPERATURES ABOVE 110 F. Mix well. Let stand for 5 minutes; stir. Set aside for use in Step 3.
2. Cream sugar, salt, and shortening in mixer bowl at medium speed.
3. Add eggs, yeast solution, water, and vanilla; mix at low speed until blended.
4. Sift together flours and milk; add to mixture. Using dough hook, mix at low speed 1 minute or until all flour mixture is incorporated into liquid. Continue mixing at medium speed 10 minutes or until dough is smooth and elastic. Dough temperature should be 78 F. to 82 F.
5. FERMENT: Cover and set in warm place, about 80 F., 1-1/2 hours or until double in bulk.
6. PUNCH: Divide into even pieces; shape each piece into a smooth ball; let rest 10 to 20 minutes.
7. MAKE-UP: Roll each piece onto a rectangular sheet, about 18 inches wide, 29 inches long, and 1/8-inch thick. Cut 6 by 9.
8. FRY: Until golden brown on underside. Turn and fry on other side. Drain on absorbent paper.
9. Sprinkle with sifted powdered sugar.

BREADS AND SWEET DOUGHS No.D 019 02

RAISED DOUGHNUTS (SWEET DOUGH MIX)

Yield 100 **Portion** 1 Doughnut

Calories	Carbohydrates	Protein	Fat	Cholesterol	Sodium	Calcium
194 cal	29 g	6 g	8 g	0 mg	323 mg	13 mg

Ingredient	Weight	Measure	Issue
SWEET DOUGH MIX	9 lbs	2 gal 1/8 qts	
YEAST,ACTIVE,DRY	3-3/4 oz	1/2 cup 1 tbsp	
EXTRACT,VANILLA	1/3 oz	1/3 tsp	
NUTMEG,GROUND	1/8 oz	1/3 tsp	
WATER	3-1/8 lbs	1 qts 2 cup	
SUGAR,POWDERED,SIFTED	1 lbs	1 qts	

Method

1. Prepare doughnuts according to directions on the container of Sweet Dough Mix.
2. When cool, roll in granulated sugar or sifted powdered sugar or in Cinnamon Sugar Filling, Recipe No. D 042 00, or dip in Vanilla Glaze, Almond Glaze, or Rum Glaze, Recipe Nos. D 046 00, D 046 01, D 046 02. Place glazed doughnuts on racks to drain.

BREADS AND SWEET DOUGHS No.D 019 03

LONGJOHNS

Yield 100   Portion 1 Each

Calories	Carbohydrates	Protein	Fat	Cholesterol	Sodium	Calcium
191 cal	26 g	4 g	8 g	11 mg	170 mg	13 mg

Ingredient	Weight	Measure	Issue
YEAST,ACTIVE,DRY	3-3/4 oz	1/2 cup 1 tbsp	
WATER,WARM	1-5/8 lbs	3 cup	
SUGAR,GRANULATED	1 lbs	2-1/4 cup	
SALT	1-1/2 oz	2-1/3 tbsp	
SHORTENING	9 oz	1-1/4 cup	
EGGS,WHOLE,FROZEN	8-5/8 oz	1 cup	
WATER,COLD	1-1/4 lbs	2-3/8 cup	
EXTRACT,VANILLA	1-3/8 oz	3 tbsp	
FLOUR,WHEAT,BREAD	3-7/8 lbs	3 qts 1 cup	
FLOUR,WHEAT,GENERAL PURPOSE	2-1/4 lbs	2 qts	
MILK,NONFAT,DRY	1-3/4 oz	3/4 cup	
NUTMEG,GROUND	1/4 oz	1 tbsp	

Method

1. Sprinkle yeast over water. DO NOT USE TEMPERATURES ABOVE 110 F. Mix well. Let stand for 5 minutes; stir. Set aside for use in Step 3.
2. Cream sugar, salt, and shortening in mixer bowl at medium speed.
3. Add eggs, yeast solution, water, and vanilla; mix at low speed until blended.
4. Sift together flours, milk, and nutmeg; add to mixture. Using dough hook, mix at low speed 1 minute or until all flour mixture is incorporated into liquid. Continue mixing at medium speed 10 minutes or until dough is smooth and elastic. Dough temperature should be between 78 F. to 82 F.
5. FERMENT: Cover and set in warm place (80 F.), 1-1/2 hours or until double in bulk.
6. PUNCH: Divide into 3 (3 lb 8 oz) pieces; shape each piece into a smooth ball; let rest 10 to 20 minutes.
7. MAKE-UP: Roll each piece into rectangular strips, 5 inches wide, 50 inches long, and 1/2-inch thick; cut into strips 1 inch wide.
8. PROOF: Place on floured sheet pan; let rise 30 minutes or until light.
9. FRY: Until golden brown on underside. Turn and fry on other side. Drain on absorbent paper.
   When cool, roll in granulated sugar or sifted powdered sugar or in Cinnamon Sugar Filling, Recipe No. D 042 00 or dip in Vanilla Glaze, Rum Glaze, Almond Glaze, Recipe Nos. D 046 00, D 046 01, D 046 02. Place glazed doughnuts on racks to drain.

BREADS AND SWEET DOUGHS No.D 019 04

CRULLERS

Yield 100   Portion   1 Doughnut

Calories	Carbohydrates	Protein	Fat	Cholesterol	Sodium	Calcium
191 cal	26 g	4 g	8 g	11 mg	170 mg	13 mg

Ingredient	Weight	Measure	Issue
YEAST,ACTIVE,DRY	3-3/4 oz	1/2 cup 1 tbsp	
WATER,WARM	1-5/8 lbs	3 cup	
SUGAR,GRANULATED	1 lbs	2-1/4 cup	
SALT	1-1/2 oz	2-1/3 tbsp	
SHORTENING	9 oz	1-1/4 cup	
EGGS,WHOLE,FROZEN	8-5/8 oz	1 cup	
WATER,COLD	1-1/4 lbs	2-3/8 cup	
EXTRACT,VANILLA	1-3/8 oz	3 tbsp	
FLOUR,WHEAT,BREAD	3-7/8 lbs	3 qts 1 cup	
FLOUR,WHEAT,GENERAL PURPOSE	2-1/4 lbs	2 qts	
MILK,NONFAT,DRY	1-3/4 oz	3/4 cup	
NUTMEG,GROUND	1/4 oz	1 tbsp	

Method

1. Sprinkle yeast over water. DO NOT USE TEMPERATURES ABOVE 110 F. Mix well. Let stand for 5 minutes; stir. Set aside for use in Step 3.
2. Cream sugar, salt, and shortening in mixer bowl at medium speed.
3. Add eggs, yeast solution, water, and vanilla; mix at low speed until blended.
4. Sift together flours, milk, and nutmeg; add to mixture. Using dough hook, mix at low speed 1 minute or until all flour mixture is incorporated into liquid. Continue mixing at medium speed 10 minutes or until dough is smooth and elastic. Dough temperature should be between 78 F. to 82 F.
5. FERMENT: Cover. Set in warm place (80 F.) 1-1/2 hours or until double in bulk.
6. PUNCH: Divide into 3 (3 lb 8 oz) pieces; shape each piece into a smooth ball; let rest 10 to 20 minutes.
7. MAKE-UP: Roll each piece into rectangular strips, 8 inches wide, 28 to 30 inches long, and 1/2-inch thick. Cut into strips 1/2-inch wide; fold in half, seal end, and twist into spiral shape.
8. PROOF: Place on floured sheet pan; let rise 30 minutes or until light.
9. FRY: Until golden brown on underside. Turn and fry on other side. Drain on absorbent paper.
   When cool, roll in granulated sugar or sifted powdered sugar or in Cinnamon Sugar Filling, Recipe No. D 042 00 or dip in Vanilla Glaze, Almond Glaze, Rum Glaze, Recipe Nos. D 046 00, D 046 01, D 046 02. Place glazed doughnuts on racks to drain.

BREADS AND SWEET DOUGHS No.D 020 00

DUMPLINGS

Yield 100 **Portion** 2 Each

Calories	Carbohydrates	Protein	Fat	Cholesterol	Sodium	Calcium
175 cal	26 g	3 g	6 g	1 mg	521 mg	73 mg

Ingredient   Weight   Measure   Issue
BISCUIT MIX   9 lbs   2 gal 1/2 qts

Method
1. Mix according to instructions on container.
2. Drop a scant 1/8-cup batter on top of simmering stew or into shallow simmering stock. Cover; cook 15 minutes. DO NOT remove cover during cooking time.

Notes
1. Shallow simmering stock should not be more than 1 inch in depth.

BREADS AND SWEET DOUGHS No.D 021 00

ENGLISH MUFFINS

Yield 100   Portion 1 Muffin

Calories	Carbohydrates	Protein	Fat	Cholesterol	Sodium	Calcium
281 cal	43 g	8 g	8 g	20 mg	130 mg	25 mg

Ingredient	**Weight**	**Measure**	**Issue**
YEAST,ACTIVE,DRY	2-7/8 oz	1/4 cup 3 tbsp	
WATER,WARM	1-1/8 lbs	2-1/4 cup	
SUGAR,GRANULATED	3/4 oz	1 tbsp	
SUGAR,GRANULATED	4 oz	1/2 cup 1 tbsp	
SALT	1 oz	1 tbsp	
SHORTENING,SOFTENED	1-1/2 lbs	3-3/8 cup	
WATER,WARM	4-1/3 lbs	2 qts 1/4 cup	
FLOUR,WHEAT,BREAD	1-3/4 lbs	1 qts 2 cup	
MILK,NONFAT,DRY	3-5/8 oz	1-1/2 cup	
FLOUR,WHEAT,BREAD	10-5/8 lbs	2 gal 3/4 qts	
EGGS,WHOLE,FROZEN	1 lbs	1-7/8 cup	

Method

1. Sprinkle yeast over water. DO NOT USE TEMPERATURES ABOVE 110 F. Mix well. Let stand 5 minutes. Add sugar; stir until dissolved. Let stand 10 minutes, then stir again. Set aside for use in Step 3.
2. Place sugar, salt, and shortening in mixer bowl. Add water; stir until shortening is melted.
3. Sift together flour and milk; add to sugar and shortening mixture. Beat at medium speed until smooth. Add yeast solution.
4. Add 1/2 of the flour mixture; mix well. Add eggs, two at a time, beating well after each addition. Add 2 remaining flour; beat to form a smooth dough.
5. FERMENT: Cover; set in a warm place (80 F.), 1-1/2 to 2 hours or until double in bulk.
6. PUNCH: Let stand 1 hour.
7. MAKE UP: Divide dough into 5 balls. Let rest 10 minutes. Roll dough to 1/2-inch thickness. Cut each dough piece into 20-4 inch circles.
8. Place cut circles in rows 4 by 6 about 1 inch apart on pans, which have been sprinkled lightly with cornmeal, about 1/2 cup per pan.
9. PROOF: At 80 F. for 45 minutes or until double in size.
   BAKE: Brown muffins on lightly greased griddle 5 minutes per side. Place browned muffins on sheet pans; bake 15 to 20

BREADS AND SWEET DOUGHS No.D 021 01

CINNAMON RAISIN ENGLISH MUFFINS

Yield 100 **Portion** 1 Muffin

Calories	Carbohydrates	Protein	Fat	Cholesterol	Sodium	Calcium
321 cal	53 g	8 g	8 g	20 mg	132 mg	34 mg

Ingredient	Weight	Measure	Issue
YEAST,ACTIVE,DRY	2-7/8 oz	1/4 cup 3 tbsp	
WATER,WARM	1-1/8 lbs	2-1/4 cup	
SUGAR,GRANULATED	3/4 oz	1 tbsp	
SUGAR,GRANULATED	4 oz	1/2 cup 1 tbsp	
SALT	1 oz	1 tbsp	
SHORTENING,SOFTENED	1-1/2 lbs	3-3/8 cup	
WATER,WARM	4-1/3 lbs	2 qts 1/4 cup	
CINNAMON,GROUND	3/4 oz	3 tbsp	
RAISINS	2-7/8 lbs	2 qts 1 cup	
FLOUR,WHEAT,BREAD	1-3/4 lbs	1 qts 2 cup	
MILK,NONFAT,DRY	3-5/8 oz	1-1/2 cup	
FLOUR,WHEAT,BREAD	10-5/8 lbs	2 gal 3/4 qts	
EGGS,WHOLE,FROZEN	1 lbs	1-7/8 cup	

Method

1. Sprinkle yeast over water. DO NOT USE TEMPERATURES ABOVE 110 F. Mix well. Let stand 5 minutes. Add sugar; stir until dissolved. Let stand 10 minutes, then stir again. Set aside for use in Step 3.
2. Place sugar, salt, and shortening in mixer bowl. Add water; stir until shortening is melted.
3. Sift together flour, cinnamon, raisins, and milk; add to sugar and shortening mixture. Beat at medium speed until smooth. Add yeast solution.
4. Add 1/2 of the flour mixture; mix well. Add eggs, two at a time, beating well after each addition. Add remaining flour; beat to form a smooth dough.
5. FERMENT: Cover and set in a warm place, about 80 F., 1-1/2 to 2 hours or until double in bulk.
6. PUNCH: Let stand 1 hour.
7. MAKE UP: Divide dough into 5 balls. Let rest 10 minutes. Roll dough to 1/2-inch thickness. Cut each dough piece into 20-4 inch circles.
8. Place cut circles in rows 4 by 6 about 1 inch apart on pans, which have been sprinkled lightly with cornmeal, using about 1/2 cup per pan.
9. PROOF: At 80 F. for 45 minutes or until double in size.
   BAKE: Brown muffins on lightly greased griddle 5 minutes per side. Place browned muffins on sheet pans; bake 15 to 20 minutes.

BREADS AND SWEET DOUGHS No.D 022 00

FRENCH TOAST

Yield 100   Portion   2 Slices

Calories	Carbohydrates	Protein	Fat	Cholesterol	Sodium	Calcium
206 cal	29 g	9 g	6 g	148 mg	324 mg	94 mg

Ingredient	Weight	Measure	Issue
WATER	5-3/4 lbs	2 qts 3 cup	
SUGAR,GRANULATED	10-5/8 oz	1-1/2 cup	
MILK,NONFAT,DRY	5-5/8 oz	2-3/8 cup	
EGGS,WHOLE,FROZEN	7-1/2 lbs	3 qts 2 cup	
BREAD,WHITE,SLICED	11 lbs	200 sl	
COOKING SPRAY,NONSTICK	2 oz	1/4 cup 1/3 tbsp	

Method

1. Place water in a mixer bowl.
2. Combine water, milk and sugar; blend well. Whip on low speed until dissolved, about 1 minute.
3. Add eggs to ingredients in mixer bowl; whip on medium speed until well blended, about 2 minutes.
4. Dip bread in egg mixture to coat both sides. DO NOT SOAK.
5. Lightly spray grill with non-stick spray. Place bread on griddle; cook on each side about 1-1/2 minutes or until golden brown.
   CCP: Internal temperature must reach 145 F. or higher for 15 seconds.

BREADS AND SWEET DOUGHS No.D 022 01

FRENCH TOAST (THICK SLICE)

Yield 100   Portion 2 Slices

Calories	Carbohydrates	Protein	Fat	Cholesterol	Sodium	Calcium
305 cal	48 g	12 g	6 g	147 mg	573 mg	104 mg

Ingredient	Weight	Measure	Issue
WATER	5-3/4 lbs	2 qts 3 cup	
MILK,NONFAT,DRY	5-5/8 oz	2-3/8 cup	
SUGAR,GRANULATED	10-5/8 oz	1-1/2 cup	
EGGS,WHOLE,FROZEN	7-1/2 lbs	3 qts 2 cup	
BREAD,FRENCH,THICK SLICE	18-3/4 lbs	200 sl	
COOKING SPRAY,NONSTICK	2 oz	1/4 cup 1/3 tbsp	

Method

1. Place water in mixer bowl.
2. Combine milk and sugar; blend well. Add to water; whip on low speed until dissolved, about 1 minute.
3. Add eggs to ingredients in mixer bowl; whip on medium speed until well blended, about 2 minutes.
4. Cut each loaf diagonally into 16 slices, 3/4 inch thick (ends removed). Dip bread in egg mixture to coat both sides. DO NOT SOAK.
5. Lightly spray griddle with non-stick spray. Place bread on griddle; cook on each side about 1-1/2 minutes or until golden brown. CCP: Internal temperature must reach 145 F. or higher for 15 seconds.

BREADS AND SWEET DOUGHS No.D 022 02

ENGLISH MUFFIN FRENCH TOAST

Yield 100   Portion   1 Muffin

Calories	Carbohydrates	Protein	Fat	Cholesterol	Sodium	Calcium
206 cal	30 g	9 g	5 g	147 mg	319 mg	139 mg

Ingredient	**Weight**	**Measure**	**Issue**
WATER	5-3/4 lbs	2 qts 3 cup	
MILK,NONFAT,DRY	5-5/8 oz	2-3/8 cup	
SUGAR,GRANULATED	10-5/8 oz	1-1/2 cup	
EGGS,WHOLE,FROZEN	7-1/2 lbs	3 qts 2 cup	
ENGLISH MUFFINS,SPLIT OR CUT	12-5/8 lbs	100 each	
COOKING SPRAY,NONSTICK	2 oz	1/4 cup 1/3 tbsp	

Method

1. Place water in mixer bowl.
2. Combine milk and sugar; blend well. Add to water; whip on low speed until dissolved, about 1 minute.
3. Add eggs to ingredients in mixer bowl; whip on medium speed until well blended, about 2 minutes.
4. Cut muffins in half; dip split muffins in batter 30 seconds. DO NOT SOAK.
5. Lightly spray griddle with non-stick spray. Place muffins on griddle, cut side down. Grill about 3 minutes; turn, grill on crust side about 1-1/2 minutes. CCP: Internal temperature must reach 145 F. or higher for 15 seconds.

BREADS AND SWEET DOUGHS No.D 022 03

FRENCH TOAST (FROZEN EGGS AND EGG WHITES)

Yield 100 **Portion** 2 Slices

Calories	Carbohydrates	Protein	Fat	Cholesterol	Sodium	Calcium
186 cal	29 g	8 g	4 g	69 mg	324 mg	85 mg

Ingredient	Weight	Measure	Issue
WATER	5-3/4 lbs	2 qts 3 cup	
MILK,NONFAT,DRY	5-5/8 oz	2-3/8 cup	
SUGAR,GRANULATED	10-5/8 oz	1-1/2 cup	
EGG WHITES,FROZEN,THAWED	3-1/2 lbs	1 qts 2-1/2 cup	
EGGS,WHOLE,FROZEN	3-1/2 lbs	1 qts 2-1/2 cup	
BREAD,WHITE,SLICED	11 lbs	200 sl	
COOKING SPRAY,NONSTICK	2 oz	1/4 cup 1/3 tbsp	

Method

1. Place water in mixing bowl.
2. Combine milk and sugar; blend well. Add to water; whip on low speed until dissolved, about 1 minute.
3. Add whole eggs and egg whites to ingredients in mixer bowl; whip on medium speed until well blended, about 2 minutes.
4. Dip bread in egg mixture to coat both sides. DO NOT SOAK.
5. Lightly spray griddle with non-stick spray. Place bread on griddle; cook on each side about 1-1/2 minutes or until golden brown.
   CCP: Internal temperature must reach 145 F. or higher for 15 seconds.

BREADS AND SWEET DOUGHS No. D 023 00

FRENCH TOAST PUFF

Yield 100   Portion 2 Halves

Calories	Carbohydrates	Protein	Fat	Cholesterol	Sodium	Calcium
284 cal	48 g	7 g	6 g	30 mg	570 mg	175 mg

Ingredient	Weight	Measure	Issue
EGGS,WHOLE,FROZEN	1-1/2 lbs	2-3/4 cup	
SUGAR,GRANULATED	1-1/4 lbs	2-3/4 cup	
SALT	1-7/8 oz	3 tbsp	
EXTRACT,VANILLA	1-1/4 oz	2-2/3 tbsp	
MILK,NONFAT,DRY	6-7/8 oz	2-7/8 cup	
WATER,WARM	7-7/8 lbs	3 qts 3 cup	
FLOUR,WHEAT,GENERAL PURPOSE	8-1/4 lbs	1 gal 3-1/2 qts	
BAKING POWDER	6-3/4 oz	3/4 cup 2 tbsp	
BREAD,WHITE,SLICE	5-1/2 lbs	100 sl	

Method

1. Combine eggs, sugar, salt, vanilla, milk, and water in mixer bowl. Beat at medium speed until well blended.
2. Add slowly flour and baking powder; mix at medium speed until smooth.
3. Cut bread in half diagonally. Dip half slices of bread in batter. Drain.
4. Fry until golden brown. Drain on absorbent paper.

Notes

1. In Step 5, serve with maple, blueberry or strawberry syrup, marmalade, jam, or jelly.
2. In Step 5, serve with well-drained canned sliced peaches, fruit cocktail or thawed, well-drained strawberries.
3. Puffs, while warm, may be rolled in Cinnamon Sugar Filling, Recipe No. D 042 00 or dusted with powdered sugar.

BREADS AND SWEET DOUGHS No.D 024 00

APPLE FRITTERS

Yield 100   **Portion** 2 Fritters

Calories	Carbohydrates	Protein	Fat	Cholesterol	Sodium	Calcium
248 cal	34 g	3 g	12 g	20 mg	273 mg	73 mg

Ingredient	Weight	Measure	Issue
FLOUR,WHEAT,GENERAL PURPOSE	4 lbs	3 qts 2-1/2 cup	
BAKING POWDER	3-1/4 oz	1/4 cup 3 tbsp	
MILK,NONFAT,DRY	3-1/4 oz	1-3/8 cup	
SALT	1-1/2 oz	2-1/3 tbsp	
SUGAR,GRANULATED	1-1/4 lbs	2-3/4 cup	
NUTMEG,GROUND	1/4 oz	1 tbsp	
CINNAMON,GROUND	1/4 oz	1 tbsp	
EGGS,WHOLE,FROZEN	1 lbs	1-7/8 cup	
WATER	3-3/4 lbs	1 qts 3-1/4 cup	
OIL,SALAD	5-3/4 oz	3/4 cup	
APPLES,CANNED,SLICED,DRAINED	6 lbs	3 qts	
SUGAR,POWDERED,SIFTED	2-1/8 lbs	2 qts	

Method

1. Sift together flour, baking powder, milk, salt, sugar, nutmeg, and cinnamon into mixer bowl.
2. Combine eggs, water, shortening or salad oil, and add to dry ingredients. Mix at low speed until well blended.
3. Drain apples and chop apples coarsely; add to batter; mix lightly.
4. Using a well rounded tablespoon, drop batter into deep fat. Fry 4 to 6 minutes. Drain on absorbent paper. Sprinkle with sifted powdered sugar.

Notes

1. In Step 3, 6 lb (7 lb 11 oz A.P.) pared, cored and diced fresh apples may be used per 100 portions.

BREADS AND SWEET DOUGHS No.D 025 00

PANCAKES

Yield 100   Portion 2 Cakes

Calories	Carbohydrates	Protein	Fat	Cholesterol	Sodium	Calcium
253 cal	41 g	7 g	6 g	53 mg	512 mg	207 mg

Ingredient	Weight	Measure	Issue
FLOUR,WHEAT,GENERAL PURPOSE	9-7/8 lbs	2 gal 1 qts	
BAKING POWDER	8-3/4 oz	1-1/8 cup	
MILK,NONFAT,DRY	13-1/4 oz	1 qts 1-1/2 cup	
SALT	1-7/8 oz	3 tbsp	
SUGAR,GRANULATED	12-1/3 oz	1-3/4 cup	
EGGS,WHOLE,FROZEN	2-2/3 lbs	1 qts 1 cup	
WATER	13 lbs	1 gal 2-1/4 qts	
OIL,SALAD	1 lbs	2 cup	
COOKING SPRAY,NONSTICK	2 oz	1/4 cup 1/3 tbsp	

Method

1. Sift together flour, baking powder, milk, salt, and sugar into mixer bowl.
2. Add eggs and water; mix at low speed about 1 minute or until blended.
3. Blend in salad oil or melted shortening about 1 minute.
4. Lightly spray griddle with non-stick spray. Pour 1/4 cup batter onto hot griddle. Cook on one side 1-1/2 to 2 minutes or until top is covered with bubbles and underside is browned. Turn; cook on other side 1-1/2 to 2 minutes.

BREADS AND SWEET DOUGHS No.D 025 01

BUTTERMILK PANCAKES (DRY BUTTERMILK)

Yield 100　　　　　　　　　　　　　　　　　　　　　**Portion** 2 Cakes

Calories	Carbohydrates	Protein	Fat	Cholesterol	Sodium	Calcium
281 cal	44 g	10 g	7 g	60 mg	478 mg	211 mg

Ingredient	Weight	Measure	Issue
FLOUR,WHEAT,GENERAL PURPOSE	9-7/8 lbs	2 gal 1 qts	
BAKING POWDER	3-7/8 oz	1/2 cup	
MILK,BUTTERMILK,DRY	2-3/8 lbs	1 qts 1/2 cup	
BAKING SODA	1 oz	2 tbsp	
SALT	1-7/8 oz	3 tbsp	
SUGAR,GRANULATED	12-1/3 oz	1-3/4 cup	
EGGS,WHOLE,FROZEN	2-2/3 lbs	1 qts 1 cup	
WATER	13 lbs	1 gal 2-1/4 qts	
OIL,SALAD	1 lbs	2 cup	
COOKING SPRAY,NONSTICK	2 oz	1/4 cup 1/3 tbsp	

Method

1. Sift together flour, baking powder, dry buttermilk, salt, sugar, and baking soda.
2. Add eggs and water; mix at low speed about 1 minute or until blended.
3. Blend in salad oil or melted shortening about 1 minute.
4. Lightly spray griddle with non-stick cooking spray. Pour 1/4 cup batter onto hot griddle. Cook on one side 1-1/2 to 2 minutes or until top is covered with bubbles and underside is browned. Turn; cook on other side 1-1/2 to 2 minutes.

BREADS AND SWEET DOUGHS No.D 025 02

BLUEBERRY PANCAKES

Yield 100   **Portion** 2 Cakes

Calories	Carbohydrates	Protein	Fat	Cholesterol	Sodium	Calcium
265 cal	43 g	8 g	7 g	53 mg	512 mg	209 mg

Ingredient	**Weight**	**Measure**	**Issue**
FLOUR,WHEAT,GENERAL PURPOSE	9-7/8 lbs	2 gal 1 qts	
BAKING POWDER	8-3/4 oz	1-1/8 cup	
MILK,NONFAT,DRY	13-1/4 oz	1 qts 1-1/2 cup	
SALT	1-7/8 oz	3 tbsp	
SUGAR,GRANULATED	12-1/3 oz	1-3/4 cup	
EGGS,WHOLE,FROZEN	2-2/3 lbs	1 qts 1 cup	
WATER	13 lbs	1 gal 2-1/4 qts	
OIL,SALAD	1 lbs	2 cup	
BLUEBERRIES,FROZEN,UNSWEETENED	5-1/8 lbs	3 qts 3 cup	
COOKING SPRAY,NONSTICK	2 oz	1/4 cup 1/3 tbsp	

Method

1. Sift together flour, baking powder, milk, salt, and sugar into mixer bowl.
2. Add eggs and water; mix at low speed about 1 minute or until blended.
3. Blend in salad oil or melted shortening about 1 minute. Use partially thawed frozen blueberries, or drain and rinse canned blueberries in cold water. Drain thoroughly and fold into batter.
4. Lightly spray non-stick cooking spray on griddle. Pour 1/4 cup batter onto hot griddle. Cook on one side 1-1/2 to 2 minutes or until top is covered with bubbles and underside is browned. Turn; cook on other side 1-1/2 to 2 minutes. Stir between batches to redistribute berries.

BREADS AND SWEET DOUGHS No.D 025 04

BUTTERMILK PANCAKES (PANCAKE MIX)

Yield 100   Portion   2 Cakes

Calories	Carbohydrates	Protein	Fat	Cholesterol	Sodium	Calcium
296 cal	48 g	7 g	8 g	14 mg	827 mg	164 mg

Ingredient
PANCAKE MIX,BUTTERMILK

Weight
12-1/2 lbs

Measure
3 gal 1-1/8 qts

Issue

Method
1 Prepare pancakes according to instructions on container.

BREADS AND SWEET DOUGHS No.D 025 05

PANCAKES (PANCAKE MIX)

Yield 100   Portion 2 Cakes

Calories	Carbohydrates	Protein	Fat	Cholesterol	Sodium	Calcium
226 cal	42 g	6 g	3 g	12 mg	716 mg	142 mg

Ingredient   **Weight**   **Measure**   **Issue**

PANCAKE MIX   13 lbs   2 gal 3-1/3 qts

Method

1 Prepare pancakes according to instructions on container.

BREADS AND SWEET DOUGHS No.D 025 06

WAFFLES, FROZEN (BROWN AND SERVE)

Yield 100   Portion 2 Each

Calories	Carbohydrates	Protein	Fat	Cholesterol	Sodium	Calcium
176 cal	27 g	4 g	6 g	22 mg	524 mg	155 mg

Ingredient	**Weight**	**Measure**	**Issue**
WAFFLES,BROWN & SERVE,FROZEN	15-3/8 lbs	200 each	

Method

1 Prepare according to instructions on container.

BREADS AND SWEET DOUGHS No.D 025 07

WAFFLES (PANCAKE MIX)

Yield 100   **Portion** 1 Each

Calories	Carbohydrates	Protein	Fat	Cholesterol	Sodium	Calcium
226 cal	42 g	6 g	3 g	12 mg	716 mg	142 mg

Ingredient	**Weight**	**Measure**	**Issue**
PANCAKE MIX	13 lbs	2 gal 3-1/3 qts	

Method

1 Prepare waffles according to instructions on container.

BREADS AND SWEET DOUGHS No.D 025 08

WAFFLES

Yield 100   Portion 1 Each

Calories	Carbohydrates	Protein	Fat	Cholesterol	Sodium	Calcium
253 cal	41 g	7 g	6 g	53 mg	512 mg	207 mg

Ingredient	Weight	Measure	Issue
FLOUR,WHEAT,GENERAL PURPOSE	9-7/8 lbs	2 gal 1 qts	
BAKING POWDER	8-3/4 oz	1-1/8 cup	
MILK,NONFAT,DRY	13-1/4 oz	1 qts 1-1/2 cup	
SALT	1-7/8 oz	3 tbsp	
SUGAR,GRANULATED	12-1/3 oz	1-3/4 cup	
EGGS,WHOLE,FROZEN	2-2/3 lbs	1 qts 1 cup	
WATER	13 lbs	1 gal 2-1/4 qts	
OIL,SALAD	1 lbs	2 cup	

Method

1. Sift together both flours, baking powder, milk, salt, and sugar into mixer bowl.
2. Add eggs and water; mix at low speed about 1 minute or until blended.
3. Blend in salad oil or melted shortening about 1 minute.
4. Pour 1/2 cup batter on preheated waffle iron. Bake until steaming stops, about 3 to 4 minutes.

BREADS AND SWEET DOUGHS No.D 025 09

WHOLE WHEAT PANCAKES

Yield 100   Portion   2 Cakes

Calories	Carbohydrates	Protein	Fat	Cholesterol	Sodium	Calcium
236 cal	37 g	8 g	7 g	53 mg	513 mg	210 mg

Ingredient	Weight	Measure	Issue
FLOUR,WHOLE WHEAT	4-1/4 lbs	1 gal	
FLOUR,WHEAT,GENERAL PURPOSE	5 lbs	1 gal 1/2 qts	
BAKING POWDER	8-3/4 oz	1-1/8 cup	
MILK,NONFAT,DRY	13-1/4 oz	1 qts 1-1/2 cup	
SALT	1-7/8 oz	3 tbsp	
SUGAR,GRANULATED	12-1/3 oz	1-3/4 cup	
EGGS,WHOLE,FROZEN	2-2/3 lbs	1 qts 1 cup	
WATER	13 lbs	1 gal 2-1/4 qts	
OIL,SALAD	1 lbs	2 cup	

Method

1. Sift together both flours, baking powder, milk, salt, and sugar into mixer bowl.
2. Add eggs and water; mix at low speed about 1 minute or until blended.
3. Blend in salad oil or melted shortening about 1 minute.
4. Pour 1/4 cup batter onto lightly greased hot griddle. Cook on one side 1-1/2 to 2 minutes or until top is covered with bubbles and underside is browned. Turn; cook on other side 1-1/2 to 2 minutes.

BREADS AND SWEET DOUGHS No.D 025 10

PANCAKES (FROZEN EGGS AND EGG WHITES)

Yield 100   Portion 2 Cakes

Calories	Carbohydrates	Protein	Fat	Cholesterol	Sodium	Calcium
247 cal	41 g	7 g	6 g	27 mg	513 mg	204 mg

Ingredient	Weight	Measure	Issue
FLOUR,WHEAT,GENERAL PURPOSE	9-7/8 lbs	2 gal 1 qts	
BAKING POWDER	8-3/4 oz	1-1/8 cup	
MILK,NONFAT,DRY	13-1/4 oz	1 qts 1-1/2 cup	
SALT	1-7/8 oz	3 tbsp	
SUGAR,GRANULATED	12-1/3 oz	1-3/4 cup	
EGGS,WHOLE,FROZEN	1-1/3 lbs	2-1/2 cup	
EGG WHITES,FROZEN,THAWED	1-1/3 lbs	2-1/2 cup	
WATER	13 lbs	1 gal 2-1/4 qts	
OIL,SALAD	1 lbs	2 cup	

Method

1 Sift together flour, baking powder, milk, salt, and sugar into mixer bowl.
2 Add eggs and water; mix at low speed about 1 minute or until blended.
3 Blend in salad oil or melted shortening about 1 minute.
4 Pour 1/4 cup batter onto lightly greased hot griddle. Cook on one side 1-1/2 to 2 minutes or until top is covered with bubbles and underside is browned. Turn; cook on other side 1-1/2 to 2 minutes.

BREADS AND SWEET DOUGHS No.D 025 11

PANCAKES (EGG SUBSTITUTE)

Yield 100   **Portion** 2 Cakes

Calories	Carbohydrates	Protein	Fat	Cholesterol	Sodium	Calcium
246 cal	41 g	8 g	6 g	1 mg	518 mg	207 mg

Ingredient   Weight   Measure   Issue

Ingredient	Weight	Measure	Issue
FLOUR,WHEAT,GENERAL PURPOSE	9-7/8 lbs	2 gal 1 qts	
BAKING POWDER	8-3/4 oz	1-1/8 cup	
MILK,NONFAT,DRY	13-1/4 oz	1 qts 1-1/2 cup	
SALT	1-7/8 oz	3 tbsp	
SUGAR,GRANULATED	12-1/3 oz	1-3/4 cup	
EGG SUBSTITUTE,PASTEURIZED	2-3/4 lbs	1 qts 1 cup	
WATER	13 lbs	1 gal 2-1/4 qts	
OIL,SALAD	1 lbs	2 cup	

Method

1. Sift together flour, baking powder, milk, salt, and sugar into mixer bowl.
2. Add egg substitute and water. Mix at low speed about 1 minute or until blended.
3. Blend in salad oil or melted shortening about 1 minute.
4. Pour 1/4 cup batter onto lightly greased hot griddle. Cook on one side 1-1/2 to 2 minutes or until top is covered with bubbles and underside is browned. Turn over and cook on other side 1-1/2 to 2 minutes.

BREADS AND SWEET DOUGHS No.D 026 00

HOT CROSS BUNS

Yield 100   Portion 1 Each

Calories	Carbohydrates	Protein	Fat	Cholesterol	Sodium	Calcium
147 cal	28 g	5 g	3 g	0 mg	270 mg	17 mg

Ingredient	Weight	Measure	Issue
YEAST,ACTIVE,DRY	2-1/4 oz	1/4 cup 1-2/3 tbsp	
WATER,WARM	3-1/8 lbs	1 qts 2 cup	
SWEET DOUGH MIX	7-1/2 lbs	1 gal 2-7/8 qts	
RAISINS	2-1/4 lbs	1 qts 3 cup	
CINNAMON,GROUND	1/2 oz	2 tbsp	
CLOVES,GROUND	<1/16th oz	<1/16th tsp	
NUTMEG,GROUND	<1/16th oz	<1/16th tsp	
COOKING SPRAY,NONSTICK	2 oz	1/4 cup 1/3 tbsp	

Method

1. Sprinkle yeast over water. DO NOT USE TEMPERATURES ABOVE 110 F. Mix well. Let stand 5 minutes; stir.
2. Add Sweet Dough Mix, raisins, cinnamon, cloves, and nutmeg.
3. Using dough hook, mix at low speed until water is absorbed. Mix at medium speed until dough is developed and cleans the bowl. Dough temperature should be 78 F. to 82 F.
4. FERMENT: Cover. Set in warm place (80 F.) 1-1/2 to 2 hours or until double in bulk.
5. PUNCH: Divide dough into 8-2 pound pieces on lightly floured work surface; shape each piece into a smooth ball. Let rest 10 to 20 minutes.
6. MAKE-UP: Roll each piece into a long rope of uniform diameter. Cut rope into pieces about 1 inch thick, weighing 1-1/2 ounces each. Shape into balls by rolling with circular motion on work surface.
7. Lightly spray pans with non-stick cooking spray. Place on pans in rows 6 by 9. Prepare 1/8 recipe Egg Wash, Recipe No. D 017 00. Brush buns in each pan with wash.
8. PROOF: At 90 F. to 100 F. about 45 minutes or until almost double in bulk.
9. BAKE: 30 minutes at 400 F. or until lightly browned. If convection oven is used, bake at 350 F. for 10 minutes on high fan, closed vent.
10. Prepare 1/8 recipe Syrup Glaze, Recipe No. D 045 00 per 100 servings. Brush buns in each pan with 1/4 cup hot glaze immediately after removal from oven.
11. When cool, prepare 1/8 recipe Decorator's Frosting, Recipe No. G 007 00 per 100 servings. Frost each bun with frosting in a cross design using a pastry bag with a small plain tip.

Notes

1. In Step 2, 1 tbsp lemon flavoring may be added per 100 servings.

BREADS AND SWEET DOUGHS No.D 027 00

KOLACHES

Yield 100   Portion   1 Roll

Calories	Carbohydrates	Protein	Fat	Cholesterol	Sodium	Calcium
240 cal	39 g	5 g	7 g	20 mg	177 mg	21 mg

Ingredient	Weight	Measure	Issue
YEAST,ACTIVE,DRY	5-1/8 oz	3/4 cup	
WATER,WARM	1-7/8 lbs	3-1/2 cup	
SUGAR,GRANULATED	7/8 oz	2 tbsp	
SUGAR,GRANULATED	1-1/4 lbs	2-3/4 cup	
SALT	1-1/2 oz	2-1/3 tbsp	
SHORTENING	1-1/3 lbs	3 cup	
EGGS,WHOLE,FROZEN	1 lbs	1-7/8 cup	
WATER	2-1/8 lbs	1 qts	
FLOUR,WHEAT,BREAD	7-1/4 lbs	1 gal 2 qts	
MILK,NONFAT,DRY	2-2/3 oz	1-1/8 cup	
CHERRY FILLING (PIE FILLING, PREPARED)	2-3/4 kg	3 unit	

Method

1. Sprinkle yeast over water. DO NOT USE TEMPERATURES ABOVE 110 F. Mix well. Let stand 5 minutes. Add sugar; stir until dissolved. Let stand 10 minutes; stir again. Set aside for use in Step 3.
2. Mix sugar, salt, and shortening in mixer bowl at medium speed 1 minute.
3. Blend in eggs, water, and yeast solution at low speed.
4. Sift flour and milk together, add to egg mixture. Mix at low speed 7 to 10 minutes or until dough is formed.
5. FERMENT: Set in warm place (80 F.) for about 1 hour.
6. PUNCH: Let rest 10 minutes. Divide dough into 2 pieces. Shape each piece into a smooth ball; let rest 10 minutes.
7. MAKE UP: Form into a rope 1-1/2 inches in diameter. Cut into 1-1/2 inch pieces. Shape into 2-ounce balls. Place 2 inches apart on greased pans. Flatten out slightly with palm of hand.
8. PROOF: About 30 minutes or until pieces are double in bulk.
9. Press down center of each piece with back of spoon. Leave a rim about 1/4-inch wide.
10. Fill center of each Kolache with about 1 ounce (2 tbsp) of Cherry Filling, Recipe No. D 041 01.
11. Brush rim with Egg Wash, Recipe No. D 017 00.
12. PROOF: 20 minutes at 350 F. or until double in bulk.
13. BAKE: At 350 F. for 25 minutes or until done. For convection oven, bake 15 minutes at 300 F.
14. If desired, cool; sprinkle with 1 lb (3 1/2 cups) sifted powder sugar or brush out edges with 1 recipe Vanilla Glaze (Recipe No. D 046 00) per 100 servings.

Notes

1. In Step 10, 7 lb (1-No. 10 cn) prepared pie filling, apple, blueberry, cherry or peach, or bakery filling, raspberry, may be used, per 100 servings.

BREADS AND SWEET DOUGHS No.D 027 01

KOLACHES (SWEET DOUGH MIX)

Yield 100   **Portion** 1 Roll

Calories	Carbohydrates	Protein	Fat	Cholesterol	Sodium	Calcium
166 cal	32 g	6 g	3 g	0 mg	325 mg	15 mg

Ingredient   Weight   Measure   Issue
SWEET DOUGH MIX   9 lbs   2 gal 1/8 qts
YEAST,ACTIVE,DRY   3-3/4 oz   1/2 cup 1 tbsp
CHERRY FILLING (PIE FILLING, PREPARED)   2-3/4 kg   3 unit

Method

1 Use sweet dough mix and active dry yeast. Prepare dough according to instructions on container.
2 PUNCH: Let rest 10 minutes. Divide dough into 2 pieces. Shape each piece into a smooth ball; let rest 10 minutes.
3 MAKE UP: Form into a rope 1-1/2 inches in diameter. Cut into 1-1/2 inch pieces. Shape into 2-ounce balls. Place 2 inches apart on greased pans. Flatten out slightly with palm of hand.
4 PROOF: About 30 minutes or until pieces are double in size.
5 Press down center of each piece with back of spoon. Leave a rim about 1/4-inch wide.
6 Fill center of each Kolache with about 1 ounce (2 tbsp) filling. Use 1 recipe Cherry Filling (Recipe No. D 041 01).
7 Brush rim with Egg Wash, Recipe No. D 017 00.
8 PROOF: 20 minutes or until double in size.
9 BAKE: 25 minutes at 350 F. or until done. For convection oven, bake 15 minutes at 300 F.
   If desired, cool; sprinkle with 1 lb (3 1/2 cups) sifted powder sugar or brush out edges with 1 recipe Vanilla Glaze (Recipe No. D 046 00) per 100 servings.

Notes

1 In Step 10, 7 lb (1-No. 10 cn) prepared pie filling, apple, blueberry, cherry or peach, or 7 lb 4 oz (7/8-No. 10 cn) bakery filling, raspberry, may be used, per 100 servings.

BREADS AND SWEET DOUGHS No.D 028 00

BRAN MUFFINS

Yield 100　　　　　　　　　　　　　　　　　**Portion** 1 Muffin

Calories	Carbohydrates	Protein	Fat	Cholesterol	Sodium	Calcium
173 cal	34 g	3 g	4 g	12 mg	240 mg	110 mg

Ingredient	Weight	Measure	Issue
APPLESAUCE,CANNED,SWEETENED	5-1/4 lbs	2 qts 1-3/8 cup	
WATER	2-1/8 lbs	1 qts	
CEREAL,ALL BRAN,BULK	1-7/8 lbs	2 qts 2 cup	
FLOUR,WHEAT,GENERAL PURPOSE	3-5/8 lbs	3 qts 1 cup	
SUGAR,GRANULATED	2-1/4 lbs	1 qts 1 cup	
BAKING POWDER	4-3/8 oz	1/2 cup 1 tbsp	
SALT	3/4 oz	1 tbsp	
CINNAMON,GROUND	1/2 oz	2 tbsp	
NUTMEG,GROUND	1/8 oz	1/3 tsp	
EGGS,WHOLE,FROZEN	9-1/2 oz	1-1/8 cup	
EGG WHITES,FROZEN,THAWED	9-1/2 oz	1-1/8 cup	
OIL,SALAD	9-5/8 oz	1-1/4 cup	
COOKING SPRAY,NONSTICK	2 oz	1/4 cup 1/3 tbsp	

Method

1. Mix applesauce with water; add to bran. Let stand for 5 minutes.
2. Sift together flour, sugar, baking powder, salt, cinnamon, and nutmeg into mixer bowl. Batter will be lumpy.
3. Add bran applesauce mixture, eggs, and salad oil or shortening; mix at low speed about 15 seconds; scrape down sides and bottom of mixer bowl. Mix until dry ingredients are moistened, about 15 seconds. DO NOT OVER MIX.
4. Lightly spray muffin cup with non-stick cooking spray. Fill each muffin cup 2/3 full.
5. Bake 25 to 30 minutes at 400 F. or until lightly brown.

BREADS AND SWEET DOUGHS No.D 028 01

RAISIN BRAN MUFFINS

Yield 100   Portion 1 Muffin

Calories	Carbohydrates	Protein	Fat	Cholesterol	Sodium	Calcium
199 cal	41 g	4 g	4 g	10 mg	240 mg	114 mg

Ingredient	Weight	Measure	Issue
APPLESAUCE,CANNED,SWEETENED	5-1/4 lbs	2 qts 1-3/8 cup	
WATER	2-1/8 lbs	1 qts	
CEREAL,ALL BRAN,BULK	1-7/8 lbs	2 qts 2 cup	
FLOUR,WHEAT,GENERAL PURPOSE	3-5/8 lbs	3 qts 1 cup	
SUGAR,GRANULATED	2-1/4 lbs	1 qts 1 cup	
BAKING POWDER	4-3/8 oz	1/2 cup 1 tbsp	
SALT	3/4 oz	1 tbsp	
CINNAMON,GROUND	1/2 oz	2 tbsp	
NUTMEG,GROUND	1/8 oz	1/3 tsp	
EGGS,WHOLE,FROZEN	8-5/8 oz	1 cup	
EGG WHITES,FROZEN,THAWED	8-1/2 oz	1 cup	
OIL,SALAD	9-5/8 oz	1-1/4 cup	
RAISINS	1-7/8 lbs	1 qts 2 cup	
COOKING SPRAY,NONSTICK	2 oz	1/4 cup 1/3 tbsp	

Method

1. Mix applesauce with water; add to bran. Let stand for 5 minutes.
2. Sift together flour, sugar, baking powder, salt, cinnamon, and nutmeg into mixer bowl. Batter will be lumpy.
3. Add bran applesauce mixture, eggs, and salad oil or shortening; mix at low speed for 15 seconds; scrape down sides and bottom of mixer bowl. Mix until dry ingredients are moistened, about 15 seconds. DO NOT OVER MIX. Fold in raisins.
4. Lightly spray each muffin cup with non-stick cooking spray. Fill each muffin cup 2/3 full.
5. Bake 25 to 30 minutes at 400 F. or until lightly brown.

BREADS AND SWEET DOUGHS No.D 028 02

BLUEBERRY BRAN MUFFINS

Yield 100   Portion   1 Muffin

Calories	Carbohydrates	Protein	Fat	Cholesterol	Sodium	Calcium
174 cal	34 g	3 g	4 g	10 mg	239 mg	110 mg

Ingredient	Weight	Measure	Issue
APPLESAUCE,CANNED,SWEETENED	5-1/4 lbs	2 qts 1-3/8 cup	
WATER	2-1/8 lbs	1 qts	
CEREAL,ALL BRAN,BULK	1-7/8 lbs	2 qts 2 cup	
FLOUR,WHEAT,GENERAL PURPOSE	3-5/8 lbs	3 qts 1 cup	
SUGAR,GRANULATED	2-1/4 lbs	1 qts 1 cup	
BAKING POWDER	4-3/8 oz	1/2 cup 1 tbsp	
SALT	3/4 oz	1 tbsp	
CINNAMON,GROUND	1/2 oz	2 tbsp	
NUTMEG,GROUND	1/8 oz	1/3 tsp	
EGGS,WHOLE,FROZEN	8-5/8 oz	1 cup	
EGG WHITES,FROZEN,THAWED	8-1/2 oz	1 cup	
OIL,SALAD	9-5/8 oz	1-1/4 cup	
BLUEBERRIES,FROZEN,UNSWEETENED	10-7/8 oz	2 cup	
COOKING SPRAY,NONSTICK	2 oz	1/4 cup 1/3 tbsp	

Method

1 Mix applesauce with water; add to bran. Let stand for 5 minutes.
2 Sift together flour, sugar, baking powder, salt, cinnamon, and nutmeg into mixer bowl. Batter will be lumpy.
3 Add bran applesauce mixture, eggs and salad oil or shortening; mix at low speed about 15 seconds. DO NOT OVER MIX. Fold in blueberries.
4 Lightly spray each muffin cup with non-stick cooking spray. Fill each muffin cup 2/3 full.
5 Bake 25 to 30 minutes at 400 F. or until lightly brown.

BREADS AND SWEET DOUGHS No.D 028 03

BANANA BRAN MUFFINS

Yield 100   Portion 1 Muffin

Calories	Carbohydrates	Protein	Fat	Cholesterol	Sodium	Calcium
181 cal	36 g	3 g	4 g	10 mg	239 mg	111 mg

Ingredient   Weight   Measure   Issue

Ingredient	Weight	Measure	Issue
APPLESAUCE,CANNED,SWEETENED	5-1/4 lbs	2 qts 1-3/8 cup	
WATER	2-1/8 lbs	1 qts	
CEREAL,ALL BRAN,BULK	1-7/8 lbs	2 qts 2 cup	
FLOUR,WHEAT,GENERAL PURPOSE	3-5/8 lbs	3 qts 1 cup	
SUGAR,GRANULATED	2-1/4 lbs	1 qts 1 cup	
BAKING POWDER	4-3/8 oz	1/2 cup 1 tbsp	
SALT	3/4 oz	1 tbsp	
CINNAMON,GROUND	1/2 oz	2 tbsp	
NUTMEG,GROUND	1/8 oz	1/3 tsp	
EGGS,WHOLE,FROZEN	8-5/8 oz	1 cup	
EGG WHITES,FROZEN,THAWED	8-1/2 oz	1 cup	
OIL,SALAD	9-5/8 oz	1-1/4 cup	
BANANA,FRESH	2 lbs		3-1/8 lbs
COOKING SPRAY,NONSTICK	2 oz	1/4 cup 1/3 tbsp	

Method

1. Mix applesauce with water; add to bran. Let stand for 5 minutes.
2. Sift together flour, sugar, baking powder, salt, cinnamon, and nutmeg into mixer bowl. Batter will be lumpy.
3. Add bran applesauce mixture, eggs and salad oil or shortening; mix at low speed about 15 seconds, scrape down sides and bottom of mixer bowl. Mix untril dry ingredients are moistened, about 15 seconds. DO NOT OVER MIX. Fold bananas into batter.
4. Lightly spray each muffin cup with non-stick cooking spray. Fill each muffin cup 2/3 full.
5. Bake 25 to 30 minutes at 400 F. or until lightly browned.

BREADS AND SWEET DOUGHS No.D 028 04

APRICOT BRAN MUFFINS

Yield 100   Portion   1 Muffin

Calories	Carbohydrates	Protein	Fat	Cholesterol	Sodium	Calcium
196 cal	40 g	4 g	4 g	10 mg	240 mg	114 mg

Ingredient	Weight	Measure	Issue
APPLESAUCE,CANNED,SWEETENED	5-1/4 lbs	2 qts 1-3/8 cup	
WATER	2-1/8 lbs	1 qts	
CEREAL,ALL BRAN,BULK	1-7/8 lbs	2 qts 2 cup	
FLOUR,WHEAT,GENERAL PURPOSE	3-5/8 lbs	3 qts 1 cup	
SUGAR,GRANULATED	2-1/4 lbs	1 qts 1 cup	
BAKING POWDER	4-3/8 oz	1/2 cup 1 tbsp	
SALT	3/4 oz	1 tbsp	
CINNAMON,GROUND	1/2 oz	2 tbsp	
NUTMEG,GROUND	1/8 oz	1/3 tsp	
EGGS,WHOLE,FROZEN	8-5/8 oz	1 cup	
EGG WHITES,FROZEN,THAWED	8-1/2 oz	1 cup	
OIL,SALAD	9-5/8 oz	1-1/4 cup	
APRICOTS,DRIED,HALVES,PITTED	1-5/8 lbs	1 qts 2 cup	
COOKING SPRAY,NONSTICK	2 oz	1/4 cup 1/3 tbsp	

Method

1. Mix applesauce with water; add to bran. Let stand for 5 minutes.
2. Sift together flour, sugar, baking powder, salt, cinnamon, and nutmeg into mixer bowl. Batter will be lumpy.
3. Add bran applesauce mixture, eggs and salad oil or shortening; mix at low speed about 15 seconds, scrape down sides and bottom of mixer bowl. Mix until dry ingredients are moistened, about 15 seconds. DO NOT OVER MIX. Fold in dried, chopped apricots.
4. Lightly spray each muffin cup with non-stick cooking spray. Fill each muffin cup 2/3 full.
5. Bake 25 to 30 minutes at 400 F. or until lightly browned.

BREADS AND SWEET DOUGHS No.D 028 05

CRANBERRY BRAN MUFFINS

Yield 100   Portion 1 Muffin

Calories	Carbohydrates	Protein	Fat	Cholesterol	Sodium	Calcium
177 cal	35 g	3 g	4 g	10 mg	239 mg	111 mg

Ingredient	Weight	Measure	Issue
APPLESAUCE,CANNED,SWEETENED	5-1/4 lbs	2 qts 1-3/8 cup	
WATER	2-1/8 lbs	1 qts	
CEREAL,ALL BRAN,BULK	1-7/8 lbs	2 qts 2 cup	
FLOUR,WHEAT,GENERAL PURPOSE	3-5/8 lbs	3 qts 1 cup	
SUGAR,GRANULATED	2-1/4 lbs	1 qts 1 cup	
BAKING POWDER	4-3/8 oz	1/2 cup 1 tbsp	
SALT	3/4 oz	1 tbsp	
CINNAMON,GROUND	1/2 oz	2 tbsp	
NUTMEG,GROUND	1/8 oz	1/3 tsp	
EGGS,WHOLE,FROZEN	8-5/8 oz	1 cup	
EGG WHITES,FROZEN,THAWED	8-1/2 oz	1 cup	
OIL,SALAD	9-5/8 oz	1-1/4 cup	
CRANBERRIES,FRESH	1-3/4 lbs	2 qts 3/8 cup	1-7/8 lbs
COOKING SPRAY,NONSTICK	2 oz	1/4 cup 1/3 tbsp	

Method

1 Mix applesauce with water; add to bran. Let stand for 5 minutes.
2 Sift together flour, sugar, baking powder, salt, cinnamon, and nutmeg into mixer bowl. Batter will be lumpy.
3 Add bran applesauce mixture, eggs and salad oil or shortening; mix at low speed about 15 seconds. DO NOT OVER MIX. Fold cranberries into batter.
4 Lightly spray each muffin cup with non-stick cooking spray. Fill each muffin cup 2/3 full.
5 Bake 25 to 30 minutes at 400 F. or until lightly browned.

BREADS AND SWEET DOUGHS No.D 029 00

MUFFINS

Yield 100   **Portion** 1 Muffin

Calories	Carbohydrates	Protein	Fat	Cholesterol	Sodium	Calcium
178 cal	30 g	4 g	5 g	34 mg	204 mg	86 mg

Ingredient	**Weight**	**Measure**	**Issue**
FLOUR,WHEAT,GENERAL PURPOSE	5 lbs	1 gal 1/2 qts	
SUGAR,GRANULATED	2-1/2 lbs	1 qts 1-5/8 cup	
MILK,NONFAT,DRY	3-5/8 oz	1-1/2 cup	
BAKING POWDER	3-7/8 oz	1/2 cup	
SALT	5/8 oz	1 tbsp	
WATER,WARM	3-2/3 lbs	1 qts 3 cup	
EGGS,WHOLE,FROZEN	1-3/4 lbs	3-1/4 cup	
APPLESAUCE,CANNED,UNSWEETENED	1-5/8 lbs	3 cup	
OIL,SALAD	11-1/2 oz	1-1/2 cup	
COOKING SPRAY,NONSTICK	1-1/2 oz	3 tbsp	

Method

1 In mixer bowl, sift together flour, sugar, milk, baking powder and salt.
2 Add warm water, eggs, applesauce, and salad oil; mix at low speed until dry ingredients are moistened about 15 seconds; scrape down sides and bottom of mixer bowl; continue to mix at low speed another 15 seconds. DO NOT OVER MIX. Batter will be lumpy.
3 Lightly spray muffin cups with non-stick cooking spray. Fill each muffin cup 2/3 full.
4 Using a convection oven, bake at 350 F. 23 to 26 minutes with open vent, fan turned off the first 10 minutes, and then low fan. Remove muffins from oven and let cool.

BREADS AND SWEET DOUGHS No.D 029 01

BLUEBERRY MUFFINS

Yield 100   Portion 1 Muffin

Calories	Carbohydrates	Protein	Fat	Cholesterol	Sodium	Calcium
187 cal	33 g	4 g	5 g	34 mg	204 mg	87 mg

Ingredient   Weight   Measure   Issue

FLOUR,WHEAT,GENERAL PURPOSE	5 lbs	1 gal 1/2 qts
SUGAR,GRANULATED	2-1/2 lbs	1 qts 1-5/8 cup
MILK,NONFAT,DRY	3-5/8 oz	1-1/2 cup
BAKING POWDER	3-7/8 oz	1/2 cup
SALT	5/8 oz	1 tbsp
WATER,WARM	3-2/3 lbs	1 qts 3 cup
EGGS,WHOLE,FROZEN	1-3/4 lbs	3-1/4 cup
APPLESAUCE,CANNED,UNSWEETENED	1-5/8 lbs	3 cup
OIL,SALAD	11-1/2 oz	1-1/2 cup
BLUEBERRIES,CANNED,DRAINED	2-1/4 lbs	1 qts
COOKING SPRAY,NONSTICK	1-1/2 oz	3 tbsp

Method

1. In mixer bowl, sift together flour, sugar, milk, baking powder and salt.
2. Add warm water, eggs, applesauce and salad oil; mix at low speed until dry ingredients are moistened about 15 seconds; scrape down sides and bottom of mixer bowl; continue to mix at low speed another 15 seconds. DO NOT OVER MIX. Batter will be lumpy.
3. Rinse blueberries, drain well. Fold into batter.
4. Lightly spray muffin cups with non-stick cooking spray. Fill each muffin cup 2/3 full.
5. Using a convection oven, bake at 350 F. 23 to 26 minutes with open vent, fan turned off the first 10 minutes, and then low fan. Remove muffins from oven and let cool. NOTES: 1. In Step 3, 2 lb A.P. (1-1/2 quarts) blueberries, frozen, IQF, thawed, may be substituted.

BREADS AND SWEET DOUGHS No.D 029 02

RAISIN MUFFINS

Yield 100 **Portion** 1 Muffin

Calories	Carbohydrates	Protein	Fat	Cholesterol	Sodium	Calcium
204 cal	37 g	4 g	5 g	34 mg	205 mg	90 mg

Ingredient	Weight	Measure	Issue
FLOUR,WHEAT,GENERAL PURPOSE	5 lbs	1 gal 1/2 qts	
SUGAR,GRANULATED	2-1/2 lbs	1 qts 1-5/8 cup	
MILK,NONFAT,DRY	3-5/8 oz	1-1/2 cup	
BAKING POWDER	3-7/8 oz	1/2 cup	
SALT	5/8 oz	1 tbsp	
WATER,WARM	3-2/3 lbs	1 qts 3 cup	
EGGS,WHOLE,FROZEN	1-3/4 lbs	3-1/4 cup	
APPLESAUCE,CANNED,UNSWEETENED	1-5/8 lbs	3 cup	
RAISINS	1-7/8 lbs	1 qts 2 cup	
OIL,SALAD	11-1/2 oz	1-1/2 cup	
COOKING SPRAY,NONSTICK	1-1/2 oz	3 tbsp	

Method

1. In mixer bowl, sift together flour, sugar, milk, baking powder and salt.
2. Add warm water, eggs, applesauce, salad oil and raisins; mix at low speed until dry ingredients are moistened about 15 seconds; scrape down sides and bottom of mixer bowl; continue to mix at low speed another 15 seconds. DO NOT OVER MIX. Batter will be lumpy.
3. Lightly spray each muffin cup with non-stick cooking spray. Fill each muffin cup 2/3 full.
4. Using a convection oven, bake at 350 F. 23 to 26 minutes with open vent, fan turned off the first 10 minutes, and then low fan. Remove muffins from oven and let cool.

BREADS AND SWEET DOUGHS No.D 029 03

BANANA MUFFINS

Yield 100   Portion   1 Muffin

Calories	Carbohydrates	Protein	Fat	Cholesterol	Sodium	Calcium
186 cal	32 g	4 g	5 g	34 mg	204 mg	87 mg

Ingredient	Weight	Measure	Issue
FLOUR,WHEAT,GENERAL PURPOSE	5 lbs	1 gal 1/2 qts	
SUGAR,GRANULATED	2-1/2 lbs	1 qts 1-5/8 cup	
MILK,NONFAT,DRY	3-5/8 oz	1-1/2 cup	
BAKING POWDER	3-7/8 oz	1/2 cup	
SALT	5/8 oz	1 tbsp	
WATER,WARM	3-2/3 lbs	1 qts 3 cup	
EGGS,WHOLE,FROZEN	1-3/4 lbs	3-1/4 cup	
APPLESAUCE,CANNED,UNSWEETENED	1-5/8 lbs	3 cup	
OIL,SALAD	11-1/2 oz	1-1/2 cup	
BANANA,FRESH,MASHED	2 lbs	1 qts	3-1/8 lbs
COOKING SPRAY,NONSTICK	1-1/2 oz	3 tbsp	

Method

1. In mixer bowl sift together flour, sugar, milk, baking powder and salt.
2. Add warm water, eggs, applesauce and salad oil; mix at low speed until dry ingredients are moistened about 15 seconds; scrape down sides and bottom of mixer bowl; continue to mix at low speed another 15 seconds. DO NOT OVER MIX. Batter will be lumpy.
3. Add mashed bananas to batter; mix at low speed another 15 seconds until blended.
4. Lightly spray each muffin cup with non-stick cooking spray. Fill each muffin cup 2/3 full.
5. Using a convection oven, bake at 350 F. 23 to 26 minutes with open vent, fan turned off the first 10 minutes, and then low fan. Remove muffins from oven and let cool.

BREADS AND SWEET DOUGHS No.D 029 04

APPLE MUFFINS

Yield 100   Portion   1 Muffin

Calories	Carbohydrates	Protein	Fat	Cholesterol	Sodium	Calcium
191 cal	34 g	4 g	5 g	34 mg	204 mg	87 mg

Ingredient	Weight	Measure	Issue
FLOUR,WHEAT,GENERAL PURPOSE	5 lbs	1 gal 1/2 qts	
SUGAR,GRANULATED	2-1/2 lbs	1 qts 1-5/8 cup	
MILK,NONFAT,DRY	3-5/8 oz	1-1/2 cup	
BAKING POWDER	3-7/8 oz	1/2 cup	
SALT	5/8 oz	1 tbsp	
WATER,WARM	3-2/3 lbs	1 qts 3 cup	
EGGS,WHOLE,FROZEN	1-3/4 lbs	3-1/4 cup	
APPLESAUCE,CANNED,UNSWEETENED	1-5/8 lbs	3 cup	
OIL,SALAD	11-1/2 oz	1-1/2 cup	
APPLES,FRESH,MEDIUM,PEELED,CORED,CHOPPED	2 lbs	1 qts 3-1/4 cup	2-1/2 lbs
SUGAR,GRANULATED	7 oz	1 cup	
CINNAMON,GROUND	1/8 oz	1/3 tsp	
COOKING SPRAY,NONSTICK	1-1/2 oz	3 tbsp	

Method

1. In mixer bowl sift together flour, sugar, milk, baking powder and salt.
2. Add warm water, eggs, applesauce and salad oil; mix at low speed until dry ingredients are moistened about 15 seconds; scrape down sides and bottom of mixer bowl; continue to mix at low speed another 15 seconds. DO NOT OVERMIX. Batter will be lumpy.
3. Fold apples into batter.
4. Mix sugar and cinnamon; sprinkle 1/2 teaspoon of cinnamon sugar mixture over each muffin.
5. Lightly spray muffin cups with non-stick cooking spray. Fill each muffin cup 2/3 full.
6. Using a convection oven, bake 23 to 26 minutes with open vent, fan turned off the first 10 minutes, and then low fan. Remove muffins from oven and let cool.

Notes

1. In Step 3, 2 lb 4 oz A.P. (1 qt-1/3 No. 10 cn) drained, chopped apple slices may be substituted.

BREADS AND SWEET DOUGHS No.D 029 05

CINNAMON CRUMB TOP MUFFINS

Yield 100   Portion 1 Muffin

Calories	Carbohydrates	Protein	Fat	Cholesterol	Sodium	Calcium
195 cal	33 g	4 g	5 g	36 mg	212 mg	90 mg

Ingredient   **Weight**   **Measure**   **Issue**

FLOUR,WHEAT,GENERAL PURPOSE   5 lbs   1 gal 1/2 qts
SUGAR,GRANULATED   2-1/2 lbs   1 qts 1-5/8 cup
MILK,NONFAT,DRY   3-5/8 oz   1-1/2 cup
BAKING POWDER   3-7/8 oz   1/2 cup
SALT   5/8 oz   1 tbsp
WATER,WARM   3-2/3 lbs   1 qts 3 cup
EGGS,WHOLE,FROZEN   1-3/4 lbs   3-1/4 cup
APPLESAUCE,CANNED,UNSWEETENED   1-5/8 lbs   3 cup
OIL,SALAD   11-1/2 oz   1-1/2 cup
COOKING SPRAY,NONSTICK   1-1/2 oz   3 tbsp
SUGAR,BROWN,PACKED   9 oz   1-3/4 cup
BUTTER   3 oz   1/4 cup 2-1/3 tbsp
FLOUR,WHEAT,GENERAL PURPOSE   1-2/3 oz   1/4 cup 2-1/3 tbsp
CINNAMON,GROUND   1/2 oz   2 tbsp

Method

1 In mixer bowl sift together flour, sugar, milk, baking powder and salt.
2 Add warm water, eggs, applesauce and salad oil; mix at low speed until dry ingredients are moistened about 15 seconds; scrape down sides and bottom of mixer bowl; continue to mix at low speed another 15 seconds. DO NOT OVERMIX. Batter will be lumpy.
3 Lightly spray muffin cups with non-stick cooking spray. Fill each muffin cup 2/3 full.
4 Mix brown sugar, butter or margarine, flour and cinnamon until mixture is crumbly. Sprinkle 1 teaspoon mixture on top of each muffin.
5 Using a convection oven, bake at 350 F. 23 to 26 minutes with open vent, fan turned on for the first 10 minutes, and then on low fan. Remove muffins from oven and cool.

BREADS AND SWEET DOUGHS No.D 029 06

CRANBERRY MUFFINS

Yield 100   Portion   1 Muffin

Calories	Carbohydrates	Protein	Fat	Cholesterol	Sodium	Calcium
182 cal	32 g	4 g	5 g	34 mg	204 mg	87 mg

Ingredient	Weight	Measure	Issue
FLOUR,WHEAT,GENERAL PURPOSE	5 lbs	1 gal 1/2 qts	
SUGAR,GRANULATED	2-1/2 lbs	1 qts 1-5/8 cup	
MILK,NONFAT,DRY	3-5/8 oz	1-1/2 cup	
BAKING POWDER	3-7/8 oz	1/2 cup	
SALT	5/8 oz	1 tbsp	
WATER,WARM	3-2/3 lbs	1 qts 3 cup	
EGGS,WHOLE,FROZEN	1-3/4 lbs	3-1/4 cup	
APPLESAUCE,CANNED,UNSWEETENED	1-5/8 lbs	3 cup	
OIL,SALAD	11-1/2 oz	1-1/2 cup	
CRANBERRIES,FRESH	2 lbs	2 qts 1-1/2 cup	2-1/8 lbs
COOKING SPRAY,NONSTICK	1-1/2 oz	3 tbsp	

Method

1. In mixer bowl sift together flour, sugar, milk, baking powder and salt.
2. Add warm water, eggs, applesauce and salad oil; mix at low speed until dry ingredients are moistened about 15 seconds; scrape down sides and bottom of mixer bowl; continue to mix at low speed another 15 seconds. DO NOT OVERMIX. Batter will be lumpy.
3. Fold cranberries into batter.
4. Lightly spray muffin cups with non-stick cooking spray. Fill each muffin cup 2/3 full.
5. Using a convection oven, bake at 350 F. 23 to 26 minutes with open vent, fan turned off the first 10 minutes, and then on low fan. Remove muffins from oven and let cool.

Notes

1. In Step 3, 2 lb 1 oz A.P. (8-1/3 cup) cranberries, brozen, IQF, thawed, may be substituted.

BREADS AND SWEET DOUGHS No.D 029 07

DATE MUFFINS

Yield 100　　　　　　　　　　　　　　　　　　　　**Portion** 1 Muffin

Calories	Carbohydrates	Protein	Fat	Cholesterol	Sodium	Calcium
196 cal	35 g	4 g	5 g	34 mg	204 mg	88 mg

Ingredient	Weight	Measure	Issue
FLOUR,WHEAT,GENERAL PURPOSE	5 lbs	1 gal 1/2 qts	
SUGAR,GRANULATED	2-1/2 lbs	1 qts 1-5/8 cup	
MILK,NONFAT,DRY	3-5/8 oz	1-1/2 cup	
BAKING POWDER	3-7/8 oz	1/2 cup	
SALT	5/8 oz	1 tbsp	
WATER,WARM	3-2/3 lbs	1 qts 3 cup	
EGGS,WHOLE,FROZEN	1-3/4 lbs	3-1/4 cup	
APPLESAUCE,CANNED,UNSWEETENED	1-5/8 lbs	3 cup	
DATES,DRIED,PITTED,CHOPPED	1-1/2 lbs		
OIL,SALAD	11-1/2 oz	1-1/2 cup	
COOKING SPRAY,NONSTICK	1-1/2 oz	3 tbsp	

Method

1 In mixer bowl, sift together flour, sugar, milk, baking powder and salt.
2 Add warm water, eggs, applesauce, salad oil and dates; mix at low speed until dry ingredients are moistened about 15 seconds; scrape down sides and bottom of mixer bowl; continue to mix at low speed another 15 seconds. DO NOT OVER MIX. Batter will be lumpy.
3 Lightly spray muffin cups with non-stick cooking spray. Fill each muffin cup 2/3 full.
4 Using a convection oven, bake at 350 F. 23 to 26 minutes with open vent, fan turned off the first 10 minutes, and then low fan. Remove muffins from oven and let cool.

BREADS AND SWEET DOUGHS No.D 029 08

NUT MUFFINS

Yield 100   Portion   1 Muffin

Calories	Carbohydrates	Protein	Fat	Cholesterol	Sodium	Calcium
272 cal	34 g	5 g	14 g	39 mg	169 mg	67 mg

Ingredient	Weight	Measure	Issue
FLOUR,WHEAT,GENERAL PURPOSE	5-1/2 lbs	1 gal 1 qts	
SUGAR,GRANULATED	2-1/4 lbs	1 qts 1 cup	
MILK,NONFAT,DRY	1-3/4 oz	3/4 cup	
BAKING POWDER	2-3/4 oz	1/4 cup 2 tbsp	
SALT	5/8 oz	1 tbsp	
EGGS,WHOLE,FROZEN	2 lbs	3-3/4 cup	
APPLESAUCE,CANNED,UNSWEETENED	1-5/8 lbs	3 cup	
WATER,WARM	2-1/8 lbs	1 qts	
OIL,SALAD	11-1/2 oz	1-1/2 cup	
PECANS,CHOPPED	3 lbs		
COOKING SPRAY,NONSTICK	2 oz	1/4 cup 1/3 tbsp	

Method

1. In mixer bowl, sift together flour, sugar, milk, baking powder and salt.
2. Add warm water, eggs, applesauce, salad oil and pecans; mix at low speed until dry ingredients are moistened about 15 seconds; scrape down sides and bottom of mixer bowl; continue to mix at low speed another 15 seconds. DO NOT OVER MIX. Batter will be lumpy.
3. Lightly spray each muffin cup with non-stick cooking spray. Fill each muffin cup 2/3 full.
4. Using a convection oven, bake at 350 F. 23 to 26 minutes with open vent, fan turned off the first 10 minutes, and then low fan. Remove muffins from oven and let cool.

BREADS AND SWEET DOUGHS No.D 029 09

OATMEAL RAISIN MUFFINS

Yield 100   Portion   1 Muffin

Calories	Carbohydrates	Protein	Fat	Cholesterol	Sodium	Calcium
216 cal	38 g	5 g	5 g	34 mg	205 mg	95 mg

Ingredient | Weight | Measure | Issue

Ingredient	Weight	Measure
FLOUR,WHEAT,GENERAL PURPOSE	3-1/2 lbs	3 qts 3/4 cup
SUGAR,GRANULATED	2-1/2 lbs	1 qts 1-5/8 cup
MILK,NONFAT,DRY	3-5/8 oz	1-1/2 cup
BAKING POWDER	3-7/8 oz	1/2 cup
SALT	5/8 oz	1 tbsp
CINNAMON,GROUND	1/4 oz	1 tbsp
CEREAL,OATMEAL,ROLLED	2 lbs	1 qts 1-3/4 cup
RAISINS	1-7/8 lbs	1 qts 2 cup
WATER,WARM	3-2/3 lbs	1 qts 3 cup
EGGS,WHOLE,FROZEN	1-3/4 lbs	3-1/4 cup
APPLESAUCE,CANNED,UNSWEETENED	1-5/8 lbs	3 cup
OIL,SALAD	11-1/2 oz	1-1/2 cup
EXTRACT,VANILLA	1-7/8 oz	1/4 cup 1/3 tbsp
COOKING SPRAY,NONSTICK	1-1/2 oz	3 tbsp

Method

1. In mixer bowl, sift together flour, sugar, milk, baking powder, salt, and cinnamon. Add rolled oats and raisins. Mix at low speed for 1 minute or until blended.
2. Add warm water, eggs, applesauce, salad oil, and vanilla; mix at low speed until dry ingredients are moistened about 15 seconds; scrape down sides and bottom of mixer bowl; continue to mix at low speed another 15 seconds. DO NOT OVER MIX. Batter will be lumpy.
3. Lightly spray muffin cups with non-stick cooking spray. Fill each muffin cup 2/3 full.
4. Using a convection oven, bake at 350 F. 23 to 26 minutes with open vent, fan turned off the first 10 minutes, and then low fan. Remove muffins from oven and let cool.

BREADS AND SWEET DOUGHS No.D 030 00

BANANA BREAD

Yield 100   Portion   1 Slice

Calories	Carbohydrates	Protein	Fat	Cholesterol	Sodium	Calcium
258 cal	33 g	6 g	12 g	37 mg	140 mg	64 mg

Ingredient	Weight	Measure	Issue
SHORTENING	1 lbs	1-1/8 cup	
SUGAR,GRANULATED	2-2/3 lbs	1 qts 2 cup	
EGGS,WHOLE,FROZEN	1-7/8 lbs	3-1/2 cup	
APPLESAUCE,CANNED,SWEETENED		1 cup	
BANANA,FRESH,MASHED	5-1/4 lbs	2 qts 2-5/8 cup	8-1/8 lbs
NUTS,UNSALTED,CHOPPED,COARSELY	2-5/8 lbs	2 qts	
FLOUR,WHEAT,GENERAL PURPOSE	3-7/8 lbs	3 qts 2 cup	
BAKING POWDER	2-2/3 oz	1/4 cup 2 tbsp	
SALT	3/8 oz	1/3 tsp	
COOKING SPRAY,NONSTICK	2 oz	1/4 cup 1/3 tbsp	

Method

1 Cream shortening and sugar in mixer bowl at medium speed 2 minutes until light and fluffy.
2 Add eggs and applesauce to mixture. Mix at medium speed 1 minute.
3 Add bananas and nuts to egg mixture. Mix at medium speed until blended.
4 Sift together flour, baking powder and salt.
5 Add dry ingredients to banana mixture; beat at low speed about 1/2 minute. Continue beating 1/2 minute longer or until blended. DO NOT OVER MIX.
6 Lightly spray each pan with non-stick cooking spray. Pour about 2 quarts of batter into each sprayed and floured loaf pan. Spread batter evenly.
7 Using a convection oven, bake at 325 F. for 70 to 75 minutes or until done on low fan, open vent.
8 Let bread cool in pans 5 minutes; then remove from pan and place on wire rack to cool completely. To enhance flavor and moistness, product may be prepared in advance. CCP: Refrigerate at 41 F. or lower overnight.

BREADS AND SWEET DOUGHS No.D 032 00

HARD ROLLS

Yield 100   **Portion** 2 Rolls

Calories	Carbohydrates	Protein	Fat	Cholesterol	Sodium	Calcium
259 cal	49 g	8 g	3 g	0 mg	425 mg	11 mg

Ingredient	Weight	Measure	Issue
YEAST,ACTIVE,DRY	2-1/2 oz	1/4 cup 2-1/3 tbsp	
WATER,WARM	1-2/3 lbs	3-1/4 cup	
WATER,COLD	6-1/4 lbs	3 qts	
EGG WHITES	8-1/2 oz	1 cup	
SUGAR,GRANULATED	3-1/2 oz	1/2 cup	
SALT	3-3/4 oz	1/4 cup 2-1/3 tbsp	
SHORTENING,SOFTENED	4-1/8 oz	1/2 cup 1 tbsp	
FLOUR,WHEAT,BREAD	14-1/2 lbs	3 gal	
COOKING SPRAY,NONSTICK	2 oz	1/4 cup 1/3 tbsp	

Method

1. Sprinkle yeast over water. DO NOT USE TEMPERATURES ABOVE 110 F. Mix well. Let stand 5 minutes; stir.
2. Place water, egg whites, sugar, salt, shortening, and flour in mixer bowl. Add yeast solution.
3. Using dough hook, mix at low speed 1 minute or until all flour mixture is incorporated into liquid. Continue mixing at medium speed 10 minutes or until dough is smooth and elastic. Dough temperature should be between 78 F. to 82 F.
4. FERMENT: Cover. Set in warm place (80 F.) about 1-1/2 hours or until double in bulk.
5. PUNCH: Divide dough into 8 2-1/2 pound pieces. Shape each piece into a smooth ball; let rest 10 to 20 minutes.
6. Roll each piece into a long rope, about 25 inches, of uniform diameter. Cut rope into pieces about 1-inch thick, weighing 1-1/2 ounces each.
7. MAKE-UP: Lightly spray sheet pans with non-stick cooking spray. Place rolls on sheet pans in rows 5 by 7 so rolls do not touch each other during proofing or baking.
8. PROOF: At 90 F. to 100 F. until double in bulk. Brush with 1 recipe hot Cornstarch Wash, Recipe No. D 040 00.
9. BAKE: 25 to 30 minutes at 400 F. or in 350 F. convection oven 15 minutes or until golden brown, on high fan, open vent. Brush with hot Cornstarch Wash, Recipe No. D 040 00 immediately after removal from oven.

BREADS AND SWEET DOUGHS No.D 033 00

HOT ROLLS

Yield 100   Portion 2 Rolls

Calories	Carbohydrates	Protein	Fat	Cholesterol	Sodium	Calcium
325 cal	56 g	9 g	7 g	0 mg	358 mg	26 mg

Ingredient	Weight	Measure	Issue
YEAST,ACTIVE,DRY	4-1/2 oz	1/2 cup 2-2/3 tbsp	
WATER,WARM	1-7/8 lbs	3-1/2 cup	
WATER,COLD	5-3/4 lbs	2 qts 3 cup	
SUGAR,GRANULATED	1-1/2 lbs	3-1/2 cup	
SALT	3-1/8 oz	1/4 cup 1-1/3 tbsp	
FLOUR,WHEAT,BREAD	14-1/2 lbs	3 gal	
MILK,NONFAT,DRY	4-1/4 oz	1-3/4 cup	
SHORTENING,SOFTENED	1-1/4 lbs	2-3/4 cup	
COOKING SPRAY,NONSTICK	1 oz	2 tbsp	

Method

1 Sprinkle yeast over water. DO NOT USE TEMPERATURES ABOVE 110 F. Mix well. Let stand 5 minutes; stir.
2 Place cold water in mixer bowl; add sugar and salt; stir until dissolved. Add yeast solution.
3 Combine flour and milk; add to liquid solution. Using dough hook, mix at low speed 1 minute or until flour mixture is incorporated into liquid.
4 Add shortening; mix at medium speed 10 minutes or until dough is smooth and elastic. Dough temperature should be between 78 F. to 82 F.
5 FERMENT: Cover. Set in warm place, about 80 F., 1-1/2 hours or until double in size.
6 PUNCH: Divide dough into 8 2 lb 14 oz pieces. MAKEUP: Shape each piece into a smooth ball; let rest 10 to 20 minutes.
7 Roll each piece into a long rope, about 32 inches, of uniform diameter. Cut rope into 25 1-3/4 oz pieces about 1-1/4 inch long. Place rolls on a lightly sprayed sheet pan.
8 PROOF: At 90 F. about 1 hour or until double in bulk.
9 BAKE: Using a 350 F. convection oven, bake for 10 to 15 minutes or until golden brown, on high fan, open vent.

BREADS AND SWEET DOUGHS No.D 033 01

HOT ROLLS (BROWN AND SERVE)

Yield 100   **Portion** 2 Rolls

Calories	Carbohydrates	Protein	Fat	Cholesterol	Sodium	Calcium
325 cal	56 g	9 g	7 g	0 mg	358 mg	26 mg

Ingredient	**Weight**	**Measure**	**Issue**
YEAST,ACTIVE,DRY	4-1/2 oz	1/2 cup 2-2/3 tbsp	
WATER,WARM	1-7/8 lbs	3-1/2 cup	
WATER,COLD	5-3/4 lbs	2 qts 3 cup	
SUGAR,GRANULATED	1-1/2 lbs	3-1/2 cup	
SALT	3-1/8 oz	1/4 cup 1-1/3 tbsp	
FLOUR,WHEAT,BREAD	14-1/2 lbs	3 gal	
MILK,NONFAT,DRY	4-1/4 oz	1-3/4 cup	
SHORTENING,SOFTENED	1-1/4 lbs	2-3/4 cup	
COOKING SPRAY,NONSTICK	1 oz	2 tbsp	

Method

1. Sprinkle yeast over water. DO NOT USE TEMPERATURES ABOVE 110 F. Mix well. Let stand 5 minutes; stir.
2. Place cold water in mixer bowl; add sugar and salt; stir until dissolved. Add yeast solution.
3. Combine flour and milk; add to liquid solution. Using dough hook, mix at low speed 1 minute or until flour mixture is incorporated into liquid.
4. Add shortening; mix at medium speed 10 minutes or until dough is smooth and elastic. Dough temperature should be between 78 F. to 82 F.
5. FERMENT: Cover. Set in warm place (80 F.) 1-1/2 hours or until double in size.
6. PUNCH: Divide dough into 8-2 lb 14 oz pieces. Shape each piece into a smooth ball; let rest 10 to 20 minutes.
7. Roll each piece into a long rope, about 32 inches, of uniform diameter about 2 inches thick. Cut rope into 25 1-3/4 oz pieces about 1-1/4 inch long.
8. MAKEUP: Shape dough pieces into balls by rolling with a circular motion. Lightly spray sheet pans with non-stick cooking
9. PROOF: At 90 F. about 30 minutes or until double in size.
10. PREBAKE: 25 minutes at 325 F. or in 300 F. convection oven for 12 to 15 minutes or until rolls begin to brown on low fan, open vent.
11. Cool on pans; wrap in aluminum foil. Refrigerate at 40 F. for up to 2 days.
12. BAKE: Bring covered rolls to room temperature about 1 hour before baking. Finish baking in 350 F. convection oven about 10 to 12 minutes or until golden brown on high fan, open vent.

BREADS AND SWEET DOUGHS No.D 033 02
HOT ROLLS (ROLL MIX)

Yield 100   **Portion** 2 Rolls

Calories	Carbohydrates	Protein	Fat	Cholesterol	Sodium	Calcium
264 cal	48 g	8 g	5 g	0 mg	416 mg	44 mg

Ingredient	Weight	Measure	Issue
ROLL,MIX	15 lbs		
YEAST,ACTIVE,DRY	6-3/4 oz	1 cup	
WATER	1-3/4 lbs	3-3/8 cup	

Method

1. Prepare dough according to instructions on container.
2. PUNCH: Divide dough into 8-2 lb 14 oz pieces. Shape each piece into a smooth ball; let rest 10 to 20 minutes.
3. Roll each piece into a long rope, about 32 inches, of uniform diameter about 2 inches thick. Cut rope into 25 1-3/4 oz pieces about 1-1/4 inch long.
4. MAKE-UP: Shape each piece into a smooth ball; let rest 10 to 20 minutes.
5. PROOF: At 90 F. about 1 hour or until double in bulk.
6. BAKE: 15 to 20 minutes at 400 F. or in 350 F. convection oven for 10 to 15 minutes or until golden brown, on high fan, open vent.

BREADS AND SWEET DOUGHS No.D 033 03

OATMEAL ROLLS

Yield 100   Portion 2 Rolls

Calories	Carbohydrates	Protein	Fat	Cholesterol	Sodium	Calcium
300 cal	46 g	8 g	9 g	0 mg	358 mg	28 mg

Ingredient	Weight	Measure	Issue
YEAST,ACTIVE,DRY	4-1/2 oz	1/2 cup 2-2/3 tbsp	
WATER,WARM	1-7/8 lbs	3-1/2 cup	
WATER,COLD	5-3/4 lbs	2 qts 3 cup	
SUGAR,GRANULATED	1-1/2 lbs	3-1/2 cup	
SALT	3-1/8 oz	1/4 cup 1-1/3 tbsp	
FLOUR,WHEAT,BREAD	9-2/3 lbs	2 gal	
CEREAL,OATMEAL,ROLLED	2 lbs	1 qts 2 cup	
MILK,NONFAT,DRY	4-1/4 oz	1-3/4 cup	
SHORTENING,SOFTENED	1-2/3 lbs	3-3/4 cup	
COOKING SPRAY,NONSTICK	1 oz	2 tbsp	

Method

1. Sprinkle yeast over water. DO NOT USE TEMPERATURES ABOVE 110 F. Mix well. Let stand 5 minutes; stir.
2. Place cold water in mixer bowl; add sugar and salt; stir until dissolved. Add yeast solution.
3. Combine flour, rolled oats and milk; add to liquid solution. Using dough hook, mix at low speed 1 minute or until flour mixture is incorporated into liquid.
4. Add shortening; mix at medium speed 10 minutes or until dough is smooth and elastic. Dough temperature should be between 78 F. to 82 F.
5. FERMENT: Cover. Set in warm place (80 F.) 1-1/2 hours or until double in size.
6. PUNCH: Divide dough into 8 2 lb 14 oz pieces. Shape each piece into a smooth ball; let rest 10 to 20 minutes.
7. Roll each piece into a long rope, about 32 inches, of uniform diameter about 2 inches thick. Cut rope into 25 1-3/4 oz pieces about 1-1/4 inches long.
8. MAKE-UP: Shape dough pieces into balls rolling with a circular motion on a worktable. Place rolls on lightly sprayed sheet pans.
9. PROOF: At 90 F. about 1 hour or until double in bulk.
   BAKE: Using a 350 F. convection oven, bake 10 to 12 minutes on high fan, open vent.

BREADS AND SWEET DOUGHS No.D 033 04

CLOVERLEAF OR TWIN ROLLS

Yield 100   Portion 2 Rolls

Calories	Carbohydrates	Protein	Fat	Cholesterol	Sodium	Calcium
357 cal	56 g	9 g	11 g	5 mg	377 mg	27 mg

Ingredient	Weight	Measure	Issue
YEAST,ACTIVE,DRY	4-1/2 oz	1/2 cup 2-2/3 tbsp	
WATER,WARM	1-7/8 lbs	3-1/2 cup	
WATER,COLD	5-3/4 lbs	2 qts 3 cup	
SUGAR,GRANULATED	1-1/2 lbs	3-1/2 cup	
SALT	3-1/8 oz	1/4 cup 1-1/3 tbsp	
FLOUR,WHEAT,BREAD	14-1/2 lbs	3 gal	
MILK,NONFAT,DRY	4-1/4 oz	1-3/4 cup	
SHORTENING,SOFTENED	1-2/3 lbs	3-3/4 cup	
BUTTER,MELTED	4 oz	1/2 cup	
BUTTER,MELTED	4 oz	1/2 cup	

Method

1 Sprinkle yeast over water. DO NOT USE TEMPERATURES ABOVE 110 F. Mix well. Let stand 5 minutes; stir.
2 Place water in mixer bowl; add sugar and salt; stir until dissolved. Add yeast solution.
3 Combine flour and milk; add to liquid solution. Using dough hook, mix at low speed 1 minute until flour mixture is incorporated into liquid.
4 Add shortening; mix at medium speed 10 minutes or until dough is smooth and elastic. Dough temperature should be between 78 F. to 82 F.
5 FERMENT: Cover. Set in warm place (80 F.) for 1-1/2 hours or until double in bulk.
6 PUNCH: Divide dough into 8 2 lb 14 oz pieces. Shape each piece into a smooth ball; let rest 10 to 20 minutes.
7 Roll each piece into a long rope, about 32 inches, of uniform diameter about 2 inches thick. Cut rope into 25 1-3/4 oz pieces about 1-1/4 inch long.
8 Divide each dough piece into thirds for cloverleaf rolls or in halves for twin rolls.
9 Shape into balls by rolling with a circular motion on work table.
10 Place in greased muffin pans. In each cup: 3 balls for cloverleaf or 2 for twin; brush with 4 ounces or 1/2 cup of melted butter or 1/3 recipe Milk Wash, Recipe No. I 004 02.
11 PROOF: At 90 F. until double in bulk.
12 BAKE: At 400 F. for 15 to 20 minutes or in a 350 F. convection oven for 10 to 15 minutes or until golden brown on high fan, open vent.
13 If desired, brush with 4 ounces or 1/2 cup of melted butter immediately after baking.

BREADS AND SWEET DOUGHS No.D 033 05

FRANKFURTER ROLLS

Yield 100   Portion   1 Roll

Calories	Carbohydrates	Protein	Fat	Cholesterol	Sodium	Calcium
227 cal	37 g	6 g	6 g	0 mg	239 mg	18 mg

Ingredient	Weight	Measure	Issue
YEAST,ACTIVE,DRY	3-3/8 oz	1/2 cup	
WATER,WARM	1-1/4 lbs	2-3/8 cup	
WATER,COLD	1 lbs	1-7/8 cup	
SUGAR,GRANULATED	1 lbs	2-3/8 cup	
SALT	2-1/8 oz	3-1/3 tbsp	
FLOUR,WHEAT,BREAD	9-2/3 lbs	2 gal	
MILK,NONFAT,DRY	2-2/3 oz	1-1/8 cup	
SHORTENING,SOFTENED	1-1/8 lbs	2-1/2 cup	
MILK AND WATER WASH		1/2 cup	

Method

1. Sprinkle yeast over water. DO NOT USE TEMPERATURES ABOVE 110 F. Mix well. Let stand 5 minutes; stir.
2. Place water in mixer bowl; add sugar and salt; stir until dissolved. Add yeast solution.
3. Combine flour and milk; add to liquid solution. Using dough hook, mix at low speed 1 minute or until flour mixture is incorporated into liquid.
4. Add shortening; mix at medium speed 10 minutes or until dough is smooth and elastic. Dough temperature should be between 78 F. to 82 F.
5. FERMENT: Cover. Set in warm place, about 180 F., 1-1/2 hours or until double in bulk.
6. PUNCH: Divide dough into 8 2 lb 14 oz pieces. Shape each piece into a smooth ball; let rest 10 to 20 minutes.
7. Roll 2-1/2-ounce pieces of dough into oblong rolls, 5 to 6 inches long.
8. Place on greased sheet pans in rows 4 by 9. Brush with 1/3 recipe Milk Wash, Recipe No. I 004 02 per 100 servings.
9. Proof at 90 F. until double in bulk.
10. Bake at 400 F. for 15 to 20 minutes or in a 350 F. convection oven for 10 to 15 minutes or until golden brown on high fan, open vent. Cool.

BREADS AND SWEET DOUGHS No.D 033 06

HAMBURGER ROLLS

Yield 100   **Portion** 1 Roll

Calories	Carbohydrates	Protein	Fat	Cholesterol	Sodium	Calcium
227 cal	37 g	6 g	6 g	0 mg	239 mg	18 mg

Ingredient	Weight	Measure	Issue
YEAST, ACTIVE, DRY	3-3/8 oz	1/2 cup	
WATER, WARM	1-1/4 lbs	2-3/8 cup	
WATER, COLD	1 lbs	1-7/8 cup	
SUGAR, GRANULATED	1 lbs	2-3/8 cup	
SALT	2-1/8 oz	3-1/3 tbsp	
FLOUR, WHEAT, BREAD	9-2/3 lbs	2 gal	
MILK, NONFAT, DRY	2-2/3 oz	1-1/8 cup	
SHORTENING, SOFTENED	1-1/8 lbs	2-1/2 cup	
MILK AND WATER WASH		1/2 cup	

Method

1. Sprinkle yeast over water. DO NOT USE TEMPERATURES ABOVE 110 F. Mix well. Let stand 5 minutes; stir.
2. Place water in mixer bowl; add sugar and salt; stir until dissolved. Add yeast solution.
3. Combine flour and milk; add to liquid solution. Using dough hook, mix at low speed 1 minute or until flour mixture is incorporated into liquid.
4. Add shortening; mix at medium speed 10 minutes or until dough is smooth and elastic. Dough temperature should be between 78 F. to 82 F.
5. FERMENT: Cover. Set in water place, about 80 F., 1-1/2 hours or until double in bulk.
6. Punch: Divide dough into 8 2 lb 14 oz pieces. Shape each piece into a smooth ball; let rest 10 to 20 minutes.
7. Shape 2-1/2 ounce pieces of dough into balls by rolling with a circular motion on work table.
8. Place on greased sheet pans in rows 4 by 6.
9. When half-proofed, flatten with hand or small can to about 1/2 inch thickness and 3-1/2 inch diameter; brush with 1/3 recipe Milk Wash, Recipe No. I 004 02 per 100 servings.
10. Proof at 90 F. until double in bulk.
11. Bake at 400 F. for 15 to 20 minutes or in 350 F. convection oven for 10 to 15 minutes or until golden brown on high fan, open vent. Cool.

BREADS AND SWEET DOUGHS No.D 033 07

PAN, CLUSTER, OR PULL APART ROLLS

Yield 100　　　　　　　　　　　　　　　　　　　Portion 2 Rolls

Calories	Carbohydrates	Protein	Fat	Cholesterol	Sodium	Calcium
357 cal	56 g	9 g	11 g	0 mg	380 mg	27 mg

Ingredient	Weight	Measure	Issue
YEAST,ACTIVE,DRY	4-1/2 oz	1/2 cup 2-2/3 tbsp	
WATER,WARM	1-7/8 lbs	3-1/2 cup	
WATER,COLD	5-3/4 lbs	2 qts 3 cup	
SUGAR,GRANULATED	1-1/2 lbs	3-1/2 cup	
SALT	3-1/8 oz	1/4 cup 1-1/3 tbsp	
FLOUR,WHEAT,BREAD	14-1/2 lbs	3 gal	
MILK,NONFAT,DRY	4-1/4 oz	1-3/4 cup	
SHORTENING,SOFTENED	1-2/3 lbs	3-3/4 cup	
MARGARINE,MELTED	4 oz	1/2 cup	
MARGARINE,MELTED	4 oz	1/2 cup	

Method

1. Sprinkle yeast over water. DO NOT USE TEMPERATURES ABOVE 110 F. Mix well. Let stand 5 minutes; stir.
2. Place water in mixer bowl; add sugar and salt; stir until dissolved. Add yeast solution.
3. Combine flour and milk; add to liquid solution. Using dough hook, mix at low speed 1 minute until flour mixture is incorporated into liquid.
4. Add shortening; mix at medium speed 10 minutes or until dough is smooth and elastic. Dough temperature should be between 78 F. to 82 F.
5. FERMENT: Cover. Set in warm place (80 F.) for 1-1/2 hours or until double in bulk.
6. PUNCH: Divide dough into about 3 pound pieces. Shape each piece into a smooth ball; let rest 10 to 20 minutes.
7. Roll each piece into a long rope, about 32 inches, of uniform diameter about 2 inches thick. Cut rope into 25 1-3/4 oz pieces about 1-1/4 inch long.
8. Shape 1-1/2 to 2-ounce dough pieces into balls by rolling with a circular motion on work table.
9. Place on greased sheet pans in rows 6 by 9. Brush with 4 ounces of melted butter or 1/4 recipe Egg Wash, Recipe No. D 017 00.
10. Proof at 90 F. until double in bulk.
11. Bake at 400 F., 15 to 20 minutes or in 350 F. convection oven 10 to 15 minutes or until golden brown on high fan, open vent.
12. Brush with 4 oz melted butter, optional, immediately after baking.

BREADS AND SWEET DOUGHS No.D 033 08

PARKER HOUSE ROLLS

Yield 100   **Portion** 2 Rolls

Calories	Carbohydrates	Protein	Fat	Cholesterol	Sodium	Calcium
357 cal	56 g	9 g	11 g	5 mg	377 mg	27 mg

Ingredient	Weight	Measure	Issue
YEAST,ACTIVE,DRY	4-1/2 oz	1/2 cup 2-2/3 tbsp	
WATER,WARM	1-7/8 lbs	3-1/2 cup	
WATER,COLD	5-3/4 lbs	2 qts 3 cup	
SUGAR,GRANULATED	1-1/2 lbs	3-1/2 cup	
SALT	3-1/8 oz	1/4 cup 1-1/3 tbsp	
FLOUR,WHEAT,BREAD	14-1/2 lbs	3 gal	
MILK,NONFAT,DRY	4-1/4 oz	1-3/4 cup	
SHORTENING,SOFTENED	1-2/3 lbs	3-3/4 cup	
BUTTER,MELTED	4 oz	1/2 cup	
BUTTER,MELTED	4 oz	1/2 cup	

Method

1. Sprinkle yeast over water. Do not use temperatures above 110 F. Mix well. Let stand 5 minutes; stir.
2. Place water in mixer bowl; add sugar and salt; stir until dissolved. Add yeast solution.
3. Combine flour and milk; add to liquid solution. Using dough hook, mix at low speed 1 minute until flour mixture is incorporated into liquid.
4. Add shortening; mix at medium speed 10 minutes or until dough is smooth and elastic. Dough temperature should be between 78 F. to 82 F.
5. FERMENT: Cover. Set in warm place, about 80 F., for 1-1/2 hours or until double in bulk.
6. PUNCH: Divide dough into about 3 pound pieces. Shape each piece into a smooth ball; let rest 10 to 20 minutes.
7. Roll each piece into a long rope, about 32 inches, of uniform diameter. Cut rope into pieces about 1-1/4 inch thick.
8. Shape 1-1/2 to 2-ounce dough pieces into balls by rolling with a circular motion on work table.
9. Cover with clean damp cloth; let rest 5 to 10 minutes.
10. Press center of each ball with a small rolling pin.
11. Brush with 4 ounces of melted butter; fold in half. Press edges together with thumb or palm of hand.
12. Place on greased sheet pans in rows 5 by 10; brush with 4 ounces of melted butter.
13. Proof at 90 F. until double in bulk.
14. Bake at 400 F., 15 to 20 minutes or in 350 F. convection oven 10 to 15 minutes or until golden brown on high fan, open vent.

BREADS AND SWEET DOUGHS No.D 033 09

POPPY SEED ROLLS

Yield 100   Portion   2 Rolls

Calories	Carbohydrates	Protein	Fat	Cholesterol	Sodium	Calcium
346 cal	56 g	9 g	9 g	0 mg	359 mg	42 mg

Ingredient	Weight	Measure	Issue
YEAST,ACTIVE,DRY	4-1/2 oz	1/2 cup 2-2/3 tbsp	
WATER,WARM	1-7/8 lbs	3-1/2 cup	
WATER,COLD	5-3/4 lbs	2 qts 3 cup	
SUGAR,GRANULATED	1-1/2 lbs	3-1/2 cup	
SALT	3-1/8 oz	1/4 cup 1-1/3 tbsp	
FLOUR,WHEAT,BREAD	14-1/2 lbs	3 gal	
MILK,NONFAT,DRY	4-1/4 oz	1-3/4 cup	
SHORTENING,SOFTENED	1-2/3 lbs	3-3/4 cup	
EGG WHITE WASH		1/2 cup	
POPPY SEEDS	3-3/4 oz	3/4 cup	

Method

1. Sprinkle yeast over water. Do not use temperatures above 110 F. Mix well. Let stand 5 minutes; stir.
2. Place water in mixer bowl; add sugar and salt; stir until dissolved. Add yeast solution.
3. Combine flour and milk; add to liquid solution. Using dough hook, mix at low speed 1 minute until flour mixture is incorporated into liquid.
4. Add shortening; mix at medium speed 10 minutes or until dough is smooth and elastic. Dough temperature should be between 78 F. to 82 F.
5. FERMENT: Cover. Set in warm place, about 80 F., for 1-1/2 hours or until double in bulk.
6. PUNCH: Divide dough into about 8 2 lb 14 oz pieces. Shape each piece into a smooth ball; let rest 10 to 20 minutes.
7. Roll each piece into a long rope, about 32 inches, of uniform diameter about 2 inches thick. Cut rope into 25 1-3/4 oz pieces about 1-1/4 inch long.
8. Shape rolls as desired.
9. Place on greased sheet pans. Brush top of rolls lightly with water or 1/6 recipe Egg White Wash, Recipe No. D 017 01. Sprinkle top of rolls with poppy seeds.
10. Proof at 90 F. until double in bulk.
11. Bake at 400 F. for 15 to 20 minutes or in a 350 F. convection oven 10 to 15 minutes or until golden brown on high fan, open vent.

BREADS AND SWEET DOUGHS No.D 033 10

SESAME SEED ROLLS

Yield 100   Portion   2 Rolls

Calories	Carbohydrates	Protein	Fat	Cholesterol	Sodium	Calcium
347 cal	56 g	9 g	9 g	0 mg	359 mg	28 mg

Ingredient	Weight	Measure	Issue
YEAST,ACTIVE,DRY	4-1/2 oz	1/2 cup 2-2/3 tbsp	
WATER,WARM	1-7/8 lbs	3-1/2 cup	
WATER,COLD	5-3/4 lbs	2 qts 3 cup	
SUGAR,GRANULATED	1-1/2 lbs	3-1/2 cup	
SALT	3-1/8 oz	1/4 cup 1-1/3 tbsp	
FLOUR,WHEAT,BREAD	14-1/2 lbs	3 gal	
MILK,NONFAT,DRY	4-1/4 oz	1-3/4 cup	
SHORTENING,SOFTENED	1-2/3 lbs	3-3/4 cup	
EGG WHITE WASH		1/2 cup	
SESAME SEEDS	3-3/4 oz	3/4 cup	

Method

1. Sprinkle yeast over water. Do not use temperatures above 110 F. Mix well. Let stand 5 minutes; stir.
2. Place water in mixer bowl; add sugar and salt; stir until dissolved. Add yeast solution.
3. Combine flour and milk; add to liquid solution. Using dough hook, mix at low speed 1 minute until flour mixture is incorporated into liquid.
4. Add shortening; mix at medium speed 10 minutes or until dough is smooth and elastic. Dough temperature should be between 78 F. to 82 F.
5. FERMENT: Cover. Set in warm place, about 80 F., for 1-1/2 hours or until double in bulk.
6. PUNCH: Divide dough into about 3 pound pieces. Shape each piece into a smooth ball; let rest 10 to 20 minutes.
7. Roll each piece into a long rope, about 32 inches, of uniform diameter about 2 inches. Cut rope into 25 1-3/4 oz pieces about 1-1/4-inch long.
8. Shape rolls as desired.
9. Place on greased sheet pans. Brush top of rolls lightly with water or 1/6 recipe Egg White Wash, Recipe No. D 017 01. Sprinkle top of rolls with sesame seeds.
10. Proof at 90 F. until double in bulk.
11. Bake at 400 F. for 15 to 20 minutes or in a 350 F. convection oven 10 to 15 minutes or until golden brown on high fan, open vent.

BREADS AND SWEET DOUGHS No.D 034 00

HOT ROLLS (SHORT-TIME FORMULA)

Yield 100   Portion 2 Rolls

Calories	Carbohydrates	Protein	Fat	Cholesterol	Sodium	Calcium
276 cal	51 g	9 g	4 g	0 mg	325 mg	30 mg

Ingredient	Weight	Measure	Issue
YEAST,ACTIVE,DRY	4-1/2 oz	1/2 cup 2-1/3 tbsp	
WATER,WARM	1-5/8 lbs	3 cup	
SUGAR,GRANULATED	1-1/8 oz	2-2/3 tbsp	
WATER	6-1/4 lbs	3 qts	
MILK,NONFAT,DRY	5-3/8 oz	2-1/4 cup	
SUGAR,GRANULATED	5-1/4 oz	3/4 cup	
FLOUR,WHEAT,BREAD	10-7/8 lbs	2 gal 1 qts	
SHORTENING,SOFTENED	9 oz	1-1/4 cup	
FLOUR,WHEAT,BREAD	3-5/8 lbs	3 qts	
SALT	2-7/8 oz	1/4 cup 2/3 tbsp	

Method

1. Sprinkle yeast over water. DO NOT USE TEMPERATURES ABOVE 110 F. Mix well. Let stand 5 minutes. Add sugar and stir until dissolved. Let stand for 10 minutes; stir. Set aside for use in Step 3.
2. Place water in mixer bowl. Add milk and sugar. Using dough hook, mix at low speed about 1 minute until blended.
3. Add flour; mix at low speed about 2 minutes or until flour is incorporated. Add shortening and yeast solution. Mix at low speed about 2 minutes until smooth.
4. Mix at medium speed 10 minutes.
5. Let rise in mixer bowl 20 minutes.
6. Sift flour and salt; add to mixture in mixer bowl. Mix at low speed 2 minutes or until flour is incorporated. Mix at medium speed 10 minutes or until smooth and elastic.
7. FERMENT: Cover. Set in warm place (80 F.) 1-1/2 hours or until double in bulk.
8. MAKE-UP: Line pans with parchment paper. Divide dough into 2 2 lb-10 oz pieces. Shape each piece into a smooth ball; let rest 15 minutes.
9. Roll each piece into a long rope, about 38 inches, of uniform diameter, about 1-1/2-inch thick. Cut rope into 25 1-2/3 inch pieces, about 1-1/3 inches long.
10. MAKE-UP: Shape each piece into a smooth ball; let rest 10 to 20 minutes.
11. PROOF: At 90 F. until double in bulk, about 45 minutes.
12. BAKE: 15 to 20 minutes at 400 F. or in 350 F. convection oven 10 to 12 minutes or until golden brown on high fan, open vent.

BREADS AND SWEET DOUGHS No.D 034 01

BROWN AND SERVE ROLLS (SHORT-TIME FORMULA)

Yield 100   **Portion** 2 Rolls

Calories	Carbohydrates	Protein	Fat	Cholesterol	Sodium	Calcium
284 cal	51 g	9 g	5 g	3 mg	335 mg	31 mg

Ingredient	Weight	Measure	Issue
YEAST,ACTIVE,DRY	4-1/2 oz	1/2 cup 2-1/3 tbsp	
WATER,WARM	1-5/8 lbs	3 cup	
SUGAR,GRANULATED	1-1/8 oz	2-2/3 tbsp	
WATER	6-1/4 lbs	3 qts	
MILK,NONFAT,DRY	5-3/8 oz	2-1/4 cup	
SUGAR,GRANULATED	5-1/4 oz	3/4 cup	
FLOUR,WHEAT,BREAD	10-7/8 lbs	2 gal 1 qts	
SHORTENING,SOFTENED	9 oz	1-1/4 cup	
FLOUR,WHEAT,BREAD	3-5/8 lbs	3 qts	
SALT	2-7/8 oz	1/4 cup 2/3 tbsp	
BUTTER,MELTED	4 oz	1/4 cup	

Method

1. Sprinkle yeast over water. DO NOT USE TEMPERATURES ABOVE 110 F. Mix well. Let stand 5 minutes. Add sugar; stir until dissolved. Let stand for 10 minutes; stir. Set aside for use in Step 3.
2. Place water in mixer bowl. Add milk and sugar. Using dough hook, mix at low speed about 1 minute until blended.
3. Add flour; mix at low speed about 2 minutes or until flour is incorporated. Add shortening and yeast solution. Mix at low speed about 2 minutes until smooth.
4. Mix at medium speed 10 minutes.
5. Let rise in mixer bowl 20 minutes.
6. Sift together flour and salt; add to mixture in mixer bowl. Mix at low speed 2 minutes or until flour is incorporated. Mix at medium speed 10 minutes or until smooth and elastic.
7. FERMENT: Cover. Set in warm place (80 F.) 1-1/2 hours or until double in bulk.
8. MAKE-UP: Line pans with parchment paper. Divide dough into 8 2 lb 10 oz pieces. Shape each piece into a smooth ball; let rest 15 minutes.
9. Roll each piece into a long rope, about 38 inches, of uniform diameter, about 1-1/2-inch thick. Cut rope into 25 1-2/3 oz pieces about 1-1/3 inches long.
10. Shape each piece into a smooth ball; let rest 10 to 20 minutes.
11. PROOF: At 90 F. until double in bulk, about 45 minutes.
12. Bake at 325 F. for 25 to 30 minutes or in 300 F. convection oven 12 to 15 minutes or until rolls begin to brown on low fan, open vent. Brush with melted margarine or butter. Cool on pans; wrap in aluminum foil. Bring covered rolls to room temperature about 1 hour before baking. Finish baking at 400 F. about 14 to 17 minutes or in 350 F. convection oven about 10 to 12 minutes or until golden brown on high fan, open vent.

BREADS AND SWEET DOUGHS No.D 034 03

WHOLE WHEAT ROLLS (SHORT-TIME FORMULA)

Yield 100   Portion   2 Rolls

Calories	Carbohydrates	Protein	Fat	Cholesterol	Sodium	Calcium
263 cal	48 g	8 g	4 g	0 mg	325 mg	30 mg

Ingredient	Weight	Measure	Issue
YEAST,ACTIVE,DRY	4-1/2 oz	1/2 cup 2-1/3 tbsp	
WATER,WARM	1-5/8 lbs	3 cup	
SUGAR,GRANULATED	1-1/8 oz	2-2/3 tbsp	
WATER	6-3/4 lbs	3 qts 1 cup	
MILK,NONFAT,DRY	5-3/8 oz	2-1/4 cup	
SUGAR,GRANULATED	7 oz	1 cup	
FLOUR,WHEAT,BREAD	3-5/8 lbs	3 qts	
FLOUR,WHEAT,BREAD	6-1/3 lbs	1 gal 1-1/4 qts	
SHORTENING,SOFTENED	9 oz	1-1/4 cup	
FLOUR,WHEAT,BREAD	3-5/8 lbs	3 qts	
SALT	2-7/8 oz	1/4 cup 2/3 tbsp	

Method

1. Sprinkle yeast over water. DO NOT USE TEMPERATURES ABOVE 110 F. Mix well. Let stand 5 minutes. Add sugar and stir until dissolved. Let stand for 10 minutes; stir. Set aside for use in Step 3.
2. Place water in mixer bowl. Add milk and sugar. Using dough hook, mix at low speed about 1 minute until blended.
3. Add flour; mix at low speed about 2 minutes or until flour is incorporated. Add shortening and yeast solution. Mix at low speed about 2 minutes until smooth.
4. Mix at medium speed 10 minutes.
5. Let rise in mixer bowl 20 minutes.
6. Sift flour and salt; add to mixture in mixer bowl. Mix at low speed 2 minutes or until flour is incorporated. Mix at medium speed 10 minutes or until smooth and elastic.
7. FERMENT: Cover. Set in warm place (80 F.) 1-1/2 hours or until double in bulk.
8. MAKE-UP: Line pans with parchment paper. Divide dough into 8 2 lb-10 oz pieces. Shape each piece into a smooth ball; let rest 15 minutes.
9. Roll each piece into a long rope, about 38 inches, of uniform diameter, about 1-1/2 inches thick. Cut rope into 25 1-2/3 oz pieces about 1-1/3 inches long.
10. Shape each piece into a smooth ball; let rest 10 to 20 minutes.
11. PROOF: At 90 F. until double in bulk, about 45 minutes.
12. BAKE: 20 to 25 minutes at 400 F. or in 350 F. convection oven 12 to 15 minutes or until golden brown on high fan, open vent.

BREADS AND SWEET DOUGHS   No.D 035 00

ONION ROLLS

Yield 100   **Portion**   2 Rolls

Calories	Carbohydrates	Protein	Fat	Cholesterol	Sodium	Calcium
312 cal	58 g	8 g	5 g	0 mg	428 mg	41 mg

Ingredient	**Weight**	**Measure**	**Issue**
YEAST,ACTIVE,DRY	3-3/8 oz	1/2 cup	
WATER,WARM	1-1/3 lbs	2-1/2 cup	
WATER	4-2/3 lbs	2 qts 1 cup	
SUGAR,GRANULATED	1-1/4 lbs	2-3/4 cup	
MILK,NONFAT,DRY	3-1/4 oz	1-3/8 cup	
SALT	3-3/4 oz	1/4 cup 2-1/3 tbsp	
ONIONS,DEHYDRATED,CHOPPED	1-5/8 lbs	3 qts 1 cup	
WATER	4-1/8 lbs	2 qts	
FLOUR,WHEAT,GENERAL PURPOSE	13-1/4 lbs	3 gal	
SHORTENING,SOFTENED	1 lbs	2-1/4 cup	

Method

1. Sprinkle yeast over water. DO NOT USE TEMPERATURES ABOVE 110 F. Mix well; let stand 5 minutes; stir. Set aside for use in Step 3.
2. Place water in mixer bowl. Add sugar, milk, and salt. Mix at low speed until smooth.
3. Soak and drain the dehydrated onions.
4. Add flour; mix at low speed. Add shortening, yeast solution, and onions; mix until well blended.
5. Mix at medium speed 15 minutes or until dough is smooth and elastic.
6. FERMENT: Cover. Set in warm place (80 F.) 2 hours or until double in bulk.
7. PUNCH: Let rest 20 minutes.
8. MAKE-UP: Shape each piece into a smooth ball; let rest 10 to 20 minutes.
9. PROOF: Until rolls are double in bulk.
10. BAKE: At 425 F. 12 to 15 minutes or until done.

BREADS AND SWEET DOUGHS No.D 035 01

ONION ROLLS (ROLL MIX)

Yield 100   **Portion** 2 Rolls

Calories	Carbohydrates	Protein	Fat	Cholesterol	Sodium	Calcium
262 cal	49 g	7 g	4 g	0 mg	376 mg	58 mg

Ingredient   Weight   Measure   Issue

Ingredient	Weight	Measure	Issue
ONIONS,DEHYDRATED,CHOPPED	1-5/8 lbs	3 qts 1 cup	
WATER	4-1/8 lbs	2 qts	
ROLL,MIX	13-1/2 lbs		
YEAST,ACTIVE,DRY	5-1/8 oz	3/4 cup	

Method

1. Soak and drain dehydrated onions. Add onions to Roll Mix and active dry yeast.
2. Prepare mix according to instructions on container.
3. PUNCH: Let rest 20 minutes.
4. MAKE-UP: Shape each piece into a smooth ball; let rest 10 to 20 minutes.
5. PROOF: Until rolls are double in bulk.
6. BAKE: At 425 F. 12 to 15 minutes or until done.

BREADS AND SWEET DOUGHS No.D 036 00

SWEET DOUGH

Yield 100   Portion   1 Roll

Calories	Carbohydrates	Protein	Fat	Cholesterol	Sodium	Calcium
201 cal	32 g	6 g	5 g	24 mg	221 mg	16 mg

Ingredient	Weight	Measure	Issue
YEAST,ACTIVE,DRY	6-3/4 oz	1 cup	
WATER,WARM	1 lbs	2 cup	
WATER	1-5/8 lbs	3 cup	
EGGS,WHOLE,FROZEN	1-1/4 lbs	2-1/4 cup	
SUGAR,GRANULATED	1-1/8 lbs	2-5/8 cup	
MILK,NONFAT,DRY	1-3/4 oz	3/4 cup	
SALT	1-7/8 oz	3 tbsp	
FLOUR,WHEAT,BREAD	7-7/8 lbs	1 gal 2-1/2 qts	
SHORTENING,SOFTENED	14-1/2 oz	2 cup	

Method

1 Sprinkle yeast over water. DO NOT USE TEMPERATURES ABOVE 110 F. Mix well. Let stand 5 minutes; stir. Set aside for use in Step 3.
2 Place water, eggs, sugar, milk, and salt in mixer bowl. Using dough hook, mix at low speed just until blended.
3 Add flour and yeast solution. Mix at low speed 1 minute or until all flour mixture is incorporated into liquid.
4 Add shortening; mix at low speed 1 minute. Continue mixing at medium speed 10 minutes or until dough is smooth and elastic. Dough temperature should be between 78 F. to 82 F.
5 FERMENT: Cover. Set in warm place (80 F.) about 1-1/2 hours or until double in bulk.
6 PUNCH: Divide dough into 3 pieces, shape into a rectangular piece. Let rest 10 to 20 minutes.

BREADS AND SWEET DOUGHS No.D 036 01

SWEET DOUGH (SWEET DOUGH MIX)

Yield 100   Portion 1 Roll

Calories	Carbohydrates	Protein	Fat	Cholesterol	Sodium	Calcium
135 cal	24 g	6 g	3 g	0 mg	323 mg	13 mg

Ingredient Weight Measure Issue

Ingredient	Weight	Measure	Issue
SWEET DOUGH MIX	9 lbs	2 gal 1/8 qts	
YEAST,ACTIVE,DRY	4-1/4 oz	1/2 cup 2 tbsp	
WATER	3-2/3 lbs	1 qts 3 cup	

Method

1 Use Sweet Dough Mix and active dry yeast. Prepare dough according to instructions on container.
2 FERMENT: Cover. Set in warm place (80 F.) about 1-1/2 hours or until double in bulk.
3 PUNCH: Divide dough into 3 pieces, let rest 10 to 20 minutes.

BREADS AND SWEET DOUGHS No.D 036 02

GLAZED ROLLS

Yield 100   **Portion** 1 Roll

Calories	Carbohydrates	Protein	Fat	Cholesterol	Sodium	Calcium
244 cal	38 g	6 g	8 g	29 mg	243 mg	17 mg

Ingredient	Weight	Measure	Issue
YEAST,ACTIVE,DRY	6-3/4 oz	1 cup	
WATER,WARM	1 lbs	2 cup	
WATER	1-5/8 lbs	3 cup	
EGGS,WHOLE,FROZEN	1-1/4 lbs	2-1/4 cup	
SUGAR,GRANULATED	1-1/8 lbs	2-5/8 cup	
MILK,NONFAT,DRY	1-3/4 oz	3/4 cup	
SALT	1-7/8 oz	3 tbsp	
FLOUR,WHEAT,BREAD	7-7/8 lbs	1 gal 2-1/2 qts	
SHORTENING,SOFTENED	14-1/2 oz	2 cup	
BUTTER	8 oz	1 cup	
VANILLA GLAZE		2-3/8 cup	

Method

1. Sprinkle yeast over water. DO NOT USE TEMPERATURES ABOVE 110 F. Mix well. Let stand 5 minutes; stir. Set aside for use in Step 3.
2. Place water, eggs, sugar, milk, and salt in mixer bowl. Using dough hook, mix at low speed just until blended.
3. Add flour and yeast solution. Mix at low speed 1 minute or until all flour mixture is incorporated into liquid.
4. Add shortening; mix at low speed 1 minute. Continue mixing at medium speed 10 minutes or until dough is smooth and elastic. Dough temperature should be between 78 F. to 82 F.
5. FERMENT: Cover. Set in a warm place (80 F.) about 1-1/2 hours or until double in bulk.
6. PUNCH: Divide dough into 3 pieces, 4 lb 5 oz each; shape into a rectangular piece. Let rest 10 to 20 minutes.
7. MAKE-UP: Roll each 4 lb 5 oz piece of dough into a long rope of uniform diameter. (If using D 036 01, Sweet Dough Mix, use 4 lb 2 oz pieces.)
8. Slice into 34 pieces, weighing 1-3/4 to 2 oz each.
9. Shape into balls by rolling with a circular motion.
10. Place on lightly greased sheet pans in rows 6 by 9.
11. Melt butter or margarine. Brush 1/2 cup on rolls in each pan.
12. PROOF: At 90 F. to 100 F. until double in bulk.
13. BAKE: At 375 F. for 20 to 25 minutes or until golden brown or in a 325 F. convection oven for 15 minutes on high fan, open vent. Cool.
14. Prepare 1 recipe Vanilla Glaze, Recipe No. D 046 00; brush about 1-1/3 cups on baked rolls in each pan for each 100 servings.

BREADS AND SWEET DOUGHS No.D 036 03

PECAN ROLLS

Yield 100 **Portion** 1 Roll

Calories	Carbohydrates	Protein	Fat	Cholesterol	Sodium	Calcium
302 cal	40 g	6 g	13 g	34 mg	261 mg	25 mg

Ingredient	Weight	Measure	Issue
YEAST,ACTIVE,DRY	6-3/4 oz	1 cup	
WATER,WARM	1 lbs	2 cup	
WATER	1-5/8 lbs	3 cup	
EGGS,WHOLE,FROZEN	1-1/4 lbs	2-1/4 cup	
SUGAR,GRANULATED	1-1/8 lbs	2-5/8 cup	
MILK,NONFAT,DRY	1-3/4 oz	3/4 cup	
SALT	1-7/8 oz	3 tbsp	
FLOUR,WHEAT,BREAD	7-7/8 lbs	1 gal 2-1/2 qts	
SHORTENING,SOFTENED	14-1/2 oz	2 cup	
PECAN TOPPING		2 qts 2 cup	
BUTTER	8 oz	1 cup	

Method

1. Sprinkle yeast over water. DO NOT USE TEMPERATURES ABOVE 110 F. Mix well. Let stand 5 minutes; stir. Set aside for use in Step 3.
2. Place water, eggs, sugar, milk, and salt in mixer bowl. Using dough hook, mix at low speed just until blended.
3. Add flour and yeast solution. Mix at low speed 1 minute or until all flour mixture is incorporated into liquid.
4. Add shortening; mix at low speed 1 minute. Continue mixing at medium speed 10 minutes or until dough is smooth and elastic. Dough temperature should be between 78 F. to 82 F.
5. FERMENT: Cover. Set in a warm place (80 F.) about 1-1/2 hours or until double in bulk.
6. PUNCH: Divide dough into 3 pieces, 4 lb 5 oz each; shape into a rectangular piece. Let rest 10 to 20 minutes.
7. Roll each 4 lb 5 oz piece of dough into a long rope of uniform diameter. (If using D 036 01, Sweet Dough Mix, use 4 lb 2 oz pieces.)
8. Slice into 34 pieces weighing 1-3/4 to 2 ounces each.
9. Shape into balls by rolling with a circular motion.
10. Prepare 1 recipe Pecan Topping, Recipe No. D 049 01 per 100 portions. Spread 1-1/4 quart in each pan.
11. Flatten balls. Place on topping mixture in rows 6 by 9.
12. Melt butter or margarine and brush 1/2 cup on rolls in each pan.
13. Proof at 90 F. to 100 F. until double in bulk.
14. Bake at 375 F. for 20 to 25 minutes or until golden brown or in 325 F. convection oven for 15 minutes on high fan, open vent.
15. Invert pans as soon as removed from oven; bottom of roll becomes top.

BREADS AND SWEET DOUGHS No. D 036 04

CINNAMON ROLLS

Yiel 100   **Portion** 1 Roll

Calories	Carbohydrates	Protein	Fat	Cholesterol	Sodium	Calcium
289 cal	47 g	6 g	9 g	34 mg	265 mg	40 mg

Ingredient	**Weight**	**Measure**	**Issue**
YEAST,ACTIVE,DRY	6-3/4 oz	1 cup	
WATER,WARM	1 lbs	2 cup	
WATER	1-5/8 lbs	3 cup	
EGGS,WHOLE,FROZEN	1-1/4 lbs	2-1/4 cup	
SUGAR,GRANULATED	1-1/8 lbs	2-5/8 cup	
MILK,NONFAT,DRY	1-3/4 oz	3/4 cup	
SALT	1-7/8 oz	3 tbsp	
FLOUR,WHEAT,BREAD	7-7/8 lbs	1 gal 2-1/2 qts	
SHORTENING,SOFTENED	14-1/2 oz	2 cup	
BUTTER	1 lbs	2 cup	
CINNAMON SUGAR FILLING		3 cup	

Method

1. Sprinkle yeast over water. DO NOT USE TEMPERATURES ABOVE 110 F. Mix well. Let stand 5 minutes; stir. Set aside for use in Step 3.
2. Place water, eggs, sugar, milk, and salt in mixer bowl. Using dough hook, mix at low speed just until blended.
3. Add flour and yeast solution. Mix at low speed 1 minute or until all flour mixture is incorporated into liquid.
4. Add shortening; mix at low speed 1 minute. Continue mixing at medium speed 10 minutes or until dough is smooth and elastic. Dough temperature should be between 78 F. to 82 F.
5. FERMENT: Cover. Set in a warm place (80 F.) about 1-1/2 hours or until double in bulk.
6. PUNCH: Divide dough into 3 pieces, 4 lb 5 oz each; shape into a rectangular piece. Let rest 10 to 20 minutes.
7. MAKE-UP: Roll each 4 lb 5 oz piece of dough into a rectangular sheet, about 18 inches wide, 36 inches long, and 1/4 inch thick. (If using D 036 01, Sweet Dough Mix, use 4 lb 2 oz pieces.)
8. Melt butter or margarine. Brush 1/2 cup on each sheet of dough. Set aside remainder for use in Step 4.
9. Prepare 1 recipe Cinnamon Sugar Filling, Recipe No. D 042 00 for 100 servings. Sprinkle 1-1/2 cups cinnamon sugar mixture over each sheet of dough.
10. Roll each piece tightly to make a long slender roll. Seal edges by pressing firmly. Elongate roll to 35 inches by rolling back and forth on work table. Brush 2 tablespoons of butter or margarine on each roll.
11. Slice each roll into 34 pieces about 1 inch wide, using dough cutter.
12. Place cut side down on lightly greased sheet pans in rows 5 by 8.
13. Proof at 90 F. to 100 F. until double in bulk.
14. Bake at 375 F. for 20 to 25 minutes or until golden brown or in 325 F. convection oven 15 minutes on high fan, open vent. Cool.
15. Glaze, if desired, with 1 recipe Vanilla Glaze, Recipe No. D 046 00 per 100 portions. Brush about 1 cup on rolls in each pan.

BREADS AND SWEET DOUGHS No.D 036 05

CINNAMON NUT ROLLS

Yield 100   Portion 1 Roll

Calories	Carbohydrates	Protein	Fat	Cholesterol	Sodium	Calcium
306 cal	38 g	7 g	15 g	34 mg	260 mg	26 mg

Ingredient	Weight	Measure	Issue
YEAST,ACTIVE,DRY	6-3/4 oz	1 cup	
WATER,WARM	1 lbs	2 cup	
WATER	1-5/8 lbs	3 cup	
EGGS,WHOLE,FROZEN	1-1/4 lbs	2-1/4 cup	
SUGAR,GRANULATED	1-1/8 lbs	2-5/8 cup	
MILK,NONFAT,DRY	1-3/4 oz	3/4 cup	
SALT	1-7/8 oz	3 tbsp	
FLOUR,WHEAT,BREAD	7-7/8 lbs	1 gal 2-1/2 qts	
SHORTENING,SOFTENED	14-1/2 oz	2 cup	
BUTTER	1 lbs	2 cup	
CINNAMON SUGAR FILLING		3 cup	
PECANS,CHOPPED	2 lbs		

Method

1. Sprinkle yeast over water. DO NOT USE TEMPERATURES ABOVE 110 F. Mix well. Let stand 5 minutes; stir. Set aside for use in Step 3.
2. Place water, eggs, sugar, milk, and salt in mixer bowl. Using dough hook, mix at low speed just until blended.
3. Add flour and yeast solution. Mix at low speed 1 minute or until all flour mixture is incorporated into liquid.
4. Add shortening; mix at low speed 1 minute. Continue mixing at medium speed 10 minutes or until dough is smooth and elastic. Dough temperature should be between 78 F. to 82 F.
5. FERMENT: Cover. Set in a warm place (80 F.) about 1-1/2 hours or until double in bulk.
6. PUNCH: Divide dough into 3 pieces, 4 lb 5 oz each; shape into a rectangular piece. Let rest 10 to 20 minutes.
7. MAKE-UP: Roll each 4 lb 5 oz piece of dough into a rectangular sheet, about 18 inches wide, 36 inches long, and 1/4 inch thick. (For D 036 01, Sweet Dough Mix, use 4 lb 2 oz pieces).
8. Melt butter or margarine. Brush 1/2 cup on each sheet of dough. Set aside remainder for use in Step 10.
9. Prepare 1 Recipe Cinnamon Sugar Nut Filling, Recipe No. D 042 01 per 100 servings. Sprinkle 1-1/2 cups cinnamon sugar mixture and 2 cups of pecans over each sheet of dough.
10. Roll each piece tightly to make a long slender roll. Seal edges by pressing firmly. Elongate roll to 35 inches by rolling back and forth on work table. Brush 2 tablespoons of butter or margarine on each roll.
11. Slice each roll into 34 pieces about 1 inch wide, using dough cutter.
12. Place cut side down on lightly greased sheet pans in rows 5 by 8.
13. Proof at 90 F. to 100 F. until double in bulk.
14. Bake at 375 F. for 20 to 25 minutes or until golden brown or in 325 F. convection oven 15 minutes on high fan, open vent. Cool.
15. Glaze, if desired, with Vanilla Glaze, Recipe No. D 046 00 per 100 servings. Brush about 1 cup on rolls in each pan.

BREADS AND SWEET DOUGHS No.D 036 06

CINNAMON RAISIN ROLLS

Yield 100   Portion 1 Roll

Calories	Carbohydrates	Protein	Fat	Cholesterol	Sodium	Calcium
298 cal	49 g	6 g	9 g	34 mg	265 mg	41 mg

Ingredient	Weight	Measure	Issue
YEAST,ACTIVE,DRY	6-3/4 oz	1 cup	
WATER,WARM	1 lbs	2 cup	
WATER	1-5/8 lbs	3 cup	
EGGS,WHOLE,FROZEN	1-1/4 lbs	2-1/4 cup	
SUGAR,GRANULATED	1-1/8 lbs	2-5/8 cup	
MILK,NONFAT,DRY	1-3/4 oz	3/4 cup	
SALT	1-7/8 oz	3 tbsp	
FLOUR,WHEAT,BREAD	7-7/8 lbs	1 gal 2-1/2 qts	
SHORTENING,SOFTENED	14-1/2 oz	2 cup	
BUTTER	1 lbs	2 cup	
CINNAMON SUGAR FILLING		3 cup	
RAISINS	10-1/4 oz	2 cup	

Method

1. Sprinkle yeast over water. DO NOT USE TEMPERATURES ABOVE 110 F. Mix well. Let stand 5 minutes; stir. Set aside for use in Step 3.
2. Place water, eggs, sugar, milk, and salt in mixer bowl. Using dough hook, mix at low speed just until blended.
3. Add flour and yeast solution. Mix at low speed 1 minute or until all flour mixture is incorporated into liquid.
4. Add shortening; mix at low speed 1 minute. Continue mixing at medium speed 10 minutes or until dough is smooth and elastic. Dough temperature should be between 78 F. to 82 F.
5. FERMENT: Cover. Set in a warm place (80 F.) about 1-1/2 hours or until double in bulk.
6. PUNCH: Divide dough into 4 pound 5 ounce pieces; shape into a rectangular piece. Let rest 10 to 20 minutes.
7. Roll each 4 pound 5 ounce piece of dough into a rectangular sheet, about 18 inches wide, 36 inches long, and 1/4 inch thick.
8. Melt butter or margarine. Brush 1/2 cup on each sheet of dough. Set aside remainder for use in Step 4.
9. Prepare 1 recipe Cinnamon Sugar Raisin Filling, Recipe No. D 042 02 per 100 portions. Sprinkle 1-1/2 cups cinnamon sugar mixture and 2 cups of raisins over each sheet of dough.
10. Roll each piece tightly to make a long slender roll. Seal edges by pressing firmly. Elongate roll to 35 inches by rolling back and forth on work table. Brush 2 tablespoons of butter or margarine on each roll.
11. Slice each roll into 34 pieces about 1 inch wide, using dough cutter.
12. Place cut side down on lightly greased sheet pans in rows 5 by 8.
13. Proof at 90 F. to 100 F. until double in bulk.
14. Bake at 375 F. for 20 to 25 minutes or until golden brown or in 325 F. convection oven 15 minutes on high fan, open vent. Cool.
15. Glaze, if desired, with Vanilla Glaze, Recipe No. D 046 00 per 100 portions. Brush about 1 cup on rolls in each pan.

BREADS AND SWEET DOUGHS No.D 036 07

BUTTERFLY ROLLS

Yield 100   **Portion** 1 Roll

Calories	Carbohydrates	Protein	Fat	Cholesterol	Sodium	Calcium
312 cal	52 g	6 g	9 g	36 mg	261 mg	19 mg

Ingredient	**Weight**	**Measure**	**Issue**
YEAST,ACTIVE,DRY	6-3/4 oz	1 cup	
WATER,WARM	1 lbs	2 cup	
WATER	1-5/8 lbs	3 cup	
EGGS,WHOLE,FROZEN	1-1/4 lbs	2-1/4 cup	
SUGAR,GRANULATED	1-1/8 lbs	2-5/8 cup	
MILK,NONFAT,DRY	1-3/4 oz	3/4 cup	
SALT	1-7/8 oz	3 tbsp	
FLOUR,WHEAT,BREAD	7-7/8 lbs	1 gal 2-1/2 qts	
SHORTENING,SOFTENED	14-1/2 oz	2 cup	
BUTTER	12 oz	1-1/2 cup	
EGG WASH		3/4 cup	
VANILLA GLAZE		2-3/4 cup	

Method

1. Sprinkle yeast over water. DO NOT USE TEMPERATURES ABOVE 110 F. Mix well. Let stand 5 minutes; stir. Set aside for use in Step 3.
2. Place water, eggs, sugar, milk, and salt in mixer bowl. Using dough hook, mix at low speed just until blended.
3. Add flour and yeast solution. Mix at low speed 1 minute or until all flour mixture is incorporated into liquid.
4. Add shortening; mix at low speed 1 minute. Continue mixing at medium speed 10 minutes or until dough is smooth and elastic. Dough temperature should be between 78 F. to 82 F.
5. FERMENT: Cover. Set in a warm place (80 F.) about 1-1/2 hours or until double in bulk.
6. PUNCH: Divide dough into 6-2 lb 2 oz pieces; shape into a rectangular piece. Let rest 10 to 20 minutes.
7. Roll each piece of dough into a rectangular sheet, about 10 inches wide, 30 inches long and 1/4 inch thick.
8. Melt butter or margarine. Brush 1/4 cup on each sheet of dough.
9. MAKE-UP: Roll each piece tightly to make long slender roll. Seal edges by pressing firmly. Elongate roll to 30 inches by rolling back and forth on work table.
10. Slice each roll into 17 pieces about 1-3/4 inches wide.
11. Press each piece firmly in center parallel to cut side of roll with back of knife or small rolling pin.
12. Place on lightly greased sheet pans in rows 4 by 8. Prepare 1/4 recipe Egg Wash, Recipe No. D 017 00 per 100 portions and brush 1/4 cup on rolls in each pan.
13. Proof at 90 F. to 100 F. until double in bulk.
14. Bake at 375 F. for 20 to 25 minutes or until golden brown or in a 325 F. convection oven for 15 minutes on high fan, open vent. Cool.
15. Glaze, if desired, with 1 recipe Vanilla Glaze, Recipe No. D 046 00 per 100 portions. Brush about 1 cup on rolls in each pan.

BREADS AND SWEET DOUGHS No. D 036 08

SUGAR ROLLS

Yield 100   Portion   1 Roll

Calories	Carbohydrates	Protein	Fat	Cholesterol	Sodium	Calcium
335 cal	56 g	6 g	10 g	36 mg	269 mg	18 mg

Ingredient	Weight	Measure	Issue
YEAST,ACTIVE,DRY	6-3/4 oz	1 cup	
WATER,WARM	1 lbs	2 cup	
WATER	1-5/8 lbs	3 cup	
EGGS,WHOLE,FROZEN	1-1/4 lbs	2-1/4 cup	
SUGAR,GRANULATED	1-1/8 lbs	2-5/8 cup	
MILK,NONFAT,DRY	1-3/4 oz	3/4 cup	
SALT	1-7/8 oz	3 tbsp	
FLOUR,WHEAT,BREAD	7-7/8 lbs	1 gal 2-1/2 qts	
SHORTENING,SOFTENED	14-1/2 oz	2 cup	
BUTTER	1 lbs	2 cup	
SUGAR,GRANULATED	14-1/8 oz	2 cup	
VANILLA GLAZE		2-3/4 cup	

Method

1 Sprinkle yeast over water. DO NOT USE TEMPERATURES ABOVE 110 F. Mix well. Let stand 5 minutes; stir. Set aside for use in Step 3.
2 Place water, eggs, sugar, milk, and salt in mixer bowl. Using dough hook, mix at low speed just until blended.
3 Add flour and yeast solution. Mix at low speed 1 minute or until all flour mixture is incorporated into liquid.
4 Add shortening; mix at low speed 1 minute. Continue mixing at medium speed 10 minutes or until dough is smooth and elastic. Dough temperature should be between 78 F. to 82 F.
5 FERMENT: Cover. Set in a warm place (80 F.) about 1-1/2 hours or until double in bulk.
6 PUNCH: Divide dough into 3 pieces, 4 lb 5 oz each; shape into a rectangular piece. Let rest 10 to 20 minutes.
7 Roll out each 4 lb 5 oz piece of dough into a rectangular sheet, about 18 inches wide, 36 inches long, and 1/4 inch thick. (If using D 036 01, Sweet Dough Mix, use 4 lb 2 oz pieces).
8 Melt butter or margarine. Brush 1/2 cup on each sheet of dough.
9 Roll each piece tightly to make a long slender roll. Seal edges by pressing firmly. Elongate roll to 35 inches by rolling back and forth on the work table. Brush 2 tbsp butter or margarine on each roll.
10 Slice each roll into 34 pieces, about 1 inch wide, using dough cutter.
11 Press cut side of each slice in 14 ounces or 2 cups granulated sugar so that surface is well coated.
12 Place sugar side up on lightly greased sheet pans in rows 5 by 8.
13 Proof at 90 F. to 100 F. until double in bulk.
14 Bake at 375 F. for 20 to 25 minutes or until golden brown or in 325 F. convection oven 15 minutes on high fan, open vent. Cool.
15 Glaze, if desired, with 1 recipe Vanilla Glaze, Recipe No. D 046 00 per 100 portions. Brush about 1 cup on rolls in each pan.

BREADS AND SWEET DOUGHS No.D 036 09

STREUSEL COFFEE CAKE

Yield 100   Portion 1 Piece

Calories	Carbohydrates	Protein	Fat	Cholesterol	Sodium	Calcium
319 cal	50 g	7 g	10 g	39 mg	274 mg	26 mg

Ingredient	Weight	Measure	Issue
YEAST,ACTIVE,DRY	6-3/4 oz	1 cup	
WATER,WARM	1 lbs	2 cup	
WATER	1-5/8 lbs	3 cup	
EGGS,WHOLE,FROZEN	1-1/4 lbs	2-1/4 cup	
SUGAR,GRANULATED	1-1/8 lbs	2-5/8 cup	
MILK,NONFAT,DRY	1-3/4 oz	3/4 cup	
SALT	1-7/8 oz	3 tbsp	
FLOUR,WHEAT,BREAD	7-7/8 lbs	1 gal 2-1/2 qts	
SHORTENING,SOFTENED	14-1/2 oz	2 cup	
EGG WASH		3/4 cup	
STREUSEL TOPPING		3 qts	
VANILLA GLAZE		2 cup	

Method

1. Sprinkle yeast over water. DO NOT USE TEMPERATURES ABOVE 110 F. Mix well. Let stand 5 minutes; stir. Set aside for use in Step 3.
2. Place water, eggs, sugar, milk, and salt in mixer bowl. Using dough hook, mix at low speed just until blended.
3. Add flour and yeast solution. Mix at low speed 1 minute or until all flour mixture is incorporated into liquid.
4. Add shortening; mix at low speed 1 minute. Continue mixing at medium speed 10 minutes or until dough is smooth and elastic. Dough temperature should be between 78 F. to 82 F.
5. FERMENT: Cover. Set in warm place (80 F.) about 1-1/2 hours or until double in bulk.
6. PUNCH: Divide dough into 2-6 lb 8 oz pieces. (If using D 036 01, Sweet Dough Mix, divide into 6 lb 4 oz pieces). Shape into a rectangular piece. Let rest 10 to 20 minutes.
7. Roll each piece of dough into a rectangular sheet, about 18 inches wide, 25 inches long and 1/2-inch thick; fit into greased sheet pans, pressing against sides; edges should be thicker than center.
8. Dock dough with fork or docker, if available.
9. Prepare 1/4 recipe Egg Wash, Recipe No. D 017 00 per 100 portions. Brush about 1/3 cup on dough in each pan. Prepare 1 recipe Streusel Topping, Recipe No. D 049 00; sprinkle 1-1/2 quart topping over dough in each pan.
10. Proof dough 20 to 35 minutes.
11. Bake at 375 F., 30 to 35 minutes or until golden brown or in 325 F. convection oven 15 minutes on high fan, open vent.
12. Prepare 2/3 recipe Vanilla Glaze, Recipe No. D 046 00 per 100 portions; drizzle about 1 cup over each cake while hot.
13. Cut 6 by 9.

BREADS AND SWEET DOUGHS No.D 036 10

SMALL COFFEE CAKE

Yield 100   Portion   1 Piece

Calories	Carbohydrates	Protein	Fat	Cholesterol	Sodium	Calcium
423 cal	81 g	6 g	9 g	36 mg	270 mg	57 mg

Ingredient	Weight	Measure	Issue
YEAST,ACTIVE,DRY	6-3/4 oz	1 cup	
WATER,WARM	1 lbs	2 cup	
WATER	1-5/8 lbs	3 cup	
EGGS,WHOLE,FROZEN	1-1/4 lbs	2-1/4 cup	
SUGAR,GRANULATED	1-1/8 lbs	2-5/8 cup	
MILK,NONFAT,DRY	1-3/4 oz	3/4 cup	
SALT	1-7/8 oz	3 tbsp	
FLOUR,WHEAT,BREAD	7-7/8 lbs	1 gal 2-1/2 qts	
SHORTENING,SOFTENED	14-1/2 oz	2 cup	
BUTTER	12 oz	1-1/2 cup	
CINNAMON SUGAR FILLING		1 qts 1/2 cup	
RAISINS	2 lbs	1 qts 2-1/4 cup	
EGG WASH		3/4 cup	
VANILLA GLAZE		2-3/4 cup	

Method

1. Sprinkle yeast over water. DO NOT USE IN TEMPERATURES ABOVE 110 F. Mix well. Let stand 5 minutes; stir. Set aside for use in Step 3.
2. Place water, eggs, sugar, milk, and salt in mixer bowl. Using dough hook, mix at low speed just until blended.
3. Add flour and yeast solution. Mix at low speed 1 minute or until all flour mixture is incorporated into liquid.
4. Add shortening; mix at low speed 1 minute. Continue mixing at medium speed 10 minutes or until dough is smooth and elastic. Dough temperature should be between 78 F. to 82 F.
5. FERMENT: Cover. Set in a warm place (80 F.) about 1-1/2 hours or until double in bulk.
6. PUNCH: Divide dough into 6-2 lb 2 oz pieces; (if using D 036 01, Sweet Dough Mix, divide into 2 lb 1 oz pieces). Shape into a rectangular piece. Let rest 10 to 20 minutes.
7. Roll each piece of dough into a rectangular sheet about 9 inches wide, 36 inches long, and 1/4 inch thick.
8. Melt butter or margarine; brush 1/4 cup on each sheet of dough. Prepare Cinnamon Sugar Filling, Recipe No. D 042 02; use 2 pounds or 6-1/4 cups of raisins; sprinkle 3/4 cup filling and 1 cup raisins over each sheet of dough.
9. Roll each piece tightly to make a long slender roll. Seal edges by pressing firmly. Elongate roll to 36 inches by rolling back and forth on work table.
10. Cut rolls into 12-inch pieces weighting about 10 ounces each.
11. Place 4 coffee cakes on each lightly greased sheet pan.
12. Make a deep 9-inch slit down the center of each piece, about 1/2 through folds of dough. Do not cut completely through all layers.
13. Prepare 1/4 recipe Egg Wash, Recipe No. D 017 00 per 100 portions. Brush about 2 teaspoons on each cake.
14. Proof at 90 F. to 100 F. until double in bulk.
15. Bake at 375 F. for 25 to 30 minutes or until golden brown or in 325 F. convection oven for 15 minutes on high fan, open vent.
16. Glaze, if desired, with 1 recipe Vanilla Glaze, Recipe No. D 046 00. Drizzle about 2 tablespoons on cakes in each pan.
17. Cut each cake into 6, 2-inch pieces.

BREADS AND SWEET DOUGHS No.D 036 11

TWIST COFFEE CAKE

Yield 100   Portion 1 Piece

Calories	Carbohydrates	Protein	Fat	Cholesterol	Sodium	Calcium
303 cal	51 g	6 g	9 g	35 mg	257 mg	29 mg

Ingredient

Ingredient	Weight	Measure	Issue
YEAST,ACTIVE,DRY	6-3/4 oz	1 cup	
WATER,WARM	1 lbs	2 cup	
WATER	1-5/8 lbs	3 cup	
EGGS,WHOLE,FROZEN	1-1/4 lbs	2-1/4 cup	
SUGAR,GRANULATED	1-1/8 lbs	2-5/8 cup	
MILK,NONFAT,DRY	1-3/4 oz	3/4 cup	
SALT	1-7/8 oz	3 tbsp	
FLOUR,WHEAT,BREAD	7-7/8 lbs	1 gal 2-1/2 qts	
SHORTENING,SOFTENED	14-1/2 oz	2 cup	
BUTTER	12 oz	1-1/2 cup	
CINNAMON SUGAR RAISIN FILLING		2-3/4 cup	
RAISINS	10-1/4 oz	2 cup	
RAISINS	10-1/4 oz	2 cup	
EGG WASH		3/4 cup	
VANILLA GLAZE		2-3/4 cup	

Method

1. Sprinkle yeast over water. DO NOT USE TEMPERATURES ABOVE 110 F. Mix well. Let stand 5 minutes; stir. Set aside for use in Step 3.
2. Place water, eggs, sugar, milk, and salt in mixer bowl. Using dough hook, mix at low speed just until blended.
3. Add flour and yeast solution. Mix at low speed 1 minute or until all flour mixture is incorporated into liquid.
4. Add shortening; mix at low speed 1 minute. Continue mixing at medium speed 10 minutes or until dough is smooth and elastic. Dough temperature should be between 78 F. to 82 F.
5. FERMENT: Cover. Set in a warm place (80 F.) about 1-1/2 hours or until double in bulk.
6. PUNCH: Divide dough into 3 pieces, 4 lb 5 oz pieces; shape into a rectangular piece. Let rest 10 to 20 minutes.
7. Roll each 4 lb 5 oz piece of dough into a rectangular sheet, about 13 inches wide, 45 inches long, and 1/4-inch thick. (If using D 036 01, Sweet Dough Mix, use 4 lb 2 oz pieces).
8. Melt butter or margarine. Brush 1/2 cup over dough in each pan. Prepare Cinnamon Sugar Raisin Filling, Recipe No. D 042 02. Sprinkle 1-1/2 cups over each sheet of dough. Sprinkle about 1 cup of raisins over center third of dough.
9. Fold 1/3 dough over center. Sprinkle 1 cup raisins on top of folded dough. Fold remaining 1/3 dough over raisins to form a strip 13 by 15 inches.
10. Cut each strip into 6-15 inch long, 2 inch wide pieces weighing about 1 pound each.
11. Slit roll down center to within 1 inch of end.
12. Twist pieces in one direction and then in opposite direction, stretching to about 19 inches.
13. Place each piece in a circle on lightly greased sheet pans; seal ends securely by fitting one end into other. Rings should not touch each other.
14. Prepare 1/4 recipe Egg Wash, Recipe No. D 017 00 per 100 portions. Brush about 2 teaspoons on each cake.
15. Proof at 90 F. to 100 F. until double in bulk.
16. Bake at 375 F. for 25 to 30 minutes or until golden brown or in 325 F. convection oven 15 minutes on high fan, open vent. Cool.
17. Glaze, if desired, with 1 recipe Vanilla Glaze, Recipe No. D 046 00 per 100 portions. Drizzle about 2/3 cup on each cake. Cut each cake into 6 pieces.

BREADS AND SWEET DOUGHS No.D 036 12

BEAR CLAWS

Yield 100   **Portion** 1 Roll

Calories	Carbohydrates	Protein	Fat	Cholesterol	Sodium	Calcium
308 cal	53 g	8 g	7 g	87 mg	254 mg	48 mg

Ingredient	Weight	Measure	Issue
YEAST,ACTIVE,DRY	6-3/4 oz	1 cup	
WATER,WARM	1 lbs	2 cup	
WATER	1-5/8 lbs	3 cup	
EGGS,WHOLE,FROZEN	1-1/4 lbs	2-1/4 cup	
SUGAR,GRANULATED	1-1/8 lbs	2-5/8 cup	
MILK,NONFAT,DRY	1-3/4 oz	3/4 cup	
SALT	1-7/8 oz	3 tbsp	
FLOUR,WHEAT,BREAD	7-7/8 lbs	1 gal 2-1/2 qts	
SHORTENING,SOFTENED	14-1/2 oz	2 cup	
CHERRY FILLING (CORNSTARCH)		2 qts 1 cup	
EGG WASH		1 gal 3/4 qts	
VANILLA GLAZE		2-3/8 cup	

Method

1. Sprinkle yeast over water. DO NOT USE TEMPERATURES ABOVE 110 F. Mix well. Let stand 5 minutes; stir. Set aside for use in Step 3.
2. Place water, eggs, sugar, milk, and salt in mixer bowl. Using dough hook, mix at low speed just until blended.
3. Add flour and yeast solution. Mix at low speed 1 minute or until all flour mixture is incorporated into liquid.
4. Add shortening; mix at low speed 1 minute. Continue mixing at medium speed 10 minutes or until dough is smooth and elastic. Dough temperature should be 78 F. to 82 F.
5. FERMENT: Cover. Set in a warm place (80 F.) about 1-1/2 hours or until double in bulk.
6. PUNCH: Divide dough into 2 pound 2 ounce pieces; shape into a rectangular piece. Let rest 10 to 20 minutes. (If using D 036 01, Sweet Dough Mix, use 2 lb 1 oz pieces.)
7. Roll each piece of dough into a rectangular sheet about 5 inches wide, 44 inches long, and 1/3-inch thick.
8. Prepare Cherry Filling, Recipe No. D 041 00, Pineapple Filling, Recipe No. D 047 00, or Nut Filling, Recipe D 043 00. Spread 1-1/2 cups cherry or pineapple or 1-1/4 cups nut filling over center of each sheet of dough.
9. Fold dough over once, lengthwise; seal along edge by pressing firmly.
10. Cut dough into 17 2-1/2-inch pieces. Make 3 cuts, 3/4-inch in depth, on sealed side of each piece to form a claw.
11. Place on lightly greased sheet pans in rows 3 by 8. Spread claws slightly. Claws should not touch each other.
12. Prepare 1/4 Recipe Egg Wash, Recipe No. D 017 00. Brush 3 tablespoons on claws in each pan.
13. Proof at 90 F. to 100 F. until double in bulk.
14. Bake at 375 F. for 20 to 25 minutes or until golden brown or in a 325 F. convection oven for 15 minutes on high fan, open vent. Cool.
15. Glaze, if desired, with 1 Recipe Vanilla Glaze, Recipe No. D 046 00. Brush about 2/3 cup over rolls in each pan.

BREADS AND SWEET DOUGHS No.D 036 13

SNAILS

Yield 100   Portion 1 Roll

Calories	Carbohydrates	Protein	Fat	Cholesterol	Sodium	Calcium
321 cal	54 g	6 g	9 g	29 mg	266 mg	20 mg

Ingredient | Weight | Measure | Issue

Ingredient	Weight	Measure	Issue
YEAST,ACTIVE,DRY	6-3/4 oz	1 cup	
WATER,WARM	1 lbs	2 cup	
WATER	1-5/8 lbs	3 cup	
EGGS,WHOLE,FROZEN	1-1/4 lbs	2-1/4 cup	
SUGAR,GRANULATED	1-1/8 lbs	2-5/8 cup	
MILK,NONFAT,DRY	1-3/4 oz	3/4 cup	
SALT	1-7/8 oz	3 tbsp	
FLOUR,WHEAT,BREAD	7-7/8 lbs	1 gal 2-1/2 qts	
SHORTENING,SOFTENED	14-1/2 oz	2 cup	
MARGARINE	12 oz	1-1/2 cup	
EGG WASH		3/4 cup	
JELLY	1-1/3 lbs	2 cup	
VANILLA GLAZE		2-1/2 cup	

Method

1. Sprinkle yeast over water. DO NOT USE TEMPERATURES ABOVE 110 F. Mix well. Let stand 5 minutes; stir. Set aside for use in Step 3.
2. Place water, eggs, sugar, milk, and salt in mixer bowl. Using dough hook, mix at low speed just until blended.
3. Add flour and yeast solution. Mix at low speed 1 minute or until all flour mixture is incorporated into liquid.
4. Add shortening; mix at low speed 1 minute. Continue mixing at medium speed 10 minutes or until dough is smooth and elastic. Dough temperature should be 78 F. to 82 F.
5. FERMENT: Cover. Set in a warm place (80 F.) about 1-1/2 hours or until double in bulk.
6. PUNCH: Divide dough into 3 pieces, 4 lb 5 oz each; shape into a rectangular piece. Let rest 10 to 20 minutes.
7. Roll each 4 lb 5 oz piece of dough into a rectangular sheet about 18 inches wide, 36 inches long, and 1/4-inch thick. (If using D 036 01, Sweet Dough Mix, use 4 lb 2 oz pieces.)
8. Melt butter or margarine. Brush 1/2 cup on each sheet of dough. Prepare 1 recipe Cinnamon Sugar Filling, Recipe No. D 042 00; sprinkle 1-1/2 cups over each sheet of dough.
9. Fold each sheet of dough in thirds lengthwise to make a strip, about 6 inches wide, 35 inches long, and 3/4 inches thick.
10. Cut strips crosswise into 34 pieces about 1-inch wide.
11. Twist pieces in one direction and then in the opposite direction. Form snails by holding one end on greased pan and winding other end around and around loosely keeping roll flat.
12. Place on lightly greased sheet pans in rows 4 by 8.
13. Prepare 1/4 Recipe Egg Wash, Recipe D 017 00; brush about 1/4 cup on snails in each pan; let rise slightly.
14. Make slight depression with back of spoon in center of each snail. Use 2 cups of jelly or jam; place about 1 teaspoon in each depression.
15. Proof at 90 F. to 100 F. until double in bulk.
16. Bake at 375 F. for 20 to 25 minutes or until golden brown or in a 325 F. convection oven for 15 minutes on high fan, open vent. Cool.
17. Glaze, if desired, with 1 Recipe Vanilla Glaze, Recipe No. D 046 00. Brush about 3/4 cup on rolls in each pan.

BREADS AND SWEET DOUGHS No.D 036 14

BOWKNOTS, FIGURE 8's, AND S SHAPES

Yield 100   Portion 1 Roll

Calories	Carbohydrates	Protein	Fat	Cholesterol	Sodium	Calcium
288 cal	52 g	6 g	6 g	29 mg	233 mg	18 mg

Ingredient	Weight	Measure	Issue
YEAST,ACTIVE,DRY	6-3/4 oz	1 cup	
WATER,WARM	1 lbs	2 cup	
WATER	1-5/8 lbs	3 cup	
EGGS,WHOLE,FROZEN	1-1/4 lbs	2-1/4 cup	
SUGAR,GRANULATED	1-1/8 lbs	2-5/8 cup	
MILK,NONFAT,DRY	1-3/4 oz	3/4 cup	
SALT	1-7/8 oz	3 tbsp	
FLOUR,WHEAT,BREAD	7-7/8 lbs	1 gal 2-1/2 qts	
SHORTENING,SOFTENED	14-1/2 oz	2 cup	
EGG WASH		3/4 cup	
VANILLA GLAZE		2-3/4 cup	

Method

1. Sprinkle yeast over water. DO NOT USE TEMPERATURES ABOVE 110 F. Mix well. Let stand 5 minutes; stir. Set aside for use in Step 3.
2. Place water, eggs, sugar, milk, and salt in mixer bowl. Using dough hook, mix at low speed just until blended.
3. Add flour and yeast solution. Mix at low speed 1 minute or until all flour mixture is incorporated into liquid.
4. Add shortening; mix at low speed 1 minute. Continue mixing at medium speed 10 minutes or until dough is smooth and elastic. Dough temperature should be 78 F. to 82 F.
5. FERMENT: Cover. Set in a warm place (80 F.) about 1-1/2 hours or until double in bulk.
6. PUNCH: Divide dough into 3 pieces, 4 lb 5 oz pieces; shape into a rectangular piece. Let rest 10 to 20 minutes.
7. Roll each 4 lb 5 oz piece of dough into a rectangular sheet about 18 inches wide, 36 inches long, and 1/4-inch thick. (If using D 036 01, Sweet Dough Mix, use 4 lb 2 oz pieces.)
8. Fold each sheet of dough in thirds lengthwise to make a strip about 6 inches wide, 35 inches long, and 3/4-inch thick.
9. Cut strips crosswise into 34 pieces about 1 inch wide.
10. Twist pieces in one direction, then in the opposite direction, stretching to about 11 inches.
11. Form into various shapes. Place on lightly greased sheet pans in rows 4 by 8.
12. Prepare 1/4 Recipe Egg Wash, Recipe No. D 017 00; brush about 1/4 cup on rolls in each pan.
13. Proof at 90 F. to 100 F. until double in bulk.
14. Bake at 375 F. for 20 to 25 minutes or until golden brown or in a 325 F. convection oven for 15 minutes on high fan, open vent. Cool.
15. Glaze, if desired, with 1 recipe Vanilla Glaze, Recipe No. D 046 00. Brush about 3/4 cup on rolls in each pan.

BREADS AND SWEET DOUGHS No.D 036 15

CINNAMON TWISTS

Yield 100   Portion 1 Roll

Calories	Carbohydrates	Protein	Fat	Cholesterol	Sodium	Calcium
368 cal	66 g	6 g	9 g	36 mg	266 mg	41 mg

Ingredient	Weight	Measure	Issue
YEAST,ACTIVE,DRY	6-3/4 oz	1 cup	
WATER,WARM	1 lbs	2 cup	
WATER	1-5/8 lbs	3 cup	
EGGS,WHOLE,FROZEN	1-1/4 lbs	2-1/4 cup	
SUGAR,GRANULATED	1-1/8 lbs	2-5/8 cup	
MILK,NONFAT,DRY	1-3/4 oz	3/4 cup	
SALT	1-7/8 oz	3 tbsp	
FLOUR,WHEAT,BREAD	7-7/8 lbs	1 gal 2-1/2 qts	
SHORTENING,SOFTENED	14-1/2 oz	2 cup	
BUTTER	12 oz	1-1/2 cup	
CINNAMON SUGAR FILLING		3 cup	
EGG WASH		3/4 cup	
VANILLA GLAZE		2-3/4 cup	

Method

1. Sprinkle yeast over water. DO NOT USE TEMPERATURES ABOVE 110 F. Mix well. Let stand 5 minutes; stir. Set aside for use in Step 3.
2. Place water, eggs, sugar, milk, and salt in mixer bowl. Using dough hook, mix at low speed just until blended.
3. Add flour and yeast solution. Mix at low speed 1 minute or until all flour mixture is incorporated into liquid.
4. Add shortening; mix at low speed 1 minute. Continue mixing at medium speed 10 minutes or until dough is smooth and elastic. Dough temperature should be 78 F. to 82 F.
5. FERMENT: Cover. Set in a warm place (80 F.) about 1-1/2 hours or until double in bulk.
6. PUNCH: Divide dough into 3 pieces, 4 lb 5 oz each; shape into a rectangular piece. Let rest 10 to 20 minutes.
7. Roll each 4 pounds 5 ounce pieces of dough into a rectangular sheet about 18 inches wide, 36 inches long, and 1/4-inch thick. If using D 036 01, use 4 lb 2 oz pieces.
8. Melt butter or margarine. Brush 1/2 cup on each sheet of dough. Prepare Cinnamon Sugar Filling, Recipe No. D 042 00; sprinkle 1-1/2 cups on each sheet of dough.
9. Fold each sheet of dough in thirds lengthwise to make a strip about 6 inches wide, 35 inches long, and 3/4-inch thick.
10. Cut strips crosswise into 34 pieces about 1 inch wide.
11. Twist pieces in one direction and then in opposite direction.
12. Place on lightly greased sheet pans in rows 4 by 8.
13. Prepare 1/4 Recipe Egg Wash, Recipe No. D 017 00; brush 1/4 cup on rolls in each pan.
14. Proof at 90 F. to 100 F. until double in bulk.
15. Bake at 375 F. for 20 to 25 minutes or in 325 F. convection oven for 15 minutes on high fan, open vent. Cool.
16. Glaze, if desired, with 1 Recipe Vanilla Glaze, Recipe No. D 046 00. Brush about 3/4 cup on rolls in each pan.

BREADS AND SWEET DOUGHS No.D 036 16

BUTTERHORNS

Yield 100   Portion   1 Roll

Calories	Carbohydrates	Protein	Fat	Cholesterol	Sodium	Calcium
311 cal	52 g	6 g	9 g	34 mg	260 mg	18 mg

Ingredient	Weight	Measure	Issue
YEAST,ACTIVE,DRY	6-3/4 oz	1 cup	
WATER,WARM	1 lbs	2 cup	
WATER	1-5/8 lbs	3 cup	
EGGS,WHOLE,FROZEN	1-1/4 lbs	2-1/4 cup	
SUGAR,GRANULATED	1-1/8 lbs	2-5/8 cup	
MILK,NONFAT,DRY	1-3/4 oz	3/4 cup	
SALT	1-7/8 oz	3 tbsp	
FLOUR,WHEAT,BREAD	7-7/8 lbs	1 gal 2-1/2 qts	
SHORTENING,SOFTENED	14-1/2 oz	2 cup	
BUTTER	12 oz	1-1/2 cup	
VANILLA GLAZE		2-3/4 cup	

Method

1. Sprinkle yeast over water. DO NOT USE TEMPERATURES ABOVE 110 F. Mix well. Let stand 5 minutes; stir. Set aside for use in Step 3.
2. Place water, eggs, sugar, milk, and salt in mixer bowl. Using dough hook, mix at low speed just until blended.
3. Add flour and yeast solution. Mix at low speed 1 minute or until all flour mixture is incorporated into liquid.
4. Add shortening; mix at low speed 1 minute. Continue mixing at medium speed 10 minutes or until dough is smooth and elastic. Dough temperature should be 78 F. to 82 F.
5. FERMENT: Cover. Set in a warm place (80 F.) about 1-1/2 hours or until double in bulk.
6. PUNCH: Divide dough into 1 pound 7 ounce pieces; shape into a rectangular piece. Let rest 10 to 20 minutes.
7. Roll each piece of dough into a rectangular sheet about 9 inches wide, 24 inches long, and about 1/4-inch thick. (For D 036 01, divide into 9-1 lb 6 oz pieces.)
8. Melt butter or margarine. Brush about 3 tablespoons on each sheet of dough.
9. Cut each strip into 12 wedges about 4 inches wide at the widest end.
10. Roll up each wedge from wide edge to point.
11. Place on lightly greased sheet pans in rows 4 by 8 with point end under roll; press firmly in place.
12. Proof at 90 F. to 100 F. until double in bulk.
13. Bake at 375 F. for 20 to 25 minutes or in a 325 F. convection oven for 15 minutes on high fan, open vent. Cool.
14. Glaze, if desired, with 1 Recipe Vanilla Glaze, Recipe No. D 046 00. Brush about 3/4 cup on rolls in each pan.

BREADS AND SWEET DOUGHS No.D 036 17

CRESCENTS

Yield 100   **Portion** 1 Roll

Calories	Carbohydrates	Protein	Fat	Cholesterol	Sodium	Calcium
311 cal	52 g	6 g	9 g	34 mg	260 mg	18 mg

Ingredient

Ingredient	Weight	Measure	Issue
YEAST,ACTIVE,DRY	6-3/4 oz	1 cup	
WATER,WARM	1 lbs	2 cup	
WATER	1-5/8 lbs	3 cup	
EGGS,WHOLE,FROZEN	1-1/4 lbs	2-1/4 cup	
SUGAR,GRANULATED	1-1/8 lbs	2-5/8 cup	
MILK,NONFAT,DRY	1-3/4 oz	3/4 cup	
SALT	1-7/8 oz	3 tbsp	
FLOUR,WHEAT,BREAD	7-7/8 lbs	1 gal 2-1/2 qts	
SHORTENING,SOFTENED	14-1/2 oz	2 cup	
BUTTER	12 oz	1-1/2 cup	
VANILLA GLAZE		2-3/4 cup	

Method

1. Sprinkle yeast over water. DO NOT USE TEMPERATURES ABOVE 110 F. Mix well. Let stand 5 minutes; stir. Set aside for use in Step 3.
2. Place water, eggs, sugar, milk, and salt in mixer bowl. Using dough hook, mix at low speed just until blended.
3. Add flour and yeast solution. Mix at low speed 1 minute or until all flour mixture is incorporated into liquid.
4. Add shortening; mix at low speed 1 minute. Continue mixing at medium speed 10 minutes or until dough is smooth and elastic. Dough temperature should be 78 F. to 82 F.
5. FERMENT: Cover. Set in a warm place (80 F.) about 1-1/2 hours or until double in bulk.
6. PUNCH: Divide dough into 9 pieces, 1 lb 7 oz pieces; shape into a rectangular piece. Let rest 10 to 20 minutes. If using D 036 01, divide into 9 1 lb 6 oz pieces.
7. Roll each piece of dough into a rectangular sheet about 9 inches wide, 24 inches long, and about 1/4-inch thick.
8. Melt butter or margarine. Brush about 3 tablespoons on each sheet of dough.
9. Cut each strip into 12 wedges about 4 inches wide at the widest end.
10. Roll up each wedge from wide edge to point.
11. Place on lightly greased sheet pans in rows 4 by 8 with point end under roll; press firmly in place.
12. Proof at 90 F. to 100 F. until double in bulk.
13. Bake at 375 F. for 20 to 25 minutes or in a 325 F. convection oven for 15 minutes on high fan, open vent. Cool.
14. Glaze, if desired, with 1 Recipe Vanilla Glaze, Recipe No. D 046 00. Brush about 3/4 cup on rolls in each pan.

BREADS AND SWEET DOUGHS No.D 037 00

QUICK COFFEE CAKE (BISCUIT MIX)

Yield 100   Portion   1 Piece

Calories	Carbohydrates	Protein	Fat	Cholesterol	Sodium	Calcium
276 cal	44 g	4 g	9 g	24 mg	443 mg	76 mg

Ingredient	Weight	Measure	Issue
FLOUR,WHEAT,GENERAL PURPOSE	1-2/3 lbs	1 qts 2 cup	
MARGARINE,SOFTENED	12 oz	1-1/2 cup	
CINNAMON,GROUND	1/4 oz	1 tbsp	
SUGAR,BROWN,PACKED	7-2/3 oz	1-1/2 cup	
BISCUIT MIX	6-3/4 lbs	1 gal 2-3/8 qts	
SUGAR,GRANULATED	1-1/2 lbs	3-1/2 cup	
MILK,NONFAT,DRY	3-5/8 oz	1-1/2 cup	
WATER	3-1/8 lbs	1 qts 2 cup	
EGGS,WHOLE,FROZEN	1-1/4 lbs	2-1/4 cup	
EXTRACT,VANILLA	7/8 oz	2 tbsp	
COOKING SPRAY,NONSTICK	2 oz	1/4 cup 1/3 tbsp	
SUGAR,POWDERED	2-1/8 lbs	2 qts	
WATER,BOILING	8-1/3 oz	1 cup	
MARGARINE,SOFTENED	2 oz	1/4 cup 1/3 tbsp	
EXTRACT,VANILLA	1/8 oz	1/8 tsp	

Method

1. TOPPING: In mixer bowl, combine flour, butter or margarine, brown sugar, cinnamon; mix at low speed 3 minutes until mixture resembles coarse cornmeal. Remove topping from mixer bowl and set aside for use in Step 6.
2. CAKE: In mixer bowl, combine Biscuit Mix, sugar and nonfat dry milk; mix at low speed 1 minute or until well blended.
3. Combine water, eggs, vanilla; add egg mixture gradually to dry mixture while mixing at low speed for 2 minutes.
4. Scrape down sides and bottom of mixer bowl; continue to mix at low speed an additional 1 minute. DO NOT OVERMIX.
5. Lightly spray pan with non-stick cooking spray. Pour 3-1/2 quarts of batter into each floured pan. Spread batter evenly.
6. Sprinkle 1 quart topping over batter in each pan.
7. Using a convection oven, bake at 325 F. for about 30 minutes on low fan, open vent. Remove cakes from oven and let cool slightly.
8. GLAZE: Combine powdered sugar, hot water, butter or margarine and vanilla; mix until smooth.
9. Drizzle about 2 cups glaze over each baked cake while cakes are still warm. Cut 6 by 9.

BREADS AND SWEET DOUGHS No.D 037 01

QUICK APPLE COFFEE CAKE (BISCUIT MIX)

Yield 100   Portion 1 Piece

Calories	Carbohydrates	Protein	Fat	Cholesterol	Sodium	Calcium
214 cal	37 g	4 g	6 g	24 mg	405 mg	76 mg

Ingredient	Weight	Measure	Issue
SUGAR,GRANULATED	1 lbs	2-1/4 cup	
CINNAMON,GROUND	1 oz	1/4 cup 1/3 tbsp	
NUTMEG,GROUND	1/8 oz	1/3 tsp	
BISCUIT MIX	6-3/4 lbs	1 gal 2-3/8 qts	
SUGAR,GRANULATED	1-1/2 lbs	3-1/2 cup	
MILK,NONFAT,DRY	3-5/8 oz	1-1/2 cup	
WATER	3-1/8 lbs	1 qts 2 cup	
EXTRACT,VANILLA	7/8 oz	2 tbsp	
EGGS,WHOLE,FROZEN	1-1/4 lbs	2-1/4 cup	
COOKING SPRAY,NONSTICK	2 oz	1/4 cup 1/3 tbsp	
APPLES,CANNED,SLICED,DRAINED	6-3/4 lbs	3 qts 1-5/8 cup	

Method

1. TOPPING: Combine sugar, cinnamon and nutmeg. Set aside for use in Steps 6 and 8.
2. Cake: In mixer bowl, combine biscuit mix, sugar and nonfat dry milk; mix at low speed 1 minute or until well blended.
3. Combine water, eggs and vanilla. Add egg mixture gradually to dry mixture while mixing at low speed for 2 minutes.
4. Scrape down sides and bottom of mixer bowl; continue to mix at low speed an additional 1 minute. DO NOT OVERMIX.
5. Lightly spray each pan with non-stick cooking spray. Pour 3-1/2 quarts of batter into each floured pan. Spread batter evenly.
6. Sprinkle 1/2 cup of topping over batter in each pan.
7. Arrange 3 pounds of apple slices evenly over batter and topping in each pan.
8. Sprinkle 3/4 cup of sugar mixture over apple slices in each pan.
9. Using a convection oven, bake 30 minutes at 325 F. on low fan, open vent.
10. Remove cakes from oven and let cool. Cut 6 by 9.

BREADS AND SWEET DOUGHS No.D 037 02

QUICK FRENCH COFFEE CAKE (BISCUIT MIX)

Yield 100　　　　　　　　　　　　　　　　　　　Portion　1 Piece

Calories	Carbohydrates	Protein	Fat	Cholesterol	Sodium	Calcium
343 cal	53 g	6 g	12 g	24 mg	444 mg	87 mg

Ingredient	Weight	Measure	Issue
FLOUR,WHEAT,GENERAL PURPOSE	1-2/3 lbs	1 qts 2 cup	
MARGARINE,SOFTENED	12 oz	1-1/2 cup	
SUGAR,BROWN,PACKED	7-2/3 oz	1-1/2 cup	
CINNAMON,GROUND	1/4 oz	1 tbsp	
BISCUIT MIX	6-3/4 lbs	1 gal 2-3/8 qts	
RAISINS	1-7/8 lbs	1 qts 2 cup	
SUGAR,GRANULATED	1-1/2 lbs	3-1/2 cup	
NUTS,UNSALTED,CHOPPED,COARSELY	1-1/2 lbs	1 qts 5/8 cup	
MILK,NONFAT,DRY	3-5/8 oz	1-1/2 cup	
NUTMEG,GROUND	1/2 oz	2 tbsp	
WATER	3-1/8 lbs	1 qts 2 cup	
EXTRACT,VANILLA	7/8 oz	2 tbsp	
EGGS,WHOLE,FROZEN	1-1/4 lbs	2-1/4 cup	
COOKING SPRAY,NONSTICK	2 oz	1/4 cup 1/3 tbsp	
SUGAR,POWDERED	2-1/8 lbs	2 qts	
WATER,BOILING	8-1/3 oz	1 cup	
MARGARINE,SOFTENED	2 oz	1/4 cup 1/3 tbsp	
EXTRACT,VANILLA	1/8 oz	1/8 tsp	

Method

1. TOPPING: In mixer bowl, combine flour, butter or margarine, brown sugar, cinnamon; mix at low speed 3 minutes until mixture resembles coarse cornmeal. Remove topping from mixer bowl and set aside for use in Step 6.
2. CAKE: In mixer bowl, combine Biscuit Mix, raisins, sugar, walnuts, nonfat dry milk and nutmeg; mix at low speed 1 minute or until well blended.
3. Combine water, eggs and vanilla. Add egg mixture gradually to dry mixture whiile mixing at low speed 2 minutes.
4. Scrape down sides and bottom of mixer bowl. Continue to mix at low speed an additional 1minute. DO NOT OVERMIX.
5. Pour 1 gallon batter into each lightly sprayed and floured pan. Spread batter evenly.
6. Sprinkle 1 quart of topping over batter in each pan.
7. Using a convection oven, bake 30 minutes on low fan, open vent. Remove cakes from oven and let cool slightly.
8. GLAZE: Combine powdered sugar, hot water, butter or margarine and vanilla; mix until smooth.
9. Drizzle 2 cups glaze over each baked cake while cakes are still warm. Cut 6 by 9.

BREADS AND SWEET DOUGHS No.D 037 03

QUICK CHERRY COFFEE CAKE (BISCUIT MIX)

Yield 100　　　　　　　　　　　　　　　　　　　Portion　1 Piece

Calories	Carbohydrates	Protein	Fat	Cholesterol	Sodium	Calcium
290 cal	48 g	4 g	9 g	24 mg	443 mg	79 mg

Ingredient	Weight	Measure	Issue
FLOUR,WHEAT,GENERAL PURPOSE	1-2/3 lbs	1 qts 2 cup	
MARGARINE,SOFTENED	12 oz	1-1/2 cup	
SUGAR,BROWN,PACKED	7-2/3 oz	1-1/2 cup	
CINNAMON,GROUND	1/4 oz	1 tbsp	
BISCUIT MIX	6-3/4 lbs	1 gal 2-3/8 qts	
SUGAR,GRANULATED	1-1/2 lbs	3-1/2 cup	
MILK,NONFAT,DRY	3-5/8 oz	1-1/2 cup	
WATER	3-1/8 lbs	1 qts 2 cup	
EGGS,WHOLE,FROZEN	1-1/4 lbs	2-1/4 cup	
EXTRACT,VANILLA	7/8 oz	2 tbsp	
COOKING SPRAY,NONSTICK	2 oz	1/4 cup 1/3 tbsp	
CHERRIES,CANNED,RED,TART,WATER PACK,INCL LIQUIDS	6-1/2 lbs	2 qts 3-7/8 cup	
SUGAR,POWDERED	2-1/8 lbs	2 qts	
WATER,BOILING	8-1/3 oz	1 cup	
MARGARINE,SOFTENED	2 oz	1/4 cup 1/3 tbsp	
EXTRACT,VANILLA	1/8 oz	1/8 tsp	

Method

1. TOPPING: In mixer bowl, combine flour, butter or margarine, brown sugar, cinnamon; mix at low speed 3 minutes until mixture resembles coarse cornmeal. Remove topping from mixer bowl and set aside for use in Step 7.
2. CAKE: In mixer bowl, combine Biscuit Mix, sugar and nonfat dry milk; mix at low speed 1 minute or until well blended.
3. Combine water, eggs and vanilla. Add egg mixture gradually to dry mixture while mixing at low speed for 2 minutes.
4. Scrape down sides and bottom of mixer bowl; continue to mix low speed an additional 1 minute. DO NOT OVERMIX.
5. Pour 3-1/2 quart batter into each lightly sprayed and floured pan. Spread batter evenly.
6. Arrange 2-1/2 pounds cherries evenly over batter in each pan.
7. Sprinkle 1 quart of topping over batter and cherries in each pan.
8. Using a convection oven, bake about 30 minutes on low fan, open vent at 325 F. Remove cakes from oven and let cool slightly.
9. GLAZE: Combine powdered sugar, hot water, butter or margarine, vanilla; mix until smooth.
   Drizzle 2 cups glaze over each baked cake while cakes are still warm. Cut 6 by 9.

BREADS AND SWEET DOUGHS No.D 037 04

QUICK ORANGE-COCONUT COFFEE CAKE (BISCUIT MIX)

Yield 100   Portion 1 Piece

Calories	Carbohydrates	Protein	Fat	Cholesterol	Sodium	Calcium
405 cal	53 g	4 g	20 g	37 mg	519 mg	66 mg

Ingredient	Weight	Measure	Issue
SUGAR,GRANULATED	1-1/2 lbs	3-1/2 cup	
MARGARINE,SOFTENED		1 cup	
BISCUIT MIX	6-3/4 lbs	1 gal 2-3/8 qts	
SUGAR,GRANULATED		2-1/4 cup	
MILK,NONFAT,DRY		1-1/2 cup	
WATER	3-1/8 lbs	1 qts 2 cup	
EGGS,WHOLE,FROZEN	1-1/4 lbs	2-1/4 cup	
EXTRACT,VANILLA	7/8 oz	2 tbsp	
ORANGE-COCONUT TOPPING		2 qts 2 cup	
COOKING SPRAY,NONSTICK	2 oz	1/4 cup 1/3 tbsp	

Method

1. TOPPING: In mixer bowl, cream sugar and butter or margarine at medium speed 2 minutes. Add coconut, orange juice, flour and orange rind; mix at low speed 2 minutes. Remove topping from mixer bowl and set aside for use in Step 6.
2. CAKE: In mixer bowl, combine bisquick mix, sugar and nonfat dry milk; mix at low speed 1 minute or until well blended.
3. Combine water, eggs and vanilla. Add egg mixture gradually to dry mixture while mixing at low speed 2 minutes.
4. Scrape down sides and bottom of mixer bowl; continue to mix low speed an additional 1 minute. DO NOT OVERMIX.
5. Lightly spray each pan with non-stick spray. Pour 3-1/2 quarts of batter into each sprayed and floured pan. Spread batter evenly.
6. Sprinkle 1 quart topping over batter in each pan.
7. Using a convection oven, bake about 30 minutes on low fan, open vent at 325 F. Remove cakes from oven and let cool slightly. Cut 6 by 9.

BREADS AND SWEET DOUGHS No.D 037 05

QUICK COFFEE CAKE

Yield 100   Portion 1 Piece

Calories	Carbohydrates	Protein	Fat	Cholesterol	Sodium	Calcium
288 cal	45 g	4 g	11 g	32 mg	246 mg	68 mg

Ingredient	Weight	Measure	Issue
FLOUR,WHEAT,GENERAL PURPOSE	1-2/3 lbs	1 qts 2 cup	
BUTTER,SOFTENED	12 oz	1-1/2 cup	
SUGAR,BROWN,PACKED	7-2/3 oz	1-1/2 cup	
CINNAMON,GROUND	1/4 oz	1 tbsp	
FLOUR,WHEAT,GENERAL PURPOSE	3-7/8 lbs	3 qts 2 cup	
SUGAR,GRANULATED	3 lbs	1 qts 2-3/4 cup	
MILK,NONFAT,DRY	3 oz	1-1/4 cup	
BAKING POWDER	2-3/4 oz	1/4 cup 2 tbsp	
SALT	1 oz	1 tbsp	
WATER	3-1/8 lbs	1 qts 2 cup	
OIL,SALAD	1-1/2 lbs	3 cup	
EGGS,WHOLE,FROZEN	1-1/4 lbs	2-1/4 cup	
EXTRACT,VANILLA	7/8 oz	2 tbsp	
SUGAR,POWDERED	2-1/8 lbs	2 qts	
WATER,BOILING	8-1/3 oz	1 cup	
BUTTER,SOFTENED	2 oz	1/4 cup 1/3 tbsp	
EXTRACT,VANILLA	1/8 oz	1/8 tsp	

Method

1. TOPPING: In mixer bowl, combine flour, butter or margarine, brown sugar, cinnamon; mix at low speed 3 minutes until mixture resembles coarse cornmeal. Remove topping from mixer bowl and set aside for use in Step 6.
2. CAKE: In mixer bowl, sift together flour, sugar and nonfat dry milk, baking powder and salt; mix at low speed 1 minute or until well blended.
3. Combine water, salad oil, eggs and vanilla. Add egg mixture gradually to dry mixture while mixing at low speed 2 minutes.
4. Scrape down sides and bottom of mixer bowl; continue to mix low speed an additional 1 minute. DO NOT OVERMIX.
5. Pour 3-1/2 quart into each lightly sprayed and floured pan. Spread batter evenly.
6. Sprinkle 1 quart of topping over batter in each pan.
7. Using a convection oven, bake on low fan, open vent at 325 F. for about 30 minutes. Remove cakes from oven and let cool slightly.
8. GLAZE: Combine powdered sugar, hot water, butter or margarine and vanilla; mix until smooth.
9. Drizzle 2 cups glaze over each baked cake while cakes are still warm. Cut 6 by 9.

BREADS AND SWEET DOUGHS No.D 038 00

TEMPURA BATTER

Yield 100   Portion 1 Gallon

Calories	Carbohydrates	Protein	Fat	Cholesterol	Sodium	Calcium
5796 cal	1069 g	204 g	66 g	2231 mg	27078 mg	3545 mg

Ingredient   **Weight**   **Measure**   **Issue**

FLOUR,WHEAT,GENERAL PURPOSE   3 lbs   2 qts 3 cup
BAKING POWDER   1-3/4 oz   1/4 cup
SALT   1-7/8 oz   3 tbsp
EGGS,WHOLE,FROZEN   1-1/8 lbs   2-1/8 cup
WATER,COLD   5-1/4 lbs   2 qts 2 cup

Method

1 Sift together flour, baking powder, and salt into mixer bowl.
2 Add water to beaten eggs.
3 Add egg mixture to dry ingredients; whip at high speed until smooth.
4 Fry in small batches. Tempura-fried foods lose crispness if allowed to stand on steam-table. DO NOT SAVE.

Notes

1 Batter may be used for Tempura Fried Shrimp, Recipe No. L 137 01 and Tempura Fried Onion Rings, Recipe No. Q 035 02.

BREADS AND SWEET DOUGHS No.D 039 00

DANISH DIAMONDS (DANISH PASTRY DOUGH)

Yield 100   **Portion** 1 Danish

Calories	Carbohydrates	Protein	Fat	Cholesterol	Sodium	Calcium
228 cal	23 g	3 g	14 g	5 mg	168 mg	15 mg

Ingredient	Weight	Measure	Issue
DANISH DOUGH,FROZEN	11 lbs	100 each	
EGG WASH	181-7/8 gm	3/4 unit	
PIE FILLING,APPLE,PREPARED	6 lbs	3 qts	
EGG WASH	181-7/8 gm	3/4 unit	

Method

1. Prepare 50 Danish squares in a batch. Thaw at room temperature 5 minutes on a lightly floured working surface. Rolling out is not necessary.
2. Prepare 1/2 Recipe (1-1/2 cups) Egg Wash, Recipe No. D 017 00. Use 3/4 cup of egg wash. Lightly brush entire surface of each square. Set aside remaining 3/4 cup egg wash for use in Step 6.
3. Place pie filling in mixer bowl. Using whip, mix on medium speed 15 seconds to break up large pieces. Place about 2 tbsp filling in center of each square. Fold lower left corner to center; fold upper right corner over top of first corner. Press firmly to seal; repeat by folding lower right corner to center; press firmly to seal. Fold upper left corner to center; press tip to seal.
4. Place squares on lightly greased pans in rows 4 by 6.
5. Brush lightly with remaining egg wash.
6. Proof at 90 F. for 30 to 45 minutes or until double in bulk.
7. Using a convection oven, bake at 325 F. for 10 minutes or until golden brown on low fan, open vent.
8. Cool. Glaze if desired, with Vanilla Glaze or Variations, Recipe Nos. D 046 00, D 046 01, D 046 02.

Notes

1. In Step 3, any type of fruit pie filling may be used.
2. In Step 3, 7 lb 11 oz of cherry, pineapple or strawberry jam may be used, per 100 portions.
3. Prepare in batches as dough becomes difficult to work with in 15 minutes.

BREADS AND SWEET DOUGHS No.D 039 01

BEAR CLAWS (DANISH PASTRY DOUGH)

Yield 100   **Portion** 1 Danish

Calories	Carbohydrates	Protein	Fat	Cholesterol	Sodium	Calcium
219 cal	20 g	3 g	15 g	5 mg	162 mg	14 mg

Ingredient	**Weight**	**Measure**	**Issue**
DANISH DOUGH,FROZEN	11 lbs	100 each	
EGG WASH	181-7/8 gm	3/4 unit	
PIE FILLING,APPLE,PREPARED	3-1/8 lbs	1 qts 2-1/4 cup	
COOKING SPRAY,NONSTICK	2 oz	1/4 cup 1/3 tbsp	
EGG WASH	181-7/8 gm	3/4 unit	

Method

1. Prepare 50 Danish squares in a batch. Thaw at room temperature 5 minutes on a lightly floured working surface. Rolling out is not necessary.
2. Prepare 1/2 Recipe Egg Wash, Recipe No. D 017 00. Use 3/4 cup of egg wash. Lightly brush entire surface of each square. Set aside remaining 3/4 cup egg wash for use in Step 6.
3. Place about 1 tablespoon of filling over half of each square. Fold in half; seal edge by pressing firmly.
4. Make 3 cuts, 3/4-inch in depth, on 4-inch sealed side of each piece to form a claw.
5. Lightly spray pans with non-stick cooking spray. Place dough on pans. Bend into slight horseshoe shape and spread claws slightly.
6. Brush lightly with remaining egg wash.
7. Proof at 90 F. to 100 F. for 30 to 45 minutes or until double in size.
8. Using a convection oven, bake at 325 F. for 10 minutes or until golden brown on low fan, open vent.
9. Cool. Glaze if desired, with Vanilla Glaze or Variations, Recipe Nos. D 046 00, D 046 01, D 046 02.

Notes

1. Prepare in batches as dough becomes difficult to work with in 15 minutes.

BREADS AND SWEET DOUGHS No.D 039 02

FRUIT TURNOVERS (FROZEN PUFF PASTRY DOUGH)

Yield 100   Portion 1 Danish

Calories	Carbohydrates	Protein	Fat	Cholesterol	Sodium	Calcium
346 cal	33 g	4 g	22 g	5 mg	155 mg	9 mg

Ingredient	Weight	Measure	Issue
PUFF PASTRY DOUGH,SQUARES,FROZEN	12-1/2 lbs	100 each	
EGG WASH	181-7/8 gm	3/4 unit	
PIE FILLING,APPLE,PREPARED	6 lbs	3 qts	
COOKING SPRAY,NONSTICK	2 oz	1/4 cup 1/3 tbsp	
EGG WASH	181-7/8 gm	3/4 unit	

Method

1 Prepare 50 Danish squares in a batch. Thaw at room temperature 5 minutes on a lightly floured working surface. Rolling out is not necessary.
2 Prepare 1/2 recipe Egg Wash (Recipe No. D 017 00). Use 3/4 cup egg wash. Lightly brush entire surface of each square. Set aside remaining 3/4 cup egg wash for use in Step 5.
3 Place about 2 tbsp filling in center of each square. Fold upper right corner over lower left corner to form a triangle. Seal by crimping edges together.
4 Make two 1-inch slits in the center.
5 Lightly spray each pan with non-stick cooking spray.   Place 24 turnovers on each pan.
6 Brush lightly with remaining egg wash.
7 Using a convection oven, bake 15 minutes in a 350 F. with low fan, open vent or until golden brown.
8 Cool. Glaze if desired, with Vanilla Glaze or Variations, Recipe Nos. D 046 00, D 046 01, D 046 02.

Notes

1 In Step 3, pie filling, prepared, fruit (apple, blueberry, cherry, or peach) may be used as filling. Place in mixer bowl. Using whip, mix on medium speed 15 seconds to break up large pieces.
2 In Step 3, 7 pounds 11 ounces of cherry, pineapple, or strawberry jam may be used, per 100 portions.
3 Prepare in batches as dough becomes difficult to work with in 15 minutes.

BREADS AND SWEET DOUGHS No.D 039 03

FRUIT PUFFS (FROZEN PUFF PASTRY DOUGH)

Yield 100   Portion 1 Danish

Calories	Carbohydrates	Protein	Fat	Cholesterol	Sodium	Calcium
357 cal	37 g	4 g	22 g	0 mg	153 mg	7 mg

Ingredient	Weight	Measure	Issue
PUFF PASTRY DOUGH,SQUARES,FROZEN	12-1/2 lbs	100 each	
WATER	12-1/2 oz	1-1/2 cup	
SUGAR,GRANULATED	1 lbs	2-1/4 cup	
PIE FILLING,APPLE,PREPARED	6 lbs	3 qts	

Method

1. Prepare 50 Danish squares in a batch. Thaw at room temperature 5 minutes on a lightly floured working surface. Rolling out is not necessary.
2. Place squares in rows 3 by 5 on pans. Brush water over each square. Sprinkle sugar over each square.
3. Place about 2 tbsp filling in center of each square. Fold lower left corner to center; fold upper right corner over top of first corner. Press firmly to seal; repeat by folding lower right corner to center; press firmly to seal. Fold upper left corner to center; press tip to seal.
4. Using a convection oven, bake in 350 F. for 15 minutes with low fan and open vent or until golden brown.
5. Cool.

Notes

1. In Step 3, pie filling, prepared, fruit (apple, blueberry, cherry, or peach) may be used as filling. Place in mixer bowl. Using whip, mix on medium speed 15 seconds to break up large pieces.
2. In Step 3, 7 lbs 11 oz cherry, pineapple or strawberry jam may be used, per 100 portions.
3. Prepare in batches as dough becomes difficult to work with in 15 minutes.

BREADS AND SWEET DOUGHS No.D 040 00

CORNSTARCH WASH

Yield 100 **Portion** 1 Quart

Calories	Carbohydrates	Protein	Fat	Cholesterol	Sodium	Calcium
122 cal	29 g	0 g	0 g	0 mg	31 mg	20 mg

Ingredient	**Weight**	**Measure**	**Issue**
CORNSTARCH	1-1/8 oz	1/4 cup 1/3 tbsp	
WATER	2-1/8 lbs	1 qts	

Method

1 Combine cornstarch and water. Bring to a boil; cook until clear.
2 Brush on bread and rolls before and immediately after baking.

Notes

1 Keep wash warm. Reheat if necessary.

BREADS AND SWEET DOUGHS No.D 041 00

CHERRY FILLING (CORNSTARCH)

Yield 100   Portion 3 Quarts

Calories	Carbohydrates	Protein	Fat	Cholesterol	Sodium	Calcium
7081 cal	1816 g	23 g	4 g	0 mg	76 mg	358 mg

Ingredient	Weight	Measure	Issue
CHERRIES,CANNED,RED,TART,WATER PACK,DRAINED	6-1/2 lbs	3 qts	
RESERVED LIQUID	1-5/8 lbs	3 cup	
CORNSTARCH	4-1/2 oz	1 cup	
SUGAR,GRANULATED	3 lbs	1 qts 2-3/4 cup	
FOOD COLOR,RED	1/8 oz	1/8 tsp	

Method

1. Drain cherries. Dissolve cornstarch in juice. Set juice and cornstarch mixture aside for use in Step 4.
2. Mash cherries with wire whip 1 minute at medium speed; combine with sugar and food coloring.
3. Bring to a boil in steam-jacketed kettle or stock pot stirring constantly to prevent scorching. Reduce heat. Simmer about 10 minutes.
4. Add reserved juice and cornstarch mixture to cherries while stirring. Cook 2 to 3 minutes until clear and thickened, stirring constantly. Remove from heat; cool.

BREADS AND SWEET DOUGHS No.D 041 01

CHERRY FILLING (PIE FILLING, PREPARED)

Yield 100   Portion 3 Quarts

Calories	Carbohydrates	Protein	Fat	Cholesterol	Sodium	Calcium
3132 cal	798 g	14 g	5 g	0 mg	245 mg	300 mg

Ingredient	**Weight**	**Measure**	**Issue**
PIE FILLING,CHERRY,PREPARED	6 lbs	3 qts	

Method

1 Mash prepared filling with a wire whip for 1 minute at medium speed.

BREADS AND SWEET DOUGHS No.D 041 03

APPLE FILLING (PIE FILLING, PREPARED)

Yield 100   Portion 3 Quarts

Calories	Carbohydrates	Protein	Fat	Cholesterol	Sodium	Calcium
2749 cal	713 g	3 g	3 g	0 mg	1197 mg	109 mg

Ingredient	Weight	Measure	Issue
PIE FILLING,APPLE,PREPARED	6 lbs	3 qts	

Method

1 Break up large pieces of prepared apple pie filling with wire whip one minute at medium speed.

BREADS AND SWEET DOUGHS No.D 041 04

BLUEBERRY FILLING (PIE FILLING, PREPARED)

Yield 100   Portion 3 Quarts

Calories	Carbohydrates	Protein	Fat	Cholesterol	Sodium	Calcium
2871 cal	754 g	0 g	0 g	0 mg	1615 mg	718 mg

Ingredient	**Weight**	**Measure**	**Issue**
PIE FILLING,BLUEBERRY,PREPARED	7 lbs	3 qts	

Method

1 Use accordingly.

BREADS AND SWEET DOUGHS No.D 041 05

RASPBERRY FILLING (PREPARED BAKERY)

Yield 100   Portion 3 Quarts

Calories	Carbohydrates	Protein	Fat	Cholesterol	Sodium	Calcium
2947 cal	774 g	0 g	0 g	0 mg	1657 mg	737 mg

Ingredient	**Weight**	**Measure**	**Issue**
RASPBERRY BAKERY FILLING	7-1/4 lbs	3 qts 3/8 cup	

Method

1 Use accordingly.

BREADS AND SWEET DOUGHS No.D 042 00

CINNAMON SUGAR FILLING

Yield 100   Portion 4-1/2 Cups

Calories	Carbohydrates	Protein	Fat	Cholesterol	Sodium	Calcium
1843 cal	480 g	1 g	1 g	0 mg	191 mg	735 mg

Ingredient   **Weight**   **Measure**   **Issue**
CINNAMON,GROUND   1 oz   1/4 cup 1/3 tbsp
SUGAR,BROWN,PACKED   1 lbs   3-1/4 cup

Method

1 Combine cinnamon and brown sugar.

Notes

1 Granulated sugar may be substituted for brown sugar.

BREADS AND SWEET DOUGHS No.D 042 01

CINNAMON SUGAR NUT FILLING

Yield 100   Portion   4-1/2 Cups

Calories	Carbohydrates	Protein	Fat	Cholesterol	Sodium	Calcium
6933 cal	646 g	232 g	433 g	0 mg	244 mg	1506 mg

Ingredient	Weight	Measure	Issue
CINNAMON,GROUND	1 oz	1/4 cup 1/3 tbsp	
SUGAR,BROWN,PACKED	1 lbs	3-1/4 cup	
NUTS,UNSALTED,CHOPPED,COARSELY	1-7/8 lbs	1 qts 2 cup	

Method

1 Combine cinnamon and brown sugar.
2 Sprinkle chopped nuts over cinnamon sugar mixture.

Notes

1 In Step 1, granulated sugar may be substituted for brown sugar.

BREADS AND SWEET DOUGHS No.D 042 02

CINNAMON SUGAR RAISIN FILLING

Yield 100   **Portion** 4-1/2 Cups

Calories	Carbohydrates	Protein	Fat	Cholesterol	Sodium	Calcium
4562 cal	1197 g	30 g	5 g	0 mg	300 mg	1179 mg

Ingredient

Ingredient	Weight	Measure	Issue
CINNAMON,GROUND	1 oz	1/4 cup 1/3 tbsp	
SUGAR,BROWN,PACKED	1 lbs	3-1/4 cup	
RAISINS	2 lbs	1 qts 2-1/4 cup	

Method

1 Combine cinnamon and brown sugar.
2 Sprinkle raisins over cinnamon sugar mixture.

Notes

1 In Step 1, granulated sugar may be substituted for brown sugar.

BREADS AND SWEET DOUGHS No.D 043 00

NUT FILLING

Yield 100   Portion   7-1/2 Cups

Calories	Carbohydrates	Protein	Fat	Cholesterol	Sodium	Calcium
10666 cal	1214 g	95 g	641 g	1118 mg	4412 mg	970 mg

Ingredient	Weight	Measure	Issue
FLOUR,WHEAT,GENERAL PURPOSE	13-1/4 oz	3 cup	
CINNAMON,GROUND	3/8 oz	1 tbsp	
SUGAR,GRANULATED	1-1/8 lbs	2-1/2 cup	
SUGAR,BROWN,PACKED	12-3/4 oz	2-1/2 cup	
BUTTER,MELTED	1-1/8 lbs	2-1/4 cup	
WALNUTS,SHELLED,CHOPPED	12-2/3 oz	3 cup	

Method

1 Sift together flour and cinnamon in mixer bowl; blend in sugars.
2 Add butter or margarine to dry ingredients; mix at low speed until well blended.
3 Add nuts, mixing at low speed. Use about 1 tbsp filling for each pastry.

BREADS AND SWEET DOUGHS No.D 044 00

OAT BRAN RAISIN MUFFINS

Yield 100   Portion   1 Muffin

Calories	Carbohydrates	Protein	Fat	Cholesterol	Sodium	Calcium
196 cal	29 g	5 g	7 g	40 mg	189 mg	88 mg

Ingredient	**Weight**	**Measure**	**Issue**
FLOUR,WHEAT,GENERAL PURPOSE	1-2/3 lbs	1 qts 2 cup	
MILK,NONFAT,DRY	5-1/8 oz	2-1/8 cup	
BAKING POWDER	2-3/4 oz	1/4 cup 2 tbsp	
SALT	5/8 oz	1 tbsp	
RAISINS	1-7/8 lbs	1 qts 2 cup	
CEREAL,OATMEAL,ROLLED	2-7/8 lbs	2 qts 1/2 cup	
CEREAL,OAT BRAN	12 oz	1 qts 2 cup	
SUGAR,BROWN,PACKED	1 lbs	3 cup	
WATER,WARM	4-2/3 lbs	2 qts 1 cup	
EGGS,WHOLE,FROZEN	2 lbs	3-3/4 cup	
OIL,SALAD	1 lbs	2 cup	
COOKING SPRAY,NONSTICK	2 oz	1/4 cup 1/3 tbsp	

Method

1 Sift together flour, milk, baking powder, and salt into mixer bowl.
2 Blend in raisins, rolled oats, oat bran, and brown sugar at low speed for 1/2 minute.
3 Add water, eggs, and oil or shortening to dry ingredients; mix at low speed until dry ingredients are moistened, about 15 seconds. DO NOT OVER MIX.
4 Lightly spray each muffin cup with non-stick cooking spray. Fill each muffin cup 2/3 full (1-No. 16 scoop).
5 Using a convection oven, bake at 350 F. for 20 minutes or until lightly browned with open vent and fan turned off first 5 minutes, then low fan.

BREADS AND SWEET DOUGHS No.D 045 00

SYRUP GLAZE

Yield 100   Portion 1 Quart

Calories	Carbohydrates	Protein	Fat	Cholesterol	Sodium	Calcium
2930 cal	790 g	1 g	1 g	0 mg	1106 mg	146 mg

Ingredient	**Weight**	**Measure**	**Issue**
SYRUP	2-1/3 lbs	3-3/8 cup	
WATER	1 lbs	2 cup	

Method

1 Combine syrup and water. Bring to a boil; boil about 5 minutes, stirring constantly.
2 Brush warm glaze over rolls or coffee cakes immediately after baking.

BREADS AND SWEET DOUGHS No.D 046 00

VANILLA GLAZE

Yield 100   Portion   2-3/4 Cups

Calories	Carbohydrates	Protein	Fat	Cholesterol	Sodium	Calcium
3125 cal	717 g	0 g	35 g	93 mg	365 mg	22 mg

Ingredient	Weight	Measure	Issue
SUGAR,POWDERED,SIFTED	1-5/8 lbs	1 qts 2 cup	
BUTTER,SOFTENED	1-1/2 oz	3 tbsp	
WATER,BOILING	6-1/4 oz	3/4 cup	
EXTRACT,VANILLA	1/4 oz	1/4 tsp	

Method

1 Combine powdered sugar, butter, boiling water, and vanilla; mix until smooth.
2 Spread glaze over baked sweet rolls or coffee cakes. Coat or dip fried doughnuts in glaze.

BREADS AND SWEET DOUGHS No.D 046 01

ALMOND GLAZE

Yield 100   Portion   2-3/4 Cups

Calories	Carbohydrates	Protein	Fat	Cholesterol	Sodium	Calcium
3115 cal	717 g	0 g	35 g	93 mg	364 mg	21 mg

Ingredient   Weight   Measure   Issue
SUGAR,POWDERED,SIFTED   1-5/8 lbs   1 qts 2 cup
BUTTER,SOFTENED   1-1/2 oz   3 tbsp
WATER,BOILING   6-1/4 oz   3/4 cup
EXTRACT,ALMOND   1/8 oz   1/8 tsp

Method

1 Combine powdered sugar, butter, boiling water, and flavoring; mix until smooth.
2 Spread glaze over baked sweet rolls or coffee cakes. Coat or dip fried doughnuts in glaze.

BREADS AND SWEET DOUGHS No.D 046 02

RUM GLAZE

Yield 100 Portion 2-3/4 Cups

Calories	Carbohydrates	Protein	Fat	Cholesterol	Sodium	Calcium
3125 cal	717 g	0 g	35 g	93 mg	365 mg	22 mg

Ingredient | Weight | Measure | Issue

Ingredient	Weight	Measure	Issue
SUGAR,POWDERED,SIFTED	1-5/8 lbs	1 qts 2 cup	
BUTTER,SOFTENED	1-1/2 oz	3 tbsp	
WATER,BOILING	6-1/4 oz	3/4 cup	
EXTRACT,RUM	1/4 oz	1/4 tsp	

Method

1 Combine powdered sugar, butter, boiling water, and flavoring; mix until smooth.
2 Spread glaze over baked sweet rolls or coffee cakes. Coat or dip fried doughnuts in glaze.

BREADS AND SWEET DOUGHS No. D 047 01

PINEAPPLE FILLING (CORNSTARCH)

Yield 100　　　　　　　　　　　　　　　　　　Portion 2-1/2 Quarts

Calories	Carbohydrates	Protein	Fat	Cholesterol	Sodium	Calcium
2942 cal	680 g	10 g	36 g	93 mg	389 mg	331 mg

Ingredient	Weight	Measure	Issue
PINEAPPLE,CANNED,CRUSHED	5 lbs	2 qts 1 cup	
BUTTER,MELTED	1-1/2 oz	3 tbsp	
SUGAR,GRANULATED	8-7/8 oz	1-1/4 cup	
CORNSTARCH	3 oz	1/2 cup 2-2/3 tbsp	
WATER	5-5/8 oz	1/2 cup 2-2/3 tbsp	

Method

1. Combine pineapple, butter, and sugar and combine over heat.
2. Dissolve cornstarch in cool water; add to hot pineapple mixture while stirring; bring to a boil; cook until thick and clear, about 5 minutes.
3. Cool slightly before using.

Notes

1. If desired, filling may be used for cake. Use 3 quarts filling for each sheet cake or 2 cups for each 9-inch layer cake.

BREADS AND SWEET DOUGHS No.D 048 00

ORANGE-COCONUT TOPPING

Yield 100   Portion 2-1/4 Quarts

Calories	Carbohydrates	Protein	Fat	Cholesterol	Sodium	Calcium
9327 cal	1063 g	42 g	581 g	497 mg	4811 mg	314 mg

Ingredient	Weight	Measure	Issue
BUTTER,SOFTENED	8 oz	1 cup	
SUGAR,GRANULATED	1 lbs	2-1/4 cup	
FLOUR,WHEAT,GENERAL PURPOSE	2-1/4 oz	1/2 cup	
JUICE,ORANGE,CANNED,UNSWEETENED	8-3/4 oz	1 cup	
ORANGE,RIND,GRATED	1-1/4 oz	1/4 cup 2-1/3 tbsp	
COCONUT,PREPARED,SWEETENED FLAKES	2-1/2 lbs	3 qts	

Method

1 Cream butter or margarine and sugar together at medium speed in mixer bowl.
2 Add flour, orange juice, orange rind, and coconut; blend.
3 Spread over sweet rolls or coffee cakes after proofing.

BREADS AND SWEET DOUGHS   No.D 049 00

STREUSEL TOPPING

Yield 100   Portion 3 Quarts

Calories	Carbohydrates	Protein	Fat	Cholesterol	Sodium	Calcium
9380 cal	1242 g	83 g	468 g	1242 mg	4892 mg	818 mg

Ingredient	Weight	Measure	Issue
FLOUR,WHEAT,GENERAL PURPOSE	1-2/3 lbs	1 qts 2 cup	
SUGAR,BROWN,PACKED	1 lbs	3-1/4 cup	
SUGAR,GRANULATED	7 oz	1 cup	
CINNAMON,GROUND	1/2 oz	2 tbsp	
BUTTER	1-1/4 lbs	2-1/2 cup	

Method

1. Place flour, sugars, and cinnamon in mixer bowl; blend thoroughly at low speed 2 minutes.
2. Add butter or margarine to dry ingredients; blend at low speed 1-1/2 to 2 minutes or until mixture resembles coarse cornmeal. DO NOT OVERMIX.
3. Sprinkle over sweet rolls and coffee cakes before baking.

Notes

1. If butter or margarine is too soft, a mass will form and mixture will not be crumbly.

BREADS AND SWEET DOUGHS   No.D 049 01

PECAN TOPPING

Yield 100   Portion   2-1/2 Quarts

Calories	Carbohydrates	Protein	Fat	Cholesterol	Sodium	Calcium
8428 cal	752 g	56 g	624 g	497 mg	2122 mg	816 mg

Ingredient
Ingredient	Weight	Measure	Issue
BUTTER	8 oz	1 cup	
SUGAR,BROWN,PACKED	1-1/3 lbs	1 qts 1/4 cup	
PECANS,CHOPPED	1-1/2 lbs		

Method

1 Combine softened butter or margarine, brown sugar, and chopped pecans.
2 Use as a topping for Pecan Rolls, Recipe No. D 036 03.

Notes

1 If butter or margarine is too soft, a mass will form and mixture will not be crumbly.

BREADS AND SWEET DOUGHS No.D 050 00
MAPLE SYRUP

Yield 100   Portion 1 Gallon

Calories	Carbohydrates	Protein	Fat	Cholesterol	Sodium	Calcium
7143 cal	1835 g	0 g	0 g	0 mg	1364 mg	1612 mg

Ingredient	Weight	Measure	Issue
SUGAR,BROWN,PACKED	4-1/8 lbs	3 qts 3/4 cup	
WATER	4-1/8 lbs	2 qts	
SALT	<1/16th oz	<1/16th tsp	
CORNSTARCH	1-1/3 oz	1/4 cup 1 tbsp	
FLAVORING,MAPLE	5/8 oz	1 tbsp	

Method

1 Combine brown sugar, water, salt, and cornstarch. Bring to a boil; reduce heat; simmer about 10 minutes or until thickened.
2 Remove from heat; add maple flavoring.

Notes

1 Hot syrup will be thin, but will thicken upon cooling.

BREADS AND SWEET DOUGHS No.D 051 00

FRYING BATTER

Yield 100 Portion 1 Gallon

Calories	Carbohydrates	Protein	Fat	Cholesterol	Sodium	Calcium
10896 cal	2068 g	350 g	114 g	2246 mg	67950 mg	3966 mg

Ingredient	Weight	Measure	Issue
FLOUR,WHEAT,GENERAL PURPOSE	5-1/2 lbs	1 gal 1 qts	
SUGAR,GRANULATED	3-1/2 oz	1/2 cup	
SALT	5-3/4 oz	1/2 cup 1 tbsp	
MILK,NONFAT,DRY	3 oz	1-1/4 cup	
BAKING POWDER	1-1/3 oz	2-2/3 tbsp	
EGGS,WHOLE,FROZEN	1-1/8 lbs	2-1/8 cup	
OIL,SALAD	1-1/4 oz	2-2/3 tbsp	
WATER	4-1/8 lbs	2 qts	

Method

1 Sift together flour, sugar, salt, milk, and baking powder into mixer bowl.
2 Combine eggs and salad oil or melted shortening; add to dry ingredients.
3 Slowly add water; beat at medium speed until smooth.

Notes

1 Batter may be used for fruits and vegetables such as apples, eggplant, and tomatoes. Moist foods should be dredged in flour before dipping into batter. When ready to fry, dip into batter; drain slightly. Fry in 350 F. to 375 F. deep fat until lightly browned.
2 Use batter the day prepared. DO NOT SAVE.

BREADS AND SWEET DOUGHS No.D 052 00

OATMEAL BREAD

Yield 100   Portion   2 Slices

Calories	Carbohydrates	Protein	Fat	Cholesterol	Sodium	Calcium
211 cal	37 g	7 g	4 g	0 mg	216 mg	21 mg

Ingredient	Weight	Measure	Issue
YEAST,ACTIVE,DRY	5-1/8 oz	3/4 cup	
WATER,WARM	1-1/3 lbs	2-1/2 cup	
WATER,COLD	3-7/8 lbs	1 qts 3-1/2 cup	
SUGAR,GRANULATED	8 oz	1-1/8 cup	
MILK,NONFAT,DRY	2-2/3 oz	1-1/8 cup	
SALT	1-7/8 oz	3 tbsp	
FLOUR,WHEAT,BREAD	8-1/2 lbs	1 gal 3 qts	
SHORTENING	7-1/4 oz	1 cup	
CEREAL,OATMEAL,ROLLED	2 lbs	1 qts 1-5/8 cup	
COOKING SPRAY,NONSTICK	2 oz	1/4 cup 1/3 tbsp	

Method

1. Sprinkle yeast over water. DO NOT USE TEMPERATURES ABOVE 110 F. Mix well. Let stand 5 minutes; stir. Set aside for use in Step 4.
2. Place water, sugar, milk, and salt in mixer bowl; blend thoroughly with a wire whip.
3. Add flour. Using dough hook, mix at low speed 1 minute or until all flour is incorporated into liquid.
4. Add yeast solution; mix at low speed 1 minute.
5. Add shortening; mix at low speed 1 minute. Continue mixing at medium speed 10 to 15 minutes or until dough is smooth and elastic. Dough temperature should be 78 F. to 82 F.
6. Add oats; mix at low speed 2 minutes. Mix at medium speed 1 minute.
7. FERMENT: Cover. Set in warm place (80 F.), 1 hour and 45 minutes or until double in bulk.
8. PUNCH: Fold sides into center and turn dough completely over. Let rest 15 minutes.
9. MAKE UP: Scale into 8 1-3/4 pound pieces; shape each piece into a smooth ball; let rest 10 minutes. Mold each piece into an oblong loaf; place each loaf seam-side down into a pan sprayed with non-stick cooking spray.
10. PROOF: At 90 F. to 100 F. about 45 minutes or until double in bulk.
11. Bake 45 to 50 minutes in 375 F. oven or in 325 F. convection oven for 30 minutes on high fan, open vent, or until done.
12. When cool, slice 25 slices, about 1/2-inch thick per loaf.

Notes

1. If using 9x4-1/2x2-3/4-inch bread pans, scale into 10 1-3/8 pound pieces; proof at 90 F. to 100 F. for 30 minutes or until double in bulk. Slice 20 slices, about 1/2-inch thick per loaf.

BREADS AND SWEET DOUGHS No.D 053 00

APPLESAUCE MUFFINS

Yield 100   Portion   1 Muffin

Calories	Carbohydrates	Protein	Fat	Cholesterol	Sodium	Calcium
181 cal	33 g	3 g	4 g	12 mg	213 mg	73 mg

Ingredient	Weight	Measure	Issue
FLOUR,WHEAT,GENERAL PURPOSE	5-3/4 lbs	1 gal 1-1/4 qts	
SUGAR,GRANULATED	2-1/4 lbs	1 qts 1 cup	
BAKING POWDER	3-7/8 oz	1/2 cup	
SALT	3/4 oz	1 tbsp	
CINNAMON,GROUND	1/2 oz	2 tbsp	
NUTMEG,GROUND	1/8 oz	1/3 tsp	
APPLESAUCE,CANNED,UNSWEETENED	5 lbs	2 qts 1-3/8 cup	
EGGS,WHOLE,FROZEN	9-5/8 oz	1-1/8 cup	
EGG WHITES,FROZEN,THAWED	9-5/8 oz	1-1/8 cup	
OIL,SALAD	10-1/4 oz	1-3/8 cup	
COOKING SPRAY,NONSTICK	2 oz	1/4 cup 1/3 tbsp	

Method

1. Sift together flour, sugar, baking powder, salt, cinnamon, and nutmeg into mixer bowl.
2. Add applesauce, eggs, egg whites, and salad oil or shortening; mix at low speed 15 seconds or until dry ingredients are moistened. DO NOT OVER MIX. Batter will be lumpy.
3. Lightly spray each muffin cup with non-stick cooking spray. Fill each muffin cup 2/3 full.
4. Bake 25 to 30 minutes in 400 F. oven or in 350 F. convection oven for 23 to 26 minutes until done, open vent, fan turned off first 10 minutes, then low fan.

BREADS AND SWEET DOUGHS No.D 053 01

APPLESAUCE RAISIN MUFFINS

Yield 100   Portion   1 Muffin

Calories	Carbohydrates	Protein	Fat	Cholesterol	Sodium	Calcium
207 cal	40 g	4 g	4 g	12 mg	214 mg	77 mg

Ingredient	Weight	Measure	Issue
FLOUR,WHEAT,GENERAL PURPOSE	5-3/4 lbs	1 gal 1-1/4 qts	
SUGAR,GRANULATED	2-1/4 lbs	1 qts 1 cup	
BAKING POWDER	3-7/8 oz	1/2 cup	
SALT	3/4 oz	1 tbsp	
CINNAMON,GROUND	1/2 oz	2 tbsp	
NUTMEG,GROUND	1/8 oz	1/3 tsp	
RAISINS	1-7/8 lbs	1 qts 2 cup	
APPLESAUCE,CANNED,UNSWEETENED	5 lbs	2 qts 1-3/8 cup	
EGGS,WHOLE,FROZEN	9-5/8 oz	1-1/8 cup	
EGG WHITES,FROZEN,THAWED	9-5/8 oz	1-1/8 cup	
OIL,SALAD	10-1/4 oz	1-3/8 cup	
COOKING SPRAY,NONSTICK	2 oz	1/4 cup 1/3 tbsp	

Method

1 Sift together flour, sugar, baking powder, salt, cinnamon, and nutmeg into mixer bowl.
2 Add applesauce, eggs, egg whites, and salad oil or shortening; mix at low speed 15 seconds until dry ingredients are moistened. Fold in raisins. DO NOT OVER MIX. Batter will be lumpy.
3 Lightly spray each muffin cup with non-stick cooking spray. Fill each muffin cup 2/3 full.
4 Bake 25 to 30 minutes in 400 F. oven or in 350 F. convection oven for 23 to 26 minutes until done, open vent, fan turned off first 10 minutes, then low fan.

BREADS AND SWEET DOUGHS No.D 053 02

APPLESAUCE ORANGE MUFFINS

Yield 100   Portion 1 Muffin

Calories	Carbohydrates	Protein	Fat	Cholesterol	Sodium	Calcium
191 cal	35 g	4 g	4 g	12 mg	213 mg	76 mg

Ingredient	Weight	Measure	Issue
FLOUR,WHEAT,GENERAL PURPOSE	5-3/4 lbs	1 gal 1-1/4 qts	
SUGAR,GRANULATED	2-1/4 lbs	1 qts 1 cup	
BAKING POWDER	3-7/8 oz	1/2 cup	
SALT	3/4 oz	1 tbsp	
CINNAMON,GROUND	1/2 oz	2 tbsp	
NUTMEG,GROUND	1/8 oz	1/3 tsp	
APPLESAUCE,CANNED,UNSWEETENED	3-1/4 lbs	1 qts 2 cup	
JUICE,ORANGE,FROZEN,CONCENTRATE,3/1,THAWED	1-7/8 lbs	3 cup	
EGGS,WHOLE,FROZEN	9-5/8 oz	1-1/8 cup	
EGG WHITES,FROZEN,THAWED	9-5/8 oz	1-1/8 cup	
OIL,SALAD	10-1/4 oz	1-3/8 cup	
COOKING SPRAY,NONSTICK	2 oz	1/4 cup 1/3 tbsp	

Method

1 Sift together flour, sugar, baking powder, salt, cinnamon, and nutmeg into mixer bowl.
2 Add applesauce, orange juice concentrate, eggs, egg whites, and salad oil or shortening; mix at low speed for 15 seconds or until dry ingredients are moistened. DO NOT OVERMIX. Batter will be lumpy.
3 Lightly spray each muffin cup with non-stick cooking spray. Fill each muffin cup 2/3 full.
4 Bake 25 to 30 minutes in 400 F. oven or in 350 F. convection oven for 23 to 26 minutes until done, open vent, fan turned off first 10 minutes, then low fan.

BREADS AND SWEET DOUGHS No.D 054 00
PINEAPPLE CARROT MUFFINS

Yield 100   Portion   1 Muffin

Calories	Carbohydrates	Protein	Fat	Cholesterol	Sodium	Calcium
145 cal	24 g	4 g	4 g	0 mg	205 mg	99 mg

Ingredient	Weight	Measure	Issue
FLOUR,WHEAT,GENERAL PURPOSE	3-1/8 lbs	2 qts 3-1/2 cup	
CEREAL,OAT BRAN	11-1/2 oz	1 qts 1-3/4 cup	
BAKING POWDER	3-1/8 oz	1/4 cup 2-2/3 tbsp	
BAKING SODA	1 oz	2 tbsp	
YOGURT,PLAIN,NONFAT	3-3/4 lbs	1 qts 3 cup	
SUGAR,BROWN,PACKED	1-1/4 lbs	1 qts	
OIL,SALAD	11-1/2 oz	1-1/2 cup	
EGG WHITES,FROZEN,THAWED	14-7/8 oz	1-3/4 cup	
PINEAPPLE,CANNED,CRUSHED,JUICE PACK,DRAINED	3-7/8 lbs	1 qts 3 cup	
CARROTS,FRESH,GRATED	1-1/2 lbs	1 qts 2-1/4 cup	1-7/8 lbs
COOKING SPRAY,NONSTICK	2 oz	1/4 cup 1/3 tbsp	

Method
1 Sift together flour, oat bran, baking powder, and baking soda. Set aside for use in Step 5.
2 Combine yogurt, brown sugar, and oil in mixer bowl. Beat at medium speed about 1 minute or until well blended.
3 Add egg whites; mix at low speed about 30 seconds.
4 Add pineapple and carrots; mix at low speed for 30 seconds.
5 Add flour mixture; mix at low speed about 15 seconds, scrape down sides and bottom of mixer bowl. Mix about 15 seconds or until ingredients are moistened. Do not overmix.
6 Lightly spray each muffin cup with non-stick cooking spray. Fill each muffin cup 2/3 full.
7 Bake 25 to 30 minutes at 400 F. or until lightly browned, or using a 350 F. convection oven, bake for 18 to 20 minutes or until lightly browned with open vent, low fan.

BREADS AND SWEET DOUGHS No.D 055 00

WHOLE WHEAT BREAD

Yield 100   **Portion** 2 Slices

Calories	Carbohydrates	Protein	Fat	Cholesterol	Sodium	Calcium
158 cal	29 g	5 g	3 g	0 mg	288 mg	25 mg

Ingredient	Weight	Measure	Issue
YEAST,ACTIVE,DRY	1-2/3 oz	1/4 cup 1/3 tbsp	
WATER,WARM	12-1/2 oz	1-1/2 cup	
WATER	4-1/8 lbs	2 qts	
MILK,NONFAT,DRY	4-1/2 oz	1-7/8 cup	
SUGAR,GRANULATED	10-5/8 oz	1-1/2 cup	
SALT	2-1/2 oz	1/4 cup 1/3 tbsp	
FLOUR,WHEAT,BREAD	4-1/4 lbs	3 qts 2 cup	
FLOUR,WHOLE WHEAT	3-1/2 lbs	3 qts 1 cup	
SHORTENING,SOFTENED	7-1/4 oz	1 cup	

Method

1. Sprinkle yeast over water. DO NOT USE TEMPERATURES ABOVE 110 F. Mix well. Let stand 5 minutes. Stir. Set aside for use in Step 4.
2. Place water, milk, sugar, and salt in mixer bowl. Using dough hook, mix at low speed about 1 minute until blended.
3. Combine flours thoroughly; add to liquid in mixer bowl. Using dough hook, mix at low speed 1 minute or until the dry ingredients are incorporated into liquid.
4. Add yeast solution; mix at low speed for one minute.
5. Add shortening; mix at low speed 1 minute. Continue mixing at medium speed for 10 to 15 minutes or until dough is smooth and elastic. Dough temperature should be 78 F. to 82 F.
6. FERMENT: Cover. Set in warm place (80 F.) 2 hours or until double in bulk.
7. PUNCH: Fold sides into center and turn dough completely over. Let rest 15 minutes.
8. PROOF: At 90 F. to 100 F. for about 1 hour or until double in size.
9. BAKE: 35 to 40 minutes at 375 F. or 30 to 35 minutes in a 325 F. convection oven until bread is done on high fan, open vent.
10. When cool, slice 25 slices, about 1/2-inch thick, per loaf.

BREADS AND SWEET DOUGHS No.D 056 00

WHOLE WHEAT BREAD (WHOLE WHEAT FLOUR SHRT TM FORM)

Yield 100   Portion 2 Slices

Calories	Carbohydrates	Protein	Fat	Cholesterol	Sodium	Calcium
158 cal	29 g	6 g	3 g	0 mg	218 mg	24 mg

Ingredient	Weight	Measure	Issue
YEAST,ACTIVE,DRY	3 oz	1/4 cup 3-1/3 tbsp	
WATER,WARM	1 lbs	2 cup	
SUGAR,GRANULATED	3/4 oz	1 tbsp	
WATER	4-1/8 lbs	2 qts	
MILK,NONFAT,DRY	3-5/8 oz	1-1/2 cup	
SUGAR,GRANULATED	5 oz	1/2 cup 3-1/3 tbsp	
FLOUR,WHEAT,BREAD	2-1/4 lbs	1 qts 3-1/2 cup	
FLOUR,WHOLE WHEAT	3-2/3 lbs	3 qts 2 cup	
SHORTENING,SOFTENED	6 oz	3/4 cup 1-1/3 tbsp	
FLOUR,WHOLE WHEAT	2-1/8 lbs	2 qts	
SALT	1-7/8 oz	3 tbsp	
COOKING SPRAY,NONSTICK	2 oz	1/4 cup 1/3 tbsp	

Method

1. Sprinkle yeast over water. DO NOT USE TEMPERATURES ABOVE 110 F. Mix well. Let stand 5 minutes. Add sugar. Stir until dissolved. Let stand 10 minutes; stir. Set aside for use in Step 3.
2. Place water in mixer bowl. Add milk, sugar, and yeast food. Using dough hook, mix at low speed until smooth.
3. Combine flours, add to bowl. Mix at low speed 2 minutes or until flour is incorporated; add shortening and yeast solution. Mix at low speed about 2 minutes until smooth.
4. Mix at medium speed 10 minutes.
5. Let rise in mixer bowl 20 minutes.
6. Sift together flour and salt; add to mixture in mixer bowl. Mix at low speed for 2 minutes or until flour is incorporated. Mix at medium speed 10 minutes or until dough is smooth and elastic.
7. FERMENT: Cover. Set in warm place (80 F.), for 25 to 30 minutes or until double in bulk.
8. MAKE UP: Scale into 8-28 ounce pieces. Roll scaled dough to pan size; place 1 loaf into each lightly greased pan.
9. PROOF: At 90 F. for 25 to 30 minutes or until double in bulk.
10. BAKE: 5 minutes at 450 F. Reduce temperature to 375 F. and bake 40 to 45 minutes or until done or in a convection oven 3 to 5 minutes on high fan, open vent. Reduce temperature to 325 F., bake 22 to 26 minutes or until done on high fan, open vent.
11. When cool, slice 25 slices, about 1/2-inch thick, per loaf.

BREADS AND SWEET DOUGHS No.D 057 00

APPLE COFFEE CAKE

Yield 100   Portion 1 Piece

Calories	Carbohydrates	Protein	Fat	Cholesterol	Sodium	Calcium
206 cal	39 g	4 g	5 g	0 mg	213 mg	65 mg

Ingredient	Weight	Measure	Issue
APPLES,CANNED,DRAINED,CHOPPED	8 lbs	1 gal	
JUICE,ORANGE	2-3/4 lbs	1 qts 1 cup	
CINNAMON,GROUND	3/4 oz	3 tbsp	
FLOUR,WHEAT,GENERAL PURPOSE	3-5/8 lbs	3 qts 1 cup	
SUGAR,GRANULATED	3 lbs	1 qts 2-3/4 cup	
FLOUR,WHOLE WHEAT	1-1/8 lbs	1 qts 1/4 cup	
MILK,NONFAT,DRY	1-3/4 oz	3/4 cup	
BAKING POWDER	2-3/4 oz	1/4 cup 2 tbsp	
SALT	5/8 oz	1 tbsp	
NUTMEG,GROUND	1/4 oz	1 tbsp	
MARGARINE,SOFTENED	1 lbs	2 cup	
WATER	2 lbs	3-3/4 cup	
EXTRACT,VANILLA	3/4 oz	1 tbsp	
EGG WHITES,FROZEN,THAWED	1-3/4 lbs	3-1/4 cup	
COOKING SPRAY,NONSTICK	2 oz	1/4 cup 1/3 tbsp	
SUGAR,BROWN,PACKED	3-7/8 oz	3/4 cup	

Method

1 Coarsely chop apples. Toss with orange juice and cinnamon. Cover.
2 Sift together flour, sugar, whole wheat flour, milk, baking powder, salt, and nutmeg into mixer bowl.
3 Add margarine, water, and vanilla to dry ingredients. Beat at low speed 1 minute until blended. Scrape down bowl; continue beating 2 minutes. Scrape down bowl.
4 Slowly add egg whites to mixture while beating at low speed 2 minutes. Scrape down bowl. Beat at medium speed 3 minutes.
5 Pour 2-1/2 quarts of batter into each lightly sprayed pan. Spread to evenly distribute batter.
6 Spread about 2-1/2 quarts apple mixture evenly over batter in each pan. Sprinkle 3 ounces or 1/3 cup of brown sugar over apples in pan.
7 Bake about 1 hour at 400 F. or until done or using a convection oven, bake at 325 F. for about 35 minutes or until done on low fan, open vent.
8 Prepare 1 recipe Vanilla Glaze, Recipe No. D 046 00. Drizzle 8 ounces or 1 cup of glaze over warm cake in each pan.
9 Cut 6 by 9.

BREADS AND SWEET DOUGHS No.D 058 00

OVEN BAKED FRENCH TOAST

Yield 100   Portion 2 Slices

Calories	Carbohydrates	Protein	Fat	Cholesterol	Sodium	Calcium
199 cal	31 g	9 g	4 g	1 mg	365 mg	99 mg

Ingredient	Weight	Measure	Issue
WATER	5-3/4 lbs	2 qts 3 cup	
EXTRACT,VANILLA	2-1/2 oz	1/4 cup 1-2/3 tbsp	
MILK,NONFAT,DRY	5-5/8 oz	2-3/8 cup	
SUGAR,GRANULATED	10-5/8 oz	1-1/2 cup	
CINNAMON,GROUND	3/8 oz	1 tbsp	
EGG SUBSTITUTE,PASTEURIZED	7-3/4 lbs	3 qts 2 cup	
BREAD,WHITE,SLICED	12 lbs	9 gal 2-7/8 qts	
COOKING SPRAY,NONSTICK	2 oz	1/4 cup 1/3 tbsp	

Method

1 Place water and vanilla in mixer bowl.
2 Combine milk, sugar, and cinnamon; blend well. Add to water; mix at low speed until dissolved or for about 1 minute.
3 Add egg substitute to ingredients in mixer bowl; mix at low speed until well blended, about 1 minute.
4 Stir egg mixture before using to redistribute cinnamon. Dip bread slices in egg mixture to coat both sides. Do not soak.
5 Lightly spray sheet pans with non-stick spray. Place dipped bread slices on pans 4 by 6.
6 Bake 20 to 25 minutes or until toast is golden brown in 450 F. oven or using a convection oven, bake at 425 F. for 12 to 14 minutes on high fan, open vent or until golden brown. Use batch method of preparation. Toast becomes tough when held more than 15 minutes.

BREADS AND SWEET DOUGHS No.D 059 00

WHOLE WHEAT ROLLS

Yield 100   Portion 2 Rolls

Calories	Carbohydrates	Protein	Fat	Cholesterol	Sodium	Calcium
261 cal	50 g	8 g	4 g	0 mg	383 mg	31 mg

Ingredient	Weight	Measure	Issue
YEAST,ACTIVE,DRY	4-1/2 oz	1/2 cup 2-2/3 tbsp	
WATER,WARM	1-7/8 lbs	3-1/2 cup	
WATER,COLD	6-3/4 lbs	3 qts 1 cup	
SUGAR,GRANULATED	1-1/2 lbs	3-3/8 cup	
SALT	3-3/8 oz	1/4 cup 1-2/3 tbsp	
FLOUR,WHOLE WHEAT	5-1/2 lbs	1 gal 1-1/4 qts	
FLOUR,WHEAT,BREAD	7-1/4 lbs	1 gal 2 qts	
MILK,NONFAT,DRY	4-1/2 oz	1-7/8 cup	
SHORTENING,SOFTENED	9 oz	1-1/4 cup	

Method

1. Sprinkle yeast over water. Do not use in temperatures above 110 F. Mix well. Let stand for 5 minutes. Stir.
2. Place water in mixer bowl; add sugar and salt; stir until dissolved. Add yeast solution.
3. Combine whole wheat flour, bread flour, and milk. Add to liquid solution. Using dough hook, mix at low speed 1 minute or until flour mixture is incorporated into liquid.
4. Add shortening; mix at medium speed 10 minutes or until dough is smooth and elastic. Dough temperature should be 78 F. to 82 F.
5. FERMENT: Cover. Set in warm place (80 F.) for 1-1/2 hours or until double in bulk.
6. PUNCH: Divide dough into 8 2-lb 14-oz pieces. Shape each piece into a smooth ball; let rest 10 to 20 minutes.
7. Roll each piece into a long rope, about 32 inches, of uniform diameter about 2 inches thick. Cut rope into 25 1-3/4 oz pieces about 1-1/4 inches long.
8. MAKE-UP: Shape into balls by rolling with a circular motion on work table.
9. PROOF: At 90 F. until double in bulk, about 1 hour.
10. BAKE: 15 to 20 minutes at 400 F., or in 350 F. convection oven for 10 to 15 minutes until golden brown, on high fan, open vent.

BREADS AND SWEET DOUGHS No.D 060 00

OATS AND FRUIT BREAKFAST SQUARES

Yield 100   Portion   1 Each

Calories	Carbohydrates	Protein	Fat	Cholesterol	Sodium	Calcium
254 cal	42 g	6 g	8 g	0 mg	116 mg	36 mg

Ingredient	Weight	Measure	Issue
FRUIT COCKTAIL,CANNED,JUICE PACK,INCL LIQUIDS	8-7/8 lbs	1 gal 1/4 qts	
FLOUR,WHEAT,GENERAL PURPOSE	2-1/4 lbs	2 qts	
CINNAMON,GROUND	1-7/8 oz	1/2 cup	
BAKING SODA	1/2 oz	1 tbsp	
MARGARINE,SOFTENED	1-1/2 lbs	3 cup	
SUGAR,BROWN,PACKED	1-5/8 lbs	1 qts 1 cup	
SUGAR,GRANULATED	1-1/8 lbs	2-5/8 cup	
RESERVED LIQUID	12-1/2 oz	1-1/2 cup	
EXTRACT,VANILLA	1-1/4 oz	2-2/3 tbsp	
EGG SUBSTITUTE,PASTEURIZED	1-1/8 lbs	2 cup	
CEREAL,OATMEAL,ROLLED	5-3/8 lbs	3 qts 3-5/8 cup	
COOKING SPRAY,NONSTICK	3/8 oz	3/8 tsp	

Method

1. Drain fruit; reserve liquid for use in Step 3 and fruit for use in Step 6.
2. Sift together flour, cinnamon, and baking soda; set aside for use in Step 5.
3. Place margarine, sugars, egg substitute, reserved liquid, and vanilla in a mixer bowl. Beat at high speed for 1 to 2 minutes or until well blended. Scrape down bowl.
4. Add oats; mix at low speed 1 minute until well blended. Scrape down bowl.
5. Add flour mixture; mix at low speed 1 to 2 minutes or until well blended. Scrape down bowl.
6. Add fruit; mix at low speed 30 seconds or until just mixed.
7. Lightly spray sheet pans. Place about 1-1/4 gallons in each sheet pan. Spread evenly.
8. Bake 35 minutes at 325 F. or until lightly browned and toothpick comes out clean on high fan, open vent.
9. Loosen from pans while still warm. Cut 6 by 9.

Notes

1. In Step 1, 4-1/4 quarts of canned, drained peaches or pears may be used for 100 portions.
2. In Step 4, a combination of 5-1/2 quarts or rolled oats and 1-1/2 quarts of oat bran cereal may be used instead of oats per 100 servings.

BREADS AND SWEET DOUGHS No.D 502 00

PUMPKIN PATCH MUFFINS

Yield 100   Portion   1 Muffin

Calories	Carbohydrates	Protein	Fat	Cholesterol	Sodium	Calcium
154 cal	25 g	3 g	5 g	0 mg	208 mg	64 mg

Ingredient	**Weight**	**Measure**	**Issue**
FLOUR,WHEAT,GENERAL PURPOSE	1-2/3 lbs	1 qts 2 cup	
FLOUR,WHOLE WHEAT	1-5/8 lbs	1 qts 2 cup	
SUGAR,GRANULATED	1-3/4 lbs	1 qts	
BAKING POWDER	2-5/8 oz	1/4 cup 1-2/3 tbsp	
SALT	7/8 oz	1 tbsp	
CINNAMON,GROUND	5/8 oz	2-2/3 tbsp	
NUTMEG,GROUND	1/2 oz	2 tbsp	
EGG SUBSTITUTE,PASTEURIZED	1-1/8 lbs	2 cup	
MILK,NONFAT,DRY	1-3/4 oz	3/4 cup	
WATER	2 lbs	3-3/4 cup	
PUMPKIN,CANNED,SOLID PACK	2-1/8 lbs	1 qts	
OIL, CANOLA	1 lbs	2 cup	
RAISINS	1-1/4 lbs	1 qts	
COOKING SPRAY,NONSTICK	2 oz	1/4 cup 1/3 tbsp	

Method

1 Combine all purpose flour, whole-wheat flour, sugar, baking powder, salt, cinnamon, and nutmeg, set aside.
2 Reconstitute milk. In a mixer bowl, combine milk, pumpkin, oil, and egg substitute, mix on low speed until blended.
3 Add flour mixture to mixer bowl; mix on low speed until dry ingredients are moistened. Fold in raisins. Do not over mix.
4 Lightly spray muffin tins with non-stick cooking spray. Fill muffin tins 2/3 full.
5 Bake at 400 F. for 15 to 20 minutes or until lightly browned.

BREADS AND SWEET DOUGHS No.D 503 00

DATE NUT BREAD

Yield 100   **Portion** 1 Slice

Calories	Carbohydrates	Protein	Fat	Cholesterol	Sodium	Calcium
189 cal	38 g	3 g	4 g	0 mg	189 mg	43 mg

Ingredient   Weight   Measure   Issue

Ingredient	Weight	Measure
WATER,ICE	4-1/8 lbs	2 qts
DATES,PIECES	4-3/4 lbs	3 qts 1/4 cup
MARGARINE	6 oz	3/4 cup
FLOUR,WHEAT,GENERAL PURPOSE	3-1/3 lbs	3 qts
FLOUR,WHOLE WHEAT	14-7/8 oz	3-1/2 cup
SUGAR,GRANULATED	1-1/2 lbs	3-3/8 cup
BAKING SODA	1-1/2 oz	3 tbsp
BAKING POWDER	1-2/3 oz	3-1/3 tbsp
ORANGE PEEL,FRESH,GRATED	1/2 oz	2-1/3 tbsp
EGG WHITES,FROZEN,THAWED	14-7/8 oz	1-3/4 cup
WALNUTS,SHELLED,HALVES AND PIECES	8-1/2 oz	2 cup
COOKING SPRAY,NONSTICK	2 oz	1/4 cup 1/3 tbsp

Method

1. In a large mixer bowl combine water, dates and margarine. Let cool 5 minutes or until the dates soften.
2. Mix together flour, whole-wheat flour, sugar, baking soda, baking powder, and orange peel.
3. Add flour mixture including egg whites to the date mixture and beat at low speed until dry ingredients have moistened. Fold in chopped walnuts.
4. Lightly spray loaf pans with non-stick cooking spray.
5. Scale 2-1/2 cups of batter into each loaf pan.
6. Bake at 350 F. for 40 to 45 minutes.

BREADS AND SWEET DOUGHS No.D 507 00

APPLESAUCE CINNAMON CRUMB TOP MUFFIN

Yield 100   **Portion** 1 Muffin

Calories	Carbohydrates	Protein	Fat	Cholesterol	Sodium	Calcium
204 cal	38 g	3 g	5 g	0 mg	217 mg	76 mg

Ingredient	Weight	Measure	Issue
FLOUR,WHEAT,GENERAL PURPOSE	5-3/4 lbs	1 gal 1-1/4 qts	
BAKING POWDER	3-7/8 oz	1/2 cup	
SALT	3/4 oz	1 tbsp	
CINNAMON,GROUND	1/2 oz	2 tbsp	
NUTMEG,GROUND	1/8 oz	1/3 tsp	
SUGAR,GRANULATED	2-1/4 lbs	1 qts 1 cup	
APPLESAUCE,CANNED,SWEETENED	5-1/4 lbs	2 qts 1-3/8 cup	
EGG SUBSTITUTE,PASTEURIZED	1-1/8 lbs	2 cup	
SHORTENING	10-7/8 oz	1-1/2 cup	
MARGARINE	3 oz	1/4 cup 2-1/3 tbsp	
FLOUR,WHEAT,GENERAL PURPOSE	1-2/3 oz	1/4 cup 2-1/3 tbsp	
SUGAR,BROWN,LIGHT	6-7/8 oz	1-3/8 cup	
COOKING SPRAY,NONSTICK	2 oz	1/4 cup 1/3 tbsp	

Method

1 Sift together flour, baking powder, salt, cinnamon, nutmeg, and sugar into mixer bowl.
2 Add applesauce, egg substitute, and salad oil or melted shortening; mix at low speed 15 seconds until dry ingredients are moistened.
3 Lightly spray each muffin cup with non-stick cooking spray. Fill each muffin cup 2/3 full. Mix softened margarine, flour, and brown sugar until crumbly. Sprinkle on top of each muffin.
4 Bake at 400 F. for 20 to 25 minutes or until lightly brown or using a convection oven, bake at 350 F. for 23 to 26 minutes open vent, turn off fan first 10 minutes, then low fan.

BREADS AND SWEET DOUGHS No.D 508 00

APPLESAUCE BLUEBERRY MUFFINS

Yield 100   Portion   1 Muffin

Calories	Carbohydrates	Protein	Fat	Cholesterol	Sodium	Calcium
195 cal	36 g	3 g	4 g	0 mg	209 mg	75 mg

Ingredient	Weight	Measure	Issue
FLOUR,WHEAT,GENERAL PURPOSE	5-3/4 lbs	1 gal 1-1/4 qts	
BAKING POWDER	3-7/8 oz	1/2 cup	
SALT	3/4 oz	1 tbsp	
CINNAMON,GROUND	1/2 oz	2 tbsp	
NUTMEG,GROUND	1/8 oz	1/3 tsp	
SUGAR,GRANULATED	2-1/4 lbs	1 qts 1 cup	
APPLESAUCE,CANNED,SWEETENED	5-1/4 lbs	2 qts 1-3/8 cup	
EGG SUBSTITUTE,PASTEURIZED	1-1/8 lbs	2 cup	
SHORTENING	10-7/8 oz	1-1/2 cup	
BLUEBERRIES,FROZEN,UNSWEETENED	2-3/8 lbs	1 qts 3 cup	
COOKING SPRAY,NONSTICK	2 oz	1/4 cup 1/3 tbsp	

Method

1 Sift together flour, baking powder, salt, cinnamon, nutmeg, and sugar into mixing bowl.
2 Add applesauce, egg substitute, and salad oil or melted shortening; mix at low speed 15 seconds until dry ingredients are moistened. Fold in blueberries. Do not overmix. Batter will be lumpy.
3 Lightly spray each muffin cup with non-stick cooking spray. Fill each muffin cup 2/3 full.
4 Bake 25 to 30 minutes at 400 F. oven or at 350 F. in a convection oven for 23 to 26 minutes or until done, open vent, fan turned off first 10 minutes, then low fan.

Notes

1 In Step 2, canned drained, rinsed blueberries 6-1/4 cups per 100 portions, may be substituted for frozen thawed blueberries.

BREADS AND SWEET DOUGHS No.D 509 00

CRAN-APPLE MUFFINS

Yield 100   **Portion** 1 Muffin

Calories	Carbohydrates	Protein	Fat	Cholesterol	Sodium	Calcium
201 cal	38 g	3 g	4 g	10 mg	209 mg	74 mg

Ingredient	Weight	Measure	Issue
FLOUR,WHEAT,GENERAL PURPOSE	5-3/4 lbs	1 gal 1-1/4 qts	
BAKING POWDER	3-7/8 oz	1/2 cup	
SALT	3/4 oz	1 tbsp	
CINNAMON,GROUND	1/2 oz	2 tbsp	
NUTMEG,GROUND	1/8 oz	1/3 tsp	
SUGAR,GRANULATED	2-1/4 lbs	1 qts 1 cup	
APPLESAUCE,CANNED,SWEETENED	5-1/4 lbs	2 qts 1-3/8 cup	
EGGS,WHOLE,FROZEN	8-5/8 oz	1 cup	
EGG WHITES,FROZEN,THAWED	8-1/2 oz	1 cup	
OIL, CANOLA	10-1/4 oz	1-3/8 cup	
CRANBERRY SAUCE,JELLIED	1-7/8 lbs	3 cup	
COOKING SPRAY,NONSTICK	2 oz	1/4 cup 1/3 tbsp	
ORANGE PEEL,FRESH,GRATED	1/8 oz	1/3 tsp	

Method

1. Sift together flour, baking powder, salt, cinnamon, nutmeg, and sugar into mixer bowl.
2. Add applesauce, eggs, egg whites, and salad oil; mix at low speed approximately 15 seconds until dry ingredients are moistened. Do not over mix. Batter will be lumpy.
3. Lightly spray each muffin cup with non-stick cooking spray. Fill each muffin cup 2/3 full. Make a well in the center of each muffin with the back of a spoon. Combine cranberry sauce and orange peel. Spoon 2 teaspoons of cranberry filling into each well.
4. Bake at 400 F. for 25 to 30 minutes or until done. Using a convection oven, bake at 350 F. for 23 to 26 minutes, open vent, fan off first 10 minutes, then low fan.

E. CEREALS AND PASTA PRODUCTS No. 0

INDEX

Card No.		Card No.	
E 001 00	Hot Oatmeal	E 010 00	Red Beans with Rice
E 001 02	Hot Farina	E 010 01	Hopping John (Black-Eye Peas with Rice)
E 002 00	Hominy Grits	E 011 00	Mexican Rice
E 002 01	Fried Hominy Grits	E 012 00	Noodles Jefferson
E 003 00	Buttered Hominy	E 013 00	Steamed Pasta
E 003 01	Fried Hominy	E 014 00	Spring Garden Rice
E 004 00	Boiled Pasta	E 015 00	Sicilian Brown Rice and Vegetables
E 004 01	Buttered Pasta	E 016 00	Islander's Rice
E 005 00	Steamed Rice	E 017 00	Mediterranean Brown Rice
E 005 01	Lyonnaise Rice	E 018 00	Spicy Brown Rice Pilaf
E 005 02	Tossed Green Rice	E 019 00	Brown Rice with Tomatoes
E 005 03	Long Grain and Wild Rice	E 020 00	Ginger Rice
E 005 04	Rice With Parmesan Cheese	E 021 00	Nutty Rice and Cheese
E 005 05	Steamed Brown Rice	E 022 00	Orzo with Lemon and Herbs
E 006 00	Steamed Rice (Steam Cooker Method)	E 023 00	Orzo, with Spinach, Tomato, and Onion
E 007 00	Pork Fried Rice	E 508 00	Southwestern Rice
E 007 02	Filipino Rice	E 510 00	Pasta Provencal
E 007 03	Shrimp Fried Rice	E 800 00	Oriental Rice
E 008 00	Rice Pilaf	E 801 00	Wild Rice
E 008 01	Orange Rice	E 803 00	Aztec Rice
E 009 00	Spanish Rice	E 804 00	Mexican Rice (Fiesta Mix)

E. CEREALS AND PASTA PRODUCTS No. 0 (1)

Card No.

E 805 00	Rice Pilaf, Using Mix
E 806 00	Georgia Rice
E 807 00	Dirty Rice

CEREALS AND PASTA PRODUCTS   No. E 001 00

HOT OATMEAL

Yield 100　　　　　　　　　　　　　　　　　　　　**Portion**   3/4 Cup

Calories	Carbohydrates	Protein	Fat	Cholesterol	Sodium	Calcium
106 cal	18 g	5 g	2 g	0 mg	216 mg	19 mg

Ingredient   Weight   Measure   Issue

CEREAL,OATMEAL,ROLLED   6 lbs   1 gal 3/8 qts
SALT   1-7/8 oz   3 tbsp
WATER,BOILING   41-3/4 lbs   5 gal

Method

1 Add cereal and salt to boiling water; stir to prevent lumping.
2 Return to a boil; reduce heat; simmer 1 to 3 minutes, stirring occasionally.
3 Turn off heat; let stand 10 minutes before serving.

CEREALS AND PASTA PRODUCTS   No. E 001 02

HOT FARINA

Yield 100   **Portion**   3/4 Cup

Calories	Carbohydrates	Protein	Fat	Cholesterol	Sodium	Calcium
104 cal	22 g	3 g	0 g	0 mg	216 mg	8 mg

Ingredient   Weight   Measure   Issue

SALT   1-7/8 oz   3 tbsp
WATER   33-1/2 lbs   4 gal
CEREAL, FARINA, DRY   6-1/4 lbs   1 gal
WATER, COLD   8-1/3 lbs   1 gal

Method

1. Add salt to hot water; bring to boil.
2. Mix cereal with cold water; pour into boiling salted water stirring constantly, until water returns to a boil. Reduce heat. Let simmer 2 to 5 minutes, stirring frequently. Turn off heat; let stand 5 minutes before serving.

CEREALS AND PASTA PRODUCTS   No. E 002 00

HOMINY GRITS

Yield 100   **Portion**   2/3 Cup

Calories	Carbohydrates	Protein	Fat	Cholesterol	Sodium	Calcium
84 cal	16 g	2 g	1 g	2 mg	107 mg	4 mg

Ingredient	Weight	Measure	Issue
WATER, BOILING	33-1/2 lbs	4 gal	
SALT	7/8 oz	1 tbsp	
BUTTER	4 oz	1/2 cup	
HOMINY GRITS, QUICK COOKING	4-1/2 lbs	3 qts 1-1/8 cup	

Method

1. Add salt and butter or margarine to boiling water.
2. Add grits gradually while stirring to prevent lumping. Bring to a boil; reduce heat; cover and cook for 5 minutes. Stir occasionally.

CEREALS AND PASTA PRODUCTS No.E 002 01

FRIED HOMINY GRITS

Yield 100   **Portion** 3 Slices

Calories	Carbohydrates	Protein	Fat	Cholesterol	Sodium	Calcium
164 cal	16 g	2 g	10 g	2 mg	107 mg	4 mg

Ingredient	Weight	Measure	Issue
WATER,BOILING	33-1/2 lbs	4 gal	
SALT	7/8 oz	1 tbsp	
BUTTER	4 oz	1/2 cup	
HOMINY GRITS,QUICK COOKING	4-1/2 lbs	3 qts 1-1/8 cup	

Method

1. Add salt and butter or margarine to boiling water.
2. Add grits gradually while stirring to prevent lumping. Bring to a boil; reduce heat; cover and cook 5 minutes. Stir occasionally.
3. Pour hot cooked grits into bread pans or in 3 steam table pans; cover and refrigerate several hours or overnight. Cut cold grits lengthwise into 3 equal strips; cut each into 1/2-inch thick slices. If slices are moist, dip in flour; fry on 400 F. preheated well-greased griddle until lightly browned, about 8 minutes per side.

CEREALS AND PASTA PRODUCTS No. E 003 00

BUTTERED HOMINY

Yield 100   **Portion** 1/3 Cup

Calories	Carbohydrates	Protein	Fat	Cholesterol	Sodium	Calcium
86 cal	12 g	1 g	4 g	7 mg	208 mg	10 mg

Ingredient	Weight	Measure	Issue
HOMINY,WHOLE,CANNED	18-7/8 lbs	3 gal 1 qts	
PEPPER,BLACK,GROUND	1/8 oz	1/3 tsp	
BUTTER	12 oz	1-1/2 cup	
PARSLEY,FRESH,BUNCH,CHOPPED	1 oz	1/2 cup	1-1/8 oz

Method

1. Drain hominy. Reserve 1 quart liquid.
2. Add pepper and reserved liquid to drained hominy in pan; heat slowly for 20 minutes.
3. Add butter or margarine.
4. Garnish with parsley or 2 tbsp paprika.

CEREALS AND PASTA PRODUCTS No. E 003 01

FRIED HOMINY

Yield 100   Portion 1/3 Cup

Calories	Carbohydrates	Protein	Fat	Cholesterol	Sodium	Calcium
134 cal	12 g	1 g	9 g	0 mg	180 mg	9 mg

Ingredient	Weight	Measure	Issue
HOMINY,WHOLE,CANNED	18-7/8 lbs	3 gal 1 qts	
SHORTENING,VEGETABLE,MELTED	1-3/4 lbs	1 qts	
PEPPER,BLACK,GROUND	1/8 oz	1/8 tsp	

Method

1 Drain hominy, discard liquid.
2 Fry hominy in melted shortening or salad oil until lightly browned. Season with black pepper.

CEREALS AND PASTA PRODUCTS   No. E 004 00

BOILED PASTA

Yield 100　　　　　　　　　　　　　　　　　　　　**Portion** 1 Cup

Calories	Carbohydrates	Protein	Fat	Cholesterol	Sodium	Calcium
207 cal	41 g	7 g	1 g	0 mg	292 mg	16 mg

Ingredient	Weight	Measure	Issue
WATER	66-7/8 lbs	8 gal	
SALT	2-1/2 oz	1/4 cup 1/3 tbsp	
OIL, SALAD	1-7/8 oz	1/4 cup 1/3 tbsp	
SPAGHETTI NOODLES, DRY	12 lbs	3 gal 1 qts	

Method

1. Add salt and salad oil to water; heat to a rolling boil.
2. Slowly add pasta while stirring constantly until water boils again. Cook according to times in Note 1; stir occasionally. DO NOT OVERCOOK.
3. Drain. Rinse with cold water; drain thoroughly.

Notes

1. Macaroni or egg noodles should cook for 8 to 10 minutes; spaghetti for 10 to 12 minutes; vermicelli for 7 to 10 minutes.
2. When held on steam table, mix 1 tablespoon salad oil with pasta in each steam table pan to prevent product from sticking together.
3. To reheat pasta before serving, place desired quantity in a wire basket; lower into boiling water 2 to 3 minutes. Drain well. Place in greased steam table pans.

CEREALS AND PASTA PRODUCTS No.E 004 01

BUTTERED PASTA

Yield 100   Portion 1 Cup

Calories	Carbohydrates	Protein	Fat	Cholesterol	Sodium	Calcium
239 cal	41 g	7 g	5 g	10 mg	329 mg	17 mg

Ingredient

Ingredient	Weight	Measure	Issue
WATER	66-7/8 lbs	8 gal	
SALT	2-1/2 oz	1/4 cup 1/3 tbsp	
OIL,SALAD	1-7/8 oz	1/4 cup 1/3 tbsp	
SPAGHETTI NOODLES,DRY	12 lbs	3 gal 1 qts	
BUTTER,MELTED	1 lbs	2 cup	

Method

1. Add salt and salad oil to water; heat to a rolling boil.
2. Slowly add pasta while stirring constantly until water boils again. Cook according to times in Note 1; stir occasionally. DO NOT OVERCOOK.
3. Drain noodles and add melted butter to pasta immediately.

Notes

1. Macaroni or egg noodles should cook for 8 to 10 minutes; spaghetti for 10 to 12 minutes; vermicelli for 7 to 10 minutes.
2. To reheat pasta before serving, place desired quantity in a wire basket; lower into boiling water 2 to 3 minutes. Drain well. Place in greased steam table pans.

CEREALS AND PASTA PRODUCTS No. E 005 00

STEAMED RICE

Yield 100   **Portion** 3/4 Cup

Calories	Carbohydrates	Protein	Fat	Cholesterol	Sodium	Calcium
148 cal	32 g	3 g	1 g	0 mg	214 mg	26 mg

Ingredient	Weight	Measure	Issue
RICE,LONG GRAIN	8-1/2 lbs	1 gal 1-1/4 qts	
WATER,COLD	23 lbs	2 gal 3 qts	
SALT	1-7/8 oz	3 tbsp	
OIL,SALAD	1-1/2 oz	3 tbsp	

Method

1. Combine rice, water, salt, and salad oil; bring to a boil. Stir occasionally.
2. Cover tightly; simmer 20 to 25 minutes. DO NOT STIR.
3. Remove from heat; transfer to shallow serving pans.

Notes

1. In Step 2, rice may be baked in a 350 F. convection oven, 35 to 40 minutes on high fan, closed vent.

CEREALS AND PASTA PRODUCTS No.E 005 01

LYONNAISE RICE

Yield 100   Portion   3/4 Cup

Calories	Carbohydrates	Protein	Fat	Cholesterol	Sodium	Calcium
164 cal	33 g	3 g	2 g	0 mg	215 mg	29 mg

Ingredient	Weight	Measure	Issue
RICE,LONG GRAIN	8-1/2 lbs	1 gal 1-1/4 qts	
WATER,COLD	23 lbs	2 gal 3 qts	
SALT	1-7/8 oz	3 tbsp	
OIL,SALAD	1-1/2 oz	3 tbsp	
ONIONS,FRESH,CHOPPED	3-1/8 lbs	2 qts 1 cup	3-1/2 lbs
OIL,SALAD	3-7/8 oz	1/2 cup	
PIMIENTO,CANNED,DRAINED,CHOPPED	13-1/2 oz	2 cup	

Method

1 Combine rice, water, salt, and salad oil; bring to a boil. Stir occasionally.
2 Cover tightly; simmer 20 to 25 minutes. DO NOT STIR. Remove from heat.
3 Saute onions in oil until tender.
4 Add sauteed onions and pimientos to cooked rice. Toss well. CCP: Hold for service at 140 F. or higher.

CEREALS AND PASTA PRODUCTS No. E 005 02

TOSSED GREEN RICE

Yield 100   **Portion**   3/4 Cup

Calories	Carbohydrates	Protein	Fat	Cholesterol	Sodium	Calcium
163 cal	33 g	3 g	2 g	0 mg	217 mg	34 mg

Ingredient | Weight | Measure | Issue

Ingredient	Weight	Measure	Issue
RICE,LONG GRAIN	8-1/2 lbs	1 gal 1-1/4 qts	
WATER,COLD	23 lbs	2 gal 3 qts	
SALT	1-7/8 oz	3 tbsp	
OIL,SALAD	1-1/2 oz	3 tbsp	
ONIONS,GREEN,FRESH,SLICED	1-1/3 lbs	1 qts 2 cup	1-1/2 lbs
PEPPERS,GREEN,FRESH,CHOPPED	2 lbs	1 qts 2 cup	2-3/8 lbs
OIL,SALAD	3-7/8 oz	1/2 cup	
PARSLEY,FRESH,BUNCH,CHOPPED	8 oz	3-3/4 cup	8-3/8 oz
PEPPER,BLACK,GROUND	1/8 oz	1/3 tsp	

Method

1 Combine rice, water, salt, and salad oil; bring to a boil. Stir occasionally.
2 Cover tightly; simmer 20 to 25 minutes. DO NOT STIR. Remove from heat.
3 Saute green onions with tops and sweet peppers in oil until tender.
4 Add to cooked rice. Add parsley and black pepper. Toss well. CCP: Hold for service at 140 F. or higher.

CEREALS AND PASTA PRODUCTS No. E 005 03

LONG GRAIN AND WILD RICE

Yield 100   **Portion** 3/4 Cup

Calories	Carbohydrates	Protein	Fat	Cholesterol	Sodium	Calcium
168 cal	34 g	7 g	1 g	0 mg	7 mg	12 mg

Ingredient	**Weight**	**Measure**	**Issue**
RICE, LONG GRAIN & WILD	10-1/8 lbs	1 gal 3-1/8 qts	
WATER, COLD	25-1/8 lbs	3 gal	
OIL, SALAD	1-1/2 oz	3 tbsp	

Method

1 Combine rice mix, water and salad oil; bring to a boil. Stir occasionally.
2 Cover tightly; simmer 20 to 25 minutes. DO NOT STIR.
3 Remove from heat; transfer to shallow serving pans. CCP: Hold for service at 140 F. or higher.

CEREALS AND PASTA PRODUCTS No. E 005 04

RICE WITH PARMESAN CHEESE

Yield 100   Portion   3/4 Cup

Calories	Carbohydrates	Protein	Fat	Cholesterol	Sodium	Calcium
187 cal	32 g	5 g	4 g	4 mg	329 mg	95 mg

Ingredient	Weight	Measure	Issue
RICE, LONG GRAIN	8-1/2 lbs	1 gal 1-1/4 qts	
WATER, COLD	23 lbs	2 gal 3 qts	
SALT	1-7/8 oz	3 tbsp	
OIL, SALAD	1-1/2 oz	3 tbsp	
MARGARINE, MELTED	8 oz	1 cup	
CHEESE, PARMESAN, GRATED	1-1/8 lbs	1 qts 1 cup	

Method

1. Combine rice, water, salt, and salad oil; bring to a boil. Stir occasionally.
2. Cover tightly; simmer 20 to 25 minutes. DO NOT STIR.
3. Remove from heat; transfer to shallow serving pans. Add melted butter to rice. Mix well to coat rice. Add grated Parmesan cheese. Toss well. CCP: Hold for service at 140 F. or higher.

CEREALS AND PASTA PRODUCTS No. E 005 05

STEAMED BROWN RICE

Yield 100   **Portion**   3/4 Cup

Calories	Carbohydrates	Protein	Fat	Cholesterol	Sodium	Calcium
168 cal	34 g	4 g	2 g	0 mg	216 mg	13 mg

Ingredient   Weight   Measure   Issue

RICE,BROWN,LONG GRAIN,RAW PARBOILED   9-3/4 lbs   1 gal 2 qts
WATER,COLD   25-1/8 lbs   3 gal
SALT   1-7/8 oz   3 tbsp
OIL,SALAD   1-1/2 oz   3 tbsp

Method

1. Combine rice, water, salt, and salad oil; bring to a boil. Stir occasionally.
2. Cover tightly; simmer for 25 minutes or until most of the water is absorbed.
3. Remove from heat; transfer to shallow serving pans. CCP: Hold for service at 140 F. or higher.

CEREALS AND PASTA PRODUCTS No. E 006 00

STEAMED RICE (STEAM COOKER METHOD)

Yield 100   **Portion**   3/4 Cup

Calories	Carbohydrates	Protein	Fat	Cholesterol	Sodium	Calcium
168 cal	34 g	4 g	2 g	0 mg	216 mg	13 mg

Ingredient

Ingredient	Weight	Measure	Issue
RICE, BROWN, LONG GRAIN, RAW PARBOILED	9-3/4 lbs	1 gal 2 qts	
WATER	25-1/8 lbs	3 gal	
SALT	1-7/8 oz	3 tbsp	
OIL, SALAD	1-1/2 oz	3 tbsp	

Method

1. Place 4-3/4 lbs rice in each pan.
2. Add 4-1/2 qts water to each pan.
3. Add 1-1/2 tbsp salt and 1-1/2 tbsp salad oil to each pan. Stir well to ensure rice is moistened.
4. Place pans in preheated steam cooker. Steam 22-27 minutes at 5 lbs PSI or 18 to 24 minutes at 15 lb PSI.

CEREALS AND PASTA PRODUCTS No.E 007 00

PORK FRIED RICE

Yield 100   **Portion** 3/4 Cup

Calories	Carbohydrates	Protein	Fat	Cholesterol	Sodium	Calcium
211 cal	29 g	8 g	6 g	55 mg	462 mg	38 mg

Ingredient	Weight	Measure	Issue
RICE,LONG GRAIN	7-1/3 lbs	1 gal 1/2 qts	
WATER,BOILING	18-3/4 lbs	2 gal 1 qts	
SALT	1-2/3 oz	2-2/3 tbsp	
OIL,SALAD	1 oz	2 tbsp	
ONIONS,FRESH,CHOPPED	2-1/2 lbs	1 qts 3 cup	2-3/4 lbs
PEPPERS,GREEN,FRESH,CHOPPED	1-1/2 lbs	1 qts 1/2 cup	1-3/4 lbs
CELERY,FRESH,CHOPPED	1-1/4 lbs	1 qts 1/2 cup	1-5/8 lbs
OIL,SALAD	5-1/8 oz	1/2 cup 2-2/3 tbsp	
EGGS,WHOLE,FROZEN	2 lbs	3-3/4 cup	
PORK,COOKED,DICED	4 lbs		
PIMIENTO,CANNED,DRAINED,CHOPPED	13-1/2 oz	2 cup	
SOY SAUCE	1 lbs	1-1/2 cup	

Method

1. Place equal amounts of rice, water, salt, and salad oil in well greased pans. Stir to combine.
2. Using a convection oven, bake at 325 F. for 30 minutes on high fan, closed vent; remove from oven. Uncover. Set aside for use in Step 4.
3. Combine onions, peppers and celery; saute in shortening or salad oil about 10 minutes or until tender.
4. Add an equal quantity of sauteed vegetables to cooked rice in each pan. Mix lightly but thoroughly.
5. Pour beaten eggs on lightly greased griddle. Cook until well done. DO NOT turn. Cut into strips; add an equal amount to rice mixture in each pan.
6. Add equal amounts of pork and pimientos to rice in each pan. Mix lightly but thoroughly.
7. Using a convection oven, bake at 350 F. for 30 minutes on high fan, closed vent. CCP: Internal temperature must reach 145 F. or higher for 15 seconds.
8. Remove from oven; blend in 1/2 cup soy sauce per pan. CCP: Hold for service at 140 F. or higher.

Notes

1. In Step 6, 4 pounds diced ham may be used per 100 servings.
2. In Step 2, rice may be prepared in small batches on 350 F. griddle or tilt frying pan. Turn occasionally until brown, 10 to 15 minutes.

CEREALS AND PASTA PRODUCTS No. E 007 02

FILIPINO RICE

Yield 100   **Portion** 3/4 Cup

Calories	Carbohydrates	Protein	Fat	Cholesterol	Sodium	Calcium
250 cal	31 g	9 g	10 g	59 mg	458 mg	25 mg

Ingredient	Weight	Measure	Issue
ONIONS,FRESH,CHOPPED	2-1/2 lbs	1 qts 3 cup	2-3/4 lbs
OIL,SALAD		2 cup	
RICE,BROWN,LONG GRAIN,DRY	8-1/2 lbs	1 gal 1-1/4 qts	
WATER	23 lbs	2 gal 3 qts	
GARLIC POWDER	<1/16th oz	<1/16th tsp	
SALT	1-2/3 oz	2-2/3 tbsp	
EGGS,WHOLE,FROZEN	2 lbs	3-3/4 cup	
PORK,COOKED,DICED	4 lbs		
SOY SAUCE	1 lbs	1-1/2 cup	

Method

1. Saute onions in a steam jacketed kettle in salad oil until light yellow.
2. Add rice; stir until well coated.
3. Add water, garlic powder, and salt to rice mixture.
4. Bring to a boil; cover; simmer 20 to 25 minutes.
5. Pour beaten eggs on lightly greased griddle. Cook until done. DO NOT turn. Cut into strips; add an equal amount to rice mixture in each pan.
6. Add an equal amount of pork to rice in each pan. Mix lightly but thoroughly.
7. Bake 45 minutes in 350 F. CCP: Internal temperature must reach 145 F. or higher for 15 seconds.
8. Remove from oven; blend in 1/2 cup soy sauce per pan. CCP: Hold for service at 140 F. or higher.

CEREALS AND PASTA PRODUCTS No.E 007 03

SHRIMP FRIED RICE

Yield 100   **Portion**   3/4 Cup

Calories	Carbohydrates	Protein	Fat	Cholesterol	Sodium	Calcium
234 cal	29 g	12 g	7 g	90 mg	502 mg	46 mg

Ingredient	Weight	Measure	Issue
RICE,LONG GRAIN	7-1/3 lbs	1 gal 1/2 qts	
WATER,BOILING	18-3/4 lbs	2 gal 1 qts	
SALT	1-2/3 oz	2-2/3 tbsp	
OIL,SALAD	1 oz	2 tbsp	
ONIONS,FRESH,CHOPPED	2-1/2 lbs	1 qts 3 cup	2-3/4 lbs
PEPPERS,GREEN,FRESH,CHOPPED	1-1/2 lbs	1 qts 1/2 cup	1-7/8 lbs
CELERY,FRESH,CHOPPED	1-1/4 lbs	1 qts 3/4 cup	1-3/4 lbs
OIL,SALAD	5-1/8 oz	1/2 cup 2-2/3 tbsp	
EGGS,WHOLE,FROZEN	2 lbs	3-3/4 cup	
COOKING SPRAY,NONSTICK	2 oz	1/4 cup 1/3 tbsp	
SHRIMP,COOKED,CHOPPED	4 lbs		
PORK,COOKED,DICED	4 lbs		
PIMIENTO,CANNED,DRAINED,CHOPPED	13-1/2 oz	2 cup	
SOY SAUCE	1 lbs	1-1/2 cup	

Method

1. Place equal amounts of rice, water, salt, and salad oil in well greased pans. Stir to combine.
2. Using a convection oven, bake at 325 F. for 30 minutes on high fan, closed vent. Remove from oven. Uncover. Set aside for use in Step 4.
3. Combine onions, peppers and celery; saute in shortening or salad oil about 10 minutes or until tender.
4. Add an equal quantity of sauteed vegetables to cooked rice in each pan. Mix lightly but thoroughly.
5. Pour beaten eggs on lightly greased griddle. Cook until well done. DO NOT TURN. Cut into strips; add an equal amount to rice mixture in each pan.
6. Add equal amounts of pork, cooked chopped shrimp and pimientos to rice in each pan. Mix lightly but thoroughly.
7. Using a convection oven, bake 45 minutes at 350 F. for 30 minutes on high fan, closed vent. CCP: Internal temperature must reach 145 F. or higher for 15 seconds.
8. Remove from oven; blend in 1/2 cup soy sauce per pan. CCP: Hold for service at 140 F. or higher.

Notes

1. In Step 6, 4 pounds diced ham may be used per 100 servings.

CEREALS AND PASTA PRODUCTS No.E 008 00

RICE PILAF

Yield 100   **Portion** 3/4 Cup

Calories	Carbohydrates	Protein	Fat	Cholesterol	Sodium	Calcium
201 cal	37 g	4 g	4 g	4 mg	927 mg	41 mg

Ingredient	Weight	Measure	Issue
BUTTER	6 oz	3/4 cup	
OIL, SALAD	5-3/4 oz	3/4 cup	
ONIONS, FRESH, CHOPPED	6-2/3 lbs	1 gal 3/4 qts	7-1/2 lbs
RICE, LONG GRAIN	9 lbs	1 gal 1-1/2 qts	
SALT	1 oz	1 tbsp	
GARLIC POWDER	3/8 oz	1 tbsp	
PEPPER, BLACK, GROUND	1/8 oz	1/8 tsp	
CHICKEN BROTH		3 gal	

Method

1. Melt butter or margarine. Add salad oil or melted shortening and onions. Stir well. Saute until onions are tender, about 5 minutes.
2. Add rice to onion mixture. Cook until rice is lightly browned, about 10 minutes, stirring constantly.
3. Place about 2 quarts onion and rice mixture into each pan.
4. Prepare broth according to recipe directions. Add salt, garlic powder and pepper; stir well. Pour 3 quarts over rice mixture in each pan; cover.
5. Using a convection oven, bake at 350 F. for 40 to 45 minutes or until tender on high fan, closed vent or until rice is tender. Stir lightly. CCP: Internal temperature must reach 145 F. or higher for 15 seconds. CCP: Hold for service at 140 F. or higher.

CEREALS AND PASTA PRODUCTS No.E 008 01

ORANGE RICE

Yield 100   **Portion** 3/4 Cup

Calories	Carbohydrates	Protein	Fat	Cholesterol	Sodium	Calcium
221 cal	42 g	4 g	4 g	4 mg	812 mg	45 mg

Ingredient	**Weight**	**Measure**	**Issue**
BUTTER	6 oz	3/4 cup	
OIL,SALAD	5-3/4 oz	3/4 cup	
ONIONS,FRESH,CHOPPED	6-2/3 lbs	1 gal 3/4 qts	7-1/2 lbs
RICE,LONG GRAIN	9 lbs	1 gal 1-1/2 qts	
JUICE,ORANGE	11 lbs	1 gal 1 qts	
CHICKEN BROTH		3 gal	

Method

1. Melt butter or margarine. Add salad oil or melted shortening and onions. Stir well. Saute until onions are tender, about 5 minutes.
2. Add rice to onion mixture. Cook until rice is lightly browned, about 10 minutes, stirring constantly.
3. Place 2 quarts of onion and rice mixture into each pan.
4. Prepare broth according to recipe directions. Add orange juice to boiling broth; stir well. Pour 3-1/4 quarts over rice mixture in each pan; cover.
5. Using a convection oven, bake at 350 F. for 40 to 45 minutes or until tender on high fan, closed vent or until rice is tender. Stir lightly. CCP: Internal temperature must reach 145 F. or higher for 15 seconds. CCP: Hold for service at 140 F. or higher.
6. May be garnished with thinly sliced oranges just before serving.

CEREALS AND PASTA PRODUCTS No.E 009 00

SPANISH RICE

Yield 100   Portion   3/4 Cup

Calories	Carbohydrates	Protein	Fat	Cholesterol	Sodium	Calcium
153 cal	31 g	4 g	2 g	2 mg	409 mg	55 mg

Ingredient	Weight	Measure	Issue
RICE,LONG GRAIN	5-3/4 lbs	3 qts 2 cup	
WATER,COLD	15-1/8 lbs	1 gal 3-1/4 qts	
OIL,SALAD	1 oz	2 tbsp	
SALT	1-1/4 oz	2 tbsp	
BACON,RAW	1-1/2 lbs		
TOMATOES,CANNED,DICED,DRAINED	19-7/8 lbs	2 gal 1 qts	
ONIONS,FRESH,CHOPPED	4-1/4 lbs	3 qts	4-2/3 lbs
PEPPERS,GREEN,FRESH,CHOPPED	2 lbs	1 qts 2 cup	2-3/8 lbs
SUGAR,GRANULATED	3-1/2 oz	1/2 cup	
SALT	1 oz	1 tbsp	
THYME,GROUND	1/3 oz	2 tbsp	
PEPPER,BLACK,GROUND	1/4 oz	1 tbsp	
GARLIC POWDER	1/4 oz	3/8 tsp	
BAY LEAF,FRESH	1/8 oz	4 each	

Method

1 Cook rice according to directions on Recipe No. E 005 00. Set aside for use in Step 4.
2 Saute bacon until crisp in steam-jacketed kettle or stock pot. Drain; discard drippings.
3 Add tomatoes, onions, peppers, sugar, salt, thyme, black pepper, garlic, and bay leaves. Stir to combine; bring to boil. Cover; reduce heat; simmer 15 minutes.
4 Add rice; stir to combine; using a convection oven, bake at 325 F. 30 minutes on high fan, closed vent. CCP: Internal temperature must reach 145 F. or higher for 15 seconds. Remove bay leaves before serving. CCP: Hold for service at 140 F. or higher.

CEREALS AND PASTA PRODUCTS No.E 010 00

RED BEANS WITH RICE

Yield 100   **Portion** 1 Cup

Calories	Carbohydrates	Protein	Fat	Cholesterol	Sodium	Calcium
225 cal	41 g	10 g	3 g	3 mg	630 mg	53 mg

Ingredient	Weight	Measure	Issue
RICE,LONG GRAIN	5-3/4 lbs	3 qts 2 cup	
WATER,COLD	15-1/8 lbs	1 gal 3-1/4 qts	
OIL,SALAD	1 oz	2 tbsp	
SALT	1-1/4 oz	2 tbsp	
BACON,SLICED,RAW	3 lbs		
ONIONS,FRESH,CHOPPED	2-1/8 lbs	1 qts 2 cup	2-1/3 lbs
BEANS,KIDNEY,DARK RED,CANNED,INCL LIQUIDS	27-1/8 lbs	3 gal	
PEPPER,BLACK,GROUND	1/4 oz	1 tbsp	
PEPPER,RED,GROUND	<1/16th oz	1/8 tsp	
GARLIC POWDER	1-1/8 oz	1/4 cup	

Method

1. Cook rice according to directions on Recipe No. E 005 00. Set aside for use in Step 6.
2. Cook bacon until crisp; drain. Set aside 2 ounces bacon fat per 100 servings for use in Step 3. Set aside bacon for use in Step 4.
3. Saute onions in bacon fat about 1 to 2 minutes or until lightly browned. Drain thoroughly.
4. Combine sauteed bacon and onions with undrained kidney beans, peppers and garlic powder.
5. Using a convection oven, bake at 325 F. for 30 minutes on high fan, closed vent. CCP: Internal temperature must reach 145 F. or higher for 15 seconds.
6. Serve 1/2 cup of beans over 1/2 cup of rice. CCP: Hold for service at 140 F. or higher.

CEREALS AND PASTA PRODUCTS No. E 010 01

HOPPING JOHN (BLACK-EYE PEAS WITH RICE)

Yield 100   Portion   2/3 Cup

Calories	Carbohydrates	Protein	Fat	Cholesterol	Sodium	Calcium
177 cal	30 g	8 g	3 g	3 mg	430 mg	32 mg

Ingredient	Weight	Measure	Issue
BACON,SLICED,RAW	3 lbs		
ONIONS,FRESH,CHOPPED	2-1/8 lbs	1 qts 2 cup	2-1/3 lbs
PEAS,BLACKEYE,CANNED,INCL LIQUIDS	27 lbs	3 gal 3/4 qts	
RICE,BROWN,LONG GRAIN,DRY	3-1/4 lbs	2 qts	
WATER	8-7/8 lbs	1 gal 1/4 qts	
PEPPER,BLACK,GROUND	1/2 oz	2 tbsp	
PEPPER,RED,GROUND	<1/16th oz	1/8 tsp	
GARLIC POWDER	2 oz	1/4 cup 3 tbsp	

Method

1. Cook bacon until crisp; drain. Set aside 2 ounces bacon fat per 100 servings, for use in Step 2; bacon for use in Step 3.
2. Saute onions in bacon fat about 1 to 2 minutes or until lightly browned. Drain thoroughly.
3. Combine undrained black-eyed peas, rice, water, sauteed onions, cooked bacon, black pepper, red pepper, and garlic. Mix well. Bring to a boil; cover tightly; reduce heat; simmer 25 minutes or until rice is tender. CCP: Internal temperature must reach 145 F. or higher for 15 seconds. CCP: Hold for service at 140 F. or higher.

CEREALS AND PASTA PRODUCTS No.E 011 00

MEXICAN RICE

Yield 100   **Portion** 3/4 Cup

Calories	Carbohydrates	Protein	Fat	Cholesterol	Sodium	Calcium
193 cal	34 g	3 g	5 g	0 mg	244 mg	37 mg

Ingredient

Ingredient	Weight	Measure	Issue
RICE,LONG GRAIN	8-1/2 lbs	1 gal 1-1/4 qts	
OIL,SALAD	1 lbs	2 cup	
ONIONS,FRESH,CHOPPED	1 lbs	3 cup	1-1/8 lbs
TOMATOES,CANNED,DICED,DRAINED	5 lbs	2 qts 1 cup	
SALT	1-7/8 oz	3 tbsp	
PEPPER,BLACK,GROUND	3/8 oz	1 tbsp	
CUMIN,GROUND	7/8 oz	1/4 cup 1/3 tbsp	
WATER	20-7/8 lbs	2 gal 2 qts	

Method

1. Place 10-1/2 cups rice, 1 cup salad oil and 1-1/2 cups onions in each pan. Stir well to coat rice.
2. Place in 400 F. oven; cook until lightly brown, about 25 minutes.
3. Combine tomatoes, salt, pepper, cumin and water.
4. Pour about 1-1/2 gallons tomato mixture over rice in each pan; stir well. Cover; return to oven; bake about 1 hour in 400 F. oven or until rice is tender.
5. Stir lightly. CCP: Internal temperature must reach 145 F. or higher for 15 seconds. CCP: Hold for service at 140 F. or higher.

Notes

1. Rice may be prepared on top of range. Follow Step 1. In Step 2, heat at medium heat until rice is lightly browned; stir occasionally. Follow Step 3. In Step 4, bring rice mixture to a boil; cover; reduce heat; cook until rice is light and fluffy. Follow Step 5.
2. Rice may be prepared in steam-jacketed kettle. In Step 1, place rice, salad oil and onions in kettle. Heat until rice is lightly browned, stirring occasionally. Omit Step 2. Follow Step 3. Add tomato mixture; bring to a boil; cover; reduce heat and cook 20 minutes at medium heat. Uncover; cook an additional 5 minutes. Omit Step 4. Follow Step 5.

CEREALS AND PASTA PRODUCTS No. E 012 00

NOODLES JEFFERSON

Yield 100   Portion   3/4 Cup

Calories	Carbohydrates	Protein	Fat	Cholesterol	Sodium	Calcium
241 cal	29 g	10 g	9 g	58 mg	509 mg	143 mg

Ingredient	Weight	Measure	Issue
WATER, WARM	50-1/8 lbs	6 gal	
SALT	1-7/8 oz	3 tbsp	
OIL, SALAD	1-1/2 oz	3 tbsp	
NOODLES, EGG	9 lbs	6 gal 2-7/8 qts	
BUTTER, MELTED	1-1/4 lbs	2-1/2 cup	
SALT	5/8 oz	1 tbsp	
PEPPER, BLACK, GROUND	1/4 oz	1 tbsp	
CHEESE, PARMESAN, GRATED	2 lbs	2 qts 1 cup	

Method

1. Add salt and oil to water; heat to a rolling boil.
2. Slowly add noodles, stirring constantly, until water boils again. Cook about 8 to 10 minutes or until tender. Drain thoroughly.
3. Add butter, salt and pepper to noodles. Stir well.
4. Add cheese; toss well. CCP: Hold for service at 140 F. or higher.

CEREALS AND PASTA PRODUCTS   No.E 013 00

STEAMED PASTA

Yield 100   **Portion** 1 Cup

Calories	Carbohydrates	Protein	Fat	Cholesterol	Sodium	Calcium
207 cal	41 g	7 g	1 g	0 mg	293 mg	17 mg

Ingredient   Weight   Measure   Issue

Ingredient	Weight	Measure	Issue
WATER	75-1/4 lbs	9 gal	
SALT	2-1/2 oz	1/4 cup 1/3 tbsp	
OIL,SALAD	1-7/8 oz	1/4 cup 1/3 tbsp	
SPAGHETTI NOODLES,DRY	12 lbs	3 gal 1 qts	

Method

1. Fill each steam table pan with 2-1/4 gallons water. Use perforated pan inside solid pan to facilitate draining.
2. Add 1 tablespoon salt and 1 tablespoon salad oil to each pan.
3. Place 3 pounds pasta in each pan. To prevent pastiness, pasta should be placed in pans just before steaming. Ensure pasta is covered with water.
4. Place pans in preheated steam cooker. Time according to type of pasta and steam cooker pressure. GUIDELINES FOR TIMING: Macaroni - 5 lb PSI, 16 minutes; 15 lb PSI, 11 minutes Noodles, Egg - 5 lb PSI, 22 minutes; 15 lb PSI, 17 minutes Spaghetti - 5 lb PSI, 20 minutes; 15 lb PSI, 15 minutes Vermicelli - 5 lb PSI, 11 minutes; 15 lb PSI, 4 minutes
5. Cooked macaroni should be rinsed in cold water and drained thoroughly to prevent sticking together. If cooked pasta is to be combined with butter or a sauce immediately, rinsing is not necessary. CCP: Hold for service at 140 F. or higher.

CEREALS AND PASTA PRODUCTS No. E 014 00

SPRING GARDEN RICE

Yield 100   Portion   3/4 Cup

Calories	Carbohydrates	Protein	Fat	Cholesterol	Sodium	Calcium
170 cal	31 g	7 g	2 g	5 mg	302 mg	160 mg

Ingredient	Weight	Measure	Issue
RICE,LONG GRAIN	6-3/4 lbs	1 gal 1/8 qts	
WATER,COLD	17-3/4 lbs	2 gal 1/2 qts	
SALT	1-1/2 oz	2-1/3 tbsp	
SQUASH,FRESH,SUMMER,SLICED	5-1/4 lbs	1 gal 1-1/4 qts	5-1/2 lbs
CARROTS,FRESH,SHREDDED	3-1/2 lbs	3 qts 2-1/2 cup	4-1/4 lbs
WATER	4-1/3 lbs	2 qts 1/4 cup	
MILK,NONFAT,DRY	4 oz	1-5/8 cup	
YOGURT,PLAIN,NONFAT	2-7/8 lbs	1 qts 1-1/4 cup	
CHEESE,PARMESAN,GRATED	1-1/4 lbs	1 qts 1-3/4 cup	
PEPPER,WHITE,GROUND	1/4 oz	1 tbsp	
GARLIC POWDER	1/8 oz	1/4 tsp	
BROCCOLI,FROZEN,SPEARS,THAWED,1-1/2""	3-1/4 lbs	2 qts	
MUSHROOMS,FRESH,WHOLE,SLICED	1-2/3 lbs	2 qts 3 cup	1-7/8 lbs
PARSLEY,FRESH,BUNCH,CHOPPED	10 oz	1 qts 3/4 cup	10-1/2 oz

Method

1 Combine rice, water and salt; bring to a boil. Stir occasionally. Cover tightly; simmer 20 minutes or until most of the water is absorbed. Remove from heat; transfer to shallow serving pans. Cover.
2 Combine squash and carrots in steam-jacketed kettle. Stir; cook 5 to 7 minutes or until tender crisp.
3 Reconstitute milk.
4 Add milk, yogurt, parmesan cheese, pepper and garlic powder to vegetables in steam-jacketed kettle. Stir well.
5 Add rice, broccoli, mushrooms, and parsley; mix lightly until all ingredients are coated with sauce. Bring to a simmer while stirring, about 5 to 7 minutes. CCP: Internal temperature must reach 145 F. or higher for 15 seconds.
6 Remove to serving pans. CCP: Hold at 140 F. or higher for service.

Notes

1 In Step 1, 7 pounds 7 ounces brown rice, 9 quarts of water and 1-1/2 ounces salt may be used per 100 servings. Follow directions on Recipe No. E 005 05, Steamed Brown Rice.
2 In Steps 1 and 2, oven method may be used; use boiling water for cold water; place 3-1/2 pounds or 2 quarts rice, 4-1/4 quarts water and 2/3 ounce or 1 tablespoon salt in each steam table pan; stir. Cover tightly; bake at 350 F. in a convection oven for 35 to 40 minutes or until most of water is absorbed on high fan, closed vent.

CEREALS AND PASTA PRODUCTS No.E 015 00

SICILIAN BROWN RICE AND VEGETABLES

Yield 100   Portion   3/4 Cup

Calories	Carbohydrates	Protein	Fat	Cholesterol	Sodium	Calcium
155 cal	29 g	6 g	2 g	4 mg	542 mg	110 mg

Ingredient	Weight	Measure	Issue
RICE,BROWN,LONG GRAIN,DRY	5-1/2 lbs	3 qts 1-3/8 cup	
WATER,COLD	13-7/8 lbs	1 gal 2-5/8 qts	
SALT	1-1/4 oz	2 tbsp	
JUICE,TOMATO,CANNED	9-1/4 lbs	1 gal 1/3 qts	
TOMATOES,CANNED,DICED,DRAINED	6-5/8 lbs	3 qts	
ONIONS,FRESH,CHOPPED	2-1/3 lbs	1 qts 2-5/8 cup	2-5/8 lbs
TOMATO PASTE,CANNED	1-1/8 lbs	2 cup	
SUGAR,BROWN,PACKED	2-1/2 oz	1/2 cup	
SALT	1 oz	1 tbsp	
BASIL,SWEET,WHOLE,CRUSHED	1-1/8 oz	1/4 cup 3-1/3 tbsp	
GARLIC POWDER	3/8 oz	1 tbsp	
OREGANO,CRUSHED	3/4 oz	1/4 cup 1-1/3 tbsp	
PEPPER,BLACK,GROUND	1/4 oz	1 tbsp	
BAY LEAF,WHOLE,DRIED	1/3 oz	10 each	
SQUASH,FRESH,SUMMER,SLICED	2-1/2 lbs	2 qts 2 cup	2-5/8 lbs
SQUASH,ZUCCHINI,FRESH,SLICED	2-1/2 lbs	2 qts 2 cup	2-5/8 lbs
CARROTS,FRESH,SHREDDED	1 lbs	1 qts 1/8 cup	1-1/4 lbs
BROCCOLI,FROZEN,SPEARS	2-3/4 lbs	2 qts	
MUSHROOMS,FRESH,WHOLE,SLICED	1-1/4 lbs	2 qts 1/8 cup	1-3/8 lbs
PARSLEY,FRESH,BUNCH,CHOPPED	8 oz	3-3/4 cup	8-3/8 oz
CHEESE,MOZZARELLA,PART SKIM,SHREDDED	1-3/4 lbs	1 qts 3 cup	

Method

1. Combine rice, water and salt; bring to a boil. Stir occasionally. Cover tightly; simmer 30 minutes or until most of the water is absorbed. Remove from heat; transfer to shallow serving pans. CCP: Cover. Hold at 140 F. or higher for use in Step 8.
2. Place tomato juice, tomatoes, onions, tomato paste, brown sugar, salt, basil, garlic powder, oregano, pepper and bay leaves in steam-jacketed kettle. Stir; bring to a boil. Reduce heat; cover; simmer 20 minutes. Remove bay leaves.
3. Stir in rice, yellow squash, zucchini and carrots. Bring to a boil; reduce heat; simmer 3 to 5 minutes or until vegetables are tender crisp. Stir occasionally.
4. Stir in broccoli, mushrooms and parsley; bring to a simmer.
5. Place 1-1/4 gallon in each steam table pan. Sprinkle 7 ounces cheese over mixture in each pan. Using a convection oven, bake at 325 F. for 12 to 15 minutes or until mixture is bubbly and cheese is melted and lightly browned on high fan, closed vent. CCP: Internal temperature must reach 145 F. or higher for 15 seconds. CCP: Hold for service at 140 F. or higher.

Notes

1. In Steps 1 and 2, oven method may be used: Use boiling water for cold water; place 2-3/4 pounds or 6-2/3 cups rice, 3-1/8 quarts boiling water, and 2-1/2 teaspoons salt in each steam table pan. Stir, cover tightly.
2. In Step 4, 2-1/2 pounds frozen summer squash and 2-1/2 pounds frozen zucchini may be used.
3. In Step 5, 1 pound canned, drained mushrooms may be used.

CEREALS AND PASTA PRODUCTS No. E 016 00

ISLANDER'S RICE

Yield 100   Portion   3/4 Cup

Calories	Carbohydrates	Protein	Fat	Cholesterol	Sodium	Calcium
149 cal	31 g	5 g	1 g	0 mg	644 mg	43 mg

Ingredient	Weight	Measure	Issue
CHICKEN BROTH		1 gal 3-1/2 qts	
BEANS,KIDNEY,DARK RED,CANNED,DRAINED	9-1/8 lbs	1 gal 1-7/8 qts	
RICE,LONG GRAIN	5-3/4 lbs	3 qts 2 cup	
ONIONS,FRESH,CHOPPED	2 lbs	1 qts 1-5/8 cup	2-1/4 lbs
GARLIC POWDER	1-1/4 oz	1/4 cup 1/3 tbsp	
THYME,GROUND	5/8 oz	1/4 cup 1/3 tbsp	
ALLSPICE,GROUND	1/3 oz	1 tbsp	
PEPPER,RED,GROUND	1/4 oz	1 tbsp	
OREGANO,CRUSHED	1/2 oz	3 tbsp	
PEPPERS,GREEN,FRESH,CHOPPED	3-5/8 lbs	2 qts 3 cup	4-3/8 lbs
PIMIENTO,CANNED,DRAINED,SLICED	1-1/2 lbs	3-1/2 cup	

Method

1. Prepare stock according to package directions.
2. Combine stock, beans, rice, onions, garlic powder, thyme, allspice, red pepper and oregano in steam-jacketed kettle or stock pot; bring to a boil. Stir occasionally.
3. Cover tightly; reduce heat; simmer 20 to 25 minutes or until most of the water is absorbed and rice is tender. Do not stir.
4. Add peppers and pimientos; stir well.
5. Transfer to serving pans. CCP: Hold for service at 140 F. or higher.

Notes

1. For vegetarian: double all ingredients; use 7-1/2 quarts vegetable stock. EACH PORTION: 1-1/2 cups.
2. OVEN METHOD: For 100 portions: Use steam table pans. Follow Step 1. In Step 2, place 4 pounds 13 ounces or 3 quarts beans, 3 pounds or 1-3/4 quarts of rice, and 1 pound or 3/4 quart onions in each pan; stir well. Combine stock with garlic powder, thyme, allspice, red pepper, and oregano; stir well. Bring to a boil. Pour 3-3/4 quarts stock mixture over rice mixture in each pan. Stir well. Omit Step 3. Cover; bake in a 350 F. convection oven for 30 minutes or until most of the water is absorbed and the rice is tender on high fan, closed vent. In Step 4, add 1-1/2 quarts peppers and 2 cups pimientos to rice mixture in each pan. Stir well to mix. Follow Step 5.

CEREALS AND PASTA PRODUCTS No. E 017 00

MEDITERRANEAN BROWN RICE

Yield 100 **Portion** 3/4 Cup

Calories	Carbohydrates	Protein	Fat	Cholesterol	Sodium	Calcium
199 cal	38 g	4 g	4 g	0 mg	699 mg	28 mg

Ingredient	Weight	Measure	Issue
OIL,SALAD	7-2/3 oz	1 cup	
ONIONS,FRESH,CHOPPED	3-1/8 lbs	2 qts 1 cup	3-1/2 lbs
RICE,BROWN,LONG GRAIN,RAW PARBOILED	8-1/8 lbs	1 gal 1 qts	
CHICKEN BROTH		2 gal 2-1/2 qts	
RAISINS,GOLDEN	1-7/8 lbs	1 qts 2 cup	
CINNAMON,GROUND	1/2 oz	2 tbsp	
ALLSPICE,GROUND	1/4 oz	1 tbsp	
CARDAMOM SEED,GROUND	1/4 oz	1 tbsp	
CILANTRO,DRY	1/4 oz	1/4 cup 1/3 tbsp	

Method

1 Heat oil in steam jacketed kettle. Add onions; cook 5 minutes or until tender, stirring occasionally.
2 Add rice; stir well until rice is coated. Stir; cook 5 minutes or until rice is lightly browned.
3 Prepare broth according to package directions. Add stock, raisins, cinnamon, allspice, and cardamom to rice. Bring to a boil; stir.
4 Reduce heat; cover tightly; simmer 25 minutes or until most of the water is absorbed. Add cilantro; mix well. CCP: Internal temperature of cooked rice mixture must reach 145 F. or higher for 15 seconds.
5 Remove from heat; transfer to shallow serving pans. Cover. CCP: Hold for service at 140 F. or higher.

Notes

1 OVEN METHOD: For 100 portions: Omit oil. Place 6-2/3 cups rice, 3-1/2 quarts boiling stock, 3 cups onions, 2 cups raisins, 2 teaspoons cinnamon, 1-1/3 teaspoons allspice, and 1-1/3 teaspoon cardamom in each steam table pan. Stir, cover tightly, bake in 350 F. convection oven 25 minutes or until most of the water is absorbed on high fan, closed vent. Fold 1/2 cup cilantro into each pan. CCP: Internal temperature of cooked rice mixture must reach 145 F. or higher for 15 seconds. CCP: Hold for service at 140 F. or higher.

CEREALS AND PASTA PRODUCTS No. E 018 00

SPICY BROWN RICE PILAF

Yield 100   Portion   3/4 Cup

Calories	Carbohydrates	Protein	Fat	Cholesterol	Sodium	Calcium
151 cal	30 g	4 g	2 g	0 mg	766 mg	36 mg

Ingredient	Weight	Measure	Issue
CHICKEN BROTH		2 gal 2-1/2 qts	
PAPRIKA,GROUND	1-1/2 oz	1/4 cup 2-2/3 tbsp	
MUSTARD,DRY	1-3/4 oz	1/4 cup 2/3 tbsp	
PEPPER,BLACK,GROUND	2/3 oz	3 tbsp	
THYME,GROUND	1/2 oz	3 tbsp	
SALT	1/2 oz	3/8 tsp	
GARLIC POWDER	1/2 oz	1 tbsp	
CUMIN,GROUND	1/3 oz	1 tbsp	
OREGANO,CRUSHED	1/2 oz	3 tbsp	
BAY LEAF,WHOLE,DRIED	1/2 oz	14 each	
PEPPER,RED,CRUSHED	<1/16th oz	1/8 tsp	
RICE,BROWN,LONG GRAIN,RAW PARBOILED	7-1/3 lbs	1 gal 1/2 qts	
COOKING SPRAY,NONSTICK	3/8 oz	3/8 tsp	
ONIONS,FRESH,CHOPPED	3-1/2 lbs	2 qts 1-7/8 cup	3-7/8 lbs
CELERY,FRESH,CHOPPED	2-1/2 lbs	2 qts 1-1/2 cup	3-3/8 lbs
PEPPERS,GREEN,FRESH,CHOPPED	2-1/2 lbs	1 qts 3-5/8 cup	3 lbs

Method

1 Prepare broth according to package directions.
2 Add paprika, mustard flour, pepper, thyme, salt, garlic powder, cumin, oregano, bay leaves, and red pepper to stock. Stir well to blend.
3 Add rice to stock in steam jacketed kettle or stock pot. Bring to a boil. Stir. Reduce heat. Cover tightly. Simmer 25 minutes or until most of the water is absorbed and rice is tender.
4 Spray steam-jacketed kettle with non-stick cooking spray. Add onions, celery, and peppers. Stir; cook 10 to 12 minutes or until vegetables are tender crisp.
5 Place approximately 8-1/2 pounds rice in each steam table pan. Add 5-1/3 cups vegetables to each pan. Mix well. CCP: Internal temperature must reach 145 F. or higher for 15 seconds. CCP: Hold for service at 140 F. or higher.

Notes

1 OVEN METHOD: For 100 portions: Follow Steps 1 and 2. Bring stock to a boil. Place 2-1/2 pounds of rice and 3-1/2 quarts stock, in each steam table pan; stir. Cover tightly; bake in 350 F. convection oven for 30 minutes or until most of the water is absorbed on high fan, closed vent. Follow Steps 4 and 5.

CEREALS AND PASTA PRODUCTS No. E 019 00

BROWN RICE WITH TOMATOES

Yield 100   **Portion** 3/4 Cup

Calories	Carbohydrates	Protein	Fat	Cholesterol	Sodium	Calcium
167 cal	35 g	4 g	1 g	0 mg	163 mg	37 mg

Ingredient	**Weight**	**Measure**	**Issue**
VEGETABLE BROTH		1 gal 3-1/2 qts	
TOMATOES,CANNED,DICED,DRAINED	13-1/4 lbs	1 gal 2 qts	
RICE,BROWN,LONG GRAIN,RAW PARBOILED	7-3/4 lbs	1 gal 3/4 qts	
ONIONS,FRESH,CHOPPED	6-1/3 lbs	1 gal 1/2 qts	7 lbs
GARLIC POWDER	2-3/8 oz	1/2 cup	
PEPPER,BLACK,GROUND	1/2 oz	2 tbsp	

Method

1. Prepare broth according to package directions in steam-jacketed kettle or stock pot.
2. Add tomatoes, brown rice, onions, garlic powder, and pepper to broth in steam-jacketed kettle or stock pot. Stir well; bring to a rolling boil, stirring occasionally. Reduce heat. Cover. Simmer 35 minutes or until most of the broth is absorbed and rice is tender. Do not stir. CCP: Internal temperature must reach 145 F. or higher for 15 seconds.
3. Stir to redistribute onions and tomatoes. Transfer to serving pans. CCP: Hold for service at 140 F. or higher.

Notes

1. Using a convection oven, bake in 2 steam table pans at 350 F. for 45 to 50 minutes on high fan, closed vent or until most of the broth is absorbed.

CEREALS AND PASTA PRODUCTS   No. E 020 00

GINGER RICE

Yield 100   Portion   3/4 Cup

Calories	Carbohydrates	Protein	Fat	Cholesterol	Sodium	Calcium
183 cal	34 g	6 g	2 g	73 mg	567 mg	43 mg

Ingredient	Weight	Measure	Issue
RICE,LONG GRAIN	8-1/2 lbs	1 gal 1-1/4 qts	
WATER,BOILING	18-3/4 lbs	2 gal 1 qts	
COOKING SPRAY,NONSTICK	1/4 oz	1/4 tsp	
EGGS,WHOLE,FROZEN	3-3/4 lbs	1 qts 3 cup	
SOY SAUCE	2-1/8 lbs	3-3/8 cup	
SUGAR,GRANULATED	1-3/4 oz	1/4 cup 1/3 tbsp	
GARLIC POWDER	1/2 oz	1 tbsp	
GINGER,GROUND	1/4 oz	1 tbsp	
PEPPER,WHITE,GROUND	1/4 oz	1 tbsp	
PEPPERS,RED FRESH,DICED	1 lbs	3 cup	1-1/4 lbs
CARROTS,FROZEN,SLICED	1 lbs	3-3/4 cup	
ONIONS,GREEN,FRESH,SLICED	1-1/8 lbs	1 qts 1-3/8 cup	1-1/3 lbs

Method

1 Place 3 pounds rice and 3 quarts water in each lightly sprayed steam table pan; stir.
2 Cover tightly. Using a convection oven, bake at 325 F. for 30 minutes on high fan, closed vent.
3 Pour eggs on lightly sprayed griddle. Cook 1-1/2 minutes or until set. Do not turn. Cut into 4-inch strips to facilitate removal. Remove immediately. Cut into 1/2-inch squares.
4 Combine soy sauce, sugar, garlic powder, white pepper, and ginger. Stir well to dissolve sugar.
5 Add 2-1/3 cups egg strips, 1-1/2 cups soy mixture, 1 cup red peppers and 1-1/4 cups of carrots to rice in each pan. Mix lightly but thoroughly.
6 Cover. CCP: Using a convection oven, bake 15 minutes on high fan, closed vent. CCP: Internal temperature must reach 145 F. or higher for 15 seconds.
7 Add 1-3/4 cups green onions to rice in each pan. Mix lightly but thoroughly. CCP: Hold for service at 140 F. or higher.

CEREALS AND PASTA PRODUCTS No.E 021 00

NUTTY RICE AND CHEESE

Yield 100   Portion 9 Ounces

Calories	Carbohydrates	Protein	Fat	Cholesterol	Sodium	Calcium
323 cal	40 g	22 g	8 g	12 mg	835 mg	289 mg

Ingredient	Weight	Measure	Issue
WATER	20-7/8 lbs	2 gal 2 qts	
SALT	1-2/3 oz	2-2/3 tbsp	
RICE,BROWN,LONG GRAIN,DRY	8-1/8 lbs	1 gal 1 qts	
CHEESE,COTTAGE,LOWFAT	14 lbs	1 gal 3 qts	
YOGURT,PLAIN,NONFAT	10-3/4 lbs	1 gal 1 qts	
EGG WHITES,FROZEN,THAWED	5 lbs	2 qts 1-3/8 cup	
ONIONS,FRESH,CHOPPED	3-7/8 lbs	2 qts 3 cup	4-1/3 lbs
ALMONDS,SLIVERED	1-3/8 lbs	1 qts 2 cup	
CHEESE,PARMESAN,GRATED	1-1/3 lbs	1 qts 2 cup	
FLOUR,WHEAT,GENERAL PURPOSE	6-5/8 oz	1-1/2 cup	
SALT	1-1/2 oz	2-1/3 tbsp	
PARSLEY,DEHYDRATED,FLAKED	1-1/4 oz	1-5/8 cup	
GARLIC POWDER	1-1/4 oz	1/4 cup 1/3 tbsp	
PEPPER,WHITE,GROUND	2/3 oz	2-2/3 tbsp	
COOKING SPRAY,NONSTICK	1/2 oz	1 tbsp	
CHEESE,PARMESAN,GRATED	7 oz	2 cup	

Method

1. Combine water, rice, and salt; bring to a boil; stir, cover tightly; simmer 25 minutes or until most of the water is absorbed.
2. Remove from heat. Transfer to sheet pans. Allow to cool 5 minutes.
3. Combine cottage cheese, yogurt, egg whites, onions, almonds, parmesan cheese, flour, salt, parsley flakes, garlic powder, and pepper in mixer bowl. Mix at low speed 1 minute. Scrape down bowl.
4. Add chilled rice to ingredients in mixer bowl. Mix at low speed 1 minute or until thoroughly blended.
5. Lightly spray steam table pans with non-stick spray. Place 12-1/4 pounds of mixture in each steam table pan. Spread evenly. Sprinkle 6 tablespoons of parmesan cheese over the top of each pan.
6. Using a convection oven, bake 55 minutes at 325 F. on high fan, open vent or until set. CCP: Internal temperature must reach 145 F. or higher for 15 seconds.
7. Cut each pan 4 by 5. CCP: Hold for service at 140 F. or higher.

CEREALS AND PASTA PRODUCTS No. E 022 00

ORZO WITH LEMON AND HERBS

Yield 100 **Portion** 3/4 Cup

Calories	Carbohydrates	Protein	Fat	Cholesterol	Sodium	Calcium
92 cal	12 g	2 g	4 g	0 mg	362 mg	19 mg

Ingredient	**Weight**	**Measure**	**Issue**
SALT	1-1/4 oz	2 tbsp	
MUSTARD,DIJON	1/2 oz	1 tbsp	
GARLIC POWDER	3/8 oz	1 tbsp	
BASIL,SWEET,WHOLE,CRUSHED	5/8 oz	1/4 cup 1/3 tbsp	
OREGANO,CRUSHED	5/8 oz	1/4 cup 1/3 tbsp	
PEPPER,BLACK,GROUND	1/4 oz	1 tbsp	
ONION POWDER	1/4 oz	1 tbsp	
JUICE,LEMON	1-1/3 lbs	2-1/2 cup	
OIL,OLIVE	11-3/8 oz	1-1/2 cup	
WATER	66-7/8 lbs	8 gal	
SALT	1-7/8 oz	3 tbsp	
OIL,SALAD	1/3 oz	1/3 tsp	
PASTA,ORZO	8-1/3 lbs	6 gal 7/8 qts	
COOKING SPRAY,NONSTICK	1-1/2 oz	3 tbsp	
ONIONS,FRESH,CHOPPED	5-1/3 lbs	3 qts 3-3/8 cup	5-7/8 lbs

Method

1 Combine salt, mustard, garlic powder, basil, oregano, pepper, and onion powder. Add lemon juice and olive oil. Stir to blend. Cover, set aside for use in Step 6.
2 Add salt and salad oil to water; heat to a rolling boil.
3 Add pasta slowly while stirring constantly until water boils again. Cook about 9 minutes or until al dente; stirring occasionally. DO NOT OVERCOOK.
4 Drain. Rinse with cold water; drain thoroughly.
5 Stir-cook onions in a lightly sprayed steam jacketed kettle or stockpot 8 to 10 minutes or until tender, stirring constantly.
6 Add the reserved lemon and herb dressing to cooked onions. Stir to blend well. Bring to a boil; reduce heat to a simmer.
7 Add the orzo to the onion and lemon mixture. Heat to a simmer while gently stirring for 1 minute to coat the orzo with the sauce. CCP: Temperature must register 145 F. or higher for 15 seconds.
8 Place 2-1/3 gallon pasta mixture in each pan. CCP: Hold for service at 140 F. or higher.

CEREALS AND PASTA PRODUCTS No.E 023 00

ORZO, WITH SPINACH, TOMATO, AND ONION

Yield 100   **Portion** 9-1/2 Ounces

Calories	Carbohydrates	Protein	Fat	Cholesterol	Sodium	Calcium
62 cal	10 g	3 g	2 g	2 mg	456 mg	104 mg

Ingredient	**Weight**	**Measure**	**Issue**
WATER	66-7/8 lbs	8 gal	
SALT	1-7/8 oz	3 tbsp	
OIL,SALAD	1 oz	2 tbsp	
PASTA,ORZO	1-2/3 lbs	1 gal 1 qts	
ONIONS,FRESH,CHOPPED	5 lbs	3 qts 2-1/8 cup	5-1/2 lbs
COOKING SPRAY,NONSTICK	2 oz	1/4 cup 1/3 tbsp	
TOMATOES,CANNED,DICED,DRAINED	13-1/4 lbs	1 gal 2 qts	
BASIL,DRIED,CRUSHED	2-1/2 oz	1 cup	
SPINACH,CHOPPED,FROZEN	4 lbs	2 qts 3-5/8 cup	
CUMIN,GROUND	7/8 oz	1/4 cup 1/3 tbsp	
PEPPER,BLACK,GROUND	2/3 oz	3 tbsp	
GARLIC POWDER	5/8 oz	2 tbsp	
SALT	1 oz	1 tbsp	
CHEESE,PARMESAN,GRATED	7 oz	2 cup	

Method

1. Add salt and salad oil to water; heat to a rolling boil.
2. Add pasta slowly while stirring constantly until water boils again. Cook about 9 minutes or until tender; stirring occasionally. DO NOT OVERCOOK.
3. Drain. Rinse with cold water; drain thoroughly. Use immediately in recipe preparation or place in shallow containers and cover.
4. Stir-cook onions in a lightly sprayed steam jacketed kettle or stockpot 8 to 10 minutes or until tender, stirring constantly.
5. Add the tomatoes, spinach, basil, salt, cumin, pepper and garlic powder, stir to combine. Bring to a boil. Cover; reduce heat; simmer for 5 minutes.
6. Add the orzo; stir to blend. Bring to a boil. Cover; reduce heat; simmer for 5 minutes. CCP: Temperature must reach 140 F. or higher for 15 seconds.
7. Place 3 gallons vegetable pasta mixture in each pan.
8. Distribute 1 cup parmesan cheese evenly over vegetable pasta mixture in each pan. CCP: Hold for service at 140 F. or higher.

CEREALS AND PASTA PRODUCTS No. E 508 00

SOUTHWESTERN RICE

Yield 100   **Portion** 3/4 Cup

Calories	Carbohydrates	Protein	Fat	Cholesterol	Sodium	Calcium
131 cal	25 g	5 g	1 g	2 mg	192 mg	64 mg

Ingredient	Weight	Measure	Issue
RICE,LONG GRAIN	5-3/4 lbs	3 qts 2 cup	
WATER	12-1/2 lbs	1 gal 2 qts	
SALT	1 oz	1 tbsp	
COOKING SPRAY,NONSTICK	2 oz	1/4 cup 1/3 tbsp	
ONIONS,FRESH,CHOPPED	11-1/4 oz	2 cup	12-1/2 oz
GARLIC POWDER	2-3/8 oz	1/2 cup	
PEPPERS,GREEN,FRESH,CHOPPED	6-5/8 oz	1-1/4 cup	8 oz
TOMATOES,CANNED,DICED,DRAINED	3 lbs	1 qts 1-1/2 cup	
PARSLEY,DEHYDRATED,FLAKED	3/8 oz	1/2 cup	
CORN,FROZEN,WHOLE KERNEL	1-1/8 lbs	3 cup	
PEPPER,BLACK,GROUND	2/3 oz	3 tbsp	
CHILI POWDER,LIGHT,GROUND	1 oz	1/4 cup 1/3 tbsp	
WORCESTERSHIRE SAUCE	4-1/4 oz	1/2 cup	
CHEESE,MONTEREY JACK,REDUCED FAT	2 lbs	2 qts	

Method

1. Combine rice, water, and salt. Bring to a boil. Cover tightly, and simmer 20 to 30 minutes.
2. Saute onions, garlic, and peppers in vegetable spray in a steam jacketed kettle. Add tomatoes, parsley, and corn. Season with pepper, chili powder, and Worcestershire sauce. Fold in cooked drained rice and thoroughly blend.
3. Divide rice in serving pans, sprinkle with cheese. Bake in 350 F. oven for 20 minutes. CCP: Internal temperature must reach 145 F. or higher for 15 seconds. CCP: Hold at 140 F. or higher for serving.

CEREALS AND PASTA PRODUCTS No. E 510 00

PASTA PROVENCAL

Yield 100   Portion 1 Cup

Calories	Carbohydrates	Protein	Fat	Cholesterol	Sodium	Calcium
295 cal	46 g	16 g	6 g	31 mg	1288 mg	212 mg

Ingredient	Weight	Measure	Issue
WATER	54-1/3 lbs	6 gal 2 qts	
SALT	1-1/2 oz	2-1/3 tbsp	
OIL,SALAD	1/3 oz	1/3 tsp	
PASTA,PENNE	10 lbs	7 gal 1-7/8 qts	
OIL,SALAD	5-3/4 oz	3/4 cup	
FLOUR,WHEAT,GENERAL PURPOSE	14-2/3 oz	3-3/8 cup	
WATER,WARM	10-1/2 lbs	1 gal 1 qts	
MILK,NONFAT,DRY	1-1/8 lbs	1 qts 3-1/2 cup	
CHICKEN BROTH		1 gal 1 qts	
CHEESE,PARMESAN,GRATED	7 oz	2 cup	
SALT	1-1/4 oz	2 tbsp	
GARLIC POWDER	1-1/4 oz	1/4 cup 1/3 tbsp	
THYME LEAVES,DRIED,GROUND	5/8 oz	1/4 cup 1/3 tbsp	
PEPPER,BLACK,GROUND	1/2 oz	2 tbsp	
BASIL,SWEET,WHOLE,CRUSHED	5/8 oz	1/4 cup 1/3 tbsp	
OREGANO,CRUSHED	7/8 oz	1/4 cup 1-2/3 tbsp	
PEPPER,RED,CRUSHED	1/8 oz	1 tbsp	
TOMATOES,CANNED,DICED,DRAINED	12-1/8 lbs	1 gal 1-1/2 qts	
BEANS,CANNELLINI,CANNED	8-1/2 lbs	3 qts 3 cup	
SPINACH,FROZEN	4 lbs	2 qts 1-1/2 cup	
ONIONS,FRESH,CHOPPED	4-3/8 lbs	3 qts 3/8 cup	4-7/8 lbs
HAM,CANNED,COOKED,DICED	4 lbs		
CARROTS,FRESH,CHOPPED	3-3/4 lbs	3 qts 1-1/4 cup	4-5/8 lbs
CELERY,FRESH,CHOPPED	2-3/4 lbs	2 qts 2-3/8 cup	3-3/4 lbs
PARSLEY,DEHYDRATED,FLAKED	3/4 oz	1 cup	

Method

1. Add salt and salad oil to water; heat to a rolling boil.
2. Add pasta slowly while stirring constantly until water boils again. Cook 7 to 9 minutes or until tender, stirring occasionally. DO NOT OVERCOOK.
3. Drain. Rinse with cold water; drain thoroughly.
4. Blend salad oil and flour together to form a roux. Using a wire whip, stir until smooth. Cook roux for 3 minutes in a steam-jacketed kettle or stockpot stirring constantly.
5. Reconstitute milk in warm water.
6. Gradually add milk and broth to roux while stirring constantly. Bring to a boil. Cover; reduce heat; simmer 5 minutes or until thickened, stirring frequently to prevent sticking.
7. Add parmesan cheese, salt, garlic powder, thyme, black pepper, basil, oregano and red pepper to thickened sauce. Stir to blend well.
8. Add tomatoes, beans, spinach, onions, ham, carrots, celery and parsley to thickened sauce. Bring to a boil. Cover; reduce heat; simmer 7 to 10 minutes until tender, stirring occasionally.
9. Add pasta to thickened sauce and vegetable mixture. Heat to a simmer while stirring for 1 minute to coat the pasta with the vegetable sauce. CCP: Temperature must register 165 F. or higher for 15 seconds.
10. Pour 3-1/8 gal pasta-vegetable mixture into 3 ungreased steam table pans; cover. CCP: Hold for service at 140 F. or higher.

F. CHEESE AND EGGS No. 0 (1)

INDEX

Card No.		Card No.	
F 001 00	Baked Macaroni and Cheese	F 010 02	Scrambled Eggs and Ham
F 002 00	Nachos	F 010 03	Scrambled Eggs (Dehydrated Egg Mix)
F 002 01	Nachos (RTU Cheese Sauce)	F 010 05	Scrambled Eggs (Frozen Eggs and Egg Whites)
F 003 00	Eggs Au Gratin (Scotch Woodcock)	F 011 00	Mushroom Quiche
F 004 00	Cooked Eggs	F 011 01	Broccoli Quiche
F 005 00	Deviled Eggs	F 011 02	Broccoli Quiche (Frozen Eggs and Egg Whites)
F 006 00	Egg Foo Young	F 011 03	Mushroom Quiche (Frozen Eggs and Egg Whites)
F 007 00	Griddle Fried Eggs		
F 008 00	Plain Omelet	F 012 00	Breakfast Burrito
F 008 01	Plain Omelet (Frozen Eggs and Egg Whites)	F 012 01	Breakfast Pita
F 008 03	Cheese Omelet	F 013 00	Veggie Egg Pocket
F 008 04	Green Pepper Omelet	F 014 00	Monterey Egg Bake
F 008 05	Ham Omelet	F 015 00	Breakfast Pizza
F 008 06	Ham and Cheese Omelet	F 015 01	Mexican Breakfast Pizza
F 008 08	Mushroom Omelet	F 015 02	Italian Breakfast Pizza
F 008 09	Onion Omelet	F 800 00	Macaroni and Cheese, Frozen
F 008 10	Western Omelet	F 801 00	Breakfast Burrito, Frozen
F 008 11	Tomato Omelet		
F 008 12	Spanish Omelet		
F 009 00	Poached Eggs		
F 010 00	Scrambled Eggs		
F 010 01	Scrambled Eggs and Cheese		

CHEESE AND EGGS No.F 001 00

BAKED MACARONI AND CHEESE

Yield 100   Portion   1 Cup

Calories	Carbohydrates	Protein	Fat	Cholesterol	Sodium	Calcium
359 cal	37 g	17 g	16 g	39 mg	721 mg	357 mg

Ingredient	Weight	Measure	Issue
MACARONI NOODLES,ELBOW,DRY	7-3/8 lbs	2 gal	
WATER,BOILING	50-1/8 lbs	6 gal	
SALT	1-2/3 oz	2-2/3 tbsp	
MILK,NONFAT,DRY	1-1/3 lbs	2 qts 1 cup	
WATER,WARM	20-7/8 lbs	2 gal 2 qts	
FLOUR,WHEAT,GENERAL PURPOSE	1-2/3 lbs	1 qts 2 cup	
WATER,COLD	2-1/8 lbs	1 qts	
SALT	1-7/8 oz	3 tbsp	
PEPPER,BLACK,GROUND	1/4 oz	1 tbsp	
CHEESE,CHEDDAR,SHREDDED	8 lbs	2 gal	
COOKING SPRAY,NONSTICK	2 oz	1/4 cup 1/3 tbsp	
BREADCRUMBS,DRY,GROUND,FINE	1-1/4 lbs	1 qts 1 cup	
MARGARINE,MELTED	10 oz	1-1/4 cup	

Method

1 Add macaroni slowly to boiling salted water; cook 8 to 10 minutes or until tender; stir occasionally to prevent sticking.
2 Drain. Set aside for use in Step 7.
3 Reconstitute milk; heat to just below boiling. DO NOT BOIL.
4 Combine flour and water to make a smooth mixture. Add mixture to hot milk, stirring constantly.
5 Add salt and pepper. Bring mixture to a boil; reduce heat; simmer 5 minutes or until thickened. Stir frequently to prevent scorching.
6 Add cheese to sauce; stir only until smooth; remove from heat.
7 Combine sauce and macaroni; mix well.
8 Lightly spray steam table pans with non-stick cooking spray. Place about 6-1/3 quart mixture in each sprayed pan.
9 Combine bread crumbs and melted butter or margarine; sprinkle 1-3/4 cup over mixture in each pan.
10 Using a convection oven, bake at 325 F. 15-20 minutes on high fan, open vent or until browned. CCP: Hold for service at 140 F. or higher.

CHEESE AND EGGS No.F 002 00

NACHOS

Yield 100   Portion   1-1/2 Ounces

Calories	Carbohydrates	Protein	Fat	Cholesterol	Sodium	Calcium
403 cal	28 g	14 g	27 g	47 mg	1259 mg	379 mg

Ingredient	Weight	Measure	Issue
PEPPERS,JALAPENOS,CANNED,CHOPPED	9-1/2 lbs	1 gal 3-7/8 qts	
WATER	1-5/8 lbs	3 cup	
RESERVED LIQUID	3-2/3 lbs	1 qts 3 cup	
CHEESE,AMERICAN,SHREDDED	11 lbs	2 gal 3 qts	
CHIPS,TORTILLA	9 lbs		

Method

1. Drain peppers. Reserve liquid from peppers. Coarsely chop peppers. Set aside for use in Step 6.
2. Combine water and reserved jalapeno liquid in steam-jacketed kettle or stock pot. Bring to a simmer. DO NOT BOIL.
3. Add cheese to hot mixture; stir constantly until melted, about 3 to 4 minutes, or until smooth and creamy. DO NOT BOIL.
4. Remove from heat; keep warm. CCP: Hold for service at 140 F. or higher.
5. Pour 2 ounces sauce over about 20 tortilla chips.
6. Sprinkle 2 teaspoons jalapeno peppers over each portion.

Notes

1. In Step 3, DO NOT use cheddar cheese. It will not produce an acceptable product.
2. In Step 3, cheese, when combined with jalapeno liquid, begins to curdle at temperatures above 170 F. to 180 F.

CHEESE AND EGGS No.F 002 01

NACHOS (RTU CHEESE SAUCE)

Yield 100   Portion   1-1/2 Ounces

Calories	Carbohydrates	Protein	Fat	Cholesterol	Sodium	Calcium
289 cal	34 g	5 g	15 g	6 mg	1028 mg	112 mg

Ingredient

Ingredient	Weight	Measure	Issue
PEPPERS,JALAPENOS,CANNED,CHOPPED	4-3/4 lbs	3 qts 3-7/8 cup	
RESERVED LIQUID	1-5/8 lbs	3 cup	
SAUCE, CHEESE, PREPARED	13-1/8 lbs	1 gal 2 qts	
CHIPS,TORTILLA	9 lbs		

Method

1. Drain peppers. Reserve liquid.
2. Combine jalapeno liquid with ready-to-use cheese sauce. Mix until smooth. Place in steam-jacketed kettle or stock pot. Heat, stirring constantly until hot, about 10 to 15 minutes. DO NOT BOIL.
3. Remove from heat; keep warm. CCP: Hold for service at 140 F. or higher.
4. Pour 2 ounces sauce over 20 tortilla chips.
5. Sprinkle 2 teaspoons jalapeno peppers over each portion.

Notes

1. Ready to use cheese sauce with jalapeno peppers may also be used.

CHEESE AND EGGS No.F 003 00

EGGS AU GRATIN (SCOTCH WOODCOCK)

Yield 100   **Portion** 2/3 Cup

Calories	Carbohydrates	Protein	Fat	Cholesterol	Sodium	Calcium
223 cal	7 g	12 g	16 g	243 mg	241 mg	179 mg

Ingredient	Weight	Measure	Issue
EGG,HARD COOKED	11 lbs	100 Eggs	
MILK,NONFAT,DRY	14-3/8 oz	1 qts 2 cup	
WATER,WARM	15-2/3 lbs	1 gal 3-1/2 qts	
BUTTER,MELTED	1-1/2 lbs	3 cup	
FLOUR,WHEAT,GENERAL PURPOSE	1-1/8 lbs	1 qts	
CHEESE,CHEDDAR,SHREDDED	3 lbs	3 qts	
BREADCRUMBS,DRY,GROUND,FINE	5-1/8 oz	1-3/8 cup	
BUTTER,MELTED	2-1/2 oz	1/4 cup 1-1/3 tbsp	

Method

1. Place eggs in baskets as needed; cover with hot water. Bring to a boil; reduce heat; simmer 10 to 15 minutes. DO NOT BOIL. Remove from water; serve immediately. CCP: All fresh shell eggs must be heated to 155 F. or higher for 15 seconds.
2. Cool; remove shells from eggs; slice eggs in half lengthwise. Arrange 100 egg halves in each steam table pan.
3. Reconstitute milk; heat to just below boiling. DO NOT BOIL.
4. Blend butter or margarine and flour together; stir until smooth. Add milk to roux, stirring constantly. Cook until thickened.
5. Add cheese to sauce; stir until cheese is melted. Stir as necessary.
6. Pour 4-3/4 quarts sauce over egg halves in each steam table pan.
7. Combine bread crumbs and butter. Sprinkle 2/3 cup buttered crumbs over mixture in each pan.
8. Using a convection oven, bake at 325 F. 10 minutes or until browned on low fan, open vent. CCP: Hold for service at 140 F. or higher.

CHEESE AND EGGS No. F 004 00

COOKED EGGS

Yield 100   Portion   2 Each

Calories	Carbohydrates	Protein	Fat	Cholesterol	Sodium	Calcium
149 cal	1 g	12 g	10 g	425 mg	126 mg	49 mg

Ingredient   **Weight**   **Measure**   **Issue**

EGGS, WHOLE, FRESH   22 lbs   200 each

Method

1. HARD COOKED EGGS: Place eggs in baskets as needed; cover with hot water. Bring to a boil; reduce heat; simmer 10 to 15 minutes. DO NOT BOIL. Remove from water; serve immediately. CCP: All fresh shell eggs must be heated to 155 F. or higher for 15 seconds.
2. SOFT COOKED EGGS: Cook individual portions. Place eggs in baskets; cover with hot water. Bring to a boil; reduce heat; simmer 4 minutes. DO NOT BOIL. Remove from water; serve immediately.

Notes

1. Remove eggs from refrigeration 30 minutes before using.
2. Eggs may be placed in perforated steamer pans and steamed to desired doneness.
3. If hard cooked eggs are to be used in salads or other dishes, plunge into cold running water immediately after cooking; add ice, if necessary, to cool eggs. CCP: Refrigerate at 41 F. or lower.
4. COLD WATER METHOD FOR COOKED EGGS: Place eggs in basket as needed; cover with cold water. Bring to a boil; reduce heat. For soft cooked eggs, simmer 1 minute. For hard cooked eggs, simmer 8 to 10 minutes. DO NOT BOIL.
5. STEAMER METHOD FOR COOKING EGGS: Grease steamer pan. Break eggs individually into a small container before dropping into greased pan. Egg depth should not exceed 2 inches. Place pan, uncovered, in steamer at 5 pound pressure for 6 to 8 minutes or 15 pound pressure for 5 to 7 minutes. Remove pan from steamer; cut eggs for easy removal. CCP: Fresh eggs must be heated to 155 F. or higher for 15 seconds. Consistency of cooked eggs can be controlled by adjusting cooking time.

CHEESE AND EGGS No. F 005 00

DEVILED EGGS

Yield 100   **Portion** 2 Halves

Calories	Carbohydrates	Protein	Fat	Cholesterol	Sodium	Calcium
115 cal	2 g	6 g	9 g	214 mg	137 mg	26 mg

Ingredient   Weight   Measure   Issue

Ingredient	Weight	Measure	Issue
EGG, HARD COOKED	11 lbs	100 Eggs	
MUSTARD, PREPARED	4-3/8 oz	1/2 cup	
PICKLE RELISH, SWEET, DRAINED	8-5/8 oz	1 cup	
SALAD DRESSING, MAYONNAISE TYPE	1-1/2 lbs	3 cup	
PAPRIKA, GROUND	1/4 oz	1 tbsp	

Method

1. Cool; remove shells from eggs; slice eggs in half lengthwise. Arrange 100 egg halves in each steam table pan. CCP: All fresh shell eggs must be heated to 155 F. or higher for 15 seconds.
2. Cool; remove shells from eggs; cut eggs in half lengthwise. Remove yolks and mash thoroughly. Set whites aside for use in Step 4.
3. Blend mustard, pickle relish and salad dressing with yolks. Mix until well blended.
4. Fill the cooked whites with yolk mixture, using 1 tablespoon filling for each egg half.
5. Sprinkle paprika on top.
6. Serve immediately or cover and refrigerate until ready to serve. CCP: Hold for service at 41 F. or lower.

CHEESE AND EGGS No.F 006 00

EGG FOO YOUNG

Yield 100   Portion 1 Omelet

Calories	Carbohydrates	Protein	Fat	Cholesterol	Sodium	Calcium
157 cal	4 g	10 g	12 g	134 mg	490 mg	27 mg

Ingredient	Weight	Measure	Issue
OIL,SALAD	7-2/3 oz	1 cup	
FLOUR,WHEAT,GENERAL PURPOSE	8-7/8 oz	2 cup	
CHICKEN BROTH		1 gal	
SOY SAUCE	10-1/8 oz	1 cup	
MOLASSES	1-1/2 oz	2 tbsp	
ONIONS,FRESH,CHOPPED	1-1/3 lbs	3-3/4 cup	1-1/2 lbs
PEPPERS,GREEN,FRESH,CHOPPED	7-7/8 oz	1-1/2 cup	9-5/8 oz
OIL,SALAD	1-7/8 oz	1/4 cup 1/3 tbsp	
CHICKEN,COOKED,DICED	4 lbs		
BEAN SPROUTS,CANNED,DRAINED	1-7/8 lbs	3 qts 2 cup	
PEPPER,BLACK,GROUND	1/8 oz	1/3 tsp	
EGGS,WHOLE,FROZEN,BEATEN	6 lbs	2 qts 3-1/4 cup	
OIL,SALAD	1 lbs	2 cup	

Method

1. Blend salad oil or shortening and flour; stir until smooth.
2. Prepare broth according to package directions. Add flour mixture to broth; mix well. Bring to a boil; reduce heat; simmer 10 minutes or until thickened.
3. Add soy sauce and molasses to sauce; simmer 5 minutes.
4. Saute onions and peppers in salad oil or olive oil until tender.
5. Combine sauteed vegetables, meat, bean sprouts, and pepper; mix well.
6. Add eggs to meat mixture; blend well.
7. Place 1/3 cup mixture on 375 F. well greased griddle; cook about 3 minutes on each side or until well done. CCP: Internal temperature must reach 165 F. or higher for 15 seconds.
8. Pour 2 tablespoons sauce over each omelet just before serving. CCP: Hold for service at 140 F. or higher.

CHEESE AND EGGS No.F 007 00

GRIDDLE FRIED EGGS

Yield 100 **Portion** 2 Each

Calories	Carbohydrates	Protein	Fat	Cholesterol	Sodium	Calcium
168 cal	1 g	12 g	12 g	425 mg	126 mg	49 mg

Ingredient	**Weight**	**Measure**	**Issue**
EGGS,WHOLE,FRESH	22 lbs	200 each	
OIL,SALAD	7-2/3 oz	1 cup	

Method

1. Break 2 eggs individually into a small bowl.
2. Fry eggs to order on a 325 F. lightly greased griddle. CCP: Internal temperature must reach 145 F. or higher for 15 seconds, 155 F. for fresh shell eggs.
3. CCP: Hold for service at 140 F. or higher.

CHEESE AND EGGS No.F 008 00

PLAIN OMELET

Yield 100   **Portion** 1 Omelet

Calories	Carbohydrates	Protein	Fat	Cholesterol	Sodium	Calcium
139 cal	1 g	11 g	10 g	392 mg	121 mg	54 mg

Ingredient   Weight   Measure   Issue

Ingredient	Weight	Measure
EGGS,WHOLE,FROZEN	20 lbs	2 gal 1-1/3 qts
COOKING SPRAY,NONSTICK	2 oz	1/4 cup 1/3 tbsp

Method

1. Place thawed eggs in mixer bowl. Using wire whip, beat just enough to thoroughly blend yolks and whites.
2. Lightly spray griddle with non-stick cooking spray. Pour 1/3 cup egg mixture for individual omelets on 325 F. griddle.
3. Cook until bottom is golden brown. DO NOT STIR. If necessary, gently lift cooked portion with a spatula to permit uncooked mixture to flow underneath. Continue cooking until eggs are set and well done. CCP: Internal temperature must reach 145 F. or higher for 15 seconds, 155 F. for fresh shell eggs.
4. Fold omelet in half or into thirds making a long oval shaped omelet. CCP: Hold for service at 140 F. or higher.

CHEESE AND EGGS No. F 008 01

PLAIN OMELET (FROZEN EGGS AND EGG WHITES)

Yield 100　　　　　　　　　　　　　　　　　　　　**Portion** 1 Omelet

Calories	Carbohydrates	Protein	Fat	Cholesterol	Sodium	Calcium
93 cal	1 g	10 g	5 g	196 mg	132 mg	30 mg

Ingredient	**Weight**	**Measure**	**Issue**
EGGS,WHOLE,FROZEN	10 lbs	1 gal 2/3 qts	
EGG WHITES,FROZEN,THAWED	10 lbs	1 gal 2/3 qts	
COOKING SPRAY,NONSTICK	2 oz	1/4 cup 1/3 tbsp	

Method

1 Thaw eggs and egg whites; place eggs in mixer bowl. Using wire whip beat just enough to thoroughly blend yolks and whites.
2 Lightly spray griddle with non-stick cooking spray. Pour 1/3 cup egg mixture for individual omelets on 325 F. griddle.
3 Cook until bottom is golden brown. DO NOT STIR. If necessary, gently lift cooked portion with a spatula to permit uncooked mixture to flow underneath. Continue cooking until eggs are set and well done. CCP: Internal temperature must reach 145 F. or higher for 15 seconds, 155 F. for fresh shell eggs.
4 Fold omelet in half or into thirds making a long oval shaped omelet. CCP: Hold for service at 140 F. or higher.

CHEESE AND EGGS No. F 008 03

CHEESE OMELET

Yield 100   **Portion** 1 Omelet

Calories	Carbohydrates	Protein	Fat	Cholesterol	Sodium	Calcium
198 cal	1 g	14 g	15 g	407 mg	212 mg	160 mg

Ingredient	Weight	Measure	Issue
EGGS,WHOLE,FROZEN	20 lbs	2 gal 1-1/3 qts	
COOKING SPRAY,NONSTICK	2 oz	1/4 cup 1/3 tbsp	
CHEESE,CHEDDAR,SHREDDED	3-1/4 lbs	3 qts 1 cup	

Method

1. Place thawed eggs in mixer bowl. Using wire whip, beat just enough to thoroughly blend.
2. Lightly spray griddle with non-stick cooking spray. Pour 1/3 cup egg mixture for individual omelets on 325 F. griddle.
3. Cook until bottom is golden brown. DO NOT STIR. If necessary, gently lift cooked portion with a spatula to permit uncooked mixture to flow underneath. Sprinkle about 2 tablespoons cheese over each omelet when partially set. Continue cooking until eggs are set and well done. CCP: Internal temperature must reach 145 F. or higher for 15 seconds, 155 F. for fresh shell eggs.
4. Fold omelet in half or into thirds making a long oval shaped omelet. CCP: Hold for service at 140 F. or higher.

CHEESE AND EGGS No.F 008 04

GREEN PEPPER OMELET

Yield 100   Portion   1 Omelet

Calories	Carbohydrates	Protein	Fat	Cholesterol	Sodium	Calcium
152 cal	3 g	11 g	10 g	392 mg	121 mg	56 mg

Ingredient	Weight	Measure	Issue
COOKING SPRAY,NONSTICK	2 oz	1/4 cup 1/3 tbsp	
PEPPERS,GREEN,FRESH,CHOPPED	7-1/8 lbs	1 gal 1-1/2 qts	8-2/3 lbs
EGGS,WHOLE,FROZEN	20 lbs	2 gal 1-1/3 qts	
COOKING SPRAY,NONSTICK	2 oz	1/4 cup 1/3 tbsp	

Method

1. Lightly spray griddle with non-stick cooking spray. Cook chopped fresh sweet peppers until tender.
2. Place thawed eggs in mixer bowl. Using wire whip, beat just enough to thoroughly blend. Lightly spray griddle with non-stick cooking spray. Pour 1/3 cup egg mixture for individual omelets on 325 F. griddle.
3. Cook until bottom is golden brown. DO NOT STIR. If necessary, gently lift cooked portion with a spatula to permit uncooked mixture to flow underneath. Sprinkle 2 tablespoons peppers over eggs when partially set. Continue cooking until eggs are set and well done. CCP: Internal temperature must reach 145 F. or higher for 15 seconds, 155 F. for fresh shell eggs.
4. Fold omelet in half or into thirds making a long oval shaped omelet. CCP: Hold for service at 140 F. or higher.

CHEESE AND EGGS No.F 008 05

HAM OMELET

Yield 100   Portion 1 Omelet

Calories	Carbohydrates	Protein	Fat	Cholesterol	Sodium	Calcium
166 cal	1 g	14 g	11 g	401 mg	352 mg	55 mg

Ingredient
EGGS,WHOLE,FROZEN
COOKING SPRAY,NONSTICK
HAM,COOKED,BONELESS

Weight
20 lbs
2 oz
4 lbs

Measure
2 gal 1-1/3 qts
1/4 cup 1/3 tbsp

Issue

Method
1 Place thawed eggs in mixer bowl. Using wire whip, beat just enough to thoroughly blend.
2 Lightly spray griddle with non-stick cooking spray. Pour 1/3 cup egg mixture for individual omelets on 325 F. griddle.
3 Cook until bottom is golden brown. DO NOT STIR. If necessary, gently lift cooked portion with a spatula to permit uncooked mixture to flow underneath.
4 Dice ham. Sprinkle 2 tablespoons ham over eggs when partially set. Continue cooking until eggs are set and well done. CCP: Internal temperature must reach 145 F. or higher for 15 seconds, 155 F. for fresh shell eggs.
5 Fold omelet in half or into thirds making a long oval shaped omelet. CCP: Hold for service at 140 F. or higher.

CHEESE AND EGGS No.F 008 06

HAM AND CHEESE OMELET

Yield 100 **Portion** 1 Omelet

Calories	Carbohydrates	Protein	Fat	Cholesterol	Sodium	Calcium
180 cal	1 g	14 g	13 g	404 mg	278 mg	103 mg

Ingredient   Weight   Measure   Issue

EGGS,WHOLE,FROZEN   20 lbs   2 gal 1-1/3 qts
COOKING SPRAY,NONSTICK   2 oz   1/4 cup 1/3 tbsp
CHEESE,CHEDDAR,SHREDDED   1-1/2 lbs   1 qts 2 cup
HAM,COOKED,BONELESS   2 lbs

Method

1. Place thawed eggs in mixer bowl. Using wire whip, beat just enough to thoroughly blend.
2. Lightly spray griddle with non-stick cooking spray. Pour 1/3 cup egg mixture for individual omelets on 325 F. griddle.
3. Cook until bottom is golden brown. DO NOT STIR. If necessary, gently lift cooked portion with a spatula to permit uncooked mixture to flow underneath.
4. Dice ham. Sprinkle about 1 tablespoon cheese and 1 tablespoon ham over eggs when partially set. Continue cooking until eggs are set and well done. CCP: Internal temperature must reach 145 F. or higher for 15 seconds, 155 F. for fresh shell eggs.
5. Fold omelet in half or into thirds making a long oval shaped omelet. CCP: Hold for service at 140 F. or higher.

CHEESE AND EGGS No.F 008 08

MUSHROOM OMELET

Yield 100   **Portion** 1 Omelet

Calories	Carbohydrates	Protein	Fat	Cholesterol	Sodium	Calcium
151 cal	3 g	11 g	10 g	392 mg	258 mg	57 mg

Ingredient | **Weight** | **Measure** | **Issue**

MUSHROOMS,CANNED,SLICED,DRAINED   7-1/8 lbs   1 gal 1-1/8 qts
COOKING SPRAY,NONSTICK   2 oz   1/4 cup 1/3 tbsp
EGGS,WHOLE,FROZEN   20 lbs   2 gal 1-1/3 qts
COOKING SPRAY,NONSTICK   2 oz   1/4 cup 1/3 tbsp

Method

1. Lightly spray griddle with non-stick cooking spray. Cook mushrooms until tender.
2. Place thawed eggs in mixer bowl. Using wire whip, beat just enough to thoroughly blend.
3. Lightly spray griddle with non-stick cooking spray. Pour 1/3 cup egg mixture for each individual omelet on 325 F. griddle.
4. Cook until bottom is golden brown. DO NOT STIR. When omelet is partially set, sprinkle about 1-1/2 tablespoon mushrooms over eggs and continue cooking until eggs are set and well done. If necessary when cooking, lift cooked portion with spatula to let uncooked mixture flow underneath. CCP: Internal temperature must reach 145 F. or higher for 15 seconds, 155 F. for fresh shell eggs.
5. Fold omelet in half or into thirds making a long oval shaped omelet. CCP: Hold for service at 140 F. or higher.

CHEESE AND EGGS No. F 008 09

ONION OMELET

Yield 100   **Portion** 1 Omelet

Calories	Carbohydrates	Protein	Fat	Cholesterol	Sodium	Calcium
150 cal	3 g	11 g	10 g	392 mg	121 mg	57 mg

Ingredient

Ingredient	Weight	Measure	Issue
ONIONS,FRESH,CHOPPED	4-1/4 lbs	3 qts	4-2/3 lbs
COOKING SPRAY,NONSTICK	2 oz	1/4 cup 1/3 tbsp	
EGGS,WHOLE,FROZEN	20 lbs	2 gal 1-1/3 qts	
COOKING SPRAY,NONSTICK	2 oz	1/4 cup 1/3 tbsp	

Method

1. Lightly spray griddle with non-stick cooking spray. Cook onions until tender.
2. Place thawed eggs in a mixer bowl. Using wire whip, beat just enough to thoroughly blend.
3. Lightly spray griddle with non-stick cooking spray. Pour 1/3 cup egg mixture for each individual omelet on 325 F. griddle.
4. Cook until bottom is golden brown. DO NOT STIR. If necessary, gently lift cooked portion with a spatula to permit uncooked portion to flow underneath. Sprinkle 1 tablespoon onions over eggs when partially set. Continue cooking until eggs are set and well done. CCP: Internal temperature must reach 145 F. or higher for 15 seconds, 155 F. for fresh shell eggs.
5. Fold omelet in half or into thirds making a long oval shaped omelet. CCP: Hold for service at 140 F. or higher.

CHEESE AND EGGS No.F 008 10

WESTERN OMELET

Yield 100   **Portion** 1 Omelet

Calories	Carbohydrates	Protein	Fat	Cholesterol	Sodium	Calcium
170 cal	4 g	13 g	11 g	396 mg	237 mg	60 mg

Ingredient / Weight / Measure / Issue

Ingredient	Weight	Measure	Issue
ONIONS,FRESH,CHOPPED	4-1/4 lbs	3 qts	4-2/3 lbs
COOKING SPRAY,NONSTICK	2 oz	1/4 cup 1/3 tbsp	
PEPPERS,GREEN,FRESH,CHOPPED	5-1/4 lbs	1 gal	6-3/8 lbs
HAM,COOKED,BONELESS	2 lbs		
EGGS,WHOLE,FROZEN	20 lbs	2 gal 1-1/3 qts	
COOKING SPRAY,NONSTICK	2 oz	1/4 cup 1/3 tbsp	

Method

1. Lightly spray griddle with non-stick cooking spray. Cook onions and peppers until tender.
2. Chop or grind ham. Combine cooked onions and peppers with chopped ham; mix thoroughly.
3. Place thawed eggs in mixer bowl. Using wire whip, beat just enough to thoroughly blend.
4. Lightly spray griddle with non-stick cooking spray. Pour 1/3 cup egg mixture for each individual omelet on griddle.
5. Cook until bottom is golden brown. DO NOT STIR. If necessary, gently lift cooked portion with a spatula to permit uncooked mixture to flow underneath. Sprinkle about 3 tablespoons onion/pepper/ham mixture over eggs when partially set. Continue cooking until eggs are set. CCP: Internal temperature must reach 145 F. or higher for 15 seconds, 155 F. for fresh shell eggs.
6. Fold omelet in half or into thirds making a long, oval shaped omelet. CCP: Hold for service at 140 F. or higher.

CHEESE AND EGGS No.F 008 11

TOMATO OMELET

Yield 100   **Portion** 1 Omelet

Calories	Carbohydrates	Protein	Fat	Cholesterol	Sodium	Calcium
145 cal	2 g	11 g	10 g	392 mg	123 mg	55 mg

Ingredient   Weight   Measure   Issue

Ingredient	Weight	Measure	Issue
EGGS,WHOLE,FROZEN	20 lbs	2 gal 1-1/3 qts	
COOKING SPRAY,NONSTICK	2 oz	1/4 cup 1/3 tbsp	
TOMATOES,FRESH,CHOPPED	6-3/4 lbs	1 gal 1/4 qts	6-7/8 lbs

Method

1. Place thawed eggs in mixer bowl. Using wire whip, beat just enough to thoroughly blend.
2. Lightly spray griddle with non-stick cooking spray. Pour 1/3 cup egg mixture for each individual omelet on 325 F. griddle.
3. Cook until bottom is golden brown. DO NOT STIR. If necessary, gently lift with a spatula to permit uncooked mixture to flow underneath. Sprinkle 2 tablespoons tomatoes over eggs when partially set. Continue cooking until eggs are set. CCP: Internal temperature must reach 145 F. or higher for 15 seconds, 155 F. for fresh shell eggs.
4. Fold omelet in half or into thirds making a long, oval shaped omelet. CCP: Hold for service at 140 F. or higher.

CHEESE AND EGGS No.F 008 12

SPANISH OMELET

Yield 100 **Portion** 1 Omelet

Calories	Carbohydrates	Protein	Fat	Cholesterol	Sodium	Calcium
183 cal	9 g	12 g	11 g	392 mg	364 mg	82 mg

Ingredient	Weight	Measure	Issue
SPANISH SAUCE		2 gal 1/4 qts	
EGGS,WHOLE,FROZEN	20 lbs	2 gal 1-1/3 qts	
COOKING SPRAY,NONSTICK	2 oz	1/4 cup 1/3 tbsp	

Method

1. Prepare 1 recipe Spanish Sauce, Recipe No. O 005 01 for use in Step 6. CCP: Hold for service at 140 F. or higher.
2. Place thawed eggs in mixer bowl. Using wire whip, beat just enough to thoroughly blend.
3. Lightly spray griddle with non-stick cooking spray. Pour 1/3 cup egg mixture for individual omelets on 325 F. griddle.
4. Cook until bottom is golden brown. DO NOT STIR. If necessary, gently lift cooked portion with a spatula to permit uncooked mixture to flow underneath. Continue cooking until eggs are set. CCP: Internal temperature must reach 145 F. or higher for 15 seconds, 155 F. for fresh shell eggs.
5. Fold omelet in half or into thirds, making a long oval shaped omelet.
6. Serve each omelet with 2 ounces of heated Spanish Sauce, Recipe No. O 005 01. CCP: Hold for service at 140 F. or higher.

CHEESE AND EGGS No. F 009 00

POACHED EGGS

Yield 100 **Portion** 2 Each

Calories	Carbohydrates	Protein	Fat	Cholesterol	Sodium	Calcium
149 cal	1 g	12 g	10 g	425 mg	127 mg	49 mg

Ingredient   **Weight**   **Measure**   **Issue**
WATER   4-1/8 lbs   2 qts
VINEGAR, DISTILLED   1 oz   2 tbsp
EGGS, WHOLE, FRESH   22 lbs   200 each

Method

1. Fill a steam table pan with water to a depth of 1 inch.
2. Add vinegar; bring to a boil; reduce to a simmer.
3. Break 2 eggs individually into a small bowl; slide gently into simmering water.
4. Cook 3 to 5 minutes or until whites are set and yolks are covered with a white film. CCP: Internal temperature must reach 155 F. or higher for 15 seconds.
5. Using a perforated skimmer, lift eggs out of pan; serve immediately. CCP: Hold for service at 140 F. or higher.

CHEESE AND EGGS No.F 010 00

SCRAMBLED EGGS

Yield 100 **Portion** 1/3 Cup

Calories	Carbohydrates	Protein	Fat	Cholesterol	Sodium	Calcium
144 cal	1 g	11 g	10 g	392 mg	121 mg	54 mg

Ingredient	Weight	Measure	Issue
EGGS,WHOLE,FROZEN	20 lbs	2 gal 1-1/3 qts	
OIL,SALAD	3-7/8 oz	1/2 cup	

Method

1 Beat eggs thoroughly.
2 Pour about 1 quart eggs on 325 F. lightly greased griddle. Cook slowly until firm, until there is no visible liquid egg, stirring occasionally. CCP: Internal temperature must reach 145 F. or higher for 15 seconds, 155 F. for fresh shell eggs. Hold for service at 140 F. or higher.

Notes

1 OVEN METHOD: Using a convection oven, bake at 350 F. 18 to 25 minutes on high fan, closed vent. After 12 minutes, stir every 5 minutes.

CHEESE AND EGGS No.F 010 01

SCRAMBLED EGGS AND CHEESE

Yield 100 **Portion** 1/3 Cup

Calories	Carbohydrates	Protein	Fat	Cholesterol	Sodium	Calcium
217 cal	1 g	15 g	16 g	411 mg	233 mg	184 mg

Ingredient	Weight	Measure	Issue
EGGS,WHOLE,FROZEN	20 lbs	2 gal 1-1/3 qts	
OIL,SALAD	3-7/8 oz	1/2 cup	
CHEESE,CHEDDAR,SHREDDED	4 lbs	1 gal	

Method

1. Beat eggs thoroughly.
2. Pour about 1 quart eggs on 325 F. lightly greased griddle. Sprinkle cheese, using about 1 cup per 1 quart of egg mixture, over partially cooked eggs. Stir gently until cheese is melted and well blended. Cook slowly until firm or until there is no visible liquid egg, stirring occasionally. CCP: Internal temperature must reach 145 F. or higher for 15 seconds, 155 F. for fresh shell eggs. Hold at 140 F. or higher.

Notes

1. OVEN METHOD: Using a 350 F. convection oven, bake 18 to 25 minutes on high fan, closed vent. After 12 minutes, stir every 5 minutes.

CHEESE AND EGGS No.F 010 02

SCRAMBLED EGGS AND HAM

Yield 100 **Portion** 1/3 Cup

Calories	Carbohydrates	Protein	Fat	Cholesterol	Sodium	Calcium
171 cal	1 g	14 g	12 g	401 mg	352 mg	55 mg

Ingredient	**Weight**	**Measure**	**Issue**
EGGS,WHOLE,FROZEN	20 lbs	2 gal 1-1/3 qts	
OIL,SALAD	3-7/8 oz	1/2 cup	
HAM,COOKED,BONELESS	4 lbs		

Method

1 Beat eggs thoroughly.
2 Pour about 1 quart eggs on 325 F. lightly greased griddle. Dice ham. Add diced ham, about 1 cup per 1 quart of egg mix, over partially cooked eggs. Stir well. Cook slowly until firm or until there is no visible liquid egg, stirring occasionally. CCP: Internal temperature must reach 145 F. or higher for 15 seconds, 155 F. for fresh shell eggs. Hold at 140 F. or higher.

Notes

1 Using a 350 F. convection oven, bake 18 to 25 minutes on high fan, closed vent. After 12 minutes, stir every 5 minutes.

CHEESE AND EGGS No.F 010 03

SCRAMBLED EGGS (DEHYDRATED EGG MIX)

Yield 100   Portion   1/3 Cup

Calories	Carbohydrates	Protein	Fat	Cholesterol	Sodium	Calcium
166 cal	1 g	12 g	12 g	451 mg	140 mg	62 mg

Ingredient	**Weight**	**Measure**	**Issue**
EGG MIX,DEHYDRATED	5-3/4 lbs	5 #3cyl	
WATER,WARM	15-2/3 lbs	1 gal 3-1/2 qts	
OIL,SALAD	3-7/8 oz	1/2 cup	

Method

1 Combine egg mix and warm water.
2 Pour about 1 quart eggs on 325 F. lightly greased griddle. Cook slowly until firm or until there is no visible liquid egg, stir occasionally. CCP: Internal temperature must reach 145 F. or higher for 15 seconds. Hold at 140 F. or higher.

Notes

1 Using a 350 F. convection oven, bake 18 to 25 minutes on high fan, closed vent. After 12 minutes, stir every 5 minutes.

CHEESE AND EGGS No.F 010 05

SCRAMBLED EGGS (FROZEN EGGS AND EGG WHITES)

Yield 100   Portion   1/3 Cup

Calories	Carbohydrates	Protein	Fat	Cholesterol	Sodium	Calcium
98 cal	1 g	10 g	6 g	196 mg	132 mg	30 mg

Ingredient	**Weight**	**Measure**	**Issue**
EGGS,WHOLE,FROZEN	10 lbs	1 gal 2/3 qts	
EGG WHITES,FROZEN,THAWED	10 lbs	1 gal 2/3 qts	
OIL,SALAD	3-7/8 oz	1/2 cup	

Method

1. Combine whole table eggs and frozen egg whites. Beat eggs thoroughly.
2. Pour about 1 quart eggs on 325 F. lightly greased griddle. Cook slowly until firm or until there is no visible liquid egg, stir occasionally. CCP: Internal temperature must reach 145 F. or higher for 15 seconds. Hold for service at 140 F. or higher.

Notes

1. Using a 350 F. convection oven, bake 18 to 25 minutes on high fan, closed vent. After 12 minutes, stir every 5 minutes.

CHEESE AND EGGS No.F 011 00

MUSHROOM QUICHE

Yield 100   Portion   4-1/2 Ounces

Calories	Carbohydrates	Protein	Fat	Cholesterol	Sodium	Calcium
199 cal	16 g	11 g	10 g	114 mg	267 mg	231 mg

Ingredient

Ingredient	Weight	Measure	Issue
MUSHROOMS,CANNED,SLICED,DRAINED	4-1/8 lbs	3 qts	
ONIONS,FRESH,CHOPPED	2-1/3 lbs	1 qts 2-5/8 cup	2-5/8 lbs
CHEESE,SWISS,SHREDDED	3-3/4 lbs	1 gal	
COOKING SPRAY,NONSTICK	2 oz	1/4 cup 1/3 tbsp	
FLOUR,WHEAT,BREAD	3-1/3 lbs	2 qts 3 cup	
MILK,NONFAT,DRY	1-1/4 oz	1/2 cup	
SALT	3/8 oz	1/3 tsp	
SUGAR,GRANULATED	1-3/4 oz	1/4 cup 1/3 tbsp	
BAKING SODA	5/8 oz	1 tbsp	
SHORTENING	7-1/4 oz	1 cup	
MILK,NONFAT,DRY	11-3/8 oz	1 qts 3/4 cup	
WATER,WARM	11-1/2 lbs	1 gal 1-1/2 qts	
EGGS,WHOLE,FROZEN	5 lbs	2 qts 1-3/8 cup	
GARLIC POWDER	3/4 oz	2-2/3 tbsp	

Method

1. Lightly spray each steam table pan with non-stick cooking spray. Combine mushrooms, onions and cheese. Spread 1-3/4 quarts evenly over bottom of each sprayed and floured pan.
2. Combine flour, milk, salt, sugar and soda in mixer bowl.
3. Cut in shortening or oil until evenly distributed and granular in appearance, about 1 minute.
4. Reconstitute milk.
5. Add eggs to milk; blend in garlic powder.
6. Add egg-milk mixture gradually to flour mixture. Scrape down bowl; beat 2 minutes at medium speed.
7. Pour about 9-1/2 cups batter over cheese and vegetable mixture in each pan. Stir gently.
8. Using a convection oven, bake at 350 F. 15 minutes on low fan, closed vent; reduce heat to 325 F.; bake an additional 30 minutes or until set and lightly browned. Let stand 10 minutes. Cut 5 by 5. CCP: Internal temperature must reach 145 F. or higher for 15 seconds. Hold for service at 140 F. or higher.

CHEESE AND EGGS No.F 011 01

BROCCOLI QUICHE

Yield 100   Portion 4-1/2 Ounces

Calories	Carbohydrates	Protein	Fat	Cholesterol	Sodium	Calcium
201 cal	16 g	12 g	10 g	114 mg	194 mg	242 mg

Ingredient	Weight	Measure	Issue
COOKING SPRAY,NONSTICK	2 oz	1/4 cup 1/3 tbsp	
ONIONS,FRESH,CHOPPED	1-1/8 lbs	3-3/8 cup	1-1/3 lbs
BROCCOLI,FROZEN,CHOPPED	6 lbs	1 gal	
CHEESE,SWISS,SHREDDED	3-3/4 lbs	1 gal	
FLOUR,WHEAT,BREAD	3-1/3 lbs	2 qts 3 cup	
MILK,NONFAT,DRY	1-1/4 oz	1/2 cup	
SALT	3/8 oz	1/3 tsp	
SUGAR,GRANULATED	1-3/4 oz	1/4 cup 1/3 tbsp	
BAKING SODA	5/8 oz	1 tbsp	
SHORTENING	7-1/4 oz	1 cup	
MILK,NONFAT,DRY	11-3/8 oz	1 qts 3/4 cup	
WATER,WARM	11-1/2 lbs	1 gal 1-1/2 qts	
EGGS,WHOLE,FROZEN	5 lbs	2 qts 1-3/8 cup	
GARLIC POWDER	3/4 oz	2-2/3 tbsp	
NUTMEG,GROUND	1/8 oz	1/3 tsp	
PEPPER,BLACK,GROUND	1/3 oz	1 tbsp	

Method

1 Lightly spray each steam table pan with non-stick cooking spray. Thaw broccoli. Combine broccoli, onions and cheese. Spread about 2 quarts mixture in each sprayed and floured pan.
2 Combine flour, milk, salt, sugar and soda in mixer bowl.
3 Cut in shortening or oil until evenly distributed and granular in appearance, about 1 minute.
4 Reconstitute milk.
5 Add eggs, nutmeg and black pepper to milk; blend in garlic powder.
6 Add egg-milk mixture gradually to flour mixture. Scrape down bowl; beat 2 minutes at medium speed.
7 Pour about 9-1/2 cups batter over cheese and vegetable mixture in each pan. Stir gently.
8 Using a convection oven, bake at 350 F. 15 minutes on low fan, closed vent; reduce temperature to 325 F.; bake an additional 30 minutes or until set and lightly browned. Let stand 10 minutes. Cut 5 by 5. CCP: Internal temperature must reach 145 F. or higher for 15 seconds. Hold for service at 140 F. or higher.

CHEESE AND EGGS No.F 011 02

BROCCOLI QUICHE (FROZEN EGGS AND EGG WHITES)

Yield 100   Portion   4-1/2 Ounces

Calories	Carbohydrates	Protein	Fat	Cholesterol	Sodium	Calcium
189 cal	16 g	11 g	9 g	64 mg	195 mg	236 mg

Ingredient	Weight	Measure	Issue
ONIONS,FRESH,CHOPPED	1-1/8 lbs	3-3/8 cup	1-1/3 lbs
BROCCOLI,FROZEN,CHOPPED	6 lbs	1 gal	
CHEESE,SWISS,SHREDDED	3-3/4 lbs	1 gal	
COOKING SPRAY,NONSTICK	2 oz	1/4 cup 1/3 tbsp	
FLOUR,WHEAT,BREAD	3-1/3 lbs	2 qts 3 cup	
MILK,NONFAT,DRY	1-1/4 oz	1/2 cup	
SALT	3/8 oz	1/3 tsp	
SUGAR,GRANULATED	1-3/4 oz	1/4 cup 1/3 tbsp	
BAKING SODA	5/8 oz	1 tbsp	
SHORTENING	7-1/4 oz	1 cup	
MILK,NONFAT,DRY	11-3/8 oz	1 qts 3/4 cup	
WATER,WARM	11-1/2 lbs	1 gal 1-1/2 qts	
EGGS,WHOLE,FROZEN	2-3/8 lbs	1 qts 1/2 cup	
EGG WHITES,FROZEN,THAWED	2-3/8 lbs	1 qts 1/2 cup	
GARLIC POWDER	3/4 oz	2-2/3 tbsp	
NUTMEG,GROUND	1/8 oz	1/3 tsp	
PEPPER,BLACK,GROUND	1/3 oz	1 tbsp	

Method

1. Lightly spray each steam table pan with non-stick cooking spray. Thaw and cut broccoli in 1/2-inch pieces. Combine broccoli, onions and cheese. Spread about 2 quarts mixture in each sprayed and floured pan.
2. Combine flour, milk, salt, sugar and soda in mixer bowl.
3. Cut in shortening or oil until evenly distributed and granular in appearance, about 1 minute.
4. Reconstitute milk.
5. Add eggs, nutmeg, and black pepper to milk; blend in garlic powder.
6. Add egg-milk mixture gradually to flour mixture. Scrape down bowl; beat 2 minutes at medium speed.
7. Pour about 9-1/2 cups batter over cheese and vegetable mixture in each pan. Stir gently.
8. Using a convection oven, bake at 325 F. for 40 minutes on low fan, closed vent or until set and lightly browned. Let stand 10 minutes. Cut 5 by 5. CCP: Internal temperature must reach 145 F. or higher for 15 seconds. Hold for service at 140 F. or higher.

CHEESE AND EGGS No.F 011 03

MUSHROOM QUICHE (FROZEN EGGS AND EGG WHITES)

Yield 100   Portion   4-1/2 Ounces

Calories	Carbohydrates	Protein	Fat	Cholesterol	Sodium	Calcium
187 cal	16 g	11 g	9 g	64 mg	269 mg	225 mg

Ingredient	Weight	Measure	Issue
MUSHROOMS,CANNED,SLICED,DRAINED	4-1/8 lbs	3 qts	
ONIONS,FRESH,CHOPPED	2-1/3 lbs	1 qts 2-5/8 cup	2-5/8 lbs
CHEESE,SWISS,SHREDDED	3-3/4 lbs	1 gal	
COOKING SPRAY,NONSTICK	2 oz	1/4 cup 1/3 tbsp	
FLOUR,WHEAT,BREAD	3-1/3 lbs	2 qts 3 cup	
MILK,NONFAT,DRY	1-1/4 oz	1/2 cup	
SALT	3/8 oz	1/3 tsp	
SUGAR,GRANULATED	1-3/4 oz	1/4 cup 1/3 tbsp	
BAKING SODA	5/8 oz	1 tbsp	
SHORTENING	7-1/4 oz	1 cup	
MILK,NONFAT,DRY	11-3/8 oz	1 qts 3/4 cup	
WATER,WARM	11-1/2 lbs	1 gal 1-1/2 qts	
EGGS,WHOLE,FROZEN	2-3/8 lbs	1 qts 1/2 cup	
EGG WHITES,FROZEN,THAWED	2-3/8 lbs	1 qts 1/2 cup	
GARLIC POWDER	3/4 oz	2-2/3 tbsp	

Method

1. Lightly spray each steam table pan with non-stick cooking spray. Combine mushrooms, onions and cheese. Spread about 2 pounds 10 ounces evenly over bottom of each sprayed and floured pan.
2. Combine flour, milk, salt, sugar and soda in mixer bowl.
3. Cut in shortening until evenly distributed and granular in appearance, about 1 minute.
4. Reconstitute milk.
5. Thaw eggs under refrigeration. Add eggs to milk; blend in garlic powder.
6. Add egg-milk mixture gradually to flour mixture. Scrape down bowl; beat 2 minutes at medium speed.
7. Pour about 9-1/2 cups of batter over cheese and vegetable mixture in each pan. Stir gently.
8. Using a convection oven, bake at 325 F. for 40 minutes on low fan, closed vent or until set and lightly browned. Let stand 10 minutes. CCP: Internal temperature must reach 145 F. or higher for 15 seconds. Cut 5 by 5.

CHEESE AND EGGS No.F 012 00

BREAKFAST BURRITO

Yield 100　　　　　　　　　　　　　　　　　　　　　　Portion　1 Each

Calories	Carbohydrates	Protein	Fat	Cholesterol	Sodium	Calcium
302 cal	26 g	16 g	14 g	167 mg	499 mg	170 mg

Ingredient	Weight	Measure	Issue
EGG WHITES,FROZEN,THAWED	7-1/2 lbs	3 qts 2 cup	
EGGS,WHOLE,FROZEN	7-1/2 lbs	3 qts 2 cup	
CHEESE,CHEDDAR,SHREDDED	2-2/3 lbs	2 qts 2-5/8 cup	
SAUSAGE,PORK,COOKED,DICED	2 lbs		
TOMATOES,FRESH,CHOPPED	2 lbs	1 qts 1 cup	2 lbs
ONIONS,FRESH,CHOPPED	1 lbs	2-5/8 cup	1 lbs
PEPPER,BLACK,GROUND	1/3 oz	1 tbsp	
OREGANO,CRUSHED	1/2 oz	3 tbsp	
COOKING SPRAY,NONSTICK	2 oz	1/4 cup 1/3 tbsp	
COOKING SPRAY,NONSTICK	2 oz	1/4 cup 1/3 tbsp	
TORTILLAS,FLOUR,8 INCH	9-1/2 lbs	100 each	

Method

1. Combine egg whites and eggs. Blend thoroughly.
2. Combine cheese, sausage, tomatoes, onions, pepper and oregano; mix thoroughly.
3. Lightly spray griddle with non-stick cooking spray. Pour about 1 quart egg mixture on 325 F. lightly sprayed griddle. Cook until partially set. Add 6 ounces cheese-sausage mixture. Cook until cheese is melted and eggs are firm. CCP: Internal temperature must reach 145 F. or higher for 15 seconds.
4. Place tortillas on lightly sprayed griddle; heat 30 seconds on each side.
5. Place about 1/2 cup cooked egg mixture in center of each tortilla; fold tortilla to cover eggs and form burrito.
6. CCP: Hold for service at 140 F. or higher.

Notes

1. In Step 2, 3-1/4 pounds (1/2 No. 10 can) of canned diced tomatoes may be used per 100 portions. Drain before using.

CHEESE AND EGGS No. F 012 01

BREAKFAST PITA

Yield 100   Portion 1 Pita

Calories	Carbohydrates	Protein	Fat	Cholesterol	Sodium	Calcium
418 cal	55 g	21 g	12 g	167 mg	801 mg	198 mg

Ingredient	Weight	Measure	Issue
EGG WHITES,FROZEN,THAWED	7-1/2 lbs	3 qts 2 cup	
EGGS,WHOLE,FROZEN	7-1/2 lbs	3 qts 2 cup	
CHEESE,CHEDDAR,SHREDDED	2-2/3 lbs	2 qts 2-5/8 cup	
SAUSAGE,PORK,COOKED,DICED	2 lbs		
TOMATOES,FRESH,CHOPPED	2 lbs	1 qts 1 cup	2 lbs
ONIONS,FRESH,CHOPPED	1 lbs	2-5/8 cup	1 lbs
PEPPER,BLACK,GROUND	1/3 oz	1 tbsp	
OREGANO,CRUSHED	1/2 oz	3 tbsp	
COOKING SPRAY,NONSTICK	2 oz	1/4 cup 1/3 tbsp	
BREAD,PITA,WHITE,8-INCH	21 lbs	100 each	

Method

1. Combine egg whites and eggs. Blend thoroughly.
2. Combine cheese, sausage, tomatoes, onions, pepper and oregano; mix thoroughly.
3. Pour about 1 quart egg mixture on lightly greased griddle. Cook until partially set. Add cheese-sausage mixture. Cook until cheese is melted and eggs are firm. CCP: Internal temperature must reach 145 F. or higher for 15 seconds.
4. Cut off top third of pita pocket and place eggs in the pocket. Place pockets on sheet pans. Using a convection oven, bake at 350 F. for 5 minutes or until warm and pliable on high fan, closed vent.
5. Place about 1/2 cup egg mixture in each pocket. CCP: Internal temperature must reach 145 F. or higher for 15 seconds. Hold for service at 140 F. or higher.

Notes

1. In Step 2, 3-1/4 pounds (1/2 No. 10 can) of canned diced tomatoes may be used per 100 portions. Drain before using.

CHEESE AND EGGS   No. F 013 00

VEGGIE EGG POCKET

Yield 100   **Portion** 1 Serving

Calories	Carbohydrates	Protein	Fat	Cholesterol	Sodium	Calcium
319 cal	45 g	20 g	6 g	5 mg	952 mg	172 mg

Ingredient	**Weight**	**Measure**	**Issue**
MUSHROOMS,CANNED,SLICED,DRAINED	3-1/8 lbs	2 qts 1 cup	
SQUASH,ZUCCHINI,FRESH,SHREDDED	3-1/4 lbs	2 qts 3-7/8 cup	3-3/8 lbs
CARROTS,FRESH,SHREDDED	4-7/8 lbs	1 gal 1 qts	6 lbs
FLOUR,WHEAT,GENERAL PURPOSE	11 oz	2-1/2 cup	
EGG SUBSTITUTE,PASTEURIZED	22-1/8 lbs	2 gal 2 qts	
SALT	5/8 oz	1 tbsp	
SALAD DRESSING,RANCH,FAT FREE	6-1/3 lbs	3 qts	
CHEESE,PARMESAN,GRATED	1 lbs	1 qts 1/2 cup	
ONIONS,FRESH,CHOPPED	2-1/4 lbs	1 qts 2-3/8 cup	2-1/2 lbs
DILL WEED,DRIED	2/3 oz	1/4 cup 2-1/3 tbsp	
PEPPER,WHITE,GROUND	1/4 oz	1 tbsp	
COOKING SPRAY,NONSTICK	2 oz	1/4 cup 1/3 tbsp	
BREAD,PITA,WHITE,8-INCH	10-1/2 lbs	50 each	

Method

1. Combine mushrooms, carrots, and zucchini. Add flour; toss lightly to coat vegetables.
2. Place egg substitute, ranch dressing, cheese, onions, dillweed, salt and pepper in mixer bowl. Using a wire whip, blend at low speed 1 minute.
3. Add vegetable mixture; mix at low speed 1 minute or until blended.
4. Lightly spray each steam table pan with non-stick cooking spray. Pour 1 gallon of egg mixture in each lightly sprayed pan.
5. Using a convection oven, bake 45-55 minutes or until eggs are set. CCP: Internal temperature must reach 145 F. or higher for 15 seconds.
6. Cut pita pockets in half. Fill each half with 3/4 cup egg mixture. Serve 1 half pocket. CCP: Hold at 140 F. higher for service.

CHEESE AND EGGS No. F 014 00

MONTEREY EGG BAKE

Yield 100 **Portion** 6 Ounces

Calories	Carbohydrates	Protein	Fat	Cholesterol	Sodium	Calcium
181 cal	14 g	19 g	6 g	5 mg	473 mg	166 mg

Ingredient	**Weight**	**Measure**	**Issue**
COOKING SPRAY,NONSTICK	2 oz	1/4 cup 1/3 tbsp	
POTATOES,WHITE,FROZEN,SHREDDED,HASHBROWN	9-1/2 lbs	1 gal 1-1/8 qts	
TOMATOES,CANNED,DICED,DRAINED	4-1/8 lbs	1 qts 3-1/2 cup	
CHEESE,CHEDDAR,LOWFAT,SHREDDED	2-1/4 lbs	2 qts 1 cup	
CHEESE,MONTEREY JACK,REDUCED FAT,SHREDDED	2-1/4 lbs	2 qts 1 cup	
PEPPERS,GREEN,FRESH,CHOPPED	2 lbs	1 qts 2 cup	2-3/8 lbs
CORN,FROZEN,WHOLE KERNEL	2 lbs	1 qts 1-1/2 cup	
PEPPERS,CHILI,GREEN,CANNED,CHOPPED,DRAINED	12-1/4 oz	2-1/2 cup	
ONIONS,GREEN,FRESH,SLICED	1-1/8 lbs	1 qts 1-3/8 cup	1-1/3 lbs
SALT	1 oz	1 tbsp	
PEPPER,WHITE,GROUND	3/8 oz	1 tbsp	
EGG SUBSTITUTE,PASTEURIZED	22-1/8 lbs	2 gal 2 qts	
WATER	3 lbs	1 qts 1-3/4 cup	
MILK,NONFAT,DRY	3 oz	1-1/4 cup	

Method

1. Lightly spray each steam table pan with non-stick cooking spray.
2. Combine potatoes, tomatoes, cheddar cheese, monterey jack cheese, green pepper, corn, green chilies, green onions, salt, and pepper; mix well.
3. Place 2-1/4 quarts of potato mixture into each steam table pan.
4. Combine egg substitute, water and nonfat dry milk; blend until mixed.
5. Pour 1-3/4 quarts of egg mixture into each steam table pan; stir to combine.
6. Using a convection oven, bake at 325 F. for 55 to 65 minutes. CCP: Internal temperature must reach 145 F. or higher for 15 seconds. Hold for service at 140 F. or higher.

CHEESE AND EGGS No.F 015 00

BREAKFAST PIZZA

Yield 100   **Portion** 1 Piece

Calories	Carbohydrates	Protein	Fat	Cholesterol	Sodium	Calcium
346 cal	44 g	24 g	7 g	12 mg	930 mg	184 mg

Ingredient	Weight	Measure	Issue
COOKING SPRAY,NONSTICK	2 oz	1/4 cup 1/3 tbsp	
DOUGH,PIZZA	16 lbs		
SAUCE,TOMATO,CANNED	4-1/3 lbs	2 qts	
BACON,TURKEY,RAW	3 lbs		
EGG SUBSTITUTE,PASTEURIZED	15-1/2 lbs	1 gal 3 qts	
SALT	1/4 oz	1/8 tsp	
PEPPER,BLACK,GROUND	1/8 oz	1/8 tsp	
CHEESE,CHEDDAR,LOWFAT,SHREDDED	6 lbs	1 gal 2 qts	
POTATOES,WHITE,FROZEN,SHREDDED,HASHBROWN	5-1/2 lbs	2 qts 3-7/8 cup	

Method

1. Lightly spray sheet pans with nonstick cooking spray.
2. Shape dough into 4-4 lb pieces. Let dough rest 15 minutes. Place dough pieces on lightly floured working surface. Roll out each piece to 1/4-inch thickness. Transfer dough to pans, pushing dough slightly up edges of pans. Gently prick dough to prevent bubbling.
3. Using a convection oven, bake 8 minutes at 450 F. on high fan, open vent until crusts are lightly browned.
4. Spread 2 cups tomato sauce evenly over crust in each pan. Set aside for use in Step 7.
5. Cook bacon until lightly browned. Drain on absorbent paper. Finely chop.
6. Add salt and pepper to eggs. Blend well. Scramble eggs until just set. Do not over cook. Pasteurized eggs will be safe at an internal temperature of 145 F. but will not set until they reach 160 F.
7. Distribute 1-1/2 quarts cheese over sauce on each crust.
8. Distribute 1-1/2 quarts scrambled eggs over cheese on each pan.
9. Distribute 1-1/4 cups bacon over eggs on each pan.
10. Distribute 1 quart shredded potatoes over bacon in each pan.
11. Using a convection oven, bake another 8 minutes or until crust is browned and hash browns begin to turn golden brown on high fan, open vent. CCP: Internal temperature must reach 145 F. or higher for 15 seconds.
12. Cut 5 by 5. CCP: Hold for service at 140 F. or higher.

CHEESE AND EGGS No.F 015 01

MEXICAN BREAKFAST PIZZA

Yield 100   Portion 1 Piece

Calories	Carbohydrates	Protein	Fat	Cholesterol	Sodium	Calcium
364 cal	50 g	26 g	6 g	6 mg	880 mg	189 mg

Ingredient	Weight	Measure	Issue
COOKING SPRAY,NONSTICK	2 oz	1/4 cup 1/3 tbsp	
DOUGH,PIZZA	16 lbs		
SAUCE,SALSA	5-3/8 lbs	2 qts 2 cup	
PEPPER,BLACK,GROUND	1/8 oz	1/8 tsp	
SALT	1/4 oz	1/8 tsp	
EGG SUBSTITUTE,PASTEURIZED	15-1/2 lbs	1 gal 3 qts	
CHEESE,MONTEREY JACK,REDUCED FAT,SHREDDED	6 lbs	1 gal 2 qts	
BEANS,BLACK,CANNED,DRAINED	5-1/3 lbs	2 qts 1-1/2 cup	
POTATOES,WHITE,FROZEN,SHREDDED,HASHBROWN	5-1/2 lbs	2 qts 3-7/8 cup	

Method

1. Lightly spray sheet pans with non-stick cooking spray.
2. Shape dough into four 4 lb pieces. Let dough rest 15 minutes. Place dough pieces on lightly floured working surface. Roll out each piece to 1/4-inch thickness. Transfer dough to pans, pushing dough slightly up edges of pans. Gently prick dough to prevent bubbling.
3. Using a convection oven, bake 8 minutes at 450 F. on high fan, open vent until crusts are lightly browned.
4. Spread 2-1/2 cups salsa evenly over crust in each pan. Set aside for use in Step 6.
5. Add salt and pepper to eggs. Blend well. Scramble eggs until just set. Do not over cook. Pasteurized eggs will be safe at an internal temperature of 145 F. but will not set until they reach 160 F.
6. Distribute 1-1/2 qt cheese over sauce on each crust.
7. Distribute 1-1/2 qt scrambled eggs over cheese on each pan.
8. Distribute 2-1/3 cup beans over eggs on each pan.
9. Distribute 1 quart shredded potatoes over beans in each pan.
10. Bake 8 minutes or until crust is browned and hash browns begin to turn golden brown on high fan, open vent. CCP: Internal temperature must reach 145 F. or higher for 15 seconds.
11. Cut 5 by 5. CCP: Hold for service at 140 F. or higher.

CHEESE AND EGGS No.F 015 02

ITALIAN BREAKFAST PIZZA

Yield 100   Portion   1 Piece

Calories	Carbohydrates	Protein	Fat	Cholesterol	Sodium	Calcium
388 cal	45 g	27 g	10 g	24 mg	798 mg	281 mg

Ingredient	**Weight**	**Measure**	**Issue**
COOKING SPRAY,NONSTICK	2 oz	1/4 cup 1/3 tbsp	
DOUGH,PIZZA	16 lbs		
SAUCE,PIZZA,CANNED	4-7/8 lbs	2 qts	
SAUSAGE LINK,TURKEY,RAW	3-1/4 lbs		
EGG SUBSTITUTE,PASTEURIZED	15-1/2 lbs	1 gal 3 qts	
BASIL,SWEET,WHOLE,CRUSHED	1/8 oz	1/3 tsp	
PEPPER,BLACK,GROUND	1/8 oz	1/8 tsp	
SALT	1/4 oz	1/8 tsp	
OREGANO,CRUSHED	1/8 oz	1/3 tsp	
CHEESE,MOZZARELLA,PART SKIM,SHREDDED	6 lbs	1 gal 2 qts	
POTATOES,WHITE,FROZEN,SHREDDED,HASHBROWN	5-1/2 lbs	2 qts 3-7/8 cup	

Method

1. Lightly spray sheet pans with nonstick cooking spray.
2. Shape dough into four 4 lb pieces. Let dough rest 15 minutes. Place dough pieces on lightly floured working surface. Roll out each piece to 1/4-inch thickness. Transfer dough to pans, pushing dough slightly up edges of pans. Gently prick dough to prevent bubbling.
3. Using a convection oven, bake 8 minutes at 450 F. on high fan, open vent until crusts are lightly browned.
4. Spread 2 cups pizza sauce evenly over crust in each pan. Set aside for use in Step 7.
5. Cook sausage until lightly browned. Drain on absorbent paper. Finely chop.
6. Add salt, pepper, oregano and basil to eggs. Blend well. Scramble eggs until just set. Do not overcook. Pasteurized eggs will be safe at an internal temperature of 145 F. but will not set until they reach 160 F.
7. Distribute 1-1/2 quart cheese over pizza sauce on each crust.
8. Distribute 1-1/2 quart scrambled eggs over cheese on each pan.
9. Distribute 1-3/4 cups sausage over scrambled eggs on each pan.
10. Distribute 1 quart shredded potatoes over sausage in each pan.
11. Using a convection oven, bake 8 minutes or until crust is browned and hash browns begin to turn golden brown on high fan, open vent. CCP: Internal temperature must reach 145 F. or higher for 15 seconds.
12. Cut 5 by 5. CCP: Hold for service at 140 F. or higher.

GUIDELINES FOR SUCCESSFUL CAKE BAKING

A. Read through entire recipe.
B. Assemble all utensils and baking pans.
   1. Preparation of Cake Pans:
      (a) Do not use warped or bent baking pans. Use only lightweight sheet pans (weighing about 4 lb) designed for baking. Shiny metal pans are best for baking cakes.
      (b) Prepare pans for baking. If cakes are to be served directly from pans, grease pans with shortening and dust with flour or spray with non-stick cooking spray. If cakes are to be removed from pans and served as layer cakes, grease and line pans with paper to ensure easy removal.
C. Check to make sure oven racks are level and in proper position for baking. Set oven thermostat to temperature specified in recipe.
D. Assemble all ingredients. Use exact ingredients specified in recipe.
   1. Preparation and Mixing of Ingredients:
      (a) The temperature of ingredients is very important in cake preparation. Shortening should be workable, neither too cold nor warm enough to liquefy. In general, all ingredients should be at room temperature unless recipe specifies otherwise. Water should be cool, and eggs should be removed from refrigeration 30 minutes before using. Eggs are easier to separate when cold but beat to greater volume when at room temperature.
      (b) Weigh or measure all ingredients accurately. Follow the mixing procedure stated on the recipe card. DO NOT overbeat or underbeat. The correct length of time for beating at each stage indicated on the recipe card should be followed <u>very closely</u>.
      (c) Whenever instructions state to add dry and liquid ingredients alternately, begin and end with dry ingredients.

GUIDELINES FOR SUCCESSFUL CAKE BAKING

2. Panning Batter
   (a) Pour the amount of batter specified in the recipe into prepared baking pans.
   (b) Spread batter evenly using a spatula.
   (c) Batter-filled baking pans should be placed immediately into a preheated oven.
3. Baking:
   (a) Space baking pans evenly in oven to allow heat to circulate around each pan. Pans SHOULD NOT touch each other or sides of oven.
   (b) To test for doneness, touch top of cake near the center. If indentation remains, the cake is not done and should be baked 3 to 5 minutes longer and tested again, or insert a toothpick near center. If clean when removed, cake is done.
   (c) When cakes are done, they should be lightly browned and beginning to shrink from sides of pans.
4. Cooling and Removing from Pans:
   (a) Remove baking pans from oven; place on racks away from drafts to cool.
   (b) Cool cakes in pans 5 to 10 minutes before removing from pans. Remove any paper liners immediately. Turn cakes right side up to cool.
   (c) Sheet cakes may be cooled in pans and frosted, or turned out onto inverted baking pans to cool before frosting.
   (d) Allow cakes to cool thoroughly before frosting.

NOTE: Use 2 lb (4-1/2 cups) shortening and 1 lb (1 qt) general purpose flour, sifted. Cream shortening and flour at medium speed in mixer bowl until smooth. (In cold weather, add 2 tbsp salad oil to the flour-shortening mixture to aid in spreading.)

BATTER CAKES
CHARACTERISTICS OF GOOD QUALITY

COLOR	Uniform color, light golden brown crust for white or yellow cake. Crusts of dark cakes may be slightly darker than inside.
SHAPE AND SIZE	Cakes should be slightly rounded on top with even height at sides. Cakes should come to slightly above top of layer or sheet pans.
CRUST	Thin tender crust with slight sheen. Flat bubbles may appear on surface and be slightly darker.
TEXTURE	Breaks easily but does not crumble. Moist but not gummy. Light, velvety, fine to medium walled cells.
FLAVOR	Determined by type of cake. Sweet, no off-flavor.

BATTER CAKES
CAUSES FOR POOR QUALITY

OUTSIDE APPEARANCE		
	Peaks	Oven too hot. Not enough liquid. Batter overmixed. Pans too close together or too close to sides of oven. Too much flour.
	Sag in center	Underbaked. Oven too cool. Too much batter in pan. Too much sugar, shortening, or leavening. Not enough eggs or flour.
CRUST		
	Too Thick	Oven too hot. Overbaked. Pan too deep. Batter overmixed.
	Cracked	Too much flour. Oven too hot. Overmixed.
	Sticky	Underbaked. High humidity. Cake placed in pastry cabinet, refrigerator, or freezer while still warm.
	Tough	Overmixed. Oven too cool. Too much flour. Not enough shortening or sugar.
	Hard	Overbaked. Pan too deep.
COLOR		
	Too Dark	Oven too hot. Too much sugar or milk solids.
	Too Light	Not enough batter in pan. Overmixed or undermixed. Underbaked.
INSIDE APPEARANCE		
	Coarse Grain	Overmixed or undermixed. Oven too cool. Too much leavening.
	Tunnels	Undermixed or overmixed. Oven too hot.

BATTER CAKES
CAUSES FOR POOR QUALITY - CONTINUED

TEXTURE	Too Dry	Overbaked. Not enough liquid or shortening. Too much flour or leavening. Omission of eggs.
	Crumbly	Not enough shortening. Too much shortening. Too much leavening. Oven too cool. Undermixed or overmixed. Not enough eggs.
	Tough	Overmixed. Too much or wrong type of flour. Not enough shortening or sugar. Oven too hot or too cool.
	Too Tender	Batter undermixed.
	Too Heavy	Too much shortening. Underbaked.
EXCESSIVE SHRINKAGE		Overmixed. Too much grease in pan. Overbaked. Not enough batter in pan.
OFF FLAVOR		Ingredients not measured accurately. Rancid pan grease. Dirty pan.
HOLLOW SPOT ON BOTTOM		Not enough liquid. Too much flour. Excess bottom heat in oven. Pan not properly prepared.
UNEVENLY BAKED		Undermixed or overmixed. Uneven or dented pan. Not panned properly. Hot or cold spots in oven. Low fan not used in convection ovens.
FALLS DURING BAKING		Overmixed. Jarred during baking. Oven too cool.
LACKS VOLUME		Not enough leavening. Undermixed or overmixed. Not enough batter in pan. Oven too hot or too cool.
CAKE STICKS TO PAN		Pan not properly prepared. Oven too cool. Cake left in pan too long. Too much liquid. Too much sugar.

G-G. DESSERTS (CAKES AND FROSTINGS No. 4

GUIDELINES FOR SCALING CAKE BATTER

9-inch Layer Pan   Pour 18 to 20 oz batter into each greased and floured layer pan. Bake 20-25 minutes. For 100 portions: Use 12 layer pans (6-2-layer cakes); cut 16 portions per cake.

16-inch Square Sheet Pan   Pour 4 to 6 lb batter into each greased and floured pan. Bake as directed on recipe card. For 100 portions: Use 3 pans; cut each cake 6 by 6.

16 by 19-inch Baking Pan (field range)   Pour 4 to 6 lb batter into each greased and floured pan. Bake as directed on recipe card. For 100 portions: Use 3 pans; cut each cake 6 by 6.

Loaf Pans (16 by 4-1/2 by 4-1/8)   Pour about 2 qt batter into each greased and floured pan. Bake 20 to 25 minutes. For 100 portions: Use 4 pans; cut 25 slices per pan.

Cupcakes   Fill each greased and floured or paper lined cup half full with batter. Bake 20 to 25 minutes. A 100-portion cake recipe will yield 13 dozen cupcakes.

GUIDELINES FOR CUTTING CAKES

There is a satisfactory method of cutting each kind of cake. The factors to keep in mind are the size and number of servings and the cutting utensil to be used. The size and number of servings depend upon the size and number of layers in the cake. A knife with a sharp straight-edged, thin blade is most suitable for cutting batter cakes. To make a clear cut, and to keep the knife blade free from frosting and cake crumbs, dip the blade into warm water before cutting portion.

The following diagrams illustrate methods of cutting cakes of various sizes and shapes. The average number of servings per cake are given.

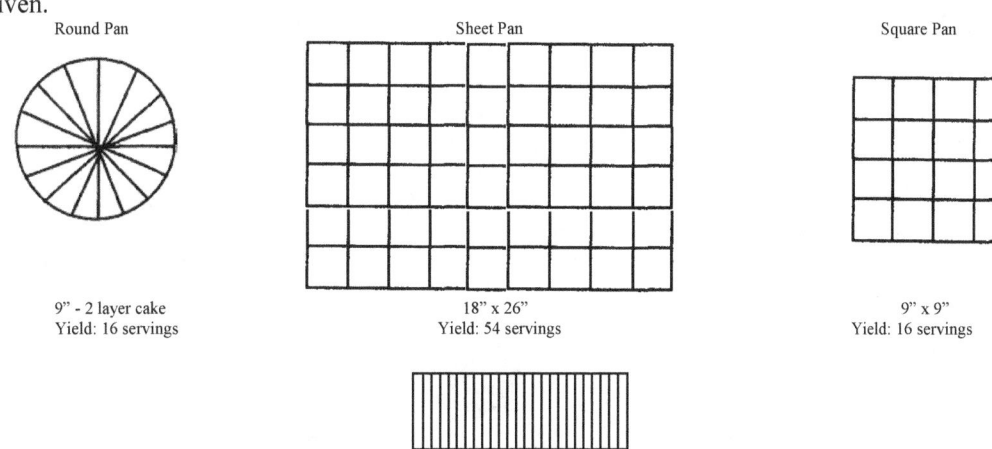

Round Pan
9" - 2 layer cake
Yield: 16 servings

Sheet Pan
18" x 26"
Yield: 54 servings

Square Pan
9" x 9"
Yield: 16 servings

16" x 4"
Yield: 25 servings

GUIDELINES FOR PREPARING FROSTINGS AND FROSTING CAKES

FROSTINGS

1. Frostings should not be so strongly flavored that they detract from the flavor of the cake. Frostings should complement the flavor of the cake.
2. If a colored frosting is desired, mix the food coloring with a small amount of the frosting and then add the cold frosting to the larger amount until the desired color is obtained. Harsh strong colors should never be used except small amounts for some specific decoration.
3. A butter cream frosting which is too thick can be thinned with a little water or milk before it is used. Care must be taken to add the liquid in very small amounts. Butter cream frosting which is too thin can be thickened by the addition of more powdered sugar. The additional powdered sugar should be mixed into the frosting until the desired consistency is reached.

FROSTING CAKES

1. Remove loose crumbs and, if necessary, trim the cake. Use a sharp knife to remove any hard or jagged edges.
2. Form layer cakes using two 9-inch layers, or a sheet cake cut in half to form 2 layers, or two sheet cakes together.
3. When frosting a layer cake, invert the bottom layer with the top side down. Place the thicker layer on the bottom. Use a spatula to spread a thin layer of frosting or filling evenly over bottom layer. (Top layer will slip if too much frosting or filling is used). Cover the top layer, top side up. Starting from the center and working outward, spread frosting on the top of the cake; then frost the sides.

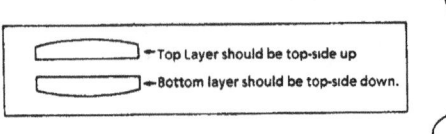

4. When frosting cupcakes, spread the specified amount of frosting on the top of the cupcake. DO NOT frost the side.

G-G. DESSERTS (CAKES AND FROSTINGS) No. 7

HIGH ALTITUDE BAKING

Since atmospheric pressure decreases as altitude increases, the requirement for baking soda also decreases. Bakery mixes are formulated for use at sea level air pressure. Follow specific high altitude instructions on the package.

When preparing cakes, hot breads, and drop cookies from basic ingredients at high altitudes, quantities of leavening agents may be adjusted as specified in the table on the back of this card.

Cakes have a tendency to stick to pans at higher altitudes; therefore the pans should be greased and dusted more heavily than those used at sea level.

Oven temperatures should be increased 25° F. at elevations of 3500 feet or more. The baking time is generally the same as at sea level; however, care should be taken to avoid overbaking since evaporation rate increases at high altitudes.

Baking powder or baking soda in recipes for cakes, hot breads, and drop cookies prepared at higher altitudes should decrease as shown on back of this card.

Amount Basic Recipe	Amounts to be Used at Higher Altitudes			
	2000 feet	**4000 feet**	**6000 feet**	**8000 feet**
1 tbsp	2-1/2 tsp	2 tsp	1-2/3 tsp	1 tsp
1-2/3 tbsp	1-2/3 tbsp	1-1/3 tbsp	1 tbsp	2 tsp
2 tbsp	1-2/3 tbsp	1-1/3 tbsp	3-1/3 tsp	2-1/2 tsp
2-1/3 tbsp	2 tbsp	1-2/3 tbsp	1-1/3 tbsp	2-2/3 tsp
3-2/3 tbsp	3 tbsp	2-2/3 tbsp	2 tbsp	1-1/3 tbsp
1/4 cup	3-1/3 tbsp	2-2/3 tbsp	2-1/3 tbsp	1-2/3 tbsp
4-2/3 tbsp	3-2/3 tbsp	3 tbsp	2-2/3 tbsp	1-2/3 tbsp
5-2/3 tbsp	4-2/3 tbsp	3-2/3 tbsp	3 tbsp	2-1/3 tbsp
6-2/3 tbsp	5-2/3 tbsp	4-2/3 tbsp	3-2/3 tbsp	2-2/3 tbsp
1/2 cup	6-2/3 tbsp	5-2/3 tbsp	4-1/3 tbsp	3-1/3 tbsp
8-2/3 tbsp	7-1/3 tbsp	6 tbsp	4-2/3 tbsp	3-1/3 tbsp
9 tbsp	7-2/3 tbsp	6-1/3 tbsp	5 tbsp	3-2/3 tbsp
11 tbsp	9-1/3 tbsp	7-2/3 tbsp	6 tbsp	4-1/3 tbsp
3/4 cup	5/8 cup	1/3 cup	6-2/3 tbsp	5 tbsp
1 cup	7/8 cup	11-1/3 tbsp	8-2/3 tbsp	6-1/3 tbsp
1-1/2 cups	1-1/4 cups	1 cup	13-1/3 tbsp	5/8 cup

G. DESSERTS (CAKE AND FROSTINGS) No. 0 (1)

INDEX

Card No.		Card No.	
G 002 00	Applesauce Cake	G 010 03	Lemon Cake (Yellow Mix)
G 003 00	Chocolate Macaroon Cake (Cake Mix)	G 010 04	Maple Nut Cake (Yellow Mix)
G 004 00	Chocolate Chip Fudge Frosting	G 010 05	Marble Cake (Mix)
G 005 00	Vanilla Frosting (Icing Mix, Vanilla, Powdered)	G 010 06	Orange Cake (Yellow Mix)
		G 011 00	Easy Chocolate Cake
G 005 01	Orange Frosting (Icing Mix, Vanilla, Powdered)	G 012 00	Devil's Food Cake
		G 012 01	Devil's Food Cake (Cake Mix)
G 006 00	Banana Cake (Cake Mix)	G 012 02	German Chocolate Cake (Mix)
G 006 01	Banana Cake (Banana Cake Mix)	G 013 00	Carrot Cake (Cake Mix)
G 007 00	Decorator's Frosting	G 014 00	Peanut Butter Cream Frosting
G 008 00	Florida Lemon Cake	G 015 00	Chocolate Fudge Frosting
G 009 00	Chocolate Frosting (Icing Mix, Chocolate Powdered)	G 016 00	Strawberry Shortcake (Biscuit Mix)
		G 016 01	Strawberry Shortcake (Cake Mix)
G 009 01	Choc Chip Frosting (Icing Mix, Chocolate Powdered)	G 017 00	Gingerbread
		G 017 01	Gingerbread (Gingerbread Cake Mix)
G 009 02	Choc Coconut Frost (Icing Mix, Chocolate Powdered)	G 018 00	Caramel Frosting
		G 019 00	Brown Sugar Frosting
G 009 03	Mocha Cream Frost (Icing Mix, Chocolate Powdered)	G 020 00	Peanut Butter Crumb Cake
		G 020 01	Peanut Butter Cake
G 010 00	Yellow Cake (Mix)	G 021 00	Pound Cake
G 010 01	Almond Cake (Yellow Mix)	G 021 01	Almond Pound Cake (Pound Cake Mix)
G 010 02	Black Walnut Cake (Yellow Mix)	G 021 02	Velvet Pound Cake (Yellow Cake Mix)

G. DESSERTS (CAKE AND FROSTINGS) No. 0 (1)

Card No.		Card No.	
G 021 03	Lemon Pound Cake (Pound Cake Mix)	G 029 03	Fruit Cocktail Upside Down Cake
G 022 00	Butter Cream Frosting	G 030 00	White Cake
G 022 01	Orange Butter Cream Frosting	G 030 01	White Cake (White Cake Mix)
G 022 02	Chocolate Butter Cream Frosting	G 030 02	Lemon Filled Cake (White Cake Mix)
G 022 03	Coconut Butter Cream Frosting	G 030 03	Raspberry Filled Cake (White Cake Mix)
G 022 04	Lemon Butter Cream Frosting	G 030 04	Strawberry Filled Cake (White Cake Mix)
G 022 05	Maple Butter Cream Frosting	G 031 00	Coconut Pecan Frosting
G 022 06	Mocha Butter Cream Frosting	G 032 00	Yellow Cake
G 023 00	Easy Vanilla Cake	G 032 01	Banana-Filled Layer Cake
G 024 00	Chocolate Glaze Frosting	G 032 02	Boston Cream Pie
G 025 00	Spice Cake	G 032 03	Marble Cake
G 025 01	Spice Cake (Yellow Cake Mix)	G 032 04	Coconut Cake
G 026 00	Cheese Cake	G 032 06	Dutch Apple Cake
G 026 01	Cheese Cake (Mix)	G 032 07	Filled Cake (Washington Pie)
G 026 02	Cheese Cake with Fruit Topping	G 032 08	Yellow Cake (Crumbs)
G 026 03	Cheese Cake Mix with Fruit Topping	G 033 00	Jelly Roll
G 026 04	Cheese Cake with Sour Cream Topping	G 034 00	Yellow Cupcakes Mix
G 026 05	Cheese Cake with Strawberries	G 034 01	Chocolate Cupcakes Mix
G 027 00	Cream Cheese Frosting	G 034 02	Spice Cake Cupcakes Mix
G 028 00	Strawberry Cake (Cake Mix)	G 034 03	Gingerbread Cupcakes Mix
G 029 00	Pineapple Upside Down Cake	G 034 04	Vanilla Cupcakes
G 029 01	Pineapple Upside Down Cake (Mix)	G 035 00	Choco-Lite Cake
G 029 02	Fruit Cocktail Upside Down Cake (Mix)	G 036 00	Lite Cheese Cake

G. DESSERTS (CAKE AND FROSTINGS) No. 0
(2)

INDEX

Card No.   Card No.

99

DESSERTS (CAKES AND FROSTINGS) No.G 002 00

APPLESAUCE CAKE

Yield 100   Portion 1 Piece

Calories	Carbohydrates	Protein	Fat	Cholesterol	Sodium	Calcium
304 cal	47 g	3 g	12 g	34 mg	172 mg	43 mg

Ingredient	Weight	Measure	Issue
FLOUR,WHEAT,GENERAL PURPOSE	3-5/8 lbs	3 qts 1 cup	
BAKING POWDER	1-3/4 oz	1/4 cup	
BAKING SODA	3/4 oz	1 tbsp	
CINNAMON,GROUND	1/2 oz	2 tbsp	
CLOVES,GROUND	1/4 oz	1 tbsp	
SALT	1/4 oz	1/8 tsp	
SUGAR,GRANULATED	2-2/3 lbs	1 qts 2 cup	
RAISINS	1-1/2 lbs	1 qts 1/2 cup	
APPLESAUCE,CANNED,UNSWEETENED	3-1/4 lbs	1 qts 2 cup	
SHORTENING	1-1/2 lbs	3-3/8 cup	
EGGS,WHOLE,FROZEN	1-3/4 lbs	3-1/4 cup	
COOKING SPRAY,NONSTICK	2 oz	1/4 cup 1/3 tbsp	

Method

1 Sift together flour, baking powder, baking soda, cinnamon, cloves, salt and sugar into mixer bowl.
2 Add raisins, applesauce and shortening to dry ingredients. Beat at low speed 1 minute, then at medium speed 2 minutes. Scrape down bowl.
3 Add eggs slowly to mixture while beating at low speed about 1 minute. Scrape down bowl. Beat at medium speed 3 minutes.
4 Lightly spray each pan with non-stick cooking spray. Pour 3-3/4 quarts batter into each sprayed and floured pan.
5 Using a convection oven, bake in 325 F. oven for 20 to 25 minutes or until done on high fan, open vent.
6 Cool; frost if desired. Cut 6 by 9.

DESSERTS (CAKES AND FROSTINGS) No.G 003 00

CHOCOLATE MACAROON CAKE (CAKE MIX)

Yield 100   Portion 1 Piece

Calories	Carbohydrates	Protein	Fat	Cholesterol	Sodium	Calcium
332 cal	51 g	4 g	13 g	30 mg	496 mg	109 mg

Ingredient	Weight	Measure	Issue
CAKE MIX,DEVILS FOOD	8-3/4 lbs		
OIL,SALAD	1 lbs	2-1/8 cup	
WATER	5-1/4 lbs	2 qts 2 cup	
EGGS,WHOLE,FROZEN	1-1/4 lbs	2-1/4 cup	
DESSERT POWDER,PUDDING,INSTANT,CHOCOLATE	1-1/2 lbs	4 cup	
COOKING SPRAY,NONSTICK	2 oz	1/4 cup 1/3 tbsp	
CAKE MIX,WHITE	2-1/2 lbs	1 qts 3-3/8 cup	
COCONUT,PREPARED,SWEETENED FLAKES	9 oz	2-3/4 cup	
WATER	1 lbs	2 cup	
VANILLA GLAZE		2-3/4 cup	

Method

1. Place Devil's Food Cake Mix, salad oil, water, eggs and dessert powder in mixer bowl. Blend at low speed until moistened, about 2 minutes. Scrape down bowl.
2. Beat at medium speed 5 to 8 minutes.
3. Pour 2 cups batter into each greased and floured pan. Set aside for use in Step 6.
4. Place White Cake Mix, coconut, and water in mixer bowl. Blend at low speed until moistened. Scrape down bowl.
5. Beat at low speed 1 minute. DO NOT OVERMIX.
6. Pour about 1 quart batter over macaroon mixture covering it completely.
7. Using a convection oven, bake 1 hour 15 minutes at 325 F. or until done on low fan, closed vent.
8. Remove from oven; cool 15 to 20 minutes; remove from pans.
9. Prepare 1 recipe of Vanilla Glaze (Recipe No. D 046 00). Drizzle 1 cup glaze over each cake.
10. Cut 20 slices per loaf.

DESSERTS (CAKES AND FROSTINGS) No.G 004 00

CHOCOLATE CHIP FUDGE FROSTING

Yield 100 **Portion** 2-1/2 Quarts

Calories	Carbohydrates	Protein	Fat	Cholesterol	Sodium	Calcium
13516 cal	2295 g	89 g	496 g	728 mg	5336 mg	2634 mg

Ingredient	Weight	Measure	Issue
CHOCOLATE,COOKING CHIPS,SEMISWEET	2-1/4 lbs	1 qts 2 cup	
BUTTER	8 oz	1 cup	
SUGAR,POWDERED,SIFTED	3-2/3 lbs	3 qts 2 cup	
MILK,NONFAT,DRY	1-3/4 oz	3/4 cup	
SALT	1/4 oz	1/8 tsp	
WATER,WARM	14-5/8 oz	1-3/4 cup	

Method

1. Melt chocolate chips and butter or margarine over very low heat. Place in mixer bowl.
2. Sift together powdered sugar, milk, and salt; add to chocolate mixture.
3. Blend in just enough water to obtain spreading consistency. Mix at medium speed 3 minutes or until smooth.
4. Spread immediately on cool cakes.

Notes

1. In Step 1, chocolate-flavored baking chips may be substituted for semi-sweet chocolate chips.
2. For 9-inch, 2-layer cakes: Spread about 1-3/4 cups frosting per cake.
3. For cupcakes: Spread about 1 tablespoon of frosting on each cupcake.

DESSERTS (CAKES AND FROSTINGS) No.G 005 00

VANILLA FROSTING (ICING MIX, VANILLA, POWDERED)

Yield 1 Portion   **Portion** 2-1/2 Quarts

Calories	Carbohydrates	Protein	Fat	Cholesterol	Sodium	Calcium
7904 cal	1808 g	6 g	94 g	0 mg	261 mg	65 mg

Ingredient	**Weight**	**Measure**	**Issue**
ICING MIX,POWDER,VANILLA	4-1/4 lbs		
WATER,WARM	12-1/2 oz	1-1/2 cup	

Method

1. Place icing mix in mixer bowl. Add hot water (120 F.) gradually while mixing at low speed. Scrape down bowl; beat at high speed 3 minutes or until thick and smooth.
2. Spread on cooled cakes.

DESSERTS (CAKES AND FROSTINGS) No.G 005 01

ORANGE FROSTING (ICING MIX, VANILLA, POWDERED)

Yield 1 Portion **Portion** 2-1/2 Quarts

Calories	Carbohydrates	Protein	Fat	Cholesterol	Sodium	Calcium
9135 cal	2122 g	25 g	97 g	0 mg	293 mg	1910 mg

Ingredient	Weight	Measure	Issue
ICING MIX,POWDER,VANILLA	4-1/4 lbs		
JUICE,ORANGE	11 oz	1-1/4 cup	
ORANGE,RIND,GRATED	2-1/2 lbs	2 qts 3-3/4 cup	
WATER,WARM	2-1/8 oz	1/4 cup 1/3 tbsp	

Method

1 Place icing mix in mixer bowl. Add orange juice, rind and hot water (120 F.) gradually while mixing at low speed. Scrape down bowl; beat at high speed 3 minutes or until thick and smooth.

2 Spread on cooled cakes.

DESSERTS (CAKES AND FROSTINGS) No.G 006 00

BANANA CAKE (CAKE MIX)

Yield 100   **Portion** 1 Piece

Calories	Carbohydrates	Protein	Fat	Cholesterol	Sodium	Calcium
216 cal	40 g	3 g	5 g	11 mg	291 mg	26 mg

Ingredient   Weight   Measure   Issue

BANANA,FRESH   3-3/4 lbs   5-3/4 lbs
CAKE MIX,YELLOW   10 lbs
OIL,SALAD   7-2/3 oz   1 cup
WATER   1 lbs   2 cup
WATER   1 lbs   2 cup
COOKING SPRAY,NONSTICK   2 oz   1/4 cup 1/3 tbsp

Method

1 Beat bananas in mixer bowl at high speed about 1 minute until smooth.
2 Add mix, contents of both soda pouches, salad oil and water to bananas. Beat at low speed 3 minutes. Scrape down bowl.
3 Add water gradually while mixing at low speed about 2 minute. Scrape down bowl. Beat at medium speed 3 minutes.
4 Lightly spray each pan with non-stick cooking spray. Pour 4-1/4 quarts of batter into each sprayed and floured pan.
5 Using a convection oven, bake at 300 F. 30 to 35 minutes or until done on low fan, open vent.
6 Cool; frost if desired.

DESSERTS (CAKES AND FROSTINGS) No.G 006 01

BANANA CAKE (BANANA CAKE MIX)

Yield 100   Portion 1 Piece

Calories	Carbohydrates	Protein	Fat	Cholesterol	Sodium	Calcium
289 cal	50 g	3 g	9 g	0 mg	304 mg	31 mg

Ingredient   **Weight**   **Measure**   **Issue**

CAKE MIX, BANANA   10 lbs   1 gal 3-3/8 qts

Method

1 Prepare mix according to instructions on container. Frost if desired.

DESSERTS (CAKES AND FROSTINGS) No.G 007 00
DECORATOR'S FROSTING

Yield 100   **Portion** 1 Quart

Calories	Carbohydrates	Protein	Fat	Cholesterol	Sodium	Calcium
4494 cal	837 g	0 g	138 g	0 mg	12 mg	11 mg

Ingredient	Weight	Measure	Issue
SUGAR,POWDERED,SIFTED	1-7/8 lbs	1 qts 3 cup	
SHORTENING	4-7/8 oz	1/2 cup 2-2/3 tbsp	
WATER	3-1/8 oz	1/4 cup 2-1/3 tbsp	
EXTRACT,VANILLA	1/4 oz	1/4 tsp	

Method

1. Cream sugar and shortening in mixer bowl 1 minute at low speed. Scrape down bowl; continue beating at medium speed 2
2. Add water and vanilla slowly to creamed mixture while beating at low speed. Scrape down bowl; continue beating at medium speed until smooth.

Notes

1. Additional water may be added to reach desired consistency.
2. This icing may be used in a pastry bag for writing and all other decorative work for cakes.
3. In Step 2, for a tinted frosting, a small amount of food coloring paste may be used.

DESSERTS (CAKES AND FROSTINGS) No. G 008 00

FLORIDA LEMON CAKE

Yield 100   Portion   1 Piece

Calories	Carbohydrates	Protein	Fat	Cholesterol	Sodium	Calcium
417 cal	52 g	4 g	22 g	53 mg	313 mg	32 mg

Ingredient	Weight	Measure	Issue
CAKE MIX,YELLOW	10 lbs		
PIE FILLING,LEMON,PREPARED	1-1/3 lbs	2-5/8 cup	
EGGS,WHOLE,FROZEN	2 lbs	3-3/4 cup	
OIL,SALAD	3-7/8 lbs	2 qts	
WATER	4-1/8 lbs	2 qts	
FLAVORING,LEMON	1-5/8 oz	3 tbsp	
COOKING SPRAY,NONSTICK	2 oz	1/4 cup 1/3 tbsp	
SUGAR,POWDERED,SIFTED	3-1/8 lbs	3 qts	
BUTTER,SOFTENED	3 oz	1/4 cup 2-1/3 tbsp	
WATER,BOILING	12-1/2 oz	1-1/2 cup	
FLAVORING,LEMON	1/2 oz	1 tbsp	

Method

1. Place cake mix and pie filling mix in mixer bowl. Blend at low speed 1 minute.
2. Add eggs; blend at low speed 1 minute. Add salad oil gradually while mixing at low speed 2 minutes. Add water and lemon flavoring while mixing; blend 3 minutes at low speed. Scrape down bowl.
3. Lightly spray each pan with non-stick cooking spray. Pour about 1-1/4 gallons batter into each sprayed and floured pan.
4. Using a convection oven, bake at 300 F. 35 to 40 minutes on low fan, open vent or until done.
5. While cake is still warm, prick entire surface with a fork.
6. Combine sugar, butter or margarine, boiling water and lemon flavoring. Mix until smooth.
7. Drizzle 2-3/4 cup glaze over each cake.
8. Cut 6 by 9.

Notes

1. In Step 3, loaf type pans may be used for sheet pans. Pour 2 quarts batter into each pan. Using a convection oven bake at 300 F. for 1 hour 15 minutes on low fan, open vent. Remove cakes from pans while still warm; prick surface with fork. Pour 1 cup glaze over each pan. Cut 20 slices per pan.

DESSERTS (CAKES AND FROSTINGS) No.G 009 00

CHOCOLATE FROSTING (ICING MIX, CHOCOLATE POWDERED)

Yield 1 Portion   **Portion** 2-1/2 Quarts

Calories	Carbohydrates	Protein	Fat	Cholesterol	Sodium	Calcium
7058 cal	1669 g	24 g	94 g	0 mg	1393 mg	209 mg

Ingredient	Weight	Measure	Issue
WATER, WARM	1 lbs	2 cup	
ICING MIX, POWDER, CHOCOLATE	4 lbs		

Method

1 Place icing mix in mixer bowl. Add hot water (120 F.) gradually while mixing at low speed. Scrape down bowl; beat at high speed 3 minutes or until thick and smooth.
2 Spread on cooled cakes.

DESSERTS (CAKES AND FROSTINGS) No.G 009 01
CHOC CHIP FROSTING (ICING MIX, CHOCOLATE POWDERED)

Yield 1 Portion **Portion** 2-1/2 Quarts

Calories	Carbohydrates	Protein	Fat	Cholesterol	Sodium	Calcium
8803 cal	1871 g	47 g	199 g	75 mg	1672 mg	859 mg

Ingredient	Weight	Measure	Issue
WATER,WARM	1 lbs	2 cup	
CHOCOLATE,COOKING CHIPS,SEMISWEET	12 oz	2 cup	
ICING MIX,POWDER,CHOCOLATE	4 lbs		

Method

1. Place icing mix in mixer bowl with chocolate chips. Add hot water (120 F.) gradually while mixing at low speed. Scrape down bowl; beat at high speed 3 minutes or until thick and smooth.
2. Spread on cooled cakes.

DESSERTS (CAKES AND FROSTINGS) No.G 009 02
CHOC COCONUT FROST (ICING MIX, CHOCOLATE POWDERED)

Yield 1 Portion **Portion** 2-1/2 Quarts

Calories	Carbohydrates	Protein	Fat	Cholesterol	Sodium	Calcium
8339 cal	1791 g	31 g	185 g	0 mg	2063 mg	247 mg

Ingredient | Weight | Measure | Issue

WATER,WARM   1 lbs   2 cup
COCONUT,PREPARED,SWEETENED FLAKES   9 oz   2-3/4 cup
ICING MIX,POWDER,CHOCOLATE   4 lbs

Method

1 Place icing mix in mixer bowl with coconut. Add hot water (120 F.) gradually while mixing at low speed. Scrape down bowl; beat at high speed 3 minutes or until thick and smooth.
2 Spread on cooled cakes.

DESSERTS (CAKES AND FROSTINGS)   No.G 009 03

MOCHA CREAM FROST (ICING MIX, CHOCOLATE POWDERED)

Yield 1 Portion   **Portion** 2-1/2 Quarts

Calories	Carbohydrates	Protein	Fat	Cholesterol	Sodium	Calcium
7084 cal	1674 g	25 g	94 g	0 mg	1397 mg	224 mg

Ingredient

Ingredient	Weight	Measure	Issue
WATER, WARM	1 lbs	2 cup	
COFFEE, INSTANT, FREEZE DRIED	3/8 oz	2 tbsp	
ICING MIX, POWDER, CHOCOLATE	4 lbs		

Method

1. Place icing mix in mixer bowl. Add hot water (120 F.) gradually while mixing at low speed and add coffee. Scrape down bowl; beat at high speed 3 minutes or until thick and smooth.
2. Spread on cooled cakes.

DESSERTS (CAKES AND FROSTINGS) No.G 010 00

YELLOW CAKE (MIX)

Yield 100   **Portion** 1 Piece

Calories	Carbohydrates	Protein	Fat	Cholesterol	Sodium	Calcium
331 cal	51 g	2 g	14 g	1 mg	318 mg	62 mg

Ingredient	Weight	Measure	Issue
CAKE MIX, YELLOW	10 lbs		

Method

1. Prepare mix according to instructions on container.
2. Cool; frost if desired

DESSERTS (CAKES AND FROSTINGS)   No. G 010 01

ALMOND CAKE (YELLOW MIX)

Yield 100 **Portion** 1 Piece

Calories	Carbohydrates	Protein	Fat	Cholesterol	Sodium	Calcium
333 cal	51 g	2 g	14 g	1 mg	318 mg	62 mg

Ingredient	Weight	Measure	Issue
CAKE MIX, YELLOW	10 lbs		
FLAVORING, ALMOND	1-7/8 oz	1/4 cup 1/3 tbsp	

Method

1. Prepare mix according to instructions on container.
2. Cool; frost if desired.

DESSERTS (CAKES AND FROSTINGS) No.G 010 02

BLACK WALNUT CAKE (YELLOW MIX)

Yield 100   Portion   1 Piece

Calories	Carbohydrates	Protein	Fat	Cholesterol	Sodium	Calcium
333 cal	51 g	2 g	14 g	1 mg	318 mg	62 mg

Ingredient   Weight   Measure   Issue

CAKE MIX,YELLOW   10 lbs
FLAVORING,BLACK WALNUT   1-7/8 oz   1/4 cup 1/3 tbsp

Method

1 Prepare mix according to instructions on container.
2 Cool; frost if desired.

DESSERTS (CAKES AND FROSTINGS) No. G 010 03
LEMON CAKE (YELLOW MIX)

Yield 100   Portion 1 Piece

Calories	Carbohydrates	Protein	Fat	Cholesterol	Sodium	Calcium
309 cal	59 g	3 g	7 g	24 mg	363 mg	31 mg

Ingredient	Weight	Measure	Issue
CAKE MIX, YELLOW	10 lbs		
FLAVORING, LEMON	2-1/8 oz	1/4 cup 1/3 tbsp	
LEMON BUTTER CREAM FROSTING		2 qts 3 cup	

Method

1. Prepare mix according to instructions on container.
2. Cool, frost if desired.

DESSERTS (CAKES AND FROSTINGS) No.G 010 04

MAPLE NUT CAKE (YELLOW MIX)

Yield 100   **Portion** 1 Piece

Calories	Carbohydrates	Protein	Fat	Cholesterol	Sodium	Calcium
344 cal	60 g	4 g	10 g	24 mg	363 mg	35 mg

Ingredient	Weight	Measure	Issue
CAKE MIX,YELLOW	10 lbs		
NUTS,UNSALTED,CHOPPED,COARSELY	1-1/4 lbs	1 qts	
FLAVORING,MAPLE	1-7/8 oz	1/4 cup 1/3 tbsp	
BUTTER CREAM FROSTING		2 qts 3 cup	

Method

1 Prepare mix according to instructions on container.
2 Cool; frost if desired.

DESSERTS (CAKES AND FROSTINGS) No. G 010 05

MARBLE CAKE (MIX)

Yield 100   **Portion** 1 Piece

Calories	Carbohydrates	Protein	Fat	Cholesterol	Sodium	Calcium
327 cal	49 g	3 g	14 g	4 mg	347 mg	88 mg

Ingredient   Weight   Measure   Issue

CAKE MIX, YELLOW   5 lbs
CAKE MIX, DEVILS FOOD   5 lbs

Method

1. Prepare mix according to instructions on container.
2. Cool; frost if desired.

DESSERTS (CAKES AND FROSTINGS) No.G 010 06

ORANGE CAKE (YELLOW MIX)

Yield 100   Portion 1 Piece

Calories	Carbohydrates	Protein	Fat	Cholesterol	Sodium	Calcium
333 cal	51 g	2 g	14 g	1 mg	318 mg	62 mg

Ingredient	Weight	Measure	Issue
CAKE MIX,YELLOW	10 lbs		
FLAVORING,ORANGE	1-7/8 oz	1/4 cup 1/3 tbsp	

Method

1 Prepare mix according to instructions on container.
2 Cool; frost if desired.

DESSERTS (CAKES AND FROSTINGS) No.G 011 00

EASY CHOCOLATE CAKE

Yield 100   Portion   1 Piece

Calories	Carbohydrates	Protein	Fat	Cholesterol	Sodium	Calcium
345 cal	56 g	3 g	13 g	0 mg	315 mg	9 mg

Ingredient	Weight	Measure	Issue
FLOUR,WHEAT,GENERAL PURPOSE	5 lbs	1 gal 1/2 qts	
SUGAR,GRANULATED	4-3/4 lbs	2 qts 2-3/4 cup	
COCOA	9-7/8 oz	3-1/4 cup	
BAKING SODA	2-1/4 oz	1/4 cup 1 tbsp	
SALT	1 oz	1 tbsp	
OIL,SALAD	1-3/4 lbs	3-3/4 cup	
VINEGAR,DISTILLED	5-5/8 oz	1/2 cup 2-2/3 tbsp	
EXTRACT,VANILLA	1-3/8 oz	3 tbsp	
WATER	5-1/4 lbs	2 qts 2 cup	
COOKING SPRAY,NONSTICK	2 oz	1/4 cup 1/3 tbsp	

Method

1 Sift together flour, sugar, cocoa, baking soda, and salt into mixer bowl.
2 Combine salad oil, vinegar and vanilla; add to dry ingredients while mixing at low speed 2 minutes.
3 Gradually add water while mixing at low speed 1 minute; scrape down bowl.
4 Mix at medium speed 2 minutes or until ingredients are well blended.
5 Lightly spray each pan with non-stick cooking spray. Pour about 3-1/2 quarts into each sprayed sheet pan.
6 Using a convection oven, bake at 325 F. for 25 minutes or until done on low fan, open vent.
7 Cool; frost if desired. Cut 6 by 9.

DESSERTS (CAKES AND FROSTINGS) No.G 012 00
DEVIL'S FOOD CAKE

Yield 100   Portion 1 Piece

Calories	Carbohydrates	Protein	Fat	Cholesterol	Sodium	Calcium
329 cal	49 g	4 g	14 g	49 mg	333 mg	32 mg

Ingredient	Weight	Measure	Issue
FLOUR,WHEAT,GENERAL PURPOSE	2-3/4 lbs	2 qts 2 cup	
SUGAR,GRANULATED	4-3/8 lbs	2 qts 2 cup	
SALT	1-1/2 oz	2-1/3 tbsp	
BAKING SODA	1-5/8 oz	3-1/3 tbsp	
COCOA	1-1/4 lbs	1 qts 2-1/2 cup	
MILK,NONFAT,DRY	4-1/4 oz	1-3/4 cup	
SHORTENING	1-3/4 lbs	1 qts	
WATER	2-5/8 lbs	1 qts 1 cup	
EGGS,WHOLE,FROZEN	2-1/2 lbs	1 qts 5/8 cup	
WATER	1-1/3 lbs	2-1/2 cup	
EXTRACT,VANILLA	7/8 oz	2 tbsp	
COOKING SPRAY,NONSTICK	2 oz	1/4 cup 1/3 tbsp	

Method

1. Sift together flour, sugar, salt, baking soda, cocoa and milk into mixer bowl.
2. Blend shortening with dry ingredients. Add water gradually; beat at low speed 2 minutes or until blended. Beat at medium speed 2 minutes. Scrape down bowl.
3. Combine eggs, water, and vanilla; add slowly to mixture while beating at low speed 1 minute. Scrape down bowl. Beat at medium speed 3 minutes.
4. Lightly spray each pan with non-stick cooking spray. Pour 4-1/2 quarts batter into each greased and floured pan. Spread evenly.
5. Using a convection oven, bake at 300 F. for 25 to 35 minutes or until done on low fan, open vent.
6. Cool; frost if desired. Cut 6 by 9.

DESSERTS (CAKES AND FROSTINGS) No. G 012 01

DEVIL'S FOOD CAKE (CAKE MIX)

Yield 100   Portion 1 Piece

Calories	Carbohydrates	Protein	Fat	Cholesterol	Sodium	Calcium
322 cal	48 g	3 g	14 g	7 mg	376 mg	115 mg

Ingredient   **Weight**   **Measure**   **Issue**

CAKE MIX, DEVILS FOOD   10 lbs

Method

1 Prepare mix according to instructions on container. Frost if desired.

DESSERTS (CAKES AND FROSTINGS) No.G 012 02
GERMAN CHOCOLATE CAKE (MIX)

Yield 100   Portion 1 Piece

Calories	Carbohydrates	Protein	Fat	Cholesterol	Sodium	Calcium
335 cal	45 g	5 g	16 g	102 mg	363 mg	64 mg

Ingredient
CAKE MIX, GERMAN CHOCOLATE
COCONUT PECAN FROSTING

Weight
10 lbs

Measure

3 qts

Issue

Method
1 Prepare mix according to instructions on container.
2 Frost if desired.

DESSERTS (CAKES AND FROSTINGS) No.G 013 00

CARROT CAKE (CAKE MIX)

Yield 100   Portion   1 Piece

Calories	Carbohydrates	Protein	Fat	Cholesterol	Sodium	Calcium
308 cal	51 g	4 g	11 g	20 mg	311 mg	93 mg

Ingredient	Weight	Measure	Issue
CAKE MIX,CARROT	10 lbs		
CREAM CHEESE FROSTING		2 qts 2 cup	

Method

1 Prepare mix according to instructions on container.
2 Cool; frost if desired.

DESSERTS (CAKES AND FROSTINGS) No.G 014 00
PEANUT BUTTER CREAM FROSTING

Yield 100   **Portion** 3 Quarts

Calories	Carbohydrates	Protein	Fat	Cholesterol	Sodium	Calcium
11838 cal	1652 g	182 g	561 g	629 mg	5645 mg	916 mg

Ingredient	Weight	Measure	Issue
PEANUT BUTTER	1-3/8 lbs	2-1/2 cup	
BUTTER,SOFTENED	10 oz	1-1/4 cup	
HONEY	15 oz	1-1/4 cup	
SUGAR,POWDERED	2-1/2 lbs	2 qts 1-5/8 cup	
MILK,NONFAT,DRY	1-5/8 oz	1/2 cup 2-2/3 tbsp	
WATER	12-1/2 oz	1-1/2 cup	
EXTRACT,VANILLA	1/4 oz	1/4 tsp	

Method

1. Cream peanut butter, butter or margarine, and honey in mixer bowl at medium speed 3 minutes.
2. Sift together powdered sugar and milk; add alternately with water and vanilla to creamed mixture while beating at low speed. Scrape down bowl; beat at medium speed 3 minutes or until smooth.
3. Spread on cool cakes.

DESSERTS (CAKES AND FROSTINGS) No. G 015 00

CHOCOLATE FUDGE FROSTING

Yield 100   **Portion** 2-1/2 Quarts

Calories	Carbohydrates	Protein	Fat	Cholesterol	Sodium	Calcium
13454 cal	2052 g	58 g	632 g	998 mg	6299 mg	749 mg

Ingredient	Weight	Measure	Issue
BUTTER	1 lbs	2 cup	
SHORTENING	8-1/8 oz	1-1/8 cup	
SUGAR,POWDERED,SIFTED	4-1/4 lbs	1 gal	
COCOA	8-1/8 oz	2-5/8 cup	
MILK,NONFAT,DRY	7/8 oz	1/4 cup 2-1/3 tbsp	
SALT	1/4 oz	1/8 tsp	
WATER,WARM	1 lbs	1-7/8 cup	
EXTRACT,VANILLA	7/8 oz	2 tbsp	

Method

1. Melt butter or margarine and shortening; pour into mixer bowl.
2. Sift together powdered sugar, cocoa, milk and salt; add to melted fats; mix at low speed until smooth.
3. Combine water and vanilla; add to mixture in bowl. Beat at medium speed until mixture obtains desired spreading consistency.
4. Spread immediately on cooled cakes.

DESSERTS (CAKES AND FROSTINGS) No.G 016 00
STRAWBERRY SHORTCAKE (BISCUIT MIX)

Yield 100 **Portion** 1 Piece

Calories	Carbohydrates	Protein	Fat	Cholesterol	Sodium	Calcium
293 cal	47 g	4 g	10 g	5 mg	550 mg	106 mg

Ingredient	Weight	Measure	Issue
BISCUIT MIX	9 lbs	2 gal 1/2 qts	
SUGAR,GRANULATED	1 lbs	2-1/4 cup	
WATER	4-2/3 lbs	2 qts 1 cup	
BUTTER,SOFTENED	6 oz	1/2 cup	
WHIPPED TOPPING (DEHYDRATED)		1 gal 2-1/4 qts	
STRAWBERRIES,FROZEN,THAWED	31-1/2 lbs	3 gal 2 qts	

Method

1. Place mix, sugar and contents of pouches in mixer bowl.
2. Blend with paddle at low speed 30 seconds.
3. Divide dough into four pieces, about 3-1/2 pounds each. Place dough on lightly floured surface; fold over 2 or 3 times; press down. Roll each piece into squares, about 16 by 16 inches and 3/8 inches thick.
4. Brush 2 pieces of dough with butter or margarine. Cut with 2-1/2 inch floured biscuit cutter.
5. Place biscuits on pans in rows 6 by 9. Brush top with remaining butter or margarine.
6. Using a convection oven, bake at 350 F. 15 minutes or until golden brown, on low fan, open vent.
7. Prepare 1-1/4 recipes Whipped Topping, Recipe No. K 002 00.
8. Place 1/4 cup strawberries on bottom half of each biscuit; top with other half. Top with 1/4 cup whipped topping.

DESSERTS (CAKES AND FROSTINGS) No.G 016 01

STRAWBERRY SHORTCAKE (CAKE MIX)

Yield 100   Portion   1 Piece

Calories	Carbohydrates	Protein	Fat	Cholesterol	Sodium	Calcium
265 cal	52 g	4 g	5 g	11 mg	304 mg	57 mg

Ingredient	Weight	Measure	Issue
CAKE MIX, YELLOW	10 lbs		
WHIPPED TOPPING (DEHYDRATED)		1 gal 2-1/4 qts	
STRAWBERRIES, FROZEN, THAWED	31-1/2 lbs	3 gal 2 qts	

Method

1. Prepare mix according to instructions on container. When cakes are cool, cut 6x9.
2. Prepare 1-1/4 recipes Whipped Topping, K 002 00.
3. Place 1/4 cup strawberries on each piece of cake. Top with 1/4 cup whipped topping.

DESSERTS (CAKES AND FROSTINGS) No.G 017 00

GINGERBREAD

Yield 100   **Portion** 1 Piece

Calories	Carbohydrates	Protein	Fat	Cholesterol	Sodium	Calcium
265 cal	42 g	3 g	10 g	24 mg	266 mg	55 mg

Ingredient	**Weight**	**Measure**	**Issue**
FLOUR,WHEAT,GENERAL PURPOSE	4-3/8 lbs	1 gal	
SUGAR,GRANULATED	3 lbs	1 qts 2-3/4 cup	
SALT	1 oz	1 tbsp	
BAKING POWDER	1-1/8 oz	2-1/3 tbsp	
BAKING SODA	1-1/3 oz	2-2/3 tbsp	
CINNAMON,GROUND	1/2 oz	2 tbsp	
GINGER,GROUND	3/4 oz	1/4 cup 1/3 tbsp	
SHORTENING	1-1/3 lbs	3 cup	
MOLASSES	2-7/8 lbs	1 qts	
EGGS,WHOLE,FROZEN	1-1/4 lbs	2-1/4 cup	
WATER,WARM	2-5/8 lbs	1 qts 1 cup	
WATER,ICE	2-5/8 lbs	1 qts 1 cup	
COOKING SPRAY,NONSTICK	2 oz	1/4 cup 1/3 tbsp	

Method

1. Sift together flour, sugar, salt, baking powder, baking soda, cinnamon, and ginger into mixer bowl.
2. Add shortening, molasses, and eggs to dry ingredients. Beat at low speed 1 minute until blended; continue beating at medium speed 2 minutes. Scrape down bowl.
3. Add water to mixture; mix at low speed only until batter is smooth.
4. Lightly spray each pan with non-stick cooking spray. Pour about 3-1/2 quarts batter into each sprayed and floured pan.
5. Using a convection oven, bake at 300 F. for 25 to 35 minutes or until done on low fan, open vent.
6. Cut 6 by 9. Serve warm if possible.

Notes

1. If desired, top each portion with 1/4 cup Whipped Topping, Recipe No. K 002 00 or 3 tablespoons Lemon Sauce, Recipe No. K 009 00 or dust with powdered sugar.

DESSERTS (CAKES AND FROSTINGS) No. G 017 01

GINGERBREAD (GINGERBREAD CAKE MIX)

Yield 100   Portion 1 Piece

Calories	Carbohydrates	Protein	Fat	Cholesterol	Sodium	Calcium
334 cal	54 g	3 g	12 g	0 mg	449 mg	64 mg

Ingredient	Weight	Measure	Issue
GINGERBREAD MIX	15 lbs		

Method

1 Prepare mix according to instructions on container. Top with whipped topping.

DESSERTS (CAKES AND FROSTINGS) No.G 018 00

CARAMEL FROSTING

Yield 100   **Portion** 2-1/2 Quarts

Calories	Carbohydrates	Protein	Fat	Cholesterol	Sodium	Calcium
12575 cal	2280 g	10 g	416 g	1121 mg	4576 mg	877 mg

Ingredient	**Weight**	**Measure**	**Issue**
BUTTER	1-1/8 lbs	2-1/4 cup	
SUGAR,BROWN,PACKED	1-1/3 lbs	1 qts 1/4 cup	
MILK,NONFAT,DRY	5/8 oz	1/4 cup 1/3 tbsp	
WATER	8-1/3 oz	1 cup	
SUGAR,POWDERED,SIFTED	3-2/3 lbs	3 qts 2 cup	

Method

1. Melt butter or margarine. Add brown sugar; mix thoroughly while mixing at low speed. Cook over low heat 2 minutes; stir constantly.
2. Combine milk and water. Add to butter mixture. Bring mixture to a boil; stir constantly. Remove from heat.
3. Pour into mixer bowl; cool 10 minutes.
4. Add powdered sugar gradually while mixing at low speed. Mix 2 minutes at medium speed until smooth.
5. Spread immediately on cooled cakes.

DESSERTS (CAKES AND FROSTINGS) No.G 019 00

BROWN SUGAR FROSTING

Yield 100   Portion   2-1/4 Quarts

Calories	Carbohydrates	Protein	Fat	Cholesterol	Sodium	Calcium
9496 cal	1997 g	12 g	185 g	502 mg	2474 mg	1336 mg

Ingredient
Ingredient	Weight	Measure	Issue
SUGAR,BROWN,PACKED	2-1/3 lbs	1 qts 3-1/4 cup	
BUTTER	8 oz	1 cup	
WATER	1 lbs	2 cup	
MILK,NONFAT,DRY	1 oz	1/4 cup 3-1/3 tbsp	
SUGAR,POWDERED,SIFTED	2-1/8 lbs	2 qts	
EXTRACT,VANILLA	7/8 oz	2 tbsp	

Method

1. Combine brown sugar, butter, or margarine, and water. Heat to boiling; cook 1 minute.
2. Remove from heat; pour into mixer bowl.
3. Sift together milk and powdered sugar; add slowly to cooked mixture while beating at low speed.
4. Add vanilla; mix at medium speed 5 minutes or until smooth and of spreading consistency.
5. Pour and spread immediately on cool cakes.

DESSERTS (CAKES AND FROSTINGS) No.G 020 00

PEANUT BUTTER CRUMB CAKE

Yield 100   Portion   1 Piece

Calories	Carbohydrates	Protein	Fat	Cholesterol	Sodium	Calcium
340 cal	52 g	6 g	13 g	11 mg	340 mg	30 mg

Ingredient   Weight   Measure   Issue

CAKE MIX,YELLOW   10 lbs
COOKING SPRAY,NONSTICK   2 oz   1/4 cup 1/3 tbsp
FLOUR,WHEAT,GENERAL PURPOSE   1-2/3 lbs   1 qts 2 cup
SUGAR,GRANULATED   2 lbs   1 qts 1/2 cup
PEANUT BUTTER   1-3/4 lbs   3 cup
MARGARINE,SOFTENED   5 oz   1/2 cup 2 tbsp

Method

1 Prepare cake mix according to instructions on container.
2 Lightly spray each pan with non-stick cooking spray. Pour 1 gallon batter into each sprayed and floured pan.
3 Combine flour, sugar, peanut butter and butter or margarine; mix at low speed 1-1/2 minutes or until crumbs are formed.
4 Sprinkle 1-1/2 quarts crumbs over batter in each pan.
5 Using a convection oven, bake at 325 F. for 30 minutes or until done on low fan, open vent.
6 Cool. Cut 6 by 9.

DESSERTS (CAKES AND FROSTINGS) No.G 020 01

PEANUT BUTTER CAKE

Yield 100   **Portion** 1 Piece

Calories	Carbohydrates	Protein	Fat	Cholesterol	Sodium	Calcium
290 cal	38 g	6 g	14 g	11 mg	345 mg	29 mg

Ingredient	**Weight**	**Measure**	**Issue**
CAKE MIX, YELLOW	10 lbs		
PEANUT BUTTER	2-1/2 lbs	1 qts 1/2 cup	
COOKING SPRAY, NONSTICK	2 oz	1/4 cup 1/3 tbsp	

Method

1. Prepare cake mix according to instructions on container. Add peanut butter.
2. Lightly spray each pan with non-stick cooking spray. Pour 4-1/2 quarts batter into each sprayed and floured pan.
3. Using a convection oven, bake at 325 F. for 30 minutes or until done on low fan, open vent.
4. Cool. Cut 6 by 9. Frost if desired.

DESSERTS (CAKES AND FROSTINGS) No.G 021 00
POUND CAKE

Yield 100   **Portion** 1 Slice

Calories	Carbohydrates	Protein	Fat	Cholesterol	Sodium	Calcium
284 cal	35 g	4 g	14 g	60 mg	186 mg	31 mg

Ingredient	Weight	Measure	Issue
FLOUR,WHEAT,GENERAL PURPOSE	4-3/8 lbs	1 gal	
SUGAR,GRANULATED	4-1/4 lbs	2 qts 1-5/8 cup	
SALT	1-1/4 oz	2 tbsp	
BAKING POWDER	3/4 oz	1 tbsp	
MILK,NONFAT,DRY	1-5/8 oz	1/2 cup 2-2/3 tbsp	
SHORTENING	2-3/4 lbs	1 qts 2 cup	
WATER	1-7/8 lbs	3-1/2 cup	
EGGS,WHOLE,FROZEN	3 lbs	1 qts 1-5/8 cup	
EXTRACT,VANILLA	1-7/8 oz	1/4 cup 1/3 tbsp	
COOKING SPRAY,NONSTICK	2 oz	1/4 cup 1/3 tbsp	

Method

1. Sift together flour, sugar, salt, baking powder, and milk into mixer bowl.
2. Add shortening and water to dry ingredients. Beat at medium speed 7 minutes. Scrape down bowl.
3. Add eggs and vanilla slowly to mixture while beating at low speed. Beat at low speed 7 minutes. Scrape down bowl.
4. Lightly spray each pan with non-stick cooking spray. Pour 2-1/2 quarts batter into each sprayed and floured pan.
5. Using a convection oven, bake at 325 F. for 1 hour 5 minutes or until done on low fan, open vent.
6. Cool; cut 6x9.

DESSERTS (CAKES AND FROSTINGS) No.G 021 01

ALMOND POUND CAKE (POUND CAKE MIX)

Yield 100   Portion 1 Piece

Calories	Carbohydrates	Protein	Fat	Cholesterol	Sodium	Calcium
201 cal	15 g	3 g	14 g	56 mg	177 mg	29 mg

Ingredient	Weight	Measure	Issue
CAKE MIX,POUND	10 lbs		
FLAVORING,ALMOND	1-7/8 oz	1/4 cup 1/3 tbsp	

Method

1 Prepare mix according to instructions on container. Add almond flavoring.

DESSERTS (CAKES AND FROSTINGS) No.G 021 02
VELVET POUND CAKE (YELLOW CAKE MIX)

Yield 100 **Portion** 1 Piece

Calories	Carbohydrates	Protein	Fat	Cholesterol	Sodium	Calcium
289 cal	42 g	3 g	12 g	26 mg	398 mg	28 mg

Ingredient	Weight	Measure	Issue
CAKE MIX,YELLOW	10 lbs		
DESSERT POWDER,PUDDING,INSTANT,VANILLA	1-1/2 lbs	3-3/4 cup	
OIL,SALAD	1 lbs	2-1/8 cup	
EGGS,WHOLE,FROZEN	12 oz	1-3/8 cup	
WATER	4-2/3 lbs	2 qts 1 cup	
FLAVORING,ALMOND	2 oz	1/4 cup 2/3 tbsp	

Method

1 Prepare mix according to instructions on container. Add dessert powder, oil, eggs, water and flavoring to cake mix.
2 Cool; cut 25 slices per loaf.

DESSERTS (CAKES AND FROSTINGS) No.G 021 03
LEMON POUND CAKE (POUND CAKE MIX)

Yield 100   Portion 1 Piece

Calories	Carbohydrates	Protein	Fat	Cholesterol	Sodium	Calcium
206 cal	15 g	3 g	15 g	54 mg	176 mg	29 mg

Ingredient

Ingredient	Weight	Measure	Issue
CAKE MIX,POUND	10 lbs		
JUICE,LEMON	2-1/8 oz	1/4 cup 1/3 tbsp	
LEMON RIND,GRATED	1/2 oz	2-2/3 tbsp	
FLAVORING,LEMON	2-1/8 oz	1/4 cup 1/3 tbsp	

Method

1 Prepare mix according to instructions on container. Add lemon juice, rind and flavoring to cake mix.

DESSERTS (CAKES AND FROSTINGS) No.G 022 00

BUTTER CREAM FROSTING

Yield 100   Portion 2-3/4 Quarts

Calories	Carbohydrates	Protein	Fat	Cholesterol	Sodium	Calcium
13120 cal	2288 g	15 g	463 g	1248 mg	7207 mg	533 mg

Ingredient

Ingredient	Weight	Measure	Issue
BUTTER,SOFTENED	1-1/4 lbs	2-1/2 cup	
SUGAR,POWDERED,SIFTED	5 lbs	1 gal 3/4 qts	
SALT	1/4 oz	1/8 tsp	
MILK,NONFAT,DRY	1 oz	1/4 cup 3-1/3 tbsp	
EXTRACT,VANILLA	7/8 oz	2 tbsp	
WATER	6-1/4 oz	3/4 cup	

Method

1. Cream butter or margarine in mixer bowl at medium speed 1 to 3 minutes or until light and fluffy.
2. Sift together powdered sugar, salt and milk; add to creamed butter or margarine.
3. Add vanilla while mixing at low speed; add just enough water to obtain a spreading consistency. Scrape down bowl. Beat at medium speed 3 to 5 minutes or until mixture is light and well blended.
4. Spread immediately on cooled cakes.

DESSERTS (CAKES AND FROSTINGS) No.G 022 01

ORANGE BUTTER CREAM FROSTING

Yield 100   **Portion** 2-3/4 Quarts

Calories	Carbohydrates	Protein	Fat	Cholesterol	Sodium	Calcium
13074 cal	2302 g	7 g	463 g	1242 mg	7042 mg	232 mg

Ingredient	Weight	Measure	Issue
BUTTER,SOFTENED	1-1/4 lbs	2-1/2 cup	
SUGAR,POWDERED,SIFTED	5 lbs	1 gal 3/4 qts	
SALT	1/4 oz	1/8 tsp	
ORANGE PEEL,FRESH,GRATED	1-1/8 oz	1/4 cup 1-2/3 tbsp	
JUICE,ORANGE	8-3/4 oz	1 cup	

Method

1. Cream butter or margarine in mixer bowl at medium speed 1 to 3 minutes or until light and fluffy.
2. Sift together powdered sugar and salt; add to creamed butter or margarine.
3. Add grated orange rind and orange juice while mixing at low speed to obtain a spreading consistency. Scrape down bowl. Beat at medium speed 3 to 5 minutes or until mixture is light and well blended.
4. Spread immediately on cooled cakes.

DESSERTS (CAKES AND FROSTINGS) No.G 022 02

CHOCOLATE BUTTER CREAM FROSTING

Yield 100　　　　　　　　　　　　　　　　　　　Portion 2-3/4 Quarts

Calories	Carbohydrates	Protein	Fat	Cholesterol	Sodium	Calcium
13907 cal	2474 g	83 g	510 g	1248 mg	7283 mg	976 mg

Ingredient	Weight	Measure	Issue
BUTTER,SOFTENED	1-1/4 lbs	2-1/2 cup	
SUGAR,POWDERED,SIFTED	5 lbs	1 gal 3/4 qts	
SALT	1/4 oz	1/8 tsp	
MILK,NONFAT,DRY	1 oz	1/4 cup 3-1/3 tbsp	
COCOA	12-1/8 oz	1 qts	
EXTRACT,VANILLA	7/8 oz	2 tbsp	
WATER,BOILING	10-1/2 oz	1-1/4 cup	

Method

1 Cream butter or margarine in mixer bowl at medium speed 1 to 3 minutes or until light and fluffy.
2 Sift together powdered sugar, salt, milk and cocoa; add to creamed butter or margarine.
3 Add vanilla while mixing at low speed; add just enough boiling water to obtain a spreading consistency. Scrape down bowl. Beat at medium speed 3 to 5 minutes or until mixture is light and well blended.
4 Spread immediately on cooled cakes.

Notes

1 Unsweetened cooking chocolate may be used. For 100 portions, melt 1 pound chocolate at low heat. Cool. Reduce butter or margarine to 1-1/2 cups. Add chocolate at end of Step 1.

DESSERTS (CAKES AND FROSTINGS) No.G 022 03

COCONUT BUTTER CREAM FROSTING

Yield 100   **Portion**   2-3/4 Quarts

Calories	Carbohydrates	Protein	Fat	Cholesterol	Sodium	Calcium
15100 cal	2476 g	27 g	603 g	1248 mg	8242 mg	592 mg

Ingredient	Weight	Measure	Issue
BUTTER,SOFTENED	1-1/4 lbs	2-1/2 cup	
SUGAR,POWDERED,SIFTED	5 lbs	1 gal 3/4 qts	
SALT	1/4 oz	1/8 tsp	
MILK,NONFAT,DRY	1 oz	1/4 cup 3-1/3 tbsp	
EXTRACT,VANILLA	7/8 oz	2 tbsp	
WATER	6-1/4 oz	3/4 cup	
COCONUT,PREPARED,SWEETENED FLAKES	9 oz	2-3/4 cup	
COCONUT,PREPARED,SWEETENED FLAKES	4-7/8 oz	1-1/2 cup	

Method

1. Cream butter or margarine in mixer bowl at medium speed 1 to 3 minutes or until light and fluffy.
2. Sift together powdered sugar, salt and milk; add to creamed butter or margarine.
3. Add vanilla while mixing at low speed; add just enough water to obtain a spreading consistency. Scrape down bowl. Beat at medium speed 3 to 5 minutes or until mixture is light and well blended. Fold in coconut.
4. Spread immediately on cooled cakes. Sprinkle additional coconut evenly over each frosted cake.

DESSERTS (CAKES AND FROSTINGS) No.G 022 04
LEMON BUTTER CREAM FROSTING

Yield 100   **Portion**   2-3/4 Quarts

Calories	Carbohydrates	Protein	Fat	Cholesterol	Sodium	Calcium
13075 cal	2294 g	16 g	463 g	1248 mg	7225 mg	572 mg

Ingredient   Weight   Measure   Issue

Ingredient	Weight	Measure
BUTTER,SOFTENED	1-1/4 lbs	2-1/2 cup
SUGAR,POWDERED,SIFTED	5 lbs	1 gal 3/4 qts
SALT	1/4 oz	1/8 tsp
MILK,NONFAT,DRY	1 oz	1/4 cup 3-1/3 tbsp
LEMON RIND,GRATED	7/8 oz	1/4 cup 1/3 tbsp
JUICE,LEMON	3-1/4 oz	1/4 cup 2-1/3 tbsp
WATER	6-1/4 oz	3/4 cup

Method

1 Cream butter or margarine in mixer bowl at medium speed 1 to 3 minutes or until light and fluffy.
2 Sift together powdered sugar, salt and milk; add to creamed butter or margarine.
3 Add grated lemon rind and lemon juice while mixing at low speed; add just enough water to obtain a spreading consistency. Scrape down bowl. Beat at medium speed 3 to 5 minutes or until mixture is light and well blended.
4 Spread immediately on cooled cakes.

DESSERTS (CAKES AND FROSTINGS) No.G 022 05

MAPLE BUTTER CREAM FROSTING

Yield 100   Portion   2-3/4 Quarts

Calories	Carbohydrates	Protein	Fat	Cholesterol	Sodium	Calcium
13194 cal	2291 g	15 g	463 g	1248 mg	7209 mg	536 mg

Ingredient	Weight	Measure	Issue
BUTTER,SOFTENED	1-1/4 lbs	2-1/2 cup	
SUGAR,POWDERED,SIFTED	5 lbs	1 gal 3/4 qts	
SALT	1/4 oz	1/8 tsp	
MILK,NONFAT,DRY	1 oz	1/4 cup 3-1/3 tbsp	
EXTRACT,VANILLA	1/2 oz	1 tbsp	
FLAVORING,MAPLE	1-3/8 oz	3 tbsp	
WATER	6-1/4 oz	3/4 cup	

Method

1 Cream butter or margarine in mixer bowl at medium speed 1 to 3 minutes or until light and fluffy.
2 Sift together powdered sugar, salt and milk; add to creamed butter or margarine.
3 Add vanilla and maple flavoring while mixing at low speed; add just enough water to obtain a spreading consistency. Scrape down bowl. Beat at medium speed 3 to 5 minutes or until mixture is light and well blended.
4 Spread immediately on cooled cakes.

DESSERTS (CAKES AND FROSTINGS) No.G 022 06

MOCHA BUTTER CREAM FROSTING

Yield 100 **Portion** 2-3/4 Quarts

Calories	Carbohydrates	Protein	Fat	Cholesterol	Sodium	Calcium
13206 cal	2332 g	28 g	478 g	1242 mg	7068 mg	315 mg

Ingredient	Weight	Measure	Issue
BUTTER,SOFTENED	1-1/4 lbs	2-1/2 cup	
SUGAR,POWDERED,SIFTED	5 lbs	1 gal 3/4 qts	
SALT	1/4 oz	1/8 tsp	
COCOA	4 oz	1-3/8 cup	
COFFEE (INSTANT)		1 cup	

Method

1. Cream butter or margarine in mixer bowl at medium speed 1 to 3 minutes or until light and fluffy.
2. Sift together powdered sugar, salt and cocoa; add to creamed butter or margarine.
3. Add double strength brewed coffee to obtain a spreading consistency. Scrape down bowl. Beat at medium speed 3 to 5 minutes or until mixture is light and well blended.
4. Spread immediately on cooled cakes.

DESSERTS (CAKES AND FROSTINGS) No.G 023 00

EASY VANILLA CAKE

Yield 100   **Portion** 1 Piece

Calories	Carbohydrates	Protein	Fat	Cholesterol	Sodium	Calcium
356 cal	58 g	3 g	13 g	0 mg	271 mg	79 mg

Ingredient	Weight	Measure	Issue
FLOUR,WHEAT,GENERAL PURPOSE	5 lbs	1 gal 1/2 qts	
OIL,SALAD	1-3/4 lbs	3-3/4 cup	
SUGAR,GRANULATED	5-1/4 lbs	3 qts	
MILK,NONFAT,DRY	2-3/8 oz	1 cup	
BAKING POWDER	3-7/8 oz	1/2 cup	
SALT	1 oz	1 tbsp	
WATER,WARM	3-1/8 lbs	1 qts 2 cup	
EGG WHITES,FROZEN,THAWED	1-3/4 lbs	3-1/4 cup	
EXTRACT,VANILLA	3-2/3 oz	1/2 cup	
COOKING SPRAY,NONSTICK	2 oz	1/4 cup 1/3 tbsp	

Method

1. Place flour in mixer bowl.
2. Gradually add oil while mixing at low speed 2 minutes. Mixture will resemble a crumbly paste.
3. Sift together sugar, milk, baking powder and salt; add to flour-oil mixture; mix at low speed 2 minutes.
4. Combine water, egg whites and vanilla; gradually add to mixture while mixing at low speed 2 minutes; scrape down bowl.
5. Mix at medium speed 2 minutes or until well blended.
6. Lightly spray each pan with non-stick cooking spray. Pour about 3-1/2 quarts batter into each sprayed pan.
7. Using a convection oven, bake at 325 F. for 35 minutes or until done on low fan, open vent.
8. Cool; frost if desired. Cut 6 by 9.

DESSERTS (CAKES AND FROSTINGS) No.G 024 00

CHOCOLATE GLAZE FROSTING

Yield 100   **Portion** 2-1/4 Cups

Calories	Carbohydrates	Protein	Fat	Cholesterol	Sodium	Calcium
2320 cal	450 g	12 g	70 g	166 mg	645 mg	99 mg

Ingredient	**Weight**	**Measure**	**Issue**
SUGAR,POWDERED	14-7/8 oz	3-1/2 cup	
COCOA	2 oz	1/2 cup 2-2/3 tbsp	
BUTTER	2-2/3 oz	1/4 cup 1-2/3 tbsp	
EXTRACT,VANILLA	1/8 oz	1/8 tsp	
WATER,BOILING	4-1/8 oz	1/2 cup	

Method

1. Sift together powdered sugar and cocoa into mixer bowl.
2. Combine butter or margarine and vanilla with sugar mixture at low speed. Add enough water to obtain spreading consistency. Beat at medium speed about 3 minutes or until smooth.
3. Spread immediately on cooled cakes.

Notes

1. In Step 1, 2-2/3 ounces unsweetened cooking chocolate may be used per 100 portions. Melt chocolate at low heat. Cool. In Step 2, reduce butter or margarine to 1-1/3 ounces or 2-2/3 tablespoons. Add cooled, melted chocolate to butter or margarine.

DESSERTS (CAKES AND FROSTINGS) No.G 025 00
SPICE CAKE

Yield 100   Portion   1 Piece

Calories	Carbohydrates	Protein	Fat	Cholesterol	Sodium	Calcium
337 cal	50 g	4 g	14 g	40 mg	320 mg	76 mg

Ingredient	Weight	Measure	Issue
FLOUR,WHEAT,GENERAL PURPOSE	4-3/8 lbs	1 gal	
SUGAR,GRANULATED	3-1/2 lbs	2 qts	
SALT	1-3/8 oz	2-1/3 tbsp	
BAKING POWDER	2-3/4 oz	1/4 cup 2 tbsp	
BAKING SODA	1/2 oz	1 tbsp	
CINNAMON,GROUND	1 oz	1/4 cup 1/3 tbsp	
CLOVES,GROUND	1/2 oz	2 tbsp	
ALLSPICE,GROUND	1/4 oz	1 tbsp	
MILK,NONFAT,DRY	3 oz	1-1/4 cup	
SHORTENING	1-7/8 lbs	1 qts 1/4 cup	
WATER	2-1/2 lbs	1 qts 5/8 cup	
EGGS,WHOLE,FROZEN	2 lbs	3-3/4 cup	
MOLASSES	8-2/3 oz	3/4 cup	
WATER	8-1/3 oz	1 cup	
EXTRACT,VANILLA	1-7/8 oz	1/4 cup 1/3 tbsp	
COOKING SPRAY,NONSTICK	2 oz	1/4 cup 1/3 tbsp	

Method

1. Sift together flour, sugar, salt, baking powder, baking soda, cinnamon, cloves, allspice and milk into mixer bowl.
2. Add shortening and water to dry ingredients. Beat at low speed 1 minute until blended. Scrape down bowl. Continue beating at medium speed 2 minutes.
3. Combine eggs, molasses, water and vanilla. Add slowly to mixture while beating at low speed. Scrape down bowl. Beat at medium speed for 3 minutes.
4. Lightly spray each pan with non-stick cooking spray. Pour 4-1/4 quarts batter into each greased and floured pan.
5. Using a convection oven, bake at 325 F. for 35 minutes or until done on low fan, open vent.
6. Cool; frost if desired. Cut 6 by 9.

DESSERTS (CAKES AND FROSTINGS) No.G 025 01

SPICE CAKE (YELLOW CAKE MIX)

Yield 100   **Portion** 1 Piece

Calories	Carbohydrates	Protein	Fat	Cholesterol	Sodium	Calcium
273 cal	52 g	3 g	7 g	11 mg	311 mg	30 mg

Ingredient	Weight	Measure	Issue
CAKE MIX,YELLOW	10 lbs		
CINNAMON,GROUND	1 oz	1/4 cup 1/3 tbsp	
CLOVES,GROUND	1/2 oz	2 tbsp	
ALLSPICE,GROUND	1/4 oz	1 tbsp	

Method

1 Prepare mix according to instructions on container. Add cinnamon, cloves and allspice. Frost if desired.

DESSERTS (CAKES AND FROSTINGS) No. G 026 00
CHEESE CAKE

Yield 100   Portion   1 Piece

Calories	Carbohydrates	Protein	Fat	Cholesterol	Sodium	Calcium
357 cal	30 g	6 g	24 g	98 mg	323 mg	53 mg

Ingredient	Weight	Measure	Issue
MARGARINE,MELTED	1-1/2 lbs	3 cup	
CRACKERS,GRAHAM,CRUMBS	3 lbs		
SUGAR,GRANULATED	12-1/3 oz	1-3/4 cup	
CHEESE,CREAM,SOFTENED,ROOM TEMPERATURE	10-1/4 lbs	1 gal 1 qts	
SUGAR,GRANULATED	3 lbs	1 qts 2-3/4 cup	
FLOUR,WHEAT,GENERAL PURPOSE	4-3/8 oz	1 cup	
MILK,NONFAT,DRY	1 oz	1/4 cup 3 tbsp	
SALT	1/4 oz	1/8 tsp	
EGGS,WHOLE,FROZEN	2-3/8 lbs	1 qts 1/2 cup	
WATER	12-1/2 oz	1-1/2 cup	
JUICE,LEMON	2-1/8 oz	1/4 cup 1/3 tbsp	
JUICE,ORANGE	2-1/4 oz	1/4 cup 1/3 tbsp	
EXTRACT,VANILLA	7/8 oz	2 tbsp	
ORANGE,RIND,GRATED	3/8 oz	2 tbsp	
LEMON RIND,GRATED	1/4 oz	1 tbsp	

Method

1. Grind graham crackers or crush on board with rolling pin. Combine butter or margarine, crumbs, and sugar in mixer bowl. Blend thoroughly at low speed, about 1 minute.
2. Press 2 quarts crumb mixture firmly in bottom of each pan. Using a convection oven, bake 3 minutes on low fan, open vent at 325 F. Cool; set aside for use in Step 8.
3. Place cream cheese in mixer bowl. Whip at medium speed until fluffy, about 3 minutes.
4. Combine sugar, flour, milk, and salt. Mix well.
5. Add to cream cheese; whip at low speed until blended, about 2 minutes. Whip at medium speed until smooth, about 1 minute.
6. Add eggs; whip at low speed 30 seconds. Whip at medium speed until smooth, about 1 minute.
7. Combine water, lemon and orange juices, vanilla, orange and lemon rinds; add to cheese mixture. Whip at low speed until well blended, about 2 minutes.
8. Spread 5-1/4 quarts cheese filling evenly over crust in each pan.
9. Using a convection oven, bake at 325 F. for 25 to 30 minutes on low fan, open vent or until filling is firm and lightly browned.
10. Refrigerate until ready to serve. Cut 6 by 9.

DESSERTS (CAKES AND FROSTINGS) No.G 026 01

CHEESE CAKE (MIX)

Yield 100   **Portion** 1 Piece

Calories	Carbohydrates	Protein	Fat	Cholesterol	Sodium	Calcium
331 cal	41 g	5 g	17 g	22 mg	440 mg	138 mg

Ingredient | Weight | Measure | Issue

Ingredient	Weight	Measure	Issue
MARGARINE,SOFTENED	1-1/2 lbs	3 cup	
CRACKERS,GRAHAM,CRUMBS	3 lbs		
SUGAR,GRANULATED	12 oz	1-3/4 cup	
CHEESECAKE MIX	8 lbs		

Method

1. Combine margarine or butter, crumbs and sugar in mixer bowl. Blend thoroughly at low speed about 1 minute.
2. Prepare mix according to instructions on container.

DESSERTS (CAKES AND FROSTINGS) No. G 026 02

CHEESE CAKE WITH FRUIT TOPPING

Yield 100   Portion 1 Piece

Calories	Carbohydrates	Protein	Fat	Cholesterol	Sodium	Calcium
432 cal	50 g	6 g	24 g	98 mg	346 mg	66 mg

Ingredient	Weight	Measure	Issue
MARGARINE,MELTED	1-1/2 lbs	3 cup	
CRACKERS,GRAHAM,CRUMBS	3 lbs		
SUGAR,GRANULATED	12-1/3 oz	1-3/4 cup	
CHEESE,CREAM,SOFTENED,ROOM TEMPERATURE	10-1/4 lbs	1 gal 1 qts	
SUGAR,GRANULATED	3 lbs	1 qts 2-3/4 cup	
FLOUR,WHEAT,GENERAL PURPOSE	4-3/8 oz	1 cup	
SALT	1/4 oz	1/8 tsp	
MILK,NONFAT,DRY	1 oz	1/4 cup 3 tbsp	
EGGS,WHOLE,FROZEN	2-3/8 lbs	1 qts 1/2 cup	
WATER	12-1/2 oz	1-1/2 cup	
JUICE,LEMON	2-1/8 oz	1/4 cup 1/3 tbsp	
JUICE,ORANGE	2-1/4 oz	1/4 cup 1/3 tbsp	
EXTRACT,VANILLA	7/8 oz	2 tbsp	
ORANGE,RIND,GRATED	3/8 oz	2 tbsp	
LEMON RIND,GRATED	1/4 oz	1 tbsp	
PIE FILLING,CHERRY,PREPARED	7-1/2 lbs	3 qts 3 cup	
PIE FILLING,BLUEBERRY,PREPARED	8-7/8 lbs	3 qts 3 cup	

Method

1 Combine butter or margarine, crumbs, and sugar in mixer bowl. Blend thoroughly at low speed, about 1 minute.
2 Press 2 quarts crumb mixture firmly in bottom of each pan. Using a convection oven, bake 3 minutes at 325 F. on low fan, open vent. Cool; set aside for use in Step 8.
3 Place cream cheese in mixer bowl. Whip at medium speed until fluffy, about 3 minutes.
4 Combine sugar, flour, milk, and salt. Mix well.
5 Add to cream cheese; whip at low speed until blended, about 2 minutes. Whip at medium speed until smooth, about 1 minute.
6 Add eggs; whip at low speed 30 seconds. Whip at medium speed until smooth, about 1 minute.
7 Combine water, lemon and orange juices, vanilla, orange and lemon rinds; add to cheese mixture. Whip at low speed until well blended, about 2 minutes.
8 Spread 5-1/4 quarts cheese filling evenly over crust in each pan.
9 Using a convection oven, bake at 325 F. 25 to 30 minutes on low fan, open vent or until firm and lightly browned.
10 Chill. Spread 7-1/2 cups canned fruit pie filling over each cake. When chilled, cut 6 by 9.

Notes

1 In Step 10, suggested fruit pie fillings include peach, apple, strawberry, or cherry.

DESSERTS (CAKES AND FROSTINGS) No.G 026 03

CHEESE CAKE MIX WITH FRUIT TOPPING

Yield 100   **Portion** 1 Piece

Calories	Carbohydrates	Protein	Fat	Cholesterol	Sodium	Calcium
366 cal	51 g	5 g	17 g	22 mg	460 mg	147 mg

Ingredient	Weight	Measure	Issue
MARGARINE,SOFTENED	1-1/2 lbs	3 cup	
CRACKERS,GRAHAM,CRUMBS	3 lbs		
SUGAR,GRANULATED	12 oz	1-3/4 cup	
CHEESECAKE MIX	8 lbs		
PIE FILLING,BLUEBERRY,PREPARED	8-3/4 lbs	3 qts 2-7/8 cup	

Method

1. Combine margarine or butter, crumbs and sugar in mixer bowl. Blend thoroughly at low speed, about 1 minute.
2. Prepare mix according to instructions on container.
3. Choice of toppings are blueberry, apple or cherry.

DESSERTS (CAKES AND FROSTINGS) No.G 026 04

CHEESE CAKE WITH SOUR CREAM TOPPING

Yield 100   Portion   1 Piece

Calories	Carbohydrates	Protein	Fat	Cholesterol	Sodium	Calcium
387 cal	35 g	6 g	25 g	103 mg	333 mg	72 mg

Ingredient	Weight	Measure	Issue
MARGARINE,MELTED	1-1/2 lbs	3 cup	
CRACKERS,GRAHAM,CRUMBS	3 lbs		
SUGAR,GRANULATED	12-1/3 oz	1-3/4 cup	
CHEESE,CREAM,SOFTENED,ROOM TEMPERATURE	10-1/4 lbs	1 gal 1 qts	
SUGAR,GRANULATED	3 lbs	1 qts 2-3/4 cup	
FLOUR,WHEAT,GENERAL PURPOSE	4-3/8 oz	1 cup	
MILK,NONFAT,DRY	1 oz	1/4 cup 3 tbsp	
SALT	1/4 oz	1/8 tsp	
EGGS,WHOLE,FROZEN	2-3/8 lbs	1 qts 1/2 cup	
WATER	12-1/2 oz	1-1/2 cup	
JUICE,ORANGE	2-1/4 oz	1/4 cup 1/3 tbsp	
JUICE,LEMON	2-1/8 oz	1/4 cup 1/3 tbsp	
EXTRACT,VANILLA	7/8 oz	2 tbsp	
ORANGE,RIND,GRATED	3/8 oz	2 tbsp	
LEMON RIND,GRATED	1/4 oz	1 tbsp	
SOUR CREAM,LOW FAT	3 lbs	1 qts 2 cup	
SUGAR,GRANULATED	12-1/3 oz	1-3/4 cup	

Method

1. Combine butter or margarine, crumbs, and sugar in mixer bowl. Blend thoroughly at low speed, about 1 minute.
2. Press 2 quarts crumb mixture firmly in bottom of each pan. Using a convection oven bake at 325 F. 3 minutes on low fan, open vent. Cool; set aside for use in Step 8.
3. Place cream cheese in mixer bowl. Whip at medium speed until fluffy, about 3 minutes.
4. Combine sugar, flour, milk, and salt. Mix well.
5. Add to cream cheese; whip at low speed until blended, about 2 minutes. Whip at medium speed until smooth, about 1 minute.
6. Add eggs; whip at low speed 30 seconds. Whip at medium speed until smooth, about 1 minute.
7. Combine water, lemon and orange juices, vanilla, orange and lemon rinds; add to cheese mixture. Whip at low speed until well blended, about 2 minutes.
8. Spread 8 pounds 5 ounces, about 5-1/4 quarts cheese filling evenly over crust in each pan.
9. Using a convection oven, bake at 325 F. 25 to 30 minutes on low fan, open vent or until firm and lightly browned.
10. Combine sour cream and last sugar. Spread about 3 cups over each baked cheese cake. Using a convection oven, bake at 325 F. 3 minutes on low fan, open vent.
11. Refrigerate until ready to serve. Cut 6 by 9.

DESSERTS (CAKES AND FROSTINGS) No.G 026 05
CHEESE CAKE WITH STRAWBERRIES

Yield 100   Portion   1 Piece

Calories	Carbohydrates	Protein	Fat	Cholesterol	Sodium	Calcium
370 cal	34 g	6 g	24 g	98 mg	324 mg	59 mg

Ingredient	Weight	Measure	Issue
MARGARINE,MELTED	1-1/2 lbs	3 cup	
CRACKERS,GRAHAM,CRUMBS	3 lbs		
SUGAR,GRANULATED	12-1/3 oz	1-3/4 cup	
CHEESE,CREAM,SOFTENED,ROOM TEMPERATURE	10-1/4 lbs	1 gal 1 qts	
SUGAR,GRANULATED	3 lbs	1 qts 2-3/4 cup	
FLOUR,WHEAT,GENERAL PURPOSE	4-3/8 oz	1 cup	
MILK,NONFAT,DRY	1 oz	1/4 cup 3 tbsp	
SALT	1/4 oz	1/8 tsp	
EGGS,WHOLE,FROZEN	2-3/8 lbs	1 qts 1/2 cup	
WATER	12-1/2 oz	1-1/2 cup	
JUICE,LEMON	2-1/8 oz	1/4 cup 1/3 tbsp	
JUICE,ORANGE	2-1/4 oz	1/4 cup 1/3 tbsp	
EXTRACT,VANILLA	7/8 oz	2 tbsp	
ORANGE,RIND,GRATED	3/8 oz	2 tbsp	
LEMON RIND,GRATED	1/4 oz	1 tbsp	
STRAWBERRIES,FROZEN,THAWED	8-3/8 lbs	3 qts 3 cup	

Method

1 Grind graham crackers or crush on board with rolling pin. Combine butter or margarine, crumbs, and sugar in mixer bowl. Blend thoroughly at low speed, about 1 minute.
2 Press 2 quarts crumb mixture firmly in bottom of each pan. Using a convection oven, bake 3 minutes on low fan, open vent at 325 F. Cool; set aside for use in Step 8.
3 Place cream cheese in mixer bowl. Whip at medium speed until fluffy, about 3 minutes.
4 Combine sugar, flour, milk, and salt. Mix well.
5 Add to cream cheese; whip at low speed until blended, about 2 minutes. Whip at medium speed until smooth, about 1 minute.
6 Add eggs; whip at low speed 30 seconds. Whip at medium speed until smooth, about 1 minute.
7 Combine water, lemon and orange juices, vanilla, orange and lemon rinds; add to cheese mixture. Whip at low speed until well blended, about 2 minutes.
8 Spread 5-1/4 quarts cheese filling evenly over crust in each pan.
9 Using a convection oven, bake at 325 F. for 25 to 30 minutes on low fan, open vent or until filling is firm and lightly browned.
10 Refrigerate until ready to serve. Place strawberries over each chilled pie. Cut 6 by 9.

DESSERTS (CAKES AND FROSTINGS) No.G 027 00

CREAM CHEESE FROSTING

Yield 100   Portion 2-1/2 Quarts

Calories	Carbohydrates	Protein	Fat	Cholesterol	Sodium	Calcium
12009 cal	1484 g	137 g	634 g	1990 mg	5378 mg	1467 mg

Ingredient	Weight	Measure	Issue
CHEESE,CREAM,SOFTENED,ROOM TEMPERATURE	4 lbs	1 qts 3-7/8 cup	
SUGAR,POWDERED,SIFTED	3-1/8 lbs	3 qts	
EXTRACT,VANILLA	7/8 oz	2 tbsp	

Method

1 Cream softened cream cheese, powdered sugar and vanilla in mixer bowl at low speed 4 minutes or until smooth and creamy.
2 Spread immediately on cooled cakes.

DESSERTS (CAKES AND FROSTINGS) No.G 028 00

STRAWBERRY CAKE (CAKE MIX)

Yield 100   Portion 1 Piece

Calories	Carbohydrates	Protein	Fat	Cholesterol	Sodium	Calcium
209 cal	38 g	2 g	5 g	0 mg	288 mg	17 mg

Ingredient

Ingredient	Weight	Measure	Issue
STRAWBERRIES,FROZEN,THAWED	2 lbs	3-1/2 cup	
DESSERT POWDER,GELATIN,STRAWBERRY	12 oz	1-1/2 cup	
CAKE MIX,WHITE	10 lbs	1 gal 3-3/8 qts	
WATER	4-1/8 lbs	2 qts	

Method

1 Thaw strawberries.
2 Prepare mix according to instructions on container. Add dessert powder and water.

DESSERTS (CAKES AND FROSTINGS) No.G 029 00

PINEAPPLE UPSIDE DOWN CAKE

Yield 100   Portion   1 Piece

Calories	Carbohydrates	Protein	Fat	Cholesterol	Sodium	Calcium
341 cal	52 g	4 g	14 g	60 mg	340 mg	93 mg

Ingredient	Weight	Measure	Issue
PINEAPPLE,CANNED,SLICES,JUICE PACK,INCL LIQUIDS	13-1/2 lbs	1 gal 2 qts	
CHERRIES,MARASCHINO,WHOLE	1 lbs	1-3/4 cup	
BUTTER,MELTED	1-1/2 lbs	3 cup	
SUGAR,BROWN,PACKED	2-1/8 lbs	1 qts 2-1/2 cup	
FLOUR,WHEAT,GENERAL PURPOSE	4 lbs	3 qts 2-1/2 cup	
SUGAR,GRANULATED	4 lbs	2 qts 1 cup	
SALT	1-1/2 oz	2-1/3 tbsp	
BAKING POWDER	3-1/4 oz	1/4 cup 3 tbsp	
MILK,NONFAT,DRY	3 oz	1-1/4 cup	
SHORTENING	1-1/2 lbs	3-3/8 cup	
WATER	2-1/3 lbs	1 qts 1/2 cup	
EGGS,WHOLE,FROZEN	2-1/4 lbs	1 qts 1/4 cup	
WATER	12-1/2 oz	1-1/2 cup	
EXTRACT,VANILLA	1-7/8 oz	1/4 cup 1/3 tbsp	

Method

1. Drain pineapple well. Drain cherries; slice in half. Set fruit aside for use in Step 3.
2. Pour 1-1/2 cups butter or margarine in each pan. Sprinkle 3-1/4 cups brown sugar evenly over butter or margarine.
3. Arrange 54 pineapple slices in rows of 6 by 9, over mixture in each pan. Place 1 cherry half, cut side up, into each pineapple slice. Set aside for use in Step 5.
4. Sift together flour, sugar, salt, baking powder, and milk into mixer bowl.
5. Add shortening and water to dry ingredients; beat at low speed 1 minute until blended. Scrape down bowl; continue beating 2 minutes.
6. Combine eggs, water, and vanilla. Add slowly to mixture while beating at low speed about 2 minutes. Scrape down bowl. Beat at medium speed 3 minutes.
7. Pour 3-1/2 quarts batter evenly over fruit in each pan.
8. Using a convection oven, bake at 325 F. 25-30 minutes on low fan, open vent or until done.
9. Remove cakes from pans while still hot. Cut 6 by 9. Serve fruit side up.

DESSERTS (CAKES AND FROSTINGS) No.G 029 01
PINEAPPLE UPSIDE DOWN CAKE (MIX)

Yield 100   Portion 1 Piece

Calories	Carbohydrates	Protein	Fat	Cholesterol	Sodium	Calcium
357 cal	60 g	3 g	13 g	26 mg	353 mg	47 mg

Ingredient | Weight | Measure | Issue

Ingredient	Weight	Measure	Issue
PINEAPPLE,CANNED,SLICES,JUICE PACK,INCL LIQUIDS	13-1/2 lbs	1 gal 2 qts	
CHERRIES,MARASCHINO,WHOLE	1 lbs	1-3/4 cup	
SUGAR,BROWN,PACKED	3 lbs	2 qts 1-3/8 cup	
BUTTER,SOFTENED	1-1/2 lbs	3 cup	
CAKE MIX,YELLOW	10 lbs		

Method

1. Drain pineapple well. Drain cherries; slice in half. Set fruit aside for use in Step 3.
2. Pour 1-1/2 cups butter or margarine in each pan. Sprinkle 3-1/4 cups brown sugar evenly over butter or margarine.
3. Arrange 54 pineapple slices, in rows 6 by 9, over mixture in each pan. Place 1 cherry half into each pineapple slice. Set aside.
4. Prepare mix according to instructions on container.
5. Pour 3-1/2 quarts batter evenly over fruit in each pan.
6. Using a convection oven, bake at 325 F. 25-30 minutes on low fan, open vent or until done.
7. Remove cakes from pans while still hot. Cut 6 by 9. Serve fruit side up.

DESSERTS (CAKES AND FROSTINGS) No.G 029 02
FRUIT COCKTAIL UPSIDE DOWN CAKE (MIX)

Yield 100   Portion   1 Piece

Calories	Carbohydrates	Protein	Fat	Cholesterol	Sodium	Calcium
278 cal	41 g	3 g	12 g	11 mg	292 mg	29 mg

Ingredient | Weight | Measure | Issue

Ingredient	Weight	Measure	Issue
FRUIT COCKTAIL,CANNED,JUICE PACK,INCL LIQUIDS	10-1/8 lbs	1 gal 7/8 qts	
CAKE MIX,YELLOW	10 lbs		

Method

1 Drain fruit cocktail well.
2 Prepare mix according to instructions on container.

DESSERTS (CAKES AND FROSTINGS) No.G 029 03

FRUIT COCKTAIL UPSIDE DOWN CAKE

Yield 100   Portion 1 Piece

Calories	Carbohydrates	Protein	Fat	Cholesterol	Sodium	Calcium
322 cal	47 g	4 g	14 g	60 mg	341 mg	87 mg

Ingredient

Ingredient	Weight	Measure	Issue
FRUIT COCKTAIL,CANNED,JUICE PACK,INCL LIQUIDS	10-1/8 lbs	1 gal 7/8 qts	
BUTTER,MELTED	1-1/2 lbs	3 cup	
SUGAR,BROWN,PACKED	2-1/8 lbs	1 qts 2-1/2 cup	
FLOUR,WHEAT,GENERAL PURPOSE	4 lbs	3 qts 2-1/2 cup	
SUGAR,GRANULATED	4 lbs	2 qts 1 cup	
SALT	1-1/2 oz	2-1/3 tbsp	
BAKING POWDER	3-1/4 oz	1/4 cup 3 tbsp	
MILK,NONFAT,DRY	3 oz	1-1/4 cup	
SHORTENING	1-1/2 lbs	3-3/8 cup	
WATER	2-1/3 lbs	1 qts 1/2 cup	
EGGS,WHOLE,FROZEN	2-1/4 lbs	1 qts 1/4 cup	
WATER	12-1/2 oz	1-1/2 cup	
EXTRACT,VANILLA	1-7/8 oz	1/4 cup 1/3 tbsp	

Method

1. Drain fruit cocktail well. Set fruit aside for use in Step 3.
2. Pour 1-1/2 cups butter or margarine in each pan. Sprinkle 3-1/4 cups brown sugar evenly over butter or margarine.
3. Spread 1-1/2 quart fruit cocktail evenly over mixture in each pan. Set aside for use in Step 5.
4. Sift together flour, sugar, salt, baking powder, and milk into mixer bowl.
5. Add shortening and water to dry ingredients; beat at low speed 1 minute until blended. Scrape down bowl; continue beating 2 minutes.
6. Combine eggs, water, and vanilla. Add slowly to mixture while beating at low speed about 2 minutes. Scrape down bowl. Beat at medium speed 3 minutes.
7. Pour 3-1/2 quarts batter evenly over fruit in each pan.
8. Using a convection oven, bake at 325 F. 25-30 minutes on low fan, open vent or until done.
9. Remove cakes from pans while still hot. Cut 6 by 9. Serve fruit side up.

DESSERTS (CAKES AND FROSTINGS) No.G 030 00

WHITE CAKE

Yield 100   **Portion** 1 Piece

Calories	Carbohydrates	Protein	Fat	Cholesterol	Sodium	Calcium
306 cal	49 g	3 g	11 g	0 mg	338 mg	89 mg

Ingredient	Weight	Measure	Issue
FLOUR,WHEAT,GENERAL PURPOSE	4 lbs	3 qts 2-1/2 cup	
SUGAR,GRANULATED	4 lbs	2 qts 1 cup	
SALT	1-1/2 oz	2-1/3 tbsp	
BAKING POWDER	4-3/8 oz	1/2 cup 1 tbsp	
MILK,NONFAT,DRY	3-1/4 oz	1-3/8 cup	
SHORTENING	1-1/2 lbs	3-3/8 cup	
WATER	2-1/4 lbs	1 qts 1/4 cup	
EGG WHITES,FROZEN,THAWED	2-3/8 lbs	1 qts 1/2 cup	
WATER	8-1/3 oz	1 cup	
EXTRACT,VANILLA	1-7/8 oz	1/4 cup 1/3 tbsp	
COOKING SPRAY,NONSTICK	2 oz	1/4 cup 1/3 tbsp	

Method

1. Sift together flour, sugar, salt, baking powder, and milk into mixer bowl.
2. Add shortening and water to dry ingredients. Beat at low speed 1 minute or until blended; continue beating at medium speed 2 minutes. Scrape down bowl.
3. Combine egg whites, water, and vanilla. Add slowly to mixture while beating at low speed. Scrape down bowl. Beat at medium speed 3 minutes.
4. Lightly spray each pan with non-stick cooking spray. Pour 1 gallon batter into each greased and floured pan.
5. Using a convection oven, bake at 300 F. for 25 to 35 minutes on low fan, open vent or until done.
6. Cool; frost if desired. Cut 6 by 9.

DESSERTS (CAKES AND FROSTINGS) No.G 030 01

WHITE CAKE (WHITE CAKE MIX)

Yield 100   Portion 1 Piece

Calories	Carbohydrates	Protein	Fat	Cholesterol	Sodium	Calcium
288 cal	50 g	2 g	9 g	0 mg	299 mg	16 mg

Ingredient | **Weight** | **Measure** | **Issue**

CAKE MIX,WHITE | 10 lbs | 1 gal 3-3/8 qts |

Method

1 Prepare mix according to instructions on container. Frost if desired.

DESSERTS (CAKES AND FROSTINGS) No.G 030 02

LEMON FILLED CAKE (WHITE CAKE MIX)

Yield 100   Portion 1 Piece

Calories	Carbohydrates	Protein	Fat	Cholesterol	Sodium	Calcium
361 cal	64 g	2 g	11 g	12 mg	370 mg	23 mg

Ingredient	Weight	Measure	Issue
CAKE MIX, WHITE	10 lbs	1 gal 3-3/8 qts	
PIE FILLING, LEMON, PREPARED	5 lbs	2 qts 2 cup	
COCONUT BUTTER CREAM FROSTING	3-1/2 kg	2-3/4 unit	

Method

1 Prepare mix according to instructions on container. Add lemon filling to cake. Frost if desired.

DESSERTS (CAKES AND FROSTINGS) No.G 030 03

RASPBERRY FILLED CAKE (WHITE CAKE MIX)

Yield 100　　　　　　　　　　　　　　　　　　**Portion** 1 Piece

Calories	Carbohydrates	Protein	Fat	Cholesterol	Sodium	Calcium
286 cal	42 g	3 g	12 g	0 mg	299 mg	28 mg

Ingredient	Weight	Measure	Issue
CAKE MIX, WHITE	10 lbs	1 gal 3-3/8 qts	
RASPBERRY BAKERY FILLING	4-1/8 lbs	1 qts 3 cup	
WHIPPED TOPPING (DEHYDRATED)		1 gal 2 qts	

Method

1 Prepare mix according to instructions on container. Add raspberry filling. Frost or top with whipped topping if desired.

DESSERTS (CAKES AND FROSTINGS) No.G 030 04

STRAWBERRY FILLED CAKE (WHITE CAKE MIX)

Yield 100   **Portion** 1 Piece

Calories	Carbohydrates	Protein	Fat	Cholesterol	Sodium	Calcium
313 cal	50 g	3 g	12 g	0 mg	296 mg	27 mg

Ingredient | Weight | Measure | Issue

CAKE MIX,WHITE   10 lbs   1 gal 3-3/8 qts
JAM,STRAWBERRY   4 lbs   1 qts 1-5/8 cup
WHIPPED TOPPING (DEHYDRATED)   1 gal 2 qts

Method

1 Prepare according to instructions on container. Add strawberry jam. Frost or top with whipped topping if desired.

DESSERTS (CAKES AND FROSTINGS) No.G 031 00

COCONUT PECAN FROSTING

Yield 100   **Portion** 3 Quarts

Calories	Carbohydrates	Protein	Fat	Cholesterol	Sodium	Calcium
14029 cal	1269 g	152 g	978 g	2326 mg	7540 mg	2852 mg

Ingredient	Weight	Measure	Issue
MILK,NONFAT,DRY	6-5/8 oz	2-3/4 cup	
WATER,WARM	2 lbs	3-3/4 cup	
BUTTER	1-1/4 lbs	2-1/2 cup	
EGGS,WHOLE,FROZEN	8-5/8 oz	1 cup	
SUGAR,GRANULATED	1-3/4 lbs	1 qts	
EXTRACT,VANILLA	5/8 oz	1 tbsp	
PECANS,CHOPPED	1 lbs		
COCONUT,PREPARED,SWEETENED FLAKES	1-1/4 lbs	1 qts 2 cup	

Method

1. Reconstitute milk.
2. Add butter, eggs and sugar to milk; blend well.
3. Cook mixture over low heat stirring constantly about 15 minutes until thickened and just begins to bubble around edge. Remove from heat.
4. Add vanilla, nuts and coconut. Stir to mix thoroughly.
5. Chill thoroughly, about 1 hour, before spreading on cooled cakes. Refrigerate cakes after frosting.

DESSERTS (CAKES AND FROSTINGS) No. G 032 00

YELLOW CAKE

Yield 100   Portion   1 Piece

Calories	Carbohydrates	Protein	Fat	Cholesterol	Sodium	Calcium
323 cal	50 g	4 g	12 g	45 mg	300 mg	75 mg

Ingredient	Weight	Measure	Issue
FLOUR,WHEAT,GENERAL PURPOSE	4-3/8 lbs	1 gal	
SUGAR,GRANULATED	4 lbs	2 qts 1 cup	
SALT	1-1/2 oz	2-1/3 tbsp	
BAKING POWDER	3-1/4 oz	1/4 cup 3 tbsp	
MILK,NONFAT,DRY	3 oz	1-1/4 cup	
SHORTENING	1-1/2 lbs	3-3/8 cup	
WATER	2-1/3 lbs	1 qts 1/2 cup	
EGGS,WHOLE,FROZEN	2-1/4 lbs	1 qts 1/4 cup	
WATER	12-1/2 oz	1-1/2 cup	
EXTRACT,VANILLA	1-7/8 oz	1/4 cup 1/3 tbsp	
COOKING SPRAY,NONSTICK	2 oz	1/4 cup 1/3 tbsp	

Method

1. Sift together flour, sugar, salt, baking powder, and milk into mixer bowl.
2. Add shortening and water to dry ingredients; beat at low speed 1 minute until blended. Scrape down bowl; continue beating 2 minutes.
3. Combine eggs, water, and vanilla. Add slowly to mixture while beating at low speed about 2 minutes. Scrape down bowl. Beat at medium speed 3 minutes.
4. Lightly spray each pan with non-stick cooking spray. Pour 3-1/2 quarts batter into each sprayed and floured pan.
5. Using a convection oven, bake at 325 F. for 30 minutes or until done on low fan, open vent.
6. Cool; frost if desired. Cut 6 by 9.

DESSERTS (CAKES AND FROSTINGS) No.G 032 01

BANANA-FILLED LAYER CAKE

Yield 100   Portion 1 Piece

Calories	Carbohydrates	Protein	Fat	Cholesterol	Sodium	Calcium
369 cal	60 g	4 g	13 g	57 mg	352 mg	80 mg

Ingredient   **Weight**   **Measure**   **Issue**

FLOUR,WHEAT,GENERAL PURPOSE   4-3/8 lbs   1 gal
SUGAR,GRANULATED   4 lbs   2 qts 1 cup
SALT   1-1/2 oz   2-1/3 tbsp
BAKING POWDER   3-1/4 oz   1/4 cup 3 tbsp
MILK,NONFAT,DRY   3 oz   1-1/4 cup
SHORTENING   1-1/2 lbs   3-3/8 cup
WATER   2-1/3 lbs   1 qts 1/2 cup
EGGS,WHOLE,FROZEN   2-1/4 lbs   1 qts 1/4 cup
WATER   12-1/2 oz   1-1/2 cup
EXTRACT,VANILLA   1-7/8 oz   1/4 cup 1/3 tbsp
COOKING SPRAY,NONSTICK   2 oz   1/4 cup 1/3 tbsp
BUTTER CREAM FROSTING   2 qts 3 cup
BANANA,FRESH,SLICED   2-1/2 lbs   1 qts 3-1/2 cup   3-7/8 lbs

Method

1. Sift together flour, sugar, salt, baking powder, and milk into mixer bowl.
2. Add shortening and water to dry ingredients; beat at low speed 1 minute until blended. Scrape down bowl; continue beating 2 minutes.
3. Combine eggs, water, and vanilla. Add slowly to mixture while beating at low speed. Scrape down bowl. Beat at medium speed 3 minutes.
4. Lightly spray each pan with non-stick cooking spray. Pour 3-1/2 quarts batter into each sprayed and floured pan.
5. Using a convection oven, bake at 325 F. for 30 minutes or until done on low fan, open vent.
6. Cool. Prepare Butter Cream Frosting, Recipe No. G 022 00. Spread frosting over 1 sheet cake. Thinly slice bananas; spread over frosting. Top with second sheet cake; spread remaining frosting evenly over sides and top of cake. Cut 4 by 25.

DESSERTS (CAKES AND FROSTINGS) No.G 032 02
BOSTON CREAM PIE

Yield 100   **Portion** 1 Slice

Calories	Carbohydrates	Protein	Fat	Cholesterol	Sodium	Calcium
330 cal	57 g	4 g	10 g	48 mg	457 mg	101 mg

Ingredient	Weight	Measure	Issue
FLOUR,WHEAT,GENERAL PURPOSE	4-3/8 lbs	1 gal	
SUGAR,GRANULATED	4 lbs	2 qts 1 cup	
SALT	1-1/2 oz	2-1/3 tbsp	
BAKING POWDER	3-1/4 oz	1/4 cup 3 tbsp	
MILK,NONFAT,DRY	3 oz	1-1/4 cup	
SHORTENING	1-1/2 lbs	3-3/8 cup	
WATER	2-1/3 lbs	1 qts 1/2 cup	
EGGS,WHOLE,FROZEN	2-1/4 lbs	1 qts 1/4 cup	
WATER	12-1/2 oz	1-1/2 cup	
EXTRACT,VANILLA	1-7/8 oz	1/4 cup 1/3 tbsp	
COOKING SPRAY,NONSTICK	2 oz	1/4 cup 1/3 tbsp	
VANILLA CREAM PUDDING (INSTANT)		1 gal 1/8 qts	
CHOCOLATE GLAZE FROSTING		1 qts 1/2 cup	
SUGAR,POWDERED	10-5/8 oz	2-1/2 cup	

Method

1. Sift together flour, sugar, salt, baking powder, and milk into mixer bowl.
2. Add shortening and water to dry ingredients; beat at low speed 1 minute until blended. Scrape down bowl; continue beating 2 minutes.
3. Combine eggs, water, and vanilla. Add slowly to mixture while beating at low speed about 2 minutes. Scrape down bowl. Beat at medium speed 3 minutes.
4. Lightly spray each pan with non-stick cooking spray. Pour 2-1/3 cups batter into each sprayed and floured 9-inch pie pan.
5. Using a convection oven, bake at 325 F. for 20 to 25 minutes or until done on low fan, open vent.
6. Cool. Split cooled cakes. Prepare Vanilla Pudding, Recipe No. J 014 00 for filling; spread 1 cup filling over bottom half of each cake. Top with other half of cake. Prepare Chocolate Glaze Frosting, Recipe No. G 024 00; spread 1/3 cup over each cake, or use powdered sugar; sprinkle 3-1/3 tablespoons over each cake. Cut 8 wedges per pie.

DESSERTS (CAKES AND FROSTINGS) No.G 032 03

MARBLE CAKE

Yield 100   Portion   1 Piece

Calories	Carbohydrates	Protein	Fat	Cholesterol	Sodium	Calcium
321 cal	50 g	4 g	13 g	47 mg	329 mg	54 mg

Ingredient	Weight	Measure	Issue
FLOUR,WHEAT,GENERAL PURPOSE	2-1/4 lbs	2 qts	
SUGAR,GRANULATED	2 lbs	1 qts 1/2 cup	
SALT	7/8 oz	1 tbsp	
BAKING POWDER	1-5/8 oz	3-1/3 tbsp	
MILK,NONFAT,DRY	1-3/4 oz	3/4 cup	
SHORTENING	10-7/8 oz	1-1/2 cup	
WATER	1-1/8 lbs	2-1/4 cup	
EGGS,WHOLE,FROZEN	1-1/8 lbs	2-1/8 cup	
WATER	6-1/4 oz	3/4 cup	
EXTRACT,VANILLA	7/8 oz	2 tbsp	
DEVIL'S FOOD CAKE (1 PIECE)	3-7/8 kg	50 unit	

Method

1. Sift together flour, sugar, salt, baking powder, and milk into mixer bowl.
2. Add shortening and water to dry ingredients; beat at low speed 1 minute until blended. Scrape down bowl; continue beating 2 minutes.
3. Combine eggs, water, and vanilla. Add slowly to mixture while beating at low speed. Scrape down bowl. Beat at medium speed 3 minutes.
4. Prepare Devil's Food Cake, Recipe Nos. G 012 00 or G 012 01.
5. Pan, alternating light and dark batters. With knife, cut carefully through batter zig-zagging to give marble effect. Using a convection oven, bake at 325 F. for 30 minutes on low fan, open vent.
6. Cool; frost if desired. Cut 6 by 9.

DESSERTS (CAKES AND FROSTINGS) No.G 032 04

COCONUT CAKE

Yield 100   Portion 1 Piece

Calories	Carbohydrates	Protein	Fat	Cholesterol	Sodium	Calcium
305 cal	42 g	4 g	14 g	52 mg	330 mg	83 mg

Ingredient	Weight	Measure	Issue
FLOUR,WHEAT,GENERAL PURPOSE	4-3/8 lbs	1 gal	
SUGAR,GRANULATED	4 lbs	2 qts 1 cup	
SALT	1-1/2 oz	2-1/3 tbsp	
BAKING POWDER	3-1/4 oz	1/4 cup 3 tbsp	
MILK,NONFAT,DRY	3 oz	1-1/4 cup	
SHORTENING	1-1/2 lbs	3-3/8 cup	
WATER	2-1/3 lbs	1 qts 1/2 cup	
EGGS,WHOLE,FROZEN	2-1/4 lbs	1 qts 1/4 cup	
WATER	12-1/2 oz	1-1/2 cup	
EXTRACT,VANILLA	1-7/8 oz	1/4 cup 1/3 tbsp	
COOKING SPRAY,NONSTICK	2 oz	1/4 cup 1/3 tbsp	
BUTTER,MELTED	12 oz	1-1/2 cup	
SUGAR,BROWN,PACKED	13-5/8 oz	2-5/8 cup	
MILK,NONFAT,DRY	7/8 oz	1/4 cup 2-1/3 tbsp	
COCONUT,PREPARED,SWEETENED FLAKES	1-5/8 lbs	2 qts	
WATER	7-1/3 oz	3/4 cup 2 tbsp	

Method

1 Sift together flour, sugar, salt, baking powder, and milk into mixer bowl.
2 Add shortening and water to dry ingredients; beat at low speed 1 minute until blended. Scrape down bowl; continue beating 2 minutes.
3 Combine eggs, water, and vanilla. Add slowly to mixture while beating at low speed. Scrape down bowl. Beat at medium speed 3 minutes.
4 Lightly spray each pan with non-stick cooking spray. Pour 3-1/2 quarts of batter into each sprayed and floured 9-inch pie pan.
5 Using a convection oven, bake at 325 F. for 25 to 30 minutes or until done on low fan, open vent.
6 Combine melted butter or margarine, brown sugar, non-fat dry milk, prepared sweetened coconut flakes, and water. As soon as cakes are removed from oven, spread about 1 quart coconut mixture over each cake. Increase oven temperature to 400 F. ; return to oven about 7 minutes or until coconut peaks are lightly browned.
7 Cool. Cut 6 by 9.

DESSERTS (CAKES AND FROSTINGS) No.G 032 06
DUTCH APPLE CAKE

Yield 100   Portion 1 Piece

Calories	Carbohydrates	Protein	Fat	Cholesterol	Sodium	Calcium
590 cal	120 g	4 g	12 g	54 mg	342 mg	79 mg

Ingredient	Weight	Measure	Issue
FLOUR,WHEAT,GENERAL PURPOSE	4-3/8 lbs	1 gal	
SUGAR,GRANULATED	4 lbs	2 qts 1 cup	
SALT	1-1/2 oz	2-1/3 tbsp	
BAKING POWDER	3-1/4 oz	1/4 cup 3 tbsp	
MILK,NONFAT,DRY	3 oz	1-1/4 cup	
SHORTENING	1-1/2 lbs	3-3/8 cup	
WATER	2-1/3 lbs	1 qts 1/2 cup	
EGGS,WHOLE,FROZEN	2-1/4 lbs	1 qts 1/4 cup	
WATER	12-1/2 oz	1-1/2 cup	
EXTRACT,VANILLA	1-7/8 oz	1/4 cup 1/3 tbsp	
PIE FILLING,APPLE,PREPARED	13 lbs	1 gal 2-1/2 qts	
VANILLA GLAZE		1 gal 2-3/4 qts	

Method

1 Sift together flour, sugar, salt, baking powder, and milk into mixer bowl.
2 Add shortening and water to dry ingredients; beat at low speed 1 minute until blended. Scrape down bowl; continue beating 2 minutes.
3 Combine eggs, water, and vanilla. Add slowly to mixture while beating at low speed. Scrape down bowl. Beat at medium speed 3 minutes.
4 Pour apple pie filling evenly over batter in each pan.
5 Using a convection oven, bake at 325 F. for 25 to 30 minutes or until done on low fan, open vent.
6 Cool. Top each portion with 1/4 cup Vanilla Glaze, Recipe No. D 046 00. Cut 6 by 9.

DESSERTS (CAKES AND FROSTINGS) No. G 032 07

FILLED CAKE (WASHINGTON PIE)

Yield 100   **Portion** 1 Slice

Calories	Carbohydrates	Protein	Fat	Cholesterol	Sodium	Calcium
308 cal	56 g	4 g	8 g	45 mg	290 mg	76 mg

Ingredient	Weight	Measure	Issue
FLOUR,WHEAT,GENERAL PURPOSE	4-3/8 lbs	1 gal	
SUGAR,GRANULATED	4 lbs	2 qts 1 cup	
SALT	1-1/2 oz	2-1/3 tbsp	
BAKING POWDER	3-1/4 oz	1/4 cup 3 tbsp	
MILK,NONFAT,DRY	3 oz	1-1/4 cup	
SHORTENING	1-1/2 lbs	3-3/8 cup	
WATER	2-1/3 lbs	1 qts 1/2 cup	
EGGS,WHOLE,FROZEN	2-1/4 lbs	1 qts 1/4 cup	
WATER	12-1/2 oz	1-1/2 cup	
EXTRACT,VANILLA	1-7/8 oz	1/4 cup 1/3 tbsp	
COOKING SPRAY,NONSTICK	2 oz		
JELLY	6 lbs	2 qts 1 cup	
SUGAR,POWDERED	10-5/8 oz	2-1/2 cup	

Method

1. Sift together flour, sugar, salt, baking powder, and milk into mixer bowl.
2. Add shortening and water to dry ingredients; beat at low speed 1 minute until blended. Scrape down bowl; continue beating 2 minutes.
3. Combine eggs, water, and vanilla. Add slowly to mixture while beating at low speed. Scrape down bowl. Beat at medium speed 3 minutes.
4. Lightly spray pie pans with non-stick cooking spray. Flour 9-inch pie pans. Pour 2-3/4 cups batter into each pan.
5. Using a convection oven, bake at 325 F. for 20 to 25 minutes or until done on low fan, open vent.
6. Cool. Split cooled cakes. Spread 3/4 cup jam or jelly over bottom half of each cake. Top with other half of cake. Sprinkle about 3-1/3 tablespoon powdered sugar over each cake. Slice each layered cake into 8 slices.

DESSERTS (CAKES AND FROSTINGS) No.G 032 08
YELLOW CAKE (CRUMBS)

Yield 100   **Portion** 1 Cup

Calories	Carbohydrates	Protein	Fat	Cholesterol	Sodium	Calcium
223 cal	34 g	4 g	8 g	45 mg	280 mg	74 mg

Ingredient	Weight	Measure	Issue
FLOUR,WHEAT,GENERAL PURPOSE	4-3/8 lbs	1 gal	
SUGAR,GRANULATED	4 lbs	2 qts 1 cup	
SALT	1-1/2 oz	2-1/3 tbsp	
BAKING POWDER	3-1/4 oz	1/4 cup 3 tbsp	
MILK,NONFAT,DRY	3 oz	1-1/4 cup	
SHORTENING	1-1/2 lbs	3-3/8 cup	
WATER	2-1/3 lbs	1 qts 1/2 cup	
EGGS,WHOLE,FROZEN	2-1/4 lbs	1 qts 1/4 cup	
WATER	12-1/2 oz	1-1/2 cup	
EXTRACT,VANILLA	1-7/8 oz	1/4 cup 1/3 tbsp	

Method

1. Sift together flour, sugar, salt, baking powder, and milk into mixer bowl.
2. Add shortening and water to dry ingredients; beat at low speed 1 minute until blended. Scrape down bowl; continue beating 2 minutes.
3. Combine eggs, water, and vanilla. Add slowly to mixture while beating at low speed about 2 minutes. Scrape down bowl. Beat at medium speed 3 minutes.
4. Pour about 7 pound 10 ounces of batter into each greased and floured pan.
5. Bake at 25 to 30 minutes or until done.
6. Cool; crumble into crumbs.

DESSERTS (CAKES AND FROSTINGS) No.G 033 00

JELLY ROLL

Yield 100   Portion 1 Slice

Calories	Carbohydrates	Protein	Fat	Cholesterol	Sodium	Calcium
240 cal	53 g	3 g	2 g	59 mg	120 mg	32 mg

Ingredient	Weight	Measure	Issue
FLOUR,WHEAT,GENERAL PURPOSE	3 lbs	2 qts 3 cup	
BAKING POWDER	1-1/8 oz	2-1/3 tbsp	
SALT	1/2 oz	3/8 tsp	
EGGS,WHOLE,FROZEN,BEATEN,ROOM TEMPERATURE	3 lbs	1 qts 1-5/8 cup	
SUGAR,GRANULATED	3 lbs	1 qts 2-3/4 cup	
WATER,WARM	1 lbs	2 cup	
EXTRACT,VANILLA	1-7/8 oz	1/4 cup 1/3 tbsp	
COOKING SPRAY,NONSTICK	2 oz	1/4 cup 1/3 tbsp	
SUGAR,POWDERED,SIFTED	12-2/3 oz	3 cup	
JELLY	8 lbs	3 qts	

Method

1. Sift together flour, baking powder and salt. Set aside for use in Step 4.
2. Combine eggs and sugar in mixer bowl. Using whip, beat at high speed 10 minutes or until mixture is light and fluffy, lemon colored, and thick enough to hold a crease.
3. Combine water and vanilla; add slowly to egg mixture while beating at low speed. Beat at low speed. DO NOT OVER MIX.
4. Add dry ingredients gradually to egg mixture while beating at low speed; beat only until ingredients are blended.
5. Lightly spray each pan with non-stick cooking spray. Pour about 2-1/4 quarts batter into each lightly sprayed, paper-lined pan.
6. Cakes should be put in oven at 5 minute intervals to allow time to roll each cake while hot. Bake 9 to 10 minutes or until done in 375 F. oven.
7. Prepare work table for rolling jelly roll while cake is baking. Place 4 sheets of paper, slightly larger than sheet pan, horizontally on work table; sprinkle generously with powdered sugar.
8. Turn baked cake upside down immediately onto paper covered with powdered sugar. Remove paper liner and pan as quickly as possible. Be careful not to tear cake. Spread 3 cups jelly evenly on each cake.
9. While cake is still hot, roll tightly, using paper to assist in shaping and molding an even roll. Cool.
10. When ready to serve, remove paper; sprinkle cake with powdered sugar. Cut 25 slices, about 1-inch thick, per roll.

DESSERTS (CAKES AND FROSTINGS) No.G 034 00

YELLOW CUPCAKES MIX

Yield 100 **Portion** 1 Cupcake

Calories	Carbohydrates	Protein	Fat	Cholesterol	Sodium	Calcium
276 cal	52 g	3 g	7 g	11 mg	311 mg	26 mg

Ingredient	Weight	Measure	Issue
CAKE MIX,YELLOW	10 lbs		
WATER	5 lbs	2 qts 1-1/2 cup	
COOKING SPRAY,NONSTICK	2 oz	1/4 cup 1/3 tbsp	

Method

1 Prepare mix according to instructions on container.
2 Lightly spray each muffin cup with non-stick cooking spray. Fill each sprayed muffin cup 2/3 full.
3 Using a convection oven, bake at 325 F. for 20 to 25 minutes or until done on low fan open vent.
4 Cool; frost or dust with powdered sugar, if desired.

DESSERTS (CAKES AND FROSTINGS)   No.G 034 01
CHOCOLATE CUPCAKES MIX

Yield 100   **Portion** 1 Cupcake

Calories	Carbohydrates	Protein	Fat	Cholesterol	Sodium	Calcium
286 cal	48 g	3 g	10 g	7 mg	376 mg	115 mg

Ingredient	Weight	Measure	Issue
CAKE MIX, DEVILS FOOD	10 lbs		
COOKING SPRAY, NONSTICK	2 oz	1/4 cup 1/3 tbsp	

Method

1. Prepare mix according to instructions on container.
2. Lightly spray each muffin cup with non-stick cooking spray. Fill each sprayed muffin cup 2/3 full.
3. Using a convection oven, bake at 325 F. for 20 to 25 minutes or until done on low fan open vent.
4. Cool; frost or dust with powdered sugar, if desired.

DESSERTS (CAKES AND FROSTINGS) No.G 034 02

SPICE CAKE CUPCAKES MIX

Yield 100   **Portion** 1 Cupcake

Calories	Carbohydrates	Protein	Fat	Cholesterol	Sodium	Calcium
278 cal	52 g	3 g	7 g	11 mg	311 mg	30 mg

Ingredient   Weight   Measure   Issue

Ingredient	Weight	Measure
CAKE MIX, YELLOW	10 lbs	
CINNAMON, GROUND	1 oz	1/4 cup 1/3 tbsp
CLOVES, GROUND	1/2 oz	2 tbsp
ALLSPICE, GROUND	1/4 oz	1 tbsp
COOKING SPRAY, NONSTICK	2 oz	1/4 cup 1/3 tbsp

Method

1. Prepare mix according to instructions on container. Add cinnamon, cloves, and allspice. Mix well.
2. Lightly spray each muffin cup with non-stick cooking spray. Fill each sprayed muffin cup 2/3 full.
3. Using a convection oven, bake at 325 F. for 20 to 25 minutes or until done on low fan open vent.
4. Cool; frost or dust with powdered sugar, if desired.

DESSERTS (CAKES AND FROSTINGS) No.G 034 03
GINGERBREAD CUPCAKES MIX

Yield 100   Portion   1 Cupcake

Calories	Carbohydrates	Protein	Fat	Cholesterol	Sodium	Calcium
298 cal	50 g	2 g	10 g	0 mg	318 mg	43 mg

Ingredient   Weight   Measure   Issue

GINGERBREAD MIX   10 lbs
COOKING SPRAY,NONSTICK   2 oz   1/4 cup 1/3 tbsp

Method

1 Prepare mix according to instructions on container.
2 Lightly spray each muffin cup with non-stick cooking spray. Fill each sprayed muffin cup 2/3 full.
3 Using a convection oven, bake at 325 F. for 20 to 25 minutes or until done on low fan open vent.
4 Cool; frost or dust with powdered sugar, if desired.

DESSERTS (CAKES AND FROSTINGS) No.G 034 04
VANILLA CUPCAKES

Yield 100   Portion   1 Cupcake

Calories	Carbohydrates	Protein	Fat	Cholesterol	Sodium	Calcium
292 cal	50 g	2 g	9 g	0 mg	299 mg	16 mg

Ingredient	Weight	Measure	Issue
CAKE MIX, WHITE	10 lbs	1 gal 3-3/8 qts	
COOKING SPRAY, NONSTICK	2 oz	1/4 cup 1/3 tbsp	

Method

1 Prepare mix according to instructions on container.
2 Lightly spray each muffin cup with non-stick cooking spray. Fill each well-greased muffin cup 2/3 full.
3 Using a convection oven, bake at 325 F. for 20 to 25 minutes or until done on low fan open vent.
4 Cool; frost or dust with powdered sugar, if desired.

DESSERTS (CAKES AND FROSTINGS) No.G 035 00
CHOCO-LITE CAKE

Yield 100   Portion 1 Piece

Calories	Carbohydrates	Protein	Fat	Cholesterol	Sodium	Calcium
225 cal	50 g	5 g	2 g	0 mg	234 mg	78 mg

Ingredient	Weight	Measure	Issue
APPLESAUCE,CANNED,UNSWEETENED	3 lbs	1 qts 1-1/2 cup	
EGG WHITES,FROZEN,THAWED	2-7/8 lbs	1 qts 1-1/2 cup	
YOGURT,VANILLA,NONFAT	1-1/8 lbs	3 cup	
WATER	12-1/2 oz	1-1/2 cup	
CHOCOLATE,COOKING,UNSWEETENED,MELTED	5-7/8 oz	1-1/4 cup	
EXTRACT,VANILLA	7/8 oz	2 tbsp	
SUGAR,GRANULATED	4-5/8 lbs	2 qts 2-1/2 cup	
FLOUR,WHEAT,GENERAL PURPOSE	3-5/8 lbs	3 qts 1 cup	
COCOA	12-1/8 oz	1 qts	
CORNSTARCH	9 oz	2 cup	
MILK,NONFAT,DRY	4 oz	1-5/8 cup	
BAKING POWDER	2-5/8 oz	1/4 cup 1-2/3 tbsp	
CINNAMON,GROUND	1 oz	1/4 cup 1/3 tbsp	
SALT	5/8 oz	1 tbsp	
BAKING SODA	2/3 oz	1 tbsp	
COOKING SPRAY,NONSTICK	2 oz	1/4 cup 1/3 tbsp	
CORN SYRUP,LIGHT	8-2/3 oz	3/4 cup	
WATER	6-1/4 oz	3/4 cup	
SUGAR,POWDERED,SIFTED	1-1/4 lbs	1 qts 1/2 cup	
COCOA	3 oz	1 cup	

Method

1. Place applesauce, egg whites, yogurt, water, melted chocolate and vanilla in mixer bowl. Mix at low speed 1 minute to blend. Mix at high speed 1 minute.
2. Sift together sugar, flour, cocoa, cornstarch, milk, baking powder, cinnamon, salt, and baking soda.
3. Add dry ingredients to mixer bowl. Mix at low speed 2 minutes. Scrape down bowl. Mix at medium speed 2 minutes or until batter is smooth.
4. Lightly spray pans with non-stick cooking spray. Pour 1 gallon batter into each pan.
5. Using a convection oven bake at 325 F. for 20-25 minutes or until done on low fan, open vent.
6. To make glaze, place syrup and water in mixer bowl. Using a wire whip, mix at low speed 1 minute.
7. Sift sugar and cocoa together.
8. Add to syrup and water mixture. Mix at low speed 1 minute; scrape bowl. Mix at high speed 2 minutes.
9. Spread 1-1/2 cups chocolate glaze over each warm cake. Cool. Cut 6 by 9.

DESSERTS (CAKES AND FROSTINGS) No.G 036 00

LITE CHEESE CAKE

Yield 100 **Portion** 1 Piece

Calories	Carbohydrates	Protein	Fat	Cholesterol	Sodium	Calcium
262 cal	44 g	9 g	6 g	4 mg	424 mg	101 mg

Ingredient	Weight	Measure	Issue
MARGARINE,MELTED	1-1/4 lbs	2-1/2 cup	
CRACKERS,GRAHAM,LOW FAT,GROUND	3 lbs		
SUGAR,GRANULATED	12-1/3 oz	1-3/4 cup	
CHEESE,CREAM,FAT FREE	10-1/4 lbs	1 gal 1 qts	
SUGAR,GRANULATED	3 lbs	1 qts 2-3/4 cup	
FLOUR,WHEAT,GENERAL PURPOSE	3-7/8 oz	3/4 cup 2 tbsp	
MILK,NONFAT,DRY	7/8 oz	1/4 cup 2 tbsp	
SALT	1/8 oz	1/8 tsp	
EGG WHITES,FROZEN,THAWED	2-2/3 lbs	1 qts 1 cup	
WATER	12-1/2 oz	1-1/2 cup	
JUICE,ORANGE,FRESH	2-1/4 oz	1/4 cup 1/3 tbsp	
JUICE,LEMON,FRESH	2-1/8 oz	1/4 cup 1/3 tbsp	
EXTRACT,VANILLA	3/4 oz	1 tbsp	
ORANGE,RIND,GRATED	1/3 oz	1 tbsp	
LEMON RIND,GRATED	1/4 oz	1 tbsp	

Method

1. Combine margarine or butter, crumbs, and sugar in mixer bowl. Blend thoroughly at low speed, about 1 minute.
2. Press about 2-1/4 quarts crumb mixture firmly into bottom of each pan. Using a convection oven, bake at 325 F. 3 minutes on low fan, open vent. Cool; set aside for use in Step 8.
3. Place cream cheese in mixer bowl. Whip at high speed until fluffy, about 3 minutes.
4. Combine sugar, flour, milk, and salt. Mix well.
5. Add to cream cheese; whip at medium speed until blended, about 2 minutes; scrape down bowl; whip at high speed until smooth, about 1 minute.
6. Add egg whites gradually while mixing at low speed 1 minute. Scrape down bowl. Whip at high speed until smooth, about 1 minute.
7. Combine water, orange and lemon juices, vanilla, orange and lemon rinds; add to cheese mixture. Whip at medium speed until well blended, about 2 minutes.
8. Pour about 1-1/4 gallons cheese filling evenly over crust in each pan. Spread evenly.
9. Using a convection oven bake at 325 F. 25 to 30 minutes or until firm and lightly browned on low fan, open vent.
10. CCP: Hold for service at 41 F. or lower. Cut 6 by 9. Cheesecake may be served with cherry or blueberry pie filling as topping.

GENERAL INFORMATION REGARDING COOKIES

TYPES:
1. <u>Sliced cookies</u> are made from a stiff dough that is generally formed into a roll, sliced, and baked on sheet pans. Care should be taken not to overmix the dough or incorporate extra flour during mixing because this will toughen the cookies. These cookies also can be rolled out and cut into squares, circles, or fancy shapes. The method of forming the dough into a roll and then slicing the roll into uniform pieces saves time and eliminate the problem of leftover dough. It is very important that the roll be uniform and that the slices be of the same thickness to ensure even baking of the cookies.
2. <u>Drop cookies</u> are made from a soft dough. A spoon or pastry bag may be used to drop the cookies onto the sheet pans. Drop cookies should all be the same size to ensure even baking.
3. <u>Bars</u> are baked and then generally cut while warm to avoid breakage. They may be formed from rolls of dough flattened in a sheet pan or from dough spread into a sheet pan before baking.
4. <u>Brownies</u> are very rich cookies. The batter is quite heavy and must be smoothed in the sheet pan to ensure an even thickness.

GUIDELINES FOR SUCCESSFUL COOKIE BAKING

1. DO NOT use warped or bent baking pans. Use only lightweight sheet pans (weighing about 4 lb) designed for baking.
2. Follow the recipe instructions regarding greasing pans as some cookies require a greased pan for baking but other cookies have enough fat in the dough to eliminate the need for greasing the pan. Heavy greasing encourages spreading of the cookies. Use cool, clean sheet pans because cookie dough will melt and spread too much if a hot sheet pan is used.

3. If cookies are to be cut into special shapes, the dough should be rolled out to 1/4 to 1/2 inch thickness on a lightly floured board, cut into the desired shapes, and baked as directed in the basic recipe. If cookie cutters are not available, an empty can of the desired size may be used. The can should have both ends removed, be thoroughly cleaned, and have the edges smoothed before it is used.
4. To cut a roll of cookie dough into even slices, it is suggested that a clean piece of wood or metal be notched according to the width desired for each cookie, and be used as a guide in slicing. For sliced cookies, a dough scraper should be used to cut the roll of cookie dough.
5. Make each cookie the same size and thickness. Space them evenly on the pan to ensure uniform baking. Cookies may be flattened with the bottom of a small can or glass dipped in sugar. Cookies may also be flattened with a fork to make a crisscross design on the top.
6. If less than a full pan of cookies is to be baked, the cookies should be spaced evenly in the center of the pan to ensure even baking.
7. Avoid overbaking cookies. Always test for doneness. Overbaked cookies become dry and lose their flavor rapidly.
8. Most cookies should be loosened from the pans and removed to other pans or racks to cool. Cookies will continue to bake if left on the hot pans and will be difficult to remove when cool.

H. DESSERTS (COOKIES) No. 0 (1)

INDEX

Card No.		Card No.	
H 001 00	Apple Cake Brownies	H 010 01	Chocolate Cookies (Chocolate Cookie Mix)
H 001 01	Apple Cake Brownies (Gingerbread Cake Mix)	H 010 02	Double Chocolate Chip Bars (Chocolate Cookie Mix)
H 002 00	Brownies		
H 002 01	Brownies (Chocolate Brownie Mix)	H 010 03	Double Chocolate Chip Cookies (Chocolate Cookie Mix)
H 002 02	Peanut Butter Brownies		
H 003 00	Butterscotch Brownies	H 011 00	Peanut Butter Cookies
H 004 00	Chewy Nut Bars	H 011 01	Peanut Butter Cookies (Sugar Cookie Mix)
H 004 01	Congo Bars	H 011 02	Peanut Butter Bars (Sugar Cookie Mix)
H 005 00	Shortbread Cookies	H 012 00	Chocolate Drop Cookies
H 006 00	Crisp Toffee Bars	H 012 01	Chocolate Drop Cookies (Chocolate Brownie Mix)
H 007 00	Oatmeal Cookies		
H 007 01	Oatmeal Chocolate Chip Cookies	H 013 00	Sugar Cookies
H 007 02	Oatmeal Nut Cookies	H 013 01	Sugar Cookies (Sugar Cookie Mix)
H 008 00	Gingerbread Cookies (Mix)	H 013 02	Snickerdoodle Cookies
H 009 00	Oatmeal Cookies (Oatmeal Cookie Mix)	H 013 03	Snickerdoodle Cookies (Sugar Cookie Mix)
H 009 01	Oatmeal Raisin Bars (Oatmeal Cookie Mix)	H 014 00	Coconut Raisin Drop Cookies
H 009 02	Oatmeal Chocolate Chip Cookies (Oatmeal Cookie Mix)	H 015 00	Crisp Drop Cookies
H 009 03	Oatmeal Raisin Cookies (Oatmeal Cookie Mix)	H 016 00	Coconut Cereal Cookies
H 009 04	Spiced Oatmeal Nut Cookies (Oatmeal Cookie Mix)	H 017 00	Hermits
		H 018 00	Raisin Nut Bars
H 010 00	Crisp Chocolate Cookies	H 018 01	Ginger Raisin Bars (Oatmeal Cookie &

H. DESSERTS (COOKIES) No. 0 (1)

Card No.   Card No.

	Gingerbread Mix)
H 019 00	Ginger Molasses Cookies (Sugar Cookie Mix)
H 019 01	Ginger Molasses Bars (Sugar Cookie Mix)
H 020 00	Chocolate Chip Cookies
H 020 01	Chocolate Chip Cookies (Sugar Cookie Mix)
H 020 02	Chocolate Chip Bars (Sugar Cookie Mix)
H 021 00	Lemon Cookies
H 021 01	Almond Cookies
H 021 02	Orange Cookies
H 021 03	Vanilla Cookies
H 022 00	Fudgy Brownies
H 023 00	Crispy Marshmallow Squares
H 024 00	Banana Split Brownies
H 025 00	Abracadabra Bars
H 800 00	Cookies, Frozen, Oatmeal Raisin
H 801 00	Cookies, Frozen, Snickerdoodle
H 802 00	Cookies, Frozen, Chocolate Chip

DESSERTS (COOKIES) No.H 001 00

APPLE CAKE BROWNIES

Yield 100 **Portion** 1 Brownie

Calories	Carbohydrates	Protein	Fat	Cholesterol	Sodium	Calcium
257 cal	36 g	4 g	11 g	24 mg	246 mg	33 mg

Ingredient	Weight	Measure	Issue
FLOUR,WHEAT,GENERAL PURPOSE	2-3/4 lbs	2 qts 2 cup	
SALT	1-1/4 oz	2 tbsp	
BAKING POWDER	1-1/8 oz	2-1/3 tbsp	
BAKING SODA	3/4 oz	1 tbsp	
CINNAMON,GROUND	1/2 oz	2 tbsp	
SHORTENING	1-5/8 lbs	3-1/2 cup	
SUGAR,GRANULATED	4-1/4 lbs	2 qts 1-5/8 cup	
EGGS,WHOLE,FROZEN,BEATEN,ROOM TEMPERATURE	1-1/4 lbs	2-1/4 cup	
EXTRACT,VANILLA	1-7/8 oz	1/4 cup 1/3 tbsp	
APPLES,CANNED,SLICED,DRAINED	6 lbs	3 qts	
NUTS,UNSALTED,CHOPPED,COARSELY	1-1/4 lbs	1 qts	
RAISINS	7-2/3 oz	1-1/2 cup	
COOKING SPRAY,NONSTICK	2 oz	1/4 cup 1/3 tbsp	

Method

1. Sift together flour, salt, baking powder, baking soda, and cinnamon. Set aside for use in Step 4.
2. Cream shortening and sugar in mixer bowl for 4 minutes at medium speed.
3. Add eggs and vanilla to creamed mixture and beat for 2 minutes at medium speed. Scrape down bowl.
4. Add dry ingredients to creamed mixture while beating at low speed.
5. Add apples, nuts and raisins to mixture. DO NOT OVERMIX. Mixture will be thick.
6. Lightly spray each pan with non-stick cooking spray. Spread one half of mixture into sprayed and floured pans.
7. Bake about 40 minutes or until done at 350 F.
8. Cool and cut 6 by 9.

Notes

1. In Step 5, 3 pound 6 ounces canned applesauce or 11 ounces canned instant applesauce rehydrated with 4-1/2 cups of water may be used per 100 portions.

DESSERTS (COOKIES) No.H 001 01

APPLE CAKE BROWNIES (GINGERBREAD CAKE MIX)

Yield 100　　　　　　　　　　　　　　　　　　**Portion** 1 Brownie

Calories	Carbohydrates	Protein	Fat	Cholesterol	Sodium	Calcium
261 cal	41 g	4 g	10 g	0 mg	299 mg	50 mg

Ingredient　　　　　　　　　　　　　　　　　　Weight　　　　Measure　　　　Issue

Ingredient	Weight	Measure
GINGERBREAD MIX	10 lbs	
APPLES,CANNED,SLICED,DRAINED	6 lbs	3 qts
NUTS,UNSALTED,CHOPPED,COARSELY	1-1/4 lbs	1 qts
RAISINS	7-2/3 oz	1-1/2 cup
COOKING SPRAY,NONSTICK	2 oz	1/4 cup 1/3 tbsp

Method

1. Use Gingerbread Mix. Prepare mix according to instructions on container.
2. Add apples, nuts and raisins to mixture. DO NOT OVERMIX. Mixture will be thick.
3. Lightly spray each pan with non-stick cooking spray. Spread one half of mixture into greased and floured pans.
4. Bake about 40 minutes or until done at 350 F.
5. Cool and cut 6 by 9.

Notes

1. In Step 2, 3 pounds 6 ounces canned applesauce or 11 ounces canned instant applesauce rehydrated with 4-1/4 cups of water may be used per 100 portions.

DESSERTS (COOKIES) No. H 002 00

BROWNIES

Yield 100   Portion   1 Brownie

Calories	Carbohydrates	Protein	Fat	Cholesterol	Sodium	Calcium
364 cal	46 g	6 g	19 g	55 mg	132 mg	45 mg

Ingredient	Weight	Measure	Issue
FLOUR,WHEAT,GENERAL PURPOSE	3 lbs	2 qts 3 cup	
SUGAR,GRANULATED	5-1/4 lbs	3 qts	
COCOA	1-1/3 lbs	1 qts 3 cup	
BAKING POWDER	1-1/8 oz	2-1/3 tbsp	
SALT	5/8 oz	1 tbsp	
SHORTENING	2-3/4 lbs	1 qts 2 cup	
EGGS,WHOLE,FROZEN	2-3/4 lbs	1 qts 1-1/4 cup	
SYRUP	1-7/8 lbs	2-5/8 cup	
EXTRACT,VANILLA	1-3/8 oz	3 tbsp	
NUTS,UNSALTED,CHOPPED,COARSELY	1-7/8 lbs	1 qts 2 cup	
COOKING SPRAY,NONSTICK	2 oz	1/4 cup 1/3 tbsp	

Method

1. Place flour, sugar, cocoa, baking powder and salt in mixer bowl; blend well at low speed for 1 minute.
2. Add shortening, eggs, syrup and vanilla to dry ingredients. Mix at low speed for 1 minute then scrape down bowl. Mix at medium speed for 2 minutes or until thoroughly blended.
3. Add nuts to batter; mix at low speed for 30 seconds.
4. Lightly spray each pan with non-stick cooking spray. Spread 4-3/4 quarts batter in sprayed pans.
5. Using a convection oven, bake for 25 to 30 minutes or until done at 325 F. on high fan, open vent. DO NOT OVERBAKE. Brownies are done when a toothpick inserted in the center of baked brownies comes out clean.
6. Cool and cut 6 by 9.

DESSERTS (COOKIES) No. H 002 01

BROWNIES (CHOCOLATE BROWNIE MIX)

Yield 100   Portion 1 Brownie

Calories	Carbohydrates	Protein	Fat	Cholesterol	Sodium	Calcium
375 cal	52 g	3 g	19 g	0 mg	206 mg	13 mg

Ingredient
BROWNIE MIX

Weight
15 lbs

Measure
2 gal 3-1/8 qts

Issue

Method
1 Prepare mix according to instructions on container.

DESSERTS (COOKIES) No. H 002 02

PEANUT BUTTER BROWNIES

Yield 100   **Portion** 1 Brownie

Calories	Carbohydrates	Protein	Fat	Cholesterol	Sodium	Calcium
377 cal	49 g	10 g	18 g	55 mg	195 mg	50 mg

Ingredient	**Weight**	**Measure**	**Issue**
FLOUR,WHEAT,GENERAL PURPOSE	3 lbs	2 qts 3 cup	
SUGAR,GRANULATED	5-1/4 lbs	3 qts	
COCOA	1-1/3 lbs	1 qts 3 cup	
BAKING POWDER	1-1/8 oz	2-1/3 tbsp	
SALT	5/8 oz	1 tbsp	
SHORTENING	1 lbs	2-1/4 cup	
PEANUT BUTTER	3 lbs	1 qts 1-1/4 cup	
EGGS,WHOLE,FROZEN	2-3/4 lbs	1 qts 1-1/4 cup	
SYRUP	1-7/8 lbs	2-5/8 cup	
EXTRACT,VANILLA	1-3/8 oz	3 tbsp	
NUTS,UNSALTED,CHOPPED,COARSELY	1-7/8 lbs	1 qts 2 cup	
COOKING SPRAY,NONSTICK	2 oz	1/4 cup 1/3 tbsp	

Method

1. Place flour, sugar, cocoa, baking powder and salt in mixer bowl; blend well at low speed for 1 minute.
2. Add shortening, peanut butter, eggs, syrup and vanilla to dry ingredients. Mix at low speed for 1 minute and scrape down bowl. Mix at medium speed for 2 minutes or until thoroughly blended.
3. Add nuts to batter and mix at low speed for 30 seconds.
4. Lightly spray each pan with non-stick cooking spray. Spread 4-1/4 quarts batter into each sprayed sheet pan.
5. Using a convection oven, bake at 325 F. for 25 to 30 minutes or until done on high fan, open vent. DO NOT OVERBAKE. Brownies are done when a toothpick inserted into center comes out clean.
6. Cool and cut 6 by 9.

DESSERTS (COOKIES) No.H 003 00

BUTTERSCOTCH BROWNIES

Yield 100   Portion 1 Brownie

Calories	Carbohydrates	Protein	Fat	Cholesterol	Sodium	Calcium
328 cal	39 g	6 g	17 g	57 mg	287 mg	108 mg

Ingredient	**Weight**	**Measure**	**Issue**
FLOUR,WHEAT,GENERAL PURPOSE	5-1/2 lbs	1 gal 1 qts	
BAKING POWDER	4-3/8 oz	1/2 cup 1 tbsp	
SALT	5/8 oz	1 tbsp	
SUGAR,BROWN,PACKED	4-1/8 lbs	3 qts 3/4 cup	
BUTTER,MELTED	1-3/4 lbs	3-1/2 cup	
EGGS,WHOLE,FROZEN	2 lbs	3-3/4 cup	
EXTRACT,VANILLA	1-7/8 oz	1/4 cup 1/3 tbsp	
NUTS,UNSALTED,CHOPPED,COARSELY	1-7/8 lbs	1 qts 2 cup	
COOKING SPRAY,NONSTICK	2 oz	1/4 cup 1/3 tbsp	

Method

1. Sift together flour, baking powder, and salt. Set aside for use in Step 3.
2. Place brown sugar in mixer bowl; add hot butter or margarine. Beat about 2 minutes at low speed until smooth and well blended.
3. Add eggs and vanilla; beat at medium speed for 8 minutes. Scrape down bowl and add dry ingredients to mixture in mixer bowl. Beat for 2 minutes at low speed or until well blended. Scrape down bowl.
4. Fold nuts into batter.
5. Lightly spray each pan with non-stick cooking spray. Spread 3-1/4 quarts batter into sprayed and floured pans.
6. Using a convection oven, bake at 300 F. 40 to 45 minutes or until done on low fan, closed vent. DO NOT OVERBAKE. Brownies are done when a toothpick inserted into center comes out clean.
7. Cut 6 by 9 while warm.

DESSERTS (COOKIES) No. H 004 00

CHEWY NUT BARS

Yield 100   **Portion** 2 Each

Calories	Carbohydrates	Protein	Fat	Cholesterol	Sodium	Calcium
225 cal	25 g	5 g	12 g	63 mg	178 mg	58 mg

Ingredient	Weight	Measure	Issue
FLOUR,WHEAT,GENERAL PURPOSE	1-7/8 lbs	1 qts 3 cup	
BAKING POWDER	1-1/8 oz	2-1/3 tbsp	
SALT	1 oz	1 tbsp	
EGGS,WHOLE,FROZEN	3-1/4 lbs	1 qts 2 cup	
SUGAR,BROWN,PACKED	3-1/2 lbs	2 qts 2-3/4 cup	
EXTRACT,VANILLA	7/8 oz	2 tbsp	
WALNUTS,SHELLED,CHOPPED	3-2/3 lbs	3 qts 2 cup	
COOKING SPRAY,NONSTICK	2 oz	1/4 cup 1/3 tbsp	

Method

1. Sift together flour, baking powder and salt. Set aside for use in Step 3.
2. Place brown sugar, eggs, and vanilla in mixer bowl. Beat at low speed for 1 minute, then at medium speed for 2 to 3 minutes or until smooth.
3. Add flour mixture; mix at low speed for 1 minute or until well blended.
4. Add nuts; mix for 1 minute at low speed.
5. Lightly spray each pan with non-stick cooking spray. Spread about 3-1/4 quarts batter into sprayed pans.
6. Using a convection oven, bake at 325 F. for 20 minutes or until done on low fan, open vent.
7. Cook; cut 6 by 18.

DESSERTS (COOKIES) No. H 004 01

CONGO BARS

Yield 100 **Portion** 2 Bars

Calories	Carbohydrates	Protein	Fat	Cholesterol	Sodium	Calcium
240 cal	31 g	4 g	12 g	41 mg	175 mg	55 mg

Ingredient	Weight	Measure	Issue
FLOUR,WHEAT,GENERAL PURPOSE	3-1/3 lbs	3 qts	
BAKING POWDER	1-1/8 oz	2-1/3 tbsp	
SALT	1 oz	1 tbsp	
EGGS,WHOLE,FROZEN	2 lbs	3-3/4 cup	
OIL,SALAD	1-1/2 lbs	3 cup	
SUGAR,BROWN,PACKED	3-1/2 lbs	2 qts 2-3/4 cup	
EXTRACT,VANILLA	7/8 oz	2 tbsp	
WALNUTS,SHELLED,CHOPPED	8-1/2 oz	2 cup	
CHOCOLATE,COOKING CHIPS,SEMISWEET	1-1/2 lbs	1 qts	
COOKING SPRAY,NONSTICK	2 oz	1/4 cup 1/3 tbsp	

Method

1. Sift together flour, baking powder and salt. Set aside for use in Step 3.
2. Place eggs, brown sugar, vanilla and oil in mixer bowl. Beat at low speed for 1 minute, then at medium speed for 2 to 3 minutes until smooth.
3. Add flour mixture; mix at low speed 1 minute or until well blended.
4. Add nuts and chocolate chips; mix for 1 minute at low speed.
5. Lightly spray each pan with non-stick cooking spray. Spread about 6 pounds 11 ounces batter into sprayed sheet pans.
6. Using a convection oven, bake at 325 F. for 25 minutes or until done on low fan, open vent.
7. Cool; cut 6 by 18.

DESSERTS (COOKIES) No. H 005 00

SHORTBREAD COOKIES

Yield 100 **Portion** 2 Cookies

Calories	Carbohydrates	Protein	Fat	Cholesterol	Sodium	Calcium
269 cal	31 g	3 g	15 g	40 mg	151 mg	9 mg

Ingredient   Weight   Measure   Issue

BUTTER, SOFTENED   4 lbs   2 qts
SUGAR, GRANULATED   2-1/4 lbs   1 qts 1 cup
FLOUR, WHEAT, GENERAL PURPOSE   6 lbs   1 gal 1-1/2 qts

Method

1. Place butter in mixer bowl; beat at medium speed until creamy.
2. Gradually add sugar; continue beating until light and fluffy, about 5 minutes.
3. Add flour; mix until blended.
4. Divide dough into 10 pieces, about 1 pound 2 ounce each. Form into rolls; chill and slice each roll into 20 pieces.
5. Place in rows, 5 by 7, on ungreased pans.
6. Bake at 350 F. for 18 minutes or until cookies are firm but not browned.

DESSERTS (COOKIES) No. H 006 00

CRISP TOFFEE BARS

Yield 100   **Portion** 2 Bars

Calories	Carbohydrates	Protein	Fat	Cholesterol	Sodium	Calcium
223 cal	21 g	4 g	14 g	26 mg	102 mg	27 mg

Ingredient	Weight	Measure	Issue
BUTTER	2-1/2 lbs	1 qts 1 cup	
SUGAR,BROWN,PACKED	1-1/8 lbs	3-3/8 cup	
EXTRACT,VANILLA	7/8 oz	2 tbsp	
FLOUR,WHEAT,GENERAL PURPOSE	3-1/3 lbs	3 qts	
CHOCOLATE,COOKING CHIPS,SEMISWEET	1-1/2 lbs	1 qts	
NUTS,UNSALTED,CHOPPED,COARSELY	1-1/4 lbs	1 qts	

Method

1. Place butter or margarine in mixer bowl; cream at medium speed for 5 minutes. Add brown sugar and vanilla; continue to beat for 5 minutes or until light and fluffy.
2. Add flour to mixture. Mix 1 minute at low speed or until thoroughly blended. Mixture will be stiff.
3. Fold chips and nuts into mixture.
4. Spread 2-3/4 quarts mixture into each ungreased pan. Press mixture evenly into pans.
5. Bake at 350 F. for 25 minutes or until lightly browned.
6. Cut 6 by 18 while still warm. When cool, remove from pans.

DESSERTS (COOKIES) No. H 007 00

OATMEAL COOKIES

Yield 100 **Portion** 2 Cookies

Calories	Carbohydrates	Protein	Fat	Cholesterol	Sodium	Calcium
296 cal	43 g	6 g	12 g	16 mg	169 mg	48 mg

Ingredient	Weight	Measure	Issue
FLOUR,WHEAT,GENERAL PURPOSE	2-1/4 lbs	2 qts	
SALT	7/8 oz	1 tbsp	
BAKING SODA	3/8 oz	3/8 tsp	
BAKING POWDER	1-1/3 oz	2-2/3 tbsp	
EGGS,WHOLE,FROZEN	12-7/8 oz	1-1/2 cup	
WATER	4-1/8 oz	1/2 cup	
EXTRACT,VANILLA	7/8 oz	2 tbsp	
SHORTENING	2 lbs	1 qts 1/2 cup	
SUGAR,GRANULATED	1-1/2 lbs	3-1/2 cup	
SUGAR,BROWN,PACKED	1-1/3 lbs	1 qts 1/4 cup	
CEREAL,OATMEAL,ROLLED	5-1/8 lbs	3 qts 3 cup	
RAISINS	1-7/8 lbs	1 qts 2 cup	
COOKING SPRAY,NONSTICK	2 oz	1/4 cup 1/3 tbsp	

Method

1. Sift together flour, salt, baking soda, and baking powder; set aside for use in Step 2.
2. Place eggs, water, vanilla, shortening, and sugars in mixer bowl. Beat at low speed for 1 to 2 minutes or until well blended. Add dry ingredients; mix at low speed for 2 to 3 minutes or until smooth.
3. Add rolled oats and raisins; mix about 1 minute.
4. Lightly spray each pan with non-stick cooking spray. Drop about 1 tablespoon dough in rows of 5 by 7, on lightly sprayed pans.
5. Using a convection oven, bake at 325 F. for 13 to 15 minutes or until lightly browned on high fan, open vent.
6. Loosen cookies from pans while still warm.

DESSERTS (COOKIES) No. H 007 01

OATMEAL CHOCOLATE CHIP COOKIES

Yield 100   Portion   2 Cookies

Calories	Carbohydrates	Protein	Fat	Cholesterol	Sodium	Calcium
322 cal	42 g	6 g	15 g	18 mg	177 mg	63 mg

Ingredient	**Weight**	**Measure**	**Issue**
FLOUR,WHEAT,GENERAL PURPOSE	2-1/4 lbs	2 qts	
SALT	7/8 oz	1 tbsp	
BAKING SODA	3/8 oz	3/8 tsp	
BAKING POWDER	1-1/3 oz	2-2/3 tbsp	
EGGS,WHOLE,FROZEN	12-7/8 oz	1-1/2 cup	
WATER	4-1/8 oz	1/2 cup	
EXTRACT,VANILLA	7/8 oz	2 tbsp	
SHORTENING	2 lbs	1 qts 1/2 cup	
SUGAR,GRANULATED	1-1/2 lbs	3-1/2 cup	
SUGAR,BROWN,PACKED	1-1/3 lbs	1 qts 1/4 cup	
CEREAL,OATMEAL,ROLLED	5-1/8 lbs	3 qts 3 cup	
CHOCOLATE,COOKING CHIPS,SEMISWEET	2-1/4 lbs	1 qts 2-1/8 cup	
COOKING SPRAY,NONSTICK	2 oz	1/4 cup 1/3 tbsp	

Method

1 Sift together flour, salt, baking soda, and baking powder; set aside for use in Step 2.
2 Place eggs, water, vanilla, shortening, and sugars in mixer bowl. Beat at low speed for 1 to 2 minutes or until well blended. Add dry ingredients; mix at low speed for 2 to 3 minutes or until smooth.
3 Add rolled oats and semisweet chocolate chips or chocolate flavored baking chips; mix about 1 minute.
4 Lightly spray each pan with non-stick cooking spray. Drop about 1 tablespoon dough in rows of 5 by 7, on lightly sprayed pans.
5 Using a convection oven, bake at 325 F. for 13 to 15 minutes or until lightly browned on high fan, open vent.
6 Loosen cookies from pans while still warm.

DESSERTS (COOKIES) No. H 007 02

OATMEAL NUT COOKIES

Yield 100   Portion 2 Cookies

Calories	Carbohydrates	Protein	Fat	Cholesterol	Sodium	Calcium
296 cal	37 g	7 g	14 g	16 mg	169 mg	47 mg

Ingredient | Weight | Measure | Issue

Ingredient	Weight	Measure
FLOUR,WHEAT,GENERAL PURPOSE	2-1/4 lbs	2 qts
SALT	7/8 oz	1 tbsp
BAKING SODA	3/8 oz	3/8 tsp
BAKING POWDER	1-1/3 oz	2-2/3 tbsp
EGGS,WHOLE,FROZEN	12-7/8 oz	1-1/2 cup
WATER	4-1/8 oz	1/2 cup
EXTRACT,VANILLA	7/8 oz	2 tbsp
SHORTENING	2 lbs	1 qts 1/2 cup
SUGAR,GRANULATED	1-1/2 lbs	3-1/2 cup
SUGAR,BROWN,PACKED	1-1/3 lbs	1 qts 1/4 cup
CEREAL,OATMEAL,ROLLED	5-1/8 lbs	3 qts 3 cup
NUTS,UNSALTED,CHOPPED,COARSELY	1 lbs	3-1/8 cup
COOKING SPRAY,NONSTICK	2 oz	1/4 cup 1/3 tbsp

Method

1. Sift together flour, salt, baking soda, and baking powder; set aside for use in Step 2.
2. Place eggs, water, vanilla, shortening, and sugars in mixer bowl. Beat at low speed for 1 to 2 minutes or until well blended. Add dry ingredients; mix at low speed for 2 to 3 minutes or until smooth.
3. Add rolled oats and unsalted nuts; mix about 1 minute.
4. Lightly spray each pan with non-stick cooking spray. Drop about 1 tablespoon dough in rows of 5 by 7, on lightly sprayed pans.
5. Using a convection oven, bake at 325 F. for 13 to 15 minutes or until lightly browned on high fan, open vent.
6. Loosen cookies from pans while still warm.

DESSERTS (COOKIES) No. H 008 00

GINGERBREAD COOKIES (MIX)

Yield 100   **Portion** 2 Cookies

Calories	Carbohydrates	Protein	Fat	Cholesterol	Sodium	Calcium
222 cal	33 g	2 g	9 g	0 mg	244 mg	25 mg

Ingredient	Weight	Measure	Issue
GINGERBREAD MIX	5 lbs		
COOKIE MIX, SUGAR	5 lbs		
SHORTENING	3-5/8 oz	1/2 cup	
WATER	1-3/8 lbs	2-5/8 cup	
COOKING SPRAY, NONSTICK	2 oz	1/4 cup 1/3 tbsp	

Method

1. Place Gingerbread Cake Mix, Sugar Cookie Mix, and shortening in mixer bowl. Mix at low speed for 1 minute.
2. Add water gradually to mixture while still beating at low speed for 1 minute until sides of bowl become clean. Scrape down bowl; mix at low speed for 1 minute.
3. Divide dough into 10 pieces, about 1 pound 2 ounce each. Form into rolls about 20 inches long; slice each roll into 20 pieces.
4. Lightly spray each pan with non-stick cooking spray. Place in rows 4 by 6 on lightly sprayed sheet pans. Flatten cookies to 1/4-inch thickness.
5. Using a convection oven, bake at 350 F. for 9 minutes or until done on low fan, open vent.
6. Loosen cookies from pans while still warm.

DESSERTS (COOKIES) No.H 009 00

OATMEAL COOKIES (OATMEAL COOKIE MIX)

Yield 100   **Portion** 2 Cookies

Calories	Carbohydrates	Protein	Fat	Cholesterol	Sodium	Calcium
253 cal	32 g	3 g	15 g	31 mg	63 mg	63 mg

Ingredient	Weight	Measure	Issue
COOKIE MIX,OATMEAL	9 lbs		
WATER	1 lbs	2 cup	
COOKING SPRAY,NONSTICK	2 oz	1/4 cup 1/3 tbsp	

Method

1. Place Oatmeal Cookie Mix and contents of soda pouches in mixer bowl. Mix to combine cookie mix and soda; add water; mix at low speed about 1 minute. Scrape down bowl once during mixing.
2. Lightly spray each pan with non-stick cooking spray. Drop about 1 level tablespoon of dough in rows, 5 by 7, on lightly sprayed pans.
3. Using a convection oven, bake at 325 F. for 12 to 14 minutes or until lightly browned on high fan, open vent.
4. Loosen cookies from pans while still warm.

DESSERTS (COOKIES) No.H 009 01

OATMEAL RAISIN BARS (OATMEAL COOKIE MIX)

Yield 100 　　　　　　　　　　　　　　　　　Portion 2 Bars

Calories	Carbohydrates	Protein	Fat	Cholesterol	Sodium	Calcium
125 cal	22 g	2 g	5 g	17 mg	34 mg	37 mg

Ingredient　　　　　　　　　　　　　　Weight　　　　Measure　　　　Issue

Ingredient	Weight	Measure
COOKIE MIX, OATMEAL	9 lbs	
RAISINS	1-1/2 lbs	1 qts 5/8 cup
WATER	1 lbs	2 cup
COOKING SPRAY, NONSTICK	2 oz	1/4 cup 1/3 tbsp

Method

1. Combine cookie mix and soda with raisins; mix until blended. Add water; mix.
2. Lightly spray each pan with non-stick cooking spray. Place about 5 pounds 11 ounces dough onto each lightly sprayed sheet pan. Roll evenly to 1/2-inch thickness with lightly floured rolling pin.
3. Using a convection oven, bake at 325 F. for 12 to 14 minutes or until lightly browned on high fan, open vent. DO NOT OVERBAKE.
4. Cut 6 by 18 while still warm.

DESSERTS (COOKIES) No.H 009 02

OATMEAL CHOCOLATE CHIP COOKIES (OATMEAL COOKIE MIX)

Yield 100   **Portion** 2 Cookies

Calories	Carbohydrates	Protein	Fat	Cholesterol	Sodium	Calcium
199 cal	21 g	2 g	14 g	18 mg	39 mg	46 mg

Ingredient	Weight	Measure	Issue
COOKIE MIX,OATMEAL	9 lbs		
CHOCOLATE,COOKING CHIPS,SEMISWEET	1-1/2 lbs	1 qts	
WATER	1 lbs	2 cup	
COOKING SPRAY,NONSTICK	2 oz	1/4 cup 1/3 tbsp	

Method

1. Combine cookie mix and soda with chocolate chips; mix until blended. Add water; mix.
2. Lightly spray each pan with non-stick cooking spray. Drop about 1 level tablespoon dough in rows, 5 by 7, on lightly sprayed pans.
3. Using a convection oven, bake at 325 F. for 12 to 14 minutes or until lightly browned on high fan, open vent.
4. Loosen cookies from pans while still warm.

DESSERTS (COOKIES) No.H 009 03

OATMEAL RAISIN COOKIES (OATMEAL COOKIE MIX)

Yield 100 **Portion** 2 Cookies

Calories	Carbohydrates	Protein	Fat	Cholesterol	Sodium	Calcium
125 cal	22 g	2 g	5 g	17 mg	34 mg	37 mg

Ingredient	Weight	Measure	Issue
COOKIE MIX,OATMEAL	9 lbs		
RAISINS	1-1/2 lbs	1 qts 5/8 cup	
WATER	1 lbs	2 cup	
COOKING SPRAY,NONSTICK	2 oz	1/4 cup 1/3 tbsp	

Method

1 Combine cookie mix and soda with raisins; mix until blended. Add water; mix.
2 Lightly spray each pan with non-stick cooking spray. Drop about 1 level tablespoon dough in rows, 5 by 7, on lightly sprayed pans.
3 Using a convection oven, bake at 325 F. for 12 to 14 minutes or until lightly browned on high fan, open vent.
4 Loosen cookies from pans while still warm.

DESSERTS (COOKIES) No.H 009 04

SPICED OATMEAL NUT COOKIES (OATMEAL COOKIE MIX)

Yield 100　　　　　　　　　　　　　　　　　　　**Portion** 2 Cookies

Calories	Carbohydrates	Protein	Fat	Cholesterol	Sodium	Calcium
172 cal	21 g	2 g	10 g	17 mg	34 mg	41 mg

Ingredient	Weight	Measure	Issue
COOKIE MIX,OATMEAL	9 lbs		
RAISINS	1 lbs	3 cup	
CINNAMON,GROUND	5/8 oz	2-2/3 tbsp	
NUTMEG,GROUND	2/3 oz	2-2/3 tbsp	
CLOVES,GROUND	1/4 oz	1 tbsp	
NUTS,UNSALTED,CHOPPED,COARSELY	8 oz	1-1/2 cup	
WATER	1 lbs	2 cup	
COOKING SPRAY,NONSTICK	2 oz	1/4 cup 1/3 tbsp	

Method

1. Combine cookie mix and soda with raisins, ground cinnamon, nutmeg, cloves, and chopped nuts; mix until blended. Add water; mix.
2. Lightly spray each pan with non-stick cooking spray. Drop about 1 level tablespoon dough in rows, 5 by 7, on lightly sprayed pans.
3. Using a convection oven, bake at 325 F. for 12 to 14 minutes or until lightly browned on high fan, open vent.
4. Loosen cookies from pans while still warm.

DESSERTS (COOKIES) No.H 010 00

CRISP CHOCOLATE COOKIES

Yield 100   **Portion** 2 Cookies

Calories	Carbohydrates	Protein	Fat	Cholesterol	Sodium	Calcium
338 cal	47 g	4 g	16 g	34 mg	167 mg	35 mg

Ingredient   Weight   Measure   Issue

Ingredient	Weight	Measure	Issue
SHORTENING	3-1/8 lbs	1 qts 3 cup	
EGGS,WHOLE,FROZEN,BEATEN,ROOM TEMPERATURE	1-3/4 lbs	3-1/4 cup	
WATER	4-1/8 oz	1/2 cup	
SUGAR,GRANULATED	5-3/4 lbs	3 qts 1 cup	
SALT	1 oz	1 tbsp	
BAKING POWDER	1-1/3 oz	2-2/3 tbsp	
COCOA	12-1/8 oz	1 qts	
FLOUR,WHEAT,GENERAL PURPOSE	5-1/2 lbs	1 gal 1 qts	

Method

1. Place ingredients in mixer bowl in order listed. Mix at low speed 1 to 2 minutes or until thoroughly blended. Scrape down bowl once during mixing.
2. Divide dough into 1 pound 10 ounce pieces. Form into rolls 2 inches thick. Wrap in waxed paper and chill at least 3 hours.
3. Slice each roll into 20 pieces. Place in rows, 5 by 7, on ungreased pans.
4. Bake about 10 minutes or until done in 350 F. oven.
5. Loosen cookies from pans while still warm.

DESSERTS (COOKIES) No. H 010 01

CHOCOLATE COOKIES (CHOCOLATE COOKIE MIX)

Yield 100   Portion   2 Cookies

Calories	Carbohydrates	Protein	Fat	Cholesterol	Sodium	Calcium
277 cal	35 g	2 g	16 g	0 mg	137 mg	9 mg

Ingredient   **Weight**   **Measure**   **Issue**

COOKIE MIX,CHOCOLATE   10 lbs

Method

1 Prepare Chocolate Cookie Mix in mixer bowl. Prepare according to instructions on container.

DESSERTS (COOKIES) No.H 010 02

DOUBLE CHOCOLATE CHIP BARS (CHOCOLATE COOKIE MIX)

Yield 100   Portion 2 Bars

Calories	Carbohydrates	Protein	Fat	Cholesterol	Sodium	Calcium
254 cal	41 g	2 g	10 g	2 mg	146 mg	28 mg

Ingredient	Weight	Measure	Issue
COOKIE MIX,CHOCOLATE	10 lbs		
WATER	2-1/3 lbs	1 qts 1/2 cup	
COOKING SPRAY,NONSTICK	2 oz	1/4 cup 1/3 tbsp	
CHOCOLATE,COOKING CHIPS,SEMISWEET	2-1/4 lbs	1 qts 2-1/8 cup	

Method

1. Place Chocolate Cookie Mix and water in mixer bowl. Beat at medium speed 1 minute. Add chocolate chips or chocolate flavored baking chips; mix at low speed. Lightly spray each pan with non-stick cooking spray. Spread 7 pounds batter in each sprayed sheet pan.
2. Bake for 25 to 30 minutes in 350 F. Cut 6 by 18 per pan while warm.

DESSERTS (COOKIES) No. H 010 03

DOUBLE CHOCOLATE CHIP COOKIES (CHOC COOKIE MIX)

Yield 100   Portion 2 Each

Calories	Carbohydrates	Protein	Fat	Cholesterol	Sodium	Calcium
254 cal	41 g	2 g	10 g	2 mg	146 mg	28 mg

Ingredient	Weight	Measure	Issue
COOKIE MIX,CHOCOLATE	10 lbs		
WATER	1-5/8 lbs	3 cup	
CHOCOLATE,COOKING CHIPS,SEMISWEET	2-1/4 lbs	1 qts 2-1/8 cup	
COOKING SPRAY,NONSTICK	2 oz	1/4 cup 1/3 tbsp	

Method

1 Place Chocolate Cookie Mix and water in mixer bowl. Mix at medium speed 1 minute. Add chocolate chips or chocolate flavored baking chips; mix on low speed. Lightly spray each pan with non-stick cooking spray. Drop by rounded tablespoon, in rows 5 by 7 on sprayed pans.
2 Bake at 375 F. for 12 to 14 minutes.
3 Loosen cookies from pans while still warm.

DESSERTS (COOKIES) No.H 011 00

PEANUT BUTTER COOKIES

Yield 100   **Portion** 2 Cookies

Calories	Carbohydrates	Protein	Fat	Cholesterol	Sodium	Calcium
257 cal	27 g	5 g	15 g	24 mg	211 mg	14 mg

Ingredient   Weight   Measure   Issue

Ingredient	Weight	Measure
SHORTENING	1-3/4 lbs	1 qts
SUGAR,GRANULATED	2 lbs	1 qts 1/2 cup
SUGAR,BROWN,PACKED	1 lbs	3-1/4 cup
EGGS,WHOLE,FROZEN	1-1/4 lbs	2-1/4 cup
EXTRACT,VANILLA	5/8 oz	1 tbsp
PEANUT BUTTER	2-1/2 lbs	1 qts 1/2 cup
FLOUR,WHEAT,GENERAL PURPOSE	3-1/3 lbs	3 qts
BAKING SODA	1-1/3 oz	2-2/3 tbsp
SALT	3/8 oz	1/3 tsp

Method

1. Place ingredients in mixer bowl in order listed. Mix at low speed 1 to 2 minutes or until smooth. Scrape down bowl once during mixing.
2. Divide dough into 10 pieces about 1 pound 3 ounces each. Form into rolls 1-3/4x20x1-1/4-inches; slice each roll into 20 pieces, about 1 ounce each.
3. Place in rows, 4 x 6, on ungreased sheet pans; using a fork, flatten to 1/4-inch thickness, forming a crisscross pattern.
4. Using a convection oven, bake at 325 F. for 10 minutes or until lightly browned on high fan, open vent.
5. Loosen cookies from pans while still warm.

DESSERTS (COOKIES) No.H 011 01

PEANUT BUTTER COOKIES (SUGAR COOKIE MIX)

Yield 100   Portion   2 Cookies

Calories	Carbohydrates	Protein	Fat	Cholesterol	Sodium	Calcium
287 cal	34 g	4 g	16 g	0 mg	245 mg	12 mg

Ingredient	**Weight**	**Measure**	**Issue**
COOKIE MIX,SUGAR	10 lbs		
WATER	1-5/8 lbs	3 cup	
PEANUT BUTTER	2-1/2 lbs	1 qts 1/2 cup	

Method

1 Prepare sugar cookie mix according to package directions.   Add water and peanut butter. Mix at low speed 1 minute. DO NOT OVERMIX.
2 Drop by slightly rounded tablespoons. Place in rows, 4 by 6, on ungreased pans; using a fork, flatten to 1/4-inch thickness, forming a crisscross pattern.
3 Using a convection oven, bake at 325 F. for 10 to 12 minutes or until lightly browned on high fan, open vent.
4 Loosen cookies from pans while still warm.

DESSERTS (COOKIES) No.H 011 02

PEANUT BUTTER BARS (SUGAR COOKIE MIX)

Yield 100   **Portion** 2 Bars

Calories	Carbohydrates	Protein	Fat	Cholesterol	Sodium	Calcium
287 cal	34 g	4 g	16 g	0 mg	245 mg	12 mg

Ingredient	Weight	Measure	Issue
COOKIE MIX,SUGAR	10 lbs		
WATER	1-1/3 lbs	2-1/2 cup	
PEANUT BUTTER	2-1/2 lbs	1 qts 1/2 cup	

Method

1. Prepare sugar cookies according to package directions. Add water and peanut butter; beat on medium speed 1 minute. DO NOT OVERMIX.
2. Spread approximately 6 pounds 14 ounces dough evenly into each pan.
3. Using a convection oven, bake at 325 F. for 20 for 25 minutes until lightly browned on low fan, closed vent. DO NOT OVERBAKE. Cut 6 by 18 while still warm.

DESSERTS (COOKIES)   No. H 012 00

CHOCOLATE DROP COOKIES

Yield 100 **Portion** 2 Cookies

Calories	Carbohydrates	Protein	Fat	Cholesterol	Sodium	Calcium
241 cal	30 g	4 g	13 g	20 mg	158 mg	27 mg

Ingredient	Weight	Measure	Issue
SHORTENING	2-1/2 lbs	1 qts 1-1/2 cup	
EGGS,WHOLE,FROZEN,BEATEN	1 lbs	1-7/8 cup	
WATER	2-1/8 lbs	1 qts	
SUGAR,BROWN,PACKED	2-3/4 lbs	2 qts 1/2 cup	
MILK,NONFAT,DRY	1-3/4 oz	3/4 cup	
FLOUR,WHEAT,GENERAL PURPOSE	4-3/8 lbs	1 gal	
BAKING SODA	2/3 oz	1 tbsp	
SALT	7/8 oz	1 tbsp	
COCOA	12-1/8 oz	1 qts	
EXTRACT,VANILLA	1-7/8 oz	1/4 cup 1/3 tbsp	
COOKING SPRAY,NONSTICK	2 oz	1/4 cup 1/3 tbsp	

Method

1. Place ingredients in mixer bowl in order listed. Mix at low speed 1 to 2 minutes or until thoroughly blended. Scrape down bowl once during mixing.
2. Lightly spray each pan with non-stick cooking spray. Drop about 2 tablespoons dough in rows, 4 x 6, on sprayed sheet pans.
3. Using a convection oven, bake at 325 F. for 12 minutes or until done on low fan, open vent.
4. Loosen cookies from pans while still warm.

DESSERTS (COOKIES) No.H 012 01

CHOCOLATE DROP COOKIES (CHOCOLATE BROWNIE MIX)

Yield 100   **Portion** 2 Cookies

Calories	Carbohydrates	Protein	Fat	Cholesterol	Sodium	Calcium
241 cal	35 g	2 g	12 g	0 mg	138 mg	9 mg

Ingredient

Ingredient	Weight	Measure	Issue
BROWNIE MIX	10 lbs	1 gal 3-3/8 qts	
WATER	1-5/8 lbs	3 cup	
COOKING SPRAY, NONSTICK	2 oz	1/4 cup 1/3 tbsp	

Method

1. Place Brownie Mix, contents of soda pouches and water in mixer bowl. Mix at medium speed 1 minute.
2. Lightly spray each pan with non-stick cooking spray. Drop about 1 tablespoon dough in rows, 4 by 6, on sprayed sheet pans.
3. Bake at 375 F. for 10 to 12 minutes or until done.
4. Loosen cookies from pans while still warm.

DESSERTS (COOKIES) No. H 013 00

SUGAR COOKIES

Yield 100   **Portion** 2 Cookies

Calories	Carbohydrates	Protein	Fat	Cholesterol	Sodium	Calcium
243 cal	40 g	3 g	8 g	20 mg	223 mg	63 mg

Ingredient	Weight	Measure	Issue
EGGS,WHOLE,FROZEN	1 lbs	1-7/8 cup	
SHORTENING	1-1/2 lbs	3-3/8 cup	
WATER	10-1/2 oz	1-1/4 cup	
EXTRACT,VANILLA	1-3/8 oz	3 tbsp	
SUGAR,GRANULATED	4-3/8 lbs	2 qts 2 cup	
FLOUR,WHEAT,GENERAL PURPOSE	5-1/4 lbs	1 gal 3/4 qts	
SALT	1 oz	1 tbsp	
BAKING POWDER	3-1/4 oz	1/4 cup 3 tbsp	
MILK,NONFAT,DRY	5/8 oz	1/4 cup 1/3 tbsp	
SUGAR,GRANULATED	5-1/4 oz	3/4 cup	
COOKING SPRAY,NONSTICK	2 oz	1/4 cup 1/3 tbsp	

Method

1. Place ingredients in mixer bowl in order listed. Beat at low speed for 1 to 2 minutes or until smooth. Scrape down bowl once during mixing.
2. Divide dough into 1-1/4 pound pieces. Roll into rolls; slice each roll into 20 pieces.
3. Lightly spray each pan with non-stick cooking spray. Dip each piece in sugar; place sugared side up in rows, 4 by 6, on sprayed sheet pans.
4. Flatten cookies to about 1/4-inch thickness.
5. Using a convection oven, bake at 350 F. for 8 to 10 minutes or until lightly browned on low fan, open vent. DO NOT OVER BAKE.
6. Loosen cookies from pans while still warm.

DESSERTS (COOKIES) No. H 013 01

SUGAR COOKIES (SUGAR COOKIE MIX)

Yield 100 **Portion** 2 Cookies

Calories	Carbohydrates	Protein	Fat	Cholesterol	Sodium	Calcium
218 cal	32 g	1 g	10 g	0 mg	191 mg	8 mg

Ingredient	Weight	Measure	Issue
COOKIE MIX, SUGAR	10 lbs		

Method

1. Prepare mix according to instructions on container. Using a convection oven, bake at 325 F. for 8 to 10 minutes on low fan, open vent.

DESSERTS (COOKIES) No.H 013 02

SNICKERDOODLE COOKIES

Yield 100   Portion   2 Cookies

Calories	Carbohydrates	Protein	Fat	Cholesterol	Sodium	Calcium
246 cal	41 g	3 g	8 g	20 mg	223 mg	68 mg

Ingredient	Weight	Measure	Issue
EGGS,WHOLE,FROZEN	1 lbs	1-7/8 cup	
SHORTENING	1-1/2 lbs	3-3/8 cup	
WATER	10-1/2 oz	1-1/4 cup	
EXTRACT,VANILLA	1-3/8 oz	3 tbsp	
SUGAR,GRANULATED	4-3/8 lbs	2 qts 2 cup	
FLOUR,WHEAT,GENERAL PURPOSE	5-1/4 lbs	1 gal 3/4 qts	
SALT	1 oz	1 tbsp	
BAKING POWDER	3-1/4 oz	1/4 cup 3 tbsp	
MILK,NONFAT,DRY	5/8 oz	1/4 cup 1/3 tbsp	
SUGAR,GRANULATED	7 oz	1 cup	
CINNAMON,GROUND	1-1/4 oz	1/4 cup 1-2/3 tbsp	
COOKING SPRAY,NONSTICK	2 oz	1/4 cup 1/3 tbsp	

Method

1. Place ingredients in mixer bowl in order listed. Beat at low speed 1 to 2 minutes or until smooth. Scrape down bowl once during mixing.
2. Divide dough into 1-1/4 pound pieces. Roll into rolls; slice each roll into 20 pieces.
3. Lightly spray each pan with non-stick cooking spray. Combine granulated sugar and ground cinnamon. Dip each piece in sugar and cinnamon mixture; place sugared side up in rows, 4 by 6, on sprayed sheet pans.
4. Flatten cookies to about 1/4-inch thickness.
5. Using a convection oven, bake at 350 F. for 8 to 10 minutes or until lightly browned on low fan, open vent.   DO NOT OVER BAKE.
6. Loosen cookies from pans while still warm.

DESSERTS (COOKIES) No. H 013 03

SNICKERDOODLE COOKIES (SUGAR COOKIE MIX)

Yield 100　　　　　　　　　　　　　　　　**Portion** 2 Cookies

Calories	Carbohydrates	Protein	Fat	Cholesterol	Sodium	Calcium
231 cal	34 g	1 g	10 g	0 mg	191 mg	12 mg

Ingredient　　　　　　　　　　　　　　Weight　　　Measure　　　　　　　Issue

COOKIE MIX, SUGAR	10 lbs		
CINNAMON, GROUND	1-1/4 oz	1/4 cup 1-2/3 tbsp	
SUGAR, GRANULATED	7 oz	1 cup	
COOKING SPRAY, NONSTICK	2 oz	1/4 cup 1/3 tbsp	

Method

1. Prepare mix according to instructions on container.
2. Combine sugar and ground cinnamon. Dip each piece in sugar and cinnamon.
3. Lightly spray cookie pans with non-stick cooking spray. Place cookies 4 by 6.
4. Using a convection oven, bake at 325 F. for 8 to 10 minutes on low fan, open vent.

DESSERTS (COOKIES) No.H 014 00

COCONUT RAISIN DROP COOKIES

Yield 100 | Portion 2 Cookies

Calories	Carbohydrates	Protein	Fat	Cholesterol	Sodium	Calcium
192 cal	25 g	3 g	9 g	8 mg	102 mg	43 mg

Ingredient	Weight	Measure	Issue
EGGS,WHOLE,FROZEN	6-3/8 oz	3/4 cup	
SHORTENING	1 lbs	2-1/4 cup	
MOLASSES	1-5/8 lbs	2-1/4 cup	
WATER	1 lbs	2 cup	
FLOUR,WHEAT,GENERAL PURPOSE	2-3/4 lbs	2 qts 2 cup	
SUGAR,GRANULATED	1 lbs	2-1/4 cup	
MILK,NONFAT,DRY	7/8 oz	1/4 cup 2-1/3 tbsp	
BAKING POWDER	3/4 oz	1 tbsp	
BAKING SODA	3/4 oz	1 tbsp	
COCONUT,PREPARED,SWEETENED FLAKES	9-7/8 oz	3 cup	
RAISINS	1 lbs	3 cup	
NUTS,UNSALTED,CHOPPED,COARSELY	1-1/4 lbs	1 qts	
COOKING SPRAY,NONSTICK	2 oz	1/4 cup 1/3 tbsp	

Method

1 Place ingredients in mixer bowl in order listed. Mix at low speed 2 minutes or until thoroughly blended.
2 Lightly spray each pan with non-stick cooking spray. Drop about 1 ounce of dough per cookie in rows, 4 by 6, on sprayed pans.
3 Bake at 375 F. for 10 minutes or until done.

DESSERTS (COOKIES) No. H 015 00

CRISP DROP COOKIES

Yield 100   Portion 2 Each

Calories	Carbohydrates	Protein	Fat	Cholesterol	Sodium	Calcium
249 cal	37 g	3 g	10 g	6 mg	233 mg	9 mg

Ingredient	Weight	Measure	Issue
FLOUR,WHEAT,GENERAL PURPOSE	5-1/2 lbs	1 gal 1 qts	
SUGAR,GRANULATED	12-1/3 oz	1-3/4 cup	
SUGAR,GRANULATED	3 lbs	1 qts 2-3/4 cup	
SYRUP	2-3/4 oz	1/4 cup 1/3 tbsp	
SHORTENING	2 lbs	1 qts 1/2 cup	
SALT	1-1/4 oz	2 tbsp	
EGGS,WHOLE,FROZEN	4-7/8 oz	1/2 cup 1 tbsp	
EXTRACT,VANILLA	7/8 oz	2 tbsp	
MILK,NONFAT,DRY	1-1/4 oz	1/2 cup	
WATER,WARM	1-1/2 lbs	2-3/4 cup	
BAKING SODA	1-1/8 oz	2-1/3 tbsp	
COOKING SPRAY,NONSTICK	2 oz	1/4 cup 1/3 tbsp	

Method

1 Sift together flour and sugar. Set aside for use in Step 4.
2 Cream sugar, syrup, shortening, salt, eggs, and vanilla at low speed 5 minutes or until light and fluffy.
3 Reconstitute milk; add soda; add to creamed mixture. Blend thoroughly.
4 Add dry ingredients to mixture; mix only until ingredients are combined. DO NOT OVERMIX.
5 Lightly spray each pan with non-stick cooking spray. Drop by tablespoons, or through size 10 plain pastry tube, in rows 5 by 7, onto lightly sprayed pans.
6 Bake at 375 F. for 14 to 16 minutes or until lightly browned.
7 Loosen cookies from pans while still warm.

DESSERTS (COOKIES) No. H 016 00

COCONUT CEREAL COOKIES

Yield 100   Portion 2 Cookies

Calories	Carbohydrates	Protein	Fat	Cholesterol	Sodium	Calcium
241 cal	31 g	3 g	12 g	20 mg	177 mg	12 mg

Ingredient	Weight	Measure	Issue
FLOUR,WHEAT,GENERAL PURPOSE	2-1/4 lbs	2 qts	
SALT	5/8 oz	1 tbsp	
BAKING SODA	1/2 oz	1 tbsp	
SHORTENING	2 lbs	1 qts 1/2 cup	
SUGAR,GRANULATED	2 lbs	1 qts 1/2 cup	
SUGAR,BROWN,PACKED	1-1/4 lbs	3-3/4 cup	
EGGS,WHOLE,FROZEN	1 lbs	1-7/8 cup	
EXTRACT,VANILLA	1/2 oz	1 tbsp	
COCONUT,PREPARED,SWEETENED FLAKES	1-1/8 lbs	1 qts 1-1/2 cup	
CEREAL,OATMEAL,ROLLED	1 lbs	3 cup	
CEREAL,CORN FLAKES,BULK	1 lbs	1 gal	

Method

1. Sift flour, salt and soda together. Set aside for use in Step 3.
2. Cream shortening and sugars in mixer bowl at low speed 1 minute. Mix at medium speed 3 minutes or until light and fluffy.
3. Add eggs and vanilla to creamed mixture. Beat at low speed 1 minute or until well blended. At low speed, add dry ingredients. Scrape bowl; mix at low speed 1 minute or until combined.
4. Add coconut and cereals to dough; mix at low speed only until ingredients are combined. Let dough stand about 30 minutes.
5. Divide dough into 10 pieces, about 1 pound 1 ounce each. Form into rolls; slice each roll into 20 pieces.
6. Place in rows, 4 by 6, on ungreased pans; flatten to 1/4-inch thickness.
7. Using a convection oven, bake at 325 F. for 8 to 10 minutes or until lightly browned on high fan, open vent.
8. Loosen cookies from pans while still warm.

Notes

1. In Step 4, other prepared cereals such as bran flakes, wheat flakes, puffed rice, puffed corn, or puffed wheat, or combination may be used for corn flakes.

DESSERTS (COOKIES) No.H 017 00

HERMITS

Yield 100   **Portion** 2 Each

Calories	Carbohydrates	Protein	Fat	Cholesterol	Sodium	Calcium
229 cal	39 g	3 g	7 g	17 mg	50 mg	22 mg

Ingredient	Weight	Measure	Issue
SUGAR,GRANULATED	2-2/3 lbs	1 qts 2 cup	
SHORTENING	1-1/3 lbs	3 cup	
BAKING SODA	1/2 oz	1 tbsp	
EGGS,WHOLE,FROZEN	14-1/4 oz	1-5/8 cup	
NUTMEG,GROUND	1/2 oz	2 tbsp	
CINNAMON,GROUND	1/2 oz	2 tbsp	
MOLASSES	1-1/8 lbs	1-1/2 cup	
WATER	8-1/3 oz	1 cup	
RAISINS	1-7/8 lbs	1 qts 2 cup	
FLOUR,WHEAT,GENERAL PURPOSE	4-2/3 lbs	1 gal 1/4 qts	
COOKING SPRAY,NONSTICK	2 oz	1/4 cup 1/3 tbsp	

Method

1. Blend sugar, shortening, baking soda, eggs, nutmeg and cinnamon in mixer bowl at low speed 1 to 2 minutes or until well blended. Scrape down bowl.
2. Add molasses, water, and raisins; mix at medium speed about 1 minute or until blended.
3. Add flour gradually; mix at low speed only until ingredients are combined.
4. Lightly spray each pan with non-stick cooking spray. Divide dough into 12 pieces, weighing about 1 pounds each; form into strips about 22 inches long. Place 3 strips on each lightly greased sheet pan. Press strips down until each is 3 inches wide, and 3/8 inches thick.
5. Using a convection oven, bake at 325 F. for 10 to 12 minutes or until done on low fan, open vent.
6. Loosen baked strips from pans while still warm; cut each strip into 16 bars.

DESSERTS (COOKIES)   No. H 018 00

RAISIN NUT BARS

Yield 100   Portion   1 Bar

Calories	Carbohydrates	Protein	Fat	Cholesterol	Sodium	Calcium
275 cal	37 g	6 g	12 g	18 mg	191 mg	30 mg

Ingredient	Weight	Measure	Issue
EGGS,WHOLE,FROZEN	12-7/8 oz	1-1/2 cup	
WATER	12-1/2 oz	1-1/2 cup	
SHORTENING	1-1/2 lbs	3-3/8 cup	
SUGAR,BROWN,PACKED	2-1/8 lbs	1 qts 2-1/2 cup	
FLOUR,WHEAT,GENERAL PURPOSE	5-1/4 lbs	1 gal 3/4 qts	
MILK,NONFAT,DRY	5/8 oz	1/4 cup 1/3 tbsp	
SALT	1 oz	1 tbsp	
BAKING SODA	3/4 oz	1 tbsp	
CINNAMON,GROUND	1/2 oz	2 tbsp	
NUTMEG,GROUND	1/8 oz	1/3 tsp	
RAISINS	1-7/8 lbs	1 qts 2 cup	
NUTS,UNSALTED,CHOPPED,COARSELY	1-7/8 lbs	1 qts 2 cup	
COOKING SPRAY,NONSTICK	2 oz	1/4 cup 1/3 tbsp	
EGGS,WHOLE,FROZEN,BEATEN	1-5/8 oz	3 tbsp	
WATER	2-1/8 oz	1/4 cup 1/3 tbsp	
SUGAR,GRANULATED	3-1/2 oz	1/2 cup	

Method

1. Place ingredients in mixer bowl in order listed. Beat at low speed 1 to 2 minutes or until thoroughly blended. Scrape down bowl once during mixing.
2. Lightly spray each pan with non-stick cooking spray. Divide dough into 1 pound 9 ounce pieces. Form into strips about 22 inches long on lightly sprayed pans. Place 3 strips per pan. Press strips down until each strip is about 4 inches wide and 3/8 inches thick.
3. Mix egg and water together. Brush top of each strip of dough with egg and water mixture.
4. Sprinkle about 2-1/2 teaspoons sugar over each strip.
5. Using a convection oven, bake at 325 F. for 10 to 12 minutes or until done on low fan, open vent.
6. While still warm, cut each strip into 12 bars, about 1-3/4 inches wide.

DESSERTS (COOKIES) No. H 018 01

GINGER RAISIN BARS (OATMEAL COOKIE & GINGRBRD MIX)

Yield 100　　　　　　　　　　　　　　　　　　　Portion 1 Bar

Calories	Carbohydrates	Protein	Fat	Cholesterol	Sodium	Calcium
100 cal	19 g	1 g	3 g	8 mg	48 mg	25 mg

Ingredient	Weight	Measure	Issue
COOKIE MIX, OATMEAL	4-1/2 lbs		
GINGERBREAD MIX	1 lbs		
WATER	1 lbs	2 cup	
RAISINS	1-7/8 lbs	1 qts 2 cup	
COOKING SPRAY, NONSTICK	2 oz	1/4 cup 1/3 tbsp	

Method

1. Prepare mix according to instructions on container.
2. Divide dough into 9 pieces, about 1-1/2 pounds each. Form strips about 22 inches long on lightly greased pans, 3 strips per pan. Press strips down until each strip is about 4 inches wide and 3/8 inch thick.
3. Using a convection oven, bake 16 to 18 minutes or until done on low fan, open vent. While still warm, cut each strip into 12 bars.

DESSERTS (COOKIES) No.H 019 00

GINGER MOLASSES COOKIES (SUGAR COOKIE MIX)

Yield 100   **Portion**   2 Cookies

Calories	Carbohydrates	Protein	Fat	Cholesterol	Sodium	Calcium
231 cal	34 g	1 g	10 g	0 mg	192 mg	15 mg

Ingredient	Weight	Measure	Issue
COOKIE MIX,SUGAR	10 lbs		
GINGER,GROUND	1-1/8 oz	1/4 cup 2-1/3 tbsp	
CINNAMON,GROUND	5/8 oz	2-2/3 tbsp	
MOLASSES	8-2/3 oz	3/4 cup	
WATER	1-5/8 lbs	3 cup	
COOKING SPRAY,NONSTICK	2 oz	1/4 cup 1/3 tbsp	

Method

1. Mix cookie mix and contents of soda pouches.
2. Add ginger, cinnamon, molasses and water. Beat at medium speed 2 minutes or until blended.
3. Lightly spray cooking pans with non-stick cooking spray. Drop by tablespoons in rows of 4 by 6, on lightly sprayed pans.
4. Bake at 375 F. for 11 to 13 minutes or until done.
5. Loosen cookies from pans while still warm.

DESSERTS (COOKIES) No.H 019 01

GINGER MOLASSES BARS (SUGAR COOKIE MIX)

Yield 100   Portion   2 Bars

Calories	Carbohydrates	Protein	Fat	Cholesterol	Sodium	Calcium
231 cal	34 g	1 g	10 g	0 mg	192 mg	15 mg

Ingredient	Weight	Measure	Issue
COOKIE MIX,SUGAR	10 lbs		
GINGER,GROUND	1-1/8 oz	1/4 cup 2-1/3 tbsp	
CINNAMON,GROUND	5/8 oz	2-2/3 tbsp	
MOLASSES	8-2/3 oz	3/4 cup	
WATER	1 lbs	2 cup	
COOKING SPRAY,NONSTICK	2 oz	1/4 cup 1/3 tbsp	

Method

1 Prepare cookie mix according to instructions on container.
2 Add ginger, cinnamon, molasses, and water. Beat at medium speed 1 minute. DO NOT OVERMIX.
3 Lightly spray pans with non-stick cooking spray. Spread dough evenly into each pan. Bake at 350 F. for 25 minutes. Cut 6 by 18 while still warm.

DESSERTS (COOKIES) No. H 020 00

CHOCOLATE CHIP COOKIES

Yield 100   Portion 2 Cookies

Calories	Carbohydrates	Protein	Fat	Cholesterol	Sodium	Calcium
266 cal	30 g	3 g	15 g	22 mg	196 mg	29 mg

Ingredient	Weight	Measure	Issue
FLOUR,WHEAT,GENERAL PURPOSE	3-5/8 lbs	3 qts 1 cup	
BAKING SODA	3/4 oz	1 tbsp	
SALT	1 oz	1 tbsp	
SHORTENING	2 lbs	1 qts 1/2 cup	
SUGAR,BROWN,PACKED	1-1/8 lbs	3-1/2 cup	
SUGAR,GRANULATED	1-1/2 lbs	3-1/2 cup	
EGGS,WHOLE,FROZEN	1 lbs	1-7/8 cup	
WATER,WARM	1 oz	2 tbsp	
EXTRACT,VANILLA	1/2 oz	1 tbsp	
CHOCOLATE,COOKING CHIPS,SEMISWEET	2-1/4 lbs	1 qts 2 cup	

Method

1. Sift together flour, baking soda, and salt. Set aside for use in Step 4.
2. Cream shortening in mixer bowl at medium speed about 1 minute. Gradually add sugars; mix at medium speed 3 minutes or until light and fluffy. Scrape down bowl.
3. Combine slightly beaten eggs and water; add gradually to creamed mixture. Blend thoroughly about 1 minute. Add vanilla. Mix thoroughly.
4. Add dry ingredients; mix only until ingredients are combined about 1 minute.
5. Add chocolate chips; mix on low speed about 1 minute or until evenly distributed.
6. Drop by tablespoons in rows, 4 by 6, on ungreased pans.
7. Using a convection oven, bake at 325 F. for 10 to 12 minutes or until lightly browned on high fan, open vent.
8. Loosen cookies from pans while still warm.

DESSERTS (COOKIES) No.H 020 01

CHOCOLATE CHIP COOKIES (SUGAR COOKIE MIX)

Yield 100   **Portion** 2 Cookies

Calories	Carbohydrates	Protein	Fat	Cholesterol	Sodium	Calcium
223 cal	32 g	1 g	10 g	0 mg	191 mg	8 mg

Ingredient	Weight	Measure	Issue
COOKIE MIX,SUGAR	10 lbs		
WATER	1-5/8 lbs	3 cup	
COOKING SPRAY,NONSTICK	2 oz	1/4 cup 1/3 tbsp	

Method

1 Prepare mix according to instructions on container. Add water.
2 Beat at medium speed 1 minute. DO NOT OVERMIX.
3 Add chocolate chips; mix on low speed about 1 minute or until evenly distributed.
4 Lightly spray sheets with non-stick cooking spray. Drop 1 tablespoon of mix onto lightly sprayed cookie sheets in rows 4 by 6.
5 Bake 12 to 14 minutes or until done. Loosen cookies from pans while still warm.

DESSERTS (COOKIES) No. H 020 02

CHOCOLATE CHIP BARS (SUGAR COOKIE MIX)

Yield 100   Portion 2 Cookies

Calories	Carbohydrates	Protein	Fat	Cholesterol	Sodium	Calcium
223 cal	32 g	1 g	10 g	0 mg	191 mg	8 mg

Ingredient	Weight	Measure	Issue
COOKIE MIX,SUGAR	10 lbs		
WATER	1-5/8 lbs	3 cup	
COOKING SPRAY,NONSTICK	2 oz	1/4 cup 1/3 tbsp	

Method

1. Prepare mix according to instructions on container. Add water.
2. Beat at medium speed 1 minute. DO NOT OVERMIX.
3. Add chocolate chips; mix on low speed about 1 minute or until evenly distributed.
4. Lightly spray sheets with non-stick cooking spray. Place dough in lightly greased sheet pans. Roll evenly into 1/2 thickness with lightly floured rolling pin.
5. Using a convection oven, bake at 325 F. for 20 to 25 minutes ot until lightly browned on low fan, open vent. DO NOT OVERBAKE. Cut 6 by 18 while still warm.

DESSERTS (COOKIES) No. H 021 00

LEMON COOKIES

Yield 100   **Portion** 2 Cookies

Calories	Carbohydrates	Protein	Fat	Cholesterol	Sodium	Calcium
310 cal	38 g	4 g	16 g	52 mg	231 mg	11 mg

Ingredient	Weight	Measure	Issue
EGGS,WHOLE,FROZEN	1-3/4 lbs	3-1/4 cup	
SHORTENING	1-3/4 lbs	1 qts	
BUTTER	1-3/4 lbs	3-1/2 cup	
FLAVORING,LEMON	1 oz	2 tbsp	
SUGAR,GRANULATED	3-1/8 lbs	1 qts 3 cup	
FLOUR,WHEAT,GENERAL PURPOSE	5-1/2 lbs	1 gal 1 qts	
SALT	1 oz	1 tbsp	
BAKING SODA	1/2 oz	1 tbsp	
SUGAR,POWDERED,SIFTED	1 lbs	1 qts	
COOKING SPRAY,NONSTICK	2 oz	1/4 cup 1/3 tbsp	

Method

1. Place ingredients in mixer bowl in order listed. Beat at low speed 1 to 2 minutes or until smooth. Scrape down bowl once during mixing.
2. Divide dough into ten 1-1/4 pound pieces. Roll into powdered sugar forming rolls 2 inches thick.
3. Lightly spray each pan with non-stick cooking spray. Slice each roll into 20 pieces. Dip top of each piece in powdered sugar; place in rows, 4 by 6 on sprayed pans. Do not flatten cookies.
4. Bake at 375 F. for 12 to 14 minutes or until done.
5. Loosen cookies from pans while still warm.

DESSERTS (COOKIES) No.H 021 01

ALMOND COOKIES

Yield 100   **Portion** 2 Cookies

Calories	Carbohydrates	Protein	Fat	Cholesterol	Sodium	Calcium
310 cal	38 g	4 g	16 g	52 mg	231 mg	11 mg

Ingredient	Weight	Measure	Issue
EGGS,WHOLE,FROZEN	1-3/4 lbs	3-1/4 cup	
SHORTENING	1-3/4 lbs	1 qts	
BUTTER	1-3/4 lbs	3-1/2 cup	
FLAVORING,ALMOND	7/8 oz	2 tbsp	
SUGAR,GRANULATED	3-1/8 lbs	1 qts 3 cup	
FLOUR,WHEAT,GENERAL PURPOSE	5-1/2 lbs	1 gal 1 qts	
SALT	1 oz	1 tbsp	
BAKING SODA	1/2 oz	1 tbsp	
SUGAR,POWDERED,SIFTED	1 lbs	1 qts	
COOKING SPRAY,NONSTICK	2 oz	1/4 cup 1/3 tbsp	

Method

1. Place ingredients in mixer bowl in order listed. Beat at low speed 1 to 2 minutes or until smooth. Scrape down bowl once during mixing.
2. Divide dough into ten 1-1/4 pound pieces. Roll into powdered sugar forming rolls 2 inches thick.
3. Lightly spray each pan with non-stick cooking spray. Slice each roll into 20 pieces. Dip top of each piece in powdered sugar; place in rows, 4 by 6 on sprayed pans. Do not flatten cookies.
4. Bake at 375 F. for 12 to 14 minutes or until done.
5. Loosen cookies from pans while still warm.

DESSERTS (COOKIES) No.H 021 02

ORANGE COOKIES

Yield 100 **Portion** 2 Cookies

Calories	Carbohydrates	Protein	Fat	Cholesterol	Sodium	Calcium
310 cal	38 g	4 g	16 g	52 mg	231 mg	11 mg

Ingredient	Weight	Measure	Issue
EGGS,WHOLE,FROZEN	1-3/4 lbs	3-1/4 cup	
SHORTENING	1-3/4 lbs	1 qts	
BUTTER	1-3/4 lbs	3-1/2 cup	
FLAVORING,ORANGE	7/8 oz	2 tbsp	
SUGAR,GRANULATED	3-1/8 lbs	1 qts 3 cup	
FLOUR,WHEAT,GENERAL PURPOSE	5-1/2 lbs	1 gal 1 qts	
SALT	1 oz	1 tbsp	
BAKING SODA	1/2 oz	1 tbsp	
ORANGE,RIND,GRATED	1 oz	1/4 cup 1 tbsp	
SUGAR,POWDERED,SIFTED	1 lbs	1 qts	
COOKING SPRAY,NONSTICK	2 oz	1/4 cup 1/3 tbsp	

Method

1. Place ingredients in mixer bowl in order listed. Add orange rind if desired (optional). Beat at low speed 1 to 2 minutes or until smooth. Scrape down bowl once during mixing.
2. Divide dough into ten 1-1/4 pound pieces. Roll into powdered sugar forming rolls 2 inches thick.
3. Lightly spray each pan with non-stick cooking spray. Slice each roll into 20 pieces. Dip top of each piece in powdered sugar; place in rows, 4 by 6 on sprayed pans. Do not flatten cookies.
4. Bake at 375 F. for 12 to 14 minutes or until done.
5. Loosen cookies from pans while still warm.

DESSERTS (COOKIES) No.H 021 03

VANILLA COOKIES

Yield 100   Portion 2 Cookies

Calories	Carbohydrates	Protein	Fat	Cholesterol	Sodium	Calcium
310 cal	38 g	4 g	16 g	52 mg	231 mg	11 mg

Ingredient	Weight	Measure	Issue
EGGS,WHOLE,FROZEN	1-3/4 lbs	3-1/4 cup	
SHORTENING	1-3/4 lbs	1 qts	
BUTTER	1-3/4 lbs	3-1/2 cup	
EXTRACT,VANILLA	7/8 oz	2 tbsp	
SUGAR,GRANULATED	3-1/8 lbs	1 qts 3 cup	
FLOUR,WHEAT,GENERAL PURPOSE	5-1/2 lbs	1 gal 1 qts	
SALT	1 oz	1 tbsp	
BAKING SODA	1/2 oz	1 tbsp	
SUGAR,POWDERED,SIFTED	1 lbs	1 qts	
COOKING SPRAY,NONSTICK	2 oz	1/4 cup 1/3 tbsp	

Method

1. Place ingredients in mixer bowl in order listed. Beat at low speed 1 to 2 minutes or until smooth. Scrape down bowl once during mixing.
2. Divide dough into ten 1-1/4 pound pieces. Roll into powdered sugar forming rolls 2 inches thick.
3. Lightly spray each pan with non-stick cooking spray. Slice each roll into 20 pieces. Dip top of each piece in powdered sugar; place in rows, 4 by 6 on sprayed pans. Do not flatten cookies.
4. Bake at 375 F. for 12 to 14 minutes or until done.
5. Loosen cookies from pans while still warm.

DESSERTS (COOKIES) No. H 022 00

FUDGY BROWNIES

Yield 100   **Portion** 1 Brownie

Calories	Carbohydrates	Protein	Fat	Cholesterol	Sodium	Calcium
232 cal	50 g	5 g	4 g	0 mg	234 mg	63 mg

Ingredient	Weight	Measure	Issue
FLOUR,WHEAT,GENERAL PURPOSE	3-1/3 lbs	3 qts	
SUGAR,GRANULATED	5-1/4 lbs	3 qts	
COCOA	1-1/2 lbs	2 qts	
BAKING POWDER	2-5/8 oz	1/4 cup 1-2/3 tbsp	
BAKING SODA	2/3 oz	1 tbsp	
SALT	3/4 oz	1 tbsp	
WATER	2-1/2 lbs	1 qts 3/4 cup	
PRUNE PUREE	3-1/3 lbs	1 qts 2 cup	
CHOCOLATE,COOKING,UNSWEETENED,MELTED	12-3/8 oz	2-5/8 cup	
EXTRACT,VANILLA	2-5/8 oz	1/4 cup 2 tbsp	
EGG WHITES,FROZEN,THAWED	2-1/2 lbs	1 qts 3/4 cup	
COOKING SPRAY,NONSTICK	2 oz	1/4 cup 1/3 tbsp	

Method

1. Sift together flour, sugar, cocoa, baking powder, baking soda, and salt. Set aside for use in Step 3.
2. Place prune puree, water, melted chocolate, and vanilla in mixer bowl; blend well at low speed for 1 minute. Add egg whites; mix at low speed for 30 seconds; scrape down bowl.
3. Add dry ingredients to mixer bowl; mix at low speed 1 minute. Scrape down bowl; mix at low speed 2 minutes or until thoroughly blended.
4. Lightly spray each pan with non-stick cooking spray. Spread 4-1/2 quarts into each lightly sprayed pan.
5. Using a convection oven, bake at 325 F. 18-20 minutes or until done on high fan, open vent. Do not over bake.
6. Cool; cut 6 by 9.

DESSERTS (COOKIES) No. H 023 00

CRISPY MARSHMALLOW SQUARES

Yield 100   Portion 2 Bars

Calories	Carbohydrates	Protein	Fat	Cholesterol	Sodium	Calcium
269 cal	52 g	2 g	6 g	0 mg	364 mg	6 mg

Ingredient	Weight	Measure	Issue
COOKING SPRAY, NONSTICK	2 oz	1/4 cup 1/3 tbsp	
MARGARINE	1-1/2 lbs	3 cup	
MARSHMALLOWS, MINIATURE	8 lbs	4 gal 2-1/8 qts	
EXTRACT, VANILLA	7/8 oz	2 tbsp	
CEREAL, RICE KRISPIES, BULK	5-7/8 lbs	5 gal	

Method

1. Lightly spray sheet pans with non-stick spray.
2. Melt margarine in steam-jacketed kettle.
3. Add marshmallows and vanilla. Stir constantly until marshmallows are completely melted, about 5 to 6 minutes.
4. Turn off heat; add cereal to marshmallow mixture; stir vigorously until cereal is well coated.
5. Turn 6 pounds 14 ounces mixture into each lightly sprayed sheet pan. Using a lightly sprayed rolling pin, roll mixture firmly to spread evenly in each pan. Cut 9 by 12. Remove from pan when cool.

DESSERTS (COOKIES) No.H 024 00

BANANA SPLIT BROWNIES

Yield 100   Portion   1 Brownie

Calories	Carbohydrates	Protein	Fat	Cholesterol	Sodium	Calcium
250 cal	53 g	3 g	4 g	0 mg	190 mg	16 mg

Ingredient | Weight | Measure | Issue

Ingredient	Weight	Measure	Issue
WATER,WARM	3-2/3 lbs	1 qts 3 cup	
BROWNIE MIX, LOWFAT CHOCOLATE	12 lbs		
BANANA,FRESH,CHOPPED	5 lbs	3 qts 3-1/8 cup	7-2/3 lbs
CHERRIES,MARASCHINO,CHOPPED	1-7/8 lbs	3-3/8 cup	
COOKING SPRAY,NONSTICK	2 oz	1/4 cup 1/3 tbsp	

Method

1. Place water in mixer bowl. Add brownie mix; mix on low speed 1 minute. Scrape down bowl. Mix on low speed 1-1/2 minutes.
2. Cut bananas 1/2 lengthwise and in 1/4 inch slices. Add bananas and cherries. Mix on low speed 15 seconds.
3. Lightly spray each sheet pan with non-stick cooking spray. Pour 4-1/2 quarts of batter into each pan. Spread evenly.
4. Using a convection oven, bake at 325 F. for 22 to 25 minutes or until done on high fan, open vent. Do not over bake.
5. Cut 6 by 9.

Notes

1. If the brownie mix package directions call for eggs, use an equal amount of egg whites. If the mix calls for oil, use an equal volume of water.

DESSERTS (COOKIES) No.H 025 00

ABRACADABRA BARS

Yield 100   Portion 2 Bars

Calories	Carbohydrates	Protein	Fat	Cholesterol	Sodium	Calcium
218 cal	42 g	3 g	4 g	0 mg	205 mg	13 mg

Ingredient	Weight	Measure	Issue
FLOUR,WHEAT,GENERAL PURPOSE	4-1/2 lbs	1 gal 1/8 qts	
BAKING SODA	1-1/3 oz	2-2/3 tbsp	
SALT	7/8 oz	1 tbsp	
CINNAMON,GROUND	1/3 oz	1 tbsp	
NUTMEG,GROUND	1/4 oz	3/8 tsp	
CLOVES,GROUND	1/4 oz	3/8 tsp	
GINGER,GROUND	1/8 oz	3/8 tsp	
SWEET POTATOES,CANNED,W/SYRUP	4-7/8 lbs	2 qts 1-3/4 cup	
SUGAR,GRANULATED	3-1/3 lbs	1 qts 3-1/2 cup	
SHORTENING	12-2/3 oz	1-3/4 cup	
EXTRACT,VANILLA	2-1/2 oz	1/4 cup 1-2/3 tbsp	
RAISINS	1-7/8 lbs	1 qts 2 cup	
COOKING SPRAY,NONSTICK	2 oz	1/4 cup 1/3 tbsp	

Method

1. Combine flour, baking soda, salt, cinnamon, nutmeg, cloves, and ginger.
2. Drain sweet potatoes, mash and set aside. Cream sugar and shortening. Add sweet potatoes and vanilla to the creamed sugar and shortening, beat on medium speed 1 minute; scrape down bowl. Beat with paddle on high speed 1 minute or until light and fluffy. Scrape down bowl.
3. Gradually add dry ingredients to sweet potato mixture, while mixing on low speed 1 minutes. Scrape down bowl; mix on medium speed 30 seconds or until just blended.
4. Fold in raisins at low speed 30 seconds.
5. Spray sheet pans very lightly with non-stick cooking spray. Using a rolling pin, spread 7 pounds 5 ounces mixture evenly in each pan.
6. Using a convection oven, bake at 325 F. 16 to 18 minutes until bars are lightly browned on low fan open vent. Cool. Cut into bars 6 by 18.

I-G. DESSERTS (PASTRY AND PIES) No. 1

MAKING ONE-CRUST PIES

BAKED PIE SHELLS

1. PREPARE AND DIVIDE DOUGH: Prepare 1/2 recipe Pie Crust (Recipe No. I-1). Divide dough into thirteen 7-1/2 oz pieces; place on lightly floured board.
2. ROLL DOUGH: Sprinkle each piece of dough lightly with flour; flatten gently. Using a floured rolling pin, roll lightly with quick strokes from center out to edge in all directions. Form a circle 1 inch larger than pie pan and about 1/8 inch thick. Shift or turn dough occasionally to prevent sticking. If edges split, pinch cracks together.
3. PLACE DOUGH IN PAN: Fold rolled dough in half; carefully place into ungreased pie pan with fold at center. Unfold and fit carefully into pie pan, being careful not to leave any air spaces between pan and dough.
4. REMOVE EXCESS DOUGH: Trim ragged edges about 1/2 inch beyond edge of pan using knife or spatula. (Incorporate excess dough into next crust, if needed). Fold extra dough back and under; crimp with the thumb and forefinger to make a high fluted edge. Dock or prick dough on bottom and sides to prevent puffing during baking. If available, place an empty pie pan inside of shell before baking to help prevent shrinking and puffing.
5. BAKE: Bake at 450° F. about 10 minutes or until golden brown or in 400° F. convection oven 8 to 10 minutes or until golden brown on high fan, open vent.
6. FILL CRUST: Fill as specified on individual recipe card.

UNBAKED SHELL

1. Follow Steps 1 through 4; omit docking or pricking of dough in Step 4.
2. Fill and bake according to instructions on specified recipe.

I-G. DESSERTS (PASTRY AND PIES) No. 2

MAKING TWO-CRUST PIES

1. PREPARE AND DIVIDE DOUGH: Prepare 1 recipe Pie Crust (Recipe No. I-1). Divide dough into 13-7-1/2 oz pieces for bottom crust and 13-7 oz pieces for top crust; place on lightly floured board.
2. ROLL DOUGH: Sprinkle each piece of dough lightly with flour; flatten gently. Using a floured rolling pin, roll lightly with quick strokes from center out to edge in all directions. Form a circle 1 inch larger than pie pan and about 1/8 inch thick. Bottom crust will be slightly thicker. Shift or turn dough occasionally to prevent sticking. If edges split, pinch cracks together.
3. BOTTOM CRUST: Fold rolled dough in half; carefully place into ungreased pie pan with fold at center. Unfold and fit carefully into pie pan, being careful not to leave any air spaces between pan and dough.
4. FILL CRUST: Fill as specified on individual recipe card.
5. TOP CRUST: Roll top crust in same manner as bottom crust. Fold in half; with knife, make several small slits near center fold to allow steam to escape during baking. Brush outer rim of bottom crust with water. Lay top crust over filling with fold at center; unfold and press edges of two crusts together lightly.
6. REMOVE EXCESS DOUGH: Trim overhanging edges of dough by using a knife or spatula. (Incorporate excess dough into next crust, if needed.) There should be little excess if skill is used in weighing and rolling dough.
7. SEAL PIE: Press edges of crust firmly together or crimp with the thumb and forefinger to make a fluted edge.
8. WASHED TOP: For a washed top, brush pies with appropriate wash as follows:
   <u>Egg and Milk Wash</u> - This wash is used for fruit pies (apple, blueberry, cherry, peach, pineapple) that are baked 30 to 35 minutes. It SHOULD NOT be used for pies requiring longer baking time as the crust will brown excessively. See Recipe No. I-4.
   <u>Egg and Water Wash</u> - This wash is used for berry and mincemeat pies that are baked 40 to 45 minutes. It SHOULD NOT be used for pies that are baked 30 to 35 minutes as the crusts will be too pale. Allow glaze to dry on crust before baking to eliminate dark spots. See Recipe No. I-4-1.
9. BAKE: Bake as specified on individual recipe card.

I. DESSERTS (PASTRY AND PIES) No. 0 (1)

INDEX

Card No.		Card No.	
I 001 00	Pie Crust	I 010 00	Apple Cobbler
I 001 01	Pie Crust (Dough Rolling Machine)	I 010 01	Peach Cobbler
I 001 02	Pie Crust (Manual Mixing Method)	I 010 02	Blueberry Cobbler
I 002 00	Graham Cracker Crust	I 010 03	Cherry Cobbler
I 002 01	Graham Cracker Crust (Preformed Crust)	I 010 04	Streusel-Topped Apple Cobbler
I 003 00	Mincemeat Pie	I 011 00	Chocolate Mousse Pie
I 004 00	Egg and Milk Wash	I 012 00	Sweet Potato Pie
I 004 01	Egg and Water Wash	I 013 00	Pumpkin Pie
I 004 02	Milk and Water Wash	I 014 00	Pineapple Pie (Canned Pineapple-Cornstarch)
I 005 00	Meringue	I 015 00	Berry Pie (Frozen Berries-Cornstarch)
I 005 01	Meringue (Dehydrated)	I 015 01	Blueberry Pie (Frozen Blueberries)
I 006 00	Vanilla Cream Pie	I 017 00	Blueberry Pie (Canned Blueberries-Cornstarch)
I 006 01	Banana Cream Pie	I 017 01	Blueberry Pie (Prepared Filling)
I 007 00	Vanilla Cream Pie (Dessert Powder, Instant)	I 019 00	Butterscotch Cream Pie (Dessert Powder, Instant)
I 007 01	Strawberry Glazed Cream Pie (Instant)		
I 007 02	Coconut Cream Pie (Instant)	I 020 00	Peach Pie (Frozen Peaches-Cornstarch)
I 007 03	Pineapple Cream Pie (Instant)	I 022 00	Cherry Pie (Canned Cherries-Cornstarch)
I 008 01	Dutch Apple Pie (Canned Apples-Cornstarch)	I 022 01	Cherry Pie (Pie Filling, Prepared)
I 008 02	French Apple Pie (Canned Apples-Cornstarch)	I 024 00	Peach Pie (Canned Peaches-Cornstarch)
I 009 00	Apple Pie (Canned Apples-Cornstarch)	I 024 01	Peach Pie (Prepared Pie Filling)
I 009 01	Apple Pie (Prepared Pie Filling)	I 026 00	Creamy Coconut Pie
I 009 02	Dutch Apple Pie (Prepared Pie Filling)	I 026 01	Creamy Banana Coconut Pie

I. DESSERTS (PASTRY AND PIES) No. 0 (1)

Card No.

I 026 02	Ambrosia Pie
I 027 00	Cherry Crumble Pie
I 028 00	Chocolate Cream Pie
I 028 01	Chocolate Cream Pie (Dessert Powder, Instant)
I 029 00	Chocolate And Vanilla Cream Pie (Instant)
I 030 00	Fried Apple Pie
I 030 01	Fried Lemon Pie
I 030 02	Fried Cherry Pie
I 030 03	Fried Peach Pie
I 030 04	Fried Blueberry Pie
I 031 00	Pecan Pie
I 031 01	Walnut Pie
I 032 00	Lemon Chiffon Pie
I 032 01	Pineapple Chiffon Pie
I 032 02	Strawberry Chiffon Pie
I 033 00	Lemon Meringue Pie
I 033 01	Lemon Meringue Pie (Pie Filling Prepared)
I 034 00	Fruit Turnovers
I 035 00	Fruit Dumplings
I 500 00	Key Lime Pie
I 800 00	Pies, Frozen
I 801 00	Elephant Ears (Frozen Puff Pastry)

DESSERTS (PASTRY AND PIES) No. I 001 00

PIE CRUST

Yield 100 **Portion** 1 Crust

Calories	Carbohydrates	Protein	Fat	Cholesterol	Sodium	Calcium
995 cal	92 g	12 g	64 g	0 mg	808 mg	19 mg

Ingredient	Weight	Measure	Issue
FLOUR,WHEAT,GENERAL PURPOSE	6-7/8 lbs	1 gal 2-1/4 qts	
SALT	1-7/8 oz	3 tbsp	
SHORTENING	3-5/8 lbs	2 qts	
WATER,COLD	2-1/8 lbs	1 qts	

Method

1. Sift together flour and salt in mixer bowl.
2. Add shortening to dry ingredients. Using pastry knife attachment, mix at low speed 30 seconds or until shortening is evenly distributed and mixture is granular in appearance.
3. Add water; mix at low speed 1 minute until dough is just formed.
4. Chill dough for at least 1 hour for ease in handling.
5. DIVIDE DOUGH: Divide dough into 13-7-1/2 oz pieces for bottom crust and 13-7 oz pieces for top crust; place on lightly floured board. ROLL DOUGH: Sprinkle each piece of dough lightly with flour; flatten gently. Using a floured rolling pin, roll lightly with quick strokes from center out to edge in all directions. Form a circle 1 inch larger than pie pan and about 1/8 inch thick. Bottom crust will be slightly thicker. Shift or turn dough occasionally to prevent sticking. If edges split, pinch cracks together. BOTTOM CRUST: Fold rolled dough in half; carefully place into ungreased pie pan with fold at center. Unfold and fit carefully into pie pan, being careful not to leave any air spaces between pan and dough. TOP CRUST: Roll top crust in same manner as bottom crust. Fold in half; with knife, make several small slits near center fold to allow steam to escape during baking. Brush outer rim of bottom crust with water. Lay top crust over filling with fold at center; unfold and press edges of two crusts together lightly. REMOVE EXCESS DOUGH: Trim overhanging edges of dough by using a knife or spatula. (Incorporate excess dough into next crust, if needed.) There should be little excess if skill is used in weighing and rolling dough. SEAL PIE: Press edges of crust firmly together or crimp with the thumb and forefinger to make a fluted edge. WASHED TOP: For a washed top, brush pies with appropriate wash as follows: Egg and Milk Wash - This wash is used for fruit pies (apple, blueberry, cherry, peach, pineapple) that are baked 30 to 35 minutes. It SHOULD NOT be used for pies requiring longer baking time as the crust will brown excessively. Egg and Water Wash - This wash is used for berry and mincemeat pies that are baked 40 to 45 minutes. It SHOULD NOT be used for pies that are baked 30 to 35 minutes as the crusts will be too pale. Allow glaze to dry on crust before baking to eliminate dark spots. BAKING INSTRUCTIONS FOR COOKED PIES: Bake as specified on individual recipe card. BAKING INSTRUCTIONS FOR UNCOOKED PIES: Bake crusts at 425 F. for about 15-18 minutes, or until light golden brown. Cool before filling. Proceed with the recipe directions.

Notes

1. Pie crust mix may be used. Omit steps 1 through 3. Follow manufacturer's directions for preparation. Follow steps 4 and 5. Quantity of pie crust mix required: 5 pounds pie crust mix yields 13-one crust pies; 10 pounds pie crust mix yields 13-two crust pies.

DESSERTS (PASTRY AND PIES) No.I 001 01

PIE CRUST (DOUGH ROLLING MACHINE)

Yield 100   Portion 1 Crust

Calories	Carbohydrates	Protein	Fat	Cholesterol	Sodium	Calcium
982 cal	87 g	14 g	65 g	0 mg	1260 mg	28 mg

Ingredient

Ingredient	Weight	Measure	Issue
FLOUR,WHEAT,BREAD	4-1/2 lbs	3 qts 3 cup	
FLOUR,WHEAT,GENERAL PURPOSE	2-1/8 lbs	1 qts 3-1/2 cup	
SALT	3 oz	1/4 cup 1 tbsp	
SUGAR,GRANULATED	1-1/3 oz	3 tbsp	
MILK,NONFAT,DRY	2/3 oz	1/4 cup 1 tbsp	
SHORTENING	3-5/8 lbs	2 qts	
WATER,COLD	1-7/8 lbs	3-1/2 cup	

Method

1. Combine sifted bread flour, sifted general purpose flour, salt, granulated sugar and nonfat dry milk in mixer bowl.
2. Add shortening to dry ingredients. Using pastry knife attachment, mix at low speed 30 seconds or until shortening is evenly distributed and mixture is granular in appearance.
3. Add water; mix at low speed 1 minute until dough is just formed.
4. Chill dough for at least 1 hour, preferably 24 hours, at 40 F. for ease in handling. Follow the equipment manufacturer's instructions for feeding/loading the dough into the machine.
5. DIVIDE DOUGH: Divide dough into 13-7-1/2 oz pieces for bottom crust and 13-7 oz pieces for top crust; place on lightly floured board. ROLL DOUGH: Sprinkle each piece of dough lightly with flour; flatten gently. Using a floured rolling pin, roll lightly with quick strokes from center out to edge in all directions. Form a circle 1 inch larger than pie pan and about 1/8 inch thick. Bottom crust will be slightly thicker. Shift or turn dough occasionally to prevent sticking. If edges split, pinch cracks together. BOTTOM CRUST: Fold rolled dough in half; carefully place into ungreased pie pan with fold at center. Unfold and fit carefully into pie pan, being careful not to leave any air spaces between pan and dough. TOP CRUST: Roll top crust in same manner as bottom crust. Fold in half; with knife, make several small slits near center fold to allow steam to escape during baking. Brush outer rim of bottom crust with water. Lay top crust over filling with fold at center; unfold and press edges of two crusts together lightly. REMOVE EXCESS DOUGH: Trim overhanging edges of dough by using a knife or spatula. (Incorporate excess dough into next crust, if needed.) There should be little excess if skill is used in weighing and rolling dough. SEAL PIE: Press edges of crust firmly together or crimp with the thumb and forefinger to make a fluted edge. WASHED TOP: For a washed top, brush pies with appropriate wash as follows: Egg and Milk Wash - This wash is used for fruit pies (apple, blueberry, cherry, peach, pineapple) that are baked 30 to 35 minutes. It SHOULD NOT be used for pies requiring longer baking time as the crust will brown excessively. Egg and Water Wash - This wash is used for berry and mincemeat pies that are baked 40 to 45 minutes. It SHOULD NOT be used for pies that are baked 30 to 35 minutes as the crusts will be too pale. Allow glaze to dry on crust before baking to eliminate dark spots. BAKING INSTRUCTIONS FOR COOKED PIES: Bake as specified on individual recipe card. BAKING INSTRUCTIONS FOR UNCOOKED PIES: Bake crusts at 425 F. for about 15-18 minutes, or until light golden brown. Cool before filling. Proceed with the recipe directions.

DESSERTS (PASTRY AND PIES) No.I 001 02

PIE CRUST (MANUAL MIXING METHOD)

Yield 100   Portion 1 Crust

Calories	Carbohydrates	Protein	Fat	Cholesterol	Sodium	Calcium
995 cal	92 g	12 g	64 g	0 mg	808 mg	19 mg

Ingredient	Weight	Measure	Issue
FLOUR,WHEAT,GENERAL PURPOSE	6-7/8 lbs	1 gal 2-1/4 qts	
SALT	1-7/8 oz	3 tbsp	
SHORTENING	3-5/8 lbs	2 qts	
WATER,COLD	2-1/8 lbs	1 qts	

Method

1. Sift together flour and salt in mixer bowl.
2. Add shortening to dry ingredients. Cut or rub shortening until evenly distributed and granular in appearance.
3. Sprinkle half of water over flour mixture and mix. Sprinkle remaining water and mix until dough is just formed.
4. Chill dough for at least 1 hour for ease in handling.
5. DIVIDE DOUGH: Divide dough into 13-7-1/2 oz pieces for bottom crust and 13-7 oz pieces for top crust; place on lightly floured board. ROLL DOUGH: Sprinkle each piece of dough lightly with flour; flatten gently. Using a floured rolling pin, roll lightly with quick strokes from center out to edge in all directions. Form a circle 1 inch larger than pie pan and about 1/8 inch thick. Bottom crust will be slightly thicker. Shift or turn dough occasionally to prevent sticking. If edges split, pinch cracks together. BOTTOM CRUST: Fold rolled dough in half; carefully place into ungreased pie pan with fold at center. Unfold and fit carefully into pie pan, being careful not to leave any air spaces between pan and dough. TOP CRUST: Roll top crust in same manner as bottom crust. Fold in half; with knife, make several small slits near center fold to allow steam to escape during baking. Brush outer rim of bottom crust with water. Lay top crust over filling with fold at center; unfold and press edges of two crusts together lightly. REMOVE EXCESS DOUGH: Trim overhanging edges of dough by using a knife or spatula. (Incorporate excess dough into next crust, if needed.) There should be little excess if skill is used in weighing and rolling dough. SEAL PIE: Press edges of crust firmly together or crimp with the thumb and forefinger to make a fluted edge. WASHED TOP: For a washed top, brush pies with appropriate wash as follows: Egg and Milk Wash - This wash is used for fruit pies (apple, blueberry, cherry, peach, pineapple) that are baked 30 to 35 minutes. It SHOULD NOT be used for pies requiring longer baking time as the crust will brown excessively. Egg and Water Wash - This wash is used for berry and mincemeat pies that are baked 40 to 45 minutes. It SHOULD NOT be used for pies that are baked 30 to 35 minutes as the crusts will be too pale. Allow glaze to dry on crust before baking to eliminate dark spots. BAKING INSTRUCTIONS FOR COOKED PIES: Bake as specified on individual recipe card. BAKING INSTRUCTIONS FOR UNCOOKED PIES: Bake crusts at 425 F. for about 15-18 minutes, or until light golden brown. Cool before filling. Proceed with the recipe directions.

DESSERTS (PASTRY AND PIES) No.I 002 00

GRAHAM CRACKER CRUST

Yield 100   **Portion** 1 Crust

Calories	Carbohydrates	Protein	Fat	Cholesterol	Sodium	Calcium
1181 cal	144 g	9 g	65 g	0 mg	1380 mg	50 mg

Ingredient	**Weight**	**Measure**	**Issue**
MARGARINE	1-7/8 lbs	3-3/4 cup	
CRACKERS,GRAHAM,CRUMBS	3-5/8 lbs		
SUGAR,GRANULATED	1-1/3 lbs	3 cup	

Method

1. Grind graham crackers or crush on board with rolling pin. Combine butter or margarine, crumbs, and sugar in mixer bowl. Mix at low speed until well blended, about 2 minutes.
2. Place about 8 ounces or 1-3/4 cups crumb mixture in each pie pan. Press firmly into an even layer against bottom and sides of each pan.
3. Chill at least 1 hour before filling is added.

Notes

1. For a firmer shell, omit Step 3; using a convection oven, bake at 325 F. for 7 minutes or until lightly browned on low fan, open vent.
2. 4 lb 1 oz (13-5 oz) preformed graham cracker crusts may be used.

DESSERTS (PASTRY AND PIES) No.I 002 01

GRAHAM CRACKER CRUST (PERFORMED CRUST)

Yield 100 **Portion** 1 Crust

Calories	Carbohydrates	Protein	Fat	Cholesterol	Sodium	Calcium
716 cal	88 g	5 g	38 g	14 mg	313 mg	57 mg

Ingredient	**Weight**	**Measure**	**Issue**
PIE CRUST PREFORMED	4 lbs		

Method

1 Use 13-5 oz preformed crusts per 100 portions.

DESSERTS (PASTRY AND PIES) No.I 003 00

MINCEMEAT PIE

Yield 100   Portion   1 Slice

Calories	Carbohydrates	Protein	Fat	Cholesterol	Sodium	Calcium
330 cal	42 g	3 g	17 g	0 mg	236 mg	11 mg

Ingredient   Weight   Measure   Issue

PIE CRUST   26 each
PIE FILLING,MINCEMEAT,CANNED   13-1/3 lbs   1 gal 2-2/3 qts
APPLES,CANNED,DRAINED,CHOPPED   4-1/2 lbs   2 qts 1 cup
SUGAR,GRANULATED   11-3/4 oz   1-5/8 cup

Method

1. DIVIDE DOUGH: Divide dough into 13-7-1/2 oz pieces for bottom crust and 13-7 oz pieces for top crust; place on lightly floured board. ROLL DOUGH: Sprinkle each piece of dough lightly with flour; flatten gently. Using a floured rolling pin, roll lightly with quick strokes from center out to edge in all directions. Form a circle 1 inch larger than pie pan and about 1/8 inch thick. Bottom crust will be slightly thicker. Shift or turn dough occasionally to prevent sticking. If edges split, pinch cracks together. BOTTOM CRUST: Fold rolled dough in half; carefully place into ungreased pie pan with fold at center. Unfold and fit carefully into pie pan, being careful not to leave any air spaces between pan and dough. FILL CRUST: Fill as specified on individual recipe card. (Step 2/3). TOP CRUST: Roll top crust in same manner as bottom crust. Fold in half; with knife, make several small slits near center fold to allow steam to escape during baking. Brush outer rim of bottom crust with water. Lay top crust over filling with fold at center; unfold and press edges of two crusts together lightly. REMOVE EXCESS DOUGH: Trim overhanging edges of dough by using a knife or spatula. (Incorporate excess dough into next crust, if needed.) There should be little excess if skill is used in weighing and rolling dough. SEAL PIE: Press edges of crust firmly together or crimp with the thumb and forefinger to make a fluted edge. WASHED TOP: For a washed top, brush pies with appropriate wash as follows: Egg and Milk Wash - This wash is used for fruit pies (apple, blueberry, cherry, peach, pineapple) that are baked 30 to 35 minutes. It SHOULD NOT be used for pies requiring longer baking time as the crust will brown excessively. Egg and Water Wash - This wash is used for berry and mincemeat pies that are baked 40 to 45 minutes. It SHOULD NOT be used for pies that are baked 30 to 35 minutes as the crusts will be too pale. Allow glaze to dry on crust before baking to eliminate dark spots. BAKING INSTRUCTIONS FOR COOKED PIES: Bake as specified on individual recipe card. BAKING INSTRUCTIONS FOR UNCOOKED PIES: Bake crusts at 425 F. for about 15-18 minutes, or until light golden brown. Cool before filling. Proceed with the recipe directions.

2. Combine mincemeat, apples, and sugar; mix until well blended.
3. Pour 3-1/2 cups filling into each unbaked pie shell. Cover with top crust. Seal edges.
4. Bake at 425 F. for 45 minutes or until lightly browned.
5. Cut 8 wedges per pie.

DESSERTS (PASTRY AND PIES) No.I 004 00

EGG AND MILK WASH

Yield 100   **Portion** 1-1/2 Cups

Calories	Carbohydrates	Protein	Fat	Cholesterol	Sodium	Calcium
91 cal	3 g	8 g	5 g	215 mg	96 mg	91 mg

Ingredient	Weight	Measure	Issue
MILK,NONFAT,DRY	1/2 oz	3 tbsp	
WATER	12-1/2 oz	1-1/2 cup	
EGGS,WHOLE,FROZEN	4-2/3 oz	1/2 cup 2/3 tbsp	

Method

1. Combine milk and water; mix until thoroughly blended.
2. Add eggs; whip until well blended.
3. Brush on pies. Allow to dry before baking. CCP: Refrigerate at 41 F. or lower until ready for use.

Notes

1. This wash will cover 13 to 15 2-crust pies that are baked 30 to 35 minutes, primarily fruit pies (apple, blueberry, cherry, peach, pineapple). It SHOULD NOT be used for pies requiring longer baking time as the crust will brown excessively.

DESSERTS (PASTRY AND PIES) No.I 004 01

EGG AND WATER WASH

Yield 100   Portion   1-1/2 Cups

Calories	Carbohydrates	Protein	Fat	Cholesterol	Sodium	Calcium
120 cal	1 g	10 g	8 g	350 mg	113 mg	51 mg

Ingredient	**Weight**	**Measure**	**Issue**
EGGS,WHOLE,FROZEN	7-5/8 oz	3/4 cup 2-1/3 tbsp	
WATER	1 lbs	2 cup	

Method

1 Combine eggs with water. Whip until well blended.
2 Brush on pies. Allow to dry before baking. CCP: Refrigerate at 41 F. or lower until ready for use.

Notes

1 Use on 2-crust pies (berry and mincemeat), bake 40 to 50 minutes. To prevent dark spots, allow wash to dry on crust before baking. This wash is used for berry and mincemeat pies. It SHOULD NOT be used for pies that are baked 30 to 35 minutes as the crusts will be too pale.

DESSERTS (PASTRY AND PIES) No.I 004 02

MILK AND WATER WASH

Yield 100 **Portion** 1-1/2 Cups

Calories	Carbohydrates	Protein	Fat	Cholesterol	Sodium	Calcium
36 cal	5 g	4 g	0 g	2 mg	59 mg	125 mg

Ingredient | **Weight** | **Measure** | **Issue**

MILK,NONFAT,DRY — 7/8 oz — 1/4 cup 2-2/3 tbsp
WATER,WARM — 14-7/8 oz — 1-3/4 cup

Method

1. Combine nonfat dry milk and warm water. Mix well.
2. Use only this wash on turnovers; allow to dry before baking. Do not use this wash on 2-crust pies.

DESSERTS (PASTRY AND PIES) No.I 005 00

MERINGUE

Yield 100   **Portion**   2-1/2 Cups

Calories	Carbohydrates	Protein	Fat	Cholesterol	Sodium	Calcium
401 cal	93 g	9 g	0 g	0 mg	406 mg	6 mg

Ingredient	**Weight**	**Measure**	**Issue**
EGG WHITES	2-3/8 lbs	1 qts 1/2 cup	
SUGAR,GRANULATED	2-2/3 lbs	1 qts 2 cup	
SALT	1/3 oz	1/4 tsp	
EXTRACT,VANILLA	1/3 oz	3/8 tsp	

Method

1. Using a whip, beat egg whites at high speed in mixer bowl until foamy, about 3 minutes.
2. Add sugar a little at a time; beat well at medium speed after each addition. Beat at high speed until stiff peaks are formed, about 6 minutes.
3. Add salt and vanilla; blend.
4. Spread about 2-1/2 cups meringue over warm pie filling, about 122 F. in each pan. Meringue should touch inner edge of crust all around and completely cover top of pie. Leave meringue somewhat rough on top.
5. Bake at 350 F. for 16 to 20 minutes or until lightly browned.   CCP: Hold for service at 41 F. or lower.

DESSERTS (PASTRY AND PIES) No.I 005 01

MERINGUE (DEHYDRATED)

Yield 100   **Portion** 2-1/2 Cups

Calories	Carbohydrates	Protein	Fat	Cholesterol	Sodium	Calcium
225 cal	56 g	2 g	0 g	0 mg	31 mg	74 mg

Ingredient	Weight	Measure	Issue
MERINGUE POWDER	3-3/8 oz	3/4 cup	
WATER,COLD	1-5/8 lbs	3 cup	
SUGAR,GRANULATED	1-1/2 lbs	3-3/8 cup	

Method

1. Add water to mixer bowl; add meringue powder.
2. Using whip, mix at low speed 1 minute or until powder is dissolved. Beat at high speed until stiff peaks form, about 5 minutes.
3. Gradually add granulated sugar beating at high speed 1 minute or until meringue is glossy.
4. Spread about 2-1/2 cups meringue over warm pie filling, about 122 F. in each pan. Meringue should touch inner edge of crust all around and completely cover top of pie. Leave meringue somewhat rough on top.
5. Bake 16 to 20 minutes at 350 F. or until lightly browned.

DESSERTS (PASTRY AND PIES) No. I 006 00

VANILLA CREAM PIE

Yield 100   Portion 1 Slice

Calories	Carbohydrates	Protein	Fat	Cholesterol	Sodium	Calcium
326 cal	38 g	4 g	17 g	46 mg	268 mg	49 mg

Ingredient	Weight	Measure	Issue
PIE CRUST		13 each	
MILK,NONFAT,DRY	10-3/8 oz	1 qts 3/8 cup	
WATER,WARM	11-7/8 lbs	1 gal 1-2/3 qts	
SUGAR,GRANULATED	1-1/2 lbs	3-3/8 cup	
SALT	3/4 oz	1 tbsp	
CORNSTARCH	13-1/2 oz	3 cup	
SUGAR,GRANULATED	1-7/8 lbs	1 qts 1/8 cup	
WATER,COLD	3-1/8 lbs	1 qts 2 cup	
EGGS,WHOLE,FROZEN	2-1/3 lbs	1 qts 3/8 cup	
MARGARINE	14-7/8 oz	1-7/8 cup	
EXTRACT,VANILLA	2-1/8 oz	1/4 cup 1 tbsp	

Method

1. PREPARE AND DIVIDE DOUGH: Prepare 1/2 recipe Pie Crust (Recipe No. I 001 00). Divide dough into 13-7-1/2 oz pieces for bottom crust; place on lightly floured board. ROLL DOUGH: Sprinkle each piece of dough lightly with flour; flatten gently. Using a floured rolling pin, roll lightly with quick strokes from center out to edge in all directions. Form a circle 1 inch larger than pie pan and about 1/8 inch thick. Shift or turn dough occasionally to prevent sticking. If edges split, pinch cracks together. BOTTOM CRUST: Fold rolled dough in half; carefully place into ungreased pie pan with fold at center. Unfold and fit carefully into pie pan, being careful not to leave any air spaces between pan and dough. REMOVE EXCESS DOUGH: Trim overhanging edges of dough by using a knife or spatula. (Incorporate excess dough into next crust, if needed.) There should be little excess if skill is used in weighing and rolling dough. BAKING INSTRUCTIONS FOR UNCOOKED PIES: Bake crusts at 425 F. for about 15-18 minutes, or until light golden brown. Cool before filling. Proceed with the recipe directions.
2. Reconstitute milk. Add sugar and salt; heat to just below boiling. DO NOT BOIL.
3. Combine cornstarch, sugar and water; stir until smooth. Add gradually to hot mixture. Cook at medium heat, stirring constantly, about 10 minutes until thickened.
4. Stir about 1 quart of hot mixture into eggs. Slowly pour egg mixture into remaining hot mixture; heat to boiling, stirring constantly. Cook 2 minutes longer. Remove from heat.
5. Add butter or margarine and vanilla; stir until well blended. Cool slightly.
6. Pour 3 cups filling into each baked pie shell. Ensure cream pie filling preparation time does not exceed 4 hours total in temperatures between 40 F. to 140 F.
7. CCP: Hold for service at 41 F. or lower.
8. Cut 8 wedges per pie. Chilled pies may be topped with Whipped Cream, Recipe No. K 001 00 or Whipped Topping Recipe No. K 002 00.

Notes

1. Filling will curdle if boiled or subjected to prolonged intense heat.

DESSERTS (PASTRY AND PIES) No. I 006 01

BANANA CREAM PIE

Yield 100   Portion 1 Slice

Calories	Carbohydrates	Protein	Fat	Cholesterol	Sodium	Calcium
292 cal	40 g	4 g	13 g	46 mg	265 mg	50 mg

Ingredient	Weight	Measure	Issue
PIE CRUST		13 each	
MILK,NONFAT,DRY	10-3/8 oz	1 qts 3/8 cup	
WATER,WARM	11-7/8 lbs	1 gal 1-2/3 qts	
SUGAR,GRANULATED	1-1/2 lbs	3-3/8 cup	
SALT	3/4 oz	1 tbsp	
CORNSTARCH	13-1/2 oz	3 cup	
SUGAR,GRANULATED	1-7/8 lbs	1 qts 1/8 cup	
WATER,COLD	3-1/8 lbs	1 qts 2 cup	
EGGS,WHOLE,FROZEN	2-1/3 lbs	1 qts 3/8 cup	
BANANA,FRESH,SLICED	7-1/2 lbs	1 gal 1-2/3 qts	11-1/2 lbs
MARGARINE	14-7/8 oz	1-7/8 cup	
EXTRACT,VANILLA	2-1/8 oz	1/4 cup 1 tbsp	

Method

1. PREPARE AND DIVIDE DOUGH: Prepare 1/2 recipe Pie Crust (Recipe No. I 001 00). Divide dough into 13-7-1/2 oz pieces for bottom crust; place on lightly floured board. ROLL DOUGH: Sprinkle each piece of dough lightly with flour; flatten gently. Using a floured rolling pin, roll lightly with quick strokes from center out to edge in all directions. Form a circle 1 inch larger than pie pan and about 1/8 inch thick. Shift or turn dough occasionally to prevent sticking. If edges split, pinch cracks together. BOTTOM CRUST: Fold rolled dough in half; carefully place into ungreased pie pan with fold at center. Unfold and fit carefully into pie pan, being careful not to leave any air spaces between pan and dough. REMOVE EXCESS DOUGH: Trim overhanging edges of dough by using a knife or spatula. (Incorporate excess dough into next crust, if needed.) There should be little excess if skill is used in weighing and rolling dough. BAKING INSTRUCTIONS FOR UNCOOKED PIES: Bake crusts at 425 F. for about 15-18 minutes, or until light golden brown. Cool before filling. Proceed with the recipe directions.
2. Reconstitute milk. Add sugar and salt; heat to just below boiling. DO NOT BOIL.
3. Combine cornstarch, sugar and water; stir until smooth. Add gradually to hot mixture. Cook at medium heat, stirring constantly, about 10 minutes until thickened.
4. Stir about 1 quart of hot mixture into eggs. Slowly pour egg mixture into remaining hot mixture; heat to boiling, stirring constantly. Cook 2 minutes longer. Remove from heat.
5. Add butter or margarine and vanilla; stir until well blended. Cool slightly. Slice bananas. Add to cooled filling. To prevent discoloration, slice bananas just before adding to filling.
6. Pour about 3-1/2 cups filling into each baked pie shell. Meringue Recipe No. I 005 00 may be spread over warm filling. Ensure cream pie filling preparation time does not exceed 4 hours total in temperatures between 40 F. to 140 F.
7. CCP: Hold for service at 41 F. or lower.
8. Cut 8 wedges per pie. Chilled pies may be topped with 1 recipe Whipped Topping, Recipe No. K 002 00.

Notes

1. Filling will curdle if boiled or subjected to prolonged intense heat.

DESSERTS (PASTRY AND PIES) No. I 007 00

VANILLA CREAM PIE (DESSERT POWDER, INSTANT)

Yield 100   Portion 1 Slice

Calories	Carbohydrates	Protein	Fat	Cholesterol	Sodium	Calcium
301 cal	43 g	3 g	13 g	1 mg	506 mg	54 mg

Ingredient	**Weight**	**Measure**	**Issue**
PIE CRUST		13 each	
MILK,NONFAT,DRY	13-3/4 oz	1 qts 1-3/4 cup	
WATER,COLD	15-1/8 lbs	1 gal 3-1/4 qts	
DESSERT POWDER,PUDDING,INSTANT,VANILLA	5-1/2 lbs	3 qts 1-1/2 cup	

Method

1. PREPARE AND DIVIDE DOUGH: Prepare 1/2 recipe Pie Crust (Recipe No. I 001 00). Divide dough into 13-7-1/2 oz pieces for bottom crust; place on lightly floured board. ROLL DOUGH: Sprinkle each piece of dough lightly with flour; flatten gently. Using a floured rolling pin, roll lightly with quick strokes from center out to edge in all directions. Form a circle 1 inch larger than pie pan and about 1/8 inch thick. Shift or turn dough occasionally to prevent sticking. If edges split, pinch cracks together. BOTTOM CRUST: Fold rolled dough in half; carefully place into ungreased pie pan with fold at center. Unfold and fit carefully into pie pan, being careful not to leave any air spaces between pan and dough. REMOVE EXCESS DOUGH: Trim overhanging edges of dough by using a knife or spatula. (Incorporate excess dough into next crust, if needed.) There should be little excess if skill is used in weighing and rolling dough. BAKING INSTRUCTIONS FOR UNCOOKED PIES: Bake crusts at 425 F. for about 15-18 minutes, or until light golden brown. Cool before filling. Proceed with the recipe directions.
2. Reconstitute milk in a large mixing bowl, with a wire whip.
3. Add dessert powder to milk and water. Using whip, blend at low speed 15 seconds or until well blended. Scrape down sides of bowl; whip at medium speed 2 minutes.
4. Pour about 3 cups filling into each baked pie shell.
5. CCP: Hold for service at 41 F. or lower.
6. Cut 8 wedges per pie. Chilled pies may be topped with Whipped Cream, Recipe No. K 001 00 or Whipped Topping Recipe No. K 002 00.

DESSERTS (PASTRY AND PIES) No. I 007 01

STRAWBERRY GLAZED CREAM PIE (INSTANT)

Yield 100   **Portion** 1 Slice

Calories	Carbohydrates	Protein	Fat	Cholesterol	Sodium	Calcium
335 cal	52 g	3 g	13 g	1 mg	508 mg	61 mg

Ingredient	**Weight**	**Measure**	**Issue**
PIE CRUST		13 each	
MILK,NONFAT,DRY	13-3/4 oz	1 qts 1-3/4 cup	
WATER,COLD	15-1/8 lbs	1 gal 3-1/4 qts	
DESSERT POWDER,PUDDING,INSTANT,VANILLA	5-1/2 lbs	3 qts 1-1/2 cup	
STRAWBERRY GLAZE TOPPING		3 qts 3 cup	

Method

1. PREPARE AND DIVIDE DOUGH: Prepare 1/2 recipe Pie Crust (Recipe No. I 001 00). Divide dough into 13-7-1/2 oz pieces for bottom crust; place on lightly floured board. ROLL DOUGH: Sprinkle each piece of dough lightly with flour; flatten gently. Using a floured rolling pin, roll lightly with quick strokes from center out to edge in all directions. Form a circle 1 inch larger than pie pan and about 1/8 inch thick. Shift or turn dough occasionally to prevent sticking. If edges split, pinch cracks together. BOTTOM CRUST: Fold rolled dough in half; carefully place into ungreased pie pan with fold at center. Unfold and fit carefully into pie pan, being careful not to leave any air spaces between pan and dough. REMOVE EXCESS DOUGH: Trim overhanging edges of dough by using a knife or spatula. (Incorporate excess dough into next crust, if needed.) There should be little excess if skill is used in weighing and rolling dough. BAKING INSTRUCTIONS FOR UNCOOKED PIES: Bake crusts at 425 F. for about 15-18 minutes, or until light golden brown. Cool before filling. Proceed with the recipe directions.
2. Reconstitute milk in a large mixing bowl with a wire whip.
3. Add dessert powder to milk and water. Using whip, blend at low speed 15 seconds or until well blended. Scrape down sides of bowl; whip at medium speed 2 minutes.
4. Pour about 3 cups filling into each baked pie shell.
5. CCP: Hold for service at 41 F. or lower.
6. Prepare 1 recipe Strawberry Glaze Topping, Recipe No. K 007 00 per 100 portions. Spread 11-1/2 ounces or 1-1/8 cups mixture over filling in each pie.
7. Cut pie into 8 wedges.

DESSERTS (PASTRY AND PIES) No. I 007 02

COCONUT CREAM PIE (INSTANT)

Yield 100   Portion   1 Slice

Calories	Carbohydrates	Protein	Fat	Cholesterol	Sodium	Calcium
355 cal	48 g	4 g	17 g	1 mg	535 mg	56 mg

Ingredient	Weight	Measure	Issue
PIE CRUST		13 each	
MILK,NONFAT,DRY	13-3/4 oz	1 qts 1-3/4 cup	
WATER,COLD	15-1/8 lbs	1 gal 3-1/4 qts	
DESSERT POWDER,PUDDING,INSTANT,VANILLA	5-1/2 lbs	3 qts 1-1/2 cup	
COCONUT,PREPARED,SWEETENED FLAKES	1-1/2 lbs	1 qts 3-1/4 cup	
COCONUT,PREPARED,SWEETENED FLAKES	14-3/4 oz	1 qts 1/2 cup	

Method

1. PREPARE AND DIVIDE DOUGH: Prepare 1/2 recipe Pie Crust (Recipe No. I 001 00). Divide dough into 13-7-1/2 oz pieces for bottom crust; place on lightly floured board. ROLL DOUGH: Sprinkle each piece of dough lightly with flour; flatten gently. Using a floured rolling pin, roll lightly with quick strokes from center out to edge in all directions. Form a circle 1 inch larger than pie pan and about 1/8 inch thick. Shift or turn dough occasionally to prevent sticking. If edges split, pinch cracks together. BOTTOM CRUST: Fold rolled dough in half; carefully place into ungreased pie pan with fold at center. Unfold and fit carefully into pie pan, being careful not to leave any air spaces between pan and dough. REMOVE EXCESS DOUGH: Trim overhanging edges of dough by using a knife or spatula. (Incorporate excess dough into next crust, if needed.) There should be little excess if skill is used in weighing and rolling dough. BAKING INSTRUCTIONS FOR UNCOOKED PIES: Bake crusts at 425 F. for about 15-18 minutes, or until light golden brown. Cool before filling. Proceed with the recipe directions.
2. Reconstitute milk in a large mixing bowl with a wire whip.
3. Add dessert powder to milk and water. Using whip, blend at low speed 15 seconds or until well blended. Scrape down sides of bowl; whip at medium speed 2 minutes.
4. Add prepared sweetened coconut flakes to filling; mix well. Pour 3-1/4 cups filling into each baked pie shell.
5. Sprinkle 1/3 cup coconut over each filled pie.
6. Cut 8 wedges per pie. CCP: Hold for service at 41 F. or lower. Chilled pies may be topped with Whipped Cream, Recipe No. K 001 00 or Whipped Topping, Recipe No. K 002 00.

DESSERTS (PASTRY AND PIES) No.I 007 03

PINEAPPLE CREAM PIE (INSTANT)

Yield 100   Portion 1 Slice

Calories	Carbohydrates	Protein	Fat	Cholesterol	Sodium	Calcium
310 cal	45 g	3 g	13 g	1 mg	506 mg	56 mg

Ingredient   **Weight**   **Measure**   **Issue**
PIE CRUST   13 each
MILK,NONFAT,DRY   13-3/4 oz   1 qts 1-3/4 cup
WATER,COLD   15-1/8 lbs   1 gal 3-1/4 qts
DESSERT POWDER,PUDDING,INSTANT,VANILLA   5-1/2 lbs   3 qts 1-1/2 cup
PINEAPPLE,CANNED,CRUSHED,JUICE PACK,DRAINED 3-1/4 lbs   1 qts 2 cup

Method

1. PREPARE AND DIVIDE DOUGH: Prepare 1/2 recipe Pie Crust (Recipe No. I 001 00). Divide dough into 13-7-1/2 oz pieces for bottom crust; place on lightly floured board. ROLL DOUGH: Sprinkle each piece of dough lightly with flour; flatten gently. Using a floured rolling pin, roll lightly with quick strokes from center out to edge in all directions. Form a circle 1 inch larger than pie pan and about 1/8 inch thick. Shift or turn dough occasionally to prevent sticking. If edges split, pinch cracks together. BOTTOM CRUST: Fold rolled dough in half; carefully place into ungreased pie pan with fold at center. Unfold and fit carefully into pie pan, being careful not to leave any air spaces between pan and dough. REMOVE EXCESS DOUGH: Trim overhanging edges of dough by using a knife or spatula. (Incorporate excess dough into next crust, if needed.) There should be little excess if skill is used in weighing and rolling dough. BAKING INSTRUCTIONS FOR UNCOOKED PIES: Bake crusts at 425 F. for about 15-18 minutes, or until light golden brown. Cool before filling. Proceed with the recipe directions.
2. Reconstitute milk in a large mixing bowl with a wire whip.
3. Add dessert powder to milk and water. Using whip, blend at low speed 15 seconds or until well blended. Scrape down sides of bowl; whip at medium speed 2 minutes.
4. Add canned, drained, crushed pineapple. Mix well. Pour about 3-1/4 cups filling into each baked pie shell.
5. Cut 8 wedges per pie. CCP: Hold for service at 41 F. or lower. Chilled pies may be topped with Whipped Cream, Recipe No. K 001 00 or Whipped Topping, Recipe No. K 002 00.

DESSERTS (PASTRY AND PIES) No. I 008 01

DUTCH APPLE PIE (CANNED APPLES-CORNSTARCH)

Yield 100   Portion 1 Slice

Calories	Carbohydrates	Protein	Fat	Cholesterol	Sodium	Calcium
358 cal	54 g	3 g	16 g	18 mg	225 mg	17 mg

Ingredient	Weight	Measure	Issue
PIE CRUST		13 each	
APPLES,CANNED,SLICED	13-7/8 lbs	1 gal 3 qts	
SUGAR,GRANULATED	3 lbs	1 qts 2-3/4 cup	
SALT	3/8 oz	1/3 tsp	
CINNAMON,GROUND	1/3 oz	1 tbsp	
NUTMEG,GROUND	3/8 oz	1 tbsp	
CORNSTARCH	7-1/2 oz	1-5/8 cup	
WATER,COLD	1-5/8 lbs	3 cup	
JUICE,LEMON	2-1/2 oz	1/4 cup 1 tbsp	
BUTTER	4 oz	1/2 cup	
STREUSEL TOPPING		3 qts 3 cup	

Method

1. PREPARE AND DIVIDE DOUGH: Prepare 1/2 recipe Pie Crust (Recipe No. I 001 00). Divide dough into 13-7 oz pieces for pie crust and place on lightly floured board. ROLL DOUGH: Sprinkle each piece of dough lightly with flour; flatten gently. Using a floured rolling pin, roll lightly with quick strokes from center out to edge in all directions. Form a circle 1 inch larger than pie pan and about 1/8 inch thick. Shift or turn dough occasionally to prevent sticking. If edges split, pinch cracks together. CRUST: Fold rolled dough in half; carefully place into ungreased pie pan with fold at center. Unfold and fit carefully into pie pan, being careful not to leave any air spaces between pan and dough. REMOVE EXCESS DOUGH: Trim overhanging edges of dough by using a knife or spatula. (Incorporate excess dough into next crust, if needed.) There should be little excess if skill is used in weighing and rolling dough. BAKING INSTRUCTIONS FOR COOKED PIES: Bake as specified on individual recipe card.
2. Drain apples; reserve juice for use in Step 3; apples for use in Step 5.
3. Take reserved juice and add water equal 1-7/8 quart per 100 portions and combine with sugar, salt, cinnamon and nutmeg; bring to a boil.
4. Combine cornstarch and water; stir until smooth. Add gradually to boiling mixture. Cook at medium heat, stirring constantly, until thick and clear. Remove from heat.
5. Fold apples, lemon juice and butter or margarine carefully into thickened mixture. Cool thoroughly.
6. Pour 2-3/4 to 3 cups filling into each unbaked pie shell.
7. Using a convection oven, bake at 375 F. for 25 minutes or until lightly browned on high fan, open vent.
8. Prepare 1-1/2 recipes No. D 049 00 Streusel Topping per 100 portions. Spread 1/3 glaze over each pie after it has cooled.
9. Cut 8 wedges per pie.

DESSERTS (PASTRY AND PIES) No. I 008 02

FRENCH APPLE PIE (CANNED APPLES-CORNSTARCH)

Yield 100   Portion 1 Slice

Calories	Carbohydrates	Protein	Fat	Cholesterol	Sodium	Calcium
417 cal	61 g	3 g	18 g	4 mg	274 mg	10 mg

Ingredient	Weight	Measure	Issue
PIE CRUST		26 each	
APPLES,CANNED,SLICED	13-7/8 lbs	1 gal 3 qts	
SUGAR,GRANULATED	3 lbs	1 qts 2-3/4 cup	
SALT	3/8 oz	1/3 tsp	
CINNAMON,GROUND	1/3 oz	1 tbsp	
NUTMEG,GROUND	3/8 oz	1 tbsp	
CORNSTARCH	7-1/2 oz	1-5/8 cup	
WATER,COLD	1-5/8 lbs	3 cup	
JUICE,LEMON	2-1/2 oz	1/4 cup 1 tbsp	
BUTTER	4 oz	1/2 cup	
VANILLA GLAZE		1 qts 1/8 cup	

Method

1. PREPARE AND DIVIDE DOUGH: Prepare 1 recipe Pie Crust (Recipe No. I 001 00). Divide dough into 13-7-1/2 oz pieces for bottom crust and 13-7 oz pieces for top crust; place on lightly floured board. ROLL DOUGH: Sprinkle each piece of dough lightly with flour; flatten gently. Using a floured rolling pin, roll lightly with quick strokes from center out to edge in all directions. Form a circle 1 inch larger than pie pan and about 1/8 inch thick. Bottom crust will be slightly thicker. Shift or turn dough occasionally to prevent sticking. If edges split, pinch cracks together. BOTTOM CRUST: Fold rolled dough in half; carefully place into ungreased pie pan with fold at center. Unfold and fit carefully into pie pan, being careful not to leave any air spaces between pan and dough. FILL CRUST: Fill as specified on individual recipe card. TOP CRUST: Roll top crust in same manner as bottom crust. Fold in half; with knife, make several small slits near center fold to allow steam to escape during baking. Brush outer rim of bottom crust with water. Lay top crust over filling with fold at center; unfold and press edges of two crusts together lightly. REMOVE EXCESS DOUGH: Trim overhanging edges of dough by using a knife or spatula. (Incorporate excess dough into next crust, if needed.) There should be little excess if skill is used in weighing and rolling dough. SEAL PIE: Press edges of crust firmly together or crimp with the thumb and forefinger to make a fluted edge. BAKING INSTRUCTIONS FOR COOKED PIES: Bake as specified on individual recipe card.
2. Drain apples; reserve juice for use in Step 3; apples for use in Step 5.
3. Take reserved juice and add water equal 1-7/8 quart per 100 portions and combine with sugar, salt, cinnamon and nutmeg; bring to a boil.
4. Combine cornstarch and water; stir until smooth. Add gradually to boiling mixture. Cook at medium heat, stirring constantly, until thick and clear. Remove from heat.
5. Fold apples, lemon juice and butter or margarine carefully into thickened mixture. Cool thoroughly.
6. Pour 2-3/4 to 3 cups filling into each unbaked pie shell. Cover with top crust. Seal edges.
7. Using a convection oven, bake at 375 F. for 25 minutes or until lightly browned on high fan, open vent.
8. Prepare 1-1/2 recipes Vanilla Glaze per 100 portions, Recipe No. D 046 00; when pies are removed and still hot, spread 1/3 glaze over each top crust.
9. Cut 8 wedges per pie.

DESSERTS (PASTRY AND PIES) No. I 009 00

APPLE PIE (CANNED APPLES-CORNSTARCH)

Yield 100   Portion 1 Slice

Calories	Carbohydrates	Protein	Fat	Cholesterol	Sodium	Calcium
370 cal	50 g	3 g	18 g	2 mg	269 mg	10 mg

Ingredient	Weight	Measure	Issue
PIE CRUST		26 each	
APPLES,CANNED,SLICED	13-7/8 lbs	1 gal 3 qts	
SUGAR,GRANULATED	3 lbs	1 qts 2-3/4 cup	
SALT	3/8 oz	1/3 tsp	
CINNAMON,GROUND	1/3 oz	1 tbsp	
NUTMEG,GROUND	3/8 oz	1 tbsp	
CORNSTARCH	7-1/2 oz	1-5/8 cup	
WATER,COLD	1-5/8 lbs	3 cup	
JUICE,LEMON	2-1/2 oz	1/4 cup 1 tbsp	
BUTTER	4 oz	1/2 cup	

Method

1 PREPARE AND DIVIDE DOUGH: Prepare 1 recipe Pie Crust (Recipe No. I 001 00). Divide dough into 13-7-1/2 oz pieces for bottom crust and 13-7 oz pieces for top crust; place on lightly floured board. ROLL DOUGH: Sprinkle each piece of dough lightly with flour; flatten gently. Using a floured rolling pin, roll lightly with quick strokes from center out to edge in all directions. Form a circle 1 inch larger than pie pan and about 1/8 inch thick. Bottom crust will be slightly thicker. Shift or turn dough occasionally to prevent sticking. If edges split, pinch cracks together. BOTTOM CRUST: Fold rolled dough in half; carefully place into ungreased pie pan with fold at center. Unfold and fit carefully into pie pan, being careful not to leave any air spaces between pan and dough. FILL CRUST: Fill as specified on individual recipe card. TOP CRUST: Roll top crust in same manner as bottom crust. Fold in half; with knife, make several small slits near center fold to allow steam to escape during baking. Brush outer rim of bottom crust with water. Lay top crust over filling with fold at center; unfold and press edges of two crusts together lightly. REMOVE EXCESS DOUGH: Trim overhanging edges of dough by using a knife or spatula. (Incorporate excess dough into next crust, if needed.) There should be little excess if skill is used in weighing and rolling dough. SEAL PIE: Press edges of crust firmly together or crimp with the thumb and forefinger to make a fluted edge. BAKING INSTRUCTIONS FOR COOKED PIES: Bake as specified on individual recipe card.

2 Drain apples; reserve juice for use in Step 3; apples for use in Step 5.

3 Take reserved juice and add water equal 1-7/8 quart per 100 portions and combine with sugar, salt, cinnamon and nutmeg; bring to a boil.

4 Combine cornstarch and water; stir until smooth. Add gradually to boiling mixture. Cook at medium heat, stirring constantly, until thick and clear. Remove from heat.

5 Fold apples, lemon juice and butter or margarine carefully into thickened mixture. Cool thoroughly.

6 Pour 2-3/4 to 3 cups filling into each unbaked pie shell. Cover with top crust. Seal edges.

7 Using a convection oven, bake at 375 F. for 25 minutes or until lightly browned on high fan, open vent.

8 Cut 8 wedges per pie.

DESSERTS (PASTRY AND PIES) No.I 009 01

APPLE PIE (PREPARED PIE FILLING)

Yield 100 **Portion** 1 Slice

Calories	Carbohydrates	Protein	Fat	Cholesterol	Sodium	Calcium
363 cal	51 g	3 g	17 g	0 mg	256 mg	9 mg

Ingredient   **Weight**   **Measure**   **Issue**

PIE CRUST   26 each
PIE FILLING,APPLE,PREPARED   22-3/4 lbs   2 gal 3-3/8 qts

Method

1. PREPARE AND DIVIDE DOUGH: Prepare 1 recipe Pie Crust (Recipe No. I 001 00). Divide dough into 13-7-1/2 oz pieces for bottom crust and 13-7 oz pieces for top crust; place on lightly floured board. ROLL DOUGH: Sprinkle each piece of dough lightly with flour; flatten gently. Using a floured rolling pin, roll lightly with quick strokes from center out to edge in all directions. Form a circle 1 inch larger than pie pan and about 1/8 inch thick. Bottom crust will be slightly thicker. Shift or turn dough occasionally to prevent sticking. If edges split, pinch cracks together. BOTTOM CRUST: Fold rolled dough in half; carefully place into ungreased pie pan with fold at center. Unfold and fit carefully into pie pan, being careful not to leave any air spaces between pan and dough. FILL CRUST: Fill as specified on individual recipe card. TOP CRUST: Roll top crust in same manner as bottom crust. Fold in half; with knife, make several small slits near center fold to allow steam to escape during baking. Brush outer rim of bottom crust with water. Lay top crust over filling with fold at center; unfold and press edges of two crusts together lightly. REMOVE EXCESS DOUGH: Trim overhanging edges of dough by using a knife or spatula. (Incorporate excess dough into next crust, if needed.) There should be little excess if skill is used in weighing and rolling dough. SEAL PIE: Press edges of crust firmly together or crimp with the thumb and forefinger to make a fluted edge. BAKING INSTRUCTIONS FOR COOKED PIES: Bake as specified on individual recipe card.
2. Use canned prepared apple pie filling.
3. Pour 3 cups filling into each unbaked pie shell. Cover with top crust. Seal edges.
4. Using a convection oven, bake at 375 F. for 25 minutes or until lightly browned on high fan, open vent.
5. Cut 8 wedges per pie.

DESSERTS (PASTRY AND PIES) No.I 009 02

DUTCH APPLE PIE (PREPARED PIE FILLING)

Yield 100   Portion 1 Slice

Calories	Carbohydrates	Protein	Fat	Cholesterol	Sodium	Calcium
335 cal	52 g	3 g	14 g	13 mg	204 mg	15 mg

Ingredient	Weight	Measure	Issue
PIE CRUST		13 each	
STREUSEL TOPPING		3 qts 1 cup	
PIE FILLING,APPLE,PREPARED	22-3/4 lbs	2 gal 3-3/8 qts	

Method

1 PREPARE AND DIVIDE DOUGH: Prepare 1/2 recipe Pie Crust (Recipe No. I 001 00). Divide dough into 13-7-1/2 oz pieces for bottom crust; place on lightly floured board. ROLL DOUGH: Sprinkle each piece of dough lightly with flour; flatten gently. Using a floured rolling pin, roll lightly with quick strokes from center out to edge in all directions. Form a circle 1 inch larger than pie pan and about 1/8 inch thick. Shift or turn dough occasionally to prevent sticking. If edges split, pinch cracks together. BOTTOM CRUST: Fold rolled dough in half; carefully place into ungreased pie pan with fold at center. Unfold and fit carefully into pie pan, being careful not to leave any air spaces between pan and dough. FILL CRUST: Fill as specified on individual recipe card. REMOVE EXCESS DOUGH: Trim overhanging edges of dough by using a knife or spatula. (Incorporate excess dough into next crust, if needed.) There should be little excess if skill is used in weighing and rolling dough. SEAL PIE: Press edges of crust firmly together or crimp with the thumb and forefinger to make a fluted edge. BAKING INSTRUCTIONS FOR COOKED PIES: Bake as specified on individual recipe card.

2 Prepare 1-1/4 recipes Streusel Topping per 100 portions, Recipe No. D 049 00.

3 Pour 3 cups filling into each unbaked pie shell. Omit top crust; sprinkle 1-1/8 cup topping over filling in each pan.

4 Using a convection oven, bake at 375 F. for 25 minutes or until lightly browned on high fan, open vent.

5 Cut 8 wedges per pie.

DESSERTS (PASTRY AND PIES) No.I 010 00

APPLE COBBLER

Yield 100 **Portion** 1 Slice

Calories	Carbohydrates	Protein	Fat	Cholesterol	Sodium	Calcium
433 cal	58 g	4 g	21 g	0 mg	311 mg	11 mg

Ingredient	Weight	Measure	Issue
PIE CRUST		32-1/2 each	
PIE FILLING,APPLE,PREPARED	24 lbs	3 gal	

Method

1. Prepare 1-1/4 recipe Pie Crust (Recipe No. I 001 00) to yield enough dough to prepare cobbler for 100 portions.
2. Divide dough into four 3-3/4 lb pieces; use 2 pieces for each sheet pan.
3. Place dough on lightly floured board; sprinkle lightly with flour; flatten gently.
4. Roll 2 pieces dough into rectangular sheets about 1/8-inch thick and large enough to fit each pan. Press dough into bottom and sides of each pan. Reserve remaining pieces for use in Step 6.
5. Pour 1-1/2 gallons filling into each pan.
6. Roll remaining pieces dough for top crusts.
7. Place top crusts carefully over filling in each pan.
8. Crimp to seal edges.
9. Cut 6 to 8 small slits, about 1/2-inch each in tops of each cobbler.
10. Using a convection oven, bake at 375 F. for 35 to 40 minutes or until lightly browned on high fan, open vent.
11. Cool; cut 6 by 9.

DESSERTS (PASTRY AND PIES) No. I 010 01

PEACH COBBLER

Yield 100   Portion   1 Serving

Calories	Carbohydrates	Protein	Fat	Cholesterol	Sodium	Calcium
484 cal	72 g	4 g	21 g	0 mg	299 mg	21 mg

Ingredient   Weight   Measure   Issue

PIE CRUST   32-1/2 each
PIE FILLING, PEACH, PREPARED   24 lbs   3 gal

Method

1. Prepare 1-1/4 Pie Crust, Recipe No. I 001 00 to yield enough dough to prepare cobbler for 100 portions.
2. Divide dough into four 3-3/4 pound pieces; use 2 pieces for each sheet pan.
3. Place dough on lightly floured board; sprinkle lightly with flour; flatten gently.
4. Roll 2 pieces dough into rectangular sheets about 1/8-inch thick and large enough to fit each pan. Press dough into bottom and sides of each pan. Reserve remaining pieces for use in Step 6.
5. Pour 1-1/2 gallons of filling into each pan.
6. Roll remaining pieces dough for top crusts.
7. Place top crusts carefully over filling in each pan.
8. Crimp to seal edges.
9. Cut 6 to 8 small slits, about 1/2-inch each, in tops of each cobbler.
10. Using a convection oven, bake at 375 F. for 35 to 40 minutes or until lightly browned on high fan, open vent.
11. Cool; cut 6 by 9.

DESSERTS (PASTRY AND PIES) No. I 010 02

BLUEBERRY COBBLER

Yield 100   Portion 1 Piece

Calories	Carbohydrates	Protein	Fat	Cholesterol	Sodium	Calcium
438 cal	60 g	4 g	21 g	0 mg	327 mg	35 mg

Ingredient	Weight	Measure	Issue
PIE CRUST		32-1/2 each	
PIE FILLING, BLUEBERRY, PREPARED	28-1/4 lbs	3 gal	

Method

1. Prepare 1-1/4 Pie Crust, Recipe No. I 001 00 to yield enough dough to prepare cobbler for 100 portions.
2. Divide dough into four 3-3/4 lb pieces; use 2 pieces for each sheet pan.
3. Place dough on lightly floured board; sprinkle lightly with flour; flatten gently.
4. Roll 2 pieces dough into rectangular sheets about 1/8-inch thick and large enough to fit each pan. Press dough into bottom and sides of each pan. Reserve remaining pieces for use in Step 6.
5. Pour 1-1/2 gallons of filling into each pan.
6. Roll remaining pieces dough for top crusts.
7. Place top crusts carefully over filling in each pan.
8. Crimp to seal edges.
9. Cut 6 to 8 small slits, about 1/2-inch each, in tops of each cobbler.
10. Using a convection oven, bake at 375 F. for 35 to 40 minutes or until lightly browned, on high fan, open vent.
11. Cool; cut 6 by 9.

DESSERTS (PASTRY AND PIES) No.I 010 03

CHERRY COBBLER

Yield 100   **Portion** 1 Piece

Calories	Carbohydrates	Protein	Fat	Cholesterol	Sodium	Calcium
449 cal	62 g	5 g	21 g	0 mg	273 mg	18 mg

Ingredient   Weight   Measure   Issue
PIE CRUST   32-1/2 each
PIE FILLING,CHERRY,PREPARED   24 lbs   3 gal

Method

1. Prepare 1-1/4 Pie Crust, Recipe No. I 001 00 to yield enough dough to prepare cobbler for 100 portions.
2. Divide dough into four 3-3/4 lb pieces; use 2 pieces for each sheet pan.
3. Place dough on lightly floured board; sprinkle lightly with flour; flatten gently.
4. Roll 2 pieces dough into rectangular sheets about 1/8 inch thick and large enough to fit each pan. Press dough into bottom and sides of each pan. Reserve remaining pieces for use in Step 6.
5. Pour 1-1/2 gallons of filling into each pan.
6. Roll remaining pieces dough for top crusts.
7. Place top crusts carefully over filling in each pan.
8. Crimp to seal edges.
9. Cut 6 to 8 small slits, about 1/2-inch each, in tops of each cobbler.
10. Using a convection oven, bake at 375 F. for 35 to 40 minutes or until lightly browned on high fan, open vent.
11. Cool; cut 6 by 9.

DESSERTS (PASTRY AND PIES) No. I 010 04

STREUSEL-TOPPED APPLE COBBLER

Yield 100   **Portion** 1 Piece

Calories	Carbohydrates	Protein	Fat	Cholesterol	Sodium	Calcium
492 cal	71 g	4 g	22 g	25 mg	303 mg	24 mg

Ingredient	Weight	Measure	Issue
PIE CRUST		19-1/2 each	
PIE FILLING,APPLE,PREPARED	24 lbs	3 gal	
STREUSEL TOPPING		1 gal 2 qts	

Method

1. Prepare 3/4 Pie Crust, Recipe No. I 001 00 to yield enough dough to prepare cobbler for 100 portions.
2. Divide dough into 2 pieces; use 1 piece for each sheet pan.
3. Place dough on lightly floured board; sprinkle lightly with flour; flatten gently.
4. Roll dough into rectangular sheets about 1/8-inch thick and large enough to fit each pan. Press dough into bottom and sides of each pan.
5. Pour 1-1/2 gallons of filling into each pan
6. Prepare 2 recipes Streusel Topping per 100 portions, Recipe No. D 049 00.
7. Spread 3 quarts topping over filling in each pan.
8. Using a convection oven, bake at 375 F. for 35 to 40 minutes or until lightly browned on high fan, open vent.
9. Cool; cut 6 by 9.

DESSERTS (PASTRY AND PIES) No. I 011 00

CHOCOLATE MOUSSE PIE

Yield 100   Portion 1 Slice

Calories	Carbohydrates	Protein	Fat	Cholesterol	Sodium	Calcium
247 cal	33 g	4 g	11 g	1 mg	377 mg	50 mg

Ingredient	**Weight**	**Measure**	**Issue**
PIE CRUST		13 each	
MILK,NONFAT,DRY	9-5/8 oz	1 qts	
WATER,COLD	10-1/2 lbs	1 gal 1 qts	
DESSERT POWDER,PUDDING,INSTANT,CHOCOLATE	3-3/4 lbs	2 qts 2 cup	
MILK,NONFAT,DRY	2-3/8 oz	1 cup	
WATER,COLD	2-1/8 lbs	1 qts	
WHIPPED TOPPING MIX,NONDAIRY,DRY	1-1/2 lbs	2 gal 1/2 qts	
SUGAR,GRANULATED	4 oz	1/2 cup 1 tbsp	
EXTRACT,VANILLA	1-3/8 oz	3 tbsp	

Method

1. PREPARE AND DIVIDE DOUGH: Prepare 1/2 recipe Pie Crust (Recipe No. I 001 00). Divide dough into 13-7-1/2 oz pieces for bottom crust; place on lightly floured board. ROLL DOUGH: Sprinkle each piece of dough lightly with flour; flatten gently. Using a floured rolling pin, roll lightly with quick strokes from center out to edge in all directions. Form a circle 1 inch larger than pie pan and about 1/8 inch thick. Shift or turn dough occasionally to prevent sticking. If edges split, pinch cracks together. BOTTOM CRUST: Fold rolled dough in half; carefully place into ungreased pie pan with fold at center. Unfold and fit carefully into pie pan, being careful not to leave any air spaces between pan and dough. REMOVE EXCESS DOUGH: Trim overhanging edges of dough by using a knife or spatula. (Incorporate excess dough into next crust, if needed.) There should be little excess if skill is used in weighing and rolling dough. BAKING INSTRUCTIONS FOR UNCOOKED PIES: Bake crusts at 425 F. for about 15-18 minutes, or until light golden brown. Cool before filling. Proceed with the recipe directions.
2. Combine milk and water in mixer bowl.
3. Add dessert powder to milk and water. Using whip, blend at low speed 15 seconds or until well blended. Scrape down bowl; whip at medium speed 2 minutes. Set aside for use in Step 7.
4. Mix milk and water in mixer bowl.
5. Add topping to milk mixture in bowl. Using whip, mix at low speed until blended.
6. Gradually add sugar and vanilla to whipped topping while mixing at low speed. Scrape down bowl. Mix at high-speed 5 minutes or until peaks are formed.
7. Add topping to pudding mixture; blend until completely mixed.
8. Pour 3-1/2 cups filling into each baked pie shell.
9. Refrigerate about 4 hours until ready to serve.
   Cut 8 wedges per pie. CCP: Hold for service at 41 F. or lower.

DESSERTS (PASTRY AND PIES) No. I 012 00

SWEET POTATO PIE

Yield 100   Portion 1 Slice

Calories	Carbohydrates	Protein	Fat	Cholesterol	Sodium	Calcium
252 cal	36 g	4 g	10 g	37 mg	221 mg	47 mg

Ingredient   Weight   Measure   Issue

Ingredient	Weight	Measure	Issue
PIE CRUST		13 each	
SWEET POTATOES,CANNED,VACUUM PACK	13-1/2 lbs	1 gal 2 qts	
EGGS,WHOLE,FROZEN	1-3/4 lbs	3-1/4 cup	
SUGAR,GRANULATED	1-1/8 lbs	2-5/8 cup	
SUGAR,BROWN,PACKED	1-1/8 lbs	3-1/2 cup	
MILK,NONFAT,DRY	5-1/8 oz	2-1/8 cup	
SALT	1/2 oz	3/8 tsp	
CINNAMON,GROUND	7/8 oz	1/4 cup	
NUTMEG,GROUND	3/8 oz	1 tbsp	
GINGER,GROUND	1/3 oz	1 tbsp	
CLOVES,GROUND	1/8 oz	3/8 tsp	
WATER,WARM	5-7/8 lbs	2 qts 3-1/4 cup	
BUTTER,MELTED	4 oz	1/2 cup	

Method

1 PREPARE AND DIVIDE DOUGH: Prepare 1/2 recipe Pie Crust (Recipe No. I 001 00). Divide dough into 13-7/1-2 oz pieces for bottom crust; place on lightly floured board. ROLL DOUGH: Sprinkle each piece of dough lightly with flour; flatten gently. Using a floured rolling pin, roll lightly with quick strokes from center out to edge in all directions. Form a circle 1 inch larger than pie pan and about 1/8 inch thick. Shift or turn dough occasionally to prevent sticking. If edges split, pinch cracks together. BOTTOM CRUST: Fold rolled dough in half; carefully place into ungreased pie pan with fold at center. Unfold and fit carefully into pie pan, being careful not to leave any air spaces between pan and dough. REMOVE EXCESS DOUGH: Trim overhanging edges of dough by using a knife or spatula. (Incorporate excess dough into next crust, if needed.) There should be little excess if skill is used in weighing and rolling dough. BAKING INSTRUCTIONS FOR UNCOOKED PIES: Bake crusts at 425 F. for about 15-18 minutes, or until light golden brown. Cool before filling. Proceed with the recipe directions.

2 Mix sweet potatoes in mixer bowl at medium speed for 5 minutes or until smooth.

3 Combine eggs, sugars, milk, salt, cinnamon, nutmeg, ginger, and cloves. Stir until well blended. Add to sweet potatoes.

4 Add water and butter or margarine to sweet potato mixture; beat at low speed until well blended.

5 Pour 2-3/4 to 3 cups filling into each unbaked pie shell.

6 Bake at 425 F. for 45 to 55 minutes or until knife inserted into filling comes out clean. Center may be soft but will set when cool.

7 Cut 8 wedges per pie. CCP: Hold for service at 41 F. or lower.

DESSERTS (PASTRY AND PIES) No. I 013 00

PUMPKIN PIE

Yield 100   Portion 1 Slice

Calories	Carbohydrates	Protein	Fat	Cholesterol	Sodium	Calcium
242 cal	35 g	4 g	10 g	46 mg	370 mg	56 mg

Ingredient   Weight   Measure   Issue

PIE CRUST		13 each
SUGAR, GRANULATED	3-5/8 lbs	2 qts 1/4 cup
SALT	1-1/8 oz	1 tbsp
FLOUR, WHEAT, GENERAL PURPOSE	6-5/8 oz	1-1/2 cup
MILK, NONFAT, DRY	8 oz	3-3/8 cup
CINNAMON, GROUND	1-1/2 oz	1/4 cup 2-1/3 tbsp
NUTMEG, GROUND	3/8 oz	1 tbsp
GINGER, GROUND	1/3 oz	1 tbsp
PUMPKIN, CANNED, SOLID PACK	10-1/2 lbs	1 gal 7/8 qts
WATER	9-3/8 lbs	1 gal 1/2 qts
EGGS, WHOLE, FROZEN	2-1/3 lbs	1 qts 3/8 cup

Method

1. PREPARE AND DIVIDE DOUGH: Prepare 1/2 recipe Pie Crust (Recipe No. I 001 00). Divide dough into 13-7-1/2 oz pieces for bottom crust; place on lightly floured board. ROLL DOUGH: Sprinkle each piece of dough lightly with flour; flatten gently. Using a floured rolling pin, roll lightly with quick strokes from center out to edge in all directions. Form a circle 1 inch larger than pie pan and about 1/8 inch thick. Shift or turn dough occasionally to prevent sticking. If edges split, pinch cracks together. BOTTOM CRUST: Fold rolled dough in half; carefully place into ungreased pie pan with fold at center. Unfold and fit carefully into pie pan, being careful not to leave any air spaces between pan and dough. REMOVE EXCESS DOUGH: Trim overhanging edges of dough by using a knife or spatula. (Incorporate excess dough into next crust, if needed.) There should be little excess if skill is used in weighing and rolling dough. BAKING INSTRUCTIONS FOR UNCOOKED PIES: Bake crusts at 425 F. for about 15-18 minutes, or until light golden brown. Cool before filling. Proceed with the recipe directions.
2. Combine sugar, salt, flour, milk, cinnamon, nutmeg and ginger in mixing bowl.
3. Add pumpkin to dry ingredients; mix at low speed until well blended. Mixture must set for one hour under refrigeration 41 F. or lower.
4. Add water and eggs; mix at low speed until well blended.
5. Pour 3-3/4 cups filling into each unbaked pie shell.
6. Bake at 375 F. for 50 to 55 minutes or until center is firm. Cool thoroughly.
7. Cut 8 wedges per pie. CCP: Hold for service at 41 F. or lower.

DESSERTS (PASTRY AND PIES) No. I 014 00

PINEAPPLE PIE (CANNED PINEAPPLE-CORNSTARCH)

Yield 100   Portion 1 Slice

Calories	Carbohydrates	Protein	Fat	Cholesterol	Sodium	Calcium
334 cal	43 g	3 g	17 g	0 mg	229 mg	8 mg

Ingredient	Weight	Measure	Issue
PIE CRUST		26 each	
PINEAPPLE,CANNED,CRUSHED,JUICE PACK,INCL LIQUIDS	3-3/4 lbs	1 qts 2-3/4 cup	
RESERVED LIQUID	4-2/3 lbs	2 qts 1 cup	
SUGAR,GRANULATED	3-1/4 lbs	1 qts 3-1/4 cup	
SALT	1/8 oz	1/8 tsp	
CORNSTARCH	8-1/2 oz	1-7/8 cup	
WATER,COLD	1-5/8 lbs	3 cup	
JUICE,LEMON	1-5/8 oz	3 tbsp	

Method

1. PREPARE AND DIVIDE DOUGH: Prepare 1 recipe Pie Crust (Recipe No. I 001 00). Divide dough into 13-7-1/2 oz pieces for bottom crust and 13-7 oz pieces for top crust; place on lightly floured board. ROLL DOUGH: Sprinkle each piece of dough lightly with flour; flatten gently. Using a floured rolling pin, roll lightly with quick strokes from center out to edge in all directions. Form a circle 1 inch larger than pie pan and about 1/8 inch thick. Bottom crust will be slightly thicker. Shift or turn dough occasionally to prevent sticking. If edges split, pinch cracks together. BOTTOM CRUST: Fold rolled dough in half; carefully place into ungreased pie pan with fold at center. Unfold and fit carefully into pie pan, being careful not to leave any air spaces between pan and dough. FILL CRUST: Fill as specified on individual recipe card. TOP CRUST: Roll top crust in same manner as bottom crust. Fold in half; with knife, make several small slits near center fold to allow steam to escape during baking. Brush outer rim of bottom crust with water. Lay top crust over filling with fold at center; unfold and press edges of two crusts together lightly. REMOVE EXCESS DOUGH: Trim overhanging edges of dough by using a knife or spatula. (Incorporate excess dough into next crust, if needed.) There should be little excess if skill is used in weighing and rolling dough. SEAL PIE: Press edges of crust firmly together or crimp with the thumb and forefinger to make a fluted edge. BAKING INSTRUCTIONS FOR COOKED PIES: Bake as specified on individual recipe card.

2. Drain pineapple; reserve juice for use in Step 3 and pineapple for use in Step 5.
3. Combine reserved juice, sugar, and salt; bring to a boil.
4. Combine cornstarch and water; stir until smooth. Add gradually to boiling mixture. Cook at medium heat, stirring constantly until thick and clear. Remove from heat.
5. Fold pineapple and lemon juice carefully into thickened mixture.
6. Pour 2-3/4 to 3 cups filling into each unbaked 9-inch pie shell. Cover with top crust. Seal edges.
7. Bake at 425 F. for 30 to 35 minutes or until lightly browned.
8. Cut 8 wedges per pie.

DESSERTS (PASTRY AND PIES) No.I 015 00

BERRY PIE (FROZEN BERRIES-CORNSTARCH)

Yield 100   Portion   1 Slice

Calories	Carbohydrates	Protein	Fat	Cholesterol	Sodium	Calcium
368 cal	48 g	4 g	18 g	4 mg	260 mg	10 mg

Ingredient	Weight	Measure	Issue
PIE CRUST		26 each	
BLUEBERRIES,FROZEN,UNSWEETENED	12-1/3 lbs	2 gal 1 qts	
SUGAR,GRANULATED	3-1/4 lbs	1 qts 3-1/4 cup	
SALT	1/3 oz	1/4 tsp	
CORNSTARCH	11-1/4 oz	2-1/2 cup	
WATER,COLD	2-1/3 lbs	1 qts 1/2 cup	
BUTTER	6 oz	3/4 cup	

Method

1 PREPARE AND DIVIDE DOUGH: Prepare 1 recipe Pie Crust (Recipe No. I 001 00). Divide dough into 13-7-1/2 oz pieces for bottom crust and 13-7 oz pieces for top crust; place on lightly floured board. ROLL DOUGH: Sprinkle each piece of dough lightly with flour; flatten gently. Using a floured rolling pin, roll lightly with quick strokes from center out to edge in all directions. Form a circle 1 inch larger than pie pan and about 1/8 inch thick. Bottom crust will be slightly thicker. Shift or turn dough occasionally to prevent sticking. If edges split, pinch cracks together. BOTTOM CRUST: Fold rolled dough in half; carefully place into ungreased pie pan with fold at center. Unfold and fit carefully into pie pan, being careful not to leave any air spaces between pan and dough. FILL CRUST: Fill as specified on individual recipe card. TOP CRUST: Roll top crust in same manner as bottom crust. Fold in half; with knife, make several small slits near center fold to allow steam to escape during baking. Brush outer rim of bottom crust with water. Lay top crust over filling with fold at center; unfold and press edges of two crusts together lightly. REMOVE EXCESS DOUGH: Trim overhanging edges of dough by using a knife or spatula. (Incorporate excess dough into next crust, if needed.) There should be little excess if skill is used in weighing and rolling dough. SEAL PIE: Press edges of crust firmly together or crimp with the thumb and forefinger to make a fluted edge. BAKING INSTRUCTIONS FOR COOKED PIES: Bake as specified on individual recipe card.

2 Thaw berries; drain; reserve juice.

3 Take reserved juice and add water to equal 6-3/4 cups per 100 portions and combine with sugar and salt; bring to a boil.

4 Combine cornstarch and water; stir until smooth. Add gradually to boiling mixture while stirring. Cook at medium heat, stirring constantly, until thick and clear. Remove from heat.

5 Fold berries and butter or margarine carefully into thickened mixture.

6 Pour 2-3/4 to 3 cups filling into each unbaked 9-inch pie shell. Cover with top crust. Seal edges.

7 Bake at 425 F. for 45 minutes or until lightly browned.

8 Cut 8 wedges per pie.

Notes

1 In Step 2, strawberries or raspberries may be used.

DESSERTS (PASTRY AND PIES) No.I 015 01

BLUEBERRY PIE (FROZEN BLUEBERRIES)

Yield 100   Portion 1 Slice

Calories	Carbohydrates	Protein	Fat	Cholesterol	Sodium	Calcium
407 cal	58 g	4 g	18 g	4 mg	261 mg	11 mg

Ingredient	Weight	Measure	Issue
PIE CRUST		26 each	
BLUEBERRIES,FROZEN,UNSWEETENED	13-1/2 lbs	2 gal 1-7/8 qts	
WATER	2-1/3 lbs	1 qts 1/2 cup	
SUGAR,GRANULATED	5-1/4 lbs	3 qts	
SALT	1/3 oz	1/4 tsp	
CORNSTARCH	11-1/4 oz	2-1/2 cup	
WATER,COLD	2-1/3 lbs	1 qts 1/2 cup	
BUTTER	6 oz	3/4 cup	

Method

1. PREPARE AND DIVIDE DOUGH: Prepare 1 recipe Pie Crust (Recipe No. I 001 00). Divide dough into 13-7-1/2 oz pieces for bottom crust and 13-7 oz pieces for top crust; place on lightly floured board. ROLL DOUGH: Sprinkle each piece of dough lightly with flour; flatten gently. Using a floured rolling pin, roll lightly with quick strokes from center out to edge in all directions. Form a circle 1 inch larger than pie pan and about 1/8 inch thick. Bottom crust will be slightly thicker. Shift or turn dough occasionally to prevent sticking. If edges split, pinch cracks together. BOTTOM CRUST: Fold rolled dough in half; carefully place into ungreased pie pan with fold at center. Unfold and fit carefully into pie pan, being careful not to leave any air spaces between pan and dough. FILL CRUST: Fill as specified on individual recipe card. TOP CRUST: Roll top crust in same manner as bottom crust. Fold in half; with knife, make several small slits near center fold to allow steam to escape during baking. Brush outer rim of bottom crust with water. Lay top crust over filling with fold at center; unfold and press edges of two crusts together lightly. REMOVE EXCESS DOUGH: Trim overhanging edges of dough by using a knife or spatula. (Incorporate excess dough into next crust, if needed.) There should be little excess if skill is used in weighing and rolling dough. SEAL PIE: Press edges of crust firmly together or crimp with the thumb and forefinger to make a fluted edge. BAKING INSTRUCTIONS FOR COOKED PIES: Bake as specified on individual recipe card.
2. Use frozen blueberries. Thawing is not necessary.
3. Combine water, sugar and salt. Bring to a boil.
4. Combine cornstarch and water; stir until smooth. Add gradually to boiling mixture. Cook at medium heat, stirring constantly, until thick and clear. Remove from heat.
5. Fold berries and butter or margarine carefully into thickened mixture.
6. Pour 3 cups filling into each unbaked 9-inch pie shell. Cover with top crust. Seal edges.
7. Bake at 425 F. for 45 minutes or until lightly browned.
8. Cut 8 wedges per pie.

DESSERTS (PASTRY AND PIES) No. I 017 00

BLUEBERRY PIE (CANNED BLUEBERRIES-CORNSTARCH)

Yield 100   Portion 1 Serving

Calories	Carbohydrates	Protein	Fat	Cholesterol	Sodium	Calcium
446 cal	71 g	4 g	17 g	0 mg	267 mg	11 mg

Ingredient	Weight	Measure	Issue
PIE CRUST		26 each	
BLUEBERRIES,CANNED,HEAVY SYRUP,INCL LIQUIDS	20-1/3 lbs	2 gal 1 qts	
RESERVED LIQUID	4-2/3 lbs	2 qts 1 cup	
SUGAR,GRANULATED	5-1/4 lbs	3 qts	
SALT	1/2 oz	3/8 tsp	
CORNSTARCH	12 oz	2-5/8 cup	
RESERVED LIQUID	2 lbs	3-3/4 cup	
JUICE,LEMON	1-5/8 oz	3 tbsp	

Method

1. PREPARE AND DIVIDE DOUGH: Prepare 1 recipe Pie Crust (Recipe No. I 001 00). Divide dough into 13-7-1/2 oz pieces for bottom crust and 13-7 oz pieces for top crust; place on lightly floured board. ROLL DOUGH: Sprinkle each piece of dough lightly with flour; flatten gently. Using a floured rolling pin, roll lightly with quick strokes from center out to edge in all directions. Form a circle 1 inch larger than pie pan and about 1/8 inch thick. Bottom crust will be slightly thicker. Shift or turn dough occasionally to prevent sticking. If edges split, pinch cracks together. BOTTOM CRUST: Fold rolled dough in half; carefully place into ungreased pie pan with fold at center. Unfold and fit carefully into pie pan, being careful not to leave any air spaces between pan and dough. FILL CRUST: Fill as specified on individual recipe card. TOP CRUST: Roll top crust in same manner as bottom crust. Fold in half; with knife, make several small slits near center fold to allow steam to escape during baking. Brush outer rim of bottom crust with water. Lay top crust over filling with fold at center; unfold and press edges of two crusts together lightly. REMOVE EXCESS DOUGH: Trim overhanging edges of dough by using a knife or spatula. (Incorporate excess dough into next crust, if needed.) There should be little excess if skill is used in weighing and rolling dough. SEAL PIE: Press edges of crust firmly together or crimp with the thumb and forefinger to make a fluted edge. BAKING INSTRUCTIONS FOR COOKED PIES: Bake as specified on individual recipe card.

2. Drain blueberries; reserve juice.
3. Combine 2-1/4 quart reserved juice, sugar, and salt; bring to a boil.
4. Combine cornstarch and 3-3/4 cups reserved juice; stir until smooth. Add gradually to boiling mixture. Cook at medium heat, stirring constantly, until thick and clear. Remove from heat.
5. Fold blueberries and lemon juice carefully into thickened mixture.
6. Pour 2-3/4 to 3 cups filling into each unbaked 9-inch pie shell. Cover with top crust. Seal edges.
7. Using a convection oven, bake at 375 F. for 20 to 25 minutes or until lightly browned on high fan, open vent.
8. Cut 8 wedges per pie.

DESSERTS (PASTRY AND PIES) No. I 017 01

BLUEBERRY PIE (PREPARED FILLING)

Yield 100 **Portion** 1 Slice

Calories	Carbohydrates	Protein	Fat	Cholesterol	Sodium	Calcium
351 cal	48 g	3 g	17 g	0 mg	262 mg	28 mg

Ingredient	Weight	Measure	Issue
PIE CRUST		26 each	
PIE FILLING,BLUEBERRY,PREPARED	22-3/4 lbs	2 gal 1-2/3 qts	

Method

1. PREPARE AND DIVIDE DOUGH: Prepare 1 recipe Pie Crust (Recipe No. I 001 00). Divide dough into 13-7-1/2 oz pieces for bottom crust and 13-7 oz pieces for top crust; place on lightly floured board. ROLL DOUGH: Sprinkle each piece of dough lightly with flour; flatten gently. Using a floured rolling pin, roll lightly with quick strokes from center out to edge in all directions. Form a circle 1 inch larger than pie pan and about 1/8 inch thick. Bottom crust will be slightly thicker. Shift or turn dough occasionally to prevent sticking. If edges split, pinch cracks together. BOTTOM CRUST: Fold rolled dough in half; carefully place into ungreased pie pan with fold at center. Unfold and fit carefully into pie pan, being careful not to leave any air spaces between pan and dough. FILL CRUST: Fill as specified on individual recipe card. TOP CRUST: Roll top crust in same manner as bottom crust. Fold in half; with knife, make several small slits near center fold to allow steam to escape during baking. Brush outer rim of bottom crust with water. Lay top crust over filling with fold at center; unfold and press edges of two crusts together lightly. REMOVE EXCESS DOUGH: Trim overhanging edges of dough by using a knife or spatula. (Incorporate excess dough into next crust, if needed.) There should be little excess if skill is used in weighing and rolling dough. SEAL PIE: Press edges of crust firmly together or crimp with the thumb and forefinger to make a fluted edge. BAKING INSTRUCTIONS FOR COOKED PIES: Bake as specified on individual recipe card.

2. Pour 3 cups filling into each unbaked 9-inch pie shell. Cover with top crust. Seal edges.
3. Using a convection oven, bake at 375 F. for 20 to 25 minutes or until lightly browned on high fan, open vent.
4. Cut 8 wedges per pie.

DESSERTS (PASTRY AND PIES) No. I 019 00

BUTTERSCOTCH CREAM PIE (DESSERT POWDER, INSTANT)

Yield 100   Portion 1 Slice

Calories	Carbohydrates	Protein	Fat	Cholesterol	Sodium	Calcium
301 cal	43 g	3 g	13 g	1 mg	492 mg	56 mg

Ingredient	Weight	Measure	Issue
PIE CRUST		13 each	
MILK,NONFAT,DRY	13-3/4 oz	1 qts 1-3/4 cup	
WATER,COLD	15-1/8 lbs	1 gal 3-1/4 qts	
DESSERT POWDER,PUDDING,INSTANT,BUTTERSCOTCH	5-1/2 lbs		

Method

1. PREPARE AND DIVIDE DOUGH: Prepare 1/2 recipe Pie Crust (Recipe No. I 001 00). Divide dough into 13-7-1/2 oz pieces for bottom crust; place on lightly floured board. ROLL DOUGH: Sprinkle each piece of dough lightly with flour; flatten gently. Using a floured rolling pin, roll lightly with quick strokes from center out to edge in all directions. Form a circle 1 inch larger than pie pan and about 1/8 inch thick. Shift or turn dough occasionally to prevent sticking. If edges split, pinch cracks together. BOTTOM CRUST: Fold rolled dough in half; carefully place into ungreased pie pan with fold at center. Unfold and fit carefully into pie pan, being careful not to leave any air spaces between pan and dough. REMOVE EXCESS DOUGH: Trim overhanging edges of dough by using a knife or spatula. (Incorporate excess dough into next crust, if needed.) There should be little excess if skill is used in weighing and rolling dough. BAKING INSTRUCTIONS FOR UNCOOKED PIES: Bake crusts at 425 F. for about 15-18 minutes, or until light golden brown. Cool before filling. Proceed with the recipe directions.
2. Reconstitute milk.
3. Add dessert powder. Using whip, blend at low speed 15 seconds or until well-blended. Scrape down sides of bowl; whip at medium speed for 2 minutes.
4. Pour 3 cups filling into each baked 9-inch pie shell.
5. Refrigerate until ready to serve. Chilled pies may be topped with Whipped Cream, Recipe No. K 001 00, or Whipped Topping, Recipe No. K 002 00.
6. Cut 8 wedges per pie. CCP: Hold for service at 41 F. or lower.

DESSERTS (PASTRY AND PIES) No.I 020 00

PEACH PIE (FROZEN PEACHES-CORNSTARCH)

Yield 100   Portion 1 Slice

Calories	Carbohydrates	Protein	Fat	Cholesterol	Sodium	Calcium
418 cal	64 g	4 g	17 g	0 mg	269 mg	9 mg

Ingredient | Weight | Measure | Issue

Ingredient	Weight	Measure	Issue
PIE CRUST		26 each	
PEACHES,FROZEN	19-7/8 lbs	2 gal 1 qts	
RESERVED LIQUID	5-1/2 lbs	2 qts 2-1/2 cup	
SUGAR,GRANULATED	3-5/8 lbs	2 qts 1/4 cup	
SALT	1/2 oz	3/8 tsp	
CORNSTARCH	10-1/8 oz	2-1/4 cup	
WATER,COLD	1-1/8 lbs	2-1/4 cup	

Method

1. PREPARE AND DIVIDE DOUGH: Prepare 1 recipe Pie Crust (Recipe No. I 001 00). Divide dough into 13-7-1/2 oz pieces for bottom crust and 13-7 oz pieces for top crust; place on lightly floured board. ROLL DOUGH: Sprinkle each piece of dough lightly with flour; flatten gently. Using a floured rolling pin, roll lightly with quick strokes from center out to edge in all directions. Form a circle 1 inch larger than pie pan and about 1/8 inch thick. Bottom crust will be slightly thicker. Shift or turn dough occasionally to prevent sticking. If edges split, pinch cracks together. BOTTOM CRUST: Fold rolled dough in half; carefully place into ungreased pie pan with fold at center. Unfold and fit carefully into pie pan, being careful not to leave any air spaces between pan and dough. FILL CRUST: Fill as specified on individual recipe card. TOP CRUST: Roll top crust in same manner as bottom crust. Fold in half; with knife, make several small slits near center fold to allow steam to escape during baking. Brush outer rim of bottom crust with water. Lay top crust over filling with fold at center; unfold and press edges of two crusts together lightly. REMOVE EXCESS DOUGH: Trim overhanging edges of dough by using a knife or spatula. (Incorporate excess dough into next crust, if needed.) There should be little excess if skill is used in weighing and rolling dough. SEAL PIE: Press edges of crust firmly together or crimp with the thumb and forefinger to make a fluted edge. BAKING INSTRUCTIONS FOR COOKED PIES: Bake as specified on individual recipe card.

2. Thaw peaches. Drain; reserve juice.
3. Combine reserved juice, sugar, and salt; bring to a boil.
4. Combine cornstarch and water; stir until smooth. Add gradually to boiling mixture. Cook at medium heat, stirring constantly, until thick and clear. Remove from heat.
5. Fold peaches carefully into thickened mixture. Cool.
6. Pour 2-3/4 to 3 cups filling into each unbaked 9-inch pie shell. Cover with top crust. Seal edges.
7. Bake at 425 F. for 30 to 35 minutes or until lightly browned.
8. Cut 8 wedges per pie.

DESKTOP DESSERTS (PASTRY AND PIES) No.I 022 00

CHERRY PIE (CANNED CHERRIES-CORNSTARCH)

Yield 100   Portion 1 Slice

Calories	Carbohydrates	Protein	Fat	Cholesterol	Sodium	Calcium
406 cal	61 g	4 g	17 g	0 mg	265 mg	16 mg

Ingredient	Weight	Measure	Issue
PIE CRUST		26 each	
CHERRIES,CANNED,RED,TART,WATER PACK,INCL LIQUIDS	19-2/3 lbs	2 gal 1 qts	
RESERVED LIQUID	2-3/4 lbs	1 qts 1-3/8 cup	
SUGAR,GRANULATED	5-1/4 lbs	3 qts	
SALT	1/2 oz	3/8 tsp	
CORNSTARCH	12 oz	2-5/8 cup	
WATER,COLD	1-1/8 lbs	2-1/4 cup	
FOOD COLOR,RED	1/4 oz	1/4 tsp	

Method

1. PREPARE AND DIVIDE DOUGH: Prepare 1 recipe Pie Crust (Recipe No. I 001 00). Divide dough into 13-7-1/2 oz pieces for bottom crust and 13-7 oz pieces for top crust; place on lightly floured board. ROLL DOUGH: Sprinkle each piece of dough lightly with flour; flatten gently. Using a floured rolling pin, roll lightly with quick strokes from center out to edge in all directions. Form a circle 1 inch larger than pie pan and about 1/8 inch thick. Bottom crust will be slightly thicker. Shift or turn dough occasionally to prevent sticking. If edges split, pinch cracks together. BOTTOM CRUST: Fold rolled dough in half; carefully place into ungreased pie pan with fold at center. Unfold and fit carefully into pie pan, being careful not to leave any air spaces between pan and dough. FILL CRUST: Fill as specified on individual recipe card. TOP CRUST: Roll top crust in same manner as bottom crust. Fold in half; with knife, make several small slits near center fold to allow steam to escape during baking. Brush outer rim of bottom crust with water. Lay top crust over filling with fold at center; unfold and press edges of two crusts together lightly. REMOVE EXCESS DOUGH: Trim overhanging edges of dough by using a knife or spatula. (Incorporate excess dough into next crust, if needed.) There should be little excess if skill is used in weighing and rolling dough. SEAL PIE: Press edges of crust firmly together or crimp with the thumb and forefinger to make a fluted edge. BAKING INSTRUCTIONS FOR COOKED PIES: Bake as specified on individual recipe card.

2. Drain cherries; reserve juice for use in Step 3 and cherries for use in Step 5.
3. Combine reserved juice, sugar, and salt; bring to a boil.
4. Combine cornstarch and water; stir until smooth. Add gradually to boiling mixture. Cook at medium heat, stirring constantly until thick and clear. Remove from heat.
5. Add red food coloring. Fold cherries carefully into thickened mixture. Cool.
6. Pour 3 cups filling into each unbaked 9-inch pie shell. Cover with top crust. Seal edges.
7. Using a convection oven, bake at 375 F. for 20 to 25 minutes or until lightly browned on high fan, open vent.
8. Cut 8 wedges per pie.

DESSERTS (PASTRY AND PIES) No. I 022 01

CHERRY PIE (PIE FILLING, PREPARED)

Yield 100   **Portion** 1 Slice

Calories	Carbohydrates	Protein	Fat	Cholesterol	Sodium	Calcium
377 cal	54 g	4 g	17 g	0 mg	219 mg	16 mg

Ingredient	Weight	Measure	Issue
PIE CRUST		26 each	
PIE FILLING,CHERRY,PREPARED	22-3/4 lbs	2 gal 3-3/8 qts	

Method

1. PREPARE AND DIVIDE DOUGH: Prepare 1 recipe Pie Crust (Recipe No. I 001 00). Divide dough into 13-7-1/2 oz pieces for bottom crust and 13-7 oz pieces for top crust; place on lightly floured board. ROLL DOUGH: Sprinkle each piece of dough lightly with flour; flatten gently. Using a floured rolling pin, roll lightly with quick strokes from center out to edge in all directions. Form a circle 1 inch larger than pie pan and about 1/8 inch thick. Bottom crust will be slightly thicker. Shift or turn dough occasionally to prevent sticking. If edges split, pinch cracks together. BOTTOM CRUST: Fold rolled dough in half; carefully place into ungreased pie pan with fold at center. Unfold and fit carefully into pie pan, being careful not to leave any air spaces between pan and dough. FILL CRUST: Fill as specified on individual recipe card. TOP CRUST: Roll top crust in same manner as bottom crust. Fold in half; with knife, make several small slits near center fold to allow steam to escape during baking. Brush outer rim of bottom crust with water. Lay top crust over filling with fold at center; unfold and press edges of two crusts together lightly. REMOVE EXCESS DOUGH: Trim overhanging edges of dough by using a knife or spatula. (Incorporate excess dough into next crust, if needed.) There should be little excess if skill is used in weighing and rolling dough. SEAL PIE: Press edges of crust firmly together or crimp with the thumb and forefinger to make a fluted edge. BAKING INSTRUCTIONS FOR COOKED PIES: Bake as specified on individual recipe card.
2. Pour 3 cups filling into each unbaked 9-inch pie shell. Cover with top crust. Seal edges.
3. Using a convection oven, bake at 375 F. for 20 to 25 minutes or until lightly browned on high fan, open vent.
4. Cut 8 wedges per pie.

DESSERTS (PASTRY AND PIES) No.I 024 00

PEACH PIE (CANNED PEACHES-CORNSTARCH)

Yield 100   Portion 1 Slice

Calories	Carbohydrates	Protein	Fat	Cholesterol	Sodium	Calcium
374 cal	54 g	4 g	17 g	0 mg	262 mg	11 mg

Ingredient	**Weight**	**Measure**	**Issue**
PIE CRUST		26 each	
PEACHES,CANNED,SLICED,JUICE PACK,INCL LIQUIDS	19-2/3 lbs	2 gal 1 qts	
RESERVED LIQUID	6-1/4 lbs	3 qts	
SUGAR,GRANULATED	3-3/4 lbs	2 qts 1/2 cup	
SALT	3/8 oz	1/3 tsp	
CORNSTARCH	9 oz	2 cup	
WATER,COLD	1-1/8 lbs	2-1/4 cup	

Method

1. PREPARE AND DIVIDE DOUGH: Prepare 1 recipe Pie Crust (Recipe No. I 001 00). Divide dough into 13-7-1/2 oz pieces for bottom crust and 13-7 oz pieces for top crust; place on lightly floured board. ROLL DOUGH: Sprinkle each piece of dough lightly with flour; flatten gently. Using a floured rolling pin, roll lightly with quick strokes from center out to edge in all directions. Form a circle 1 inch larger than pie pan and about 1/8 inch thick. Bottom crust will be slightly thicker. Shift or turn dough occasionally to prevent sticking. If edges split, pinch cracks together. BOTTOM CRUST: Fold rolled dough in half; carefully place into ungreased pie pan with fold at center. Unfold and fit carefully into pie pan, being careful not to leave any air spaces between pan and dough. FILL CRUST: Fill as specified on individual recipe card. TOP CRUST: Roll top crust in same manner as bottom crust. Fold in half; with knife, make several small slits near center fold to allow steam to escape during baking. Brush outer rim of bottom crust with water. Lay top crust over filling with fold at center; unfold and press edges of two crusts together lightly. REMOVE EXCESS DOUGH: Trim overhanging edges of dough by using a knife or spatula. (Incorporate excess dough into next crust, if needed.) There should be little excess if skill is used in weighing and rolling dough. SEAL PIE: Press edges of crust firmly together or crimp with the thumb and forefinger to make a fluted edge. BAKING INSTRUCTIONS FOR COOKED PIES: Bake as specified on individual recipe card.

2. Drain peaches; reserve juice for use in Step 3; peaches for use in Step 5.
3. Combine reserved juice, sugar, and salt; bring to a boil.
4. Combine cornstarch and water; stir until smooth. Add gradually to boiling mixture. Cook at medium heat, stirring constantly until thick and clear. Remove from heat.
5. Fold peaches carefully into thickened mixture. Cool.
6. Pour about 3 cups filling into each unbaked 9-inch pie shell. Cover with top crust. Seal edges.
7. Using a convection oven, bake at 375 F. for 20 to 25 minutes or until lightly browned on high fan, open vent.
8. Cut 8 wedges per pie.

DESSERTS (PASTRY AND PIES) No.I 024 01

PEACH PIE (PREPARED PIE FILLING)

Yield 100   Portion   1 Slice

Calories	Carbohydrates	Protein	Fat	Cholesterol	Sodium	Calcium
410 cal	64 g	4 g	17 g	0 mg	245 mg	19 mg

Ingredient   **Weight**   **Measure**   **Issue**

PIE CRUST   26 each
PIE FILLING,PEACH,PREPARED   22-3/4 lbs   2 gal 3-3/8 qts

Method

1. PREPARE AND DIVIDE DOUGH: Prepare 1 recipe Pie Crust (Recipe No. I 001 00). Divide dough into 13-7-1/2 oz pieces for bottom crust and 13-7 oz pieces for top crust; place on lightly floured board. ROLL DOUGH: Sprinkle each piece of dough lightly with flour; flatten gently. Using a floured rolling pin, roll lightly with quick strokes from center out to edge in all directions. Form a circle 1 inch larger than pie pan and about 1/8 inch thick. Bottom crust will be slightly thicker. Shift or turn dough occasionally to prevent sticking. If edges split, pinch cracks together. BOTTOM CRUST: Fold rolled dough in half; carefully place into ungreased pie pan with fold at center. Unfold and fit carefully into pie pan, being careful not to leave any air spaces between pan and dough. FILL CRUST: Fill as specified on individual recipe card. TOP CRUST: Roll top crust in same manner as bottom crust. Fold in half; with knife, make several small slits near center fold to allow steam to escape during baking. Brush outer rim of bottom crust with water. Lay top crust over filling with fold at center; unfold and press edges of two crusts together lightly. REMOVE EXCESS DOUGH: Trim overhanging edges of dough by using a knife or spatula. (Incorporate excess dough into next crust, if needed.) There should be little excess if skill is used in weighing and rolling dough. SEAL PIE: Press edges of crust firmly together or crimp with the thumb and forefinger to make a fluted edge. BAKING INSTRUCTIONS FOR COOKED PIES: Bake as specified on individual recipe card.
2. Pour 3 cups filling into each unbaked 9-inch pie shell. Cover with top crust. Seal edges.
3. Using a convection oven, bake at 375 F. for 20 to 25 minutes or until lightly browned on high fan, open vent.
4. Cut 8 wedges per pie.

DESSERTS (PASTRY AND PIES) No. I 026 00

CREAMY COCONUT PIE

Yield 100   Portion 1 Slice

Calories	Carbohydrates	Protein	Fat	Cholesterol	Sodium	Calcium
296 cal	23 g	4 g	21 g	23 mg	214 mg	43 mg

Ingredient	Weight	Measure	Issue
PIE CRUST		13 each	
MILK,NONFAT,DRY	3-5/8 oz	1-1/2 cup	
WATER,COLD	3-7/8 lbs	1 qts 3-1/2 cup	
MILK,NONFAT,DRY	2-3/8 oz	1 cup	
WATER,WARM	3 lbs	1 qts 1-5/8 cup	
CHEESE,CREAM,SOFTENED,ROOM TEMPERATURE	4-1/2 lbs	2 qts 3/4 cup	
SUGAR,GRANULATED	8 oz	1-1/8 cup	
COCONUT,PREPARED,SWEETENED FLAKES	3-1/8 lbs	3 qts 3 cup	
FLAVORING,ALMOND	1-3/8 oz	3 tbsp	
WHIPPED TOPPING MIX,NONDAIRY,DRY	5-2/3 oz	2 qts	

Method

1. PREPARE AND DIVIDE DOUGH: Prepare 1/2 recipe Pie Crust (Recipe No. I 001 00). Divide dough into 13-7-1/2 oz pieces for bottom crust; place on lightly floured board. ROLL DOUGH: Sprinkle each piece of dough lightly with flour; flatten gently. Using a floured rolling pin, roll lightly with quick strokes from center out to edge in all directions. Form a circle 1 inch larger than pie pan and about 1/8 inch thick. Shift or turn dough occasionally to prevent sticking. If edges split, pinch cracks together. BOTTOM CRUST: Fold rolled dough in half; carefully place into ungreased pie pan with fold at center. Unfold and fit carefully into pie pan, being careful not to leave any air spaces between pan and dough. REMOVE EXCESS DOUGH: Trim overhanging edges of dough by using a knife or spatula. (Incorporate excess dough into next crust, if needed.) There should be little excess if skill is used in weighing and rolling dough. BAKING INSTRUCTIONS FOR UNCOOKED PIES: Bake crusts at 425 F. for about 15-18 minutes, or until light golden brown. Cool before filling. Proceed with the recipe directions.
2. Combine milk and water in mixer bowl. CCP: Refrigerate at 41 F. or lower for use in Step 5.
3. Combine 2nd milk and 2nd water in mixer bowl.
4. Combine cream cheese, sugar, coconut and almond flavoring with milk in mixer bowl. Whip at low speed 1 minute; scrape down sides of bowl. Whip 3 minutes at low speed, or until well blended. Set aside for use in Step 6.
5. Place cold milk and water (from Step 2) in mixer bowl. Add topping. Blend 3 minutes at low speed. Scrape down sides of bowl. Whip at high speed about 5 to 10 minutes or until stiff peaks are formed.
6. Add whipped topping to cream cheese mixture. Blend at low speed 1 minute; scrape down sides of bowl. Blend at low speed 1 minute or until smooth.
7. Pour 4-2/3 cups filling into each crust.
8. Toasted coconut, chopped unsalted nuts, or chopped maraschino cherries may be sprinkled over pies before placing in freezer. Place pies in freezer 4 hours or until firm.
9. Let pies stand at room temperature 5 minutes before cutting. Cut 8 wedges per pie. CCP: Hold for service at 41 F. or lower.

Notes

1. 4 pound and 1 ounce preformed, graham cracker pie crusts may be used per 100 servings.

DESSERTS (PASTRY AND PIES) No.I 026 01

CREAMY BANANA COCONUT PIE

Yield 100   Portion 1 Slice

Calories	Carbohydrates	Protein	Fat	Cholesterol	Sodium	Calcium
307 cal	26 g	4 g	21 g	23 mg	214 mg	44 mg

Ingredient	Weight	Measure	Issue
PIE CRUST		13 each	
MILK,NONFAT,DRY	3-5/8 oz	1-1/2 cup	
WATER,COLD	3-7/8 lbs	1 qts 3-1/2 cup	
MILK,NONFAT,DRY	2-3/8 oz	1 cup	
WATER,WARM	3 lbs	1 qts 1-5/8 cup	
CHEESE,CREAM,SOFTENED,ROOM TEMPERATURE	4-1/2 lbs	2 qts 3/4 cup	
SUGAR,GRANULATED	8 oz	1-1/8 cup	
COCONUT,PREPARED,SWEETENED FLAKES	3-1/8 lbs	3 qts 3 cup	
BANANA,FRESH	3 lbs		4-5/8 lbs
WHIPPED TOPPING MIX,NONDAIRY,DRY	5-2/3 oz	2 qts	

Method

1. PREPARE AND DIVIDE DOUGH: Prepare 1/2 recipe Pie Crust (Recipe No. I 001 00). Divide dough into 13-7-1/2 oz pieces for bottom crust; place on lightly floured board. ROLL DOUGH: Sprinkle each piece of dough lightly with flour; flatten gently. Using a floured rolling pin, roll lightly with quick strokes from center out to edge in all directions. Form a circle 1 inch larger than pie pan and about 1/8 inch thick. Shift or turn dough occasionally to prevent sticking. If edges split, pinch cracks together. BOTTOM CRUST: Fold rolled dough in half; carefully place into ungreased pie pan with fold at center. Unfold and fit carefully into pie pan, being careful not to leave any air spaces between pan and dough. REMOVE EXCESS DOUGH: Trim overhanging edges of dough by using a knife or spatula. (Incorporate excess dough into next crust, if needed.) There should be little excess if skill is used in weighing and rolling dough. BAKING INSTRUCTIONS FOR UNCOOKED PIES: Bake crusts at 425 F. for about 15-18 minutes, or until light golden brown. Cool before filling. Proceed with the recipe directions.
2. Combine milk and water in mixer bowl. CCP: Refrigerate at 41 F. or lower for use in Step 5.
3. Combine 2nd milk and 2nd water in mixer bowl.
4. Combine cream cheese, sugar, coconut and peeled ripe bananas with milk in mixer bowl. Whip at low speed 1 minute; scrape down sides of bowl. Whip 3 minutes at low speed, or until well blended. Set aside for use in Step 6.
5. Place cold milk and water (from Step 2) in mixer bowl. Add topping. Blend 3 minutes at low speed. Scrape down sides of bowl. Whip at high speed about 5 to 10 minutes or until stiff peaks are formed.
6. Add whipped topping to cream cheese mixture. Blend at low speed 1 minute; scrape down sides of bowl. Blend at low speed 1 minute or until smooth.
7. Pour 1-1/4 quart filling into each 9-inch pie crust.
8. Toasted coconut, chopped unsalted nuts, or chopped maraschino cherries may be sprinkled over pies before placing in freezer. Place pies in freezer 4 hours or until firm.
9. Let pies stand at room temperature 5 minutes before cutting. Cut 8 wedges per pie. CCP: Hold for service at 41 F. or lower.

Notes

1. 13 5-ounce pie crusts, preformed, graham cracker pie crusts, may be used per 100 portions.

DESSERTS (PASTRY AND PIES) No.I 026 02

AMBROSIA PIE

Yield 100   Portion 1 Slice

Calories	Carbohydrates	Protein	Fat	Cholesterol	Sodium	Calcium
314 cal	28 g	4 g	21 g	23 mg	214 mg	44 mg

Ingredient	Weight	Measure	Issue
PIE CRUST		13 each	
MILK,NONFAT,DRY	3-5/8 oz	1-1/2 cup	
WATER,COLD	3-7/8 lbs	1 qts 3-1/2 cup	
JUICE,ORANGE	5-1/2 lbs	2 qts 2 cup	
MILK,NONFAT,DRY	2-3/8 oz	1 cup	
CHEESE,CREAM	4-1/2 lbs	2 qts 3/4 cup	
SUGAR,GRANULATED	1 lbs	2-1/4 cup	
COCONUT,PREPARED,SWEETENED FLAKES	3-1/8 lbs	3 qts 3 cup	
FOOD COLOR,YELLOW	1/4 oz	1/4 tsp	
FOOD COLOR,RED	1/8 oz	1/8 tsp	
WHIPPED TOPPING MIX,NONDAIRY,DRY	5-2/3 oz	2 qts	

Method

1. PREPARE AND DIVIDE DOUGH: Prepare 1/2 recipe Pie Crust (Recipe No. I 001 00). Divide dough into 13-7-1/2 oz pieces for bottom crust; place on lightly floured board. ROLL DOUGH: Sprinkle each piece of dough lightly with flour; flatten gently. Using a floured rolling pin, roll lightly with quick strokes from center out to edge in all directions. Form a circle 1 inch larger than pie pan and about 1/8 inch thick. Shift or turn dough occasionally to prevent sticking. If edges split, pinch cracks together. BOTTOM CRUST: Fold rolled dough in half; carefully place into ungreased pie pan with fold at center. Unfold and fit carefully into pie pan, being careful not to leave any air spaces between pan and dough. REMOVE EXCESS DOUGH: Trim overhanging edges of dough by using a knife or spatula. (Incorporate excess dough into next crust, if needed.) There should be little excess if skill is used in weighing and rolling dough. BAKING INSTRUCTIONS FOR UNCOOKED PIES: Bake crusts at 425 F. for about 15-18 minutes, or until light golden brown. Cool before filling. Proceed with the recipe directions.
2. Combine milk and water in mixer bowl. CCP: Refrigerate at 41 F. or lower for use in Step 5.
3. Combine nonfat dry milk with orange juice.
4. Combine cream cheese, sugar, and coconut with milk in mixer bowl. Add yellow and red food coloring. Whip at low speed 1 minute; scrape down sides of bowl. Whip 3 minutes at low speed, or until well blended. Set aside for use in Step 6.
5. Place cold milk and water (from Step 2) in mixer bowl. Add topping; blend 3 minutes at low speed. Scrape down sides of bowl. Whip at high speed about 5 to 10 minutes or until stiff peaks are formed.
6. Add whipped topping to cream cheese mixture. Blend at low speed 1 minute; scrape down sides of bowl. Blend at low speed 1 minute or until smooth.
7. Pour 4-2/3 cups filling into each 9-inch pie crust.
8. Toasted coconut, chopped unsalted nuts, or chopped maraschino cherries may be sprinkled over pies before placing in freezer. Place pies in freezer 4 hours or until firm.
9. Let pies stand at room temperature 5 minutes before cutting. Cut 8 wedges per pie. CCP: Hold for service at 41 F. or lower.

Notes

1. 13-5 ounce pie crusts, preformed, graham cracker pie crusts, may be used.

DESSERTS (PASTRY AND PIES) No.I 027 00

CHERRY CRUMBLE PIE

Yield 100   Portion   1 Slice

Calories	Carbohydrates	Protein	Fat	Cholesterol	Sodium	Calcium
456 cal	77 g	4 g	16 g	0 mg	185 mg	16 mg

Ingredient   Weight   Measure   Issue

Ingredient	Weight	Measure
FLOUR,WHEAT,GENERAL PURPOSE	8 lbs	1 gal 3-1/4 qts
SALT	1-1/2 oz	2-1/3 tbsp
SUGAR,GRANULATED	4-3/8 lbs	2 qts 2 cup
SHORTENING	3-1/8 lbs	1 qts 3 cup
CHERRIES,CANNED,RED,TART,WATER PACK,INCL LIQUIDS	19-2/3 lbs	2 gal 1 qts
SUGAR,GRANULATED	1-1/2 lbs	3-3/8 cup
SUGAR,GRANULATED	2-1/4 lbs	1 qts 1 cup
CORNSTARCH	7-7/8 oz	1-3/4 cup
SALT	1/8 oz	1/8 tsp
WATER,COLD	1 lbs	2 cup
MARGARINE	3 oz	1/4 cup 2-1/3 tbsp
FOOD COLOR,RED	1/4 oz	1/4 tsp

Method

1. Mix flour, salt, sugar, and shortening in a mixer bowl 1 minute at low speed to form a crumbly mixture.
2. Place 1-1/2 cups of mixture in each pan; press firmly into an even layer against bottom and sides of pan. Set remaining crumb mixture aside for use in Step 5.
3. Drain cherries. Set aside juice for use in Step 7.
4. Combine cherries and sugar. Spread 2 cups mixture over crumbs in each pan.
5. Spread 1 cup reserved crumb mixture over cherries in each 9-inch pan.
6. Using a convection oven, bake 35 to 40 minutes at 350 F. or until done on low fan, open vent.
7. Take reserved juice add water to equal 1 gallon per 100 portions and combine with sugar; bring to a boil.
8. Combine cornstarch, salt, and water; stir until smooth. Add gradually to boiling mixture. Stir until well blended; cook at medium heat about 5 minutes. Add margarine or butter and food coloring.
9. Pour 1-1/2 cups of sauce over each baked pie.
10. Cool; cut 8 wedges per pie.

DESSERTS (PASTRY AND PIES) No. I 028 00

CHOCOLATE CREAM PIE

Yield 100   Portion 1 Slice

Calories	Carbohydrates	Protein	Fat	Cholesterol	Sodium	Calcium
333 cal	43 g	5 g	16 g	31 mg	257 mg	60 mg

Ingredient	Weight	Measure	Issue
PIE CRUST		13 each	
MILK,NONFAT,DRY	13-1/4 oz	1 qts 1-1/2 cup	
WATER,WARM	14-1/8 lbs	1 gal 2-3/4 qts	
SUGAR,GRANULATED	1-7/8 lbs	1 qts 1/4 cup	
SALT	3/4 oz	1 tbsp	
CORNSTARCH	1-1/8 lbs	1 qts	
SUGAR,GRANULATED	1-7/8 lbs	1 qts 1/4 cup	
COCOA	9-7/8 oz	3-1/4 cup	
WATER,COLD	1-1/3 lbs	2-1/2 cup	
EGGS,WHOLE,FROZEN	1-1/2 lbs	2-7/8 cup	
MARGARINE	10-5/8 oz	1-3/8 cup	
EXTRACT,VANILLA	1 oz	2-1/3 tbsp	

Method

1. PREPARE AND DIVIDE DOUGH: Prepare 1/2 recipe Pie Crust (Recipe No. I 001 00). Divide dough into 13-7-1/2 oz pieces for bottom crust; place on lightly floured board. ROLL DOUGH: Sprinkle each piece of dough lightly with flour; flatten gently. Using a floured rolling pin, roll lightly with quick strokes from center out to edge in all directions. Form a circle 1 inch larger than pie pan and about 1/8 inch thick. Shift or turn dough occasionally to prevent sticking. If edges split, pinch cracks together. BOTTOM CRUST: Fold rolled dough in half; carefully place into ungreased pie pan with fold at center. Unfold and fit carefully into pie pan, being careful not to leave any air spaces between pan and dough. REMOVE EXCESS DOUGH: Trim overhanging edges of dough by using a knife or spatula. (Incorporate excess dough into next crust, if needed.) There should be little excess if skill is used in weighing and rolling dough. BAKING INSTRUCTIONS FOR UNCOOKED PIES: Bake crusts at 425 F. for about 15-18 minutes, or until light golden brown. Cool before filling. Proceed with the recipe directions.
2. Reconstitute milk. Add sugar and salt; heat to just below boiling. DO NOT BOIL.
3. Combine cornstarch, sugar, cocoa, and water; stir until smooth. Add gradually to hot mixture. Cook at medium heat, stirring constantly, about 10 minutes until thickened.
4. Stir 1 quart of hot mixture into eggs. Slowly pour egg mixture into remaining hot mixture; heat to boiling stirring constantly. Cook 2 minutes longer. Remove from heat.
5. Add margarine or butter and vanilla; stir until well blended. Cool slightly.
6. Pour 3 cups of filling into each 9-inch baked pie shell. Meringue, Recipe No. I 005 00 or I 005 01 may be spread over chilled filling, about 50 F. Ensure cream pie filling preparation time does not exceed 4 hours total in temperatures between 40 F. to 140 F.
7. Refrigerate until ready to serve. CCP: Hold for service at 41 F. or lower.
8. Cut 8 wedges per pie. Chilled pies may be topped with Whipped Cream, Recipe No. K 001 00 or Whipped Topping, Recipe No. K 002 00.

Notes

1. Filling will curdle or scorch if boiled or subjected to prolonged intense heat.

DESSERTS (PASTRY AND PIES) No. I 028 01
CHOCOLATE CREAM PIE (DESSERT POWDER, INSTANT)

Yield 100   Portion 1 Slice

Calories	Carbohydrates	Protein	Fat	Cholesterol	Sodium	Calcium
331 cal	50 g	4 g	14 g	1 mg	620 mg	64 mg

Ingredient	Weight	Measure	Issue
PIE CRUST		13 each	
MILK,NONFAT,DRY	15 oz	1 qts 2-1/4 cup	
WATER,COLD	16-3/4 lbs	2 gal	
DESSERT POWDER,PUDDING,INSTANT,CHOCOLATE	7-1/2 lbs	1 gal 1 qts	

Method

1. PREPARE AND DIVIDE DOUGH: Prepare 1/2 recipe Pie Crust (Recipe No. I 001 00). Divide dough into 13-7-1/2 oz pieces for bottom crust; place on lightly floured board. ROLL DOUGH: Sprinkle each piece of dough lightly with flour; flatten gently. Using a floured rolling pin, roll lightly with quick strokes from center out to edge in all directions. Form a circle 1 inch larger than pie pan and about 1/8 inch thick. Shift or turn dough occasionally to prevent sticking. If edges split, pinch cracks together. BOTTOM CRUST: Fold rolled dough in half; carefully place into ungreased pie pan with fold at center. Unfold and fit carefully into pie pan, being careful not to leave any air spaces between pan and dough. REMOVE EXCESS DOUGH: Trim overhanging edges of dough by using a knife or spatula. (Incorporate excess dough into next crust, if needed.) There should be little excess if skill is used in weighing and rolling dough. BAKING INSTRUCTIONS FOR UNCOOKED PIES: Bake crusts at 425 F. for about 15-18 minutes, or until light golden brown. Cool before filling. Proceed with the recipe directions.
2. Combine nonfat dry milk and cold water, 50 F. in mixer bowl. Add dessert powder pudding, instant, chocolate to milk and water.
3. Using whip, blend at low speed for 15 seconds or until well blended.
4. Scrape down sides of bowl; whip at medium speed 2 minutes.
5. Pour 3 cups filling into each baked 9-inch pie shell. Meringue, Recipe No. I 005 00 or I 005 01 may be spread over chilled filling, about 50 F. Ensure cream pie filling preparation time does not exceed 4 hours total in temperatures between 40 F. to 140 F.
6. Refrigerate until ready to serve.
7. Chilled pies may be topped with Whipped Cream, Recipe No. K 001 00 or Whipped Topping, Recipe No. K 002 00. Cut 8 wedges per pie. CCP: Hold for service at 41 F. or lower.

Notes

1. Filling will curdle or scorch if boiled or subjected to prolonged intense heat.

DESSERTS (PASTRY AND PIES) No.I 029 00

CHOCOLATE AND VANILLA CREAM PIE (INSTANT)

Yield 100   Portion 1 Slice

Calories	Carbohydrates	Protein	Fat	Cholesterol	Sodium	Calcium
290 cal	40 g	4 g	13 g	1 mg	446 mg	58 mg

Ingredient	Weight	Measure	Issue
PIE CRUST		13 each	
MILK,NONFAT,DRY	14-3/8 oz	1 qts 2 cup	
WATER,COLD	15-2/3 lbs	1 gal 3-1/2 qts	
DESSERT POWDER,PUDDING,INSTANT,CHOCOLATE	2-1/4 lbs	1 qts 2 cup	
DESSERT POWDER,PUDDING,INSTANT,VANILLA	2-1/2 lbs	1 qts 2 cup	
WHIPPED TOPPING MIX,NONDAIRY,DRY	1 oz	1-1/2 cup	
SUGAR,GRANULATED	5/8 oz	1 tbsp	
EXTRACT,VANILLA	5/8 oz	1 tbsp	

Method

1. PREPARE AND DIVIDE DOUGH: Prepare 1/2 recipe Pie Crust (Recipe No. I 001 00). Divide dough into 13-7-1/2 oz pieces for bottom crust; place on lightly floured board. ROLL DOUGH: Sprinkle each piece of dough lightly with flour; flatten gently. Using a floured rolling pin, roll lightly with quick strokes from center out to edge in all directions. Form a circle 1 inch larger than pie pan and about 1/8 inch thick. Shift or turn dough occasionally to prevent sticking. If edges split, pinch cracks together. BOTTOM CRUST: Fold rolled dough in half; carefully place into ungreased pie pan with fold at center. Unfold and fit carefully into pie pan, being careful not to leave any air spaces between pan and dough. REMOVE EXCESS DOUGH: Trim overhanging edges of dough by using a knife or spatula. (Incorporate excess dough into next crust, if needed.) There should be little excess if skill is used in weighing and rolling dough. BAKING INSTRUCTIONS FOR UNCOOKED PIES: Bake crusts at 425 F. for about 15-18 minutes, or until light golden brown. Cool before filling. Proceed with the recipe directions.
2. Combine milk and water in mixer bowl.
3. Pour 3-1/2 quarts chilled milk into mixer bowl; add dessert powder. Using whip, blend at low speed 15 seconds or until well blended. Scrape down sides of bowl. Whip at medium speed 2 minutes or until smooth.
4. Pour 1-1/3 cups filling into each baked pie shell.
5. Pour 1 gallon chilled milk into mixer bowl; add dessert powder. Using whip, blend 15 seconds at low speed or until well blended. Scrape down sides of bowl. Whip at medium speed 2 minutes or until smooth. Set aside for use in Step 7.
6. Pour 1-1/2 cups chilled milk into mixer bowl; add topping, sugar and vanilla. Whip at low speed 3 minutes or until blended. Scrape down sides of bowl. Whip at high speed until stiff.
7. Fold whipped topping into vanilla pie filling. Spread 1-3/4 cups over chocolate filling in each baked pie shell.
8. Refrigerate at least 1 hour or until ready to serve.
9. Cut 8 wedges per pie. CCP: Hold for service at 41 F. or lower. Chilled pies may be topped with Whipped Cream, Recipe No. K 001 00 or Whipped Topping, Recipe No. K 002 00.

Notes

1. 1 pound 5 ounces canned dessert topping and frozen bakery products, may be used. Omit Step 6.

DESSERTS (PASTRY AND PIES) No.I 030 00

FRIED APPLE PIE

Yield 100   **Portion** 1 Pie

Calories	Carbohydrates	Protein	Fat	Cholesterol	Sodium	Calcium
366 cal	52 g	5 g	16 g	0 mg	340 mg	73 mg

Ingredient	Weight	Measure	Issue
FLOUR,WHEAT,GENERAL PURPOSE	9-7/8 lbs	2 gal 1 qts	
MILK,NONFAT,DRY	2-2/3 oz	1-1/8 cup	
BAKING POWDER	3-1/4 oz	1/4 cup 3 tbsp	
SALT	1-7/8 oz	3 tbsp	
SHORTENING	1-1/3 lbs	3 cup	
WATER	2-7/8 lbs	1 qts 1-1/2 cup	
PIE FILLING,APPLE,PREPARED	14 lbs	1 gal 3 qts	

Method

1. Sift together flour, milk, baking powder, and salt into mixer bowl.
2. Blend shortening into dry ingredients at low speed until mixture resembles coarse crumbs.
3. Add water; mix at low speed only enough to form soft dough.
4. On lightly floured board, roll dough into a rectangular sheet, about 1/8-inch thick. Cut into 6 circles.
5. Place 1/4 cup filling in the center of each circle. Wash edges of circle with water. Fold over to form a half circle; seal edges with a fork.
6. Fry pies, a few at a time, 2 minutes on one side, turn and fry 2 minutes on other side until golden brown. Drain on absorbent paper.

Notes

1. Pie crust mix may be used. Omit steps 1 through 3. Use 6 pounds 14 ounces of pie crust mix. Follow manufacturer's directions for mixing. Follow Steps 4 through 6.

DESSERTS (PASTRY AND PIES) No.I 030 01

FRIED LEMON PIE

Yield 100   **Portion** 1 Pie

Calories	Carbohydrates	Protein	Fat	Cholesterol	Sodium	Calcium
349 cal	47 g	5 g	16 g	0 mg	338 mg	76 mg

Ingredient	Weight	Measure	Issue
FLOUR,WHEAT,GENERAL PURPOSE	9-7/8 lbs	2 gal 1 qts	
MILK,NONFAT,DRY	2-2/3 oz	1-1/8 cup	
BAKING POWDER	3-1/4 oz	1/4 cup 3 tbsp	
SALT	1-7/8 oz	3 tbsp	
SHORTENING	1-1/3 lbs	3 cup	
WATER	2-7/8 lbs	1 qts 1-1/2 cup	
PIE FILLING,LEMON,PREPARED	14 lbs	1 gal 3 qts	

Method

1. Sift together flour, milk, baking powder, and salt into mixer bowl.
2. Blend shortening into dry ingredients at low speed until mixture resembles coarse crumbs.
3. Add water; mix at low speed only enough to form soft dough.
4. On lightly floured board, roll dough into a rectangular sheet, about 1/8-inch thick. Cut into 6 circles.
5. Place 1/4 cup filling in the center of each circle. Wash edges of circle with water. Fold over to form a half circle; seal edges with a fork.
6. Fry pies, a few at a time, 2 minutes on one side, turn and fry 2 minutes on other side until golden brown. Drain on absorbent paper.

Notes

1. Pie crust mix may be used. Omit steps 1 through 3. Use 6 pounds 14 ounces of pie crust mix. Follow manufacturer's directions for mixing. Follow Steps 4 through 6.

DESSERTS (PASTRY AND PIES) No.I 030 02

FRIED CHERRY PIE

Yield 100   **Portion** 1 Pie

Calories	Carbohydrates	Protein	Fat	Cholesterol	Sodium	Calcium
375 cal	54 g	5 g	16 g	0 mg	318 mg	78 mg

Ingredient	Weight	Measure	Issue
FLOUR,WHEAT,GENERAL PURPOSE	9-7/8 lbs	2 gal 1 qts	
MILK,NONFAT,DRY	2-2/3 oz	1-1/8 cup	
BAKING POWDER	3-1/4 oz	1/4 cup 3 tbsp	
SALT	1-7/8 oz	3 tbsp	
SHORTENING	1-1/3 lbs	3 cup	
WATER	2-7/8 lbs	1 qts 1-1/2 cup	
PIE FILLING,CHERRY,PREPARED	14 lbs	1 gal 3 qts	

Method

1. Sift together flour, milk, baking powder, and salt into mixer bowl.
2. Blend shortening into dry ingredients at low speed until mixture resembles coarse crumbs.
3. Add water; mix at low speed only enough to form soft dough.
4. On lightly floured board, roll dough into a rectangular sheet, about 1/8-inch thick. Cut into 6 circles.
5. Place 1/4 cup filling in the center of each circle. Wash edges of circle with water. Fold over to form a half circle; seal edges with a fork.
6. Fry pies, a few at a time, 2 minutes on one side, turn and fry 2 minutes on other side until golden brown. Drain on absorbent paper.

Notes

1. Pie crust mix may be used. Omit steps 1 through 3. Use 6 pounds 14 ounces of pie crust mix. Follow manufacturer's directions for mixing. Follow Steps 4 through 6.

DESSERTS (PASTRY AND PIES) No.I 030 03

FRIED PEACH PIE

Yield 100   Portion   1 Pie

Calories	Carbohydrates	Protein	Fat	Cholesterol	Sodium	Calcium
395 cal	59 g	5 g	16 g	0 mg	333 mg	79 mg

Ingredient	Weight	Measure	Issue
FLOUR,WHEAT,GENERAL PURPOSE	9-7/8 lbs	2 gal 1 qts	
MILK,NONFAT,DRY	2-2/3 oz	1-1/8 cup	
BAKING POWDER	3-1/4 oz	1/4 cup 3 tbsp	
SALT	1-7/8 oz	3 tbsp	
SHORTENING	1-1/3 lbs	3 cup	
WATER	2-7/8 lbs	1 qts 1-1/2 cup	
PIE FILLING,PEACH,PREPARED	14 lbs	1 gal 3 qts	

Method

1 Sift together flour, milk, baking powder, and salt into mixer bowl.
2 Blend shortening into dry ingredients at low speed until mixture resembles coarse crumbs.
3 Add water; mix at low speed only enough to form soft dough.
4 On lightly floured board, roll dough into a rectangular sheet, about 1/8-inch thick. Cut into 6 circles.
5 Place 1/4 cup filling in the center of each circle. Wash edges of circle with water. Fold over to form a half circle; seal edges with a fork.
6 Fry pies, a few at a time, 2 minutes on one side, turn and fry 2 minutes on other side until golden brown. Drain on absorbent paper.

Notes

1 Pie crust mix may be used. Omit steps 1 through 3. Use 6 pounds 14 ounces of pie crust mix. Follow manufacturer's directions for mixing. Follow Steps 4 through 6.

DESSERTS (PASTRY AND PIES) No.I 030 04

FRIED BLUEBERRY PIE

Yield 100 **Portion** 1 Pie

Calories	Carbohydrates	Protein	Fat	Cholesterol	Sodium	Calcium
358 cal	50 g	5 g	16 g	0 mg	344 mg	85 mg

Ingredient	Weight	Measure	Issue
FLOUR,WHEAT,GENERAL PURPOSE	9-7/8 lbs	2 gal 1 qts	
MILK,NONFAT,DRY	2-2/3 oz	1-1/8 cup	
BAKING POWDER	3-1/4 oz	1/4 cup 3 tbsp	
SALT	1-7/8 oz	3 tbsp	
SHORTENING	1-1/3 lbs	3 cup	
WATER	2-7/8 lbs	1 qts 1-1/2 cup	
PIE FILLING,BLUEBERRY,PREPARED	14 lbs	1 gal 2 qts	

Method

1. Sift together flour, milk, baking powder, and salt into mixer bowl.
2. Blend shortening into dry ingredients at low speed until mixture resembles coarse crumbs.
3. Add water; mix at low speed only enough to form soft dough.
4. On lightly floured board, roll dough into a rectangular sheet, about 1/8-inch thick. Cut into 6 circles.
5. Place 1/4 cup filling in the center of each circle. Wash edges of circle with water. Fold over to form a half circle; seal edges with a fork.
6. Fry pies, a few at a time, 2 minutes on one side, turn and fry 2 minutes on other side until golden brown. Drain on absorbent paper.

Notes

1. Pie crust mix may be used. Omit steps 1 through 3. Use 6 pounds 14 ounces of pie crust mix. Follow manufacturer's directions for mixing. Follow Steps 4 through 6.

DESSERTS (PASTRY AND PIES) No. I 031 00

PECAN PIE

Yield 100   Portion 1 Slice

Calories	Carbohydrates	Protein	Fat	Cholesterol	Sodium	Calcium
504 cal	77 g	6 g	21 g	126 mg	396 mg	25 mg

Ingredient	Weight	Measure	Issue
PIE CRUST		13 each	
EGGS,WHOLE,FROZEN	6 lbs	2 qts 3-1/4 cup	
SUGAR,GRANULATED	4-7/8 lbs	2 qts 3 cup	
BUTTER,MELTED	12 oz	1-1/2 cup	
CORN SYRUP,LIGHT	11-5/8 lbs	1 gal	
EXTRACT,VANILLA	1-7/8 oz	1/4 cup 1/3 tbsp	
SALT	1-1/2 oz	2-1/3 tbsp	
PECANS,CHOPPED	2-1/2 lbs		

Method

1 PREPARE AND DIVIDE DOUGH: Prepare 1/2 recipe Pie Crust (Recipe No. I 001 00). Divide dough into 13-7-1/2 oz pieces for bottom crust; place on lightly floured board. ROLL DOUGH: Sprinkle each piece of dough lightly with flour; flatten gently. Using a floured rolling pin, roll lightly with quick strokes from center out to edge in all directions. Form a circle 1 inch larger than pie pan and about 1/8 inch thick. Shift or turn dough occasionally to prevent sticking. If edges split, pinch cracks together. BOTTOM CRUST: Fold rolled dough in half; carefully place into ungreased pie pan with fold at center. Unfold and fit carefully into pie pan, being careful not to leave any air spaces between pan and dough. REMOVE EXCESS DOUGH: Trim overhanging edges of dough by using a knife or spatula. (Incorporate excess dough into next crust, if needed.) There should be little excess if skill is used in weighing and rolling dough.

2 Place eggs in mixer bowl; add sugar gradually while beating at low speed. Add butter or margarine; mix thoroughly.

3 Add corn syrup, vanilla, and salt; beat at low speed until smooth.

4 Place 3/4 cup pecans into each unbaked pie shell.

5 Pour 2-3/4 cups filling over pecans in each 9-inch pie pan.

6 Bake at 350 F. for 35 minutes or until filling is set. DO NOT OVERBAKE.

7 Refrigerate until ready to serve.

8 Cut 8 wedges per pie. CCP: Hold for service at 41 F. or lower.

DESSERTS (PASTRY AND PIES) No. I 031 01

WALNUT PIE

Yield 100   **Portion** 1 Slice

Calories	Carbohydrates	Protein	Fat	Cholesterol	Sodium	Calcium
502 cal	76 g	6 g	21 g	126 mg	397 mg	32 mg

Ingredient	Weight	Measure	Issue
PIE CRUST		13 each	
EGGS,WHOLE,FROZEN	6 lbs	2 qts 3-1/4 cup	
SUGAR,GRANULATED	4-7/8 lbs	2 qts 3 cup	
BUTTER,MELTED	12 oz	1-1/2 cup	
CORN SYRUP,LIGHT	11-5/8 lbs	1 gal	
EXTRACT,VANILLA	1-7/8 oz	1/4 cup 1/3 tbsp	
SALT	1-1/2 oz	2-1/3 tbsp	
WALNUTS,SHELLED,CHOPPED	2-1/2 lbs	2 qts 1-1/2 cup	

Method

1 PREPARE AND DIVIDE DOUGH: Prepare 1/2 recipe Pie Crust (Recipe No. I 001 00). Divide dough into 13-7-1/2 oz pieces for bottom crust; place on lightly floured board. ROLL DOUGH: Sprinkle each piece of dough lightly with flour; flatten gently. Using a floured rolling pin, roll lightly with quick strokes from center out to edge in all directions. Form a circle 1 inch larger than pie pan and about 1/8 inch thick. Shift or turn dough occasionally to prevent sticking. If edges split, pinch cracks together. BOTTOM CRUST: Fold rolled dough in half; carefully place into ungreased pie pan with fold at center. Unfold and fit carefully into pie pan, being careful not to leave any air spaces between pan and dough. REMOVE EXCESS DOUGH: Trim overhanging edges of dough by using a knife or spatula. (Incorporate excess dough into next crust, if needed.) There should be little excess if skill is used in weighing and rolling dough.

2 Place eggs in mixer bowl; add sugar gradually while beating at low speed. Add butter or margarine; mix thoroughly.

3 Add corn syrup, vanilla, and salt; beat at low speed until smooth.

4 Place 3/4 cup chopped walnuts into each unbaked pie shell.

5 Pour 2-3/4 cups filling over walnuts in each 9-inch pie pan.

6 Bake at 350 F. for 35 minutes or until filling is set. DO NOT OVERBAKE.

7 Refrigerate until ready to serve.

8 Cut 8 wedges per pie. CCP: Hold for service at 41 F. or lower.

DESSERTS (PASTRY AND PIES) No. I 032 00
LEMON CHIFFON PIE

Yield 100   Portion 1 Slice

Calories	Carbohydrates	Protein	Fat	Cholesterol	Sodium	Calcium
216 cal	30 g	3 g	10 g	0 mg	151 mg	10 mg

Ingredient	Weight	Measure	Issue
PIE CRUST		13 each	
DESSERT POWDER,GELATIN,LEMON	3-1/4 lbs	1 qts 2-1/2 cup	
SUGAR,GRANULATED	5-1/4 oz	3/4 cup	
WATER,BOILING	5-1/2 lbs	2 qts 2-1/2 cup	
WATER,COLD	3-1/8 lbs	1 qts 2 cup	
JUICE,LEMON	12-7/8 oz	1-1/2 cup	
WATER,COLD	1-1/2 lbs	2-7/8 cup	
WHIPPED TOPPING MIX,NONDAIRY,DRY	12 oz	1 gal 1/4 qts	
MILK,NONFAT,DRY	1-1/3 oz	1/2 cup 1 tbsp	
SUGAR,GRANULATED	2-2/3 oz	1/4 cup 2-1/3 tbsp	
EXTRACT,VANILLA	3/4 oz	1 tbsp	
LEMON RIND,GRATED	7/8 oz	1/4 cup 1/3 tbsp	

Method

1. PREPARE AND DIVIDE DOUGH: Prepare 1/2 recipe Pie Crust (Recipe No. I 001 00). Divide dough into 13-7/1-2 oz pieces for bottom crust; place on lightly floured board. ROLL DOUGH: Sprinkle each piece of dough lightly with flour; flatten gently. Using a floured rolling pin, roll lightly with quick strokes from center out to edge in all directions. Form a circle 1 inch larger than pie pan and about 1/8 inch thick. Shift or turn dough occasionally to prevent sticking. If edges split, pinch cracks together. BOTTOM CRUST: Fold rolled dough in half; carefully place into ungreased pie pan with fold at center. Unfold and fit carefully into pie pan, being careful not to leave any air spaces between pan and dough. REMOVE EXCESS DOUGH: Trim overhanging edges of dough by using a knife or spatula. (Incorporate excess dough into next crust, if needed.) There should be little excess if skill is used in weighing and rolling dough. BAKING INSTRUCTIONS FOR UNCOOKED PIES: Bake crusts at 425 F. for about 15-18 minutes, or until light golden brown. Cool before filling. Proceed with the recipe directions.
2. Dissolve gelatin and sugar in boiling water; add cold water. Mix until well blended.
3. Add juice to gelatin mixture; mix until blended.
4. Refrigerate until gelatin is thickened but not firm.
5. Pour cold water into chilled mixer bowl; add topping, milk, sugar, and vanilla. Using whip, beat at low speed 3 minutes or until well blended. Scrape down whip and bowl. Whip at high speed 5 to 10 minutes or until mixture forms stiff peaks. Set aside for use in Step 7.
6. Using whip, beat thickened gelatin at high speed 10 minutes or until foamy and soft peaks form.
7. Fold whipped topping and lemon rind into gelatin. Mix carefully at low speed until well blended.
8. Pour 1-1/4 quart filling into each baked pie shell.
9. Refrigerate about 2 hours or until set. CCP: Hold for service at 41 F. or lower.
10. Cut 8 wedges per pie.

Notes

1. In Step 5, 2 pound 10 ounces of canned dessert topping and frozen bakery products may be used for all ingredients, per 100 servings.

DESSERTS (PASTRY AND PIES) No. I 032 01
PINEAPPLE CHIFFON PIE

Yield 100   **Portion** 1 Slice

Calories	Carbohydrates	Protein	Fat	Cholesterol	Sodium	Calcium
216 cal	30 g	3 g	10 g	0 mg	149 mg	10 mg

Ingredient	Weight	Measure	Issue
PIE CRUST		13 each	
DESSERT POWDER,GELATIN,LEMON	3-1/4 lbs	1 qts 2-1/2 cup	
WATER	4-2/3 lbs	2 qts 1 cup	
WATER,COLD	1-1/2 lbs	2-7/8 cup	
MILK,NONFAT,DRY	1-1/3 oz	1/2 cup 1 tbsp	
SUGAR,GRANULATED	2-2/3 oz	1/4 cup 2-1/3 tbsp	
EXTRACT,VANILLA	3/4 oz	1 tbsp	
PINEAPPLE,CANNED,CRUSHED,JUICE PACK,DRAINED	2-1/2 lbs	1 qts 1/2 cup	
WHIPPED TOPPING MIX,NONDAIRY,DRY	12 oz	1 gal 1/4 qts	

Method

1 PREPARE AND DIVIDE DOUGH: Prepare 1/2 recipe Pie Crust (Recipe No. I 001 00). Divide dough into 13-7-1/2 oz pieces for bottom crust; place on lightly floured board. ROLL DOUGH: Sprinkle each piece of dough lightly with flour; flatten gently. Using a floured rolling pin, roll lightly with quick strokes from center out to edge in all directions. Form a circle 1 inch larger than pie pan and about 1/8 inch thick. Shift or turn dough occasionally to prevent sticking. If edges split, pinch cracks together. BOTTOM CRUST: Fold rolled dough in half; carefully place into ungreased pie pan with fold at center. Unfold and fit carefully into pie pan, being careful not to leave any air spaces between pan and dough. REMOVE EXCESS DOUGH: Trim overhanging edges of dough by using a knife or spatula. (Incorporate excess dough into next crust, if needed.) There should be little excess if skill is used in weighing and rolling dough. BAKING INSTRUCTIONS FOR UNCOOKED PIES: Bake crusts at 425 F. for about 15-18 minutes, or until light golden brown. Cool before filling. Proceed with the recipe directions.
2 Dissolve gelatin in boiling water; add cold water. Mix until well blended.
3 Refrigerate until gelatin is thickened but not firm.
4 Pour cold water into chilled mixer bowl; add topping, milk, sugar, and vanilla. Using whip, beat at low speed 3 minutes or until well blended. Scrape down whip and bowl. Whip at high speed 5 to 10 minutes or until mixture forms stiff peaks. Set aside for use in Step 7.
5 Using whip, beat thickened gelatin at high speed 10 minutes or until foamy and soft peaks form.
6 Fold whipped topping and drained pineapple into gelatin. Mix carefully at low speed until well blended.
7 Pour 5-3/4 cups filling into each baked pie shell.
8 Refrigerate about 2 hours or until set. CCP: Hold for service at 41 F. or lower.
9 Cut 8 wedges per pie.

Notes

1 In Step 5, 2 pound 10 ounces of canned dessert topping and frozen bakery products may be used for all ingredients, per 100 servings.

DESSERTS (PASTRY AND PIES) No. I 032 02

STRAWBERRY CHIFFON PIE

Yield 100   Portion 1 Slice

Calories	Carbohydrates	Protein	Fat	Cholesterol	Sodium	Calcium
209 cal	28 g	3 g	10 g	0 mg	145 mg	13 mg

Ingredient	Weight	Measure	Issue
PIE CRUST		13 each	
DESSERT POWDER,GELATIN,STRAWBERRY	2 lbs	1 qts 1-1/2 cup	
WATER,BOILING	5-1/2 lbs	2 qts 2-1/2 cup	
WATER,COLD	4-2/3 lbs	2 qts 1 cup	
WATER,COLD	1-1/2 lbs	2-7/8 cup	
MILK,NONFAT,DRY	1-1/3 oz	1/2 cup 1 tbsp	
SUGAR,GRANULATED	2-2/3 oz	1/4 cup 2-1/3 tbsp	
EXTRACT,VANILLA	3/4 oz	1 tbsp	
WHIPPED TOPPING MIX,NONDAIRY,DRY	12 oz	1 gal 1/4 qts	
STRAWBERRIES,FROZEN,THAWED	5 lbs	2 qts 1 cup	

Method

1. PREPARE AND DIVIDE DOUGH: Prepare 1/2 recipe Pie Crust (Recipe No. I 001 00). Divide dough into 13-7-1/2 oz pieces for bottom crust; place on lightly floured board. ROLL DOUGH: Sprinkle each piece of dough lightly with flour; flatten gently. Using a floured rolling pin, roll lightly with quick strokes from center out to edge in all directions. Form a circle 1 inch larger than pie pan and about 1/8 inch thick. Shift or turn dough occasionally to prevent sticking. If edges split, pinch cracks together. BOTTOM CRUST: Fold rolled dough in half; carefully place into ungreased pie pan with fold at center. Unfold and fit carefully into pie pan, being careful not to leave any air spaces between pan and dough. REMOVE EXCESS DOUGH: Trim overhanging edges of dough by using a knife or spatula. (Incorporate excess dough into next crust, if needed.) There should be little excess if skill is used in weighing and rolling dough. BAKING INSTRUCTIONS FOR UNCOOKED PIES: Bake crusts at 425 F. for about 15-18 minutes, or until light golden brown. Cool before filling. Proceed with the recipe directions.
2. Dissolve gelatin in boiling water; add cold water. Mix until well blended.
3. Refrigerate until gelatin is thickened but not firm.
4. Pour cold water into chilled mixer bowl; add topping, milk, sugar, and vanilla. Using whip, beat at low speed 3 minutes or until well blended. Scrape down whip and bowl. Whip at high speed for 5 to 10 minutes or until mixture forms stiff peaks. Set aside for use in Step 7.
5. Using whip, beat thickened gelatin at high speed for 10 minutes or until foamy and soft peaks form.
6. Fold whipped topping and thawed, drained strawberries into gelatin. Mix carefully at low speed until well blended.
7. Pour 5-3/4 cups filling into each baked pie shell.
8. Refrigerate 2 hours or until set. Keep refrigerated until ready to serve.
9. Cut 8 wedges per pie. CCP: Hold for service at 41 F. or lower.

Notes

1. In Step 5, 2 pound 10 ounces of canned dessert topping and frozen bakery products may be used for all ingredients, per 100 servings.

DESSERTS (PASTRY AND PIES) No. I 033 00

LEMON MERINGUE PIE

Yield 100   **Portion** 1 Slice

Calories	Carbohydrates	Protein	Fat	Cholesterol	Sodium	Calcium
327 cal	53 g	3 g	12 g	39 mg	317 mg	11 mg

Ingredient	**Weight**	**Measure**	**Issue**
PIE CRUST		13 each	
SUGAR,GRANULATED	7 lbs	1 gal	
SALT	1-3/8 oz	2-1/3 tbsp	
LEMON RIND,GRATED	2-1/2 oz	3/4 cup	
WATER	9-3/8 lbs	1 gal 1/2 qts	
CORNSTARCH	1-3/8 lbs	1 qts 1 cup	
WATER,COLD	2-1/3 lbs	1 qts 1/2 cup	
EGGS,WHOLE,FROZEN,BEATEN	1-5/8 lbs	3 cup	
BUTTER	12 oz	1-1/2 cup	
JUICE,LEMON	2-1/8 lbs	1 qts	
FOOD COLOR,YELLOW	<1/16th oz	2 drop	
MERINGUE	532 gm	7-1/2 unit	

Method

1. PREPARE AND DIVIDE DOUGH: Prepare 1/2 recipe Pie Crust (Recipe No. I 001 00). Divide dough into 13-7-1/2 oz pieces for bottom crust; place on lightly floured board. ROLL DOUGH: Sprinkle each piece of dough lightly with flour; flatten gently. Using a floured rolling pin, roll lightly with quick strokes from center out to edge in all directions. Form a circle 1 inch larger than pie pan and about 1/8 inch thick. Shift or turn dough occasionally to prevent sticking. If edges split, pinch cracks together. BOTTOM CRUST: Fold rolled dough in half; carefully place into ungreased pie pan with fold at center. Unfold and fit carefully into pie pan, being careful not to leave any air spaces between pan and dough. REMOVE EXCESS DOUGH: Trim overhanging edges of dough by using a knife or spatula. (Incorporate excess dough into next crust, if needed.) There should be little excess if skill is used in weighing and rolling dough. BAKING INSTRUCTIONS FOR UNCOOKED PIES: Bake crusts at 425 F. for about 15-18 minutes, or until light golden brown. Cool before filling. Proceed with the recipe directions.
2. Combine sugar, salt, lemon rind, and water. Bring to a boil.
3. Combine cornstarch and water; stir until smooth. Add gradually to boiling mixture; cook at medium heat, stirring constantly until thick and clear.
4. Stir about 1 quart hot mixture into eggs. Slowly pour egg mixture into remaining hot mixture, stirring constantly. Cook at medium heat; stirring frequently, until mixture returns to a boil. Remove from heat.
5. Add butter or margarine, lemon juice, and food coloring; stir until well blended. Cool slightly.
6. Pour 2-3/4 to 3 cups filling into each baked 9-inch pie shell.
7. Prepare 1 recipe Meringue, Recipe No. I 005 00 or I 005 01 per 100 portions. Spread 2-1/2 cups completely over warm filling, about 122 F., in each pan. Meringue should touch inner edge of crust all around and completely cover top of pie. Leave meringue somewhat rough on top.
8. Bake at 350 F. for 15 to 20 minutes or until lightly browned.
9. Refrigerate until ready to serve.
10. Cut 8 wedges per pie. CCP: Hold for service at 41 F. or lower.

DESSERTS (PASTRY AND PIES) No.I 033 01

LEMON MERINGUE PIE (PIE FILLING PREPARED)

Yield 100　　　　　　　　　　　　　　　　　　　　　　**Portion** 1 Slice

Calories	Carbohydrates	Protein	Fat	Cholesterol	Sodium	Calcium
213 cal	33 g	2 g	9 g	0 mg	156 mg	11 mg

Ingredient	**Weight**	**Measure**	**Issue**
PIE CRUST		13 each	
PIE FILLING,LEMON,PREPARED	21 lbs	2 gal 2-1/2 qts	
MERINGUE	532 gm	7-1/2 unit	

Method

1 PREPARE AND DIVIDE DOUGH: Prepare 1/2 recipe Pie Crust (Recipe No. I 001 00). Divide dough into 13-7-1/2 oz pieces for bottom crust; place on lightly floured board. ROLL DOUGH: Sprinkle each piece of dough lightly with flour; flatten gently. Using a floured rolling pin, roll lightly with quick strokes from center out to edge in all directions. Form a circle 1 inch larger than pie pan and about 1/8 inch thick. Shift or turn dough occasionally to prevent sticking. If edges split, pinch cracks together. BOTTOM CRUST: Fold rolled dough in half; carefully place into ungreased pie pan with fold at center. Unfold and fit carefully into pie pan, being careful not to leave any air spaces between pan and dough. REMOVE EXCESS DOUGH: Trim overhanging edges of dough by using a knife or spatula. (Incorporate excess dough into next crust, if needed.) There should be little excess if skill is used in weighing and rolling dough. BAKING INSTRUCTIONS FOR UNCOOKED PIES: Bake crusts at 425 F. for about 15-18 minutes, or until light golden brown. Cool before filling. Proceed with the recipe directions.

2 Heat filling to 122 F. ; pour about 3-1/4 cups of filling into each baked 9-inch pie shell.

3 Prepare Meringue, Recipe No. I 005 00. Spread 2-1/2 cups completely over warm filling, about 122 F., in each 9-inch pie pan. Meringue should touch inner edge of crust all around and completely cover top of pie. Leave meringue somewhat rough on top.

4 Bake at 350 F. for 15 to 20 minutes or until lightly browned.

5 Refrigerate until ready to serve.

6 Cut 8 wedges per pie. CCP: Hold for service at 41 F. or lower.

DESSERTS (PASTRY AND PIES) No. I 034 00

FRUIT TURNOVERS

Yield 100   **Portion** 1 Turnover

Calories	Carbohydrates	Protein	Fat	Cholesterol	Sodium	Calcium
315 cal	38 g	3 g	17 g	0 mg	236 mg	11 mg

Ingredient	**Weight**	**Measure**	**Issue**
PIE CRUST		26 each	
PIE FILLING, APPLE, PREPARED	12 lbs	1 gal 2 qts	
MILK AND WATER WASH		3 cup	

Method

1. Prepare Pie Crust, Recipe No. I 001 00 to yield enough dough to prepare cobbler for 100 portions. Divide dough into 8 pieces.
2. Place dough on lightly floured board; sprinkle each piece lightly with flour; flatten gently. Roll dough into 18 by 24-inch rectangular sheet about 1/8-inch thick. Cut into twelve 6-inch squares. Brush edges of each square with water.
3. Place 1/4 cup of fruit filling in the center of each square. Fold opposite corner of dough together forming a triangle. Seal by crimping edges.
4. Make 2-1/2 inch slits near the center fold to allow steam to escape during baking.
5. Place 12 turnovers on each lightly greased sheet pan.
6. Brush top of each turnover with Milk and Water wash. Allow to dry before baking. See Recipe No. I 004 02. Do not use Egg and Milk wash or Egg and Water wash for turnovers. The egg and milk will cause the turnovers to brown excessively and egg and water wash will cause turnovers to be too pale in color.
7. Bake at 425 F. for 20 minutes or until lightly browned.

DESSERTS (PASTRY AND PIES) No.I 035 00

FRUIT DUMPLINGS

Yield 100 **Portion** 1 Each

Calories	Carbohydrates	Protein	Fat	Cholesterol	Sodium	Calcium
378 cal	44 g	4 g	21 g	0 mg	287 mg	8 mg

Ingredient	Weight	Measure	Issue
PIE CRUST	7-1/4 kg	32-1/2 unit	
PIE FILLING,APPLE,PREPARED	12 lbs	1 gal 2 qts	

Method

1. Prepare Pie Crust, Recipe No. I 001 00 to yield enough dough to prepare cobbler for 100 portions. Divide dough into 8 pieces.
2. Place dough on lightly floured board; sprinkle each piece lightly with flour; flatten gently. Roll dough into 18x24-inch rectangular sheet, about 1/8-inch thick. Cut into 12, 6-inch squares. Brush edges of each square with water.
3. Place 1/4 cup of fruit filling in the center or each pastry square. Bring points of pastry up over filling. Seal edges tightly.
4. Place 12 dumplings on each sheet pan.
5. Bake at 425 F. 20 minutes or until lightly browned.
6. Serve with dessert sauce. See Recipe Section K.

DESSERTS (PASTRY AND PIES) No.I 500 00

KEY LIME PIE

Yield 100   Portion 1 Slice

Calories	Carbohydrates	Protein	Fat	Cholesterol	Sodium	Calcium
337 cal	60 g	4 g	10 g	73 mg	177 mg	20 mg

Ingredient	Weight	Measure	Issue
COOKIES,CHOCOLATE,CRUSHED	5 lbs	1 gal 1-1/8 qts	
SHORTENING	10-7/8 oz	1-1/2 cup	
EGGS,WHOLE,FRESH	3-3/4 lbs	34 each	
JUICE,LIME	1 lbs	2 cup	
LIMES,FRESH	14-1/4 oz	6 each	
FLOUR,WHEAT,GENERAL PURPOSE	1-1/4 lbs	1 qts 1/2 cup	
SUGAR,GRANULATED	7 lbs	1 gal	
MARGARINE	8 oz	1 cup	
WATER	10-1/2 lbs	1 gal 1 qts	
SUGAR,GRANULATED	1-1/3 lbs	3 cup	

Method

1. Crush chocolate wafer cookies to equal 5 quarts. In a mixer, combine the crushed cookies and shortening. Mix on low speed for 3 minutes. Divide among large sheet pans. Press crust evenly into bottom of pans.
2. Separate egg yolks from whites. In a heavy saucepan or steam kettle, beat egg yolks with lime juice and 2 tablespoon lime zest. Place over low heat.
3. Beat in flour, sugar, margarine, and water, alternating each ingredient so as to maintain a smooth consistency. Cook stirring constantly for 3 minutes on medium low heat. CCP: Internal temperature must reach 155 F. or higher for 15 seconds. Pour into the cookie crust.
4. In a mixer, beat egg whites until stiff, but not dry. Gradually add sugar and whip for 3 minutes. Spread over filling. Bake at 450 F. for 10 minutes or until meringue is brown.
5. Chill for 1 hour before serving. Cut 6 by 9. CCP: Hold for service at 41 F. or lower.

J. DESSERTS (PUDDING AND OTHER DESSERTS) No. 0 (1)
INDEX

Card No.		Card No.	
J 001 01	Apple Crisp (Pie Filling & Cookie Mix)	J 006 06	Spiced Fruit Cup
J 002 00	Vanilla Soft Serve Ice Cream (Dehy)	J 006 07	Mandarin Orange and Pineapple Fruit Cup
J 002 01	Chocolate Soft Serve Ice Cream (Dehy)	J 007 00	Fruit Gelatin
J 002 02	Chocolate Milk Shake (Dehy Mix)	J 007 01	Banana Gelatin
J 002 03	Strawberry Soft Serve Ice Cream (Dehy)	J 007 02	Fruit Flavored Gelatin
J 002 04	Vanilla Milk Shake (Dehy Mix)	J 007 03	Fruit Gelatin (Crushed Ice Method)
J 003 00	Baked Apples	J 007 04	Strawberry Gelatin
J 003 01	Baked Apples with Raisin Nut Filling	J 007 05	Peach Gelatin
J 003 02	Baked Apples with Raisin Coconut Filling	J 008 00	Peach Crisp
J 004 00	Vanilla Soft Serve Ice Cream (Liquid Mix)	J 008 01	Cherry Crisp (Pie Filling Cookie Mix)
J 004 01	Strawberry Soft Serve Ice Cream (Liquid Mix)	J 008 02	Cherry Crisp
J 004 02	Vanilla Milk Shake (Liquid Mix)	J 008 03	Peach Crisp (Pie Filling Cookie Mix)
J 004 03	Chocolate Milk Shake (Liquid Mix)	J 008 04	Blueberry Crisp (Pie Filling Cookie Mix)
J 004 04	Chocolate Soft Serve Ice Cream (Liquid Mix)	J 010 01	Apple Crunch (Apple Pie Filling)
J 005 00	Fluffy Fruit Cup	J 010 02	Blueberry Crunch (Blueberry Pie Filling)
J 005 01	Yogurt Fruit Cup	J 010 03	Cherry Crunch (Cherry Pie Filling)
J 006 00	Fruit Cup	J 010 05	Peach Crunch (Peach Pie Filling)
J 006 01	Ambrosia	J 011 00	Banana Split
J 006 02	Banana Fruit Cup	J 012 00	Vanilla Soft Serve Yogurt (Dehydrated)
J 006 03	Melon Fruit Cup	J 012 01	Chocolate Soft Serve Yogurt (Dehydrated)
J 006 04	Strawberry Fruit Cup	J 013 00	Tapioca Pudding
J 006 05	Fruit Cocktail Fruit Cup	J 014 00	Vanilla Cream Pudding (Instant)

J. DESSERTS (PUDDING AND OTHER DESSERTS) No. 0 (1)

Card No.

J 014 01	Banana Cream Pudding (Instant)
J 014 02	Coconut Cream Pudding (Instant)
J 014 03	Pineapple Cream Pudding (Instant)
J 014 04	Butterscotch Cream Pudding (Instant)
J 014 05	Chocolate Cream Pudding (Instant)
J 015 00	Baked Rice Pudding
J 015 01	Baked Rice Pudding (Frozen Eggs and Egg Whites)
J 016 00	Bread Pudding
J 016 01	Chocolate Chip Bread Pudding
J 016 02	Coconut Bread Pudding
J 017 00	Cream Puffs
J 017 01	Eclairs
J 018 00	Vanilla Cream Pudding
J 018 01	Chocolate Cream Pudding
J 020 00	Creamy Rice Pudding
J 021 00	Fluffy Pineapple Rice Cup
J 022 00	Breakfast Bread Pudding
J 023 00	Baked Cinnamon Apple Slices
J 500 00	Bread Pudding with Hard Sauce
J 504 00	Baked Bananas

DESSERTS (PUDDINGS AND OTHER DESSERTS)   No. J 001 01

APPLE CRISP (PIE FILLING & COOKIE MIX)

Yield 100   **Portion** 1 Piece

Calories	Carbohydrates	Protein	Fat	Cholesterol	Sodium	Calcium
190 cal	34 g	1 g	7 g	12 mg	103 mg	30 mg

Ingredient Weight Measure Issue

Ingredient	Weight	Measure
PIE FILLING, APPLE, PREPARED	18 lbs	2 gal 1 qts
COOKIE MIX, OATMEAL	6-3/4 lbs	
MARGARINE, SOFTENED	1 lbs	2 cup

Method

1. Place 10-1/2 pounds of the pie filling in each pan.
2. Combine oatmeal cookie mix with margarine.
3. Sprinkle 3 pounds 13 ounces of oatmeal-margarine mixture evenly over apples, in each pan.
4. Using a convection oven, bake at 350 F. for 30 minutes or until top is bubbling and lightly browned on low fan, open vent.
5. Cut 6 by 9. Serve with serving spoon or spatula.

DESSERTS (PUDDINGS AND OTHER DESSERTS) No.J 002 00
VANILLA SOFT SERVE ICE CREAM (DEHY)

Yield 100 **Portion** 3/4 Cup

Calories	Carbohydrates	Protein	Fat	Cholesterol	Sodium	Calcium
166 cal	40 g	1 g	0 g	1 mg	71 mg	29 mg

Ingredient	Weight	Measure	Issue
ICE MILK-MILKSHAKE,DEHYDRATED,VAN	10 lbs		
WATER	20-7/8 lbs	2 gal 2 qts	

Method

1. Stir dehydrated mix into water. Mix thoroughly with wire whip or mixer. Cover container.
2. Chill 4 to 24 hours in refrigerator to 35 F. to 40 F.
3. Stir until smooth. Pour mixture into top hopper of soft serve ice cream freezer; start dasher motor; turn on refrigeration according to manufacturer's directions. Freeze to a temperature of 18 F. to 22 F., about 10 minutes or until product can be drawn with a stiff consistency that will hold a peak.

DESSERTS (PUDDINGS AND OTHER DESSERTS) No. J 002 01
CHOCOLATE SOFT SERVE ICE CREAM (DEHY)

Yield 100   **Portion** 3/4 Cup

Calories	Carbohydrates	Protein	Fat	Cholesterol	Sodium	Calcium
166 cal	40 g	1 g	1 g	1 mg	168 mg	29 mg

Ingredient | Weight | Measure | Issue

ICE MILK-MILKSHAKE,DEHYDRATED,CHOC   10 lbs
WATER   20-7/8 lbs   2 gal 2 qts

Method

1. Stir dehydrated mix into water. Mix thoroughly with wire whip or mixer. Cover container.
2. Chill 4 to 24 hours in refrigerator until 35 F. to 40 F.
3. Stir until smooth. Pour mixture into top hopper of soft serve ice cream freezer; start dasher motor; turn on refrigeration according to manufacturer's directions. Freeze to a temperature of 18 F. to 22 F., about 10 minutes or until product can be drawn with a stiff consistency that will hold a peak.

DESSERTS (PUDDINGS AND OTHER DESSERTS)   No. J 002 02

CHOCOLATE MILK SHAKE (DEHY MIX)

Yield 100   Portion 8 Ounces

Calories	Carbohydrates	Protein	Fat	Cholesterol	Sodium	Calcium
166 cal	40 g	1 g	1 g	1 mg	169 mg	29 mg

Ingredient   Weight   Measure   Issue

ICE MILK-MILKSHAKE,DEHYDRATED,CHOC   10 lbs
WATER   25-1/8 lbs   3 gal

Method

1 Stir dehydrated mix into water. Mix thoroughly with wire whip or mixer. Cover container.
2 Chill 4 to 24 hours in refrigerator until 35 F. to 40 F.
3 Stir until smooth. Pour mixture into top hopper of milk shake mix machine, according to manufacturer's directions. Freeze to a temperature of 27 F. to 30 F., about 10 minutes.

DESSERTS (PUDDINGS AND OTHER DESSERTS) No.J 002 03
STRAWBERRY SOFT SERVE ICE CREAM (DEHY)

Yield 100   Portion   3/4 Cup

Calories	Carbohydrates	Protein	Fat	Cholesterol	Sodium	Calcium
176 cal	43 g	1 g	0 g	1 mg	71 mg	33 mg

Ingredient	**Weight**	**Measure**	**Issue**
ICE MILK-MILKSHAKE,DEHYDRATED,VAN	10 lbs		
WATER	17-1/4 lbs	2 gal 1/4 qts	
STRAWBERRIES,FROZEN,THAWED	6-1/2 lbs	2 qts 3-1/2 cup	
FOOD COLOR,RED	1/8 oz	1/8 tsp	

Method

1. Stir dehydrated mix into water. Mix thoroughly with wire whip or mixer. Cover container.
2. Chill 4 to 24 hours in refrigerator to 35 F. to 40 F. Crush strawberries; red food coloring may be added.
3. Stir until smooth. Pour mixture into top hopper of soft serve ice cream freezer; remove mix feed and air control units. Start dasher motor; turn on refrigeration according to manufacturer's directions. Stir occasionally. Freeze to a temperature of 18 F. to 22 F., about 10 minutes or until product can be drawn with a stiff consistency that will hold a peak.

DESSERTS (PUDDINGS AND OTHER DESSERTS)   No. J 002 04

VANILLA MILK SHAKE (DEHY MIX)

Yield 100   Portion 1 Cup

Calories	Carbohydrates	Protein	Fat	Cholesterol	Sodium	Calcium
166 cal	40 g	1 g	0 g	1 mg	72 mg	29 mg

Ingredient	Weight	Measure	Issue
ICE MILK-MILKSHAKE,DEHYDRATED,VAN	10 lbs		
WATER	25-1/8 lbs	3 gal	

Method

1 Stir dehydrated mix into water. Mix thoroughly with wire whip or mixer. Cover container.
2 Chill 4 to 24 hours in refrigerator to 35 F. to 40 F.
3 Stir until smooth. Pour mixture into top hopper of milk shake machine, according to manufacturer's directions; freeze to a temperature of 27 F. to 30 F.

DESSERTS (PUDDINGS AND OTHER DESSERTS) No. J 003 00
BAKED APPLES

Yield 100   Portion 1 Serving

Calories	Carbohydrates	Protein	Fat	Cholesterol	Sodium	Calcium
207 cal	51 g	0 g	1 g	2 mg	34 mg	11 mg

Ingredient	Weight	Measure	Issue
APPLES,COOKING,FRESH,UNPEELED	28-1/8 lbs	100 each	33-1/8 lbs
SUGAR,GRANULATED	7 lbs	1 gal	
CINNAMON,GROUND	1/8 oz	1/3 tsp	
SALT	1/4 oz	1/8 tsp	
WATER,ICE	5-1/4 lbs	2 qts 2 cup	
BUTTER	4 oz	1/2 cup	

Method

1. Score apples once around middle to prevent bursting. Place apples on pans.
2. Mix sugar, cinnamon and salt thoroughly.
3. Combine with water and butter or margarine. Pour 1-1/2 quarts of syrup over apples in each pan.
4. Using a convection oven, bake at 325 F. for 30 minutes or until tender on low fan, closed vent. Baste occasionally.
5. Serve each apple with 2 tablespoons syrup.

DESSERTS (PUDDINGS AND OTHER DESSERTS)   No.J 003 01

BAKED APPLES WITH RAISIN NUT FILLING

Yield 100   **Portion** 1 Serving

Calories	Carbohydrates	Protein	Fat	Cholesterol	Sodium	Calcium
264 cal	58 g	1 g	5 g	2 mg	35 mg	16 mg

Ingredient	**Weight**	**Measure**	**Issue**
APPLES,COOKING,FRESH,UNPEELED	28-1/8 lbs	100 each	33-1/8 lbs
RAISINS	1-1/2 lbs	1 qts 1/2 cup	
PECANS,CHOPPED	1-1/4 lbs		
SUGAR,GRANULATED	7 lbs	1 gal	
CINNAMON,GROUND	1/8 oz	1/3 tsp	
SALT	1/4 oz	1/8 tsp	
WATER,ICE	5-1/4 lbs	2 qts 2 cup	
BUTTER	4 oz	1/2 cup	

Method

1. Score apples once around middle to prevent bursting. Place apples on pans.
2. Mix raisins with finely chopped, unsalted nuts. Fill cavity in center of each apple with 1-2/3 tablespoons of mixture.
3. Mix sugar, cinnamon and salt thoroughly.
4. Combine with water and butter or margarine. Pour 1-1/2 quart syrup over apples in each pan.
5. Using a convection oven, bake at 325 F. for 30 minutes or until tender on low fan, closed vent, basting occasionally.
6. Serve each apple with 2 tablespoon of syrup.

Notes

1. In Step 4, baking time will vary depending on variety and size of apples.

DESSERTS (PUDDINGS AND OTHER DESSERTS) No. J 003 02
BAKED APPLES WITH RAISIN COCONUT FILLING

Yield 100   Portion 1 Serving

Calories	Carbohydrates	Protein	Fat	Cholesterol	Sodium	Calcium
241 cal	58 g	0 g	2 g	2 mg	42 mg	14 mg

Ingredient	Weight	Measure	Issue
APPLES,COOKING,FRESH,UNPEELED	28-1/8 lbs	100 each	33-1/8 lbs
RAISINS	1-1/2 lbs	1 qts 1/2 cup	
COCONUT,PREPARED,SWEETENED FLAKES	9-7/8 oz	3 cup	
SUGAR,GRANULATED	7 lbs	1 gal	
CINNAMON,GROUND	1/8 oz	1/3 tsp	
SALT	1/4 oz	1/8 tsp	
WATER,ICE	5-1/4 lbs	2 qts 2 cup	
BUTTER	4 oz	1/2 cup	

Method

1. Score apples once around middle to prevent bursting. Place apples on pans.
2. Mix raisins with prepared, sweetened, flaked coconut. Fill cavity in center of each apple with 1 tablespoon of mixture.
3. Mix sugar, cinnamon and salt thoroughly.
4. Combine with water and butter or margarine. Pour 1-1/2 quart syrup over apples in each pan.
5. Using a convection oven, bake at 325 F. for 30 minutes or until tender on low fan, closed vent, basting occasionally.
6. Serve each apple with 2 tablespoons of syrup.

DESSERTS (PUDDINGS AND OTHER DESSERTS)   No. J 004 00

VANILLA SOFT SERVE ICE CREAM (LIQUID MIX)

Yield 100   **Portion** 3/4 Cup

Calories	Carbohydrates	Protein	Fat	Cholesterol	Sodium	Calcium
101 cal	16 g	3 g	3 g	0 mg	0 mg	84 mg

Ingredient   **Weight**   **Measure**   **Issue**

ICE MILK MIX,LIQ,VAN,CHILLED   29-1/4 lbs   3 gal 2 qts

Method

1 Pour mix into top hopper of soft serve ice cream freezer; start dasher motor; turn on refrigeration according to manufacturer's directions.
2 Freeze to a temperature of 18 F. to 22 F., about 10 minutes or until product can be drawn with a stiff consistency that will hold a peak.

DESSERTS (PUDDINGS AND OTHER DESSERTS) No.J 004 01
STRAWBERRY SOFT SERVE ICE CREAM (LIQUID MIX)

Yield 100   **Portion** 3/4 Cup

Calories	Carbohydrates	Protein	Fat	Cholesterol	Sodium	Calcium
111 cal	18 g	3 g	3 g	0 mg	1 mg	89 mg

Ingredient	Weight	Measure	Issue
FOOD COLOR,RED	1/8 oz	1/8 tsp	
ICE MILK MIX,LIQ,VAN,CHILLED	29-1/4 lbs	3 gal 2 qts	
STRAWBERRIES,FROZEN,THAWED	6-1/2 lbs	2 qts 3-1/2 cup	

Method

1. Pour mix into top hopper of soft serve ice cream freezer; start dasher motor; turn on refrigeration according to manufacturer's directions. Crush strawberries; drain. Red food coloring may be added.
2. Add strawberry mixture to soft serve mixture. Freeze to a temperature of 18 F. to 22 F., about 10 minutes or until product can be drawn with a stiff consistency that will hold a peak.

Notes

1. While drawing ice cream, strawberries must be stirred up occasionally from the bottom of freezer hopper.

DESSERTS (PUDDINGS AND OTHER DESSERTS)   No.J 004 02

VANILLA MILK SHAKE (LIQUID MIX)

Yield 100   **Portion** 1 Cup

Calories	Carbohydrates	Protein	Fat	Cholesterol	Sodium	Calcium
115 cal	18 g	3 g	4 g	0 mg	1 mg	97 mg

Ingredient	Weight	Measure	Issue
ICE MILK MIX,LIQ,VAN,CHILLED	33-3/8 lbs	3 gal	
WATER,COLD	8-1/3 lbs	1 gal	

Method

1. Combine liquid milk shake mix and cold water.
2. Pour sufficient amount into top hopper of soft serve ice cream freezer; start dasher motor; turn on refrigeration. Prepare according to manufacturer's directions; freeze to a temperature of 27 F. to 30 F.

DESSERTS (PUDDINGS AND OTHER DESSERTS) No. J 004 03
CHOCOLATE MILK SHAKE (LIQUID MIX)

Yield 100   Portion   1 Cup

Calories	Carbohydrates	Protein	Fat	Cholesterol	Sodium	Calcium
130 cal	20 g	4 g	4 g	0 mg	0 mg	108 mg

Ingredient	Weight	Measure	Issue
ICE MILK,MIX,LIQ,CHOC,CHILLED	37-5/8 lbs	4 gal 2 qts	

Method

1 Pour mix into top hopper of soft serve ice cream freezer; start dasher motor; turn on refrigeration according to manufacturer's directions.
2 Freeze to a temperature of 27 F. to 30 F.

DESSERTS (PUDDINGS AND OTHER DESSERTS) No.J 004 04
CHOCOLATE SOFT SERVE ICE CREAM (LIQUID MIX)

Yield 100 **Portion** 3/4 Cup

Calories	Carbohydrates	Protein	Fat	Cholesterol	Sodium	Calcium
101 cal	16 g	3 g	3 g	0 mg	0 mg	84 mg

Ingredient	**Weight**	**Measure**	**Issue**
ICE MILK,MIX,LIQ,CHOC,CHILLED	29-1/4 lbs	3 gal 2 qts	

Method

1 Pour mix into top hopper of soft serve ice cream freezer; start dasher motor; turn on refrigeration according to manufacturer's directions.

2 Freeze to a temperature of 18 F. to 22 F., about 10 minutes, or until product can be drawn with a stiff consistency that will hold a peak.

DESSERTS (PUDDINGS AND OTHER DESSERTS) No.J 005 00
FLUFFY FRUIT CUP

Yield 100   Portion   1/2 Cup

Calories	Carbohydrates	Protein	Fat	Cholesterol	Sodium	Calcium
82 cal	20 g	1 g	0 g	0 mg	5 mg	20 mg

Ingredient

Ingredient	Weight	Measure	Issue
PINEAPPLE,CANNED,CHUNKS,JUICE PACK,DRAINED	7-7/8 lbs	1 gal 1/2 qts	
CHERRIES,MARASCHINO,WHOLE	1-2/3 lbs	3 cup	
ORANGE,FRESH,SECTIONS	4 lbs	2 qts 2-1/8 cup	5-1/2 lbs
GRAPES,FRESH,CUT IN HALVES	2-7/8 lbs	2 qts 1/8 cup	3 lbs
BANANA,FRESH,SLICED	4-1/3 lbs	3 qts 1-1/8 cup	6-2/3 lbs
MARSHMALLOWS,MINIATURE	1 lbs	2 qts 1 cup	
WATER,COLD	1 lbs	2 cup	
WHIPPED TOPPING MIX,NONDAIRY,DRY	1-3/8 oz	2 cup	
MILK,NONFAT,DRY	7/8 oz	1/4 cup 2-1/3 tbsp	
SUGAR,GRANULATED	1/2 oz	1 tbsp	
EXTRACT,VANILLA	1/2 oz	1 tbsp	

Method

1 Drain pineapple. Drain cherries; cut into halves.
2 Combine pineapple, cherries, oranges, grapes, bananas and marshmallows; mix well. Set aside for use in Step 4.
3 Pour cold water into mixer bowl; add topping, milk, sugar and vanilla. Whip at low speed for 3 minutes or until thoroughly blended.
4 Fold mixed fruit into whipped topping. Mix carefully until thoroughly blended.
5 Refrigerate until ready to serve. CCP: Hold for service at 41 F. or lower.

DESSERTS (PUDDINGS AND OTHER DESSERTS)   No. J 005 01
YOGURT FRUIT CUP

Yield 100 **Portion** 1/2 Cup

Calories	Carbohydrates	Protein	Fat	Cholesterol	Sodium	Calcium
90 cal	21 g	2 g	0 g	1 mg	16 mg	50 mg

Ingredient

Ingredient	Weight	Measure	Issue
PINEAPPLE,CANNED,CHUNKS,JUICE PACK,DRAINED	7-7/8 lbs	1 gal 1/2 qts	
CHERRIES,MARASCHINO,WHOLE	1-2/3 lbs	3 cup	
ORANGE,FRESH,SECTIONS	4 lbs	2 qts 2-1/8 cup	5-1/2 lbs
GRAPES,FRESH,CUT IN HALVES	2-7/8 lbs	2 qts 1/8 cup	3 lbs
BANANA,FRESH,SLICED	4-1/3 lbs	3 qts 1-1/8 cup	6-2/3 lbs
MARSHMALLOWS,MINIATURE	1 lbs	2 qts 1 cup	
YOGURT,PLAIN,LOWFAT	4 lbs	1 qts 3-1/2 cup	

Method

1. Drain pineapple. Drain cherries; cut into halves.
2. Combine pineapple, cherries, oranges, grapes, bananas and marshmallows; mix well. Set aside for use in Step 3.
3. Fold yogurt into mixed fruit. Mix lightly until just combined.
4. Refrigerate until ready to serve. CCP: Hold for service at 41 F. or lower.

DESSERTS (PUDDINGS AND OTHER DESSERTS) No. J 006 00
FRUIT CUP

Yield 100   Portion   1/2 Cup

Calories	Carbohydrates	Protein	Fat	Cholesterol	Sodium	Calcium
61 cal	16 g	1 g	0 g	0 mg	3 mg	15 mg

Ingredient	Weight	Measure	Issue
PEACHES,CANNED,SLICED	6-1/2 lbs	3 qts	
PEARS,CANNED,SLICES	6-1/2 lbs	3 qts	
PINEAPPLE,CANNED,CHUNKS,JUICE PACK,INCL LIQUIDS	6-5/8 lbs	3 qts	
ORANGE,FRESH,CHOPPED	3 lbs	1 qts 3-7/8 cup	4-1/8 lbs
APPLES,FRESH,MEDIUM,UNPEELED,DICED	3-1/2 lbs	3 qts 1/8 cup	4-1/8 lbs

Method

1 Drain peaches and pears. Reserve juices. Cut fruit into 3/4-inch pieces.
2 Combine pineapple, peaches, pears, oranges, apples and juices from all fruit. Mix thoroughly.
3 Cover; CCP: Hold for service at 41 F. or lower.

DESSERTS (PUDDINGS AND OTHER DESSERTS)   No. J 006 01
AMBROSIA

Yield 100   **Portion**   1/2 Cup

Calories	Carbohydrates	Protein	Fat	Cholesterol	Sodium	Calcium
85 cal	18 g	1 g	2 g	0 mg	15 mg	16 mg

Ingredient	Weight	Measure	Issue
PEACHES,CANNED,SLICED,JUICE PACK,INCL LIQUIDS	6-1/2 lbs	3 qts	
PEARS,CANNED,JUICE PACK,SLICES,INCL LIQUID	6-1/2 lbs	3 qts	
PINEAPPLE,CANNED,CRUSHED,JUICE PACK,INCL LIQUIDS	6-5/8 lbs	3 qts	
ORANGE,FRESH,CHOPPED	3-1/8 lbs	1 qts 3-7/8 cup	4-1/4 lbs
APPLES,FRESH,MEDIUM,UNPEELED,DICED	3-1/3 lbs	3 qts 1/8 cup	3-7/8 lbs
COCONUT,PREPARED,SWEETENED FLAKES	1 lbs	1 qts 1 cup	

Method

1 Drain peaches and pears. Reserve juices. Cut fruit into 3/4-inch pieces.
2 Combine pineapple, peaches, pears, oranges, apples and sweetened coconut flakes and juices from all fruit. Mix thoroughly.
3 Cover. CCP: Hold for service at 41 F. or lower.

DESSERTS (PUDDINGS AND OTHER DESSERTS)   No. J 006 02
BANANA FRUIT CUP

Yield 100 **Portion** 1/2 Cup

Calories	Carbohydrates	Protein	Fat	Cholesterol	Sodium	Calcium
82 cal	21 g	1 g	0 g	0 mg	1 mg	14 mg

Ingredient	Weight	Measure	Issue
BANANA,FRESH,SLICED	11-1/2 lbs	2 gal 2/3 qts	17-2/3 lbs
PINEAPPLE,CANNED,CHUNKS,JUICE PACK,INCL LIQUIDS	6-5/8 lbs	3 qts	
ORANGE,FRESH,CHOPPED	3-1/8 lbs	1 qts 3-7/8 cup	4-1/4 lbs
APPLES,FRESH,MEDIUM,UNPEELED,DICED	3-1/3 lbs	3 qts 1/8 cup	3-7/8 lbs

Method

1. Combine bananas, pineapple, oranges and apples. Mix thoroughly.
2. Cover. CCP: Hold for service at 41 F. or lower.

DESSERTS (PUDDINGS AND OTHER DESSERTS) No.J 006 03
MELON FRUIT CUP

Yield 100   **Portion** 1/2 Cup

Calories	Carbohydrates	Protein	Fat	Cholesterol	Sodium	Calcium
54 cal	14 g	1 g	0 g	0 mg	3 mg	14 mg

Ingredient

Ingredient	Weight	Measure	Issue
PEACHES,CANNED,SLICED,JUICE PACK,INCL LIQUIDS	6-1/2 lbs	3 qts	
PEARS,CANNED,JUICE PACK,SLICES,INCL LIQUID	6-1/2 lbs	3 qts	
WATERMELON,FRESH,DICED	7 lbs	1 gal 1-1/4 qts	13-1/2 lbs
ORANGE,FRESH,SECTIONS,PEELED,DICED	3-1/8 lbs	1 qts 3-7/8 cup	10-7/8 each
APPLES,FRESH,MEDIUM,UNPEELED,DICED	3-1/3 lbs	3 qts 1/8 cup	3-7/8 lbs

Method

1. Drain peaches and pears. Reserve juices. Cut fruit into 3/4 inch pieces.
2. Seed melon. Combine melon with oranges, peaches, pears, apples and juices from fruit. Mix thoroughly.
3. Cover; CCP: Hold for service at 41 F. or lower.

DESSERTS (PUDDINGS AND OTHER DESSERTS) No. J 006 04
STRAWBERRY FRUIT CUP

Yield 100   **Portion** 1/2 Cup

Calories	Carbohydrates	Protein	Fat	Cholesterol	Sodium	Calcium
58 cal	14 g	1 g	0 g	0 mg	3 mg	21 mg

Ingredient | Weight | Measure | Issue

Ingredient	Weight	Measure	Issue
PEACHES,CANNED,SLICED,JUICE PACK,INCL LIQUIDS	6-1/2 lbs	3 qts	
PINEAPPLE,CANNED,CHUNKS,JUICE PACK,INCL LIQUIDS	6-5/8 lbs	3 qts	
ORANGE,FRESH,SECTIONS,PEELED,DICED	3-1/8 lbs	2 qts	11 each
STRAWBERRIES,FRESH,SLICED	8-3/4 lbs	1 gal 2 qts	1 gal 2-3/8 qts
KIWIFRUIT,FRESH,CHOPPED	2-7/8 lbs	1 qts 3-1/4 cup	3-1/4 lbs

Method

1. Drain peaches. Reserve juices. Cut fruit into 3/4-inch pieces.
2. Combine pineapple, peaches, oranges and juices from all fruit.
3. Slice strawberries into quarters. Combine strawberries with fruit mixture; mix thoroughly. Cut kiwi into 3/8-inch slices. Garnish with kiwifruit. Place 1 slice kiwifruit on each portion.
4. Cover; CCP: Hold for service at 41 F. or lower.

DESSERTS (PUDDINGS AND OTHER DESSERTS) No. J 006 05
FRUIT COCKTAIL FRUIT CUP

Yield 100   **Portion** 1/2 Cup

Calories	Carbohydrates	Protein	Fat	Cholesterol	Sodium	Calcium
58 cal	15 g	1 g	0 g	0 mg	4 mg	14 mg

Ingredient	Weight	Measure	Issue
ORANGE,FRESH,CHOPPED	3-1/8 lbs	1 qts 3-7/8 cup	4-1/4 lbs
FRUIT COCKTAIL,CANNED,JUICE PACK,INCL LIQUIDS	20-1/4 lbs	2 gal 1-2/3 qts	
APPLES,FRESH,MEDIUM,UNPEELED,DICED	3-1/3 lbs	3 qts 1/8 cup	3-7/8 lbs

Method

1. Quickly combine apples and oranges with canned fruit cocktail to prevent discoloration; mix thoroughly.
2. Cover; CCP: Hold for service at 41 F. or lower.

DESSERTS (PUDDINGS AND OTHER DESSERTS) No. J 006 06
SPICED FRUIT CUP

Yield 100   Portion   1/2 Cup

Calories	Carbohydrates	Protein	Fat	Cholesterol	Sodium	Calcium
58 cal	15 g	0 g	0 g	0 mg	3 mg	18 mg

Ingredient	Weight	Measure	Issue
FRUIT COCKTAIL,CANNED,JUICE PACK,INCL LIQUIDS	12-1/2 lbs	1 gal 2 qts	
CINNAMON,GROUND	1/8 oz	1/8 tsp	
NUTMEG,GROUND	1/8 oz	1/3 tsp	
SUGAR,BROWN,PACKED	8-1/2 oz	1-5/8 cup	
APPLES,FRESH,MEDIUM,UNPEELED,DICED	4 lbs	3 qts 2-1/2 cup	4-3/4 lbs
ORANGE,FRESH,SECTIONS,PEELED,DICED	5-1/4 lbs	3 qts 1-3/8 cup	18-1/3 each

Method

1. Drain fruit cocktail and reserve juice for Step 2. Combine drained juice with ground cinnamon, ground nutmeg, and packed brown sugar. Bring to a boil; reduce heat; simmer 5 minutes. Chill.
2. Combine fruit cocktail, apples and oranges. Pour chilled syrup over fruits; mix lightly.
3. Cover; CCP: Hold for service at 41 F. or lower.

DESSERTS (PUDDINGS AND OTHER DESSERTS) No.J 006 07

MANDARIN ORANGE AND PINEAPPLE FRUIT CUP

Yield 100   Portion   1/2 Cup

Calories	Carbohydrates	Protein	Fat	Cholesterol	Sodium	Calcium
74 cal	19 g	1 g	0 g	0 mg	4 mg	20 mg

Ingredient

Ingredient	Weight	Measure	Issue
PINEAPPLE,CANNED,CHUNKS,JUICE PACK,DRAINED	20-1/4 lbs	1 gal 3-3/8 qts	
ORANGES,MANDARIN,CANNED,DRAINED	15-1/4 lbs	1 gal 3 qts	
CHERRIES,MARASCHINO,HALVES	1-1/8 lbs	2 cup	

Method

1. Combine pineapple and mandarin oranges.
2. Top each portion with 1/2 a maraschino cherry, if desired.
3. Cover; CCP: Hold for service at 41 F. or lower.

DESSERTS (PUDDINGS AND OTHER DESSERTS) No. J 007 00
FRUIT GELATIN

Yield 100   **Portion** 2/3 Cup

Calories	Carbohydrates	Protein	Fat	Cholesterol	Sodium	Calcium
116 cal	28 g	2 g	0 g	0 mg	65 mg	8 mg

Ingredient	**Weight**	**Measure**	**Issue**
FRUIT COCKTAIL,CANNED,JUICE PACK,INCL LIQUIDS	12-1/2 lbs	1 gal 2 qts	
DESSERT POWDER,GELATIN,STRAWBERRY	5-1/8 lbs	2 qts 2-1/2 cup	
WATER,BOILING	12-1/2 lbs	1 gal 2 qts	
RESERVED LIQUID	6-1/4 lbs	3 qts	
WATER,COLD	6-1/4 lbs	3 qts	

Method

1 Drain fruit; reserve juice for use in Step 3 and fruit for use in Step 5.
2 Dissolve gelatin in boiling water.
3 Add juice and water; stir to mix well.
4 Pour about 1 gallon into each pan. Chill until slightly thickened.
5 Fold an equal quantity of fruit into gelatin in each pan. Chill until firm. CCP: Hold for service at 41 F. or lower.

Notes

1 In Step 1, 2 No. 10 cans of the following canned fruit may be used per 100 servings: Canned Fruit Cocktail, Canned Mixed Fruit Chunks, Canned Peaches, quarters or slices, Canned Pears, quarters or slices, Canned Pineapple, chunks or tidbits.

DESSERTS (PUDDINGS AND OTHER DESSERTS)   No. J 007 01
BANANA GELATIN

Yield 100   Portion   2/3 Cup

Calories	Carbohydrates	Protein	Fat	Cholesterol	Sodium	Calcium
130 cal	32 g	2 g	0 g	0 mg	64 mg	6 mg

Ingredient | Weight | Measure | Issue

Ingredient	Weight	Measure	Issue
DESSERT POWDER,GELATIN,STRAWBERRY	5-1/8 lbs	2 qts 2-1/2 cup	
WATER,BOILING	12-1/2 lbs	1 gal 2 qts	
WATER,COLD	12-1/2 lbs	1 gal 2 qts	
BANANA,FRESH,SLICED	9-3/4 lbs	1 gal 3-3/8 qts	15 lbs

Method

1 Dissolve gelatin in boiling water.
2 Add cold water; stir to mix well.
3 Pour 1 gallon into each pan. Chill until slightly thickened.
4 Fold 2-1/4 quarts of banana into gelatin in each pan. Chill until firm.

DESSERTS (PUDDINGS AND OTHER DESSERTS) No. J 007 02
FRUIT FLAVORED GELATIN

Yield 100   **Portion** 1/2 Cup

Calories	Carbohydrates	Protein	Fat	Cholesterol	Sodium	Calcium
90 cal	21 g	2 g	0 g	0 mg	63 mg	3 mg

Ingredient

Ingredient	Weight	Measure	Issue
DESSERT POWDER,GELATIN,STRAWBERRY	5-1/8 lbs	2 qts 2-1/2 cup	
WATER,BOILING	12-1/2 lbs	1 gal 2 qts	
WATER,COLD	12-1/2 lbs	1 gal 2 qts	

Method

1. Dissolve gelatin in boiling water.
2. Add water; stir to mix well.
3. Pour 1 gallon into each steam table pan. Chill until firm.

DESSERTS (PUDDINGS AND OTHER DESSERTS) No.J 007 03

FRUIT GELATIN (CRUSHED ICE METHOD)

Yield 100 **Portion** 2/3 Cup

Calories	Carbohydrates	Protein	Fat	Cholesterol	Sodium	Calcium
116 cal	28 g	2 g	0 g	0 mg	65 mg	7 mg

Ingredient

Ingredient	Weight	Measure	Issue
FRUIT COCKTAIL,CANNED,JUICE PACK,INCL LIQUIDS	12-1/2 lbs	1 gal 2 qts	
DESSERT POWDER,GELATIN,STRAWBERRY	5-1/8 lbs	2 qts 2-1/2 cup	
RESERVED LIQUID	9-3/8 lbs	1 gal 1/2 qts	
ICE CUBES	12-1/2 lbs	3 gal 3-5/8 qts	

Method

1. Drain fruit; reserve juice for use in Step 2 and fruit for use in Step 3.
2. Dissolve gelatin in boiling water and juice.
3. Crush the ice. Add crushed ice, stirring constantly until ice is melted and gelatin begins to thicken. Add fruit; stir until blended; pour into pans. Chill until firm.

Notes

1. In Step 1, 2 No. 10 cans of the following canned fruit may be used per 100 servings: Canned Fruit Cocktail, Canned Mixed Fruit Chunks, Canned Peaches, quarters or slices, Canned Pears, quarters or slices, Canned Pineapple, chunks or tidbits.

DESSERTS (PUDDINGS AND OTHER DESSERTS) No.J 007 04
STRAWBERRY GELATIN

Yield 100   **Portion** 2/3 Cup

Calories	Carbohydrates	Protein	Fat	Cholesterol	Sodium	Calcium
117 cal	28 g	2 g	0 g	0 mg	64 mg	14 mg

Ingredient   Weight   Measure   Issue

Ingredient	Weight	Measure	Issue
DESSERT POWDER,GELATIN,STRAWBERRY	5-1/8 lbs	2 qts 2-1/2 cup	
WATER,BOILING	14-5/8 lbs	1 gal 3 qts	
STRAWBERRIES,FROZEN,THAWED	16-7/8 lbs	1 gal 3-1/2 qts	
JUICE,LEMON	4-1/3 oz	1/2 cup	

Method

1. Dissolve strawberry flavored gelatin in boiling water.
2. Add strawberries and lemon juice to gelatin. Stir until strawberries are completely thawed and separated.
3. Pour 5-1/2 quarts of gelatin mixture into each pan. Chill until firm.

DESSERTS (PUDDINGS AND OTHER DESSERTS)   No. J 007 05

PEACH GELATIN

Yield 100   **Portion** 2/3 Cup

Calories	Carbohydrates	Protein	Fat	Cholesterol	Sodium	Calcium
160 cal	39 g	2 g	0 g	0 mg	66 mg	4 mg

Ingredient | Weight | Measure | Issue

Ingredient	Weight	Measure	Issue
DESSERT POWDER,GELATIN,ORANGE	5-1/8 lbs	2 qts 2-1/2 cup	
WATER,BOILING	14-5/8 lbs	1 gal 3 qts	
PEACHES,FROZEN	16-1/2 lbs	1 gal 3-1/2 qts	

Method

1. Dissolve orange flavored gelatin in boiling water.
2. Add partially thawed sliced or quartered peaches to orange flavored gelatin. Stir peaches until thawed and separated.
3. Pour 5-1/2 quarts into each pan. Chill until firm.

DESSERTS (PUDDINGS AND OTHER DESSERTS) No.J 008 00
PEACH CRISP

Yield 100   Portion 1 Piece

Calories	Carbohydrates	Protein	Fat	Cholesterol	Sodium	Calcium
203 cal	32 g	2 g	8 g	0 mg	211 mg	23 mg

Ingredient	Weight	Measure	Issue
PEACHES,CANNED,SLICED	19-2/3 lbs	2 gal 1 qts	
COOKING SPRAY,NONSTICK	2 oz	1/4 cup 1/3 tbsp	
SUGAR,GRANULATED	1 lbs	2-1/4 cup	
FLOUR,WHEAT,GENERAL PURPOSE	6-5/8 oz	1-1/2 cup	
SALT	1/4 oz	1/8 tsp	
CINNAMON,GROUND	1/4 oz	1 tbsp	
NUTMEG,GROUND	1/8 oz	1/3 tsp	
FLOUR,WHEAT,GENERAL PURPOSE	1-3/8 lbs	1 qts 1 cup	
BAKING POWDER	1/4 oz	1/4 tsp	
BAKING SODA	1/4 oz	1/4 tsp	
SALT	5/8 oz	1 tbsp	
CEREAL,OATMEAL,ROLLED	1 lbs	2-7/8 cup	
SUGAR,BROWN,PACKED	1-2/3 lbs	1 qts 1-3/8 cup	
MARGARINE,SOFTENED	2 lbs	1 qts	

Method

1. Drain fruit; reserve juice.
2. Spray each pan with non-stick cooking spray. Arrange about 3 quarts of peaches in each pan. Pour 3 cups reserve juice over peaches in each pan.
3. Combine sugar, flour, salt, cinnamon, and nutmeg; sprinkle about 2 cups evenly over peaches in each pan. Stir lightly to moisten flour mixture.
4. Combine flour, baking powder, baking soda, salt, rolled oats, brown sugar and margarine; mix only until blended.
5. Sprinkle 2-1/2 quarts of mixture over the fruit in each pan.
6. Using a convection oven, bake at 350 F. for 30 minutes or until top is lightly browned on low fan, open vent.
7. Cut 6 by 9 and serve with serving spoon or spatula.

DESSERTS (PUDDINGS AND OTHER DESSERTS) No. J 008 01
CHERRY CRISP (PIE FILLING COOKIE MIX)

Yield 100   **Portion** 1 Piece

Calories	Carbohydrates	Protein	Fat	Cholesterol	Sodium	Calcium
215 cal	41 g	1 g	6 g	8 mg	69 mg	30 mg

Ingredient   Weight   Measure   Issue

Ingredient	Weight	Measure
PIE FILLING,CHERRY,PREPARED	24-1/2 lbs	3 gal 1/4 qts
COOKING SPRAY,NONSTICK	2 oz	1/4 cup 1/3 tbsp
COOKIE MIX,OATMEAL	4-1/2 lbs	
MARGARINE,SOFTENED	1 lbs	2 cup

Method

1. Pour 5-1/2 quarts of prepared pie filling into each pan.
2. Combine cookie mix and margarine. Sprinkle half of mixture evenly over cherries in each pan.
3. Using a convection oven, bake at 350 F. for 30 minutes or until top is lightly browned on low fan, open vent.
4. Cut 6 by 9. Serve with serving spoon or spatula.

DESSERTS (PUDDINGS AND OTHER DESSERTS) No.J 008 02
CHERRY CRISP

Yield 100   Portion 1 Piece

Calories	Carbohydrates	Protein	Fat	Cholesterol	Sodium	Calcium
232 cal	39 g	2 g	8 g	0 mg	209 mg	30 mg

Ingredient	Weight	Measure	Issue
CHERRIES,CANNED,RED,TART,WATER PACK,INCL LIQUIDS	24-1/2 lbs	2 gal 3-1/4 qts	
COOKING SPRAY,NONSTICK	2 oz	1/4 cup 1/3 tbsp	
SUGAR,GRANULATED	2 lbs	1 qts 1/2 cup	
FLOUR,WHEAT,GENERAL PURPOSE	6-5/8 oz	1-1/2 cup	
SALT	1/4 oz	1/8 tsp	
CINNAMON,GROUND	1/4 oz	1 tbsp	
NUTMEG,GROUND	1/8 oz	1/3 tsp	
FLOUR,WHEAT,GENERAL PURPOSE	1-3/8 lbs	1 qts 1 cup	
BAKING POWDER	1/4 oz	1/4 tsp	
BAKING SODA	1/4 oz	1/4 tsp	
SALT	5/8 oz	1 tbsp	
CEREAL,OATMEAL,ROLLED	1 lbs	2-7/8 cup	
SUGAR,BROWN,PACKED	1-2/3 lbs	1 qts 1-3/8 cup	
MARGARINE,SOFTENED	2 lbs	1 qts	

Method

1 Drain fruit; reserve juice for use in Step 2.
2 Lightly spray pans with non-stick cooking spray. Arrange about 3 quarts of cherries in each sprayed pan. Pour 3 cups reserve juice over cherries in each pan.
3 Combine sugar, flour, salt, cinnamon, and nutmeg; sprinkle about 2 cups evenly over cherries in each pan. Stir lightly to moisten flour mixture.
4 Combine flour, baking powder, baking soda, salt, rolled oats, brown sugar, margarine; mix only until blended.
5 Sprinkle 2-1/2 quart mixture over fruit in each pan.
6 Using a convection oven, bake at 350 F. for 30 minutes or until top is lightly browned on low fan, open vent.
7 Cut 6 by 9 and serve with serving spoon or spatula.

DESSERTS (PUDDINGS AND OTHER DESSERTS) No.J 008 03
PEACH CRISP (PIE FILLING COOKIE MIX)

Yield 100   Portion 1 Piece

Calories	Carbohydrates	Protein	Fat	Cholesterol	Sodium	Calcium
250 cal	51 g	1 g	6 g	8 mg	96 mg	33 mg

Ingredient

Ingredient	Weight	Measure	Issue
PIE FILLING,PEACH,PREPARED	24-1/2 lbs	3 gal 1/4 qts	
COOKING SPRAY,NONSTICK	2 oz	1/4 cup 1/3 tbsp	
COOKIE MIX,OATMEAL	4-1/2 lbs		
MARGARINE,SOFTENED	1 lbs	2 cup	

Method

1 Lightly spray each pan with non-stick cooking spray. Pour about 5-1/2 quarts of pie filling into each sprayed pan.
2 Combine canned oatmeal cookie mix with softened margarine; mix until crumbly.
3 Sprinkle 2-1/2 quarts of mixture over fruit in each pan.
4 Using a convection oven, bake at 350 F. for 30 minutes or until top is lightly browned on low fan, open vent.
5 Cut 6 by 9 and serve with serving spoon or spatula.

DESSERTS (PUDDINGS AND OTHER DESSERTS)   No. J 008 04

BLUEBERRY CRISP (PIE FILLING COOKIE MIX)

Yield 100   **Portion** 1 Serving

Calories	Carbohydrates	Protein	Fat	Cholesterol	Sodium	Calcium
219 cal	35 g	1 g	10 g	8 mg	158 mg	44 mg

Ingredient   Weight   Measure   Issue

Ingredient	Weight	Measure	Issue
PIE FILLING, BLUEBERRY, PREPARED	24-1/2 lbs	2 gal 2-3/8 qts	
COOKING SPRAY, NONSTICK	2 oz	1/4 cup 1/3 tbsp	
COOKIE MIX, OATMEAL	4-1/2 lbs		
MARGARINE, SOFTENED	2 lbs	1 qts	

Method

1. Lightly spray each pan with non-stick cooking spray. Pour about 5-1/2 quarts of pie filling into each sprayed pan.
2. Combine canned oatmeal cookie mix with margarine; mix until crumbly.
3. Sprinkle 2-1/2 quarts of mixture over fruit in each pan.
4. Using a convection oven, bake at 350 F. for 30 minutes or until top is lightly browned on low fan, open vent.
5. Cut 6 by 9 and serve with serving spoon or spatula.

DESSERTS (PUDDINGS AND OTHER DESSERTS) No. J 010 01
APPLE CRUNCH (APPLE PIE FILLING)

Yield 100   **Portion** 1 Piece

Calories	Carbohydrates	Protein	Fat	Cholesterol	Sodium	Calcium
231 cal	42 g	2 g	7 g	6 mg	236 mg	18 mg

Ingredient	Weight	Measure	Issue
PIE FILLING,APPLE,PREPARED	18 lbs	2 gal 1 qts	
COOKING SPRAY,NONSTICK	2 oz	1/4 cup 1/3 tbsp	
JUICE,LEMON	3-1/4 oz	1/4 cup 2-1/3 tbsp	
CAKE MIX,YELLOW	5 lbs		
COCONUT,PREPARED,SWEETENED FLAKES	1 lbs	1 qts 1 cup	
MARGARINE,SOFTENED	1 lbs	2 cup	

Method

1. Spray each pan with non-stick cooking spray. Spread 4-1/2 quarts filling in each sprayed sheet pan. Sprinkle 3 tablespoons of lemon juice on top of mixture in each pan.
2. Combine cake mix and coconut; add margarine; mix until crumbly.
3. Sprinkle 2-3/4 quarts of mixture over each pan.
4. Using a convection oven, bake at 325 F. for 30 minutes or until lightly brown on low fan, open vent.
5. Cut 6 by 9.

Notes

1. In Step 2, 1 pound chopped unsalted nuts may be substituted for coconut per 100 servings.

DESSERTS (PUDDINGS AND OTHER DESSERTS)   No. J 010 02
BLUEBERRY CRUNCH (BLUEBERRY PIE FILLING)

Yield 100   Portion 1 Serving

Calories	Carbohydrates	Protein	Fat	Cholesterol	Sodium	Calcium
235 cal	43 g	2 g	7 g	6 mg	249 mg	36 mg

Ingredient	Weight	Measure	Issue
PIE FILLING,BLUEBERRY,PREPARED	21-1/4 lbs	2 gal 1 qts	
COOKING SPRAY,NONSTICK	2 oz	1/4 cup 1/3 tbsp	
JUICE,LEMON	3-1/4 oz	1/4 cup 2-1/3 tbsp	
CAKE MIX,YELLOW	5 lbs		
COCONUT,PREPARED,SWEETENED FLAKES	1 lbs	1 qts 1 cup	
MARGARINE,SOFTENED	1 lbs	2 cup	

Method

1. Spray each pan with non-stick cooking spray. Spread 4-1/2 quarts of pie filling into each sprayed sheet pan. Sprinkle 3 tablespoons of lemon juice on top of mixture in each pan.
2. Combine cake mix and coconut; add margarine; mix until crumbly.
3. Sprinkle 2-3/4 quarts of mixture over each pan.
4. Using a convection oven, bake at 325 F. for 30 minutes or until lightly browned on low fan, open vent.
5. Cut 6 by 9.

Notes

1. In Step 2, 1 pound chopped unsalted nuts may be substituted for coconut, per 100 servings.

DESSERTS (PUDDINGS AND OTHER DESSERTS) No.J 010 03
CHERRY CRUNCH (CHERRY PIE FILLING)

Yield 100   Portion 1 Piece

Calories	Carbohydrates	Protein	Fat	Cholesterol	Sodium	Calcium
243 cal	44 g	2 g	7 g	6 mg	207 mg	24 mg

Ingredient	Weight	Measure	Issue
PIE FILLING,CHERRY,PREPARED	18 lbs	2 gal 1 qts	
COOKING SPRAY,NONSTICK	2 oz	1/4 cup 1/3 tbsp	
JUICE,LEMON	3-1/4 oz	1/4 cup 2-1/3 tbsp	
CAKE MIX,YELLOW	5 lbs		
COCONUT,PREPARED,SWEETENED FLAKES	1 lbs	1 qts 1 cup	
MARGARINE,SOFTENED	1 lbs	2 cup	

Method

1 Spray each pan with non-stick cooking spray. Spread 4-1/2 quarts of pie filling into each sprayed sheet pan. Sprinkle 3 tablespoons of lemon juice on top of mixture in each pan.
2 Combine cake mix and coconut; add margarine; mix until crumbly.
3 Sprinkle 2-3/4 quarts of mixture over each pan.
4 Using a convection oven, bake at 325 F. for 30 minutes on low fan, open vent or until lightly browned.
5 Cut 6 by 9.

Notes

1 In Step 2, 1 pound chopped unsalted nuts may be substituted for coconut, per 100 servings.

DESSERTS (PUDDINGS AND OTHER DESSERTS) No. J 010 05
PEACH CRUNCH (PEACH PIE FILLING)

Yield 100 **Portion** 1 Piece

Calories	Carbohydrates	Protein	Fat	Cholesterol	Sodium	Calcium
269 cal	52 g	2 g	7 g	6 mg	227 mg	26 mg

Ingredient	Weight	Measure	Issue
PIE FILLING,PEACH,PREPARED	18 lbs	2 gal 1 qts	
COOKING SPRAY,NONSTICK	2 oz	1/4 cup 1/3 tbsp	
JUICE,LEMON	3-1/4 oz	1/4 cup 2-1/3 tbsp	
CAKE MIX,YELLOW	5 lbs		
COCONUT,PREPARED,SWEETENED FLAKES	1 lbs	1 qts 1 cup	
MARGARINE,SOFTENED	1 lbs	2 cup	

Method

1. Spray each pan with non-stick cooking spray. Spread 4-1/2 quart filling in each sprayed sheet pan. Sprinkle 3 tablespoons of lemon juice on top of mixture in each pan.
2. Combine cake mix and coconut; add margarine and butter; mix until crumbly.
3. Sprinkle 2-3/4 quarts of mixture over each pan.
4. Using a convection oven bake at 325 F. for 30 minutes or until lightly browned on low fan, open vent.
5. Cut 6 by 9.

Notes

1. In Step 2, 1 pound chopped unsalted nuts may be used for coconut per 100 servings.

ESSERTS (PUDDINGS AND OTHER DESSERTS) No.J 011 00
BANANA SPLIT

Yield 100   Portion 1 Each

Calories	Carbohydrates	Protein	Fat	Cholesterol	Sodium	Calcium
361 cal	53 g	5 g	16 g	30 mg	110 mg	132 mg

Ingredient	Weight	Measure	Issue
BANANA,FRESH	13 lbs		20 lbs
JUICE,ORANGE	1-1/8 lbs	2 cup	
ICE CREAM,VANILLA	15-1/8 lbs	3 gal 1 qts	
ICE CREAM TOPPING,FUDGE	8-5/8 lbs	3 qts 1 cup	
WHIPPED TOPPING,12 OZ CAN	1-1/4 lbs	2 qts	
PECANS,CHOPPED	8 oz		
CHERRIES,MARASCHINO,SLICED	1-1/8 lbs	2 cup	

Method

1 Peel and slice bananas lengthwise into quarters; place on pan.
2 Pour juice over bananas; cover with waxed paper; refrigerate until ready to serve.
3 Make banana splits to order. Place 1/2 cup ice cream in soup bowl. Drain 2 banana quarters; place 1 on each side of ice cream. Ladle 2 tablespoons of topping over ice cream. Top with 1 tablespoon whipped topping, 1 teaspoon chopped pecans and 1/2 maraschino cherry.

Notes

1 In Step 3, Chocolate Sauce, Recipe No. K 005 00, or Butterscotch, Fudge, Marshmallow, Pineapple, or Strawberry Topping, or Whipped Topping, Recipe No. K 002 00 may be used.

DESSERTS (PUDDINGS AND OTHER DESSERTS) No. J 012 00

VANILLA SOFT SERVE YOGURT (DEHYDRATED)

Yield 100 **Portion** 3/4 Cup

Calories	Carbohydrates	Protein	Fat	Cholesterol	Sodium	Calcium
166 cal	40 g	1 g	0 g	1 mg	71 mg	29 mg

Ingredient	Weight	Measure	Issue
YOGURT MIX,DEHYDRATED,VANILLA	10 lbs		
WATER	10 lbs		

Method

1. Stir dehydrated mix into water. Mix thoroughly with wire whip or mixer. Cover container.
2. Chill 4 to 24 hours in refrigerator to 35 F. to 40 F.
3. Stir until smooth. Pour mixture into top hopper to soft serve ice cream freezer; start dasher motor; turn on refrigeration according to manufacturer's directions. Freeze to temperature of 18 F. to 22 F., about 10 minutes, or until product can be drawn with a stiff consistency that will hold a peak.

DESSERTS (PUDDINGS AND OTHER DESSERTS) No.J 012 01
CHOCOLATE SOFT SERVE YOGURT (DEHYDRATED)

Yield 100   Portion   3/4 Cup

Calories	Carbohydrates	Protein	Fat	Cholesterol	Sodium	Calcium
166 cal	40 g	1 g	1 g	1 mg	169 mg	30 mg

Ingredient   Weight   Measure   Issue

YOGURT MIX,DEHYDRATED,CHOCOLATE   10 lbs
WATER   10 lbs

Method

1. Stir dehydrated mix into water. Mix thoroughly with wire whip or mixer. Cover container.
2. Chill 4 to 24 hours in refrigerator to 35 F. to 40 F.
3. Stir until smooth. Pour mixture into top hopper to soft serve ice cream freezer; start dasher motor; turn on refrigeration according to manufacturer's directions. Freeze to temperature of 18 F. to 22 F., about 10 minutes, or until product can be drawn with a stiff consistency that will hold a peak.

DESSERTS (PUDDINGS AND OTHER DESSERTS) No. J 013 00
TAPIOCA PUDDING

Yield 100   **Portion** 1/2 Cup

Calories	Carbohydrates	Protein	Fat	Cholesterol	Sodium	Calcium
119 cal	21 g	3 g	3 g	36 mg	139 mg	92 mg

Ingredient	Weight	Measure	Issue
MILK,NONFAT,DRY	1-1/2 lbs	2 qts 2 cup	
WATER,WARM	23-1/2 lbs	2 gal 3-1/4 qts	
BUTTER	8 oz	1 cup	
TAPIOCA,QUICK-COOKING	14-1/3 oz	2-5/8 cup	
SUGAR,GRANULATED	3 lbs	1 qts 2-3/4 cup	
SALT	5/8 oz	1 tbsp	
EGGS,WHOLE,FROZEN	1-1/2 lbs	2-7/8 cup	
EXTRACT,VANILLA	1-3/8 oz	3 tbsp	

Method

1. Reconstitute milk. Reserve 2 cups for use in Step 3.
2. Heat remaining milk in steam jacketed kettle or stock pot to a boil. Add butter or margarine.
3. Combine reserved milk with tapioca, sugar, salt, and eggs.
4. Add tapioca mixture to hot milk in steam-jacketed kettle or stock pot. Bring to just a boil; reduce heat; cook without boiling, stirring occasionally until slightly thickened, about 5 minutes. The mixture will be thin. Turn off heat; cool in kettle 15 to 20 minutes.
5. Add vanilla; blend well. Pour 1 gallon into each pan. Cover surface of pudding with waxed paper. Refrigerate until ready to serve. Mixture will thicken as it cools. CCP: Hold for service at 41 F. or lower.

Notes

1. Garnish with Whipped Topping, Recipe No. K 002 00 and maraschino cherry half (optional).

DESSERTS (PUDDINGS AND OTHER DESSERTS) No.J 014 00

VANILLA CREAM PUDDING (INSTANT)

Yield 100   Portion 1/2 Cup

Calories	Carbohydrates	Protein	Fat	Cholesterol	Sodium	Calcium
136 cal	32 g	2 g	0 g	1 mg	503 mg	77 mg

Ingredient	Weight	Measure	Issue
MILK,NONFAT,DRY	1-1/3 lbs	2 qts 3/4 cup	
WATER,COLD	23 lbs	2 gal 3 qts	
DESSERT POWDER,PUDDING,INSTANT,VANILLA	6-7/8 lbs	1 gal 1/4 qts	

Method

1. Reconstitute milk. Chill to 50 F. Place in mixer bowl.
2. Add dessert powder. Using whip, blend at low speed 15 seconds or until well blended. Scrape sides and bottom of bowl; whip at medium speed 2 minutes or until smooth.
3. Pour 4-1/2 quarts pudding into each pan. Cover surface of pudding with waxed paper.
4. Refrigerate at least 1 hour or until ready to serve. Pudding may be garnished with well-drained fruit or whipped topping.
   CCP: Hold for service at 41 F. or lower.

DESSERTS (PUDDINGS AND OTHER DESSERTS) No. J 014 01
BANANA CREAM PUDDING (INSTANT)

Yield 100 **Portion** 1/2 Cup

Calories	Carbohydrates	Protein	Fat	Cholesterol	Sodium	Calcium
134 cal	32 g	2 g	0 g	1 mg	403 mg	63 mg

Ingredient	Weight	Measure	Issue
MILK,NONFAT,DRY	1 lbs	1 qts 3 cup	
WATER,COLD	18-1/4 lbs	2 gal 3/4 qts	
DESSERT POWDER,PUDDING,INSTANT,VANILLA	5-1/2 lbs	3 qts 1-1/2 cup	
BANANA,FRESH,SLICED	6 lbs	1 gal 1/2 qts	9-1/4 lbs

Method

1. Reconstitute milk. Chill to 50 F. Place in mixer bowl.
2. Add dessert powder. Using whip, blend at low speed 15 seconds or until well blended. Scrape sides and bottom of bowl; whip at medium speed 2 minutes or until smooth.
3. Pour 3-2/3 quarts pudding into each pan. Fold 1-1/2 quarts of banana into each pan. Cover surface of pudding with waxed paper.
4. Refrigerate at least 1 hour or until ready to serve. Pudding may be garnished with well-drained fruit or whipped topping. CCP: Hold for service at 41 F. or lower.

Notes

1. To prevent discoloration, slice bananas just before adding to pudding.

DESSERTS (PUDDINGS AND OTHER DESSERTS) No.J 014 02
COCONUT CREAM PUDDING (INSTANT)

Yield 100　　　　　　　　　　　　　　　　　　Portion　1/2 Cup

Calories	Carbohydrates	Protein	Fat	Cholesterol	Sodium	Calcium
178 cal	36 g	2 g	3 g	1 mg	525 mg	78 mg

Ingredient	Weight	Measure	Issue
MILK,NONFAT,DRY	1-1/3 lbs	2 qts 3/4 cup	
WATER,COLD	23 lbs	2 gal 3 qts	
DESSERT POWDER,PUDDING,INSTANT,VANILLA	6-7/8 lbs	1 gal 1/4 qts	
COCONUT,PREPARED,SWEETENED FLAKES	1-7/8 lbs	2 qts 1 cup	

Method

1. Reconstitute milk. Chill to 50 F. Place in mixer bowl.
2. Add dessert powder. Using whip, blend at low speed 15 seconds or until well blended. Scrape sides and bottom of bowl; whip at medium speed 2 minutes or until smooth.
3. Pour 4-1/2 quarts pudding into each pan. Fold coconut into pudding. Cover surface of pudding with waxed paper.
4. Refrigerate at least 1 hour or until ready to serve. Pudding may be garnished with well drained fruit or whipped topping. CCP: Hold for service at 41 F. or lower.

DESSERTS (PUDDINGS AND OTHER DESSERTS) No.J 014 03
PINEAPPLE CREAM PUDDING (INSTANT)

Yield 100   **Portion** 1/2 Cup

Calories	Carbohydrates	Protein	Fat	Cholesterol	Sodium	Calcium
128 cal	31 g	2 g	0 g	1 mg	403 mg	66 mg

Ingredient | Weight | Measure | Issue

Ingredient	Weight	Measure	Issue
MILK,NONFAT,DRY	1 lbs	1 qts 3 cup	
WATER,COLD	18-1/4 lbs	2 gal 3/4 qts	
DESSERT POWDER,PUDDING,INSTANT,VANILLA	5-1/2 lbs	3 qts 1-1/2 cup	
PINEAPPLE,CANNED,CRUSHED,JUICE PACK,DRAINED	7-1/4 lbs	1 #10cn	

Method

1. Reconstitute milk. Chill to 50 F. Place in mixer bowl.
2. Add vanilla dessert powder. Using whip, blend at low speed 15 seconds or until well blended. Scrape sides and bottom of bowl; whip at medium speed 2 minutes or until smooth.
3. Pour 3-2/3 quarts of pudding into each pan. Fold drained pineapple into pudding. Cover surface of pudding with waxed paper.
4. Refrigerate at least 1 hour or until ready to serve. Pudding may be garnished with well-drained fruit or whipped topping. CCP: Hold for service at 41 F. or lower.

DESSERTS (PUDDINGS AND OTHER DESSERTS)   No.J 014 04

BUTTERSCOTCH CREAM PUDDING (INSTANT)

Yield 100   **Portion**   1/2 Cup

Calories	Carbohydrates	Protein	Fat	Cholesterol	Sodium	Calcium
136 cal	32 g	2 g	0 g	1 mg	485 mg	79 mg

Ingredient
	Weight	Measure	Issue
MILK,NONFAT,DRY	1-1/3 lbs	2 qts 3/4 cup	
WATER,COLD	23 lbs	2 gal 3 qts	
DESSERT POWDER,PUDDING,INSTANT,BUTTERSCOTCH	6-7/8 lbs		

Method

1. Reconstitute milk. Chill to 50 F. Place in mixer bowl.
2. Add butterscotch dessert powder. Using whip, blend at low speed 15 seconds or until well blended. Scrape sides and bottom of bowl; whip at medium speed 2 minutes or until smooth.
3. Pour 4-1/2 quarts pudding into each pan. Cover surface of pudding with waxed paper.
4. Refrigerate at least 1 hour or until ready to serve. Pudding may be garnished with well-drained fruit or whipped topping. CCP: Hold for service at 41 F. or lower.

DESSERTS (PUDDINGS AND OTHER DESSERTS) No.J 014 05
CHOCOLATE CREAM PUDDING (INSTANT)

Yield 100   **Portion** 1/2 Cup

Calories	Carbohydrates	Protein	Fat	Cholesterol	Sodium	Calcium
128 cal	30 g	3 g	1 g	1 mg	465 mg	81 mg

Ingredient   Weight   Measure   Issue

Ingredient	Weight	Measure	Issue
MILK,NONFAT,DRY	1-1/3 lbs	2 qts 3/4 cup	
WATER,COLD	22-1/4 lbs	2 gal 2-2/3 qts	
DESSERT POWDER,PUDDING,INSTANT,CHOCOLATE	6-5/8 lbs	1 gal 3/8 qts	

Method

1. Reconstitute milk. Chill to 50 F. Place in mixer bowl.
2. Add chocolate dessert powder. Using whip, blend at low speed 15 seconds or until well blended. Scrape sides and bottom of bowl; whip at medium speed 2 minutes or until smooth.
3. Pour 4-1/2 quarts pudding into each pan. Cover surface of pudding with waxed paper.
4. Refrigerate at least 1 hour or until ready to serve. Pudding may be garnished with well-drained fruit or whipped topping. CCP: Hold for service at 41 F. or lower.

DESSERTS (PUDDINGS AND OTHER DESSERTS) No.J 015 00
BAKED RICE PUDDING

Yield 100   **Portion**   1/2 Cup

Calories	Carbohydrates	Protein	Fat	Cholesterol	Sodium	Calcium
173 cal	30 g	4 g	4 g	48 mg	156 mg	62 mg

Ingredient	Weight	Measure	Issue
RICE,LONG GRAIN	3-1/4 lbs	2 qts	
WATER,COLD	12-1/2 lbs	1 gal 2 qts	
SALT	3/4 oz	1 tbsp	
MILK,NONFAT,DRY	10-3/4 oz	1 qts 1/2 cup	
WATER,WARM	11-1/2 lbs	1 gal 1-1/2 qts	
EGGS,WHOLE,FROZEN	2-3/8 lbs	1 qts 1/2 cup	
MARGARINE,MELTED	12 oz	1-1/2 cup	
SUGAR,GRANULATED	2 lbs	1 qts 1/2 cup	
EXTRACT,VANILLA	1-3/8 oz	3 tbsp	
CINNAMON,GROUND	1/4 oz	1 tbsp	
NUTMEG,GROUND	1/8 oz	1/8 tsp	
COOKING SPRAY,NONSTICK	2 oz	1/4 cup 1/3 tbsp	
RAISINS	1-7/8 lbs	1 qts 2 cup	

Method

1. Combine rice, water, and salt. Bring to a boil, stirring occasionally. Reduce heat; cover tightly; simmer 20 to 25 minutes or until water is absorbed.
2. Reconstitute milk; add eggs, margarine or butter, sugar, vanilla, cinnamon and nutmeg; blend thoroughly.
3. Spray each pan with non-stick cooking spray. Place 1-3/4 quarts of cooked, cooled rice and 1-1/2 cup of raisins in each sprayed pan. Blend thoroughly.
4. Pour 2 quarts egg mixture over rice-raisin mixture in each pan.
5. Using a convection oven, bake at 325 F. 30 to 35 minutes or until lightly browned on low fan, open vent and a knife inserted in center comes out clean.
6. Cover, refrigerate until ready to serve. CCP: Hold for service at 41 F. or lower.
7. Cut 4 by 6.

DESSERTS (PUDDINGS AND OTHER DESSERTS) No.J 015 01
BAKED RICE PUDDING (FROZEN EGGS AND EGG WHITES)

Yield 100   Portion 1/2 Cup

Calories	Carbohydrates	Protein	Fat	Cholesterol	Sodium	Calcium
168 cal	30 g	4 g	4 g	24 mg	157 mg	59 mg

Ingredient	Weight	Measure	Issue
RICE,LONG GRAIN	3-1/4 lbs	2 qts	
WATER,COLD	12-1/2 lbs	1 gal 2 qts	
SALT	3/4 oz	1 tbsp	
MILK,NONFAT,DRY	10-3/4 oz	1 qts 1/2 cup	
WATER,WARM	11-1/2 lbs	1 gal 1-1/2 qts	
EGGS,WHOLE,FROZEN	1-1/4 lbs	2-1/4 cup	
EGG WHITES	1-1/4 lbs	2-1/4 cup	
MARGARINE,MELTED	12 oz	1-1/2 cup	
SUGAR,GRANULATED	2 lbs	1 qts 1/2 cup	
EXTRACT,VANILLA	1-3/8 oz	3 tbsp	
CINNAMON,GROUND	1/4 oz	1 tbsp	
NUTMEG,GROUND	1/8 oz	1/8 tsp	
COOKING SPRAY,NONSTICK	2 oz	1/4 cup 1/3 tbsp	
RAISINS	1-7/8 lbs	1 qts 2 cup	

Method

1. Combine rice, water, and salt. Bring to a boil, stirring occasionally. Reduce heat; cover tightly; simmer 20 to 25 minutes or until water is absorbed.
2. Reconstitute milk; add eggs, egg whites, margarine or butter, sugar, vanilla, cinnamon and nutmeg; blend thoroughly.
3. Spray each pan with non-stick cooking spray. Place 1-3/4 quarts cooked cooled rice and 1-1/2 cups raisins in each sprayed pan. Blend thoroughly.
4. Pour 2 quarts egg mixture over rice-raisin mixture in each pan.
5. Using a convection oven, bake 30 to 35 minutes in 325 F. oven or until lightly browned on low fan, open vent and a knife inserted in center comes out clean.
6. Cover, refrigerate until ready to serve. CCP: Hold for service at 41 F. or lower.
7. Cut 4 by 6.

DESSERTS (PUDDINGS AND OTHER DESSERTS) No.J 016 00
BREAD PUDDING

Yield 100   **Portion** 2/3 Cup

Calories	Carbohydrates	Protein	Fat	Cholesterol	Sodium	Calcium
205 cal	34 g	5 g	6 g	30 mg	310 mg	93 mg

Ingredient	Weight	Measure	Issue
BREAD,WHITE,CUBED	4-1/8 lbs	3 gal 1-1/2 qts	
COOKING SPRAY,NONSTICK	2 oz	1/4 cup 1/3 tbsp	
MARGARINE,MELTED	1 lbs	2 cup	
EGGS,WHOLE,FROZEN	1-1/2 lbs	2-3/4 cup	
EGG WHITES,FROZEN,THAWED	1-1/2 lbs	2-3/4 cup	
SUGAR,GRANULATED	2-2/3 lbs	1 qts 2 cup	
SALT	1 oz	1 tbsp	
NUTMEG,GROUND	1/4 oz	1 tbsp	
EXTRACT,VANILLA	1-7/8 oz	1/4 cup 1/3 tbsp	
MILK,NONFAT,DRY	1 lbs	1 qts 3 cup	
WATER,WARM	18-3/4 lbs	2 gal 1 qts	
RAISINS	2-7/8 lbs	2 qts 1 cup	

Method

1 Spray each pan with non-stick cooking spray. Place 4-1/2 quarts bread in each sprayed steam table pan. Pour margarine or butter over bread cubes and toss lightly. Toast in oven until light brown.
2 Add sugar, salt, nutmeg, and vanilla to eggs; blend thoroughly.
3 Reconstitute milk; combine with egg mixture. Pour 1 gallon over bread cubes in each pan.
4 Add 3 cups raisins to each pan.
5 Bake at 350 F. for 15 minutes: stir to distribute the raisins. Bake 45 minutes or until firm.
6 Cover; refrigerate until ready to serve. CCP: Hold for service at 41 F. or lower.
7 Cut 4 by 8.

DESSERTS (PUDDINGS AND OTHER DESSERTS) No. J 016 01
CHOCOLATE CHIP BREAD PUDDING

Yield 100   Portion   2/3 Cup

Calories	Carbohydrates	Protein	Fat	Cholesterol	Sodium	Calcium
218 cal	30 g	6 g	9 g	32 mg	318 mg	106 mg

Ingredient	Weight	Measure	Issue
BREAD,WHITE,CUBED	4-1/8 lbs	3 gal 1-1/2 qts	
COOKING SPRAY,NONSTICK	2 oz	1/4 cup 1/3 tbsp	
MARGARINE,MELTED	1 lbs	2 cup	
EGGS,WHOLE,FROZEN	1-1/2 lbs	2-3/4 cup	
EGG WHITES	1-1/2 lbs	2-3/4 cup	
SUGAR,GRANULATED	2-2/3 lbs	1 qts 2 cup	
SALT	1 oz	1 tbsp	
NUTMEG,GROUND	1/4 oz	1 tbsp	
EXTRACT,VANILLA	1-7/8 oz	1/4 cup 1/3 tbsp	
MILK,NONFAT,DRY	1 lbs	1 qts 3 cup	
WATER,WARM	18-3/4 lbs	2 gal 1 qts	
CHOCOLATE,COOKING CHIPS,SEMISWEET	2-1/4 lbs	1 qts 2-1/8 cup	

Method

1. Spray each pan with non-stick cooking spray. Place 4-1/2 quarts bread in each sprayed steam table pan. Pour margarine or butter over bread cubes, toss lightly. Toast in oven until light brown.
2. Add sugar, salt, nutmeg, and vanilla to eggs; blend thoroughly.
3. Reconstitute milk; combine with egg mixture. Pour 1 gallon over bread cubes in each pan.
4. Add 12 ounces of chocolate chips to each pan.
5. Bake 1 hour or until firm in 350 F. oven.
6. Cover; CCP: Hold for service at 41 F. or lower.
7. Cut 4 by 8.

DESSERTS (PUDDINGS AND OTHER DESSERTS) No.J 016 02
COCONUT BREAD PUDDING

Yield 100   Portion 2/3 Cup

Calories	Carbohydrates	Protein	Fat	Cholesterol	Sodium	Calcium
208 cal	28 g	5 g	8 g	30 mg	331 mg	88 mg

Ingredient | Weight | Measure | Issue

Ingredient	Weight	Measure	Issue
COOKING SPRAY,NONSTICK	2 oz	1/4 cup 1/3 tbsp	
BREAD,WHITE,CUBED	4-1/8 lbs	3 gal 1-1/2 qts	
COCONUT,PREPARED,SWEETENED FLAKES	1-7/8 lbs	2 qts 1 cup	
MARGARINE,MELTED	1 lbs	2 cup	
EGGS,WHOLE,FROZEN	1-1/2 lbs	2-3/4 cup	
EGG WHITES	1-1/2 lbs	2-3/4 cup	
SUGAR,GRANULATED	2-2/3 lbs	1 qts 2 cup	
SALT	1 oz	1 tbsp	
NUTMEG,GROUND	1/4 oz	1 tbsp	
EXTRACT,VANILLA	1-7/8 oz	1/4 cup 1/3 tbsp	
MILK,NONFAT,DRY	1 lbs	1 qts 3 cup	
WATER,WARM	18-3/4 lbs	2 gal 1 qts	

Method

1. Lightly spray each pan with non-stick cooking spray. Place 4-1/2 quarts bread in each pan. Pour margarine over bread cubes; toss flaked coconut with bread cubes. Toast in oven until lightly brown.
2. Add sugar, salt, nutmeg, and vanilla to eggs; blend thoroughly.
3. Reconstitute milk; combine with egg mixture. Pour 1 gallon over bread cubes in each pan.
4. Bake 1 hour or until firm in 350 F. oven.
5. Cover; CCP: Hold for service at 41 F. or lower.
6. Cut 4 by 8.

DESSERTS (PUDDINGS AND OTHER DESSERTS) No. J 017 00

CREAM PUFFS

Yield 100   **Portion** 1 Each

Calories	Carbohydrates	Protein	Fat	Cholesterol	Sodium	Calcium
139 cal	10 g	3 g	10 g	90 mg	121 mg	14 mg

Ingredient	Weight	Measure	Issue
BUTTER	2 lbs	1 qts	
WATER, BOILING	4-1/8 lbs	2 qts	
FLOUR, WHEAT, GENERAL PURPOSE	2-3/4 lbs	2 qts 2 cup	
SALT	1/4 oz	1/8 tsp	
EGGS, WHOLE, FROZEN	3-5/8 lbs	1 qts 2-5/8 cup	
COOKING SPRAY, NONSTICK	2 oz	1/4 cup 1/3 tbsp	

Method

1. Combine butter or margarine and water; bring to a boil.
2. Add flour and salt all at once, stirring rapidly. Cook 2 minutes or until mixture leaves sides of pan and forms a ball.
3. Remove from heat; place in mixer bowl. Cool slightly.
4. Add eggs, while beating at high speed, using a flat paddle. Beat until mixture is thick and shiny.
5. Spray each pan with non-stick cooking spray. Drop 2-1/2 tablespoons of batter in rows, 2 inches apart on sprayed pans.
6. Bake 10 minutes at 400 F.; reduce oven temperature to 350 F.; bake 30 minutes longer or until firm. Turn off oven.
7. Open oven door slightly; leave puffs in oven 8 to 10 minutes to dry out after baking. Shells should be slightly moist inside.
8. Using a pastry tube, fill shells. See Note 1.
9. CCP: Hold for service at 41 F. or lower.

Notes

1. Fill shells with 2/3 recipe Vanilla Cream Pudding Recipe No. J 014 00, 1 recipe Whipped Topping Recipe No. K 002 00, or commercial prepared hard ice cream may be used. Fill shells with 1/3 cup filling. Sprinkle with sifted powdered sugar or cover with Chocolate Glaze Frosting, Recipe No. G 024 00.

DESSERTS (PUDDINGS AND OTHER DESSERTS)   No. J 017 01
ECLAIRS

Yield 100   **Portion** 1 Each

Calories	Carbohydrates	Protein	Fat	Cholesterol	Sodium	Calcium
139 cal	10 g	3 g	10 g	90 mg	121 mg	14 mg

Ingredient	Weight	Measure	Issue
BUTTER	2 lbs	1 qts	
WATER, BOILING	4-1/8 lbs	2 qts	
FLOUR, WHEAT, GENERAL PURPOSE	2-3/4 lbs	2 qts 2 cup	
SALT	1/4 oz	1/8 tsp	
EGGS, WHOLE, FROZEN	3-5/8 lbs	1 qts 2-5/8 cup	
COOKING SPRAY, NONSTICK	2 oz	1/4 cup 1/3 tbsp	

Method

1. Combine butter and water; bring to a boil.
2. Add flour and salt all at once stirring rapidly. Cook 2 minutes or until mixture leaves the sides of the pan and forms a ball.
3. Remove from heat; place in mixer bowl. Cool slightly.
4. Add eggs, while beating at high speed, using a flat paddle. Beat until mixture is thick and shiny.
5. Spray each pan with non-stick cooking spray. Use a pastry bag or drop 2-1/2 tablespoons of batter 2 to 6 inches apart on sprayed pans; spread each mound into a 1x4-1/2 inch rectangle, rounding sides or piling batter on top.
6. Bake at 400 F. for 10 minutes; reduce oven temperature to 350 F.; bake 30 minutes longer or until firm. Turn off oven.
7. Open oven door slightly; leave puffs in oven 8 to 10 minutes to dry out after baking. Shells should be slightly moist inside.
8. Using a pastry tube, fill shells. See Note 1.
9. Refrigerate filled shells until served.

Notes

1. Fill shells with 2/3 recipe Vanilla Cream Pudding Recipe No. J 014 00, 1 recipe Whipped Topping Recipe No. K 002 00, or commercial prepared hard ice cream may be used. Fill shells with 1/3 cup filling. Sprinkle with sifted powdered sugar or cover with Chocolate Glaze Frosting, Recipe No. G 024 00.

DESSERTS (PUDDINGS AND OTHER DESSERTS) No.J 018 00
VANILLA CREAM PUDDING

Yield 100   **Portion** 1/2 Cup

Calories	Carbohydrates	Protein	Fat	Cholesterol	Sodium	Calcium
154 cal	25 g	3 g	5 g	58 mg	193 mg	58 mg

Ingredient	Weight	Measure	Issue
MILK,NONFAT,DRY	13-3/4 oz	1 qts 1-3/4 cup	
WATER,WARM	15-2/3 lbs	1 gal 3-1/2 qts	
SUGAR,GRANULATED	2 lbs	1 qts 1/2 cup	
SALT	1 oz	1 tbsp	
CORNSTARCH	1-1/8 lbs	1 qts	
SUGAR,GRANULATED	2 lbs	1 qts 1/2 cup	
WATER	5-1/4 lbs	2 qts 2 cup	
EGGS,WHOLE,FROZEN	2-3/8 lbs	1 qts 1/2 cup	
BUTTER	1 lbs	2 cup	
EXTRACT,VANILLA	2-3/4 oz	1/4 cup 2-1/3 tbsp	

Method

1. Reconstitute milk. Add sugar and salt. Heat to just below boiling. DO NOT BOIL.
2. Combine cornstarch, sugar, and water; stir until smooth. Add gradually to hot mixture. Cook at medium heat, stirring constantly, about 10 minutes or until thickened.
3. Stir 1 quart of hot mixture into eggs. Slowly pour egg mixture into remaining hot milk mixture; heat to boiling, stirring constantly. Cook about 2 minutes longer. Remove from heat.
4. Add butter or margarine and vanilla; stir until well blended.
5. Pour 1 gallon of pudding into each pan. Cover surface of pudding with waxed paper.
6. Refrigerate until ready to serve. CCP: Hold for service at 41 F. or lower.

Notes

1. Pudding will curdle if boiled or subjected to prolonged intense heat.

DESSERTS (PUDDINGS AND OTHER DESSERTS) No. J 018 01
CHOCOLATE CREAM PUDDING

Yield 100 **Portion** 1/2 Cup

Calories	Carbohydrates	Protein	Fat	Cholesterol	Sodium	Calcium
181 cal	36 g	2 g	4 g	11 mg	180 mg	56 mg

Ingredient	Weight	Measure	Issue
MILK,NONFAT,DRY	13-3/4 oz	1 qts 1-3/4 cup	
WATER,WARM	15-2/3 lbs	1 gal 3-1/2 qts	
SUGAR,GRANULATED	4 lbs	2 qts 1 cup	
SALT	1 oz	1 tbsp	
COCOA	12-1/8 oz	1 qts	
CORNSTARCH	1-1/8 lbs	1 qts	
SUGAR,GRANULATED	2 lbs	1 qts 1/2 cup	
WATER	5-1/4 lbs	2 qts 2 cup	
BUTTER	1 lbs	2 cup	
EXTRACT,VANILLA	2-3/4 oz	1/4 cup 2-1/3 tbsp	

Method

1. Reconstitute milk. Add sugar and salt. Heat to just below boiling. DO NOT BOIL.
2. Combine cocoa with cornstarch, sugar, and water; stir until smooth. Add gradually to hot mixture. Cook at medium heat stirring constantly, about 10 minutes or until thickened.
3. Add butter or margarine and vanilla; stir until well blended.
4. Pour 1 gallon of pudding into each pan. Cover surface of pudding with waxed paper.
5. Refrigerate until ready to serve. CCP: Hold for service at 41 F. or lower.

Notes

1. Pudding will curdle if boiled or subjected to prolonged intense heat.

DESSERTS (PUDDINGS AND OTHER DESSERTS) No. J 020 00

CREAMY RICE PUDDING

Yield 100   **Portion**   1/2 Cup

Calories	Carbohydrates	Protein	Fat	Cholesterol	Sodium	Calcium
170 cal	30 g	3 g	4 g	49 mg	254 mg	63 mg

Ingredient	**Weight**	**Measure**	**Issue**
RICE, LONG GRAIN	2-2/3 lbs	1 qts 2-1/2 cup	
WATER, BOILING	6-3/4 lbs	3 qts 1 cup	
SALT	5/8 oz	1 tbsp	
SUGAR, GRANULATED	2 lbs	1 qts 1/2 cup	
CORNSTARCH	7-7/8 oz	1-3/4 cup	
MILK, NONFAT, DRY	12 oz	1 qts 1 cup	
SALT	1 oz	1 tbsp	
CINNAMON, GROUND	1/8 oz	1/8 tsp	
NUTMEG, GROUND	1/8 oz	1/8 tsp	
WATER, WARM	7-1/3 lbs	3 qts 2 cup	
EGGS, WHOLE, FROZEN	2 lbs	3-3/4 cup	
WATER, BOILING	6-1/4 lbs	3 qts	
BUTTER	14 oz	1-3/4 cup	
EXTRACT, VANILLA	1-3/8 oz	3 tbsp	
RAISINS	1-7/8 lbs	1 qts 2 cup	
CINNAMON, GROUND	1/8 oz	1/3 tsp	

Method

1 Cook rice in boiling, salted water 20 to 25 minutes or until tender. Cover; set aside for use in Step 6.
2 In a steam jacketed kettle, combine sugar, cornstarch, milk, salt, cinnamon, and nutmeg; mix until well blended.
3 Add water to dry mixture; stir until smooth.
4 Add eggs; blend well.
5 Slowly add water to egg mixture, stirring with a wire whip. Cook until thickened, stirring constantly.
6 Turn off heat; add cooked rice, butter or margarine, vanilla, and raisins.
7 Pour 1 gallon of pudding into each pan.
8 Sprinkle cinnamon or nutmeg over pudding in each pan.
9 Cover surface of pudding with waxed paper. CCP: Hold for service at 41 F. or lower.

Notes

1 Pudding may be served hot. Omit Step 9.

DESSERTS (PUDDINGS AND OTHER DESSERTS) No. J 021 00
FLUFFY PINEAPPLE RICE CUP

Yield 100   **Portion** 1/2 Cup

Calories	Carbohydrates	Protein	Fat	Cholesterol	Sodium	Calcium
140 cal	22 g	2 g	6 g	0 mg	56 mg	22 mg

Ingredient	Weight	Measure	Issue
WATER,COLD	3-1/8 lbs	1 qts 2 cup	
RICE,LONG GRAIN	1-1/4 lbs	3 cup	
SALT	1/4 oz	1/8 tsp	
OIL,SALAD	1/2 oz	1 tbsp	
PINEAPPLE,CANNED,CRUSHED	6-5/8 lbs	3 qts	
CHERRIES,MARASCHINO,CHOPPED,DRAINED	8-7/8 oz	1 cup	
RESERVED LIQUID	3-7/8 lbs	1 qts 3-1/2 cup	
WHIPPED TOPPING MIX,NONDAIRY,DRY	2 lbs	2 gal 3-1/4 qts	
MILK,NONFAT,DRY	3-1/4 oz	1-3/8 cup	
EXTRACT,VANILLA	1-7/8 oz	1/4 cup 1/3 tbsp	
MARSHMALLOWS,MINIATURE	1-1/4 lbs	2 qts 3 cup	
COCONUT,PREPARED,SWEETENED FLAKES	1 lbs	1 qts 1 cup	

Method

1. Combine water, rice, salt and salad oil; bring to a boil. Stir occasionally.
2. Cover tightly; simmer 20 to 25 minutes. DO NOT STIR.
3. Remove from heat and refrigerate for use in Step 5.
4. Drain pineapple; reserve juice for use in Step 6.
5. Combine rice, pineapple and cherries. Refrigerate for use in Step 7.
6. Pour reserved juice and water into mixer bowl; add topping, milk and vanilla. Using whip at low speed, whip 3 minutes or until thoroughly blended. Scrape down bowl. Whip at high speed 5 to 10 minutes or until stiff peaks form.
7. Combine rice mixture and marshmallows and coconut. Mix thoroughly. Fold in whipped topping. Mix lightly.
8. Refrigerate until ready to serve. CCP: Hold for service at 41 F. or lower.

DESSERTS (PUDDINGS AND OTHER DESSERTS) No. J 022 00
BREAKFAST BREAD PUDDING

Yield 100   **Portion** 2/3 Cup

Calories	Carbohydrates	Protein	Fat	Cholesterol	Sodium	Calcium
206 cal	39 g	7 g	3 g	1 mg	300 mg	99 mg

Ingredient	Weight	Measure	Issue
COOKING SPRAY,NONSTICK	2 oz	1/4 cup 1/3 tbsp	
PEACHES,CANNED,QUARTERS,DICED,DRAINED	8-3/4 lbs	1 gal	
BREAD,WHITE,CUBED	4-1/8 lbs	3 gal 1-1/2 qts	
MILK,NONFAT,DRY	15 oz	1 qts 2-1/4 cup	
EGG SUBSTITUTE,PASTEURIZED	3 lbs	1 qts 1-1/2 cup	
WATER,WARM	15-2/3 lbs	1 gal 3-1/2 qts	
SUGAR,BROWN,PACKED	1-3/8 lbs	1 qts 3/8 cup	
EXTRACT,VANILLA	1-7/8 oz	1/4 cup 1/3 tbsp	
SALT	7/8 oz	1 tbsp	
CINNAMON,GROUND	1/4 oz	1 tbsp	
GINGER,GROUND	1/8 oz	1/3 tsp	
CEREAL,GRANOLA,TOASTED OAT MIX,LOW FAT	4-5/8 lbs	1 gal 3/4 qts	

Method

1. Lightly spray steam table pans with non-stick cooking spray. Place 1 quart peaches and 3-1/2 quarts bread in each pan. Mix lightly.
2. Reconstitute milk; add egg substitute, brown sugar, vanilla, salt, cinnamon, and ginger to milk, blend thoroughly.
3. Pour 2-1/2 quarts egg mixture over bread mixture in each pan.
4. Evenly distribute 4-3/4 cups granola on top of each pan.
5. Using a convection oven, bake 30 minutes at 325 F. or until lightly browned and a knife inserted in center comes out clean on low fan, open vent. CCP: Internal temperature must reach 145 F. or higher for 15 seconds.
6. CCP: Hold for service at 140 F. or higher.
7. Cut 4 by 6.

DESSERTS (PUDDINGS AND OTHER DESSERTS)   No. J 023 00

BAKED CINNAMON APPLE SLICES

Yield 100 **Portion** 1/2 Cup

Calories	Carbohydrates	Protein	Fat	Cholesterol	Sodium	Calcium
106 cal	26 g	0 g	1 g	0 mg	4 mg	8 mg

Ingredient	**Weight**	**Measure**	**Issue**
APPLES,CANNED,SLICED	27-3/4 lbs	3 gal 2 qts	
EXTRACT,VANILLA	2-1/2 oz	1/4 cup 1-2/3 tbsp	
SUGAR,GRANULATED	3-1/2 oz	1/2 cup	
CINNAMON,GROUND	1/2 oz	2 tbsp	
NUTMEG,GROUND	1/8 oz	1/4 tsp	
SUGAR,GRANULATED	14-1/8 oz	2 cup	
CINNAMON,GROUND	1/4 oz	1 tbsp	

Method

1. Blend sugar, cinnamon, and nutmeg. Combine with apples and vanilla. Place 3-1/3 quarts mixture in each pan.
2. Blend 2nd sugar and cinnamon. Sprinkle 1/2 cup evenly over apples in each pan.
3. Using a convection oven, bake at 375 F. for 20 minutes or until mixture begins to simmer and sugar begins to brown on high fan, open vent. CCP: Hold at 140 F. or higher for service.

DESSERTS (PUDDINGS AND OTHER DESSERTS) No. J 500 00
BREAD PUDDING WITH HARD SAUCE

Yield 100   Portion   1 Piece

Calories	Carbohydrates	Protein	Fat	Cholesterol	Sodium	Calcium
243 cal	43 g	5 g	6 g	1 mg	206 mg	54 mg

Ingredient	**Weight**	**Measure**	**Issue**
BREAD,WHITE,CUBED		3 gal 3 qts	
APPLES,COOKING,FRESH,PARED,CHOPPED	1-2/3 lbs	1 qts 3 cup	2-1/8 lbs
RAISINS	1-7/8 lbs	1 qts 2 cup	
EGG SUBSTITUTE,PASTEURIZED	3-1/3 lbs	1 qts 2 cup	
MILK,NONFAT,DRY	3-5/8 oz	1-1/2 cup	
WATER	3-7/8 lbs	1 qts 3-1/2 cup	
MARGARINE	1 lbs	2 cup	
SUGAR,GRANULATED	3-1/2 lbs	2 qts	
NUTMEG,GROUND	1/2 oz	2 tbsp	
EXTRACT,VANILLA	7/8 oz	2 tbsp	
CINNAMON,GROUND	1/2 oz	2 tbsp	
WATER	8-1/3 oz	1 cup	
SUGAR,GRANULATED	1-3/4 lbs	1 qts	
FLAVORING,RUM	2-3/4 oz	1/4 cup 2-1/3 tbsp	
MARGARINE	4 oz	1/2 cup	
EGG SUBSTITUTE,PASTEURIZED	11-3/4 oz	1-3/8 cup	

Method

1. Preheat oven to 350 F. Place bread in steam table pans.
2. Combine apples and raisins. Divide apples and raisins evenly among pans.
3. Reconstitute milk. Combine margarine, egg substitute, sugar, nutmeg, vanilla, cinnamon, and milk. Pour over bread and fruit. Fold lightly. Bake 20 to 30 minutes until set.
4. In medium saucepan, heat water, sugar, and extract until sugar is dissolved. Add margarine a little at a time until melted and combined. Temper the eggs with hot mixture, then add eggs. Stir and heat until sauce thickens slightly. Pour sauce over pudding. CCP: Internal temperature must reach 145 F. or higher for 15 seconds. Hold for service at 140 F. or higher.

DESSERTS (PUDDINGS AND OTHER DESSERTS) No. J 504 00
BAKED BANANAS

Yield 100   Portion   1/2 Cup

Calories	Carbohydrates	Protein	Fat	Cholesterol	Sodium	Calcium
169 cal	44 g	1 g	0 g	0 mg	4 mg	13 mg

Ingredient | Weight | Measure | Issue

Ingredient	Weight	Measure	Issue
SUGAR,BROWN,LIGHT	1-3/8 lbs	1 qts 3/8 cup	
WATER	2-1/8 lbs	1 qts	
HONEY	3 lbs	1 qts	
BANANA,FRESH	25 lbs		38-1/2 lbs

Method

1. Heat brown sugar, water, and honey in a saucepan over low heat until sugar is dissolved, about 5 minutes.
2. Cut bananas in half crosswise. Place 25 halves into each steam table pan. Pour 3/4 cup of syrup over each pan of bananas.
3. Using a convection oven, bake at 350 F. for 10 minutes until lightly browned.
4. Serve with sauce. CCP: Hold for service at 140 F. or higher.

K. DESSERTS (SAUCES AND TOPPINGS) No. 0

INDEX

Card No.

K 001 00	Whipped Cream
K 002 00	Whipped Topping (Dehydrated)
K 002 01	Whipped Topping (Frozen)
K 003 00	Rum Sauce
K 004 00	Cherry Sauce
K 005 00	Chocolate Sauce
K 005 01	Chocolate Coconut Sauce
K 005 02	Chocolate Marshmallow Sauce
K 005 03	Chocolate Nut Sauce
K 005 04	Chocolate Mint Sauce
K 006 00	Cherry Jubilee Sauce
K 007 00	Strawberry Glaze Topping

DESSERTS (SAUCES AND TOPPINGS) No.K 001 00

WHIPPED CREAM

Yield 100   **Portion** 2 Tablespoons

Calories	Carbohydrates	Protein	Fat	Cholesterol	Sodium	Calcium
72 cal	2 g	0 g	7 g	26 mg	7 mg	12 mg

Ingredient   Weight   Measure   Issue

Ingredient	Weight	Measure	Issue
CREAM,WHIPPING,COLD	4-1/4 lbs	2 qts	
SUGAR,POWDERED,SIFTED	5-1/4 oz	1-1/4 cup	
EXTRACT,VANILLA	7/8 oz	2 tbsp	

Method

1. Pour cream into chilled mixer bowl. Using whip at medium speed, whip 1 gallon of cream 3 to 7 minutes or until slightly thickened.
2. Gradually add sugar and vanilla. Whip 7 to 8 minutes or until stiff. DO NOT OVER WHIP.
3. Cover; refrigerate until ready to serve. CCP: Hold for service at 41 F. or lower.

DESSERTS (SAUCES AND TOPPINGS) No.K 002 00

WHIPPED TOPPING (DEHYDRATED)

Yield 100   **Portion** 3 Tablespoons

Calories	Carbohydrates	Protein	Fat	Cholesterol	Sodium	Calcium
29 cal	3 g	0 g	2 g	0 mg	8 mg	7 mg

Ingredient	Weight	Measure	Issue
WATER,COLD	2 lbs	3-3/4 cup	
WHIPPED TOPPING MIX,NONDAIRY,DRY	1 lbs	1 gal 1-5/8 qts	
MILK,NONFAT,DRY	1-5/8 oz	1/2 cup 2-2/3 tbsp	
EXTRACT,VANILLA	7/8 oz	2 tbsp	

Method

1. Place cold water in mixer bowl; add topping, milk, and vanilla. Using whip at low speed, whip 3 minutes or until well blended. Scrape down bowl.
2. Whip at high speed 5 to 10 minutes or until stiff peaks form. Cover; refrigerate until ready to serve. CCP: Hold for service at 41 F. or lower.

Notes

1. When topping is used for icing cakes, fold 2 cups sifted powdered sugar into whipped topping.

DESSERTS (SAUCES AND TOPPINGS) No.K 002 01

WHIPPED TOPPING (FROZEN)

Yield 100 **Portion** 3 Tablespoons

Calories	Carbohydrates	Protein	Fat	Cholesterol	Sodium	Calcium
14 cal	1 g	0 g	1 g	0 mg	1 mg	0 mg

Ingredient	Weight	Measure	Issue
WHIPPED TOPPING,FROZEN,NONDAIRY	1 lbs	1 qts 2 cup	

Method

1 Thaw topping in chilled mixer bowl. Using whip at medium speed, whip topping 10 to 20 minutes or until stiff peaks form. Cover; refrigerate until ready to serve. CCP: Hold for service at 41 F. or lower.

Notes

1 When topping is used for icing cakes, fold 2 cups sifted powdered sugar into whipped topping.

DESSERTS (SAUCES AND TOPPINGS) No.K 003 00
RUM SAUCE

Yield 100   **Portion** 2 Tablespoons

Calories	Carbohydrates	Protein	Fat	Cholesterol	Sodium	Calcium
130 cal	12 g	0 g	9 g	25 mg	101 mg	18 mg

Ingredient	**Weight**	**Measure**	**Issue**
BUTTER	2-1/2 lbs	1 qts 1 cup	
SUGAR,BROWN,PACKED	2-3/4 lbs	2 qts 1/2 cup	
MILK,NONFAT,DRY	1-1/4 oz	1/2 cup	
WATER	1 lbs	2 cup	
FLAVORING,RUM	1-7/8 oz	1/4 cup 1/3 tbsp	

Method

1 Melt butter or margarine; add brown sugar. Cook on low heat for 2 minutes, stirring constantly.
2 Reconstitute milk; add to sugar mixture. Cook, stirring constantly, until mixture comes to a boil.
3 Remove immediately from heat; cool 10 minutes.
4 Add rum flavoring; stir until well blended.

DESSERTS (SAUCES AND TOPPINGS) No.K 004 00
CHERRY SAUCE

Yield 100 Portion 2-1/2 Tablespoons

Calories	Carbohydrates	Protein	Fat	Cholesterol	Sodium	Calcium
46 cal	12 g	0 g	0 g	0 mg	4 mg	4 mg

Ingredient	Weight	Measure	Issue
PIE FILLING,CHERRY,PREPARED	8-3/4 lbs	1 gal 3/8 qts	
WATER	8-1/3 oz	1 cup	

Method

1 Combine pie filling with water in mixer bowl and mix well.

DESSERTS (SAUCES AND TOPPINGS) No. K 005 00

CHOCOLATE SAUCE

Yield 100   Portion 2 Tablespoons

Calories	Carbohydrates	Protein	Fat	Cholesterol	Sodium	Calcium
83 cal	16 g	1 g	3 g	6 mg	31 mg	19 mg

Ingredient	Weight	Measure	Issue
MILK,NONFAT,DRY	4-1/4 oz	1-3/4 cup	
WATER,WARM	3-1/8 lbs	1 qts 2 cup	
SUGAR,GRANULATED	3 lbs	1 qts 2-3/4 cup	
COCOA	9-1/8 oz	3 cup	
WATER,COLD	1 lbs	2 cup	
BUTTER	10 oz	1-1/4 cup	
EXTRACT,VANILLA	1/2 oz	1 tbsp	

Method

1. Reconstitute milk. Set aside for use in Step 3.
2. Mix sugar and cocoa with water to form a paste. Bring to a boil, stirring constantly; cool slightly.
3. Add milk stirring constantly. Bring to a boil; cook 3 minutes. Remove from heat immediately.
4. Add butter or margarine and vanilla; stir. Serve warm or at room temperature.

Notes

1. In Step 2, for 100 portions, 1 pound unsweetened, cooking chocolate may be used for cocoa. In Step 4, reduce butter or margarine to 1/4 cup. Add chocolate with butter or margarine.

DESSERTS (SAUCES AND TOPPINGS) No. K 005 01

CHOCOLATE COCONUT SAUCE

Yield 100 **Portion** 2 Tablespoons

Calories	Carbohydrates	Protein	Fat	Cholesterol	Sodium	Calcium
109 cal	18 g	1 g	4 g	6 mg	45 mg	20 mg

Ingredient	Weight	Measure	Issue
MILK,NONFAT,DRY	4-1/4 oz	1-3/4 cup	
WATER,WARM	3-1/8 lbs	1 qts 2 cup	
SUGAR,GRANULATED	3 lbs	1 qts 2-3/4 cup	
COCOA	9-1/8 oz	3 cup	
WATER,COLD	1 lbs	2 cup	
BUTTER	10 oz	1-1/4 cup	
EXTRACT,VANILLA	1/2 oz	1 tbsp	
COCONUT,PREPARED,SWEETENED FLAKES	1-1/8 lbs	1 qts 1-1/2 cup	

Method

1. Reconstitute milk. Set aside for use in Step 3.
2. Mix sugar and cocoa with water to form a paste. Bring to a boil, stirring constantly; cool slightly.
3. Add milk stirring constantly. Bring to a boil; cook 3 minutes. Remove from heat immediately.
4. Add butter or margarine and vanilla; stir.
5. Just before serving, add sweetened, flaked coconut to sauce and mix well.

Notes

1. In Step 2, for 100 portions, 1 pound unsweetened, cooking chocolate may be used for cocoa. In Step 4, reduce butter or margarine to 1/4 cup. Add chocolate with butter or margarine.

DESSERTS (SAUCES AND TOPPINGS) No. K 005 02

CHOCOLATE MARSHMALLOW SAUCE

Yield 100   Portion   2 Tablespoons

Calories	Carbohydrates	Protein	Fat	Cholesterol	Sodium	Calcium
97 cal	19 g	1 g	3 g	6 mg	33 mg	19 mg

Ingredient	Weight	Measure	Issue
MILK,NONFAT,DRY	4-1/4 oz	1-3/4 cup	
WATER,WARM	3-1/8 lbs	1 qts 2 cup	
SUGAR,GRANULATED	3 lbs	1 qts 2-3/4 cup	
COCOA	9-1/8 oz	3 cup	
WATER,COLD	1 lbs	2 cup	
BUTTER	10 oz	1-1/4 cup	
EXTRACT,VANILLA	1/2 oz	1 tbsp	
MARSHMALLOWS,MINIATURE	1 lbs	2 qts 1 cup	

Method

1. Reconstitute milk. Set aside for use in Step 3.
2. Mix sugar and cocoa with water to form a paste. Bring to a boil, stirring constantly; cool slightly.
3. Add milk stirring constantly. Bring to a boil; cook 3 minutes. Remove from heat immediately.
4. Add butter or margarine and vanilla; stir.
5. Just before serving, add miniature marshmallows to sauce and mix well.

Notes

1. In Step 2, for 100 portions, 1 pound unsweetened, cooking chocolate may be used for cocoa. In Step 4, reduce butter or margarine to 1/4 cup. Add chocolate with butter or margarine.

DESSERTS (SAUCES AND TOPPINGS) No. K 005 03

CHOCOLATE NUT SAUCE

Yield 100   Portion 2 Tablespoons

Calories	Carbohydrates	Protein	Fat	Cholesterol	Sodium	Calcium
109 cal	16 g	2 g	5 g	6 mg	32 mg	23 mg

Ingredient	Weight	Measure	Issue
MILK,NONFAT,DRY	4-1/4 oz	1-3/4 cup	
WATER,WARM	3-1/8 lbs	1 qts 2 cup	
SUGAR,GRANULATED	3 lbs	1 qts 2-3/4 cup	
COCOA	9-1/8 oz	3 cup	
WATER,COLD	1 lbs	2 cup	
BUTTER	10 oz	1-1/4 cup	
EXTRACT,VANILLA	1/2 oz	1 tbsp	
NUTS,UNSALTED,CHOPPED,COARSELY	1 lbs	3-1/8 cup	

Method

1. Reconstitute milk. Set aside for use in Step 3.
2. Mix sugar and cocoa with water to form a paste. Bring to a boil, stirring constantly; cool slightly.
3. Add milk stirring constantly. Bring to a boil; cook 3 minutes. Remove from heat immediately.
4. Add butter or margarine and vanilla; stir.
5. Just before serving, add chopped unsalted nuts to sauce and mix well.

Notes

1. In Step 2, for 100 portions, 1 pound unsweetened, cooking chocolate may be used for cocoa. In Step 4, reduce butter or margarine to 1/4 cup. Add chocolate with butter or margarine.

DESSERTS (SAUCES AND TOPPINGS) No.K 005 04

CHOCOLATE MINT SAUCE

Yield 100 **Portion** 2 Tablespoons

Calories	Carbohydrates	Protein	Fat	Cholesterol	Sodium	Calcium
83 cal	16 g	1 g	3 g	6 mg	31 mg	19 mg

Ingredient	Weight	Measure	Issue
MILK,NONFAT,DRY	4-1/4 oz	1-3/4 cup	
WATER,WARM	3-1/8 lbs	1 qts 2 cup	
SUGAR,GRANULATED	3 lbs	1 qts 2-3/4 cup	
COCOA	9-1/8 oz	3 cup	
WATER,COLD	1 lbs	2 cup	
BUTTER	10 oz	1-1/4 cup	
FLAVORING,PEPPERMINT	1/2 oz	1 tbsp	

Method

1. Reconstitute milk. Set aside for use in Step 3.
2. Mix sugar and cocoa with water to form a paste. Bring to a boil, stirring constantly; cool slightly.
3. Add milk stirring constantly. Bring to a boil; cook 3 minutes. Remove from heat immediately.
4. Add butter or margarine and peppermint flavoring; stir. Serve warm or at room temperature.

Notes

1. In Step 2, for 100 portions, 1 pound unsweetened, cooking chocolate may be used for cocoa. In Step 4, reduce butter or margarine to 1/4 cup. Add chocolate with butter or margarine.

DESSERTS (SAUCES AND TOPPINGS) No.K 006 00
CHERRY JUBILEE SAUCE

Yield 100   Portion   1/4 Cup

Calories	Carbohydrates	Protein	Fat	Cholesterol	Sodium	Calcium
86 cal	22 g	0 g	0 g	0 mg	25 mg	6 mg

Ingredient	Weight	Measure	Issue
CHERRIES,CANNED,DARK,SWEET,PITTED,INCL LIQUIDS	13-3/8 lbs	1 gal 2 qts	
CORNSTARCH	3-3/8 oz	3/4 cup	
SALT	1/4 oz	1/8 tsp	
SUGAR,GRANULATED	1-3/4 lbs	1 qts	
FLAVORING,BRANDY	1-3/8 oz	3 tbsp	

Method

1 Drain cherries; set aside for use in Step 5. Take cherry juice and add water to equal 1 gallon per 100 portions.
2 Combine cornstarch, salt, and sugar. Add liquid; mix well.
3 Cook over medium heat until mixture comes to a boil.
4 Reduce heat; continue cooking slowly, stirring occasionally until sauce is thick and clear.
5 Remove from heat; add brandy flavoring and cherries.
6 Serve warm or cold.

DESSERTS (SAUCES AND TOPPINGS) No. K 007 00

STRAWBERRY GLAZE TOPPING

Yield 100 **Portion** 2-1/2 Tablespoons

Calories	Carbohydrates	Protein	Fat	Cholesterol	Sodium	Calcium
36 cal	9 g	0 g	0 g	0 mg	2 mg	7 mg

Ingredient

Ingredient	Weight	Measure	Issue
STRAWBERRIES,FROZEN,THAWED	9 lbs	1 gal	
CORNSTARCH	7-1/2 oz	1-5/8 cup	
SUGAR,GRANULATED	12-1/3 oz	1-3/4 cup	
RESERVED LIQUID	4-2/3 lbs	2 qts 1 cup	

Method

1. Drain strawberries. Set juice aside for use in Step 2; berries for use in Step 3.
2. Combine cornstarch, sugar and strawberry juice. Bring to a boil. Cook at medium heat, stirring constantly until thick and clear. Remove from heat.
3. Fold strawberries per 100 portions into thickened mixture.
4. Chill topping.

TIMETABLES FOR ROASTING TURKEYS (UNSTUFFED)

Weight of Turkeys	Oven Temperature	Cooking Time (hours)	Convection Oven Temperature	Convection Oven Time (hours)
8 to 12 lb.	325° F.	3 to 4	300° F.	2-1/4 to 3
12 to 16 lb.	325° F.	3-1/2 to 4-1/2	300° F.	2-3/4 to 3-1/2
16 to 20 lb.	325° F.	4 to 5	300° F.	3 to 3-3/4
20 to 24 lb.	325° F.	4-1/2 to 5-1/2	300° F.	3-1/2 to 4-1/4

For best result in slicing, allow to stand 30 minutes.

SERVINGS PER TURKEY

Ready-to-Cook Weight	Number of Servings
8 to 12 lb.	10 to 20
12 to 15 lb.	20 to 32
16 to 24 lb.	33 to 50

(about 2 servings per pound)

Disclaimer:
Time is approximate (16-20 minutes per pound)
Cook to internal temperature as recommend by HACCP regulations. Place thermometer in the spot located between the thigh and breast.
Do **NOT** cook stuffed birds.

L. MEAT, FISH AND POULTRY No. 0(1)

INDEX

Card No. .... Card No.

Card No.		Card No.	
L 001 01	Grilled or Oven Fried Bacon (Precooked Bacon)	L 012 00	Country Style Steak
L 002 00	Oven Fried Bacon	L 013 00	Pepper Steak
L 002 02	Grilled Bacon	L 013 01	Oriental Pepper Steak
L 002 03	Grilled or Oven Fried Canadian Bacon	L 014 00	Ground Beef Cordon Bleu
L 003 00	Chicken Enchiladas (Canned Chicken)	L 015 00	Steak Smothered with Onions
L 003 01	Chicken Enchiladas (Cooked Diced)	L 015 01	Steak Strips Smothered with Onions
L 004 00	Roast Rib of Beef	L 016 00	Swiss Steak with Tomato Sauce
L 004 01	Steamship Round Of Beef (Round, Bone-In)	L 016 01	Swiss Steak with Brown Gravy
L 004 02	Steamship Round Of Beef (Round, Boneless)	L 016 03	Swiss Steak with Tomato Soup
L 004 03	Roast Rib Of Beef (Boneless Ribeye Roll)	L 016 04	Swiss Steak with Mushroom Gravy
L 005 00	Roast Beef	L 017 00	Braised Beef and Noodles
L 005 01	Roast Beef (Precooked)	L 017 01	Braised Beef Cubes
L 006 00	Sukiyaki	L 018 00	Barbecued Beef Cubes
L 007 00	Grilled Steak	L 018 01	Barbecued Beef Cubes (Canned Beef)
L 007 01	Grilled Tenderloin Steak	L 019 00	Stuffed Flounder Creole
L 008 00	Teriyaki Steak	L 020 00	Beef and Corn Pie
L 009 00	Spinach Lasagna	L 020 01	Turkey Corn Pie
L 010 00	Beef Pot Roast	L 021 00	Beef Pot Pie with Biscuit Topping
L 010 01	Ginger Pot Roast	L 021 01	Beef Pot Pie with Pie Crust Topping
L 010 02	Yankee Pot Roast	L 022 00	Beef Stew
L 011 00	Simmered Beef	L 022 01	Beef Stew (Canned)

L. MEAT, FISH AND POULTRY No. 0(1)

Card No. .... Card No.

Card No.	Item	Card No.	Item
L 022 02	El Rancho Stew	L 031 01	Cheese Ravioli (Frozen)
L 023 00	Caribbean Chicken Breast (Breast Boneless)	L 031 02	Beef Ravioli (Canned in Tomato Sauce)
L 024 00	Stuffed Cabbage Rolls	L 032 00	Parmesan Fish
L 024 01	Stuffed Cabbage Rolls (Tomato Soup)	L 033 00	Roast Beef Hash
L 024 02	Stuffed Cabbage Rolls (Ground Turkey)	L 033 01	Roast Beef Hash (Canned)
L 025 00	Lasagna	L 033 02	Roast Beef Hash (Canned Beef Chunks)
L 025 01	Lasagna (Ground Turkey)	L 034 00	Tacos (Ground Beef)
L 025 02	Lasagna (Frozen)	L 034 01	Tacos (Ground Turkey)
L 025 03	Lasagna (Canned Pizza Sauce)	L 035 00	Meat Loaf
L 026 00	Baked Breaded Clam Strips	L 035 01	Turkey Loaf
L 026 01	French Fried Breaded Clam Strips	L 035 02	Tomato Meat Loaf
L 027 00	Beef Balls Stroganoff	L 035 03	Cajun Meat Loaf
L 027 01	Turkey Balls Stroganoff	L 036 00	Minced Beef
L 028 00	Chili Con Carne	L 037 00	Salisbury Steak
L 028 02	Chili Macaroni	L 037 02	Grilled Hamburger Steak
L 028 03	Chili Con Carne (Ground Turkey)	L 038 00	Spaghetti with Meat Sauce (Ground Turkey)
L 028 04	Chili Macaroni (Ground Turkey)	L 038 01	Spaghetti with Meat Sauce (Ground Beef)
L 029 00	Beef Porcupines	L 038 02	Spaghetti with Meat Sauce, RTU (Ground Turkey)
L 029 01	Turkey Porcupines		
L 030 00	Creamed Ground Beef	L 038 03	Spaghetti with Meat Sauce, RTU (Ground Beef)
L 030 01	Creamed Ground Turkey	L 039 00	Spaghetti with Meatballs (Ground Turkey)
L 031 00	Beef Ravioli (Frozen)	L 039 01	Spaghetti with Meatballs (Ground Beef)

L. MEAT, FISH AND POULTRY No. 0(2)

INDEX

Card No. .... Card No.

L 040 00	Stuffed Green Peppers (Ground Beef)	L 052 00	Creamed Chipped Beef
L 040 01	Stuffed Green Peppers (Frozen)	L 053 00	Beef Stroganoff
L 040 02	Stuffed Green Peppers (Ground Turkey)	L 053 01	Beef Stroganoff (Cream of Mushroom Soup)
L 041 00	Swedish Meatballs (Ground Beef)	L 053 02	Hamburger Stroganoff
L 041 01	Swedish Meatballs (Ground Turkey)	L 053 03	Ground Turkey Stroganoff
L 042 00	Chili Conquistador (Ground Beef)	L 053 04	Beef Stroganoff (Fajita Strips)
L 042 01	Chili Conquistador (Ground Turkey)	L 054 00	Steak Ranchero
L 043 00	Beef Fajitas (Fajita Strips)	L 055 00	Beef Cordon Bleu
L 043 01	Chicken Fajitas (Fajita Strips)	L 056 00	Southern Fried Catfish Fillets
L 043 02	Turkey Fajitas	L 057 00	Tamale Pie (Ground Beef)
L 044 00	Turkey Curry	L 057 01	Hot Tamales with Chili Gravy
L 045 00	Stuffed Beef Rolls	L 057 02	Tamale Pizza
L 045 01	Beef Brogul	L 058 00	Chili and Macaroni (Canned Chili Con Carne)
L 046 00	Beef and Bean Tostadas	L 059 00	Chili Con Carne (with Beans)
L 047 00	Beef Pie with Biscuit Topping (Canned Beef)	L 060 00	Hamburger Parmesan
L 048 00	Baked Chicken and Rice (Cooked Diced)	L 061 00	Texas Hash (Ground Beef)
L 048 01	Baked Chicken and Rice (Canned Chicken)	L 061 01	Texas Hash (Ground Turkey)
L 049 00	Turkey Cutlet	L 062 00	Yakisoba (Beef and Spaghetti)
L 050 00	Chalupa	L 062 01	Hamburger Yakisoba (Ground Beef)
L 051 00	Chicken Parmesan (Precooked Fillet)	L 062 02	Turkey Yakisoba
L 051 01	Chicken Parmesan (Breast Boneless)	L 063 00	Enchiladas (Ground Beef)

L. MEAT, FISH AND POULTRY No. 0(2)

Card No.		Card No.	
L 063 01	Enchiladas (Frozen)	L 074 00	Chilies Rellenos
L 063 02	Enchiladas (Ground Turkey)	L 075 00	Broccoli, Cheese, and Rice
L 064 00	Creole Macaroni (Ground Beef)	L 076 00	Beef Manicotti (Cannelloni)
L 064 01	Creole Macaroni (Ground Turkey)	L 076 01	Cheese Manicotti
L 065 00	Hungarian Goulash	L 077 00	Savory Roast Lamb
L 066 00	Sauerbraten	L 078 00	Chicken Adobo (8 Pc)
L 067 00	Glazed Ham Loaf	L 079 00	Sweet and Sour Pork Chops
L 068 00	Scalloped Ham and Noodles	L 079 01	Sweet and Sour Chicken (8 Pc)
L 069 00	Baked Ham	L 079 02	Sweet and Sour Chicken (Cooked Diced)
L 069 01	Grilled Ham Steak	L 080 00	Pork Chop Suey
L 070 00	Barbecued Ham Steak	L 080 01	Shrimp Chop Suey
L 070 01	Barbecued Ham Steak (Canned Ham)	L 081 00	Roast Pork
L 071 00	Baked Canned Ham	L 081 01	Roast Pork Tenderloin
L 071 01	Baked Ham Steak (Canned Ham)	L 081 02	Barbecued Pork Loin
L 071 02	Grilled Ham Steak (Canned Ham)	L 082 00	Sweet and Sour Pork
L 071 03	Grilled Ham Slice (Canned Ham)	L 083 00	Creole Pork Chops
L 072 00	Baked Ham, Macaroni, and Tomatoes (Canned Ham)	L 083 01	Barbecued Pork Chops
		L 084 00	Baked Stuffed Pork Chops
L 072 01	Baked Luncheon Meat, Macaroni, and Cheese	L 084 01	Pork Chops with Apple Rings
L 072 02	Baked Ham, Macaroni and Tomatoes (Canned Chunks)	L 085 00	Braised Pork Chops
		L 085 01	Grilled Pork Chops
L 073 00	Scalloped Ham and Potatoes (Canned Ham)	L 085 02	Pork Chops with Mushroom Gravy

L. MEAT, FISH AND POULTRY No. 0(3)

INDEX

Card No.		Card No.	
L 086 01	Creole Pork Steaks (Frozen Breaded Pork Steaks)	L 100 00	Simmered Pork Hocks (Ham Hocks)
L 086 02	Breaded Pork Steaks (Frozen)	L 101 00	Italian Style Veal Steaks
L 086 03	Pork Schnitzel (Frozen Breaded Pork Steaks)	L 102 00	Veal Paprika Steak
L 087 00	Pork Chops Mexicana	L 103 00	Veal Parmesan
L 088 00	Grilled Polish Sausage	L 103 01	Veal Steak
L 088 01	Baked Italian Sausage (Hot or Sweet)	L 104 00	Jaegerschnitzel
L 088 02	Grilled Frankfurters	L 105 00	Veal Cubes Parmesan
L 088 03	Grilled Bratwurst	L 106 00	Roast Veal
L 088 05	Simmered Knockwurst	L 106 01	Roast Veal with Herbs
L 089 00	Grilled Sausage Patties	L 107 00	Braised Liver with Onions
L 089 02	Grilled Sausage Patties (Preformed)	L 107 01	Grilled Liver
L 091 00	Grilled Sausage Links (Cooked Pork and Beef)	L 108 00	Breaded Liver
L 092 00	Barbecued Spareribs	L 108 01	Breaded Liver with Onion and Mushroom Gravy
L 093 00	Braised Spareribs	L 109 00	Oven Fried Chicken Fillets (3 Oz)
L 093 01	Spareribs and Sauerkraut	L 109 01	Fried Chicken Fillets (3 Oz)
L 094 00	Sweet and Sour Spareribs	L 109 02	Oven Fried Chicken Fillets (5 Oz)
L 095 00	Cantonese Spareribs	L 109 03	Fried Chicken Fillets (5 Oz)
L 096 00	Roast Fresh Ham	L 109 04	Oven Fried Chicken Fillet Nuggets
L 097 00	Shrimp Jambalaya	L 109 05	Fried Chicken Fillet Nuggets
L 099 00	Pork Adobo	L 110 00	Corned Beef Hash
		L 110 01	Corned Beef Hash (Canned)

L. MEAT, FISH AND POULTRY No. 0(3)

Card No. .... Card No.

Card No.	Name	Card No.	Name
L 111 00	New England Boiled Dinner	L 119 07	Cajun Baked Fish
L 111 01	New England Boiled Dinner (Precooked Frozen Beef)	L 120 00	Baked Stuffed Fish
L 112 00	Simmered Corned Beef	L 121 00	Shrimp Scampi
L 112 01	Apple Glazed Corned Beef	L 122 00	Pan Fried Fish
L 112 02	Baked Corned Beef (Precooked Frozen)	L 122 01	Tempura Fish
L 113 00	Baked Frankfurters with Sauerkraut	L 122 02	Deep Fat Fried Fish
L 113 01	Baked Knockwurst with Sauerkraut	L 123 00	Oven Fried Fish
L 114 00	Teriyaki Chicken (8 Pc)	L 124 00	Baked Fish Portions
L 114 01	Teriyaki Chicken (Thighs)	L 124 01	Baked Fish Portions (Batter Dipped)
L 115 00	Spicy Baked Fish	L 124 02	French Fried Fish Portions
L 116 00	Macaroni Tuna Salad	L 124 03	French Fried Fish Portions (Batter Dip)
L 116 01	Chicken Rotini Salad (Canned Chicken)	L 124 04	Fish and Chips
L 116 02	Chicken Rotini Salad (Cooked Diced)	L 124 05	Baked Fish Nuggets
L 117 01	Grilled Luncheon Meat	L 124 06	French Fried Fish Nuggets
L 119 00	Baked Fish	L 125 00	Chipper Fish
L 119 01	Baked Fish with Garlic Butter	L 126 00	Fried Oysters
L 119 02	Onion-Lemon Baked Fish	L 126 01	Fried Oysters (Breaded, Frozen)
L 119 03	Lemon Baked Fish	L 127 00	Boiled Lobster, Whole
L 119 04	Herbed Baked Fish	L 127 01	Boiled Lobster Tail, Frozen
L 119 05	Mustard-Dill Baked Fish	L 127 03	Boiled Crab Legs, Alaskan King, Frozen
L 119 06	Fish Amandine	L 127 04	Boiled Shrimp, Frozen
		L 128 00	Salmon Cakes

INDEX

Card No.		Card No.	
L 129 00	Salmon Loaf	L 142 01	Rock Cornish Hens with Syrup Glaze
L 130 00	Scalloped Salmon and Peas	L 142 02	Herbed Cornish Hens
L 131 00	Chopstick Tuna	L 143 00	Baked Chicken (8 Pc)
L 132 00	Tuna Salad	L 143 01	Mexican Baked Chicken (8 Pc)
L 132 01	Salmon Salad (Canned Salmon)	L 143 02	Herbed Baked Chicken (8 Pc)
L 133 00	Baked Tuna and Noodles	L 143 03	Baked Chicken (Breast Boneless)
L 133 01	Baked Tuna and Noodles (Cream of Mushroom Soup)	L 143 04	Mexican Baked Chicken (Breast Boneless)
		L 143 05	Herbed Baked Chicken (Breast Boneless)
L 134 00	Fried Scallops	L 144 00	Baked Turkey and Noodles
L 135 00	Creole Scallops	L 144 01	Baked Chicken and Noodles (Canned Chicken)
L 135 01	Creole Fish	L 144 03	Baked Chicken and Noodles (Cooked Diced)
L 135 02	Creole Fish Fillets	L 145 00	Chicken Vega (8 Pc)
L 136 00	Creole Shrimp	L 146 00	Barbecued Chicken (8 Pc)
L 137 00	French Fried Shrimp	L 146 01	Barbecued Chicken (Breast Boneless)
L 137 01	Tempura Shrimp	L 147 00	Chicken a La King (Cooked Diced)
L 137 02	French Fried Shrimp (Breaded, Frozen)	L 147 01	Chicken a La King (Canned Chicken)
L 138 00	Shrimp Curry	L 147 02	Turkey a La King
L 139 00	Shrimp Salad	L 148 00	Chicken Cacciatore (8 Pc)
L 140 00	Seafood Newburg	L 148 01	Chicken Cacciatore (Cooked Diced)
L 141 00	Crab Cakes	L 149 00	Baked Chicken and Gravy (8 Pc)
L 142 00	Honey Glazed Rock Cornish Hens	L 149 01	Baked Chicken with Mushroom Gravy (8 Pc)

L. MEAT, FISH AND POULTRY No. 0(4)

Card No. .... Card No.

L 149 02	Baked Chicken with Mushroom Gravy (8 Pc Cnd Soup)	L 158 00	Savory Baked Chicken (8 Pc)
L 150 00	Turkey Pot Pie	L 158 01	Savory Baked Chicken (Thighs)
L 150 01	Chicken Pot Pie (Canned Chicken)	L 159 00	Szechwan Chicken (8 Pc)
L 150 03	Chicken Pot Pie (Cooked Diced)	L 159 01	Szechwan Chicken (Breast Boneless)
L 151 00	Chicken Salad (Cooked Diced)	L 160 00	Chicken Chow Mein (Cooked Diced)
L 151 01	Chicken Salad (Canned Chicken)	L 160 01	Chicken Chow Mein (Canned Chicken)
L 151 02	Turkey Salad (Boneless, Frozen)	L 161 00	Roast Turkey
L 152 00	Chicken Tetrazzini (Canned Chicken)	L 162 00	Roast Turkey (Boneless Turkey)
L 152 01	Tuna Tetrazzini (Canned Tuna)	L 162 01	Roast Turkey With Barbecue Sauce
L 152 02	Chicken Tetrazzini (Cooked Diced)	L 163 00	Turkey Nuggets
L 153 00	Chinese Five-Spice Chicken (8 Pc)	L 164 00	Roast Duck
L 154 00	Creole Chicken (8 Pc)	L 164 01	Hawaiian Baked Duck
L 154 01	Creole Chicken (Cooked Diced)	L 164 02	Roast Duck With Apple Jelly Glaze
L 155 00	Fried Chicken (8 Pc)	L 164 03	Honey Glazed Duck
L 155 01	Southern Fried Chicken (8 Pc)	L 165 00	Pizza
L 155 02	Fried Chicken (Precooked Breaded, Frozen For Deep Fat Fry)	L 165 01	Pizza (Thick Crust)
		L 165 02	Mushroom, Green Pepper and Onion Pizza
L 156 00	Oven Baked Chicken (8 Pc)	L 165 03	Hamburger Pizza
L 156 01	Fried Chicken (Precooked, Breaded Chicken, Frozen For Oven)	L 165 04	Pepperoni, Green Pepper, and Mushroom Pizza
		L 165 05	Pepperoni Pizza
		L 165 06	Pizza (Roll Mix)
L 157 00	Pineapple Chicken (8 Pc)	L 165 07	Pork or Italian Sausage Pizza

INDEX

Card No.		Card No.	
L 165 08	French Bread Pizza	L 178 01	Tropical Chicken Salad (Canned Chicken)
L 165 09	Sausage, Green Pepper, and Onion Pizza	L 179 00	Honey Ginger Chicken (Breast Boneless)
L 165 10	Pizza (Pourable Pizza Crust)	L 180 00	Turkey Sausage Patties
L 166 00	Pizza (12 Inch Frozen Crust)	L 181 00	Chicken in Orange Sauce (Breast Boneless)
L 167 00	Chuck Wagon Stew (Beans with Beef)	L 182 00	Fiesta Chicken (Fajita Strips)
L 168 00	Baked Scallops	L 183 00	Buffalo Chicken (8 Pc)
L 169 00	Baked Whole Trout	L 184 00	Grilled Turkey Patties (Ground Turkey)
L 169 01	Baked Trout Fillets	L 185 00	Caribbean Catfish
L 170 00	Chili (without Beans)	L 185 01	Caribbean Flounder
L 171 00	Cheese Pita Pizza	L 186 00	Baked Yogurt Chicken (Breast Boneless)
L 171 01	Mushroom, Onion, and Green Pepper Pita Pizza	L 187 00	Hot and Spicy Chicken (8 Pc)
L 172 00	Beef Stew (Canned Beef Chunks)	L 188 00	Turkey Fingers
L 173 00	Cheese Tortellini Marinara	L 189 00	Italian Broccoli Pasta
L 173 01	Spinach Tortellini Marinara (Frozen)	L 190 00	Cranberry Glazed Chicken (Breast Boneless)
L 173 02	Cheese Tortellini Marinara (Dehydrated)	L 191 00	Chicken & Italian Vegetable Pasta (Fajita Strips)
L 174 00	Rice Frittata	L 192 00	Honey Lemon Chicken Breast (Breast Boneless)
L 175 00	Potato Frittata	L 193 00	Cajun Roast Beef
L 176 00	Vegetable Stuffed Peppers	L 193 01	Cajun Roast Tenderloin of Beef
L 177 00	Bombay Chicken (8 Pc)	L 194 00	Tropical Baked Pork Chops
L 177 01	Bombay Chicken (Breast Boneless)	L 195 00	Teriyaki Beef Strips
L 178 00	Tropical Chicken Salad (Cooked Diced)	L 195 01	Teriyaki Beef Strips (Fajita Strips)

L. MEAT, FISH AND POULTRY No. 0(5)

Card No. .... Card No.

L 196 00	Southwestern Sweet Potatoes, Black Beans, and Corn	L 221 00	Turkey Divan
L 196 01	Southwestern Sweet Potatoes, Black Beans, and Corn (Canned)	L 222 00	Spicy Italian Pork Chops
		L 223 00	Lime Chicken Soft Tacos (Fajita Strips)
		L 224 00	Sausage, Beans and Greens
L 197 00	Dijon Baked Pork Chops	L 225 00	Orange & Rosemary Honey Glazed Pork Chops
L 198 00	Greek Lemon Turkey Pasta	L 500 00	Russian Turkey Stew
L 200 00	Grilled Turkey Sausage Links	L 501 00	Pasta Primavera
L 201 00	Tamale Pie (Turkey)	L 502 00	Fish Florentine
L 202 00	Oriental Tuna Patties	L 503 00	Jamaican Rum Chicken (Breast Boneless)
L 203 00	Vegetable Curry with Rice	L 504 00	Baked Fish Scandia
L 204 00	Turkey Peach Pasta Salad (Entree)	L 506 00	Thai Beef Salad
L 205 00	Italian Rice and Beef	L 507 00	Vegetarian Burrito
L 206 00	Bayou Chicken (Breast Boneless)	L 508 00	Vegetable Lasagna
L 207 00	Southwestern Shrimp Linguine	L 510 00	Tuna Plate Trio
L 208 00	Pasta Toscano	L 512 00	Grilled Turkey Sausage Patty (Pre-Made)
L 209 00	Seafood Stew	L 515 00	Oven Fried Turkey Bacon
L 210 00	Sante Fe Glazed Chicken (Breast Boneless)	L 523 00	Mambo Pork Roast
L 212 00	White Bean Chicken Chili (Cooked Diced)	L 524 00	White Fish with Mushrooms
L 213 00	Chicken Briyani (Cooked Diced)	L 800 00	Turkey Polynesian
L 216 00	Cheddar Chicken and Broccoli (Cooked Diced)	L 802 00	Angel Hair Pasta, Filipino Style with Shrimp
L 217 00	Asian Barbecue Turkey	L 803 00	Oven Roasted Turkey, Precooked
L 219 00	Lemon N' Herb Turkey Fillets	L 804 00	Lasagna (Frozen)

INDEX

Card No.		Card No.	
L 805 00	Mexican Turkey Pasta	L 827 00	Spaghetti & Meat Sauce (Precooked Ground Beef)
L 806 00	Basil Baked Fish Portions		
L 807 00	Tuna Noodle Casserole, Frozen	L 827 01	Spaghetti & Meat Balls (Precooked Meatballs)
L 808 00	Turkey Tetrazzini, Frozen	L 828 00	Baked Flounder Fillets with Lemon Pepper
L 809 00	Shepherd's Pie	L 829 00	Hunter Style Turkey Stew
L 810 00	Beef Stir Fry	L 831 00	Beef Stroganoff, Frozen
L 811 00	Indonesian Style Beef Over Noodles	L 832 00	Honey Glazed Chicken (Breast Boneless)
L 812 00	Hot & Spicy Chicken Wings	L 833 00	Rosemary Turkey Roast
L 813 00	Mambo Pork Roast Using Precooked Pork	L 834 00	Swedish Meatballs (Precooked Meatballs)
L 814 00	Kielbasa with Sauerkraut and Apples	L 835 00	Mexican Pepper Steak
L 816 00	Tarragon Chicken & Rice (Fajita Strips)	L 836 00	St Louis Style BBQ Pork Ribs, Precooked
L 817 00	Cajun Roast Beef (Precooked Roast Beef)	L 837 00	Meatloaf (Precooked)
L 818 00	Baked Tandouri Chicken (Breast Boneless)	L 837 01	Cajun Meatloaf (Precooked)
L 819 00	Baked Ham and Spaghetti Pie	L 838 00	Pork Tenderloin, Precooked
L 820 00	Cantonese BBQ Pork Ribs, Precooked	L 839 00	Chicken Cordon Bleu
L 821 00	Herb Turkey Roast w/Tomato Gravy Precooked Turkey	L 840 00	Blackened Fish
		L 841 00	Manicotti, Frozen
L 822 00	Beef and Bean Burritos, Frozen	L 842 00	Salisbury Steak in Gravy, Frozen
L 825 00	Corned Beef And Cabbage (Precooked Corned Beef)	L 843 00	Cabbage Rolls, Stuffed, Frozen
		L 844 00	Jerked Roast Turkey
L 826 00	Savory Baked Chicken (Breast Boneless)	L 845 00	Lemon Pepper Catfish

MEAT, FISH, AND POULTRY No.L 001 01

GRILLED OR OVEN FRIED BACON (PRECOOKED BACON)

Yield 100   Portion 2 Slices

Calories	Carbohydrates	Protein	Fat	Cholesterol	Sodium	Calcium
29 cal	0 g	2 g	2 g	4 mg	81 mg	1 mg

Ingredient   **Weight**   **Measure**   **Issue**

BACON,COOKED   4 lbs

Method

1. Place bacon on 350 F. griddle. Heat 5 minutes until crisp but not brittle turning once after 3 minutes.
2. Drain on absorbent paper. CCP: Hold for service at 140 F. or higher.

Notes

1. Precooked bacon may be oven fried. Using a convection oven, bake 4 to 5 minutes at 375 F. or until slightly crisp on high fan, closed vent.

MEAT, FISH, AND POULTRY No. L 002 00

OVEN FRIED BACON

Yield 100   **Portion** 2 Slices

Calories	Carbohydrates	Protein	Fat	Cholesterol	Sodium	Calcium
88 cal	0 g	5 g	8 g	13 mg	243 mg	2 mg

Ingredient   Weight   Measure   Issue
BACON, SLICED, RAW   12 lbs

Method

1. Arrange slices in rows, 2-1/2 pounds per pan, down the length of 18x26 sheet pan, with fat edges slightly overlapping lean edges.
2. Using a convection oven, bake 25 minutes at 325 F. on high fan, open vent. Drain excess fat. Bake an additional 5 to 10 minutes or until bacon is slightly crisp. DO NOT OVERCOOK.
3. Drain thoroughly. Place on absorbent paper or in perforated steam table pan. CCP: Hold for service at 140 F. or higher.

MEAT, FISH, AND POULTRY No.L 002 02

GRILLED BACON

Yield 100 **Portion** 2 Slices

Calories	Carbohydrates	Protein	Fat	Cholesterol	Sodium	Calcium
88 cal	0 g	5 g	8 g	13 mg	243 mg	2 mg

Ingredient   **Weight**   **Measure**   **Issue**
BACON,SLICED,RAW   12 lbs

Method

1. Place bacon slices on 350 F. griddle. Grill approximately 5 minutes turning once after 3 minutes, until slightly crisp. Remove excess fat as it accumulates on griddle.
2. Drain thoroughly. Place on absorbent paper or in perforated steam table pan. CCP: Hold for service at 140 F. or higher.

MEAT. FISH. AND POULTRY No.L 002 03

GRILLED OR OVEN FRIED CANADIAN BACON

Yield 100   **Portion** 2 Slices

Calories	Carbohydrates	Protein	Fat	Cholesterol	Sodium	Calcium
29 cal	0 g	4 g	1 g	9 mg	245 mg	2 mg

Ingredient   Weight   Measure   Issue

BACON,CANADIAN,SLICED,1 OZ   12-1/2 lbs

Method

1. Grill bacon on lightly greased 350 F. griddle about 1 minute on each side.
2. Drain thoroughly. Place on absorbent paper or in perforated steam table pan. CCP: Hold for service at 140 F. or higher.

Notes

1. Canadian bacon may be oven fried. Using a convection oven, bake at 350 F. for 6 to 8 minutes on high fan, open vent.

MEAT, FISH, AND POULTRY No.L 003 00

CHICKEN ENCHILADAS (CANNED CHICKEN)

Yield 100 **Portion** 2 Enchiladas

Calories	Carbohydrates	Protein	Fat	Cholesterol	Sodium	Calcium
412 cal	34 g	32 g	16 g	71 mg	2091 mg	137 mg

Ingredient	**Weight**	**Measure**	**Issue**
COOKING SPRAY,NONSTICK	1/4 oz	1/4 tsp	
ONIONS,FRESH,CHOPPED	5 lbs	3 qts 2-1/8 cup	5-1/2 lbs
SAUCE,ENCHILADA,CANNED	41-1/2 lbs	4 gal 3-1/2 qts	
CHILI POWDER,LIGHT,GROUND	5-1/4 oz	1-1/4 cup	
PEPPER,RED,GROUND	1 oz	1/4 cup 1-2/3 tbsp	
GARLIC POWDER	1 oz	3-1/3 tbsp	
CHICKEN,BONED,CANNED,PIECES	23-3/4 lbs	2 gal 3-1/2 qts	
TORTILLAS,WHEAT,6 INCH	8-1/2 lbs		
CHEESE,CHEDDAR,LOWFAT,SHREDDED	4 lbs	1 gal	

Method

1. Lightly spray kettle or stock pot with non-stick cooking spray. Stir-cook onions in a lightly sprayed steam jacketed kettle or stock pot 5 minutes or until tender.
2. Combine onions, 6-1/4 qt enchilada sauce, chili powder, red pepper, and garlic powder. Blend well. Gently fold in chicken.
3. Spread 1-1/4 cup enchilada sauce in each sheet pan.
4. Place 1/3 cup (1-No. 12 scoop) of chicken filling in center of each tortilla. Roll tortilla tightly around filling. Place 3 rows seam-side down in each sheet pan (about 50 per pan).
5. Pour remaining enchilada sauce evenly over enchiladas in each pan.
6. Using a convection oven, bake 25 minutes at 300 F. on high fan, closed vent. CCP: Internal temperature must reach 165 F. or higher for 15 seconds.
7. Sprinkle 1 lb (1qt) cheese over enchiladas in each pan. Bake 3 minutes to melt cheese. CCP: Hold for service at 140 F. or higher.

MEAT, FISH, AND POULTRY   No. L 003 01

CHICKEN ENCHILADAS (COOKED DICED)

Yield 100   Portion   2 Enchiladas

Calories	Carbohydrates	Protein	Fat	Cholesterol	Sodium	Calcium
533 cal	48 g	44 g	17 g	105 mg	1769 mg	149 mg

Ingredient	Weight	Measure	Issue
COOKING SPRAY,NONSTICK	1/4 oz	1/4 tsp	
ONIONS,FRESH,CHOPPED	5 lbs	3 qts 2-1/8 cup	5-1/2 lbs
SAUCE,ENCHILADA,CANNED	41-1/2 lbs	4 gal 3-1/2 qts	
CHILI POWDER,LIGHT,GROUND	5-1/4 oz	1-1/4 cup	
PEPPER,RED,GROUND	1 oz	1/4 cup 1-2/3 tbsp	
GARLIC POWDER	1 oz	3-1/3 tbsp	
CHICKEN,COOKED,DICED	25 lbs		
TORTILLAS,WHEAT,6 INCH	14-1/8 lbs	200 each	
CHEESE,CHEDDAR,LOWFAT,SHREDDED	4 lbs	1 gal	

Method

1. Lightly spray kettle or stock pot with non-stick cooking spray. Stir-cook onions in a lightly sprayed steam jacketed kettle or stockpot 5 minutes or until tender, let cool.
2. Combine onions, 6-1/4 qt enchilada sauce, chili powder, red pepper, and garlic powder. Blend well. Gently fold in chicken; cover.
3. Spread 1-1/4 cup enchilada sauce in each sheet pan.
4. Place 1/3 cup of chicken filling in center of each tortilla. Roll tortilla tightly around filling. Place 3 rows seam-side down in each sheet pan (about 50 per pan).
5. Pour remaining enchilada sauce evenly over enchiladas in each pan.
6. Using a convection oven, bake 25 minutes at 300 F. on high fan, closed vent. CCP: Internal temperature must reach 165 F. or higher for 15 seconds.
7. Sprinkle 1 lb (1 qt) cheese over enchiladas in each pan. Bake 3 minutes to melt cheese. CCP: Hold for service at 140 F. or higher.

MEAT, FISH, AND POULTRY No.L 004 00

ROAST RIB OF BEEF

Yield 100   **Portion** 6 Ounces

Calories	Carbohydrates	Protein	Fat	Cholesterol	Sodium	Calcium
743 cal	0 g	67 g	50 g	222 mg	161 mg	24 mg

Ingredient   Weight   Measure   Issue

BEEF,RIBEYE,PERFECT CHOICE,RAW   75 lbs
PEPPER,BLACK,GROUND   1/2 oz   2 tbsp

Method

1. Rub each roast with pepper.
2. Place roasts in 18x24 roasting pans. DO NOT ADD WATER. DO NOT COVER. Insert meat thermometer in center of roasts; DO NOT touch bone with thermometer.
3. Using a convection oven, roast 3 to 4 hours at 300 F. on low fan, closed vent or until roast reaches desired degree of doneness. CCP: Internal temperature must reach 145 F. or higher for 15 seconds.
4. Let roast stand about 20 minutes before slicing. CCP: Hold for service at 140 F. or higher.

Notes

1. Remove roasts from oven when meat thermometer registers 140 F. for rare, 160 F. for medium, and 170 F. for well done.
2. 50 pounds beef rib may be used per 100 portions. EACH PORTION: 4 oz.

MEAT, FISH, AND POULTRY No. L 004 01

STEAMSHIP ROUND OF BEEF (ROUND, BONE-IN)

Yield 100   Portion 6 Ounces

Calories	Carbohydrates	Protein	Fat	Cholesterol	Sodium	Calcium
470 cal	0 g	71 g	18 g	216 mg	115 mg	12 mg

Ingredient	Weight	Measure	Issue
BEEF,ROUND,BOTTOM,LEAN,RAW	75 lbs		
PEPPER,BLACK,GROUND	1/2 oz	2 tbsp	

Method

1. Use bone-in rounds. Rub each roast with pepper.
2. Place roasts in 18x24 roasting pans. DO NOT ADD WATER. DO NOT COVER. Insert meat thermometer in center of roasts; DO NOT touch bone with thermometer.
3. Using a convection oven, roast at 300 F. about 3 hours on high fan, closed vent and last 4 hours on low fan, closed vent, or until roast reaches desired degree of doneness. CCP: Internal temperature must reach 145 F. or higher for 15 seconds.
4. Let roast stand about 20 minutes before slicing. CCP: Hold for service at 140 F. or higher.

Notes

1. Remove roasts from oven when meat thermometer registers 140 F. for rare, 160 F. for medium, and 170 F. for well done.
2. 50 pounds bone-in rounds may be used per 100 portions. EACH PORTION: 4 ounces.

MEAT, FISH, AND POULTRY No.L 004 02

STEAMSHIP ROUND OF BEEF (ROUND, BONELESS)

Yield 100   Portion 6 Ounces

Calories	Carbohydrates	Protein	Fat	Cholesterol	Sodium	Calcium
407 cal	0 g	62 g	16 g	187 mg	99 mg	10 mg

Ingredient	Weight	Measure	Issue
BEEF,ROUND,BOTTOM,LEAN,RAW	65 lbs		
PEPPER,BLACK,GROUND	1/2 oz	2 tbsp	

Method

1. Use boneless rounds or racks. Rub each roast with pepper.
2. Place roasts in 18x24 roasting pans. DO NOT ADD WATER. DO NOT COVER. Insert meat thermometer in center of roasts.
3. Using a convection oven, roast at 300 F. about 3 hours on high fan, closed vent and last 2 hours on low fan, closed vent or until roast reaches desired degree of doneness. CCP: Internal temperature must reach 145 F. or higher for 15 seconds.
4. Let roast stand about 20 minutes before slicing. CCP: Hold for service at 140 F. or higher.

Notes

1. Remove roasts from oven when meat thermometer registers 140 F. for rare, 160 F. for medium, and 170 F. for well done.
2. 40 pounds boneless rounds may be used per 100 portions. EACH PORTION: 4 ounces.

MEAT. FISH. AND POULTRY   No.L 004 03

ROAST RIB OF BEEF (BONELESS RIBEYE ROLL)

Yield 100   Portion   6 Ounces

Calories	Carbohydrates	Protein	Fat	Cholesterol	Sodium	Calcium
675 cal	0 g	57 g	48 g	195 mg	137 mg	23 mg

Ingredient	Weight	Measure	Issue
BEEF,RIBEYE ROLL,RAW	65 lbs		
PEPPER,BLACK,GROUND	1/2 oz	2 tbsp	

Method

1. Use boneless ribeye rolls. Rub each roast with pepper.
2. Place roasts in roasting pans. DO NOT ADD WATER. DO NOT COVER. Insert meat thermometer in center of roasts.
3. Using a convection oven, roast about 2 to 3 hours at 300 F. on high fan, closed vent or until roast reaches desired degree of doneness. CCP: Internal temperature must reach 145 F. or higher for 15 seconds.
4. Let roast stand about 20 minutes before slicing. CCP: Hold for service at 140 F. or higher.

Notes

1. Remove roasts from oven when meat thermometer registers 140 F. for rare, 160 F. for medium, and 170 F. for well done.
2. 40 pound boneless ribeye rolls may be used. EACH PORTION: 4 Ounces.

MEAT, FISH, AND POULTRY No.L 005 00

ROAST BEEF

Yield 100　　　　　　　　　　　　　　　　　　　　**Portion** 4 Ounces

Calories	Carbohydrates	Protein	Fat	Cholesterol	Sodium	Calcium
276 cal	0 g	39 g	12 g	112 mg	86 mg	11 mg

Ingredient	Weight	Measure	Issue
PEPPER,BLACK,GROUND	1/2 oz	2 tbsp	
BEEF,OVEN ROAST,TEMPERED	40 lbs		

Method

1. Place roasts fat side up in 18x20 roasting in pans according to size without crowding. Sprinkle with pepper.
2. Insert meat thermometer into center of thickest part of main muscle. DO NOT ADD WATER. DO NOT COVER.
3. Using a convection oven, roast 1 hour 45 minutes at 325 F., depending on size of roasts. Roast to desired degree of doneness. CCP: Internal temperature must reach 155 F. or higher for 15 seconds.
4. Let stand 20 minutes before slicing. CCP: Hold for service at 140 F. or higher.

Notes

1. 26 pounds of precooked roast beef may be used.
2. Frozen roasts will require 1 hour or longer cooking time.
3. Remove roasts from oven when meat thermometer registers 140 F. for rare; 160 F. for medium; and 170 F. for well done.
4. Internal temperature will rise about 10 degrees during 20 minute standing period.

MEAT, FISH, AND POULTRY   No. L 005 01

ROAST BEEF (PRECOOKED)

Yield 100   Portion 4 Ounces

Calories	Carbohydrates	Protein	Fat	Cholesterol	Sodium	Calcium
249 cal	0 g	35 g	11 g	101 mg	78 mg	9 mg

Ingredient   Weight   Measure   Issue

BEEF, OVEN ROAST, PRE COOKED   26 lbs

Method

1 Thaw beef. CCP: Hold for service at 140 F. or higher.

MEAT, FISH, AND POULTRY No.L 006 00

SUKIYAKI

Yield 100   **Portion** 1 Cup

Calories	Carbohydrates	Protein	Fat	Cholesterol	Sodium	Calcium
219 cal	9 g	27 g	8 g	70 mg	770 mg	48 mg

Ingredient

Ingredient	Weight	Measure	Issue
BEEF,OVEN ROAST,TEMPERED	25 lbs		
SOY SAUCE	2-1/2 lbs	1 qts	
SUGAR,GRANULATED	7 oz	1 cup	
PEPPER,BLACK,GROUND	1/4 oz	1 tbsp	
MUSHROOMS,CANNED,DRAINED	13-3/4 oz	2-1/2 cup	
COOKING SPRAY,NONSTICK	2 oz	1/4 cup 1/3 tbsp	
CELERY,FRESH,SLICED	8 lbs	1 gal 3-5/8 qts	11 lbs
ONIONS,FRESH,SLICED	5 lbs	1 gal 7/8 qts	5-1/2 lbs
PEPPERS,GREEN,FRESH,JULIENNE	3 lbs	2 qts 1-1/8 cup	3-2/3 lbs
ONIONS,GREEN,FRESH,SLICED	5 lbs	1 gal 1-2/3 qts	5-1/2 lbs
BEAN SPROUTS,CANNED,DRAINED	3-1/4 lbs	2 qts 3-1/2 cup	

Method

1. Slice beef into 1/8-inch thick slices. Cut slices into strips 2 inches long and 1/2-inch wide. Set aside for use in Step 5.
2. Combine soy sauce, sugar, pepper and mushrooms. Set aside for use in Step 8.
3. Lightly spray steam-jacketed kettle or stock pot.
4. Add celery; saute 1-1/2 minutes, stirring constantly.
5. Add beef strips; continue stir frying 1-1/2 minutes.
6. Add onions; stir-fry 1-1/2 minutes.
7. Add green peppers; stir-fry 1 minute.
8. Add mushroom sauce mixture, green onions and bean sprouts; stir-fry 30 seconds. Remove from heat. CCP: Internal temperature must reach 155 F. or higher for 15 seconds.
9. CCP: Hold at 140 F. or higher for service.

MEAT. FISH. AND POULTRY No.L 007 00

GRILLED STEAK

Yield 100   **Portion** 1 Steak

Calories	Carbohydrates	Protein	Fat	Cholesterol	Sodium	Calcium
433 cal	0 g	45 g	27 g	144 mg	101 mg	18 mg

Ingredient   **Weight**   **Measure**   **Issue**

SHORTENING,VEGETABLE,MELTED   14-1/2 oz   2 cup
BEEF LOIN,STRIP STEAK,BONELESS,RAW,SIRLOIN,LEAN   47 lbs

Method

1 Preheat grill; lightly grease with shortening.
2 Grill steaks to desired degree of doneness: SIRLOIN: Rare - 6 minutes; Medium - 7-1/2 minutes; Well done - 9-1/2 minutes; RIBEYE: Rare - 3-1/2 minutes; Medium - 4 minutes; Well Done - 5 minutes; STRIP LOIN: Rare - 5 minutes; Medium - 6 minutes; Well done - 7 minutes. CCP: Internal temperature must reach 145 F. or higher for 15 seconds.

Notes

1 Do not hold steaks in ovens, warming cabinets, or on grills after cooking. This will cause steaks to dry out and be tough.
2 Steaks may be prepared in convection oven. Arrange in rows 3 by 5 on rack. Place racks on sheet pans. DO NOT TURN STEAKS. Cook in 400 F. oven to desired degree of doneness.

MEAT, FISH, AND POULTRY No.L 007 01

GRILLED TENDERLOIN STEAK

Yield 100   Portion   1 Steak

Calories	Carbohydrates	Protein	Fat	Cholesterol	Sodium	Calcium
436 cal	0 g	38 g	30 g	129 mg	88 mg	12 mg

Ingredient   Weight   Measure   Issue
BEEF,TENDERLOIN,RAW   44 lbs

Method

1. Use thawed beef tenderloin. Trim excess fat to 1/4-inch and slice tenderloins into 6 ounce steaks, about 3/4 inch thick. Grill on 400 F. griddle for 3 to 6 minutes for rare, 4 to 7 minutes for medium and 5 to 9 minutes for well done. CCP: Internal temperature must reach 145 F. or higher for 15 seconds. Hold for service at 140 F. or higher.

Notes

1. The narrow tail section may be butterflied or flattened to produce steaks of more uniform thickness. The cooking time varies due to size variations of tenderloins.

MEAT, FISH, AND POULTRY No. L 008 00

TERIYAKI STEAK

Yield 100　　　　　　　　　　　　　　　　　　　**Portion** 1 Steak

Calories	Carbohydrates	Protein	Fat	Cholesterol	Sodium	Calcium
434 cal	6 g	48 g	23 g	144 mg	1551 mg	31 mg

Ingredient	Weight	Measure	Issue
BEEF LOIN,STRIP STEAK,BONELESS,RAW,SIRLOIN,LEAN	47 lbs		
JUICE,PINEAPPLE,CANNED,UNSWEETENED	5 lbs	2 qts 1 cup	
SOY SAUCE	5-3/4 lbs	2 qts 1 cup	
WATER	11 lbs	1 gal 1-1/4 qts	
GINGER,GROUND	3-3/8 oz	1-1/8 cup	
GARLIC POWDER	7/8 oz	3 tbsp	
PEPPER,BLACK,GROUND	1-1/3 oz	1/4 cup 2-1/3 tbsp	
COOKING SPRAY,NONSTICK	2 oz	1/4 cup 1/3 tbsp	

Method

1 Arrange 25 steaks in each 18x24 roasting pan.
2 Combine pineapple juice, soy sauce, water, ginger, garlic and pepper. Pour 2-1/4 quarts sauce over steaks in each pan. Cover; CCP: Marinate under refrigeration at 41 F. or lower for 3 hours, turning steaks after 1-1/2 hours. Drain. Bring marinade to a boil. CCP: Internal temperature must reach 165 F. or higher for 15 seconds.
3 Preheat griddle; spray lightly with cooking spray. Grill steaks on each side to desired degree of doneness turning frequently. CCP: Internal temperature must reach 145 F. or higher for 15 seconds.
4 Serve with 1/4 cup sauce. CCP: Hold for service at 140 F. or higher.

MEAT, FISH, AND POULTRY No.L 009 00

SPINACH LASAGNA

Yield 100 **Portion** 9-1/2 Ounces

Calories	Carbohydrates	Protein	Fat	Cholesterol	Sodium	Calcium
370 cal	45 g	25 g	12 g	89 mg	1142 mg	425 mg

Ingredient	Weight	Measure	Issue
ONIONS,FRESH,CHOPPED	3-1/8 lbs	2 qts 1 cup	3-1/2 lbs
COOKING SPRAY,NONSTICK	2 oz	1/4 cup 1/3 tbsp	
TOMATOES,CANNED,CRUSHED,INCL LIQUIDS	26-1/2 lbs	3 gal	
TOMATO PASTE,CANNED	8-1/8 lbs	3 qts 2 cup	
WATER	4-1/8 lbs	2 qts	
BAY LEAF,WHOLE,DRIED	1/8 oz	4 each	
GARLIC POWDER	5/8 oz	2 tbsp	
OREGANO,CRUSHED	1/3 oz	2 tbsp	
BASIL,DRIED,CRUSHED	1/3 oz	2 tbsp	
THYME,GROUND	1/3 oz	2 tbsp	
PEPPER,RED,GROUND	<1/16th oz	1/8 tsp	
SUGAR,GRANULATED	3-1/2 oz	1/2 cup	
SALT	1-7/8 oz	3 tbsp	
SPINACH,CHOPPED,FROZEN	15 lbs	2 gal 2-7/8 qts	
EGGS,WHOLE,FROZEN	3 lbs	1 qts 1-5/8 cup	
NUTMEG,GROUND	1/8 oz	1/3 tsp	
CHEESE,COTTAGE,LOWFAT	11 lbs	1 gal 1-1/2 qts	
CHEESE,MOZZARELLA,SHREDDED	6 lbs	1 gal 2 qts	
CHEESE,PARMESAN,GRATED	14-1/8 oz	1 qts	
NOODLES,LASAGNA,UNCOOKED	6 lbs	1 gal 2-1/2 qts	
CHEESE,PARMESAN,GRATED	5-1/4 oz	1-1/2 cup	

Method

1. Lightly spray steam jacketed kettle and saute onions.
2. Combine sauteed onions with tomatoes, tomato paste, water, bay leaves, garlic, oregano, basil, thyme, pepper, sugar, and salt; mix well.
3. Bring to a boil; reduce heat; simmer 1 hour or until thickened, stirring occasionally. Remove bay leaves. CCP: Hold at 140 F. or higher for use in Step 8.
4. Drain spinach. Press out excess water. Set aside for use in Step 7.
5. Add nutmeg to eggs; blend well.
6. Combine eggs with cheese; mix well.
7. Stir spinach into egg-cheese mixture. Mix lightly but thoroughly; place in shallow steam table pans.
8. PANNING INSTRUCTIONS: Arrange in layers in each pan. During panning, remove small amounts of filling from refrigeration at a time. Ensure entire panning procedure does not exceed 3 hours total time between temperatures of 40 F. to 140 F. Progressive preparation and immediate baking of the product will ensure food safety. Layer: 1. 2 cups sauce 2. Noodles, flat and in rows 3. 5-1/2 cups chilled spinach-cheese filling 4. 1 quart sauce 5. Noodles, flat and in rows 6. 5-1/2 cups chilled spinach-cheese filling 7. Noodles, flat and in rows 8. 1-1/2 quarts sauce Sprinkle with parmesan cheese.
9. Cover. Using a convection oven, bake 1-1/4 hours at 300 F. Remove cover; bake 10 to 15 minutes. CCP: Internal temperature must reach 145 F. or higher for 15 seconds.
10. Cut 5 by 4. CCP: Hold for service at 140 F. or higher.

MEAT, FISH, AND POULTRY No. L 010 00

BEEF POT ROAST

Yield 100   **Portion** 3-1/2 Ounces

Calories	Carbohydrates	Protein	Fat	Cholesterol	Sodium	Calcium
406 cal	6 g	35 g	26 g	114 mg	411 mg	18 mg

Ingredient	Weight	Measure	Issue
BEEF,POT ROAST,RAW	40 lbs		
WATER,BOILING	8-1/3 lbs	1 gal	
SALT	3 oz	1/4 cup 1 tbsp	
PEPPER,BLACK,GROUND	1/2 oz	2 tbsp	
ONIONS,FRESH,SLICED	3 lbs	3 qts	3-3/8 lbs
GARLIC POWDER	1/3 oz	1 tbsp	
FLOUR,WHEAT,GENERAL PURPOSE	1-1/4 lbs	1 qts 1/2 cup	
WATER,COLD	3-1/8 lbs	1 qts 2 cup	
RESERVED STOCK	10-1/2 lbs	1 gal 1 qts	

Method

1. Place roasts in stock pot or steam-jacketed kettle; brown on all sides; add water.
2. Add salt, pepper, onions and garlic. Cover. Simmer 3-1/2 to 4-1/2 hours or until tender. CCP: Internal temperature must reach 145 F. or higher for 15 seconds. Remove scum as it rises to the surface during cooking. Remove cooked beef. Skim off excess fat from stock. Reserve stock for use in Step 4.
3. Let roast stand 20 minutes; slice 1/8-inch thick. CCP: Hold for service at 140 F. or higher.
4. Combine flour and water, stirring until smooth; add to stock, stirring constantly. Cook 10 minutes or until slightly thickened. Remove bay leaves before serving. CCP: Temperature must reach 165 F. or higher for 15 seconds.
5. Serve sauce with sliced meat. CCP: Hold for service at 140 F. or higher.

MEAT, FISH, AND POULTRY No.L 010 01

GINGER POT ROAST

Yield 100   Portion   3-1/2 Ounces

Calories	Carbohydrates	Protein	Fat	Cholesterol	Sodium	Calcium
415 cal	8 g	36 g	26 g	114 mg	475 mg	27 mg

Ingredient	**Weight**	**Measure**	**Issue**
BEEF,POT ROAST,RAW	40 lbs		
WATER,BOILING	8-1/3 lbs	1 gal	
SALT	3 oz	1/4 cup 1 tbsp	
ONIONS,FRESH,CHOPPED	3 lbs	2 qts 1/2 cup	3-1/3 lbs
TOMATOES,CANNED,DICED,INCL LIQUIDS	6-3/8 lbs	2 qts 3-1/8 cup	
GINGER,GROUND	1/2 oz	2-2/3 tbsp	
THYME,GROUND	<1/16th oz	1/8 tsp	
BAY LEAF,WHOLE,DRIED	1/8 oz	4 lf	
PEPPER,BLACK,GROUND	1/2 oz	2 tbsp	
GARLIC POWDER	1/3 oz	1 tbsp	
FLOUR,WHEAT,GENERAL PURPOSE	1-1/4 lbs	1 qts 1/2 cup	
WATER,COLD	3-1/8 lbs	1 qts 2 cup	
RESERVED STOCK	16-3/4 lbs	2 gal	

Method

1. Place roasts in stock pot or steam-jacketed kettle; brown on all sides; add water.
2. Add salt, pepper, chopped onions, diced tomatoes, ground ginger, ground thyme, bay leaves and garlic to roasts. Cover. Simmer 3-1/2 to 4-1/2 hours or until tender. CCP: Internal temperature must reach 145 F. or higher for 15 seconds. Remove scum as it rises to the surface during cooking. Remove cooked beef. Skim off excess fat from stock. Reserve stock for use in Step 4.
3. Let roast stand 20 minutes; slice 1/8-inch thick.
4. Combine flour and water, stirring until smooth; add to stock, stirring constantly. Cook 10 minutes or until slightly thickened. Remove bay leaves before serving. CCP: Temperature must reach 165 F. or higher for 15 seconds.
5. Serve sauce with sliced meat. CCP: Hold for service at 140 F. or higher.

MEAT, FISH, AND POULTRY No.L 010 02

YANKEE POT ROAST

Yield 100   Portion 3-1/2 Ounces

Calories	Carbohydrates	Protein	Fat	Cholesterol	Sodium	Calcium
419 cal	9 g	36 g	26 g	114 mg	479 mg	30 mg

Ingredient

Ingredient	Weight	Measure	Issue
BEEF,POT ROAST,RAW	40 lbs		
WATER,BOILING	8-1/3 lbs	1 gal	
SALT	3 oz	1/4 cup 1 tbsp	
PEPPER,BLACK,GROUND	1/2 oz	2 tbsp	
ONIONS,FRESH,SLICED	3 lbs	2 qts 3-7/8 cup	3-1/3 lbs
GARLIC POWDER	1/3 oz	1 tbsp	
CARROTS,FRESH,CHOPPED	2 lbs	1 qts 3-1/8 cup	2-1/2 lbs
PARSLEY,FRESH,BUNCH,CHOPPED	2 oz	3/4 cup 3 tbsp	2-1/8 oz
TOMATOES,CANNED,DICED,INCL LIQUIDS	6-3/8 lbs	2 qts 3-1/8 cup	
ALLSPICE,GROUND	1/4 oz	1 tbsp	
BAY LEAF,WHOLE,DRIED	1/8 oz	4 lf	
THYME,GROUND	<1/16th oz	1/8 tsp	
VINEGAR,DISTILLED	8-1/3 oz	1 cup	
FLOUR,WHEAT,GENERAL PURPOSE	1-1/4 lbs	1 qts 1/2 cup	
WATER,COLD	3-1/8 lbs	1 qts 2 cup	
RESERVED STOCK	16-3/4 lbs	2 gal	

Method

1. Place roasts in stock pot or steam-jacketed kettle; brown on all sides; add water.
2. Add salt, pepper, onions, garlic, diced fresh carrots, chopped fresh parsley, canned tomatoes, ground allspice, bay leaves, ground thyme, and vinegar to roasts. Cover. Simmer 3-1/2 to 4-1/2 hours or until tender. CCP: Internal temperature must reach 145 F. or higher for 15 seconds. Remove scum as it rises to the surface during cooking. Remove cooked beef. Skim off excess fat from stock and reserve stock for use in Step 4.
3. Let roast stand 20 minutes; slice 1/8-inch thick.
4. Combine flour and water until smooth; add to stock, stirring constantly. Cook 10 minutes or until slightly thickened. Remove bay leaves before serving. CCP: Temperature must reach 165 F. or higher for 15 seconds.
5. Serve sauce with sliced meat. CCP: Hold for service at 140 F. or higher.

MEAT, FISH, AND POULTRY No.L 011 00

SIMMERED BEEF

Yield 100   **Portion** 4 Ounces

Calories	Carbohydrates	Protein	Fat	Cholesterol	Sodium	Calcium
397 cal	4 g	35 g	26 g	114 mg	416 mg	34 mg

Ingredient	**Weight**	**Measure**	**Issue**
BEEF,POT ROAST,RAW	40 lbs		
WATER,BOILING	33-1/2 lbs	4 gal	
CARROTS,FRESH,CHOPPED	2 lbs	1 qts 3-1/8 cup	2-1/2 lbs
CELERY,FRESH,CHOPPED	2 lbs	1 qts 3-1/2 cup	2-3/4 lbs
ONIONS,FRESH,CHOPPED	4 lbs	2 qts 3-3/8 cup	4-1/2 lbs
BAY LEAF,WHOLE,DRIED	1/8 oz	4 each	
CLOVES,WHOLE	4-2/3 oz	20 each	
SALT	2-7/8 oz	1/4 cup 2/3 tbsp	
PEPPER,BLACK,GROUND	1/4 oz	1 tbsp	

Method

1. Place roasts in stock pot or steam-jacketed kettle; brown on all sides; add water to cover.
2. Add carrots, celery, onions, bay leaves, cloves, salt and pepper.
3. Simmer 2-1/2 to 3 hours or until tender. DO NOT BOIL OR OVERCOOK. CCP: Internal temperature must reach 145 F. or higher for 15 seconds. Remove scum as it rises to the surface during cooking. Remove cooked beef; remove bay leaves and cloves.
4. Let roast stand 20 minutes before slicing. CCP: Hold for service at 140 F. or higher.

MEAT, FISH, AND POULTRY No.L 012 00

COUNTRY STYLE STEAK

Yield 100 **Portion** 6-1/2 Ounces

Calories	Carbohydrates	Protein	Fat	Cholesterol	Sodium	Calcium
393 cal	14 g	39 g	19 g	137 mg	624 mg	36 mg

Ingredient	Weight	Measure	Issue
BEEF,SWISS STEAK,LEAN,RAW,THAWED	37-1/2 lbs		
FLOUR,WHEAT,GENERAL PURPOSE	2-1/4 lbs	2 qts	
SALT	3-3/4 oz	1/4 cup 2-1/3 tbsp	
PEPPER,BLACK,GROUND	1/4 oz	1 tbsp	
MILK,NONFAT,DRY	3-1/4 oz	1-3/8 cup	
WATER	3-7/8 lbs	1 qts 3-1/2 cup	
EGGS,WHOLE,FROZEN	1-1/2 lbs	2-3/4 cup	
BREADCRUMBS	2-5/8 lbs	2 qts 3 cup	
SALT	5/8 oz	1 tbsp	
PEPPER,BLACK,GROUND	1/4 oz	1 tbsp	
OIL, CANOLA	1-7/8 lbs	1 qts	

Method

1. Dredge steaks in mixture of flour, salt, and pepper; shake off excess.
2. Reconstitute milk; add eggs; blend thoroughly.
3. Combine bread crumbs, salt, and pepper.
4. Dip steaks in egg and milk mixture; then in seasoned bread crumbs.
5. Brown steaks 1-1/2 minutes on each side on 350 F. well greased griddle.
6. Overlap steaks in lightly greased 18x24 roasting pans. Cover pans tightly.
7. Using a convection oven, bake 1-1/2 hours at 325 F. or until steaks are tender. CCP: Internal temperature must reach 145 F. or higher for 15 seconds. Hold for service at 140 F. or higher.

MEAT. FISH. AND POULTRY No.L 013 00

PEPPER STEAK

Yield 100   Portion 5-1/2 Ounces

Calories	Carbohydrates	Protein	Fat	Cholesterol	Sodium	Calcium
225 cal	7 g	30 g	8 g	86 mg	443 mg	16 mg

Ingredient	Weight	Measure	Issue
BEEF,SWISS STEAK,LEAN,RAW,THAWED	30 lbs		
COOKING SPRAY,NONSTICK	2 oz	1/4 cup 1/3 tbsp	
WATER	8-1/3 lbs	1 gal	
TOMATO PASTE,CANNED	2 lbs	3-1/2 cup	
SOY SAUCE	1-1/4 lbs	2 cup	
SUGAR,GRANULATED	1-3/4 oz	1/4 cup 1/3 tbsp	
PEPPER,BLACK,GROUND	1/3 oz	1 tbsp	
GARLIC POWDER	1/2 oz	1 tbsp	
CORNSTARCH	4-1/2 oz	1 cup	
WATER,COLD	2-1/8 lbs	1 qts	
PEPPERS,GREEN,FRESH,CHOPPED	8 lbs	1 gal 2-1/8 qts	9-3/4 lbs
ONIONS,FRESH,CHOPPED	2-3/4 lbs	1 qts 3-3/4 cup	3 lbs

Method

1 Lightly spray griddle with non-stick cooking spray. Cut steaks into 1/2-inch strips; brown strips 5 minutes on 350 F. griddle turning frequently.
2 Place strips in each roasting pan.
3 Combine water, tomato paste, soy sauce, sugar, pepper, and garlic powder. Blend well. Bring to a boil.
4 Dissolve cornstarch in water; stir until smooth; add to sauce mixture. Cook until thickened, about 3 minutes, stirring constantly.
5 Pour sauce evenly over beef strips in each pan. Cover. Bake in a convection oven at 325 F. for 1-1/2 hours on high fan, closed vent.
6 Add 4 lbs (4-3/4 quart) peppers and 1 lb 5 oz (1 quart) onions to each pan. Stir to distribute vegetables. Cover; bake 20 minutes or until beef is tender. CCP: Internal temperature must reach 145 F. or higher for 15 seconds. Hold for service at 140 F. or higher.

MEAT. FISH. AND POULTRY No.L 013 01

ORIENTAL PEPPER STEAK

Yield 100   Portion   5-1/2 Ounces

Calories	Carbohydrates	Protein	Fat	Cholesterol	Sodium	Calcium
227 cal	8 g	30 g	8 g	86 mg	463 mg	18 mg

Ingredient	Weight	Measure	Issue
BEEF,SWISS STEAK,LEAN,RAW,THAWED	30 lbs		
COOKING SPRAY,NONSTICK	2 oz	1/4 cup 1/3 tbsp	
WATER	8-1/3 lbs	1 gal	
TOMATO PASTE,CANNED	2 lbs	3-1/2 cup	
SOY SAUCE	1-1/4 lbs	2 cup	
SUGAR,GRANULATED	1-3/4 oz	1/4 cup 1/3 tbsp	
PEPPER,BLACK,GROUND	1/3 oz	1 tbsp	
GARLIC POWDER	1/2 oz	1 tbsp	
CORNSTARCH	4-1/2 oz	1 cup	
WATER,COLD	2-1/8 lbs	1 qts	
BEAN SPROUTS,CANNED,DRAINED	3-1/4 lbs	2 qts 3-3/4 cup	
PEPPERS,GREEN,FRESH,CHOPPED	8 lbs	1 gal 2-1/8 qts	9-3/4 lbs
ONIONS,FRESH,CHOPPED	2-3/4 lbs	1 qts 3-3/4 cup	3 lbs

Method

1. Lightly spray griddle with non-stick cooking spray. Cut steaks into 1/2-inch strips; brown strips 5 minutes on 350 F. griddle turning frequently.
2. Place strips in roasting pans.
3. Combine water, tomato paste, soy sauce, sugar, pepper, and garlic powder. Blend well. Bring to a boil.
4. Dissolve cornstarch in water; stir until smooth; add to sauce mixture. Cook until thickened, about 3 minutes, stirring constantly.
5. Pour sauce evenly over beef strips in each pan. Cover. Using a convection oven, bake at 325 F. for 1-1/2 hours on high fan, closed vent.
6. Add 4 lbs (4-3/4 quart) peppers and 1 lb 5 oz (1 quart) onions to each pan. Add drained bean sprouts. Stir to distribute vegetables. Cover; bake 20 minutes or until beef is tender. CCP: Internal temperature must reach 145 F. or higher for 15 seconds. Hold for service at 140 F. or higher.

MEAT, FISH, AND POULTRY No.L 014 00

GROUND BEEF CORDON BLEU

Yield 100   Portion 5 Ounces

Calories	Carbohydrates	Protein	Fat	Cholesterol	Sodium	Calcium
377 cal	7 g	36 g	22 g	124 mg	766 mg	296 mg

Ingredient	Weight	Measure	Issue
BREAD,WHITE,CUBED	2-1/2 lbs	2 gal 1/8 qts	
GARLIC POWDER	1/3 oz	1 tbsp	
WATER	4-2/3 lbs	2 qts 1 cup	
BEEF,GROUND,BULK,RAW,90% LEAN	22-1/2 lbs		
ONIONS,FRESH,CHOPPED	1-3/8 lbs	1 qts	1-5/8 lbs
EGGS,WHOLE,FROZEN	6-3/8 oz	3/4 cup	
SALT	3 oz	1/4 cup 1 tbsp	
PEPPER,BLACK,GROUND	1/4 oz	1 tbsp	
CHEESE,SWISS,SLICED	6-1/4 lbs	100 sl	
HAM,COOKED,BONELESS,SLICED	3-3/4 lbs	100 sl	

Method

1. Combine bread, garlic, and water. Let stand 10 minutes until water is absorbed.
2. Add ground beef, onions, eggs, salt, and pepper; mix well. Shape into 200, 2-1/2 ounce patties. Flatten patties to 3-1/2 inch diameter.
3. Cut cheese slices in half. Place halved cheesed slices on 100 patties. Place another halved cheese slice on top of ham. Fold ham around cheese. Fold ham and cheese no larger than 3 by 3 inches to fit inside patties and ensure a good seal. Add remaining ham slices. Place remaining patties on top; enclose securely by sealing edges together.
4. Using a convection oven, bake in 350 F. for 15 minutes or until done on high fan, closed vent. CCP: Internal temperature must reach 155 F. or higher for 15 seconds. CCP: Hold for service at 140 F. or higher.

MEAT. FISH. AND POULTRY No.L 015 00

STEAK SMOTHERED WITH ONIONS

Yield 100   Portion   3-1/2 Ounces

Calories	Carbohydrates	Protein	Fat	Cholesterol	Sodium	Calcium
329 cal	8 g	37 g	16 g	108 mg	297 mg	25 mg

Ingredient	Weight	Measure	Issue
BEEF,SWISS STEAK,LEAN,RAW,THAWED	37-1/2 lbs		
OIL,SALAD	1-1/2 lbs	3 cup	
ONIONS,FRESH,SLICED	20 lbs	4 gal 3-3/4 qts	22-1/4 lbs
SALT	1 oz	1 tbsp	
PEPPER,BLACK,GROUND	1/8 oz	1/3 tsp	
BEEF BROTH		2 qts	

Method

1. Brown steaks on 350 F. well greased griddle; 1 minute on each side.
2. Place steaks in roasting pans.
3. Evenly distribute onions over steaks in each pan, approximately 7 quarts per pan.
4. Prepare stock according to directions. Add salt and pepper; stir.
5. Pour 1 quart stock over steaks in each pan. Cover pan.
6. Using a convection oven, bake 1-1/2 hours at 325 F. or until tender on closed vent, high fan. CCP: Internal temperature must reach 145 F. or higher for 15 seconds. CCP: Hold for service at 140 F. or higher.

MEAT, FISH, AND POULTRY No. L 015 01

STEAK STRIPS SMOTHERED WITH ONIONS

Yield 100   Portion 3/4 Cup

Calories	Carbohydrates	Protein	Fat	Cholesterol	Sodium	Calcium
291 cal	8 g	30 g	15 g	86 mg	286 mg	24 mg

Ingredient

Ingredient	Weight	Measure	Issue
BEEF,SWISS STEAK,LEAN,RAW,THAWED	30 lbs		
OIL,SALAD	1-2/3 lbs	3 cup	
ONIONS,FRESH,SLICED	20 lbs	4 gal 3-3/4 qts	22-1/4 lbs
SALT	1 oz	1 tbsp	
PEPPER,BLACK,GROUND	1/8 oz	1/3 tsp	
BEEF BROTH		2 qts	

Method

1. Slice each steak into thin strips, 1/2-inch wide. Brown steaks on 350 F. well greased griddle; 1 minute on each side.
2. Place steaks in roasting pans.
3. Evenly distribute onions over steaks in each pan, approximately 7 quarts per pan.
4. Prepare stock according to recipe directions. Add salt and pepper; stir.
5. Pour 1 quart over steaks in each pan. Cover pan.
6. Using a convection oven, bake at 325 F. for 1-1/2 hours or until tender on closed vent, high fan. CCP: Internal temperature must reach 145 F. or higher for 15 seconds. CCP: Hold for service at 140 F. or higher.

MEAT, FISH, AND POULTRY No.L 016 00

SWISS STEAK WITH TOMATO SAUCE

Yield 100   Portion 7-1/2 Ounces

Calories	Carbohydrates	Protein	Fat	Cholesterol	Sodium	Calcium
328 cal	8 g	37 g	16 g	108 mg	388 mg	27 mg

Ingredient	Weight	Measure	Issue
BEEF,SWISS STEAK,LEAN,RAW,THAWED	37-1/2 lbs		
OIL,SALAD	1-1/2 lbs	3 cup	
BEEF BROTH		1 qts 1 cup	
SALT	1 oz	1 tbsp	
PEPPER,BLACK,GROUND	1/2 oz	2 tbsp	
GARLIC POWDER	1/8 oz	1/4 tsp	
WORCESTERSHIRE SAUCE	6-1/3 oz	3/4 cup	
ONIONS,FRESH,CHOPPED	3-1/8 lbs	2 qts 1 cup	3-1/2 lbs
PEPPERS,GREEN,FRESH,CHOPPED	2 lbs	1 qts 2 cup	2-3/8 lbs
TOMATOES,CANNED,DICED,INCL LIQUIDS	13-3/4 lbs	1 gal 2 qts	
FLOUR,WHEAT,GENERAL PURPOSE	8-7/8 oz	2 cup	
WATER	1 lbs	2 cup	

Method

1. Brown steaks on 325 F. well greased griddle.
2. Overlap steaks in roasting pans.
3. Prepare broth according to package directions.
4. Add salt, pepper, garlic, Worcestershire sauce, onions, peppers and tomatoes to stock. Stir well. Heat to boiling.
5. Pour about 4-1/2 quarts sauce over steaks in each pan. Cover.
6. Using a convection oven, bake at 325 F. for 2 hours or until tender on high fan, closed vent. CCP: Internal temperature must reach 145 F. or higher for 15 seconds. Skim off excess fat. Place steaks in 4 steam table roasting pans. Place sauce in steam-jacketed kettle or stock pot.
7. Mix flour and water to make a smooth paste; add to sauce. Cook 2 minutes or until thickened stirring constantly.
8. Pour 8-1/2 cups sauce over steaks in each pan. CCP: Hold for service at 140 F. or higher.

MEAT, FISH, AND POULTRY No. L 016 01

SWISS STEAK WITH BROWN GRAVY

Yield 100   Portion   7-1/2 Ounces

Calories	Carbohydrates	Protein	Fat	Cholesterol	Sodium	Calcium
329 cal	7 g	37 g	16 g	108 mg	545 mg	16 mg

Ingredient	Weight	Measure	Issue
BEEF,SWISS STEAK,LEAN,RAW,THAWED	37-1/2 lbs		
OIL,SALAD	1-1/2 lbs	3 cup	
ONIONS,FRESH,CHOPPED	3-1/8 lbs	2 qts 1 cup	3-1/2 lbs
BEEF BROTH		2 gal	
PEPPER,BLACK,GROUND	1/2 oz	2 tbsp	
GARLIC POWDER	1/8 oz	1/4 tsp	
WORCESTERSHIRE SAUCE	6-1/3 oz	3/4 cup	
ONIONS,FRESH,CHOPPED	3-1/8 lbs	2 qts 1 cup	3-1/2 lbs
FLOUR,WHEAT,GENERAL PURPOSE	1-1/8 lbs	1 qts	
WATER	2-1/8 lbs	1 qts	

Method

1. Grill steaks on well greased griddle 5 minutes on one side and then 4 minutes in the other.
2. Evenly layer 25 steaks into each ungreased steam table pan.
3. Cook onions in a lightly sprayed steam-jacketed kettle or stock pot 8 to 10 minutes, stirring constantly.
4. Prepare beef broth according to instructions on package.
5. Add broth, pepper, garlic powder, Worcestershire sauce to cooked onions; stir to blend. Bring to a boil; reduce heat to simmer.
6. Blend flour and cold water to make a slurry. Add slurry to broth and onions, stirring constantly. Bring to a boil. Cover; reduce heat; simmer 2 minutes or until thickened, stirring frequently.
7. Pour gravy evenly over steaks in each pan.
8. Using a convection oven, bake 2 hours at 325 F. or until tender on high fan, closed vent. CCP: Internal temperature must reach 145 F. or higher for 15 seconds. Hold at 140 F. or higher for service.

MEAT. FISH. AND POULTRY No.L 016 03

SWISS STEAK WITH TOMATO SOUP

Yield 100   Portion   7-1/2 Ounces

Calories	Carbohydrates	Protein	Fat	Cholesterol	Sodium	Calcium
339 cal	9 g	37 g	17 g	108 mg	489 mg	17 mg

Ingredient	Weight	Measure	Issue
BEEF,SWISS STEAK,LEAN,RAW,THAWED	37-1/2 lbs		
OIL,SALAD	1-1/2 lbs	3 cup	
SOUP,CONDENSED,TOMATO	12-1/2 lbs	1 gal 1-5/8 qts	
WATER	5-3/4 lbs	2 qts 3 cup	
SALT	1 oz	1 tbsp	
PEPPER,BLACK,GROUND	1/2 oz	2 tbsp	
GARLIC POWDER	1/8 oz	1/4 tsp	
ONIONS,FRESH,CHOPPED	3-1/8 lbs	2 qts 1 cup	3-1/2 lbs
PEPPERS,GREEN,FRESH,CHOPPED	2 lbs	1 qts 2 cup	2-3/8 lbs

Method

1. Brown steaks on 325 F. well greased griddle.
2. Overlap steaks in roasting pans.
3. Mix tomato soup with water.
4. Add salt, pepper, garlic, onions and sweet peppers to tomato soup. Stir to mix well. Heat to boiling.
5. Pour about 6-1/4 quarts sauce over steaks in each pan. Cover.
6. Using a convection oven, bake 2 hours at 325 F. on high fan, closed vent or until steaks are tender. CCP: Internal temperature must reach 145 F. or higher for 15 seconds. Skim off excess fat. Remove steaks to steam table roasting pans. Place sauce in steam-jacketed kettle or stock pot. Heat to boiling.
7. Pour 8-1/2 cups sauce over steaks in each pan. CCP: Hold for service at 140 F. or higher.

MEAT, FISH, AND POULTRY No.L 016 04

SWISS STEAK WITH MUSHROOM GRAVY

Yield 100   Portion   7-1/2 Ounces

Calories	Carbohydrates	Protein	Fat	Cholesterol	Sodium	Calcium
338 cal	6 g	37 g	18 g	108 mg	451 mg	25 mg

Ingredient	Weight	Measure	Issue
BEEF,SWISS STEAK,LEAN,RAW,THAWED	37-1/2 lbs		
OIL,SALAD	1 lbs	3 cup	
SOUP,CONDENSED,CREAM OF MUSHROOM	12-1/2 lbs	1 gal 1-5/8 qts	
WATER	8-1/3 lbs	1 gal	
PEPPER,BLACK,GROUND	1/2 oz	2 tbsp	
GARLIC POWDER	1/8 oz	1/4 tsp	
ONIONS,FRESH,CHOPPED	3-1/8 lbs	2 qts 1 cup	3-1/2 lbs

Method

1. Brown steaks on 325 F. well greased griddle.
2. Overlap steaks in roasting pans.
3. Mix soup with water.
4. Add pepper, garlic, and onions to soup. Stir to mix well. Heat to boiling.
5. Pour 5-1/4 quarts sauce over steaks in each pan. Cover.
6. Using a convection oven, bake 2-1/2 hours at 325 F. on high fan, closed vent or until steaks are tender. CCP: Internal temperature must reach 145 F. or higher for 15 seconds. Skim off excess fat. Place sauce in steam jacketed kettle or stock pot. Remove steaks to steam table roasting pans. Place sauce in steam jacketed kettle and heat to boiling.
7. Pour 8-1/2 cups sauce over steaks in each pan. CCP: Hold for service at 140 F. or higher.

MEAT, FISH, AND POULTRY No.L 017 00

BRAISED BEEF AND NOODLES

Yield 100   Portion   1-1/4 Cups

Calories	Carbohydrates	Protein	Fat	Cholesterol	Sodium	Calcium
294 cal	21 g	26 g	11 g	81 mg	716 mg	27 mg

Ingredient | Weight | Measure | Issue

Ingredient	Weight	Measure	Issue
BEEF,DICED,LEAN,RAW	30 lbs		
WATER	14-5/8 lbs	1 gal 3 qts	
ONIONS,FRESH,SLICED	4 lbs	1 gal	4-1/2 lbs
CATSUP	2-1/8 lbs	1 qts	
PEPPER,BLACK,GROUND	2/3 oz	3 tbsp	
THYME,GROUND	1/2 oz	3 tbsp	
GARLIC POWDER	3/8 oz	1 tbsp	
BAY LEAF,WHOLE,DRIED	1/4 oz	6 each	
SALT	3-3/8 oz	1/4 cup 1-2/3 tbsp	
NOODLES,EGG	3-1/2 lbs	2 gal 2-1/2 qts	
WATER,BOILING	58-1/2 lbs	7 gal	
SALT	1-1/2 oz	2-1/3 tbsp	
FLOUR,WHEAT,GENERAL PURPOSE	1-1/2 lbs	1 qts 1-1/2 cup	
WATER,COLD	3-1/8 lbs	1 qts 2 cup	

Method

1. Place beef, water, onions, catsup, pepper, thyme, garlic powder, bay leaves and salt in steam-jacketed kettle or stock pot. Bring to a boil; reduce heat; cover; simmer about 2 hours or until tender. Skim off excess fat. Remove bay leaves.
2. Add noodles to boiling salted water; return to a boil; cook 8 to 10 minutes or until tender; drain thoroughly.
3. Combine flour and water to make smooth mixture; stir into beef mixture. Blend well. Return to boil. Reduce heat; cook 10 minutes or until thickened. CCP: Internal temperature must reach 145 F. or higher for 15 seconds.
4. Add cooked noodles to beef mixture. Stir well. CCP: Hold for service at 140 F. or higher.

MEAT, FISH, AND POULTRY No.L 017 01

BRAISED BEEF CUBES

Yield 100   Portion   6-1/2 Ounces

Calories	Carbohydrates	Protein	Fat	Cholesterol	Sodium	Calcium
223 cal	7 g	24 g	10 g	66 mg	428 mg	14 mg

Ingredient	Weight	Measure	Issue
BEEF,DICED,LEAN,RAW	30 lbs		
WATER	10-1/2 lbs	1 gal 1 qts	
ONIONS,FRESH,SLICED	4 lbs	1 gal	4-1/2 lbs
PEPPER,BLACK,GROUND	2/3 oz	3 tbsp	
THYME,GROUND	1/2 oz	3 tbsp	
GARLIC POWDER	3/8 oz	1 tbsp	
BAY LEAF,WHOLE,DRIED	1/4 oz	6 each	
SALT	3-3/8 oz	1/4 cup 1-2/3 tbsp	
WATER,COLD	3-1/8 lbs	1 qts 2 cup	
FLOUR,WHEAT,GENERAL PURPOSE	1-1/2 lbs	1 qts 1-1/2 cup	

Method

1. Place beef, water, onions, pepper, thyme, garlic powder, bay leaves and salt in steam-jacketed kettle or stock pot. Bring to a boil; reduce heat; cover; simmer about 2 hours or until tender. Skim off excess fat. Remove bay leaves.
2. Combine flour and water to make smooth mixture; stir into beef mixture. Blend well. Return to boil. Reduce heat; cook 10 minutes or until thickened. CCP: Internal temperature must reach 145 F. or higher for 15 seconds.
3. CCP: Hold for service at 140 F. or higher.

MEAT, FISH, AND POULTRY No.L 018 00

BARBECUED BEEF CUBES

Yield 100 **Portion** 6-1/2 Ounces

Calories	Carbohydrates	Protein	Fat	Cholesterol	Sodium	Calcium
307 cal	29 g	25 g	11 g	66 mg	1238 mg	44 mg

Ingredient	Weight	Measure	Issue
BARBECUE SAUCE		3 gal 1 qts	
WATER	10-1/2 lbs	1 gal 1 qts	
BEEF,DICED,LEAN,RAW	30 lbs		

Method

1 Prepare 2 recipes Barbecue Sauce, Recipe No. O 002 00. DO NOT COOK. Add water. Stir or utilize prepared BBQ Sauce.
2 Cook beef in steam-jacketed kettle or stock pot 15 minutes, uncovered, stirring constantly.
3 Cover; cook 15 minutes.
4 Add barbecue sauce mixture; cover; simmer 1 hour or until tender. CCP: Internal temperature must reach 145 F. or higher for 15 seconds. CCP: Hold for service at 140 F. or higher.

MEAT, FISH, AND POULTRY No.L 018 01

BARBECUED BEEF CUBES (CANNED BEEF)

Yield 100   **Portion** 6-1/2 Ounces

Calories	Carbohydrates	Protein	Fat	Cholesterol	Sodium	Calcium
287 cal	18 g	27 g	12 g	73 mg	817 mg	29 mg

Ingredient	Weight	Measure	Issue
BARBECUE SAUCE		2 gal 1/3 qts	
BEEF,CANNED,CHUNKS,W/NATURAL JUICE,DRAINED	20-1/2 lbs	4 gal 2-3/4 qts	

Method

1. Prepare 1-1/3 recipes Barbecue Sauce, Recipe No. O 002 00. Bring to a boil; reduce heat. Simmer 25 minutes or utilize prepared BBQ sauce.
2. Drain beef. Add beef chunks to barbecue sauce. Mix well. Cook 15 minutes, or until beef is heated thoroughly. CCP: Internal temperature must reach 145 F. or higher for 15 seconds. Hold for service at 140 F. or higher.

MEAT, FISH, AND POULTRY No. L 019 00

STUFFED FLOUNDER CREOLE

Yield 100　　　　　　　　　　　　　　　　　　Portion 4-1/2 Ounces

Calories	Carbohydrates	Protein	Fat	Cholesterol	Sodium	Calcium
306 cal	30 g	32 g	6 g	97 mg	387 mg	63 mg

Ingredient	Weight	Measure	Issue
CREOLE SAUCE		2 gal 1/2 qts	
CELERY,FRESH,CHOPPED	12-2/3 oz	3 cup	1-1/8 lbs
ONIONS,FRESH,CHOPPED	1-5/8 lbs	1 qts 1/2 cup	1-3/4 lbs
BUTTER,MELTED	12 oz	1-1/2 cup	
CRACKER CRUMBS	5-7/8 lbs	1 gal 1-3/4 qts	
PEPPER,BLACK,GROUND	1/4 oz	3/8 tsp	
THYME,GROUND	1/3 oz	2 tbsp	
SHRIMP,COOKED	2 lbs		
WATER	2-1/8 lbs	1 qts	
FISH,FLOUNDER/SOLE FILLET,RAW	30 lbs		

Method

1. Prepare 1 Creole Sauce, Recipe No. O 005 00 or utilize prepared Creole Sauce. CCP: Hold at 140 F. or higher for use in Step 8.
2. Saute celery and onions in melted butter or margarine until tender.
3. Combine cracker crumbs, pepper, and thyme; add to vegetables. Add shrimp to vegetable crumb mixture.
4. Add water to vegetable-crumb-shrimp mixture; toss mixture but do not pack.
5. Separate fillets. Place 1/4 cup vegetable-crumb-shrimp mixture on each fillet; roll fillets using toothpicks to hold together.
6. Place 25 rolled fillets in each greased steam table pan, in rows 3 by 8.
7. Bake 20 minutes at 375 F. Remove from oven.
8. Cover fish in each pan with 2 quarts hot Creole Sauce.
9. Bake 5 to 10 minutes or until thoroughly heated. CCP: Internal temperature must reach 165 F. or higher for 15 seconds. CCP: Hold for service at 140 F. or higher.

MEAT, FISH, AND POULTRY No. L 020 00

BEEF AND CORN PIE

Yield 100   Portion   1-1/2 Cups

Calories	Carbohydrates	Protein	Fat	Cholesterol	Sodium	Calcium
372 cal	18 g	33 g	19 g	113 mg	674 mg	46 mg

Ingredient	Weight	Measure	Issue
BEEF,GROUND,BULK,RAW,90% LEAN	30 lbs		
ONIONS,FRESH,CHOPPED	3-1/8 lbs	2 qts 1 cup	3-1/2 lbs
PEPPERS,GREEN,FRESH,CHOPPED	2 lbs	1 qts 2 cup	2-3/8 lbs
SALT	2-1/2 oz	1/4 cup 1/3 tbsp	
PEPPER,BLACK,GROUND	1/2 oz	2 tbsp	
GARLIC POWDER	3/4 oz	2-2/3 tbsp	
CORN,CANNED,WHOLE KERNEL,INCL LIQUIDS	13-1/2 lbs	1 gal 2 qts	
MASHED POTATOES (INSTANT)		4 gal 1/2 qts	

Method

1 Cook beef with onions and peppers until beef loses its pink color, stirring to break apart. Drain or skim off excess fat.
2 Add salt, pepper and garlic. Mix well.
3 Place 10 pounds of beef mixture in each roasting pan.
4 Spread 2 quarts corn with liquid on top of beef mixture in each pan.
5 Prepare 1 Recipe Mashed Potatoes, Recipe No. Q 057 00. Spread 5-1/2 quarts mashed potatoes over beef mixture and corn in each pan.
6 Using a convection oven, bake 20 minutes at 300 F. 20 minutes on high fan, open vent until potatoes are evenly browned. CCP: Internal temperature must reach 155 F. or higher for 15 seconds.
7 Cut 5 by 7. CCP: Hold for service at 140 F. or higher.

MEAT, FISH, AND POULTRY No.L 020 01

TURKEY CORN PIE

Yield 100   Portion   1-1/2 Cups

Calories	Carbohydrates	Protein	Fat	Cholesterol	Sodium	Calcium
284 cal	18 g	28 g	12 g	92 mg	731 mg	66 mg

Ingredient	Weight	Measure	Issue
TURKEY,GROUND,90% LEAN,RAW	30 lbs		
ONIONS,FRESH,CHOPPED	3-1/8 lbs	2 qts 1 cup	3-1/2 lbs
PEPPERS,GREEN,FRESH,CHOPPED	2-1/8 lbs	1 qts 2 cup	2-5/8 lbs
SALT	2-1/2 oz	1/4 cup 1/3 tbsp	
PEPPER,BLACK,GROUND	1/2 oz	2 tbsp	
GARLIC POWDER	1-1/8 oz	1/4 cup	
CORN,CANNED,WHOLE KERNEL,INCL LIQUIDS	13-1/2 lbs	1 gal 2 qts	
MASHED POTATOES (INSTANT)		4 gal 1 qts	

Method

1 Cook turkey with onions and peppers until turkey loses its pink color. Drain or skim off excess fat.
2 Add salt, pepper and garlic powder. Mix well.
3 Place 10 pounds turkey mixture in each roasting pan.
4 Spread 2 quarts corn with liquid on top of turkey mixture in each pan.
5 Prepare 1 Recipe Mashed Potatoes, Recipe No. Q 057 00. Spread 5-1/2 quarts mashed potatoes over turkey mixture and corn in each pan.
6 Using a convection oven, bake 20 minutes at 300 F. on high fan, open vent, or until potatoes are evenly browned. CCP: Internal temperature must reach 165 F. or higher for 15 seconds.
7 Cut 5 by 7. CCP: Hold for service at 140 F. or higher.

MEAT. FISH. AND POULTRY No.L 021 00

BEEF POT PIE WITH BISCUIT TOPPING

Yield 100   Portion 1 Cup

Calories	Carbohydrates	Protein	Fat	Cholesterol	Sodium	Calcium
412 cal	41 g	28 g	15 g	66 mg	825 mg	141 mg

Ingredient	Weight	Measure	Issue
BEEF,DICED,LEAN,RAW	30 lbs		
ONIONS,FRESH,CHOPPED	3-1/2 lbs	2 qts 2 cup	3-7/8 lbs
WATER	25-1/8 lbs	3 gal	
JUICE,TOMATO,CANNED	12-1/3 lbs	1 gal 1-3/4 qts	
SALT	1-7/8 oz	3 tbsp	
PEPPER,BLACK,GROUND	1/2 oz	2 tbsp	
CARROTS,FRESH,SLICED	6 lbs	1 gal 1-1/3 qts	7-1/3 lbs
POTATOES,FRESH,CHOPPED	9 lbs	1 gal 2-5/8 qts	11-1/8 lbs
FLOUR,WHEAT,GENERAL PURPOSE	11 oz	2-1/2 cup	
WATER	2-1/8 lbs	1 qts	
BAKING POWDER BISCUITS		100 each	

Method

1. Cook beef and onions in a steam-jacketed kettle about 5 minutes.
2. Add water, tomato juice, salt, and pepper to meat. Bring to a boil; reduce heat; cover; simmer 1 hour 15 minutes.
3. Add carrots; cover; simmer 10 minutes.
4. Add potatoes, cover; simmer 20 minutes or until vegetables are tender.
5. Combine flour and water; add to meat and vegetable mixture while stirring; simmer 5 minutes or until thickened, stirring constantly.
6. Place 7 quarts mixture in each steam table pan. CCP: Hold for service at 140 F. or higher.
7. Prepare Baking Powder Biscuits, D 001 01. Place 25 biscuits on top of hot mixture in each pan.
8. Using a convection oven, bake at 400 F. for 10 to 15 minutes or until biscuits are lightly browned. CCP: Hold for service at 140 F. or higher.

MEAT, FISH, AND POULTRY No.L 021 01

BEEF POT PIE WITH PIE CRUST TOPPING

Yield 100   Portion 1 Cup

Calories	Carbohydrates	Protein	Fat	Cholesterol	Sodium	Calcium
332 cal	24 g	26 g	15 g	66 mg	515 mg	28 mg

Ingredient | Weight | Measure | Issue

Ingredient	Weight	Measure	Issue
BEEF,DICED,LEAN,RAW	30 lbs		
ONIONS,FRESH,CHOPPED	3-1/2 lbs	2 qts 2 cup	3-7/8 lbs
WATER	25-1/8 lbs	3 gal	
JUICE,TOMATO,CANNED	12-1/3 lbs	1 gal 1-3/4 qts	
SALT	1-7/8 oz	3 tbsp	
PEPPER,BLACK,GROUND	1/2 oz	2 tbsp	
CARROTS,FRESH,SLICED	6 lbs	1 gal 1-1/3 qts	7-1/3 lbs
POTATOES,FRESH,CHOPPED	9 lbs	1 gal 2-5/8 qts	11-1/8 lbs
FLOUR,WHEAT,GENERAL PURPOSE	11 oz	2-1/2 cup	
WATER	2-1/8 lbs	1 qts	
FLOUR,WHEAT,GENERAL PURPOSE	1-7/8 lbs	1 qts 3 cup	
SALT	1/3 oz	1/4 tsp	
SHORTENING	14-1/2 oz	2 cup	
WATER,COLD	8-1/3 oz	1 cup	

Method

1. Cook beef and onions in a steam-jacketed kettle about 5 minutes.
2. Add water, tomato juice, salt and pepper to meat. Bring to a boil; reduce heat; cover; simmer 1 hour 15 minutes.
3. Add carrots; cover; simmer 10 minutes.
4. Add potatoes, cover; simmer 20 minutes or until vegetables are tender.
5. Combine flour and water; add to meat and vegetable mixture while stirring; simmer 5 minutes or until thickened, stirring constantly.
6. Place 7 quarts mixture in each steam table pan. CCP: Hold for service at 140 F. or higher.
7. Sift flour and salt together in a mixing bowl.
8. Add shortening to dry ingredients. Using a pastry knife attachment, mix at low speed 30 seconds or until shortening is evenly distributed and mixture is granular in appearance.
9. Add water; mix at low speed 1 minute until dough is just formed. Chill dough at least 1 hour for ease in handling.
10. Divide dough into 4-1 pound balls. Roll each ball into a rectangle about 18x10 inches, about 1/8-inch thick. Cut each rectangle into 25 pieces about 3-1/2x2 inches. Place 25 pieces on top of hot, 180 F., meat mixture in each pan. Using a convection oven, bake at 400 F. 25 to 30 minutes or until lightly browned on low fan, open vent. CCP: Internal temperature must reach 155 F. or higher for 15 seconds. Hold for service at 140 F. or higher.

MEAT, FISH, AND POULTRY No.L 022 00

BEEF STEW

Yield 100   Portion 1-1/4 Cups

Calories	Carbohydrates	Protein	Fat	Cholesterol	Sodium	Calcium
286 cal	22 g	25 g	11 g	66 mg	593 mg	44 mg

Ingredient	Weight	Measure	Issue
BEEF,DICED,LEAN,RAW	30 lbs		
WATER	16-3/4 lbs	2 gal	
TOMATOES,CANNED,DICED,DRAINED	6-5/8 lbs	3 qts	
SALT	4-1/4 oz	1/4 cup 3 tbsp	
PEPPER,BLACK,GROUND	1/2 oz	2 tbsp	
GARLIC POWDER	5/8 oz	2 tbsp	
THYME,GROUND	1/4 oz	1 tbsp	
BAY LEAF,WHOLE,DRIED	1/8 oz	4 lf	
CARROTS,FRESH,SLICED	8 lbs	1 gal 3-1/8 qts	9-3/4 lbs
CELERY,FRESH,SLICED	4-1/4 lbs	1 gal	5-7/8 lbs
ONIONS,FRESH,QUARTERED	3 lbs	2 qts 3-7/8 cup	3-1/3 lbs
POTATOES,FRESH,CHOPPED	10-1/3 lbs	1 gal 3-1/2 qts	12-3/4 lbs
FLOUR,WHEAT,GENERAL PURPOSE	1-1/4 lbs	1 qts 1/2 cup	
WATER,COLD	3-1/8 lbs	1 qts 2 cup	

Method

1. Place beef, water, tomatoes, salt, pepper, garlic, thyme and bay leaves in steam-jacketed kettle or stock pot. Bring to a boil; reduce heat; cover. Simmer 1 hour 40 minutes or until tender.
2. Add carrots to beef mixture. Cover; simmer 15 minutes.
3. Add celery, onions, and potatoes. Stir. Cover; simmer 20 minutes or until vegetables are tender.
4. Remove bay leaves. Combine flour and water. Add to stew while stirring. Cook 5 minutes or until thickened. CCP: Internal temperature must reach 145 F. or higher for 15 seconds. Hold for service at 140 F. or higher.

Notes

1. In Step 2, 2 No. 10 canned carrots, drained or 8 pounds frozen carrots may be used per 100 servings.

MEAT, FISH, AND POULTRY No. L 022 01

BEEF STEW (CANNED)

Yield 100 **Portion** 1-1/4 Cups

Calories	Carbohydrates	Protein	Fat	Cholesterol	Sodium	Calcium
286 cal	21 g	15 g	16 g	49 mg	1240 mg	36 mg

Ingredient
BEEF STEW,CANNED,W/VEGETABLES

Weight
67 lbs

Measure

Issue

Method

1. Heat to a serving temperature. CCP: Internal temperature must reach 165 F. or higher for 15 seconds. Hold for service at 140 F. or higher.

MEAT, FISH, AND POULTRY No.L 022 02

EL RANCHO STEW

Yield 100 **Portion** 1 Cup

Calories	Carbohydrates	Protein	Fat	Cholesterol	Sodium	Calcium
278 cal	20 g	25 g	11 g	66 mg	396 mg	26 mg

Ingredient	Weight	Measure	Issue
BEEF,DICED,LEAN,RAW	30 lbs		
WATER	8-1/3 lbs	1 gal	
SALT	3 oz	1/4 cup 1 tbsp	
PEPPER,BLACK,GROUND	1/2 oz	2 tbsp	
CARROTS,FRESH,SLICED	8 lbs	1 gal 3-1/8 qts	9-3/4 lbs
ONIONS,FRESH,QUARTERED	4 lbs	3 qts 3-3/4 cup	4-1/2 lbs
PEAS,GREEN,FROZEN	2 lbs	1 qts 2-1/4 cup	
POTATOES,FRESH,CHOPPED	10 lbs	1 gal 3-1/4 qts	12-1/3 lbs
FLOUR,WHEAT,GENERAL PURPOSE	1-1/8 lbs	1 qts	
WATER,COLD	2-1/8 lbs	1 qts	

Method

1. Place beef, water, salt and pepper in steam-jacketed kettle or stock pot. Bring to a boil; reduce heat; cover. Simmer 1 hour 40 minutes or until tender.
2. Add carrots to beef mixture. Cover; simmer 15 minutes.
3. Add onions and potatoes. Stir. Cover; simmer 20 minutes or until vegetables are tender. Add frozen peas. Simmer 10 minutes or until peas are tender.
4. Combine flour and water. Add to stew while stirring. Cook 5 minutes or until thickened. CCP: Internal temperature must reach 145 F. or higher for 15 seconds. Hold for service at 140 F. or higher.

Notes

1. 3 pounds drained, canned peas may be used per 100 portions. Add canned peas after thickening.

MEAT. FISH. AND POULTRY No.L 023 00

CARIBBEAN CHICKEN BREAST (BREAST BONELESS)

Yield 100 **Portion** 5 Ounces

Calories	Carbohydrates	Protein	Fat	Cholesterol	Sodium	Calcium
202 cal	7 g	32 g	4 g	88 mg	212 mg	24 mg

Ingredient	Weight	Measure	Issue
CHICKEN,BREAST,BNLS/SKNLS,5 OZ	31-1/4 lbs		
JUICE,LEMON	2-1/8 lbs	1 qts	
HONEY	1-3/8 lbs	1-7/8 cup	
CHICKEN BROTH		1 cup	
PAPRIKA,GROUND	3-7/8 oz	1 cup	
GARLIC POWDER	1-5/8 oz	1/4 cup 1-2/3 tbsp	
SALT	1 oz	1 tbsp	
LEMON RIND,GRATED	1 oz	1/4 cup 1-1/3 tbsp	
GINGER,GROUND	1 oz	1/4 cup 1-2/3 tbsp	
PEPPER,RED,GROUND	3/8 oz	2 tbsp	
OREGANO,CRUSHED	3/4 oz	1/4 cup 1-1/3 tbsp	
COOKING SPRAY,NONSTICK	1-1/2 oz	3 tbsp	

Method

1. Wash chicken thoroughly under cold running water. Drain well. Remove excess fat. Place chicken in each roasting pan; cover.
2. Combine lemon juice, honey, chicken broth, paprika, garlic powder, salt, lemon rind, ginger, red pepper, and oregano. Mix well.
3. Pour marinade evenly over chicken in each roasting pan; cover. CCP: Marinate under refrigeration at 41 F. or lower for 45 minutes.
4. Place chicken breasts on each lightly sprayed sheet pan. Lightly spray chicken with cooking spray. CCP: Refrigerate remaining marinade at 41 F. or lower for use in Step 6.
5. Using a convection oven, bake at 325 F. for 12-14 minutes on high fan, open vent. CCP: Internal temperature must reach 165 F. or higher for 15 seconds.
6. Bring reserved marinade to a boil. Cover; reduce heat; simmer 2 minutes. CCP: Temperature must reach 165 F. or higher for 15 seconds.
7. Transfer chicken to steam table pans. Discard chicken drippings.
8. Pour approximately 1 cup marinade evenly over chicken in each pan. Discard any unused marinade. CCP: Hold for service at 140 F. or higher.

MEAT, FISH, AND POULTRY No.L 024 00
STUFFED CABBAGE ROLLS

Yield 100	Portion 2 Rolls

Calories	Carbohydrates	Protein	Fat	Cholesterol	Sodium	Calcium
325 cal	25 g	27 g	13 g	85 mg	751 mg	80 mg

Ingredient	Weight	Measure	Issue
BEEF BROTH		2 qts	
TOMATO PASTE,CANNED	5-3/4 lbs	2 qts 2 cup	
SUGAR,GRANULATED	1-1/2 lbs	3-1/2 cup	
JUICE,LEMON	1-1/8 lbs	2 cup	
CABBAGE,GREEN,FRESH,HEAD	24 lbs	9 gal 2-7/8 qts	30 lbs
WATER,BOILING	25-1/8 lbs	3 gal	
SALT	1/2 oz	3/8 tsp	
BEEF,GROUND,BULK,RAW,90% LEAN	24 lbs		
STEAMED RICE		3 qts	
ONIONS,FRESH,CHOPPED	3-1/2 lbs	2 qts 2 cup	3-7/8 lbs
CATSUP	1 lbs	2 cup	
WORCESTERSHIRE SAUCE	8-1/2 oz	1 cup	
PEPPER,BLACK,GROUND	2/3 oz	3 tbsp	
GARLIC POWDER	1/8 oz	1/8 tsp	
SALT	1-7/8 oz	3 tbsp	

Method

1. Prepare broth according to package directions. Blend in tomato paste, sugar and lemon juice. Set aside for use in Step 8.
2. Add cabbage to boiling salted water in steam-jacketed kettle or stock pot; cover; cook 10 minutes or until leaves are pliable.
3. Drain well; separate 200 leaves; remove larger ribs; set aside for use in Step 6.
4. Shred remaining cabbage coarsely. Set aside for use in Step 7.
5. Combine beef, cooked rice, onions, catsup, Worcestershire sauce, salt, pepper, and garlic. Mix lightly but thoroughly.
6. Place 1/4 cup meat mixture on each cabbage leaf. Fold sides of leaf over mixture; roll tightly.
7. Place 25 cabbage rolls seam side down in each pan. Spread shredded cabbage evenly over rolls in each steam table pan.
8. Pour 2-1/2 cups sauce over cabbage in each pan.
9. Using a convection oven, bake 1 hour at 325 F. on high fan, closed vent or until cabbage is tender and beef is done. CCP: Internal temperature must reach 155 F. or higher for 15 seconds. Skim off excess fat. CCP: Hold for service at 140 F. or higher.

MEAT, FISH, AND POULTRY No.L 024 01

STUFFED CABBAGE ROLLS (TOMATO SOUP)

Yield 100   Portion 2 Rolls

Calories	Carbohydrates	Protein	Fat	Cholesterol	Sodium	Calcium
322 cal	23 g	27 g	14 g	85 mg	697 mg	75 mg

Ingredient	Weight	Measure	Issue
SOUP,CONDENSED,TOMATO	11-1/8 lbs	1 gal 1 qts	
JUICE,LEMON	1-1/8 lbs	2 cup	
SUGAR,GRANULATED	14-1/8 oz	2 cup	
CABBAGE,GREEN,FRESH,HEAD	24 lbs	9 gal 2-7/8 qts	30 lbs
WATER,BOILING	25-1/8 lbs	3 gal	
SALT	1/2 oz	3/8 tsp	
BEEF,GROUND,BULK,RAW,90% LEAN	24 lbs		
STEAMED RICE		2 qts 2 cup	
ONIONS,FRESH,CHOPPED	3-1/2 lbs	2 qts 2 cup	3-7/8 lbs
CATSUP	1 lbs	2 cup	
WORCESTERSHIRE SAUCE	8-1/2 oz	1 cup	
SALT	1-7/8 oz	3 tbsp	
PEPPER,BLACK,GROUND	2/3 oz	3 tbsp	
GARLIC POWDER	1/8 oz	1/8 tsp	

Method

1. Blend tomato soup, lemon juice, and sugar. Set aside for use in Step 8.
2. Add cabbage to boiling salted water in steam-jacketed kettle or stock pot; cover; cook 10 minutes or until leaves are pliable.
3. Drain well; separate 200 leaves; remove larger ribs; set aside for use in Step 6.
4. Shred remaining cabbage coarsely. Set aside for use in Step 7.
5. Combine beef, cooked rice, onions, catsup, Worcestershire sauce, salt, pepper, and garlic powder. Mix lightly but thoroughly.
6. Place 1/4 cup meat mixture on each cabbage leaf. Fold sides of leaf over mixture; roll tightly.
7. Place 25 cabbage rolls seam side down in each 12x20x2-1/2 steam table pan. Spread shredded cabbage evenly over rolls in each pan.
8. Pour 2-1/2 cups sauce over cabbage rolls in each pan.
9. Using a convection oven, bake 1 hour at 325 F. on high fan, closed vent or until cabbage is tender. Skim off excess fat, CCP: Internal temperature must reach 155 F. or higher for 15 seconds. CCP: Hold for service at 140 F. or higher.

MEAT, FISH, AND POULTRY No.L 024 02

STUFFED CABBAGE ROLLS (GROUND TURKEY)

Yield 100   Portion 2 Rolls

Calories	Carbohydrates	Protein	Fat	Cholesterol	Sodium	Calcium
258 cal	26 g	23 g	8 g	68 mg	681 mg	97 mg

Ingredient	Weight	Measure	Issue
WATER	4-1/8 lbs	2 qts	
TOMATO PASTE,CANNED	5-3/4 lbs	2 qts 2 cup	
SUGAR,GRANULATED	1-1/2 lbs	3-1/2 cup	
JUICE,LEMON	1-1/8 lbs	2 cup	
CABBAGE,GREEN,FRESH,HEAD	24 lbs	9 gal 2-7/8 qts	30 lbs
WATER,BOILING	25-1/8 lbs	3 gal	
SALT	1/2 oz	3/8 tsp	
STEAMED RICE		3 qts	
ONIONS,FRESH,CHOPPED	3-1/2 lbs	2 qts 2 cup	3-7/8 lbs
TURKEY,GROUND,90% LEAN,RAW	24 lbs		
CATSUP	1 lbs	2 cup	
WORCESTERSHIRE SAUCE	8-1/2 oz	1 cup	
PARSLEY,FRESH,BUNCH	4-7/8 oz	2 cup	5-1/8 oz
SALT	1-7/8 oz	3 tbsp	
PEPPER,BLACK,GROUND	2/3 oz	3 tbsp	
GARLIC POWDER	1/3 oz	1 tbsp	

Method

1. Blend water, tomato paste, sugar and lemon juice.
2. Add cabbage to boiling salted water in steam-jacketed kettle or stock pot; cover; cook 10 minutes or until leaves are pliable.
3. Drain well; separate 200 leaves; remove larger ribs; set aside for use in Step 6.
4. Shred remaining cabbage coarsely. Set aside for use in Step 7.
5. Combine turkey, cooked rice, onions, catsup, Worcestershire sauce, salt, pepper and garlic powder. Add parsley. Mix lightly but thoroughly.
6. Place 1/4 cup meat mixture on each cabbage leaf. Fold sides of leaf over mixture; roll tightly.
7. Place 25 cabbage rolls seam side down in each steam table pan. Spread shredded cabbage evenly over rolls in each pan.
8. Pour 2-1/2 cups sauce over cabbage rolls in each pan.
9. Using a convection oven, bake 1 hour at 325 F. on high fan, closed vent. CCP: Internal temperature must reach 165 F. or higher for 15 seconds. Skim off excess fat. CCP: Hold for service at 140 F. or higher.

MEAT, FISH, AND POULTRY No.L 025 00

LASAGNA

Yield 100   **Portion** 9.5 Ounces

Calories	Carbohydrates	Protein	Fat	Cholesterol	Sodium	Calcium
403 cal	35 g	33 g	14 g	131 mg	963 mg	289 mg

Ingredient	Weight	Measure	Issue
BEEF,GROUND,BULK,RAW,90% LEAN	12 lbs		
TOMATOES,CANNED,DICED,DRAINED	5 lbs	2 qts 1 cup	
TOMATO PASTE,CANNED	7-3/4 lbs	3 qts 1-1/2 cup	
WATER	5-1/4 lbs	2 qts 2 cup	
ONIONS,FRESH,CHOPPED	4-1/4 lbs	3 qts	4-2/3 lbs
SUGAR,GRANULATED	5-1/4 oz	3/4 cup	
SALT	1-7/8 oz	3 tbsp	
BASIL,SWEET,WHOLE,CRUSHED	7/8 oz	1/4 cup 1-2/3 tbsp	
GARLIC POWDER	5/8 oz	2 tbsp	
OREGANO,CRUSHED	7/8 oz	1/4 cup 1-2/3 tbsp	
THYME,GROUND	1/3 oz	2 tbsp	
PEPPER,BLACK,GROUND	1/4 oz	1 tbsp	
PEPPER,RED,GROUND	<1/16th oz	1/8 tsp	
EGGS,WHOLE,FROZEN	3-5/8 lbs	1 qts 2-3/4 cup	
CHEESE,COTTAGE,LOWFAT	11 lbs	1 gal 1-1/2 qts	
CHEESE,MOZZARELLA,PART SKIM,SHREDDED	3-3/4 lbs	3 qts 3 cup	
CHEESE,PARMESAN,GRATED	14-1/8 oz	1 qts	
PARSLEY,DEHYDRATED,FLAKED	1/4 oz	1/4 cup 2-1/3 tbsp	
NOODLES,LASAGNA,UNCOOKED	6 lbs	1 gal 2-1/2 qts	
CHEESE,PARMESAN,GRATED	5-1/4 oz	1-1/2 cup	

Method

1. Cook beef until beef loses its pink color, stirring to break apart. Drain or skim off excess fat.
2. Add tomatoes, tomato paste, water, onions, sugar, salt, basil, garlic powder, oregano, thyme, black pepper, and red pepper. Blend well; simmer 1 hour.
3. Combine eggs, cheeses, and parsley. Mix well; place in pans; cover.
4. PANNING INSTRUCTIONS: Arrange in layers in each pan. During panning remove small amounts of filling from refrigeration at a time. Ensure entire panning procedure does not exceed 4 hours total time between temperatures of 40 F. to 140 F. Progressive preparation and immediate baking of the product will ensure food safety. Layer: 1. 2-1/2 cups meat sauce 2. Noodles, flat and in rows 3. 3-1/2 cups chilled filling 4. 1 quart meat sauce 5. Noodles, flat and in rows 6. 3-1/2 cups chilled filling 7. 1 quart meat sauce 8. Noodles, flat and in rows 9. 1-1/4 quart meat sauce 10. Sprinkle with parmesan cheese.
5. Cover. Using a convection oven, bake at 300 F. for 55 minutes on high fan, closed vent. Uncover; bake 5 minutes. CCP: Internal temperature must reach 155 F. or higher for 15 seconds.
6. Let stand 10 to 15 minutes before cutting to allow cheeses to firm. Cut 4 by 5. CCP: Hold for service at 140 F. or higher.

MEAT, FISH, AND POULTRY No.L 025 01

LASAGNA (GROUND TURKEY)

Yield 100   Portion   9-1/2 Ounces

Calories	Carbohydrates	Protein	Fat	Cholesterol	Sodium	Calcium
396 cal	40 g	33 g	12 g	126 mg	1077 mg	320 mg

Ingredient	Weight	Measure	Issue
TURKEY,GROUND,90% LEAN,RAW	13-1/4 lbs		
TOMATOES,CANNED,DICED,DRAINED	19-7/8 lbs	2 gal 1 qts	
TOMATO PASTE,CANNED	7-3/4 lbs	3 qts 1-1/2 cup	
WATER	5-1/4 lbs	2 qts 2 cup	
ONIONS,FRESH,CHOPPED	4-1/4 lbs	3 qts	4-2/3 lbs
SUGAR,GRANULATED	5-1/4 oz	3/4 cup	
SALT	1-7/8 oz	3 tbsp	
BASIL,SWEET,WHOLE,CRUSHED	7/8 oz	1/4 cup 1-2/3 tbsp	
GARLIC POWDER	5/8 oz	2 tbsp	
OREGANO,CRUSHED	7/8 oz	1/4 cup 1-2/3 tbsp	
THYME,GROUND	1/3 oz	2 tbsp	
PEPPER,BLACK,GROUND	1/4 oz	1 tbsp	
PEPPER,RED,GROUND	<1/16th oz	1/8 tsp	
EGGS,WHOLE,FROZEN	3-5/8 lbs	1 qts 2-3/4 cup	
CHEESE,COTTAGE,LOWFAT	11 lbs	1 gal 1-1/2 qts	
CHEESE,MOZZARELLA,PART SKIM,SHREDDED	3-3/4 lbs	3 qts 3 cup	
CHEESE,PARMESAN,GRATED	14-1/8 oz	1 qts	
PARSLEY,DEHYDRATED,FLAKED	1/4 oz	1/4 cup 2-1/3 tbsp	
NOODLES,LASAGNA,UNCOOKED	6 lbs	1 gal 2-1/2 qts	
CHEESE,PARMESAN,GRATED	5-1/4 oz	1-1/2 cup	

Method

1. Cook turkey until turkey loses its pink color, stirring to break apart. Drain or skim off excess fat.
2. Add tomatoes, tomato paste, water, onions, sugar, salt, basil, garlic powder, oregano, thyme, black pepper and red pepper. Blend well; simmer 1 hour.
3. Combine eggs, cheeses, and parsley. Mix well; place in pans; cover.
4. PANNING INSTRUCTIONS: Arrange in layers in each pan. During panning remove small amounts of filling from refrigeration at a time. Ensure entire panning procedure does not exceed 4 hours total time between temperatures of 40 F. to 140 F. Progressive preparation and immediate baking of the product will ensure food safety. Layer: 1. 2-1/2 cups meat sauce 2. Noodles, flat and in rows 3. 3-1/2 cups chilled filling 4. 1 quart meat sauce 5. Noodles, flat and in rows 6. 3-1/2 cups chilled filling 7. 1 quart meat sauce 8. Noodles, flat and in rows 9. 1-1/4 quart meat sauce. 10. Sprinkle with parmesan cheese.
5. Cover. Using a convection oven, bake at 300 F. for 55 minutes on high fan, closed vent. Uncover; bake 5 minutes. CCP: Internal temperature must reach 165 F. or higher for 15 seconds.
6. Let stand 10 to 15 minutes before cutting to allow cheeses to firm. Cut 4 by 5. CCP: Hold for service at 140 F. or higher.

MEAT, FISH, AND POULTRY No. L 025 02

LASAGNA (FROZEN)

Yield 100   Portion 9 Ounces

Calories	Carbohydrates	Protein	Fat	Cholesterol	Sodium	Calcium
312 cal	27 g	21 g	13 g	36 mg	760 mg	272 mg

Ingredient   **Weight**   **Measure**   **Issue**

LASAGNA, WITH MEAT & SAUCE, FROZEN   50 lbs

Method

1 Follow manufacturer's directions for heating and serving. CCP: Internal temperature must reach 165 F. or higher for 15 seconds. CCP: Hold at 140 F. or higher for service.

MEAT, FISH, AND POULTRY No.L 025 03

LASAGNA (CANNED PIZZA SAUCE)

Yield 100   Portion   9 Ounces

Calories	Carbohydrates	Protein	Fat	Cholesterol	Sodium	Calcium
454 cal	40 g	35 g	16 g	136 mg	757 mg	356 mg

Ingredient	Weight	Measure	Issue
BEEF,GROUND,BULK,RAW,90% LEAN	12 lbs		
SAUCE,PIZZA,CANNED	37-1/3 lbs	3 gal 3-1/2 qts	
ONIONS,FRESH,CHOPPED	4-1/4 lbs	3 qts	4-2/3 lbs
SUGAR,GRANULATED	2-1/3 oz	1/4 cup 1-2/3 tbsp	
BASIL,SWEET,WHOLE,CRUSHED	5/8 oz	1/4 cup 1/3 tbsp	
OREGANO,CRUSHED	5/8 oz	1/4 cup 1/3 tbsp	
GARLIC POWDER	1/3 oz	1 tbsp	
PEPPER,BLACK,GROUND	1/4 oz	1 tbsp	
THYME,GROUND	1/8 oz	1 tbsp	
PEPPER,RED,GROUND	<1/16th oz	1/8 tsp	
EGGS,WHOLE,FROZEN	3-5/8 lbs	1 qts 2-3/4 cup	
CHEESE,COTTAGE,LOWFAT	11 lbs	1 gal 1-1/2 qts	
CHEESE,MOZZARELLA,PART SKIM,SHREDDED	3-3/4 lbs	3 qts 3 cup	
CHEESE,PARMESAN,GRATED	14-1/8 oz	1 qts	
PARSLEY,DEHYDRATED,FLAKED	1/4 oz	1/4 cup 2-1/3 tbsp	
NOODLES,LASAGNA,UNCOOKED	6 lbs	1 gal 2-1/2 qts	
CHEESE,PARMESAN,GRATED	5-1/4 oz	1-1/2 cup	

Method

1. Cook beef until beef loses its pink color, stirring to break apart. Drain or skim off excess fat.
2. Stir onions, sugar, basil, oregano, garlic powder, black pepper, thyme and red pepper into pizza sauce.
3. Add meat to pizza sauce. Simmer 20 minutes. Skim off excess fat.
4. Combine eggs, cheeses, and parsley. Mix well; place in shallow pans; cover.
5. PANNING INSTRUCTIONS: Arrange in layers in each pan. During panning remove small amounts of filling from refrigeration at a time. Ensure entire panning procedure does not exceed 4 hours total time between temperatures of 40 F. to 140 F. Progressive preparation and immediate baking of the product will ensure food safety. Layer: 1. 2-1/2 cups meat sauce 2. Noodles, flat and in rows 3. 3-1/2 cups chilled filling 4. 1 quart meat sauce 5. Noodles, flat and in rows 6. 3-1/2 cups chilled filling 7. 1 quart meat sauce 8. Noodles, flat and in rows 9. 1-1/4 quart meat sauce Sprinkle with parmesan cheese.
6. Cover. Using a convection oven, bake at 300 F. for 1 hour on high fan, closed vent. Uncover; bake 10 to 15 minutes. CCP: Internal temperature must reach 155 F. for 15 seconds.
7. Let stand 10 to 15 minutes before cutting to allow cheeses to firm. Cut 4 by 5. CCP: Hold for service at 140 F. or higher.

MEAT. FISH. AND POULTRY   No.L 026 00

BAKED BREADED CLAM STRIPS

Yield 100   **Portion**   3 Ounces

Calories	Carbohydrates	Protein	Fat	Cholesterol	Sodium	Calcium
285 cal	27 g	12 g	14 g	13 mg	623 mg	36 mg

Ingredient   **Weight**   **Measure**   **Issue**

CLAM STRIPS,BREADED,FROZEN   25 lbs

Method

1 Place 3 pounds 2 ounces clam strips on sheet pans.
2 Using a convection oven, bake 8 to 10 minutes at 375 F. or until golden brown on high fan, open vent. CCP: Internal temperature must reach 145 F. or higher for 15 seconds.

Notes

1 In Step 2, DO NOT over cook or over brown; clams will be tough and rubbery.
2 Prepare clams in small batches. Clams tend to become soggy if held for an extended period.

MEAT, FISH, AND POULTRY No.L 026 01

FRENCH FRIED BREADED CLAM STRIPS

Yield 100　　　　　　　　　　　　　　　　　　　Portion 3 Ounces

Calories	Carbohydrates	Protein	Fat	Cholesterol	Sodium	Calcium
325 cal	27 g	12 g	19 g	13 mg	623 mg	36 mg

Ingredient　　　　　　　　　　　　　　　Weight　　　　Measure　　　　Issue
CLAM STRIPS,BREADED,FROZEN　　　　　25 lbs

Method
1. Fry in 350 F. deep fat about 1 minute or until golden brown. Drain well in basket or on absorbent paper. CCP: Internal temperature must reach 145 F. or higher for 15 seconds.

Notes
1. Prepare clams in small batches. Clams tend to become soggy if held for an extended period.

MEAT, FISH, AND POULTRY No.L 027 00

BEEF BALLS STROGANOFF

Yield 100   Portion   3-1/2 Ounces

Calories	Carbohydrates	Protein	Fat	Cholesterol	Sodium	Calcium
268 cal	13 g	23 g	13 g	87 mg	668 mg	59 mg

Ingredient	Weight	Measure	Issue
MILK,NONFAT,DRY	3 oz	1-1/4 cup	
WATER,WARM	2 lbs	3-3/4 cup	
EGGS,WHOLE,FROZEN	8-5/8 oz	1 cup	
PARSLEY,FRESH,BUNCH,CHOPPED	1-5/8 oz	3/4 cup	1-2/3 oz
SALT	1-1/4 oz	2 tbsp	
PEPPER,BLACK,GROUND	1/2 oz	2 tbsp	
THYME,GROUND	1/8 oz	1 tbsp	
PEPPER,RED,GROUND	1/8 oz	1/3 tsp	
GARLIC POWDER	1/8 oz	1/8 tsp	
BREAD,WHITE,STALE,SLICED	2-3/4 lbs	2 gal 7/8 qts	
BEEF,GROUND,BULK,RAW,90% LEAN	20 lbs		
ONIONS,FRESH,CHOPPED	11-1/4 oz	2 cup	12-1/2 oz
FLOUR,WHEAT,GENERAL PURPOSE	1-1/8 lbs	1 qts	
WATER	3-1/8 lbs	1 qts 2 cup	
BEEF BROTH		1 gal 1-1/2 qts	
PAPRIKA,GROUND	1 oz	1/4 cup 1/3 tbsp	
PEPPER,BLACK,GROUND	1/8 oz	1/4 tsp	
MUSHROOMS,CANNED,STEMS & PIECES,INCL LIQUIDS	3-1/2 lbs	2 qts 2-1/8 cup	
SOUR CREAM,LOW FAT	3 lbs	1 qts 2 cup	

Method

1. Reconstitute milk.
2. Blend in eggs, milk, parsley, salt, black pepper, thyme, red pepper, and garlic powder.
3. Place bread in mixer; mix at medium speed 5 minutes or until coarse crumbs are formed. Pour milk mixture over bread in mixer; mix lightly at low speed 1/2 minute; let stand 10 minutes.
4. Add beef and onions to bread mixture. Mix at low speed 1 minute. Do not over mix.
5. Shape into 300 balls weighing 1-1/3 ounce each; place 100 meat balls on each sheet pan.
6. Using a convection oven, bake at 350 F. for 8 to 10 minutes on high fan, closed vent or until browned.
7. Combine flour and water, stirring until smooth.
8. Prepare broth according to package directions. Bring to a boil, reduce heat; gradually add flour mixture stirring constantly. Add paprika and pepper. Return to a boil; reduce heat; simmer 5 to 10 minutes or until thickened.
9. Add mushrooms. Stir well. Turn off heat. Remove 1 quart sauce. Stir into sour cream until smooth; combine with remaining sauce; stir until smooth.
10. Place 100 meatballs in each steam table pan. Pour 3 quarts sauce over beef balls in each pan. Cover.
11. Using a convection oven, bake at 300 F. for 15 minutes on high fan, closed vent. CCP: Internal temperature must reach 155 F. or higher for 15 seconds. CCP: Hold for service at 140 F. or higher.

MEAT, FISH, AND POULTRY No.L 027 01

TURKEY BALLS STROGANOFF

Yield 100   Portion   3-1/2 Ounces

Calories	Carbohydrates	Protein	Fat	Cholesterol	Sodium	Calcium
208 cal	13 g	20 g	8 g	72 mg	733 mg	74 mg

Ingredient	Weight	Measure	Issue
MILK,NONFAT,DRY	3 oz	1-1/4 cup	
WATER,WARM	2 lbs	3-3/4 cup	
EGGS,WHOLE,FROZEN	8-5/8 oz	1 cup	
PARSLEY,FRESH,BUNCH,CHOPPED	1-5/8 oz	3/4 cup	1-2/3 oz
SALT	1-1/4 oz	2 tbsp	
PEPPER,BLACK,GROUND	1/2 oz	2 tbsp	
THYME,GROUND	1/8 oz	1 tbsp	
PEPPER,RED,GROUND	1/8 oz	1/3 tsp	
GARLIC POWDER	1/8 oz	1/8 tsp	
BREAD,WHITE,STALE,SLICED	2-3/4 lbs	2 gal 7/8 qts	
TURKEY,GROUND,90% LEAN,RAW	20 lbs		
ONIONS,FRESH,CHOPPED	11-1/4 oz	2 cup	12-1/2 oz
FLOUR,WHEAT,GENERAL PURPOSE	1-1/8 lbs	1 qts	
WATER	3-1/8 lbs	1 qts 2 cup	
CHICKEN BROTH		1 gal 1-1/2 qts	
PAPRIKA,GROUND	1 oz	1/4 cup 1/3 tbsp	
PEPPER,BLACK,GROUND	1/8 oz	1/4 tsp	
MUSHROOMS,CANNED,SLICED,INCL LIQUIDS	3-1/2 lbs	2 qts 2-1/8 cup	
SOUR CREAM,LOW FAT	3 lbs	1 qts 2 cup	

Method

1. Reconstitute milk.
2. Blend in eggs, milk, parsley, salt, black pepper, thyme, red pepper, and garlic powder.
3. Place bread in mixer; mix at medium speed 5 minutes or until coarse crumbs are formed. Pour milk mixture over bread in mixer; mix lightly at low speed 1/2 minute; let stand 10 minutes.
4. Add turkey and onions to bread mixture. Mix at low speed 1 minute. Do not over mix.
5. Shape into balls weighing 1-1/3 ounce each; place 100 meat balls on each sheet pan.
6. Using a convection oven, bake at 350 F. for 8 to 10 minutes on high fan, closed vent. Discard drippings.
7. Combine flour and water, stirring until smooth.
8. Prepare broth according to package directions. Bring to a boil, reduce heat; gradually add flour mixture stirring constantly. Add paprika and pepper. Return to boil; reduce heat; simmer 5 to 10 minutes or until thickened.
9. Add mushrooms. Stir well. Turn off heat. Remove 1 quart sauce. Stir into sour cream until smooth; combine with remaining sauce; stir until smooth.
10. Place 100 turkey balls in each steam table pan. Pour 3 quarts sauce over turkey balls in each pan. Cover.
11. Using a convection oven, bake at 300 F. for 15 minutes on high fan, closed vent. CCP: Internal temperature must reach 165 F. or higher for 15 seconds. Hold for service at 140 F. or higher.

MEAT, FISH, AND POULTRY No. L 028 00

CHILI CON CARNE

Yield 100   **Portion** 1 Cup

Calories	Carbohydrates	Protein	Fat	Cholesterol	Sodium	Calcium
286 cal	30 g	24 g	9 g	50 mg	912 mg	76 mg

Ingredient	Weight	Measure	Issue
BEEF,GROUND,BULK,RAW,90% LEAN	14 lbs		
CHILI POWDER,DARK,GROUND	8-1/2 oz	2 cup	
CUMIN,GROUND	1-2/3 oz	1/2 cup	
PAPRIKA,GROUND	2 oz	1/2 cup	
SALT	1-7/8 oz	3 tbsp	
GARLIC POWDER	7/8 oz	3 tbsp	
PEPPER,RED,GROUND	3/8 oz	2 tbsp	
BEANS,KIDNEY,DARK RED,CANNED,INCL LIQUIDS	33-7/8 lbs	3 gal 3 qts	
RESERVED LIQUID	8-1/3 lbs	1 gal	
WATER	8-1/3 lbs	1 gal	
TOMATOES,CANNED,DICED,DRAINED	6-5/8 lbs	3 qts	
TOMATO PASTE,CANNED	2 lbs	3-1/2 cup	
ONIONS,FRESH,CHOPPED	3-1/8 lbs	2 qts 1 cup	3-1/2 lbs

Method

1. Place beef in steam-jacketed kettle; cook in its own juice until it loses its pink color, stirring to break apart. Drain or skim off excess fat.
2. Combine chili powder, cumin, paprika, salt, garlic powder and red pepper. Stir into cooked beef.
3. Drain beans; reserve beans for use in Step 4. Combine bean liquid with hot water to make 2 gallons; reserve for use in Step 4.
4. Add beans, tomatoes, tomato paste, and onions to cooked beef; stir well. Add reserved bean liquid and hot water to the beef mixture; stir. Bring to a boil; cover; reduce heat; simmer 1 hour. DO NOT BOIL. Stir occasionally. CCP: Internal temperature must reach 155 F. or higher for 15 seconds. Hold for service at 140 F. or higher.

Notes

1. In Step 3, 10 pounds dry, kidney, pinto or white, beans may be used; pick over beans, removing discolored beans and foreign matter. Wash beans thoroughly. Cover with 6 gallons water; boil 2 minutes, turn off heat. Cover; let soak 1 hour. Bring beans to a boil; add 1-2/3 ounces or 2-2/3 tablespoons salt. Cover; simmer 1-1/2 hours or until tender. If necessary, add more water to keep beans covered. Drain beans; reserve beans for use in Step 4. Combine bean liquid with hot water to make 2 gallon; reserve for use in Step 4.

MEAT. FISH. AND POULTRY No.L 028 02

CHILI MACARONI

Yield 100   Portion   1-1/4 Cups

Calories	Carbohydrates	Protein	Fat	Cholesterol	Sodium	Calcium
330 cal	42 g	22 g	9 g	50 mg	501 mg	61 mg

Ingredient	Weight	Measure	Issue
BEEF,GROUND,BULK,RAW,90% LEAN	14 lbs		
CHILI POWDER,DARK,GROUND	8-1/2 oz	2 cup	
CUMIN,GROUND	1-2/3 oz	1/2 cup	
PAPRIKA,GROUND	2 oz	1/2 cup	
SALT	1-7/8 oz	3 tbsp	
GARLIC POWDER	7/8 oz	3 tbsp	
PEPPER,RED,GROUND	3/8 oz	2 tbsp	
WATER	75-1/4 lbs	9 gal	
MACARONI NOODLES,ELBOW,DRY	9 lbs	2 gal 1-3/4 qts	
TOMATOES,CANNED,DICED,DRAINED	12-3/4 lbs	1 gal 1-7/8 qts	
TOMATO PASTE,CANNED	4 lbs	1 qts 3 cup	
ONIONS,FRESH,CHOPPED	3-1/8 lbs	2 qts 1 cup	3-1/2 lbs
WATER	16-3/4 lbs	2 gal	

Method

1 Place beef in steam-jacketed kettle; cook in its own juice until it loses its pink color, stirring to break apart. Drain or skim off excess fat.
2 Combine chili powder, cumin, paprika, salt, garlic powder and red pepper. Stir into cooked beef.
3 Prepare macaroni. See Recipe No. E 004 00.
4 Combine diced tomatoes, tomato paste, chopped onions and water to meat; bring to a simmer; cover; cook 30 minutes. DO NOT BOIL. Stir occasionally. Add cooked macaroni, combine thoroughly. CCP: Internal temperature must reach 155 F. or higher for 15 seconds. Hold for service at 140 F. or higher.

MEAT. FISH. AND POULTRY No. L 028 03

CHILI CON CARNE (GROUND TURKEY)

Yield 100   Portion 1 Cup

Calories	Carbohydrates	Protein	Fat	Cholesterol	Sodium	Calcium
255 cal	30 g	22 g	6 g	45 mg	972 mg	85 mg

Ingredient	Weight	Measure	Issue
TURKEY,GROUND,90% LEAN,RAW	16 lbs		
CHILI POWDER,DARK,GROUND	8-1/2 oz	2 cup	
CUMIN,GROUND	1-2/3 oz	1/2 cup	
PAPRIKA,GROUND	2 oz	1/2 cup	
SALT	1-7/8 oz	3 tbsp	
GARLIC POWDER	7/8 oz	3 tbsp	
PEPPER,RED,GROUND	3/8 oz	2 tbsp	
BEANS,KIDNEY,DARK RED,CANNED,INCL LIQUIDS	33-7/8 lbs	3 gal 3 qts	
RESERVED LIQUID	8-1/3 lbs	1 gal	
WATER		1 gal	
TOMATOES,CANNED,DICED,INCL LIQUIDS	6-7/8 lbs	3 qts	
TOMATO PASTE,CANNED	2 lbs	3-1/2 cup	
ONIONS,FRESH,CHOPPED	3-1/8 lbs	2 qts 1 cup	3-1/2 lbs

Method

1. Place turkey in steam-jacketed kettle; cook in its own juice until it loses its pink color, stirring to break apart. Drain or skim off excess fat.
2. Combine chili powder, cumin, paprika, salt, garlic powder and red pepper. Stir into cooked turkey.
3. Drain beans; reserve beans for use in Step 4. Combine bean liquid with hot water to make 2 gallons; reserve for use in Step 4.
4. Add beans, tomatoes, tomato paste, and onions to cooked turkey; stir well. Add reserved bean liquid and hot water to the beef mixture; stir. Bring to a boil; cover; reduce heat; simmer 1 hour. DO NOT BOIL. Stir occasionally. CCP: Internal temperature must reach 165 F. or higher for 15 seconds. Hold for service at 140 F. or higher.

Notes

1. In Step 3, 10 pounds dry, kidney, pinto or white, beans may be used; pick over beans, removing discolored beans and foreign matter. Wash beans thoroughly. Cover with 6 gallons water; boil 2 minutes, turn off heat. Cover; let soak 1 hour. Bring beans to a boil; add 1-2/3 ounces or 2-2/3 tablespoons salt. Cover; simmer 1-1/2 hours or until tender. If necessary, add more water to keep beans covered. Drain beans; reserve beans for use in Step 4. Combine bean liquid with hot water to make 2 gallons; reserve for use in Step 4.

MEAT. FISH. AND POULTRY No.L 028 04

CHILI MACARONI (GROUND TURKEY)

Yield 100　　　　　　　　　　　　　　　　　　　Portion 1-1/4 Cups

Calories	Carbohydrates	Protein	Fat	Cholesterol	Sodium	Calcium
300 cal	42 g	21 g	6 g	45 mg	533 mg	72 mg

Ingredient	Weight	Measure	Issue
TURKEY,GROUND,90% LEAN,RAW	16 lbs		
CHILI POWDER,DARK,GROUND	8-1/2 oz	2 cup	
CUMIN,GROUND	1-2/3 oz	1/2 cup	
PAPRIKA,GROUND	2 oz	1/2 cup	
SALT	1-7/8 oz	3 tbsp	
GARLIC POWDER	7/8 oz	3 tbsp	
PEPPER,RED,GROUND	3/8 oz	2 tbsp	
MACARONI NOODLES,ELBOW,DRY	9 lbs	2 gal 1-3/4 qts	
WATER,BOILING	75-1/4 lbs	9 gal	
TOMATOES,CANNED,DICED,DRAINED	12-3/4 lbs	1 gal 1-7/8 qts	
TOMATO PASTE,CANNED	4 lbs	1 qts 3 cup	
ONIONS,FRESH,CHOPPED	3-1/8 lbs	2 qts 1 cup	3-1/2 lbs
WATER	16-3/4 lbs	2 gal	

Method

1 Place turkey in steam-jacketed kettle; cook in its own juice until it loses its pink color, stirring to break apart. Drain or skim off excess fat.
2 Combine chili powder, cumin, paprika, salt, garlic powder and red pepper. Stir into cooked turkey.
3 Prepare macaroni. See Recipe No. E 004 00.
4 Combine diced tomatoes, tomato paste, chopped onions and water to meat; bring to a simmer; cover; cook 30 minutes. DO NOT BOIL. Stir occasionally. Add cooked macaroni, combine thoroughly. CCP: Internal temperature must reach 165 F. or higher for 15 seconds. CCP: Hold for service at 140 F. or higher.

MEAT, FISH, AND POULTRY No.L 029 00

BEEF PORCUPINES

Yield 100　　　　　　　　　　　　　　　　　　　　**Portion** 5 Ounces

Calories	Carbohydrates	Protein	Fat	Cholesterol	Sodium	Calcium
350 cal	23 g	27 g	16 g	85 mg	891 mg	33 mg

Ingredient	Weight	Measure	Issue
TOMATO SAUCE		2 gal 2-1/2 qts	
RICE, LONG GRAIN	2-7/8 lbs	1 qts 3 cup	
WATER	7-1/3 lbs	3 qts 2 cup	
SALT	1 oz	1 tbsp	
BEEF, GROUND, BULK, RAW, 90% LEAN	24 lbs		
ONIONS, FRESH, CHOPPED	1-5/8 lbs	1 qts 1/2 cup	1-3/4 lbs
PEPPERS, GREEN, FRESH, CHOPPED	1-1/2 lbs	1 qts 1/2 cup	1-3/4 lbs
PEPPER, BLACK, GROUND	1/4 oz	1 tbsp	
SALT	2-1/2 oz	1/4 cup 1/3 tbsp	
GARLIC POWDER	1/3 oz	1 tbsp	
WORCESTERSHIRE SAUCE	2-1/8 oz	1/4 cup 1/3 tbsp	

Method

1. Prepare 1-1/2 recipes tomato sauce. See Recipe No. O 015 00 or use prepared tomato sauce. Set aside for use in Step 6.
2. Cook rice according to directions in Recipe No. E 005 00. Cool.
3. Thoroughly combine cooled rice with ground beef, onions, peppers, salt, garlic powder and Worcestershire Sauce. DO NOT OVERMIX.
4. Shape into 200 balls weighing about 3-2/3 ounces each.
5. Place an equal quantity of balls on each steam table pan. Using a convection oven, bake at 325 F. at 15 minutes on high fan, closed vent, or until brown.
6. Place approximately 40 meatballs in each steam table pan. Pour 8-1/3 cups sauce over balls in each pan.
7. Cover, using a convection oven, bake 30 minutes at 325 F. on high fan, closed vent or until thoroughly heated. CCP: Internal temperature must reach 155 F. or higher for 15 seconds. Hold for service at 140 F. or higher.

MEAT, FISH, AND POULTRY No.L 029 01

TURKEY PORCUPINES

Yield 100   Portion   5 Ounces

Calories	Carbohydrates	Protein	Fat	Cholesterol	Sodium	Calcium
218 cal	14 g	21 g	8 g	68 mg	596 mg	37 mg

Ingredient	Weight	Measure	Issue
TOMATO SAUCE		2 gal 2-1/2 qts	
RICE,LONG GRAIN	2-7/8 lbs	1 qts 3 cup	
WATER	7-1/3 lbs	3 qts 2 cup	
SALT	1 oz	1 tbsp	
TURKEY,GROUND,90% LEAN,RAW	24 lbs		
ONIONS,FRESH,CHOPPED	1-5/8 lbs	1 qts 1/2 cup	1-3/4 lbs
PEPPERS,GREEN,FRESH,CHOPPED	1-1/2 lbs	1 qts 1/2 cup	1-3/4 lbs
PEPPER,BLACK,GROUND	1/4 oz	1 tbsp	
SALT	2-1/2 oz	1/4 cup 1/3 tbsp	
GARLIC POWDER	1/3 oz	1 tbsp	
WORCESTERSHIRE SAUCE	2-1/8 oz	1/4 cup 1/3 tbsp	

Method

1. Prepare 1-1/2 recipes Tomato Sauce. See Recipe No. O 015 00 or utilize prepared sauce. Set aside for use in Step 6.
2. Cook rice according to directions in Recipe No. E 005 00. Cool.
3. Thoroughly combine cooled rice with ground turkey, onions, peppers, salt, garlic and Worcestershire Sauce. DO NOT
4. Shape into 200 balls weighing about 3-2/3 oz each.
5. Place an equal quantity of balls on each steam table pan. Using a convection oven, bake 15 minutes at 325 F. or until brown. Drain or skim off excess fat.
6. Place approximately 40 meatballs in each steam table pan. Pour 8-1/3 cups sauce over balls in each pan.
7. Cover; using a convection oven, bake 30 minutes at 325 F. on high fan, closed vent or until thoroughly heated. CCP: Internal temperature must reach 165 F. or higher for 15 seconds. Hold for service at 140 F. or higher.

MEAT, FISH, AND POULTRY No.L 030 00

CREAMED GROUND BEEF

Yield 100					Portion 5-1/2 Ounces

Calories	Carbohydrates	Protein	Fat	Cholesterol	Sodium	Calcium
214 cal	9 g	21 g	10 g	65 mg	283 mg	81 mg

Ingredient | Weight | Measure | Issue

Ingredient	Weight	Measure	Issue
BEEF,GROUND,BULK,RAW,90% LEAN	18 lbs		
ONIONS,FRESH,CHOPPED	1 lbs	3 cup	1-1/8 lbs
FLOUR,WHEAT,GENERAL PURPOSE	1-2/3 lbs	1 qts 2 cup	
SALT	1-7/8 oz	3 tbsp	
PEPPER,BLACK,GROUND	1/4 oz	1 tbsp	
WATER,WARM	22 lbs	2 gal 2-1/2 qts	
WORCESTERSHIRE SAUCE	2-1/8 oz	1/4 cup 1/3 tbsp	
MILK,NONFAT,DRY	1-1/4 lbs	2 qts 1/2 cup	

Method

1. Cook beef in steam jacketed kettle or roasting pan until beef loses its pink color, stirring to break apart. Drain or skim off excess fat. Add onions; stir-cook 3 minutes.
2. Combine flour, salt and pepper. Sprinkle evenly over beef and onions. Mix thoroughly; cook about 5 minutes until flour is absorbed.
3. Reconstitute milk. Blend Worcestershire sauce into milk. Add to meat mixture.
4. Heat to a simmer, stirring frequently. Cook 10 minutes or until thickened. CCP: Internal temperature must reach 155 F. or higher for 15 seconds. Hold for service at 140 F. or higher.

MEAT, FISH, AND POULTRY   No.L 030 01

CREAMED GROUND TURKEY

Yield 100   Portion   5-1/2 Ounces

Calories	Carbohydrates	Protein	Fat	Cholesterol	Sodium	Calcium
160 cal	10 g	18 g	6 g	52 mg	455 mg	93 mg

Ingredient	Weight	Measure	Issue
TURKEY,GROUND,90% LEAN,RAW	18 lbs		
ONIONS,FRESH,CHOPPED	1 lbs	3 cup	1-1/8 lbs
FLOUR,WHEAT,GENERAL PURPOSE	1-2/3 lbs	1 qts 2 cup	
SALT	3-1/8 oz	1/4 cup 1-1/3 tbsp	
GARLIC POWDER	5/8 oz	2 tbsp	
PEPPER,BLACK,GROUND	3/8 oz	1 tbsp	
MILK,NONFAT,DRY	1-1/4 lbs	2 qts 1/2 cup	
WATER,WARM	22 lbs	2 gal 2-1/2 qts	
WORCESTERSHIRE SAUCE	6-1/3 oz	3/4 cup	

Method

1. Cook turkey in steam-jacketed kettle or roasting pan until turkey loses its pink color, stirring to break apart. CCP: Temperature must reach 165 F. or higher. Drain or skim off excess fat. Add onions; stir-cook 3 minutes.
2. Combine flour, salt, garlic powder, and pepper. Sprinkle evenly over turkey and onion mixture. Mix thoroughly; cook about 5 minutes until flour is absorbed.
3. Reconstitute milk. Add to mixture.
4. Add Worcestershire sauce. Heat to a simmer, stirring frequently. Simmer 10 minutes until thickened. CCP: Internal temperature must reach 165 F. or higher for 15 seconds. CCP: Hold at 140 F. or higher for service.

MEAT, FISH, AND POULTRY No.L 031 00

BEEF RAVIOLI (FROZEN)

Yield 100 **Portion** 8 Ounces

Calories	Carbohydrates	Protein	Fat	Cholesterol	Sodium	Calcium
330 cal	39 g	15 g	12 g	81 mg	754 mg	141 mg

Ingredient	Weight	Measure	Issue
WATER	83-5/8 lbs	10 gal	
RAVIOLI,W/O SAUCE,FROZEN	27-1/4 lbs	3 gal 2-1/2 qts	
SAUCE,PIZZA,CANNED	28-7/8 lbs	3 gal	

Method

1. Heat water to a boil.
2. Place ravioli in boiling water. Cook 12 to 15 minutes or until tender. Drain. CCP: Internal temperature must reach 165 F. or higher for 15 seconds. Hold for service at 140 F. or higher.
3. Bring sauce to a boil. Serve over hot ravioli. Each portion is 4 Ravioli (5 ounces) with 1/2 cup sauce (3-1/2 ounces).

MEAT, FISH, AND POULTRY No.L 031 01

CHEESE RAVIOLI (FROZEN)

Yield 100   **Portion** 8 Ounces

Calories	Carbohydrates	Protein	Fat	Cholesterol	Sodium	Calcium
181 cal	27 g	7 g	5 g	7 mg	713 mg	113 mg

Ingredient   Weight   Measure   Issue

Ingredient	Weight	Measure	Issue
WATER,BOILING	83-5/8 lbs	10 gal	
RAVIOLI,CHEESE,W/O SAUCE,FROZEN	27-1/4 lbs	3 gal 1-5/8 qts	
SAUCE,PIZZA,CANNED	28-7/8 lbs	3 gal	

Method

1. Heat water to a boil.
2. Place ravioli in boiling water. Cook 12 to 15 minutes or until tender. Drain. CCP: Internal temperature must reach 165 F. or higher for 15 seconds. Hold for service at 140 F. or higher.
3. Bring sauce to a boil. Serve over hot ravioli. Each portion is 4 Ravioli (5 ounces) with 1/2 cup sauce (3-1/2 ounces).

MEAT, FISH, AND POULTRY No. L 031 02

BEEF RAVIOLI (CANNED IN TOMATO SAUCE)

Yield 100   Portion 1 Cup

Calories	Carbohydrates	Protein	Fat	Cholesterol	Sodium	Calcium
230 cal	37 g	8 g	5 g	15 mg	1178 mg	20 mg

Ingredient   **Weight**   **Measure**   **Issue**

RAVIOLI,BEEF,W/MEAT SAUCE,CANNED   54 lbs   6 gal 3 qts

Method

1. Heat canned beef ravioli in tomato sauce to a boil. CCP: Internal temperature must reach 165 F. or higher for 15 seconds. Hold for service at 140 F. or higher.

MEAT. FISH. AND POULTRY No.L 032 00

PARMESAN FISH

Yield 100 Portion 4 Ounces

Calories	Carbohydrates	Protein	Fat	Cholesterol	Sodium	Calcium
212 cal	1 g	32 g	8 g	88 mg	395 mg	226 mg

Ingredient	Weight	Measure	Issue
FISH,FLOUNDER/SOLE FILLET,RAW	30 lbs		
CHEESE,PARMESAN,GRATED	3-1/8 lbs	3 qts 2 cup	
PARSLEY,DEHYDRATED,FLAKED	1-2/3 oz	2-1/4 cup	
PAPRIKA,GROUND	2/3 oz	2-2/3 tbsp	
OREGANO,CRUSHED	1/3 oz	2 tbsp	
PEPPER,BLACK,GROUND	3/8 oz	1 tbsp	
BASIL,DRIED,CRUSHED	1/8 oz	1 tbsp	
WATER,WARM	1 lbs	1-7/8 cup	
MILK,NONFAT,DRY	7/8 oz	1/4 cup 2-1/3 tbsp	
COOKING SPRAY,NONSTICK	2 oz	1/4 cup 1/3 tbsp	
BUTTER,MELTED	8 oz	1 cup	

Method

1. Separate fillets or steak; cut into 4-1/2 oz portions.
2. Combine cheese, parsley, paprika, oregano, pepper and basil. Blend thoroughly.
3. Reconstitute milk; dip fish into milk; drain.
4. Lightly spray pans with non-stick cooking spray. Dredge fish in cheese mixture; shake off excess. Arrange fish in single layers on pans.
5. Drizzle about 1/4 cup butter or margarine over fish in each pan.
6. Using a convection oven, bake at 325 F. for 15-20 minutes or until lightly browned. CCP: Internal temperature must reach 145 F. or higher for 15 seconds. Hold for service at 140 F. or higher.

MEAT, FISH, AND POULTRY No.L 033 00

ROAST BEEF HASH

Yield 100   **Portion** 1/2 Cup

Calories	Carbohydrates	Protein	Fat	Cholesterol	Sodium	Calcium
159 cal	12 g	15 g	6 g	39 mg	315 mg	13 mg

Ingredient	Weight	Measure	Issue
BEEF,OVEN ROAST,PRE COOKED	10 lbs		
SHORTENING	3-5/8 oz	1/2 cup	
ONIONS,FRESH,CHOPPED	2-1/2 lbs	1 qts 3 cup	2-3/4 lbs
PEPPERS,GREEN,FRESH,CHOPPED	1-1/2 lbs	1 qts 1/2 cup	1-3/4 lbs
POTATOES,FRESH,PEELED,CUBED	10 lbs	1 gal 3-1/4 qts	12-1/3 lbs
WATER,BOILING	14-5/8 lbs	1 gal 3 qts	
SALT	1/4 oz	1/8 tsp	
WATER	1-5/8 lbs	3 cup	
CATSUP	14-1/8 oz	1-5/8 cup	
SALT	1-7/8 oz	3 tbsp	
PEPPER,BLACK,GROUND	1/8 oz	1/4 tsp	
GARLIC POWDER	1/4 oz	1/2 tsp	
COOKING SPRAY,NONSTICK	2 oz	1/4 cup 1/3 tbsp	

Method

1. Chop beef finely. Set aside for use in Step 5.
2. Saute onions, and peppers in shortening or salad oil for 10 minutes or until tender. Stir frequently.
3. Place potatoes in boiling salted water. Return to boil; reduce heat; cook 10 minutes or until tender. Drain. Set aside for use in Step 5.
4. Blend water, catsup, salt, pepper and garlic powder.
5. Combine beef, sauteed vegetables, potatoes and catsup mixture. Mix thoroughly.
6. Lightly spray steam table pans with non-stick cooking spray. Place about 6-1/2 qts beef mixture into each steam table pan.
7. Using a convection oven, bake at 325 F. for 25 minutes or until lightly browned. CCP: Internal temperature must reach 145 F. or higher for 15 seconds. Hold for service at 140 F. or higher.

MEAT. FISH. AND POULTRY No.L 033 01

ROAST BEEF HASH (CANNED)

Yield 100 Portion 1/2 Cup

Calories	Carbohydrates	Protein	Fat	Cholesterol	Sodium	Calcium
200 cal	12 g	11 g	12 g	38 mg	411 mg	22 mg

Ingredient	Weight	Measure	Issue
BEEF,ROAST,HASH,CANNED	27 lbs	3 gal 1 qts	

Method

1. Heat Roast Beef Hash according to directions on container. CCP: Internal temperature must reach 165 F. or higher for 15 seconds. Hold for service at 140 F. or higher.

MEAT, FISH, AND POULTRY   No.L 033 02

ROAST BEEF HASH (CANNED BEEF CHUNKS)

Yield 100   Portion   1/2 Cup

Calories	Carbohydrates	Protein	Fat	Cholesterol	Sodium	Calcium
196 cal	12 g	17 g	9 g	46 mg	322 mg	12 mg

Ingredient	Weight	Measure	Issue
BEEF,CANNED,CHUNKS,W/NATURAL JUICE,DRAINED	12-3/4 lbs	2 gal 3-5/8 qts	
ONIONS,FRESH,CHOPPED	2-1/2 lbs	1 qts 3 cup	2-3/4 lbs
PEPPERS,GREEN,FRESH,CHOPPED	1-1/2 lbs	1 qts 1/2 cup	1-3/4 lbs
OIL,SALAD	3-7/8 oz	1/2 cup	
POTATOES,FRESH,CHOPPED	10 lbs	1 gal 3-1/4 qts	12-1/3 lbs
WATER,BOILING	14-5/8 lbs	1 gal 3 qts	
SALT	1/4 oz	1/8 tsp	
RESERVED LIQUID	1-5/8 lbs	3 cup	
CATSUP	14-1/8 oz	1-5/8 cup	
SALT	1-7/8 oz	3 tbsp	
PEPPER,BLACK,GROUND	1/8 oz	1/4 tsp	
GARLIC POWDER	1/4 oz	1/2 tsp	
COOKING SPRAY,NONSTICK	2 oz	1/4 cup 1/3 tbsp	

Method

1. Drain beef chunks. Chop fine.
2. Saute onions and peppers in salad oil for 10 minutes or until tender. Stir frequently.
3. Place potatoes in boiling salted water. Return to boil; reduce heat; cook 10 minutes or until tender. Drain. Set aside reserved liquid.
4. Blend reserved liquid, catsup, salt, pepper and garlic powder.
5. Combine beef, sauteed vegetables, potatoes and catsup mixture. Mix thoroughly.
6. Lightly spray steam table pan with non-stick cooking spray. Place 6-1/2 quarts beef mixture into each steam table pan.
7. Using a convection oven, bake at 325 F. for 25 minutes or until lightly browned. CCP: Internal temperature must reach 165 F. or higher for 15 seconds. Hold for service at 140 F. or higher.

MEAT, FISH, AND POULTRY No.L 034 00

TACOS (GROUND BEEF)

Yield 100   Portion 2 Tacos

Calories	Carbohydrates	Protein	Fat	Cholesterol	Sodium	Calcium
443 cal	21 g	32 g	26 g	106 mg	810 mg	259 mg

Ingredient	Weight	Measure	Issue
TACO SAUCE		3 qts 2 cup	
BEEF,GROUND,BULK,RAW,90% LEAN	22 lbs		
SALT	2-1/3 oz	1/4 cup	
PEPPER,RED,GROUND	<1/16th oz	1/8 tsp	
CUMIN,GROUND	1/4 oz	1 tbsp	
GARLIC POWDER	1/2 oz	1 tbsp	
CHILI POWDER,LIGHT,GROUND	2-1/8 oz	1/2 cup	
FLOUR,WHEAT,GENERAL PURPOSE	10-1/4 oz	2-3/8 cup	
SHELLS,TACO,CORN	4-2/3 lbs	200 each	
CHEESE,CHEDDAR,GRATED	6 lbs	1 gal 2 qts	
LETTUCE,ICEBERG,FRESH,CHOPPED	5-7/8 lbs	3 gal	6-1/4 lbs
ONIONS,FRESH,CHOPPED	3-1/8 lbs	2 qts 1 cup	3-1/2 lbs

Method

1 Prepare 1 recipe Taco Sauce Recipe No. O 007 00 or utilize prepared Taco Sauce.
2 Cook beef until beef loses its pink color; stir to break apart. Drain fat.
3 Combine salt, red pepper, cumin, garlic, chili powder, and flour; add to beef. Saute 5 minutes. CCP: Internal temperature must reach 155 F. or higher for 15 seconds. Hold at 140 F. or higher for use in Step 6.
4 Arrange taco shells on sheet pans. Using a convection oven, bake 2 to 3 minutes at 325 F. on high fan, open vent until just heated.
5 Place 1/4 cup meat filling in each taco; line up next to each other in steam table pan. CCP: Hold for service at 140 F. or higher.
6 Just before serving, top each taco with 2 tablespoons cheese, 2-1/3 tablespoons lettuce, 2 teaspoons onions, and 1 tablespoon taco sauce.

MEAT, FISH, AND POULTRY No.L 034 01

TACOS (GROUND TURKEY)

Yield 100						Portion 2 Tacos

Calories	Carbohydrates	Protein	Fat	Cholesterol	Sodium	Calcium
377 cal	21 g	28 g	21 g	90 mg	847 mg	273 mg

Ingredient	Weight	Measure	Issue
TACO SAUCE		3 qts 2 cup	
TURKEY,GROUND,90% LEAN,RAW	22 lbs		
SALT	2-1/3 oz	1/4 cup	
PEPPER,RED,GROUND	<1/16th oz	1/8 tsp	
CUMIN,GROUND	1/4 oz	1 tbsp	
GARLIC POWDER	1/2 oz	1 tbsp	
FLOUR,WHEAT,GENERAL PURPOSE	10-1/4 oz	2-3/8 cup	
CHILI POWDER,LIGHT,GROUND	2-1/8 oz	1/2 cup	
SHELLS,TACO,CORN	4-2/3 lbs	200 each	
CHEESE,CHEDDAR,GRATED	6 lbs	1 gal 2 qts	
LETTUCE,ICEBERG,FRESH,CHOPPED	5-7/8 lbs	3 gal	6-1/4 lbs
ONIONS,FRESH,CHOPPED	3-1/8 lbs	2 qts 1 cup	3-1/2 lbs

Method

1. Prepare Taco Sauce, Recipe No. O 007 00 or utilize prepared Taco Sauce.
2. Cook turkey until turkey loses its pink color; stir to break apart. Drain fat.
3. Combine salt, red pepper, cumin, garlic, chili powder, and flour; add to turkey. Saute 5 minutes. CCP: Internal temperature must reach 165 F. or higher for 15 seconds. Hold at 140 F. or higher.
4. Arrange taco shells on sheet pans. Using a convection oven, bake 2 to 3 minutes at 325 F. on high fan, open vent until just heated.
5. Place 1/4 cup turkey filling in each taco; line up next to each other in steam table pan. CCP: Hold for service at 140 F. or higher.
6. Just before serving, top each taco with 2 tablespoons cheese, 2-1/3 tablespoons lettuce, 2 teaspoons onions, and 1 tablespoon taco sauce.

MEAT, FISH, AND POULTRY No.L 035 00

MEAT LOAF

Yield 100 **Portion** 6 Ounces

Calories	Carbohydrates	Protein	Fat	Cholesterol	Sodium	Calcium
343 cal	11 g	33 g	18 g	154 mg	648 mg	48 mg

Ingredient	Weight	Measure	Issue
BEEF,GROUND,BULK,RAW,90% LEAN	30 lbs		
BREADCRUMBS	3-3/4 lbs	1 gal	
SALT	3-3/4 oz	1/4 cup 2-1/3 tbsp	
PEPPER,BLACK,GROUND	1/4 oz	1 tbsp	
GARLIC POWDER	1/3 oz	1 tbsp	
MILK,NONFAT,DRY	2-3/8 oz	1 cup	
WATER	2-7/8 lbs	1 qts 1-1/2 cup	
CELERY,FRESH,CHOPPED	1 lbs	3-3/4 cup	1-3/8 lbs
ONIONS,FRESH,CHOPPED	1 lbs	2-7/8 cup	1-1/8 lbs
PEPPERS,GREEN,FRESH,CHOPPED	1 lbs	3 cup	1-1/4 lbs
EGGS,WHOLE,FROZEN	2-3/8 lbs	1 qts 1/2 cup	
JUICE,TOMATO,CANNED	3-1/8 lbs	1 qts 1-3/4 cup	

Method

1 Combine beef with bread crumbs, salt, pepper and garlic; mix until well blended.
2 Reconstitute milk.
3 Add milk, celery, onions, sweet peppers, eggs, and tomato juice. Mix lightly but thoroughly. DO NOT OVERMIX.
4 Place 11 pounds 6 ounces meat mixture into each steam table pan and divide into 2 loaves per pan.
5 Using a convection oven, bake 1 hour 15 minutes at 300 F. CCP: Internal temperature must reach 155 F. or higher for 15 seconds. Skim off excess fat and liquid during cooking.
6 Let stand 20 minutes before slicing. Cut 13 slices per loaf. CCP: Hold for service at 140 F. or higher.

MEAT, FISH, AND POULTRY No. L 035 01

TURKEY LOAF

Yield 100   Portion 6 Ounces

Calories	Carbohydrates	Protein	Fat	Cholesterol	Sodium	Calcium
253 cal	11 g	28 g	11 g	132 mg	699 mg	67 mg

Ingredient	Weight	Measure	Issue
TURKEY,GROUND,90% LEAN,RAW	30 lbs		
BREADCRUMBS	3-3/4 lbs	1 gal	
SALT	3-3/4 oz	1/4 cup 2-1/3 tbsp	
PEPPER,BLACK,GROUND	1/4 oz	1 tbsp	
GARLIC POWDER	1/3 oz	1 tbsp	
MILK,NONFAT,DRY	2-3/8 oz	1 cup	
WATER	2-7/8 lbs	1 qts 1-1/2 cup	
CELERY,FRESH,CHOPPED	1 lbs	3-3/4 cup	1-3/8 lbs
ONIONS,FRESH,CHOPPED	1 lbs	3 cup	1-1/8 lbs
PEPPERS,GREEN,FRESH,CHOPPED	1 lbs	3 cup	1-1/4 lbs
EGGS,WHOLE,FROZEN	2-3/8 lbs	1 qts 1/2 cup	
JUICE,TOMATO,CANNED	3-1/8 lbs	1 qts 1-3/4 cup	

Method

1. Combine turkey with bread crumbs, salt, pepper and garlic; mix until well blended.
2. Reconstitute milk.
3. Add milk, celery, onions, sweet peppers, eggs, and tomato juice. Mix lightly but thoroughly. DO NOT OVERMIX.
4. Place 11 pounds 6 ounces meat mixture into each steam table pan and divide into 2 loaves per pan.
5. Using a convection oven, bake 1 hour 15 minutes at 325 F. on high fan, closed vent. Skim off excess fat and liquid during cooking period. CCP: Internal temperature must reach 165 F. or higher for 15 seconds.
6. Let stand 20 minutes before slicing. Cut 13 slices per loaf. CCP: Hold for service at 140 F. or higher.

MEAT, FISH, AND POULTRY No.L 035 02

TOMATO MEAT LOAF

Yield 100   Portion 6 Ounces

Calories	Carbohydrates	Protein	Fat	Cholesterol	Sodium	Calcium
372 cal	16 g	34 g	18 g	154 mg	717 mg	51 mg

Ingredient	Weight	Measure	Issue
BEEF,GROUND,BULK,RAW,90% LEAN	30 lbs		
BREADCRUMBS	3-3/4 lbs	1 gal	
SALT	2-1/2 oz	1/4 cup 1/3 tbsp	
PEPPER,BLACK,GROUND	1/4 oz	1 tbsp	
GARLIC POWDER	1/3 oz	1 tbsp	
MILK,NONFAT,DRY	2-3/8 oz	1 cup	
WATER	2-7/8 lbs	1 qts 1-1/2 cup	
CELERY,FRESH,CHOPPED	1 lbs	3-3/4 cup	1-3/8 lbs
ONIONS,FRESH,CHOPPED	4 oz	1/2 cup 3-1/3 tbsp	4-1/2 oz
PEPPERS,GREEN,FRESH,CHOPPED	1 lbs	3 cup	1-1/4 lbs
EGGS,WHOLE,FROZEN	2-3/8 lbs	1 qts 1/2 cup	
SOUP,CONDENSED,TOMATO	5-1/8 lbs	2 qts 1-1/4 cup	
WORCESTERSHIRE SAUCE	1-5/8 oz	3 tbsp	
WATER	1-1/3 lbs	2-1/2 cup	
SOUP,CONDENSED,TOMATO	5-1/8 lbs	1-1/2 #3cyl	

Method

1 Combine beef with bread crumbs, salt, pepper and garlic; mix until well blended.
2 Reconstitute milk.
3 Add milk, celery, onions, sweet peppers, eggs, tomato soup, and Worcestershire sauce. Mix lightly but thoroughly. DO NOT OVERMIX.
4 Place 11 pounds 6 ounces meat mixture into each steam table pan and divide into 2 loaves per pan.
5 Using a convection oven, bake 1 hour 15 minutes at 300 F. on high fan, closed vent. CCP: Internal temperature must reach 155 F. or higher for 15 seconds. Skim off excess fat and liquid during cooking period.
6 Combine tomato soup and water. Bring to a boil. Let meat loaf stand 20 minutes before slicing. Cut 13 slices per loaf. Pour tomato soup mixture evenly over baked meat loaf slices. CCP: Hold for service at 140 F. or higher.

MEAT, FISH, AND POULTRY No.L 035 03

CAJUN MEAT LOAF

Yield 100 Portion 6 Ounces

Calories	Carbohydrates	Protein	Fat	Cholesterol	Sodium	Calcium
403 cal	23 g	35 g	19 g	154 mg	989 mg	85 mg

Ingredient	Weight	Measure	Issue
BEEF,GROUND,BULK,RAW,90% LEAN	30 lbs		
BREADCRUMBS	3-3/4 lbs	1 gal	
SALT	3-3/4 oz	1/4 cup 2-1/3 tbsp	
PEPPER,BLACK,GROUND	7/8 oz	1/4 cup 1/3 tbsp	
GARLIC POWDER	2/3 oz	2-1/3 tbsp	
PEPPER,RED,GROUND	3/8 oz	2 tbsp	
OREGANO,CRUSHED	1/3 oz	2 tbsp	
BASIL,SWEET,WHOLE,CRUSHED	1/3 oz	2 tbsp	
THYME,GROUND	1/3 oz	2 tbsp	
ONION POWDER	1/2 oz	2 tbsp	
MILK,NONFAT,DRY	2-3/8 oz	1 cup	
WATER	2-7/8 lbs	1 qts 1-1/2 cup	
CELERY,FRESH,CHOPPED	1 lbs	3-3/4 cup	1-3/8 lbs
ONIONS,FRESH,CHOPPED	1 lbs	3 cup	1-1/8 lbs
PEPPERS,GREEN,FRESH,CHOPPED	1 lbs	3 cup	1-1/4 lbs
EGGS,WHOLE,FROZEN	2-3/8 lbs	1 qts 1/2 cup	
CATSUP	3-1/8 lbs	1 qts 2 cup	
WORCESTERSHIRE SAUCE	8-1/2 oz	1 cup	
CREOLE SAUCE		2 gal 1/2 qts	

Method

1 Combine beef with bread crumbs, salt, pepper, garlic powder, red pepper, oregano, basil, thyme, and onion powder; mix until well blended.
2 Reconstitute milk.
3 Add milk, celery, onions, sweet peppers, eggs, catsup, and Worcestershire sauce. Mix lightly but thoroughly. DO NOT OVERMIX.
4 Place 11 pounds 6 ounces meat mixture into each steam table pan and divide into 2 loaves per pan.
5 Using a convection oven, bake 1 hour 15 minutes at 300 F. on high fan, closed vent. CCP: Internal temperature must reach 155 F. or higher for 15 seconds. Skim off excess fat and liquid during cooking period.
6 Let stand 20 minutes before slicing. Cut 13 slices per loaf. CCP: Hold for service at 140 F. or higher.
7 Serve with Cajun Creole Sauce, Recipe No. O 005 02.

MEAT, FISH, AND POULTRY No.L 036 00

MINCED BEEF

Yield 100   Portion 5 Ounces

Calories	Carbohydrates	Protein	Fat	Cholesterol	Sodium	Calcium
249 cal	15 g	23 g	11 g	71 mg	301 mg	40 mg

Ingredient

Ingredient	Weight	Measure	Issue
BEEF,GROUND,BULK,RAW,90% LEAN	20 lbs		
ONIONS,FRESH,CHOPPED	2-1/8 lbs	1 qts 2 cup	2-1/3 lbs
FLOUR,WHEAT,GENERAL PURPOSE	2-1/8 lbs	1 qts 3-1/2 cup	
TOMATOES,CANNED,CRUSHED,INCL LIQUIDS	19-7/8 lbs	2 gal 1 qts	
MACE,GROUND	3/8 oz	2 tbsp	
SALT	1-1/4 oz	2 tbsp	
PEPPER,BLACK,GROUND	1/8 oz	1/3 tsp	

Method

1. Cook beef with onions until beef loses its pink color, stirring to break apart. Drain or skim off excess fat.
2. Sprinkle flour over beef; continue cooking until flour is absorbed.
3. Add tomatoes, mace or nutmeg, salt and pepper. Stir to mix well. Simmer 10 to 15 minutes. CCP: Internal temperature must reach 155 F. or higher for 15 seconds. Hold for service at 140 F. or higher. Minced beef may be served over toast, biscuits, rice or pasta.

MEAT, FISH, AND POULTRY No.L 037 00

SALISBURY STEAK

Yield 100   Portion 4.5 Ounces

Calories	Carbohydrates	Protein	Fat	Cholesterol	Sodium	Calcium
327 cal	12 g	31 g	16 g	119 mg	514 mg	49 mg

Ingredient	Weight	Measure	Issue
MILK,NONFAT,DRY	3-1/4 oz	1-3/8 cup	
WATER,WARM	3-7/8 lbs	1 qts 3-1/2 cup	
BREADCRUMBS	4-3/4 lbs	1 gal 1 qts	
BEEF,GROUND,BULK,RAW,90% LEAN	28 lbs		
ONIONS,FRESH,CHOPPED	3 lbs	2 qts 1/2 cup	3-1/3 lbs
EGGS,WHOLE,FROZEN	1 lbs	1-7/8 cup	
SALT	3 oz	1/4 cup 1 tbsp	
PEPPER,BLACK,GROUND	1/4 oz	1 tbsp	
WORCESTERSHIRE SAUCE	2-7/8 oz	1/4 cup 1-2/3 tbsp	

Method

1. Reconstitute milk.
2. Add milk to bread; let stand 5 minutes.
3. Combine bread mixture with beef, onions, eggs, salt, pepper, and Worcestershire sauce; mix thoroughly.
4. Shape into steaks about 1 inch thick by 4 inches weighing 6 ounces.
5. Place on sheet pans; using a convection oven, bake at 325 F. on high fan, open vent for 20-25 minutes or until well done. CCP: Internal temperature must reach 155 F. or higher for 15 seconds. Hold for service at 140 F. or higher.

Notes

1. Salisbury steak may be grilled. Lightly spray griddle with non-stick cooking spray. Cook patties on lightly sprayed 350 F. griddle. Grill 8 minutes on each side or until steaks are well done. CCP: Internal temperature must reach 155 F. or higher for 15 seconds. Hold for service at 140 F. or higher.

MEAT, FISH, AND POULTRY No.L 037 02

GRILLED HAMBURGER STEAK

Yield 100   **Portion** 4-1/2 Ounces

Calories	Carbohydrates	Protein	Fat	Cholesterol	Sodium	Calcium
345 cal	0 g	38 g	20 g	133 mg	79 mg	11 mg

Ingredient	Weight	Measure	Issue
BEEF,GROUND,BULK,RAW,90% LEAN	37-1/2 lbs		
COOKING SPRAY,NONSTICK	2 oz	1/4 cup 1/3 tbsp	

Method

1. Shape into steaks about 1 inch thick by 4 inches weighing 6 ounces each.
2. Lightly spray griddle with non-stick cooking spray. Grill steaks on 350 F. griddle for 9 minutes on each side or until well done. CCP: Internal temperature must reach 155 F. or higher for 15 seconds. Hold for service at 140 F. or higher.

MEAT, FISH, AND POULTRY No.L 038 00

SPAGHETTI WITH MEAT SAUCE (GROUND TURKEY)

Yield 100   Portion 1 Cup

Calories	Carbohydrates	Protein	Fat	Cholesterol	Sodium	Calcium
405 cal	63 g	25 g	7 g	51 mg	1422 mg	102 mg

Ingredient	**Weight**	**Measure**	**Issue**
TURKEY,GROUND,90% LEAN,RAW	18 lbs		
TOMATOES,CANNED,DICED,INCL LIQUIDS	27-5/8 lbs	3 gal	
TOMATO PASTE,CANNED	11-1/2 lbs	1 gal 1 qts	
WATER	6-1/4 lbs	3 qts	
ONIONS,FRESH,CHOPPED	4-1/4 lbs	3 qts	4-2/3 lbs
SUGAR,GRANULATED	7 oz	1 cup	
SALT	3-3/8 oz	1/4 cup 1-2/3 tbsp	
GARLIC POWDER	1-5/8 oz	1/4 cup 1-2/3 tbsp	
BASIL,DRIED,CRUSHED	7/8 oz	1/4 cup 1-2/3 tbsp	
THYME,GROUND	1/2 oz	3 tbsp	
OREGANO,CRUSHED	7/8 oz	1/4 cup 1-2/3 tbsp	
PEPPER,BLACK,GROUND	1/4 oz	1 tbsp	
BAY LEAF,WHOLE,DRIED	1/4 oz	8 each	
WATER,BOILING	83-5/8 lbs	10 gal	
SPAGHETTI NOODLES,DRY	12 lbs	3 gal 1 qts	
SALT	2-1/2 oz	1/4 cup 1/3 tbsp	

Method

1. Cook turkey in steam-jacketed kettle or stock pot until turkey loses its pink color. Stir. Drain or skim off excess fat.
2. Add tomatoes, tomato paste, water, onions, sugar, salt, garlic powder, basil, thyme, oregano, pepper, and bay leaves to turkey. Mix well.
3. Bring to a boil; reduce heat; cover; cook at low heat about 1 hour, stirring frequently. CCP: Internal temperature must reach 165 F. or higher for 15 seconds.
4. Remove bay leaves before serving. CCP: Hold for service at 140 F. or higher.
5. Add salt to boiling water. Slowly add spaghetti while stirring constantly until water boils again. Cook 10 to 12 minutes or until tender, stirring occasionally. Drain thoroughly.

MEAT, FISH, AND POULTRY No.L 038 01

SPAGHETTI WITH MEAT SAUCE (GROUND BEEF)

Yield 100   Portion   1 Cup

Calories	Carbohydrates	Protein	Fat	Cholesterol	Sodium	Calcium
441 cal	63 g	26 g	10 g	57 mg	1388 mg	90 mg

Ingredient	Weight	Measure	Issue
BEEF,GROUND,BULK,RAW,90% LEAN	16 lbs		
TOMATOES,CANNED,DICED,INCL LIQUIDS	27-5/8 lbs	3 gal	
TOMATO PASTE,CANNED	11-1/2 lbs	1 gal 1 qts	
WATER	6-1/4 lbs	3 qts	
ONIONS,FRESH,CHOPPED	4-1/4 lbs	3 qts	4-2/3 lbs
SUGAR,GRANULATED	7 oz	1 cup	
SALT	3-3/8 oz	1/4 cup 1-2/3 tbsp	
GARLIC POWDER	1-5/8 oz	1/4 cup 1-2/3 tbsp	
BASIL,DRIED,CRUSHED	7/8 oz	1/4 cup 1-2/3 tbsp	
THYME,GROUND	1/2 oz	3 tbsp	
OREGANO,CRUSHED	7/8 oz	1/4 cup 1-2/3 tbsp	
PEPPER,BLACK,GROUND	1/4 oz	1 tbsp	
BAY LEAF,WHOLE,DRIED	1/4 oz	8 each	
WATER,BOILING	83-5/8 lbs	10 gal	
SPAGHETTI NOODLES,DRY	12 lbs	3 gal 1 qts	
SALT	2-1/2 oz	1/4 cup 1/3 tbsp	

Method

1 Cook beef in steam-jacketed kettle or stock pot until beef loses its pink color, stirring to break apart. Drain or skim off excess fat.
2 Add tomatoes, tomato paste, water, onions, sugar, salt, garlic powder, basil, thyme, oregano, pepper, and bay leaves to beef. Mix well.
3 Bring to a boil; reduce heat; cover; cook at low heat about 1 hour, stirring frequently. CCP: Internal temperature must reach 155 F. or higher for 15 seconds.
4 Remove bay leaves before serving. CCP: Hold for service at 140 F. or higher.
5 Add salt to boiling water. Slowly add spaghetti while stirring constantly until water boils again. Cook 10 to 12 minutes or until tender, stirring occasionally. Drain thoroughly.

MEAT. FISH. AND POULTRY No.L 038 02

SPAGHETTI WITH MEAT SAUCE, RTU (GROUND TURKEY)

Yield 100   Portion   1 Cup

Calories	Carbohydrates	Protein	Fat	Cholesterol	Sodium	Calcium
433 cal	55 g	24 g	12 g	51 mg	1127 mg	74 mg

Ingredient	**Weight**	**Measure**	**Issue**
TURKEY,GROUND,90% LEAN,RAW	18 lbs		
SAUCE,SPAGHETTI,CANNED,RTU	46-1/3 lbs	5 gal 1 qts	
SALT	2-1/3 oz	1/4 cup	
WATER,BOILING	83-5/8 lbs	10 gal	
SPAGHETTI NOODLES,DRY	12 lbs	3 gal 1 qts	

Method

1. Cook turkey in steam-jacketed kettle or stock pot until turkey loses its pink color, stirring to break apart.
2. Add spaghetti sauce to meat; stir well to distribute meat.
3. Cook at medium heat until sauce comes to a boil; reduce heat, simmer 5 minutes to blend flavors, stirring as necessary. CCP: Internal temperature must reach 165 F. or higher for 15 seconds. Hold for service at 140 F. or higher.
4. Add salt to boiling water. Slowly add spaghetti while stirring constantly until water boils again. Cook 10 to 12 minutes or until tender, stirring occasionally. Drain thoroughly.

MEAT, FISH, AND POULTRY No.L 038 03

SPAGHETTI WITH MEAT SAUCE, RTU (GROUND BEEF)

Yield 100   Portion 1 Cup

Calories	Carbohydrates	Protein	Fat	Cholesterol	Sodium	Calcium
469 cal	55 g	26 g	15 g	57 mg	1092 mg	62 mg

Ingredient | Weight | Measure | Issue

Ingredient	Weight	Measure	Issue
BEEF,GROUND,BULK,RAW,90% LEAN	16 lbs		
SAUCE,SPAGHETTI,CANNED,RTU	46-1/3 lbs	5 gal 1 qts	
SALT	2-1/3 oz	1/4 cup	
WATER,BOILING	83-5/8 lbs	10 gal	
SPAGHETTI NOODLES,DRY	12 lbs	3 gal 1 qts	

Method

1 Cook beef in steam-jacketed kettle or stock pot until beef loses its pink color, stirring to break apart.
2 Add spaghetti sauce to meat; stir well to distribute meat.
3 Cook at medium heat until sauce comes to a boil; reduce heat, simmer 5 minutes to blend flavors, stirring as necessary. CCP: Internal temperature must reach 155 F. or higher for 15 seconds. Hold for service at 140 F. or higher.
4 Add salt to boiling water. Slowly add spaghetti while stirring constantly until water boils again. Cook 10 to 12 minutes or until tender, stirring occasionally. Drain thoroughly.

MEAT, FISH, AND POULTRY No.L 039 00

SPAGHETTI WITH MEATBALLS (GROUND TURKEY)

Yield 100   Portion 1 Serving

Calories	Carbohydrates	Protein	Fat	Cholesterol	Sodium	Calcium
443 cal	66 g	28 g	8 g	72 mg	1630 mg	110 mg

Ingredient	Weight	Measure	Issue
TOMATOES,CANNED,DICED,INCL LIQUIDS	26-1/2 lbs	2 gal 3-1/2 qts	
TOMATO PASTE,CANNED	9-1/4 lbs	1 gal	
WATER	8-1/3 lbs	1 gal	
ONIONS,FRESH,CHOPPED	3-1/8 lbs	2 qts 1 cup	3-1/2 lbs
SUGAR,GRANULATED	7 oz	1 cup	
SALT	2-1/2 oz	1/4 cup 1/3 tbsp	
GARLIC POWDER	1 oz	3-1/3 tbsp	
BASIL,SWEET,WHOLE,CRUSHED	5/8 oz	1/4 cup 1/3 tbsp	
THYME,GROUND	1/3 oz	2 tbsp	
OREGANO,CRUSHED	5/8 oz	1/4 cup 1/3 tbsp	
PEPPER,RED,GROUND	1/8 oz	1/4 tsp	
BAY LEAF,WHOLE,DRIED	3/8 oz	12 lf	
TURKEY,GROUND,90% LEAN,RAW	20 lbs		
ONIONS,FRESH,CHOPPED	2-3/8 lbs	1 qts 2-3/4 cup	2-2/3 lbs
BREADCRUMBS,DRY,GROUND,FINE	2-3/8 lbs	2 qts 2 cup	
EGGS,WHOLE,FROZEN	12-7/8 oz	1-1/2 cup	
SALT	3 oz	1/4 cup 1 tbsp	
PEPPER,BLACK,GROUND	1/4 oz	1 tbsp	
SALT	2-1/2 oz	1/4 cup 1/3 tbsp	
SPAGHETTI NOODLES,DRY	12 lbs	3 gal 1 qts	
WATER,BOILING	66-7/8 lbs	8 gal	

Method

1 Combine tomatoes, tomato paste, water, onions, sugar, salt, garlic powder, basil, thyme, oregano, red pepper, and bay leaves; mix well. Bring to a boil; reduce heat; simmer 1 hour or until thickened, stirring occasionally. Remove bay leaves.
2 Combine turkey, onions, bread crumbs, eggs, salt, and pepper; mix lightly but thoroughly.
3 Shape into 300 - 1-1/3 ounce balls. Place 100 balls in each pan.
4 Using a convection oven, bake 12-14 minutes at 350 F. on high fan, closed vent or until browned. CCP: Internal temperature must reach 165 F. or higher for 15 seconds. Discard fat. Remove to serving pan. CCP: Hold for service at 140 F. or higher.
5 Add salt to boiling water. Slowly add spaghetti while stirring constantly until water boils again. Cook about 10 to 12 minutes or until tender, stirring occasionally. Do not overcook. Drain thoroughly.
6 EACH PORTION: 3 meatballs, 3/4 cup sauce, and 1 cup spaghetti.

MEAT, FISH, AND POULTRY No.L 039 01

SPAGHETTI WITH MEATBALLS (GROUND BEEF)

Yield 100   Portion   1 Cup

Calories	Carbohydrates	Protein	Fat	Cholesterol	Sodium	Calcium
500 cal	66 g	31 g	13 g	87 mg	1590 mg	97 mg

Ingredient	Weight	Measure	Issue
TOMATOES,CANNED,DICED,INCL LIQUIDS	26-1/2 lbs	2 gal 3-1/2 qts	
TOMATO PASTE,CANNED	9-1/4 lbs	1 gal	
WATER	8-1/3 lbs	1 gal	
ONIONS,FRESH,CHOPPED	3-1/8 lbs	2 qts 1 cup	3-1/2 lbs
SUGAR,GRANULATED	7 oz	1 cup	
SALT	2-1/2 oz	1/4 cup 1/3 tbsp	
GARLIC POWDER	1 oz	3-1/3 tbsp	
BASIL,SWEET,WHOLE,CRUSHED	5/8 oz	1/4 cup 1/3 tbsp	
THYME,GROUND	1/3 oz	2 tbsp	
OREGANO,CRUSHED	5/8 oz	1/4 cup 1/3 tbsp	
PEPPER,RED,GROUND	1/8 oz	1/4 tsp	
BAY LEAF,WHOLE,DRIED	3/8 oz	12 lf	
BEEF,GROUND,BULK,RAW,90% LEAN	20 lbs		
ONIONS,FRESH,CHOPPED	2-1/3 lbs	1 qts 2-5/8 cup	2-5/8 lbs
BREADCRUMBS	2-1/8 lbs	2 qts 1 cup	
EGGS,WHOLE,FROZEN	12-7/8 oz	1-1/2 cup	
SALT	3 oz	1/4 cup 1 tbsp	
PEPPER,BLACK,GROUND	1/4 oz	1 tbsp	
WATER,BOILING	66-7/8 lbs	8 gal	
SALT	2-1/2 oz	1/4 cup 1/3 tbsp	
SPAGHETTI NOODLES,DRY	12 lbs	3 gal 1 qts	

Method

1 Combine tomatoes, tomato paste, water, onions, sugar, salt, garlic powder, basil, thyme, oregano, red pepper and bay leaves; mix well.
2 Bring to a boil; reduce heat; simmer 1 hour or until thickened, stirring occasionally. Remove bay leaves.
3 Combine beef, onions, bread crumbs, eggs, salt, and pepper; mix lightly but thoroughly.
4 Shape into 300 1-1/3 ounce balls. Place 100 balls on each pan.
5 Using a convection oven, bake 10-12 minutes at 350 F. on high fan, closed vent or until browned. CCP: Internal temperature must reach 155 F. or higher for 15 seconds. Discard fat. Remove to serving pan. CCP: Hold for service at 140 F. or higher.
6 Add salt to boiling water. Slowly add spaghetti while stirring constantly until water boils again. Cook about 10 to 12 minutes or until tender, stirring occasionally. Do not overcook. Drain thoroughly.
7 EACH PORTION: 3 meatballs, 3/4 cup sauce, 1 cup spaghetti.

MEAT, FISH, AND POULTRY No.L 040 00

STUFFED GREEN PEPPERS (GROUND BEEF)

Yield 100   Portion 1 Half

Calories	Carbohydrates	Protein	Fat	Cholesterol	Sodium	Calcium
342 cal	24 g	27 g	15 g	85 mg	960 mg	36 mg

Ingredient	Weight	Measure	Issue
TOMATO SAUCE		1 gal 2-1/2 qts	
PEPPERS,GREEN,FRESH	17-1/4 lbs	3 gal 1-1/8 qts	21 lbs
WATER,BOILING	8-1/3 lbs	1 gal	
STEAMED RICE		1 gal 2 qts	
BEEF,GROUND,BULK,RAW,90% LEAN	24 lbs		
ONIONS,FRESH,CHOPPED	2-7/8 lbs	2 qts 1/4 cup	3-1/4 lbs
SALT	5-1/8 oz	1/2 cup	
PEPPER,BLACK,GROUND	1/8 oz	1/3 tsp	
WORCESTERSHIRE SAUCE	12-2/3 oz	1-1/2 cup	
WATER	2-1/8 lbs	1 qts	
WATER	2-1/8 lbs	1 qts	

Method

1. Prepare Tomato Sauce, Recipe No O 015 00.
2. Cut each pepper in half lengthwise; remove core.
3. Place peppers in boiling water. Return to a boil; cook 1 minute. Drain well. Set aside for use in Step 6.
4. Prepare rice according to Recipe No. E 005 00.
5. Combine cooked rice, ground beef, onions, salt, pepper, Worcestershire sauce and water with 2 quarts tomato sauce. DO NOT OVERMIX.
6. Fill each pepper with 3/4 cup beef mixture. Place filled peppers in roasting pans.
7. Pour 1 cup water around peppers in each pan.
8. Pour remaining sauce over peppers in each pan. Cover pans.
9. Bake about 1-1/2 hours at 350 F. or until tender. CCP: Internal temperature must reach 155 F. or higher for 15 seconds. Hold at 140 F. or higher for service.

MEAT, FISH, AND POULTRY No. L 040 01

STUFFED GREEN PEPPERS (FROZEN)

Yield 100 **Portion** 1 Pepper

Calories	Carbohydrates	Protein	Fat	Cholesterol	Sodium	Calcium
260 cal	26 g	16 g	10 g	50 mg	1429 mg	43 mg

Ingredient	Weight	Measure	Issue
PEPPERS,STUFFED	50 lbs		
SAUCE,TOMATO,CANNED	14 lbs	1 gal 2-1/2 qts	

Method

1 Pour tomato sauce evenly over peppers. Follow manufacturer's directions for cooking stuffed peppers. CCP: Internal temperature must reach 165 F. or higher for 15 seconds. Hold at 140 F. or higher.

MEAT, FISH, AND POULTRY No. L 040 02

STUFFED GREEN PEPPERS (GROUND TURKEY)

Yield 100   Portion   1 Half

Calories	Carbohydrates	Protein	Fat	Cholesterol	Sodium	Calcium
269 cal	24 g	23 g	10 g	68 mg	1000 mg	51 mg

Ingredient	Weight	Measure	Issue
TOMATO SAUCE		1 gal 2-1/2 qts	
PEPPERS,GREEN,FRESH	17-1/4 lbs	3 gal 1-1/8 qts	21 lbs
WATER,BOILING	8-1/3 lbs	1 gal	
STEAMED RICE		1 gal 2 qts	
TURKEY,GROUND,90% LEAN,RAW	24 lbs		
ONIONS,FRESH,CHOPPED	2-7/8 lbs	2 qts 1/4 cup	3-1/4 lbs
SALT	5-1/8 oz	1/2 cup	
PEPPER,BLACK,GROUND	1/8 oz	1/3 tsp	
WORCESTERSHIRE SAUCE	12-2/3 oz	1-1/2 cup	
WATER	2-1/8 lbs	1 qts	
WATER	2-1/8 lbs	1 qts	

Method

1. Prepare Tomato Sauce, Recipe No O 015 00.
2. Cut each pepper in half lengthwise; remove core.
3. Place peppers in boiling water. Return to a boil; cook 1 minute. Drain well. Set aside for use in Step 6.
4. Prepare rice according to Recipe No. E 005 00.
5. Combine cooked rice, ground turkey, onions, salt, pepper, Worcestershire sauce and water with 2 quarts tomato sauce. DO NOT OVERMIX.
6. Fill each pepper with 3/4 cup turkey mixture. Place filled peppers in roasting pans.
7. Pour 1 cup water around peppers in each pan.
8. Pour remaining sauce over peppers in each pan. Cover pans.
9. Bake about 1-1/2 hours at 350 F. or until tender. CCP: Internal temperature must reach 165 F. or higher for 15 seconds. Hold at 140 F. or higher for service.

MEAT, FISH, AND POULTRY No.L 041 00

SWEDISH MEATBALLS (GROUND BEEF)

Yield 100   Portion 3-1/2 Ounces

Calories	Carbohydrates	Protein	Fat	Cholesterol	Sodium	Calcium
244 cal	11 g	23 g	12 g	84 mg	753 mg	30 mg

Ingredient | Weight | Measure | Issue

Ingredient	Weight	Measure	Issue
BREAD,WHITE,SLICED	2 lbs	1 gal 2-1/2 qts	
MILK,NONFAT,DRY	1-3/4 oz	3/4 cup	
WATER,WARM	2 lbs	3-3/4 cup	
EGGS,WHOLE,FROZEN	10-3/4 oz	1-1/4 cup	
SALT	1-1/4 oz	2 tbsp	
NUTMEG,GROUND	1/8 oz	1/3 tsp	
PEPPER,BLACK,GROUND	1/8 oz	1/4 tsp	
ALLSPICE,GROUND	1/8 oz	1/4 tsp	
BEEF,GROUND,BULK,RAW,90% LEAN	20 lbs		
ONIONS,FRESH,CHOPPED	11-1/4 oz	2 cup	12-1/2 oz
BEEF BROTH		2 gal 1/2 qts	
FLOUR,WHEAT,GENERAL PURPOSE	1-3/8 lbs	1 qts 1 cup	
WATER	2-5/8 lbs	1 qts 1 cup	
NUTMEG,GROUND	1/4 oz	1 tbsp	
PEPPER,BLACK,GROUND	1/4 oz	1 tbsp	
GARLIC POWDER	1/8 oz	1/8 tsp	
PAPRIKA,GROUND	1/4 oz	1 tbsp	

Method

1. Place bread in mixer bowl; mix at medium speed 5 minutes or until crumbs are formed.
2. Reconstitute milk.
3. Blend in eggs, salt, nutmeg, pepper, and allspice. Pour over bread; mix at low speed 1/2 minute; let stand 10 minutes.
4. Add beef and onions to bread mixture. Mix at low speed 1 minute. Do not over mix.
5. Shape into 300 balls weighing 1-1/3 ounces; place 100 meatballs on each sheet pan.
6. Using a convection oven, bake at 350 F. on high fan, closed vent 8-10 minutes or until browned and done. CCP: Internal temperature must reach 155 F. or higher for 15 seconds. Remove meatballs to steam table pans. CCP: Hold at 140 F. or higher for use in Step 10.
7. Prepare stock according to package directions.
8. Combine flour and water, stirring until smooth; add to stock, bring to a boil; reduce heat; simmer 10 minutes or until thickened, stirring constantly.
9. Add nutmeg, pepper, and garlic powder; stir well.
10. Pour 2-3/4 quarts gravy over meatballs in each pan.
11. Using a convection oven, bake at 350 F. 15 minutes or until heated thoroughly on high fan, closed vent. CCP: Internal temperature must reach 155 F. or higher for 15 seconds.
12. Sprinkle each pan with 1 teaspoon paprika before serving. CCP: Hold for service at 140 F.

MEAT, FISH, AND POULTRY No.L 041 01

SWEDISH MEATBALLS (GROUND TURKEY)

Yield 100   Portion   3-1/2 Ounces

Calories	Carbohydrates	Protein	Fat	Cholesterol	Sodium	Calcium
186 cal	11 g	19 g	7 g	70 mg	917 mg	48 mg

Ingredient	Weight	Measure	Issue
BREAD,WHITE,SLICED	2 lbs	1 gal 2-1/2 qts	
MILK,NONFAT,DRY	1-3/4 oz	3/4 cup	
WATER,WARM	2 lbs	3-3/4 cup	
EGGS,WHOLE,FROZEN	10-3/4 oz	1-1/4 cup	
SALT	1-1/4 oz	2 tbsp	
NUTMEG,GROUND	1/8 oz	1/3 tsp	
PEPPER,BLACK,GROUND	1/8 oz	1/4 tsp	
ALLSPICE,GROUND	1/8 oz	1/4 tsp	
TURKEY,GROUND,90% LEAN,RAW	20 lbs		
ONIONS,FRESH,CHOPPED	11-1/4 oz	2 cup	12-1/2 oz
CHICKEN BROTH		2 gal 1-3/4 qts	
FLOUR,WHEAT,GENERAL PURPOSE	1-3/8 lbs	1 qts 1 cup	
WATER	2-5/8 lbs	1 qts 1 cup	
NUTMEG,GROUND	1/4 oz	1 tbsp	
PEPPER,BLACK,GROUND	1/4 oz	1 tbsp	
GARLIC POWDER	1/8 oz	1/8 tsp	
PAPRIKA,GROUND	1/4 oz	1 tbsp	

Method

1. Place bread in mixer bowl; mix at medium speed 5 minutes or until crumbs are formed.
2. Reconstitute milk.
3. Blend in egg substitute, salt, nutmeg, pepper and allspice. Pour over bread; mix at low speed 1/2 minute; let stand 10 minutes.
4. Add turkey and onions to bread mixture. Mix at low speed 1 minute. Do not overmix.
5. Shape into balls weighing 1-1/3 ounces each; place 100 meatballs on each sheet pan.
6. Using a convection oven, bake 8-10 minutes at 350 F. on high fan, closed vent or until browned and done. Remove meatballs to steam table pans. Set aside for use in Step 10.
7. Prepare stock according to package directions.
8. Combine flour and water, stirring until smooth; add to stock, bring to a boil; reduce heat; simmer 10 minutes or until thickened, stirring constantly.
9. Add nutmeg, pepper, and garlic powder; stir well.
10. Pour 3-1/2 quarts gravy over meatballs in each pan.
11. Using a convection oven, bake at 350 F. 15 minutes or until heated thoroughly on high fan, closed vent. CCP: Internal temperature must reach 165 F. or higher for 15 seconds.
12. Sprinkle each pan with 1 teaspoon paprika before serving. CCP: Hold for service at 140 F. or higher.

MEAT, FISH, AND POULTRY No. L 042 00

CHILI CONQUISTADOR (GROUND BEEF)

Yield 100   **Portion** 8-1/2 Ounces

Calories	Carbohydrates	Protein	Fat	Cholesterol	Sodium	Calcium
462 cal	45 g	29 g	18 g	86 mg	908 mg	74 mg

Ingredient	Weight	Measure	Issue
RICE, LONG GRAIN	2-1/8 lbs	1 qts 1-1/4 cup	
WATER, COLD	5-3/4 lbs	2 qts 3 cup	
SALT	1/2 oz	3/8 tsp	
BEEF, GROUND, BULK, RAW, 90% LEAN	24 lbs		
TOMATOES, CANNED, DICED, DRAINED	19-7/8 lbs	2 gal 1 qts	
ONIONS, FRESH, CHOPPED	2-1/8 lbs	1 qts 2 cup	2-1/3 lbs
CHILI POWDER, DARK, GROUND	5-5/8 oz	1-3/8 cup	
SALT	1-7/8 oz	3 tbsp	
GARLIC POWDER	1/3 oz	1 tbsp	
PEPPER, RED, GROUND	<1/16th oz	1/8 tsp	
CORN BREAD MIX	9 lbs	1 gal 2-2/3 qts	

Method

1 Combine rice, water, and salt. Bring to a boil; stir occasionally.
2 Cover tightly; simmer 20 to 25 minutes. Do not stir.
3 Cook beef until beef loses its pink color, stirring to break apart. Drain or skim off excess fat.
4 Add tomatoes, onions, chili powder, salt, garlic powder and red pepper to meat mixture; stir until blended; heat to simmer.
5 Combine rice with chili mixture; mix well. Place 5-3/4 quarts mixture in each pan.
6 Prepare corn bread mix according to instructions on container.
7 Spread 1-3/4 quarts corn bread batter over chili mixture in each pan.
8 Using a convection oven, bake 30 minutes at 375 F. on high fan, open vent or until corn bread is golden brown and done. CCP: Internal temperature must reach 155 F. or higher for 15 seconds.
9 Cut 5 by 5. CCP: Hold at 140 F. or higher for service.

MEAT, FISH, AND POULTRY No. L 042 01

CHILI CONQUISTADOR (GROUND TURKEY)

Yield 100   Portion 8-1/2 Ounces

Calories	Carbohydrates	Protein	Fat	Cholesterol	Sodium	Calcium
390 cal	45 g	25 g	13 g	68 mg	949 mg	89 mg

Ingredient	Weight	Measure	Issue
RICE, LONG GRAIN	2-1/8 lbs	1 qts 1-1/4 cup	
WATER, COLD	5-3/4 lbs	2 qts 3 cup	
SALT	1/2 oz	3/8 tsp	
TURKEY, GROUND, 90% LEAN, RAW	24 lbs		
TOMATOES, CANNED, DICED, DRAINED	19-7/8 lbs	2 gal 1 qts	
ONIONS, FRESH, CHOPPED	2-1/8 lbs	1 qts 2 cup	2-1/3 lbs
CHILI POWDER, DARK, GROUND	5-5/8 oz	1-3/8 cup	
SALT	1-7/8 oz	3 tbsp	
GARLIC POWDER	1/3 oz	1 tbsp	
PEPPER, RED, GROUND	<1/16th oz	1/8 tsp	
CORN BREAD MIX	9 lbs	1 gal 2-2/3 qts	

Method

1. Combine rice, water, and salt. Bring to a boil; stir occasionally.
2. Cover tightly; simmer 20 to 25 minutes. Do not stir. CCP: Hold for 140 F. or higher.
3. Cook turkey until turkey loses its pink color. Stir to break apart. CCP: Internal temperature must reach 165 F. or higher for 15 seconds. Drain or skim off excess fat.
4. Add tomatoes, onions, chili powder, salt, garlic powder and red pepper to meat mixture; stir until blended; heat to simmer.
5. Combine rice with chili mixture; mix well. Place mixture evenly in each steam table pan.
6. Prepare corn bread mix according to instructions on container.
7. Spread corn bread batter evenly over chili mixture in each pan.
8. Using a convection oven, bake for 30 minutes at 375 F. on high fan, open vent or until corn bread is golden brown and done. CCP: Internal temperature must reach 165 F. or higher for 15 seconds.
9. Cut 5 by 5. CCP: Hold for service at 140 F. or higher.

MEAT, FISH, AND POULTRY No.L 043 00

BEEF FAJITAS (FAJITA STRIPS)

Yield 100   **Portion** 2 Fajitas

Calories	Carbohydrates	Protein	Fat	Cholesterol	Sodium	Calcium
458 cal	59 g	27 g	13 g	51 mg	1081 mg	148 mg

Ingredient	**Weight**	**Measure**	**Issue**
JUICE,LIME	1-1/2 lbs	3 cup	
SALT	3 oz	1/4 cup 1 tbsp	
GARLIC POWDER	2-3/8 oz	1/2 cup	
ONION POWDER	1-1/8 oz	1/4 cup 1 tbsp	
PEPPER,BLACK,GROUND	3/4 oz	3-1/3 tbsp	
CUMIN,GROUND	1/3 oz	1 tbsp	
PEPPER,RED,GROUND	1/4 oz	1 tbsp	
TOMATOES,CANNED,CRUSHED,DRAINED	7-1/4 lbs	1 #10cn	
BEEF,FAJITA STRIPS	18 lbs		
TORTILLAS,FLOUR,8 INCH	19-1/8 lbs	200 each	
COOKING SPRAY,NONSTICK	2 oz	1/4 cup 1/3 tbsp	
ONIONS,FRESH,1/4"" STRIPS	5-1/8 lbs	1 gal 1 qts	5-5/8 lbs
PEPPERS,GREEN,FRESH,JULIENNE	5 lbs	3 qts 3-1/4 cup	6-1/8 lbs
COOKING SPRAY,NONSTICK	2 oz	1/4 cup 1/3 tbsp	
SALSA		3 qts 2 cup	

Method

1. Combine lime juice, salt, garlic powder, onion powder, black pepper, cumin, tomatoes and red pepper. Stir well to blend.
2. Pour mixture over beef strips. Mix thoroughly to evenly distribute seasonings around all surfaces of beef. Cover. CCP: Marinate under refrigeration at 41 F. or lower for 45 minutes for use in Step 5.
3. Wrap tortillas in foil; place in a 150 F. oven or in a warmer for 15 minutes or until tortillas are soft and pliable.
4. Lightly spray griddle with non-stick cooking spray. Grill onions and peppers 6 to 8 minutes while tossing intermittently; lightly spray with cooking spray as needed.
5. Lightly spray griddle with non-stick cooking spray. Grill beef strips 3 to 4 minutes or until lightly browned while tossing intermittently; lightly spray with cooking spray as needed. CCP: Internal temperature must reach 145 F. or higher for 15 seconds.
6. Place 6 to 7 cooked fajita strips (3 oz.), 3 tbsp onion/sweet pepper mixture in center of each tortilla. Roll tortilla tightly around mixture. Secure tortilla with a toothpick.
7. Serve with 2 tbsp of salsa. Use batch preparation methods to prevent the fajitas from getting soggy. CCP: Hold for service at 140 F. or higher.

MEAT, FISH, AND POULTRY No. L 043 01

CHICKEN FAJITAS (FAJITA STRIPS)

Yield 100　　　　　　　　　　　　　　　　　　　　Portion 2 Fajitas

Calories	Carbohydrates	Protein	Fat	Cholesterol	Sodium	Calcium
449 cal	56 g	32 g	10 g	65 mg	985 mg	144 mg

Ingredient	Weight	Measure	Issue
JUICE,LIME	1-1/4 lbs	2-1/2 cup	
SALT	2-1/3 oz	1/4 cup	
GARLIC POWDER	1-1/4 oz	1/4 cup 1/3 tbsp	
ONION POWDER	7/8 oz	1/4 cup	
PEPPER,BLACK,GROUND	7/8 oz	1/4 cup	
CUMIN,GROUND	1/4 oz	1 tbsp	
PEPPER,RED,GROUND	1/8 oz	3/8 tsp	
CHICKEN,FAJITA STRIPS	23 lbs		
TORTILLAS,FLOUR,8 INCH	19-1/8 lbs	200 each	
COOKING SPRAY,NONSTICK	2 oz	1/4 cup 1/3 tbsp	
ONIONS,FRESH,1/4"" STRIPS	5-1/8 lbs	1 gal 1 qts	5-5/8 lbs
PEPPERS,GREEN,FRESH,JULIENNE	5 lbs	3 qts 3-1/4 cup	6-1/8 lbs
COOKING SPRAY,NONSTICK	2 oz	1/4 cup 1/3 tbsp	
SALSA		3 qts 2 cup	

Method

1. Combine lime juice, salt, garlic powder, onion powder, black pepper, cumin, and red pepper. Stir well to blend.
2. Pour mixture over chicken strips. Mix thoroughly to evenly distribute seasonings around all surfaces of chicken. Cover. CCP: Marinate under refrigeration at 41 F. or lower for 45 minutes for use in Step 5.
3. Wrap tortillas in foil; place in a 150 F. oven or in a warmer for 15 minutes or until tortillas are soft and pliable.
4. Lightly spray griddle with non-stick cooking spray. Grill onions and peppers 6 to 8 minutes while tossing intermittently; lightly spray with cooking spray as needed.
5. Lightly spray griddle with non-stick cooking spray. Grill chicken strips 5 to 7 minutes or until lightly browned while tossing intermittently; lightly spray with cooking spray as needed. CCP: Internal temperature must register 165 F. or higher for 15 seconds. Hold at 140 F. or higher for use in Step 6.
6. Place 6 to 7 cooked fajita strips (3 oz.), 3 tbsp onion/sweet pepper mixture in center of each tortilla. Roll tortilla tightly around mixture. Secure tortilla with a toothpick.
7. Serve with 2 tbsp of salsa. Use batch preparation methods to prevent the fajitas from getting soggy. CCP: Hold for service at 140 F. or higher.

MEAT, FISH, AND POULTRY No.L 043 02

TURKEY FAJITAS

Yield 100   **Portion** 2 Fajitas

Calories	Carbohydrates	Protein	Fat	Cholesterol	Sodium	Calcium
491 cal	59 g	30 g	15 g	65 mg	1620 mg	170 mg

Ingredient	Weight	Measure	Issue
JUICE,LIME	1-1/4 lbs	2-1/2 cup	
SALT	2-1/3 oz	1/4 cup	
GARLIC POWDER	2 oz	1/4 cup 3 tbsp	
ONION POWDER	7/8 oz	1/4 cup	
PEPPER,BLACK,GROUND	5/8 oz	2-2/3 tbsp	
CUMIN,GROUND	1/4 oz	1 tbsp	
PEPPER,RED,GROUND	1/8 oz	3/8 tsp	
TURKEY,BNLS,WHITE AND DARK MEAT	26 lbs		
TORTILLAS,FLOUR,8 INCH	19-1/8 lbs	200 each	
ONIONS,FRESH,1/4"" STRIPS	5-1/8 lbs	1 gal 1 qts	5-5/8 lbs
PEPPERS,GREEN,FRESH,JULIENNE	5 lbs	3 qts 3-1/4 cup	6-1/8 lbs
COOKING SPRAY,NONSTICK	2 oz		
COOKING SPRAY,NONSTICK	2 oz		
SALSA		3 qts 2 cup	

Method

1. Combine lime juice, salt, garlic powder, onion powder, black pepper, cumin, and red pepper. Stir well to blend.
2. Cut turkey into 1/4 inch thick slices. Cut slices into 3/8 inch strips, 2 to 3 inches long.
3. Pour marinade mixture over turkey strips. Mix thoroughly to evenly distribute seasonings around all surfaces of turkey. Cover. CCP: Marinate under refrigeration at 41 F. or lower for 45 minutes for use in Step 6.
4. Wrap tortillas in foil; place in a 150 F. oven or in a warmer for 15 minutes or until tortillas are soft and pliable.
5. Lightly spray griddle with non-stick cooking spray. Grill onions and peppers 6 to 8 minutes while tossing intermittently; lightly spray with cooking spray as needed.
6. Lightly spray griddle with non-stick cooking spray. Grill turkey strips 5 to 7 minutes or until lightly browned while tossing intermittently; lightly spray with cooking spray as needed. CCP: Internal temperature must register 165 F. or higher for 15 seconds. CCP: Hold at 140 F. or higher for use in Step 7.
7. Place 6 to 7 cooked fajita strips (3 oz), 3 tbsp onion/sweet pepper mixture in center of each tortilla. Roll tortilla tightly around mixture. Secure tortilla with a toothpick. Batch preparation methods should be used to prevent the fajitas from getting soggy.
8. Serve with 2 tbsp of salsa. CCP: Hold for service at 140 F. or higher.

MEAT. FISH. AND POULTRY No.L 044 00

TURKEY CURRY

Yield 100   Portion 7 Ounces

Calories	Carbohydrates	Protein	Fat	Cholesterol	Sodium	Calcium
250 cal	27 g	15 g	10 g	39 mg	1475 mg	60 mg

Ingredient	Weight	Measure	Issue
TURKEY,BNLS,WHITE AND DARK MEAT,DICED	15-1/2 lbs		
WATER	50-1/8 lbs	6 gal	
SALT	5-3/4 oz	1/2 cup 1 tbsp	
BAY LEAF,WHOLE,DRIED	1/3 oz	9 lf	
COOKING SPRAY,NONSTICK	2 oz	1/4 cup 1/3 tbsp	
ONIONS,FRESH,CHOPPED	4-1/4 lbs	3 qts	4-3/4 lbs
CELERY,FRESH,CHOPPED	6 lbs	1 gal 1-2/3 qts	8-1/4 lbs
GARLIC POWDER	1/8 oz	1/4 tsp	
FLOUR,WHEAT,GENERAL PURPOSE	1-1/8 lbs	1 qts	
SALT	1 oz	1 tbsp	
PEPPER,BLACK,GROUND	1/8 oz	1/3 tsp	
CURRY POWDER	2-2/3 oz	3/4 cup	
GINGER,GROUND	1/4 oz	1 tbsp	
HOT SAUCE	1/3 oz	1/3 tsp	
WORCESTERSHIRE SAUCE	4-1/4 oz	1/2 cup	
CHICKEN BROTH		1 gal	
APPLES,FRESH,MEDIUM,PEELED,CORED,CHOPPED	8 lbs	1 gal 3-1/4 qts	10-1/4 lbs
RAISINS	1-7/8 lbs	1 qts 2 cup	
COCONUT,PREPARED,SWEETENED FLAKES	2-1/2 lbs	3 qts	

Method

1. Place turkey in stock pot or steam jacketed kettle; add water, salt and bay leaves. Bring to a boil; reduce heat; simmer turkey in 6 gallons water 3 to 4 hours or until tender. CCP: Internal temperature must reach 165 F. or higher for 15 seconds. Drain. Cool. Dice cooked turkey. CCP: Refrigerate at 41 F. or lower for use in Step 6.
2. Lightly spray stock pot or steam jacketed kettle with non-stick cooking spray. Cook onions, celery, and garlic until tender in stock pot or steam-jacketed kettle.
3. Add flour, salt, pepper, curry powder, ginger, hot sauce and Worcestershire sauce to vegetable mixture; stir to blend.
4. Prepare chicken broth according to package directions; add gradually to vegetables and spices stirring constantly. Cook until thickened, about 1 minute.
5. Add apples and raisins. Cook 10 minutes or until apples are tender.
6. Add coconut and turkey to apple-vegetable mixture. Heat 20 minutes. CCP: Internal temperature must reach 165 F. or higher for 15 seconds.
7. CCP: Serve immediately or hold for service at 140 F. or higher.

MEAT, FISH, AND POULTRY No.L 045 00

STUFFED BEEF ROLLS

Yield 100   Portion 1 Roll

Calories	Carbohydrates	Protein	Fat	Cholesterol	Sodium	Calcium
373 cal	24 g	30 g	16 g	76 mg	790 mg	38 mg

Ingredient	Weight	Measure	Issue
BEEF,OVEN ROAST,TEMPERED	27 lbs		
COOKING SPRAY,NONSTICK	2 oz	1/4 cup 1/3 tbsp	
CELERY,FRESH,CHOPPED	2-1/2 lbs	2 qts 1-1/2 cup	3-3/8 lbs
ONIONS,FRESH,CHOPPED	1 lbs	2-7/8 cup	1-1/8 lbs
BREAD,WHITE,CUBED	3-2/3 lbs	3 gal	
BEEF BROTH		3 qts	
SEASONING,POULTRY	1/8 oz	1/3 tsp	
PEPPER,BLACK,GROUND	1/8 oz	1/4 tsp	
FLOUR,WHEAT,GENERAL PURPOSE	2-1/4 lbs	2 qts	
COOKING SPRAY,NONSTICK	2 oz	1/4 cup 1/3 tbsp	
BROWN GRAVY		1 gal 3-3/4 qts	

Method

1. Slice beef into 4 ounce slices, 1/4 inch thick.
2. Lightly spray pan with non-stick cooking spray. Cook celery and onions until tender.
3. Add celery and onions to cubed bread; toss lightly.
4. Prepare stock according to package directions; add poultry seasoning and pepper. Add to bread mixture; mix thoroughly. DO NOT OVERMIX.
5. Place 1/4 cup stuffing in the center of each beef slice; roll tightly around stuffing.
6. Lightly spray griddle with non-stick cooking spray. Dredge beef rolls in flour; grill on 350 F. griddle 3 to 5 minutes or until browned on all sides. Place 25 beef rolls in each pan.
7. Prepare 1-1/4 recipes Brown Gravy, Recipe No. O 016 00.
8. Pour about 1-3/4 quarts gravy over beef rolls in each steam table pan.
9. Cover. Using a convection oven, bake for 45 minutes at 300 F. on high fan, closed vent. CCP: Internal temperature must reach 165 F. or higher for 15 seconds. Hold for service at 140 F. or higher.

MEAT, FISH, AND POULTRY No.L 045 01

BEEF BROGUL

Yield 100   Portion   7 Ounces

Calories	Carbohydrates	Protein	Fat	Cholesterol	Sodium	Calcium
416 cal	34 g	35 g	16 g	81 mg	1174 mg	153 mg

Ingredient	Weight	Measure	Issue
BEEF,OVEN ROAST,TEMPERED	27 lbs		
MUSHROOMS,CANNED,STEMS & PIECES,CHOPPED,DRAINED	3 lbs	2 qts 3/4 cup	
CELERY,FRESH,CHOPPED	2-1/2 lbs	2 qts 1-1/2 cup	3-3/8 lbs
SHORTENING	2-3/8 oz	1/4 cup 1-2/3 tbsp	
ONIONS,FRESH,CHOPPED	1 lbs	2-7/8 cup	1-1/8 lbs
CHEESE,PARMESAN,GRATED	1-1/3 lbs	1 qts 2 cup	
BREAD,WHITE,SLICED	3-2/3 lbs	3 gal	
BEEF BROTH		3 qts	
SEASONING,POULTRY	1/8 oz	1/3 tsp	
PEPPER,BLACK,GROUND	1/8 oz	1/4 tsp	
SHORTENING	14-1/2 oz	2 cup	
FLOUR,WHEAT,GENERAL PURPOSE	2-1/4 lbs	2 qts	
TOMATO PASTE,CANNED	18-1/2 lbs	2 gal	
OREGANO,CRUSHED	3/8 oz	2-2/3 tbsp	
BASIL,SWEET,WHOLE,CRUSHED	1/4 oz	1 tbsp	
THYME,GROUND	1/8 oz	1 tbsp	
WATER,COLD	6-1/4 lbs	3 qts	
WATER,BOILING	10-1/2 lbs	1 gal 1 qts	

Method

1. Slice beef into 4 ounce slices, 1/4 inch thick.
2. Chop canned, drained mushrooms; saute with celery and onions.
3. Add grated Parmesan cheese to mixture, and add to cubed bread; toss lightly.
4. Prepare stock according to package directions. Add poultry seasoning and pepper. Add to bread mixture; mix lightly but thoroughly. DO NOT OVERMIX.
5. Place 1/3 cup, stuffing in center of each beef slice; roll tightly around stuffing.
6. Dredge beef rolls in flour; grill on well greased griddle at 350 F. for 3 to 5 minutes or until browned on all sides. Place 25 beef rolls in each pan.
7. Combine tomato paste and crushed oregano, crushed sweet basil and ground thyme. Add cold water; mix until smooth. Add to boiling water, stirring constantly. Cook at medium heat until sauce comes to a boil. Simmer 1 minute, stirring as necessary.
8. Pour 2-1/4 quarts sauce over beef rolls in each pan.
9. Cover. Using a convection oven, bake at 300 F. for 45 minutes or until tender on high fan, closed vent. CCP: Internal temperature must reach 165 F. or higher for 15 seconds. Hold for service at 140 F. or higher.

Notes

1. In Step 1, beef, boneless, frozen, top round will provide the most uniform slice and portion.

MEAT, FISH, AND POULTRY No.L 046 00

BEEF AND BEAN TOSTADAS

Yield 100 **Portion** 2 Tostadas

Calories	Carbohydrates	Protein	Fat	Cholesterol	Sodium	Calcium
539 cal	45 g	32 g	26 g	102 mg	835 mg	386 mg

Ingredient	Weight	Measure	Issue
BEEF,GROUND,BULK,RAW,90% LEAN	14-1/2 lbs		
SALT	1-1/4 oz	2 tbsp	
PEPPER,RED,GROUND	<1/16th oz	1/8 tsp	
CUMIN,GROUND	1/8 oz	3/8 tsp	
GARLIC POWDER	1/8 oz	1/4 tsp	
CHILI POWDER,LIGHT,GROUND	1-3/8 oz	1/4 cup 1-2/3 tbsp	
FLOUR,WHEAT,GENERAL PURPOSE	6-5/8 oz	1-1/2 cup	
BEANS,REFRIED	20 lbs	2 gal 1 qts	
LETTUCE,ICEBERG,FRESH,CHOPPED	5-3/8 lbs	2 gal 3-1/8 qts	5-3/4 lbs
TOMATOES,FRESH,CHOPPED	7-1/8 lbs	1 gal 1/2 qts	7-1/4 lbs
ONIONS,FRESH,CHOPPED	2 lbs	1 qts 1-1/2 cup	2-1/8 lbs
TORTILLAS,CORN,6 INCH	11-1/2 lbs	200 each	
CHEESE,CHEDDAR,GRATED	6-1/2 lbs	1 gal 2-1/2 qts	
SAUCE,SALSA	3-3/4 lbs	1 qts 3 cup	
SOUR CREAM	6-3/8 lbs	3 qts 1/2 cup	

Method

1. Cook beef until beef loses its pink color; stir to break apart. Drain fat. CCP: Internal temperature must reach 155 F. or higher for 15 seconds. Combine flour, chili powder, salt, garlic powder, cumin and red pepper. Add to beef. Cook 5 minutes, stirring occasionally. CCP: Hold at 140 F. or higher for use in Step 5.
2. Place refried beans in a steam jacketed kettle or stockpot. Cover; heat slowly for 15 to 20 minutes or until steaming, stirring frequently to prevent sticking.
3. Shred lettuce and chop tomatoes and onions; cover.
4. Place tortillas in rows 4 by 6 on sheet pans. Bake 6 to 8 minutes or until tortillas are lightly toasted or browned on low fan, open vent.
5. Use batch preparation methods when assembling tostadas. Tostadas may be served with 1 recipe Guacamole (Recipe No. M 052 00) per 100 portions. Follow assembly instructions. Arrange each tostada as follows: 1. One tostada shell 2. 2 tbsp refried beans, spread evenly 3. 2 tbsp taco filling, spread evenly 4. 2 tbsp shredded cheese 5. 2 tbsp shredded lettuce 6. 1 tbsp chopped tomatoes 7. 1 tsp chopped onions 8. 1 tbsp salsa 9. 1 tbsp sour cream

MEAT, FISH, AND POULTRY No. L 047 00

BEEF PIE WITH BISCUIT TOPPING (CANNED BEEF)

Yield 100   Portion 1 Cup

Calories	Carbohydrates	Protein	Fat	Cholesterol	Sodium	Calcium
583 cal	42 g	42 g	27 g	105 mg	625 mg	93 mg

Ingredient	Weight	Measure	Issue
CARROTS,FRESH,SLICED	5 lbs	1 gal 1/2 qts	6-1/8 lbs
ONIONS,FRESH,QUARTERED	4-5/8 lbs	1 gal 5/8 qts	5-1/8 lbs
POTATOES,FRESH,PEELED,CUBED	6-1/4 lbs	1 gal 5/8 qts	7-3/4 lbs
WATER	16-3/4 lbs	2 gal	
BEEF,CANNED,CHUNKS,W/NATURAL JUICE,DRAINED	29 lbs	6 gal 2-1/2 qts	
COOKING SPRAY,NONSTICK	2 oz	1/4 cup 1/3 tbsp	
SHORTENING,VEGETABLE,MELTED	14-1/2 oz	2 cup	
FLOUR,WHEAT,GENERAL PURPOSE	1-2/3 lbs	1 qts 2 cup	
RESERVED STOCK	20-7/8 lbs	2 gal 2 qts	
PEPPER,BLACK,GROUND	1/8 oz	1/8 tsp	
PEAS,GREEN,CANNED,DRAINED	6-1/8 lbs	1 gal 1/8 qts	
BAKING POWDER BISCUITS (BISCUIT MIX) (1 BISC)	3-5/8 kg	100 unit	

Method

1 Simmer carrots 10 to 15 minutes. Add onions and potatoes. Cook 20 minutes or until just tender.
2 Drain vegetables. Reserve liquid for use in Step 5; vegetables for use in Step 6.
3 Drain beef chunks; reserve juices for use in Step 5.
4 Lightly spray pan with non-stick cooking spray. Combine shortening or salad oil and flour; brown lightly on low heat.
5 Add beef juices, vegetable liquid or water gradually. Cook 15 minutes or until thickened. Stir constantly. Add pepper.
6 Add beef; cook until simmering. Add vegetables and simmer. CCP: Internal temperature must reach 165 F. or higher for 15 seconds. Stew must reach 180 F. or raw dough on bottom of biscuits will result.
7 Pour about 6-1/2 quarts meat mixture into each pan.
8 Add 3 cups peas to each pan. Stir lightly.
9 Prepare 1 recipe Baking Powder Biscuits, Recipe No. D 001 01. Place 25 biscuits on top of mixture in each pan.
10 Using a convection oven, bake at 400 F. for 10-15 minutes on low fan, open vent or until biscuits are browned. (Stew must reach 180 F. or raw dough on bottom of biscuits will result.) CCP: Internal temperature must reach 155 F. or higher for 15 seconds. Hold for service at 140 F. or higher.

MEAT, FISH, AND POULTRY No.L 048 00

BAKED CHICKEN AND RICE (COOKED DICED)

Yield 100   Portion 1 Cup

Calories	Carbohydrates	Protein	Fat	Cholesterol	Sodium	Calcium
341 cal	34 g	28 g	9 g	79 mg	1282 mg	69 mg

Ingredient

Ingredient	Weight	Measure	Issue
CHICKEN BROTH		3 gal	
WATER	15-2/3 lbs	1 gal 3-1/2 qts	
RICE,LONG GRAIN	5-3/4 lbs	3 qts 2 cup	
SALT	3-1/8 oz	1/4 cup 1-1/3 tbsp	
PEPPER,BLACK,GROUND	1/3 oz	1 tbsp	
GARLIC POWDER	1/3 oz	1 tbsp	
WATER,WARM	6 lbs	2 qts 3-1/2 cup	
MILK,NONFAT,DRY	5-3/8 oz	2-1/4 cup	
WATER,COLD	4-1/8 lbs	2 qts	
FLOUR,WHEAT,GENERAL PURPOSE	2-1/4 lbs	2 qts	
CHICKEN,COOKED,DICED	18 lbs		
BREADCRUMBS,DRY,GROUND,FINE	1-3/8 lbs	1 qts 2 cup	
BUTTER,MELTED	9 oz	1-1/8 cup	
PAPRIKA,GROUND	3/4 oz	3 tbsp	

Method

1. Combine broth, water, rice, salt, pepper, and garlic powder in a steam jacketed kettle or stockpot; bring to a boil. Cover tightly; reduce heat; simmer 20 minutes. Do not stir. There will be excess liquid in cooked rice.
2. Reconstitute milk in warm water. Stir milk into cooked rice.
3. Blend flour and cold water together to make a smooth slurry. Add slurry to rice mixture stirring constantly. Bring to a boil. Cover; reduce heat; simmer 10 minutes or until thickened, stirring frequently to prevent sticking.
4. Stir chicken gently into thickened rice mixture.
5. Pour 1-3/4 gal chicken and rice mixture into each ungreased steam table pan.
6. Combine crumbs, paprika, and margarine or butter. Sprinkle 1-1/2 cups crumb mixture evenly over chicken and rice in each pan.
7. Using a convection oven, bake 25 minutes or until browned at 325 F., on high fan, open vent. CCP: Internal temperature must reach 165 F. or higher for 15 seconds. Hold for service at 140 F. or higher.

MEAT. FISH. AND POULTRY No.L 048 01

BAKED CHICKEN AND RICE (CANNED CHICKEN)

Yield 100   Portion 1 Cup

Calories	Carbohydrates	Protein	Fat	Cholesterol	Sodium	Calcium
323 cal	34 g	21 g	11 g	56 mg	1400 mg	70 mg

Ingredient	Weight	Measure	Issue
CHICKEN,BONED,CANNED,PIECES	18 lbs	1 gal 3-1/8 qts	
CHICKEN BROTH		3 gal	
WATER,COLD	15-2/3 lbs	1 gal 3-1/2 qts	
RICE,LONG GRAIN	5-3/4 lbs	3 qts 2 cup	
SALT	1-1/4 oz	2 tbsp	
PEPPER,BLACK,GROUND	1/3 oz	1 tbsp	
GARLIC POWDER	1/3 oz	1 tbsp	
MILK,NONFAT,DRY	5-3/8 oz	2-1/4 cup	
WATER,WARM	4-1/8 lbs	2 qts	
WATER,COLD	4-1/8 lbs	2 qts	
FLOUR,WHEAT,GENERAL PURPOSE	1-7/8 lbs	1 qts 3 cup	
BREADCRUMBS,DRY,GROUND,FINE	1-7/8 lbs	2 qts	
BUTTER,MELTED	1 lbs	2 cup	
PAPRIKA,GROUND	3/4 oz	3 tbsp	

Method

1. Cut chicken into 1-inch pieces.
2. Combine broth, water, rice, salt, pepper, and garlic powder in a steam jacketed kettle or stockpot; bring to a boil. Cover tightly; reduce heat; simmer 20 minutes. Do not stir. There will be excess liquid in cooked rice.
3. Reconstitute milk in warm water. Stir milk into cooked rice.
4. Blend flour and cold water together to make a smooth slurry. Add slurry to rice mixture stirring constantly. Bring to a boil. Cover; reduce heat; simmer 10 minutes or until thickened, stirring frequently to prevent sticking.
5. Stir chicken gently into thickened rice mixture.
6. Pour 1-3/4 gal chicken and rice mixture into each ungreased steam table pan.
7. Combine crumbs, paprika and butter or margarine. Sprinkle 1-1/2 cups crumb mixture evenly over chicken and rice in each pan.
8. Using a convection oven, bake for 25 minutes at 325 F. or until lightly browned on high fan, open vent. CCP: Internal temperature must reach 165 F. or higher for 15 seconds. Hold for service at 140 F. or higher.

MEAT, FISH, AND POULTRY No.L 049 00

TURKEY CUTLET

Yield 100 **Portion** 4-1/2 Ounces

Calories	Carbohydrates	Protein	Fat	Cholesterol	Sodium	Calcium
319 cal	21 g	30 g	12 g	81 mg	987 mg	73 mg

Ingredient	Weight	Measure	Issue
TURKEY,BNLS,WHITE AND DARK MEAT	32-1/2 lbs		
FLOUR,WHEAT,GENERAL PURPOSE	2-1/4 lbs	2 qts	
SEASONING,POULTRY	1/2 oz	1/4 cup 1/3 tbsp	
PEPPER,BLACK,GROUND	1/8 oz	1/3 tsp	
BREADCRUMBS,DRY,GROUND,FINE	4-1/4 lbs	1 gal 1/2 qts	
PAPRIKA,GROUND	1 oz	1/4 cup 1/3 tbsp	
EGG WHITES,FROZEN,THAWED	2-1/2 lbs	1 qts 5/8 cup	
COOKING SPRAY,NONSTICK	2 oz	1/4 cup 1/3 tbsp	

Method

1. Thaw turkey under refrigeration at 41 F. or lower.
2. Dredge slices in mixture of flour, pepper and poultry seasoning; shake off excess.
3. Combine bread crumbs and paprika.
4. Dip floured slices into egg whites. Dredge in seasoned bread crumbs until well coated; shake off excess.
5. Lightly spray sheet pans with non-stick cooking spray. Place 17 cutlets in a single layer on each pan, spray breasts lightly with cooking spray.
6. Using a convection oven, bake at 325 F. on high fan, closed vent for 15 minutes. Turn cutlets, bake 15 minutes more or until golden brown. CCP: Internal temperature must reach 165 F. or higher for 15 seconds. Hold for service at 140 F. or higher.

MEAT, FISH, AND POULTRY No. L 050 00

CHALUPA

Yield 100 **Portion** 1 Cup

Calories	Carbohydrates	Protein	Fat	Cholesterol	Sodium	Calcium
380 cal	27 g	36 g	14 g	98 mg	543 mg	91 mg

Ingredient	Weight	Measure	Issue
BEANS,PINTO,DRY	8-1/2 lbs	1 gal 1 qts	
WATER,COLD	25-1/8 lbs	3 gal	
PORK CUBES,RAW	32 lbs		
WATER	41-3/4 lbs	5 gal	
ONIONS,FRESH,CHOPPED	1-3/4 lbs	1 qts 1 cup	2 lbs
GARLIC POWDER	1/2 oz	1 tbsp	
SALT	3-3/4 oz	1/4 cup 2-1/3 tbsp	
CHILI POWDER,DARK,GROUND	5-1/4 oz	1-1/4 cup	
CUMIN,GROUND	2-1/4 oz	1/2 cup 2-2/3 tbsp	
OREGANO,CRUSHED	3-3/4 oz	1-1/2 cup	
PEPPERS,JALAPENOS,CANNED,CHOPPED	2-2/3 oz	1/2 cup 1 tbsp	

Method

1. Pick over beans, removing discolored beans and foreign matter. Wash beans thoroughly.
2. Cover with water; bring to a boil; boil 2 minutes; turn off heat.
3. Cover; let soak 1 hour. Drain beans.
4. Combine pork, water, onions, garlic, salt, chili powder, cumin, oregano, and jalapeno peppers with beans in steam-jacketed kettle or stock pot; simmer 1-1/2 to 2 hours or until beans are tender. DO NOT COVER. Stir occasionally. CCP: Internal temperature must reach 145 F. or higher for 15 seconds. Hold at 140 F. or higher for service.

Notes

1. Chalupas can be served with shredded lettuce, chopped onions, chopped tomatoes, sour cream.

MEAT, FISH, AND POULTRY No. L 051 00

CHICKEN PARMESAN (PRECOOKED FILLET)

Yield 100 Portion 7 Ounces

Calories	Carbohydrates	Protein	Fat	Cholesterol	Sodium	Calcium
559 cal	25 g	29 g	38 g	77 mg	1224 mg	181 mg

Ingredient	Weight	Measure	Issue
CHICKEN FILLET,BREADED,PRECOOKED,FROZEN,5 OZ	32 lbs		
CHEESE,MOZZARELLA	3 lbs	2 qts 2-3/8 cup	
SAUCE,PIZZA,CANNED	16-7/8 lbs	1 gal 3 qts	
CHEESE,PARMESAN,GRATED	7 oz	2 cup	

Method

1. Place 15 fillets on each sheet pan. Using a convection oven, bake at 375 F. 12 to 14 minutes on high fan, closed vent or until thoroughly heated.
2. Cut cheese slices in half. Place 1/2 slice cheese on each fillet.
3. Heat sauce to a simmer. Pour about 1 quart over each sheet pan.
4. Sprinkle about 4-1/2 tablespoons parmesan cheese over fillets in each pan.
5. Using a convection oven, bake at 375 F. 5-10 minutes or until cheese is melted on high fan, closed vent. CCP: Hold for service at 140 F. or higher.

MEAT. FISH. AND POULTRY No.L 051 01

CHICKEN PARMESAN (BREAST BONELESS)

Yield 100　　　　　　　　　　　　　　　　　　　　Portion　5 Ounces

Calories	Carbohydrates	Protein	Fat	Cholesterol	Sodium	Calcium
261 cal	7 g	38 g	8 g	100 mg	319 mg	178 mg

Ingredient	Weight	Measure	Issue
CHICKEN,BREAST,BNLS/SKNLS,5 OZ	31-1/4 lbs		
COOKING SPRAY,NONSTICK	2 oz	1/4 cup 1/3 tbsp	
CHEESE,MOZZARELLA,PART SKIM	3 lbs		
SAUCE,PIZZA,CANNED	16-7/8 lbs	1 gal 3 qts	
BASIL,SWEET,WHOLE,CRUSHED	1/2 oz	3-1/3 tbsp	
PEPPER,BLACK,GROUND	1/4 oz	1 tbsp	
OREGANO,CRUSHED	1/2 oz	3-1/3 tbsp	
GARLIC POWDER	1/8 oz	1/4 tsp	
CHEESE,PARMESAN,GRATED	7 oz	2 cup	

Method

1. Wash chicken breasts thoroughly under cold running water. Drain well.
2. Place chicken breasts in each lightly sprayed sheet pan.
3. Lightly spray chicken breasts in each pan with cooking spray.
4. Using a convection oven, bake 8 to 10 minutes at 325 F. on high fan, closed vent.
5. Place 1 oz mozzarella cheese on each chicken breast.
6. Add herbs to sauce; stir. Ladle about 1/4 cup of sauce over each chicken breast.
7. Sprinkle about 1/3 cup parmesan cheese evenly over chicken breasts in each pan.
8. Using a convection oven, bake at 325 F. an additional 4-6 minutes cheese melts on high fan, closed vent. CCP: Internal temperature must reach 165 F. or higher for 15 seconds. Hold for service at 140 F. or higher.

MEAT, FISH, AND POULTRY No.L 052 00

CREAMED CHIPPED BEEF

Yield 100 **Portion** 6 Ounces

Calories	Carbohydrates	Protein	Fat	Cholesterol	Sodium	Calcium
169 cal	12 g	13 g	7 g	15 mg	1219 mg	110 mg

Ingredient	Weight	Measure	Issue
BEEF,CHIPPED,DRIED,CHOPPED	7 lbs		
WATER,WARM	8-1/3 lbs	1 gal	
MILK,NONFAT,DRY	1-3/4 lbs	3 qts	
WATER,WARM	31-1/3 lbs	3 gal 3 qts	
MARGARINE,SOFTENED	1-1/2 lbs	3-1/8 cup	
FLOUR,WHEAT,GENERAL PURPOSE	2-1/4 lbs	2 qts	
PEPPER,BLACK,GROUND	1/2 oz	2 tbsp	

Method

1 Separate dried beef slices, cut into 1-inch slices.
2 Place beef in 190 F. water. Soak 5 minutes. Drain thoroughly.
3 Reconstitute milk. Heat to just below boiling. DO NOT BOIL.
4 Combine butter or margarine with flour and pepper; add to milk, stirring constantly. Cook 5 minutes until thickened.
5 Add beef to sauce; blend well. CCP: Internal temperature must reach 145 F. or higher for 15 seconds. Hold for service at 140 F. or higher.

MEAT. FISH. AND POULTRY No.L 053 00

BEEF STROGANOFF

Yield 100   Portion 6 Ounces

Calories	Carbohydrates	Protein	Fat	Cholesterol	Sodium	Calcium
250 cal	8 g	31 g	9 g	93 mg	844 mg	48 mg

Ingredient	Weight	Measure	Issue
BEEF,SWISS STEAK,LEAN,RAW,THAWED	30 lbs		
COOKING SPRAY,NONSTICK	2 oz	1/4 cup 1/3 tbsp	
MUSHROOMS,CANNED,STEMS & PIECES,INCL LIQUIDS	3-1/8 lbs	2 qts 1 cup	
ONIONS,FRESH,CHOPPED	3-1/2 lbs	2 qts 2 cup	3-7/8 lbs
BEEF BROTH		1 gal 1 qts	
FLOUR,WHEAT,GENERAL PURPOSE	1-1/8 lbs	1 qts	
SALT	3-3/4 oz	1/4 cup 2-1/3 tbsp	
PAPRIKA,GROUND	1-1/4 oz	1/4 cup 1-1/3 tbsp	
PEPPER,BLACK,GROUND	1/4 oz	1 tbsp	
GARLIC POWDER	1/3 oz	1 tbsp	
MILK,NONFAT,DRY	2-3/8 oz	1 cup	
WATER,WARM	5 lbs	2 qts 1-1/2 cup	
SOUR CREAM,LOW FAT	4 lbs	2 qts	

Method

1. Slice beef into strips about 1/2 inch wide. Lightly spray griddle with cooking spray. Grill beef strips 3 to 4 minutes or until lightly browned while tossing intermittently.
2. Drain mushrooms. Reserve 1 quart mushroom liquid. Set mushrooms aside for use in Step 7.
3. Cook onions in a lightly sprayed steam-jacketed kettle or stock pot 8 to 10 minutes, stirring constantly.
4. Add beef broth to cooked onions; stir to blend. Bring to a boil; reduce heat to a simmer.
5. Blend flour and mushroom liquid, stirring to make a slurry. Add slurry to broth and onions, stirring constantly. Bring to a boil. Cover; reduce heat ; simmer 10 minutes or until thickened, stirring frequently.
6. Reconstitute milk with warm water. Add salt, paprika, pepper and garlic powder; stir milk mixture into thickened broth. Bring to a boil. Cover; reduce heat; simmer 2 minutes.
7. Stir beef strips and mushrooms gently into thickened sauce. Heat to a simmer. CCP: Temperature must reach 145 F. or higher for 15 seconds. Remove from heat.
8. Carefully blend sour cream with 1 quart of sauce mixture. Combine with remaining sauce mixture. Mix well.
9. Pour beef stroganoff into each ungreased pan. CCP: Hold for service at 140 F. or higher. Serve with Boiled Noodles or Steamed Rice.

MEAT, FISH, AND POULTRY No. L 053 01

BEEF STROGANOFF (CREAM OF MUSHROOM SOUP)

Yield 100   Portion 3/4 Cup

Calories	Carbohydrates	Protein	Fat	Cholesterol	Sodium	Calcium
292 cal	8 g	31 g	14 g	93 mg	570 mg	53 mg

Ingredient	Weight	Measure	Issue
MUSHROOMS,CANNED,SLICED,INCL LIQUIDS	3-1/8 lbs	2 qts 1 cup	
SOUP,CONDENSED,CREAM OF MUSHROOM	14-3/8 lbs	1 gal 2-1/2 qts	
PAPRIKA,GROUND	1-1/4 oz	1/4 cup 1-1/3 tbsp	
PEPPER,BLACK,GROUND	1/8 oz	1/8 tsp	
GARLIC POWDER	1/4 oz	1/3 tsp	
BEEF,SWISS STEAK,LEAN,RAW,THAWED	30 lbs		
COOKING SPRAY,NONSTICK	2 oz	1/4 cup 1/3 tbsp	
ONIONS,FRESH,CHOPPED	3-1/2 lbs	2 qts 2 cup	3-7/8 lbs
SOUR CREAM,LOW FAT	4 lbs	2 qts	

Method

1. Drain mushrooms; reserve liquid for use in Step 2 and mushrooms for Step 6.
2. Combine cream of mushroom soup with paprika, pepper, and garlic powder; stir well; add reserved mushroom liquid; stir well.
3. Slice beef into strips about 1/2-inch wide.
4. Spray griddle with non-stick cooking spray. Brown strips 5 minutes turning frequently.
5. Place about 11 pounds 3 ounces strips in each pan.
6. Add about 1 quart mushrooms and 1-1/4 quarts onions to meat in each pan; stir well.
7. Add about 1 gallon sauce to meat in each pan. Stir well.
8. Cover; Using a convection oven, bake 1 hour 15 minutes at 325 F. or until tender on high fan, closed vent. CCP: Internal temperature must reach 145 F. or higher for 15 seconds.
9. Remove from oven. Skim off excess fat.
10. Add 1 quart sour cream to each pan, stirring to blend. Heat. CCP: Hold for service at 140 F. or higher. Serve with Boiled Noodles or Steamed Rice.

Notes

1. In Step 5, 33-3/4 pounds beef fajita strips may be used.

MEAT. FISH. AND POULTRY No.L 053 02

HAMBURGER STROGANOFF

Yield 100   Portion 6 Ounces

Calories	Carbohydrates	Protein	Fat	Cholesterol	Sodium	Calcium
330 cal	8 g	32 g	18 g	113 mg	862 mg	52 mg

Ingredient	Weight	Measure	Issue
BEEF,GROUND,BULK,RAW,90% LEAN	30 lbs		
ONIONS,FRESH,CHOPPED	3-1/2 lbs	2 qts 2 cup	3-7/8 lbs
MUSHROOMS,CANNED,STEMS & PIECES,INCL LIQUIDS	3-1/8 lbs	2 qts 1 cup	
BEEF BROTH		1 gal 1 qts	
FLOUR,WHEAT,GENERAL PURPOSE	1-1/8 lbs	1 qts	
SALT	3-3/4 oz	1/4 cup 2-1/3 tbsp	
PAPRIKA,GROUND	1-1/4 oz	1/4 cup 1-1/3 tbsp	
PEPPER,BLACK,GROUND	1/4 oz	1 tbsp	
GARLIC POWDER	1/3 oz	1 tbsp	
MILK,NONFAT,DRY	2-3/8 oz	1 cup	
WATER,WARM	5 lbs	2 qts 1-1/2 cup	
SOUR CREAM,LOW FAT	4 lbs	2 qts	

Method

1. Cook beef in a steam jacketed kettle or stock pot for 10 minutes, stirring to break apart.
2. Drain mushrooms. Reserve 1 quart mushroom liquid. Set mushrooms aside for use in Step 7.
3. Add beef broth to cooked onions and beef; stir to blend. Bring to a boil; reduce heat to a simmer.
4. Blend flour and mushroom liquid, stirring to make a slurry. Add slurry to broth and onions, stirring constantly. Bring to a boil. Cover; reduce heat; simmer 10 minutes or until thickened, stirring frequently.
5. Reconstitute milk with warm water. Add salt, paprika, pepper and garlic powder; stir milk mixture into thickened broth. Bring to a boil. Cover; reduce heat; simmer 2 minutes.
6. Stir beef, onions and mushrooms gently into thickened sauce. Heat to a simmer. CCP: Temperature must reach 155 F. or higher for 15 seconds. Remove from heat.
7. Carefully blend sour cream with 1 quart of sauce mixture. Combine with remaining sauce mixture. Mix well.
8. Pour stroganoff into each ungreased pan. CCP: Hold for service at 140 F. or higher. Serve with Boiled Noodles or Steamed Rice.

MEAT, FISH, AND POULTRY No.L 053 03

GROUND TURKEY STROGANOFF

Yield 100   **Portion** 6 Ounces

Calories	Carbohydrates	Protein	Fat	Cholesterol	Sodium	Calcium
244 cal	8 g	27 g	11 g	91 mg	913 mg	71 mg

Ingredient	Weight	Measure	Issue
TURKEY,GROUND,90% LEAN,RAW	30 lbs		
COOKING SPRAY,NONSTICK	2 oz	1/4 cup 1/3 tbsp	
MUSHROOMS,CANNED,STEMS & PIECES,INCL LIQUIDS	3-1/8 lbs	2 qts 1 cup	
ONIONS,FRESH,CHOPPED	3-1/2 lbs	2 qts 2 cup	3-7/8 lbs
BEEF BROTH		1 gal 1 qts	
FLOUR,WHEAT,GENERAL PURPOSE	1-1/8 lbs	1 qts	
SALT	3-3/4 oz	1/4 cup 2-1/3 tbsp	
PAPRIKA,GROUND	1-1/4 oz	1/4 cup 1-1/3 tbsp	
PEPPER,BLACK,GROUND	1/4 oz	1 tbsp	
GARLIC POWDER	1/3 oz	1 tbsp	
MILK,NONFAT,DRY	2-3/8 oz	1 cup	
WATER,WARM	5 lbs	2 qts 1-1/2 cup	
SOUR CREAM,LOW FAT	4 lbs	2 qts	

Method

1. Lightly spray griddle with cooking spray. Grill turkey 3 to 4 minutes or until lightly browned while tossing intermittently.
2. Drain mushrooms. Reserve 1 quart mushroom liquid. Set mushrooms aside for use in Step 7.
3. Cook onions in a lightly sprayed steam-jacketed kettle or stock pot 8 to 10 minutes, stirring constantly.
4. Add beef broth to cooked onions; stir to blend. Bring to a boil; reduce heat to a simmer.
5. Blend flour and mushroom liquid, stirring to make a slurry. Add slurry to broth and onions, stirring constantly. Bring to a boil. Cover; reduce heat ; simmer 10 minutes or until thickened, stirring frequently.
6. Reconstitute milk with warm water. Add salt, paprika, pepper and garlic powder; stir milk mixture into thickened broth. Bring to a boil. Cover; reduce heat; simmer 2 minutes.
7. Stir turkey and mushrooms gently into thickened sauce. Heat to a simmer. CCP: Temperature must reach 165 F. or higher for 15 seconds. Remove from heat.
8. Carefully blend sour cream with 1 quart of sauce mixture. Combine with remaining sauce mixture. Mix well.
9. Pour turkey stroganoff into each ungreased pan. CCP: Hold for service at 140 F. or higher. Serve with Boiled Noodles or Steamed Rice.

MEAT. FISH. AND POULTRY No.L 053 04

BEEF STROGANOFF (FAJITA STRIPS)

Yield 100   Portion   6 Ounces

Calories	Carbohydrates	Protein	Fat	Cholesterol	Sodium	Calcium
291 cal	8 g	35 g	12 g	103 mg	871 mg	51 mg

Ingredient	Weight	Measure	Issue
BEEF,FAJITA STRIPS	33-3/4 lbs		
COOKING SPRAY,NONSTICK	2 oz	1/4 cup 1/3 tbsp	
MUSHROOMS,CANNED,STEMS & PIECES,INCL LIQUIDS	3-1/8 lbs	2 qts 1 cup	
ONIONS,FRESH,CHOPPED	3-1/2 lbs	2 qts 2 cup	3-7/8 lbs
BEEF BROTH		1 gal 1 qts	
FLOUR,WHEAT,GENERAL PURPOSE	1-1/8 lbs	1 qts	
SALT	3-3/4 oz	1/4 cup 2-1/3 tbsp	
PAPRIKA,GROUND	1-1/4 oz	1/4 cup 1-1/3 tbsp	
PEPPER,BLACK,GROUND	1/4 oz	1 tbsp	
GARLIC POWDER	1/3 oz	1 tbsp	
MILK,NONFAT,DRY	2-3/8 oz	1 cup	
WATER,WARM	5 lbs	2 qts 1-1/2 cup	
SOUR CREAM,LOW FAT	4 lbs	2 qts	

Method

1. Lightly spray griddle with cooking spray. Grill beef strips 3 to 4 minutes or until lightly browned while tossing intermittently.
2. Drain mushrooms. Reserve 1 quart mushroom liquid. Set mushrooms aside for use in Step 7.
3. Cook onions in a lightly sprayed steam-jacketed kettle or stock pot 8 to 10 minutes, stirring constantly.
4. Add beef broth to cooked onions; stir to blend. Bring to a boil; reduce heat to a simmer.
5. Blend flour and mushroom liquid, stirring to make a slurry. Add slurry to broth and onions, stirring constantly. Bring to a boil. Cover; reduce heat; simmer 10 minutes or until thickened, stirring frequently.
6. Reconstitute milk with warm water. Add salt, paprika, pepper and garlic powder; stir milk mixture into thickened broth. Bring to a boil. Cover; reduce heat; simmer 2 minutes.
7. Stir beef strips and mushrooms gently into thickened sauce. Heat to a simmer. CCP: Temperature must reach 145 F. or higher for 15 seconds. Remove from heat.
8. Carefully blend sour cream with 1 quart of sauce mixture. Combine with remaining sauce mixture. Mix well.
9. Pour beef stroganoff into each ungreased pan. CCP: Hold for service at 140 F. or higher. Serve with Boiled Noodles or Steamed Rice.

MEAT. FISH. AND POULTRY No.L 054 00

STEAK RANCHERO

Yield 100   Portion   4-1/2 Ounces

Calories	Carbohydrates	Protein	Fat	Cholesterol	Sodium	Calcium
296 cal	12 g	37 g	10 g	108 mg	432 mg	18 mg

Ingredient

Ingredient	Weight	Measure	Issue
BEEF,SWISS STEAK,LEAN,RAW,THAWED	37-1/2 lbs		
COOKING SPRAY,NONSTICK	3/4 oz	1 tbsp	
PEPPERS,GREEN,FRESH,CHOPPED	4 lbs	3 qts	4-3/4 lbs
ONIONS,FRESH,CHOPPED	3-1/8 lbs	2 qts 1 cup	3-1/2 lbs
SALT	1-7/8 oz	3 tbsp	
CHILI POWDER,DARK,GROUND	2-1/8 oz	1/2 cup	
PAPRIKA,GROUND	1 oz	1/4 cup 1/3 tbsp	
GARLIC POWDER	5/8 oz	2 tbsp	
CUMIN,GROUND	1/8 oz	1/3 tsp	
SOUP,CONDENSED,TOMATO	6-1/4 lbs	2 qts 3-1/4 cup	
WATER	10-1/2 lbs	1 gal 1 qts	
FLOUR,WHEAT,GENERAL PURPOSE	8 oz	1-3/4 cup	
WATER,COLD	1 lbs	2 cup	

Method

1. Lightly spray griddle with non-stick cooking spray. Grill steaks 5 minutes on one side and 4 minutes on the other side.
2. Evenly shingle 25 steaks into each ungreased steam table pan.
3. Cook onions and peppers in a lightly sprayed steam jacketed kettle or stock pot 8 to 10 minutes stirring constantly.
4. Add water, tomato soup, salt, chili powder, paprika, cumin and garlic powder to cooked onions and peppers; stir to blend. Bring to a boil; reduce heat to a simmer.
5. Blend flour and cold water stirring to make a slurry. Add slurry to tomato soup mixture stirring constantly to make Ranchero Sauce. Bring to a boil. Cover; reduce heat; simmer 2 minutes or until thickened, stirring frequently.
6. Pour Ranchero Sauce evenly over steaks in each pan.
7. Cover. Using a convection oven, bake 2 hours at 325 F. or until tender, on high fan, closed vent. CCP: Internal temperature must reach 145 F. or higher for 15 seconds. Hold for service at 140 F. or higher.

Notes

1. In Step 1, 31-1/8 lb (100-5 oz.) cube steaks may be substituted. In Step 7, reduce baking time to 1-1/2 hours.

MEAT, FISH, AND POULTRY No. L 055 00

BEEF CORDON BLEU

Yield 100   Portion 5 Ounces

Calories	Carbohydrates	Protein	Fat	Cholesterol	Sodium	Calcium
369 cal	9 g	32 g	22 g	128 mg	727 mg	113 mg

Ingredient	Weight	Measure	Issue
BEEF,OVEN ROAST,TEMPERED	25 lbs		
HAM,COOKED,1 OZ SLICE	3-1/8 lbs		
CHEESE,AMERICAN,SLICED	2-5/8 lbs		
POTATO,WHITE,INSTANT,GRANULES	6-3/4 oz	1 qts	
MILK,NONFAT,DRY	3-1/4 oz	1-3/8 cup	
WATER	3-7/8 lbs	1 qts 3-1/2 cup	
EGGS,WHOLE,FROZEN	2 lbs	3-3/4 cup	
BREADCRUMBS	2-7/8 lbs	3 qts	
SALT	3 oz	1/4 cup 1 tbsp	
PEPPER,BLACK,GROUND	1/4 oz	1 tbsp	
SHORTENING	1-3/4 lbs	1 qts	

Method

1. Slice beef into 1/4-inch thick slices, 4 ounces per slice.
2. Slice ham and cheese in 1/2. Place 1/2 slice ham and 1/2 slice cheese on each slice of beef.
3. Fold beef slice in half, enclosing ham and cheese. Pound edges of beef together to seal.
4. Dredge beef in instant potato granules. Set aside for use in Step 6.
5. Reconstitute milk; combine with eggs.
6. Dip beef in milk and egg mixture. Drain.
7. Dredge in mixture of bread crumbs, salt, and pepper; shake off excess.
8. Fry on griddle at 350 F. for 3 minutes on each side or until golden brown. CCP: Internal temperature must reach 145 F. or higher for 15 seconds. CCP: Hold for service at 140 F. or higher.

MEAT. FISH. AND POULTRY No.L 056 00

SOUTHERN FRIED CATFISH FILLETS

Yield 100   Portion   4 Ounces

Calories	Carbohydrates	Protein	Fat	Cholesterol	Sodium	Calcium
250 cal	10 g	21 g	13 g	68 mg	248 mg	11 mg

Ingredient   Weight   Measure   Issue

Ingredient	Weight	Measure	Issue
FISH,CATFISH,FILLET	30 lbs		
CORN MEAL	1-7/8 lbs	1 qts 2 cup	
FLOUR,WHEAT,GENERAL PURPOSE	1-1/8 lbs	1 qts	
SALT	1-1/2 oz	2-1/3 tbsp	
PEPPER,BLACK,GROUND	1/3 oz	1 tbsp	

Method

1. Separate fillets; cut into 4-1/2 ounce portions, if necessary.
2. Dredge fish in mixture of cornmeal, flour, salt and pepper; shake off excess.
3. Fry at 365 F. about 4 minutes or until golden brown. CCP: Internal temperature must reach 145 F. or higher for 15 seconds.
4. Drain well in basket or on absorbent paper. CCP: Hold for service at 140 F. or higher.

MEAT, FISH, AND POULTRY No.L 057 00

TAMALE PIE (GROUND BEEF)

Yield 100   **Portion** 9-1/2 Ounces

Calories	Carbohydrates	Protein	Fat	Cholesterol	Sodium	Calcium
385 cal	36 g	24 g	16 g	71 mg	514 mg	152 mg

Ingredient	Weight	Measure	Issue
CORN MEAL	6-1/2 lbs	1 gal 1-1/3 qts	
WATER,BOILING	25-1/8 lbs	3 gal	
BEEF,GROUND,BULK,RAW,90% LEAN	16 lbs		
ONIONS,FRESH,CHOPPED	2 lbs	1 qts 1-5/8 cup	2-1/4 lbs
PEPPERS,GREEN,FRESH,CHOPPED	1 lbs	3 cup	1-1/4 lbs
TOMATOES,CANNED,CRUSHED,INCL LIQUIDS	13-1/4 lbs	1 gal 2 qts	
CORN,CANNED,WHOLE KERNEL,DRAINED	6-3/4 lbs	1 gal 2/3 qts	
OLIVES,RIPE,PITTED,SLICED,INCL LIQUIDS	3-1/2 lbs	3 qts	
CHILI POWDER,DARK,GROUND	4-1/4 oz	1 cup	
CUMIN,GROUND	1-1/4 oz	1/4 cup 2-1/3 tbsp	
GARLIC POWDER	1 oz	3-1/3 tbsp	
SALT	7/8 oz	1 tbsp	
PEPPER,RED,GROUND	1/4 oz	1 tbsp	
COOKING SPRAY,NONSTICK	2 oz	1/4 cup 1/3 tbsp	
CHEESE,CHEDDAR,GRATED	3 lbs	3 qts	

Method

1. Mix cornmeal; chili powder and salt together; gradually stir into boiling water. Bring to a boil.
2. Reduce heat; simmer 25 minutes, stirring frequently until a stiff paste is formed. Set aside for use in Step 5.
3. Cook beef with onions and peppers until beef loses its pink color, stirring to break apart. Drain or skim off excess fat.
4. Add tomatoes, corn, olives, chili powder, cumin, salt, garlic powder, and red pepper to beef mixture; simmer 15 minutes, stirring frequently.
5. Lightly spray each pan with non-stick cooking spray. Spread 2-1/3 cup cornmeal paste over bottom and sides of each pan to form a thin crust.
6. Pour 2 quarts meat mixture over crust in each pan.
7. Spread 4-2/3 cups cornmeal paste evenly over meat mixture in each pan.
8. Using a convection oven, bake at 325 F. 50 to 60 minutes on low fan, open vent; remove from oven. CCP: Internal temperature must reach 155 F. or higher for 15 seconds.
9. Sprinkle 1-1/2 cups cheese evenly over each pan.
10. Let stand 10 to 15 minutes to allow filling to firm and cheese to melt. CCP: Hold for service at 140 F. or higher. Cut 3 by 4.

MEAT, FISH, AND POULTRY No.L 057 01

HOT TAMALES WITH CHILI GRAVY

Yield 100   Portion 2 Each

Calories	Carbohydrates	Protein	Fat	Cholesterol	Sodium	Calcium
284 cal	16 g	12 g	20 g	47 mg	778 mg	51 mg

Ingredient   **Weight**   **Measure**   **Issue**
TAMALE,BEEF   28-2/3 lbs
CHILI GRAVY   1 gal 2-1/4 qts

Method

1 Use canned beef tamales or frozen beef tamales. Heat according to directions on container. CCP: Internal temperature must reach 165 F. or higher for 15 seconds.
2 Serve with Chili Gravy, Recipe No. O 016 03.

MEAT, FISH, AND POULTRY No. L 057 02

TAMALE PIZZA

Yield 100   Portion   5 Ounces

Calories	Carbohydrates	Protein	Fat	Cholesterol	Sodium	Calcium
336 cal	22 g	22 g	18 g	71 mg	676 mg	155 mg

Ingredient	Weight	Measure	Issue
CORN BREAD MIX	4-1/2 lbs	3 qts 1-3/8 cup	
COOKING SPRAY,NONSTICK	2 oz	1/4 cup 1/3 tbsp	
BEEF,GROUND,BULK,RAW,90% LEAN	16 lbs		
ONIONS,FRESH,CHOPPED	2-1/8 lbs	1 qts 2 cup	2-1/3 lbs
PEPPERS,GREEN,FRESH,CHOPPED	14-1/2 oz	2-3/4 cup	1-1/8 lbs
TOMATOES,CANNED,DICED,DRAINED	13-1/4 lbs	1 gal 2 qts	
GARLIC POWDER	1-1/8 oz	1/4 cup	
SALT	1 oz	1 tbsp	
CHILI POWDER,DARK,GROUND	4-1/4 oz	1 cup	
PEPPER,RED,GROUND	1/8 oz	1/4 tsp	
CUMIN,GROUND	5/8 oz	3 tbsp	
OLIVES,RIPE,PITTED,SLICED,INCL LIQUIDS	3-1/8 lbs	2 qts 2-3/4 cup	
CHEESE,CHEDDAR,GRATED	3 lbs	3 qts	

Method

1. Use canned cornbread mix. Prepare according to directions on container.
2. Spread 1-3/4 cups corn bread batter in a thin layer in each greased pan.
3. Cook beef with onions and peppers until beef loses its pink color, stirring to break apart. CCP: Internal temperature must reach 155 F. or higher for 15 seconds. Drain or skim off excess fat.
4. Add tomatoes, garlic, salt, chili powder, red pepper, cumin, and olives to beef mixture; simmer 15 minutes, stirring frequently.
5. Cover batter with 1-1/2 quart meat filling.
6. Sprinkle 1-1/2 cups cheese evenly over each pan. Using a convection oven, bake at 375 F. 15 minutes on low fan, open vent.
7. Let stand 10 to 15 minutes to allow filling to firm and cheese to melt.
8. Cut 3x4. CCP: Hold for service at 140 F. or higher.

Notes

1. 9-inch pie pans may be used. In each pan, use 1 cup corn bread batter, 3-2/3 cups meat filling and 7/8 cup shredded cheese. Cut into 8 wedges. EACH PORTION: 1 wedge or 5 ounces.

MEAT, FISH, AND POULTRY No.L 058 00

CHILI AND MACARONI (CANNED CHILI CON CARNE)

Yield 100   Portion   1-1/2 Cups

Calories	Carbohydrates	Protein	Fat	Cholesterol	Sodium	Calcium
420 cal	49 g	24 g	14 g	36 mg	1116 mg	216 mg

Ingredient	Weight	Measure	Issue
MACARONI NOODLES,ELBOW,DRY	7-3/8 lbs	2 gal	
SALT	1-2/3 oz	2-2/3 tbsp	
WATER,BOILING	43-7/8 lbs	5 gal 1 qts	
ONIONS,FRESH,CHOPPED	8-1/2 lbs	1 gal 2 qts	9-3/8 lbs
SHORTENING	7-1/4 oz	1 cup	
CHILI CON CARNE,CANNED,NO BEANS	33-3/4 lbs	3 gal 3 qts	
TOMATOES,CANNED,INCL LIQUIDS	13-3/4 lbs	1 gal 2 qts	
WATER	4-1/8 lbs	2 qts	
PEPPER,BLACK,GROUND	2/3 oz	3 tbsp	
CHILI POWDER,DARK,GROUND	1 oz	1/4 cup 1/3 tbsp	
CUMIN,GROUND	1/4 oz	1 tbsp	
CHEESE,CHEDDAR,GRATED	4 lbs	1 gal	

Method

1 Add macaroni slowly to salted water; bring to a boil, stirring occasionally. Boil 10 to 15 minutes. Drain; set aside for use in Step 5.
2 Saute onions in shortening for 10 minutes or until tender.
3 Heat chili to boiling. Reduce heat; skim off excess fat.
4 Add macaroni, onions, tomatoes, water, salt, pepper, chili powder and cumin; mix thoroughly.
5 Pour 2-1/4 gallons chili mixture into each pan.
6 Sprinkle 1 quart cheese over mixture in each pan.
7 Using a convection oven, bake 20 to 30 minutes at 350 F. on high fan, closed vent or until cheese is lightly browned and mixture is thoroughly heated. CCP: Internal temperature must reach 165 F. or higher for 15 seconds. Hold for service at 140 F. or higher.

MEAT, FISH, AND POULTRY   No.L 059 00

CHILI CON CARNE (WITH BEANS)

Yield 100   **Portion** 1 Cup

Calories	Carbohydrates	Protein	Fat	Cholesterol	Sodium	Calcium
222 cal	29 g	16 g	5 g	13 mg	866 mg	66 mg

Ingredient	**Weight**	**Measure**	**Issue**
ONIONS,FRESH,CHOPPED	4-1/4 lbs	3 qts	4-2/3 lbs
PEPPERS,GREEN,FRESH,CHOPPED	4 lbs	3 qts	4-3/4 lbs
CHILI CON CARNE,CANNED,NO BEANS	27 lbs	3 gal	
BEANS,KIDNEY,DARK RED,CANNED,DRAINED	18-3/4 lbs	3 gal	
CHILI POWDER,DARK,GROUND	1/2 oz	2 tbsp	

Method

1. Cook onions and peppers until onions are transparent in steam jacketed kettle or stock pot.
2. Remove excess solid fat from surface of chili con carne cans. Add chili con carne, beans and chili powder to onions and peppers.
3. Cover; bring to a boil; reduce heat; uncover; simmer 10 to 15 minutes or until thoroughly heated stirring frequently. CCP: Internal temperature must reach 165 F. or higher for 15 seconds. Hold for service at 140 F. or higher.

MEAT, FISH, AND POULTRY No.L 060 00

HAMBURGER PARMESAN

Yield 100   Portion 5 Ounces

Calories	Carbohydrates	Protein	Fat	Cholesterol	Sodium	Calcium
334 cal	16 g	28 g	17 g	101 mg	451 mg	203 mg

Ingredient   **Weight**   **Measure**   **Issue**

PIZZA SAUCE   1 gal
BEEF,GROUND,BULK,RAW,90% LEAN   18-3/4 lbs
FLOUR,WHEAT,GENERAL PURPOSE   1-3/8 lbs   1 qts 1 cup
MILK,NONFAT,DRY   1-3/4 oz   3/4 cup
WATER,WARM   2 lbs   3-3/4 cup
EGGS,WHOLE,FROZEN   9-5/8 oz   1-1/8 cup
BREADCRUMBS,DRY,GROUND,FINE   2-3/8 lbs   2 qts 2 cup
CHEESE,PARMESAN,GRATED   2-2/3 oz   3/4 cup
CHEESE,MOZZARELLA,SLICED   6-1/4 lbs   1 gal 2-1/4 qts

Method

1. Prepare Pizza Sauce, Recipe No. O 012 00. Set aside for use in Step 8.
2. Dredge patties in flour, shake off excess.
3. Reconstitute milk; and eggs. Stir to blend well.
4. Dip patties in milk and egg mixture. Drain well.
5. Combine crumbs and cheese. Dredge patties in crumb-cheese mixture; shake off excess.
6. Fry 2-1/2 minutes in 350 F. deep fat or until evenly browned. Drain well in basket or on absorbent paper. CCP: Internal temperature must reach 155 F. or higher for 15 seconds.
7. Place 20 patties in each sheet pan. Top each patty with 1 slice cheese.
8. Pour about 2 tablespoons of sauce evenly over patties in each pan.
9. Using a convection oven, bake at 325 F. 4 to 5 minutes or cheese is melted and patties are cooked, on high fan, closed vent. CCP: Hold at 140 F. or higher for service.

MEAT. FISH. AND POULTRY No.L 061 00

TEXAS HASH (GROUND BEEF)

Yield 100   Portion   1 Cup

Calories	Carbohydrates	Protein	Fat	Cholesterol	Sodium	Calcium
325 cal	24 g	27 g	13 g	85 mg	320 mg	57 mg

Ingredient	Weight	Measure	Issue
BEEF,GROUND,BULK,RAW,90% LEAN	24 lbs		
TOMATOES,CANNED,CRUSHED,INCL LIQUIDS	19-7/8 lbs	2 gal 1 qts	
ONIONS,FRESH,CHOPPED	5-1/4 lbs	3 qts 3 cup	5-7/8 lbs
PEPPERS,GREEN,FRESH,CHOPPED	4 lbs	3 qts	4-3/4 lbs
RICE,LONG GRAIN	3-2/3 lbs	2 qts 1 cup	
WATER	1 lbs	2 cup	
CHILI POWDER,DARK,GROUND	3-1/8 oz	3/4 cup	
SALT	1-1/4 oz	2 tbsp	
PEPPER,BLACK,GROUND	1/2 oz	2 tbsp	

Method

1. Cook beef until beef loses its pink color; stirring to break apart. Drain or skim off excess fat.
2. Add tomatoes, onions, peppers, rice, water, chili powder, salt and pepper to beef. Mix thoroughly. Heat to a simmer.
3. Place about 5 3/4 qts mixture into each steam table pan.
4. Cover pans; bake at 375 F. for 1 hour or until rice is tender. CCP: Internal temperature must reach 155 F. or higher for 15 seconds. Hold for service at 140 F. or higher.

MEAT. FISH. AND POULTRY No.L 061 01

TEXAS HASH (GROUND TURKEY)

Yield 100   Portion   1 Cup

Calories	Carbohydrates	Protein	Fat	Cholesterol	Sodium	Calcium
255 cal	24 g	23 g	8 g	68 mg	362 mg	73 mg

Ingredient

Ingredient	Weight	Measure	Issue
TURKEY,GROUND,90% LEAN,RAW	24 lbs		
TOMATOES,CANNED,CRUSHED,INCL LIQUIDS	19-7/8 lbs	2 gal 1 qts	
ONIONS,FRESH,CHOPPED	5-1/4 lbs	3 qts 3 cup	5-7/8 lbs
PEPPERS,GREEN,FRESH,CHOPPED	4 lbs	3 qts	4-3/4 lbs
RICE,LONG GRAIN	3-2/3 lbs	2 qts 1 cup	
WATER	1 lbs	2 cup	
CHILI POWDER,DARK,GROUND	3-1/8 oz	3/4 cup	
SALT	1-1/4 oz	2 tbsp	
PEPPER,BLACK,GROUND	1/2 oz	2 tbsp	

Method

1. Cook turkey until turkey loses its pink color; stirring to break apart. Drain or skim off excess fat.
2. Add tomatoes, onions, peppers, rice, water, chili powder, salt and pepper to turkey. Mix thoroughly. Heat to a simmer.
3. Place about 5 3/4 qts mixture into each pan.
4. Cover pans; using a convection oven, bake at 325 F. 1 hour or until rice is tender on high fan, closed vent. CCP: Internal temperature must reach 165 F. or higher for 15 seconds. Hold for service at 140 F. or higher.

MEAT, FISH, AND POULTRY No.L 062 00

YAKISOBA (BEEF AND SPAGHETTI)

Yield 100   Portion 1 Cup

Calories	Carbohydrates	Protein	Fat	Cholesterol	Sodium	Calcium
329 cal	31 g	30 g	8 g	70 mg	1002 mg	27 mg

Ingredient	Weight	Measure	Issue
BEEF,OVEN ROAST,TEMPERED	25 lbs		
WATER,BOILING	50-1/8 lbs	6 gal	
SALT	1-2/3 oz	2-2/3 tbsp	
SPAGHETTI NOODLES,DRY	8 lbs	2 gal 5/8 qts	
COOKING SPRAY,NONSTICK	1 oz	2 tbsp	
ONIONS,FRESH,CHOPPED	4-1/4 lbs	3 qts	4-3/4 lbs
PEPPERS,GREEN,FRESH,CHOPPED	4-1/4 lbs	3 qts 1 cup	5-1/4 lbs
WATER	4-1/8 lbs	2 qts	
SOY SAUCE	1-5/8 lbs	2-1/2 cup	
SALT	3-1/8 oz	1/4 cup 1-1/3 tbsp	
GARLIC POWDER	1 oz	3-1/3 tbsp	
GINGER,GROUND	1 oz	1/4 cup 1-1/3 tbsp	
PEPPER,BLACK,GROUND	3/8 oz	1 tbsp	
ONIONS,GREEN,FRESH,CHOPPED	11-1/8 oz	3-1/2 cup	

Method

1. Trim excess fat from roast. Slice beef into thin slices, 1/4-inch or less. Cut slices into strips 2-1/4x2-inches.
2. Add salt to boiling water. Slowly add spaghetti while stirring constantly until water boils again. Cook spaghetti in water until tender, about 10 to 12 minutes, stirring occasionally. Do not over cook. Drain thoroughly.
3. Spray steam-jacketed kettle or tilt fry pan with non-stick cooking spray. Cook beef 3 to 4 minutes or until beef begins to lose red color, stirring constantly.
4. Add onions and peppers; cook 4 minutes or until beef is done and vegetables are tender-crisp, stirring constantly. CCP: Internal temperature of beef must reach 145 F. or higher for 15 seconds.
5. Combine water, soy sauce, garlic powder, ginger, and pepper. Add to meat mixture. Stir to distribute seasonings.
6. Add spaghetti and green onions; stir until thoroughly mixed. CCP: Internal temperature must reach 145 F. or higher for 15 seconds. Hold at 140 F. or higher for service.

Notes

1. In Step 1, 18 pounds 12 ounces of ready-to-use beef cut for fajitas may be used per 100 portions. Cut into 2-inch pieces.

MEAT, FISH, AND POULTRY No. L 062 01

HAMBURGER YAKISOBA (GROUND BEEF)

Yield 100 **Portion** 1 Cup

Calories	Carbohydrates	Protein	Fat	Cholesterol	Sodium	Calcium
393 cal	24 g	35 g	16 g	106 mg	813 mg	28 mg

Ingredient	Weight	Measure	Issue
SPAGHETTI NOODLES,DRY	6 lbs	1 gal 2-1/2 qts	
WATER,BOILING	33-1/2 lbs	4 gal	
SALT	1-1/4 oz	2 tbsp	
BEEF,GROUND,BULK,RAW,90% LEAN	30 lbs		
PEPPERS,GREEN,FRESH,JULIENNE	5-1/4 lbs	1 gal	6-3/8 lbs
ONIONS,FRESH,CHOPPED	4-1/4 lbs	3 qts	4-3/4 lbs
SOY SAUCE	1-1/4 lbs	2 cup	
SALT	2-1/2 oz	1/4 cup 1/3 tbsp	
GARLIC POWDER	3/4 oz	2-2/3 tbsp	
GINGER,GROUND	3/4 oz	1/4 cup 1/3 tbsp	
PEPPER,BLACK,GROUND	1/3 oz	1 tbsp	
WATER	4-1/8 lbs	2 qts	
ONIONS,GREEN,FRESH,SLICED	12-1/3 oz	3-1/2 cup	13-3/4 oz

Method

1. Cook spaghetti in salted water until tender, 10 to 12 minutes.
2. Cook beef in steam-jacketed kettle or stock pot until beef loses its pink color, stirring to break apart. Drain and skim off excess fat.
3. Combine beef with onions, peppers, soy sauce, salt, garlic powder, ginger, and pepper. Cook until onions are tender, about 10 minutes. CCP: Internal temperature must reach 155 F. or higher for 15 seconds.
4. Add water, green onions, and spaghetti. Mix thoroughly. Heat to serving temperature. CCP: Internal temperature must reach 155 F. or higher for 15 seconds. Hold for service at 140 F. or higher.

MEAT, FISH, AND POULTRY No.L 062 02

TURKEY YAKISOBA

Yield 100 Portion 1 Cup

Calories	Carbohydrates	Protein	Fat	Cholesterol	Sodium	Calcium
315 cal	24 g	31 g	10 g	90 mg	871 mg	47 mg

Ingredient	Weight	Measure	Issue
SPAGHETTI NOODLES,DRY	6 lbs	1 gal 2-1/2 qts	
WATER,BOILING	33-1/2 lbs	4 gal	
SALT	1-1/4 oz	2 tbsp	
TURKEY,GROUND,90% LEAN,RAW	32 lbs		
ONIONS,FRESH,CHOPPED	4-1/4 lbs	3 qts	4-2/3 lbs
PEPPERS,GREEN,FRESH,CHOPPED	5-1/4 lbs	1 gal	6-3/8 lbs
SOY SAUCE	1-1/4 lbs	2 cup	
SALT	2-1/2 oz	1/4 cup 1/3 tbsp	
GARLIC POWDER	3/4 oz	2-2/3 tbsp	
GINGER,GROUND	3/4 oz	1/4 cup 1/3 tbsp	
PEPPER,BLACK,GROUND	1/3 oz	1 tbsp	
WATER	4-1/8 lbs	2 qts	
ONIONS,GREEN,FRESH,CHOPPED	12-1/3 oz	3-1/2 cup	13-3/4 oz

Method

1. Cook spaghetti in salted water until tender, 10 to 12 minutes.
2. Cook turkey in steam-jacketed kettle or stock pot until turkey loses its pink color, stirring to break apart. Drain and skim off excess fat.
3. Combine turkey with onions, peppers, soy sauce, salt, garlic powder, ginger, and pepper. Cook until onions are tender, about 10 minutes. CCP: Internal temperature must reach 165 F. or higher for 15 seconds.
4. Add water, green onions, and spaghetti. Mix thoroughly. Heat to serving temperature. CCP: Internal temperature must reach 165 F. or higher for 15 seconds. Hold for service at 140 F. or higher.

MEAT, FISH, AND POULTRY No.L 063 00

ENCHILADAS (GROUND BEEF)

Yield 100   Portion 2 Enchiladas

Calories	Carbohydrates	Protein	Fat	Cholesterol	Sodium	Calcium
440 cal	34 g	27 g	22 g	83 mg	690 mg	246 mg

Ingredient	Weight	Measure	Issue
FLOUR,WHEAT,GENERAL PURPOSE	1-1/4 lbs	1 qts 1/2 cup	
SHORTENING	1 lbs	2-1/4 cup	
TOMATO PASTE,CANNED	2-1/4 lbs	3-7/8 cup	
CHILI POWDER,DARK,GROUND	4-1/2 oz	1 cup	
CUMIN,GROUND	1 oz	1/4 cup 2/3 tbsp	
BEEF BROTH		2 qts 1-1/4 cup	
PEPPER,BLACK,GROUND	<1/16th oz	1/8 tsp	
BEEF,GROUND,BULK,RAW,90% LEAN	18 lbs		
ONIONS,FRESH,CHOPPED	2-1/2 lbs	1 qts 3-1/8 cup	2-3/4 lbs
CHILI POWDER,DARK,GROUND	4-1/4 oz	1 cup	
SALT	1-7/8 oz	3 tbsp	
PEPPER,RED,GROUND	1/2 oz	2-2/3 tbsp	
GARLIC POWDER	1/3 oz	1 tbsp	
TORTILLAS,CORN,6 INCH	11-1/2 lbs	200 each	
CHEESE,CHEDDAR,SHREDDED	4 lbs	1 gal	
ONIONS,FRESH,CHOPPED	1-7/8 lbs	1 qts 1-3/8 cup	2-1/8 lbs

Method

1. Blend together melted shortening or salad oil and sifted general purpose flour until smooth. Cook at low heat 2 minutes. Add canned tomato paste, chili powder, ground cumin; blend well.
2. Prepare beef broth following package directions. Add stock to roux, stirring constantly. Bring to a boil; reduce heat; simmer 10 minutes or until thickened, stirring constantly. Add pepper. Stir to blend.
3. Cook beef until beef loses its pink color, stirring to break apart. Drain or skim off excess fat.
4. Add 2 quarts gravy, onions, chili powder, salt, red pepper, and garlic powder to beef. Blend well. CCP: Hold at 140 F. or higher for use in Step 7. Spread 2 cups gravy in each pan.
5. Wrap tortillas in foil; place in 150 F. oven or in a warmer for 15 minutes or until warm and pliable.
6. Place 3 tablespoons meat filling in center of each tortilla. Roll tightly around filling; place seam-side down in pan, 50 per pan.
7. Pour 1-1/4 quarts gravy evenly over enchiladas in each pan.
8. Using a convection oven, bake 18-20 minutes in 325 F. oven or until thoroughly heated. CCP: Internal temperature must reach 155 F. or higher for 15 seconds. Remove from oven.
9. Sprinkle 1 quart cheese and 1-1/3 cups onions over enchiladas in each pan.
10. Heat in oven 3 minutes to melt cheese. CCP: Hold for service at 140 F. or higher for 15 seconds.

MEAT, FISH, AND POULTRY   No. L 063 01

ENCHILADAS (FROZEN)

Yield 100   **Portion** 2 Enchiladas

Calories	Carbohydrates	Protein	Fat	Cholesterol	Sodium	Calcium
416 cal	37 g	19 g	22 g	51 mg	1669 mg	347 mg

Ingredient	Weight	Measure	Issue
ENCHILADAS,FROZEN	50 lbs		
CHEESE,CHEDDAR,LOWFAT,SHREDDED	4 lbs	1 gal	
ONIONS,FRESH,CHOPPED	2-1/8 lbs	1 qts 2 cup	2-1/3 lbs

Method

1. Follow manufacturer's directions on container for heating frozen enchiladas. CCP: Internal temperature must reach 165 F. or higher for 15 seconds.
2. Sprinkle an equal quantity of cheese and onions over enchiladas in each sheet pan. Using a convection oven, bake 3 minutes at 300 F. on high fan, closed vent to melt cheese. CCP: Hold at 140 F. or higher for service.

MEAT, FISH, AND POULTRY No. L 063 02

ENCHILADAS (GROUND TURKEY)

Yield 100   **Portion** 2 Enchiladas

Calories	Carbohydrates	Protein	Fat	Cholesterol	Sodium	Calcium
329 cal	22 g	23 g	18 g	70 mg	692 mg	213 mg

Ingredient	Weight	Measure	Issue
FLOUR,WHEAT,GENERAL PURPOSE	1-1/4 lbs	1 qts 1/2 cup	
SHORTENING	1 lbs	2-1/4 cup	
TOMATO PASTE,CANNED	2-1/4 lbs	3-7/8 cup	
CHILI POWDER,DARK,GROUND	4-1/2 oz	1 cup	
CUMIN,GROUND	1 oz	1/4 cup 2/3 tbsp	
CHICKEN BROTH		2 qts 1-1/4 cup	
PEPPER,BLACK,GROUND	<1/16th oz	1/8 tsp	
TURKEY,GROUND,90% LEAN,RAW	18 lbs		
ONIONS,FRESH,CHOPPED	2-1/2 lbs	1 qts 3-1/8 cup	2-3/4 lbs
CHILI POWDER,DARK,GROUND	4-1/4 oz	1 cup	
SALT	1-7/8 oz	3 tbsp	
PEPPER,RED,GROUND	1/2 oz	2-2/3 tbsp	
GARLIC POWDER	1/3 oz	1 tbsp	
TORTILLAS,CORN,6 INCH	5-3/4 lbs	100 each	
CHEESE,CHEDDAR,SHREDDED	4 lbs	1 gal	
ONIONS,FRESH,CHOPPED	1-7/8 lbs	1 qts 1-3/8 cup	2-1/8 lbs

Method

1. Blend together melted shortening or salad oil and sifted general purpose flour until smooth. Cook at low heat 2 minutes. Add canned tomato paste, chili powder, ground cumin; blend well.
2. Prepare chicken broth following package directions. Add stock to roux, stirring constantly. Bring to a boil; reduce heat; simmer 10 minutes or until thickened, stirring constantly. Add pepper. Stir to blend.
3. Cook turkey until turkey loses its pink color, stirring to break apart. Drain or skim off excess fat.
4. Add 2 quarts gravy, onions, chili powder, salt, red pepper, and garlic powder to turkey. Blend well. CCP: Hold at 140 F. or higher for use in Step 7. Spread 2 cups gravy in each pan.
5. Wrap tortillas in foil; place in 150 F. oven or in a warmer for 15 minutes or until warm and pliable.
6. Place 3 tablespoons meat filling in center of each tortilla. Roll tightly around filling; place seam-side down in pan, 50 per pan.
7. Pour 1-1/4 quarts gravy evenly over enchiladas in each pan.
8. Using a convection oven, bake 18-20 minutes in 325 F. oven or until thoroughly heated. CCP: Internal temperature must reach 165 F. or higher for 15 seconds. Remove from oven.
9. Sprinkle 1 quart cheese and 1-1/3 cups onions over enchiladas in each pan.
10. Heat in oven 3 minutes to melt cheese. CCP: Hold for service at 140 F. or higher.

MEAT, FISH, AND POULTRY No. L 064 00

CREOLE MACARONI (GROUND BEEF)

Yield 100   Portion   1 Cup

Calories	Carbohydrates	Protein	Fat	Cholesterol	Sodium	Calcium
267 cal	32 g	18 g	8 g	43 mg	869 mg	69 mg

Ingredient	Weight	Measure	Issue
WATER	12-1/2 lbs	1 gal 2 qts	
TOMATO PASTE,CANNED	7-3/4 lbs	3 qts 1-1/2 cup	
TOMATOES,CANNED,DICED,DRAINED	6-5/8 lbs	3 qts	
PEPPERS,GREEN,FRESH,CHOPPED	1-1/2 lbs	1 qts 1/2 cup	1-3/4 lbs
ONIONS,FRESH,CHOPPED	1-1/4 lbs	3-1/2 cup	1-3/8 lbs
SUGAR,GRANULATED	5-1/4 oz	3/4 cup	
SALT	3-3/8 oz	1/4 cup 1-2/3 tbsp	
PEPPER,BLACK,GROUND	3/8 oz	1 tbsp	
GARLIC POWDER	1/3 oz	1 tbsp	
BASIL,SWEET,WHOLE,CRUSHED	3/8 oz	2-2/3 tbsp	
PEPPER,RED,GROUND	1/4 oz	1 tbsp	
THYME,GROUND	1/8 oz	1 tbsp	
MACARONI NOODLES,ELBOW,DRY	6 lbs	1 gal 2-1/2 qts	
SALT	1 oz	1 tbsp	
WATER,BOILING	33-1/2 lbs	4 gal	
BEEF,GROUND,BULK,RAW,90% LEAN	11 lbs		
CHEESE,AMERICAN,SHREDDED	1 lbs	1 qts	

Method

1. Combine water, tomato paste, tomatoes, peppers, onions, sugar, salt, black pepper, garlic powder, basil, red pepper, and thyme in steam-jacketed kettle or stock pot. Mix well; bring to a boil; reduce heat; cover; simmer 10 to 15 minutes or until thickened.
2. Add macaroni to salted water; bring to a boil stirring constantly. Cook 10 minutes, stirring occasionally; drain well. Do not overcook.
3. Brown beef until beef loses its pink color. Drain or skim off excess fat.
4. Combine beef, tomato sauce mixture, and macaroni. Mix well.
5. Pour about 8-1/4 quarts macaroni mixture in each pan.
6. Sprinkle 1-1/3 cups cheese over macaroni mixture in each pan.
7. Using a convection oven, bake 20 minutes at 325 F. on high fan, closed vent or until mixture is bubbling and cheese is melted. CCP: Internal temperature must reach 155 F. or higher for 15 seconds. Hold for service at 140 F. or higher.

MEAT, FISH, AND POULTRY No.L 064 01

CREOLE MACARONI (GROUND TURKEY)

Yield 100   **Portion** 1 Cup

Calories	Carbohydrates	Protein	Fat	Cholesterol	Sodium	Calcium
242 cal	32 g	16 g	6 g	38 mg	646 mg	77 mg

Ingredient	**Weight**	**Measure**	**Issue**
WATER	12-1/2 lbs	1 gal 2 qts	
TOMATO PASTE,CANNED	8-1/8 lbs	3 qts 2 cup	
TOMATOES,CANNED,DICED,DRAINED	6-5/8 lbs	3 qts	
PEPPERS,GREEN,FRESH,CHOPPED	1-1/2 lbs	1 qts 1/2 cup	1-3/4 lbs
ONIONS,FRESH,CHOPPED	1-3/8 lbs	1 qts	1-5/8 lbs
SUGAR,GRANULATED	5-1/4 oz	3/4 cup	
SALT	1 oz	1 tbsp	
PEPPER,BLACK,GROUND	1/2 oz	2 tbsp	
GARLIC POWDER	1/3 oz	1 tbsp	
BASIL,SWEET,WHOLE,CRUSHED	1/2 oz	3 tbsp	
PEPPER,RED,GROUND	1/4 oz	1 tbsp	
THYME,GROUND	1/8 oz	1 tbsp	
MACARONI NOODLES,ELBOW,DRY	6 lbs	1 gal 2-1/2 qts	
WATER,BOILING	33-1/2 lbs	4 gal	
SALT	1 oz	1 tbsp	
TURKEY,GROUND,90% LEAN,RAW	12 lbs		
CHEESE,AMERICAN,SHREDDED	1 lbs	1 qts	

Method

1. Combine water, tomato paste, tomatoes, peppers, onions, sugar, salt, black pepper, garlic powder, basil, red pepper, and thyme in steam-jacketed kettle or stock pot. Mix well; bring to a boil; reduce heat; cover; simmer 10 to 15 minutes or until thickened.
2. Add macaroni to salted water; bring to a boil stirring constantly. Cook 10 minutes, stirring occasionally; drain well. Do not overcook.
3. Brown turkey until turkey loses its pink color, stirring to break apart. Drain or skim off excess fat.
4. Combine turkey, tomato sauce mixture, and macaroni. Mix well.
5. Pour about 8-1/4 quarts macaroni mixture in each pan.
6. Sprinkle 1-1/3 cups cheese over macaroni mixture in each pan.
7. Using a convection oven, bake 20 minutes at 325 F. on high fan, closed vent or until mixture is bubbling and cheese is melted. CCP: Internal temperature must reach 165 F. or higher for 15 seconds. Hold for service at 140 F. or higher.

MEAT, FISH, AND POULTRY No. L 065 00

HUNGARIAN GOULASH

Yield 100 **Portion** 6-1/2 Ounces

Calories	Carbohydrates	Protein	Fat	Cholesterol	Sodium	Calcium
382 cal	37 g	30 g	12 g	104 mg	637 mg	40 mg

Ingredient	Weight	Measure	Issue
BEEF,DICED,LEAN,RAW	30 lbs		
WATER	10-1/2 lbs	1 gal 1 qts	
ONIONS,FRESH,CHOPPED	7-1/2 lbs	1 gal 1-1/3 qts	8-1/3 lbs
PAPRIKA,GROUND	3-7/8 oz	1 cup	
SALT	2-1/2 oz	1/4 cup 1/3 tbsp	
GARLIC POWDER	1-1/4 oz	1/4 cup 1/3 tbsp	
PEPPER,BLACK,GROUND	2/3 oz	3 tbsp	
THYME,GROUND	1/3 oz	2 tbsp	
WATER,COLD	2-1/8 lbs	1 qts	
FLOUR,WHEAT,GENERAL PURPOSE	1-1/8 lbs	1 qts	
WATER,BOILING	100-1/3 lbs	12 gal	
NOODLES,EGG	9 lbs	6 gal 2-7/8 qts	
SALT	2-1/2 oz	1/4 cup 1/3 tbsp	

Method

1 Place beef, hot water, onions, paprika, salt, garlic powder, pepper and thyme in steam jacketed kettle. Bring to a boil. Reduce heat; cover; simmer about 2 hours or until beef is tender. Skim excess fat. CCP: Internal temperature must reach 145 F. or higher for 15 seconds.
2 Combine cold water and flour to make a smooth mixture. Stir in beef mixture until well blended. Return to boil; reduce heat; cook 10 minutes or until thickened. CCP: Hold for service at 140 F. or higher.
3 Add noodles to boiling salted water, while stirring constantly. Cook 8-10 minutes. Stir occasionally. Do not overcook. Drain thoroughly.
4 Serve 3/4 cup (6 oz) goulash with 1 cup noodles.

MEAT, FISH, AND POULTRY No. L 066 00

SAUERBRATEN

Yield 100　　　　　　　　　　　　　　　　　　Portion 3-1/2 Ounces

Calories	Carbohydrates	Protein	Fat	Cholesterol	Sodium	Calcium
467 cal	19 g	36 g	27 g	114 mg	491 mg	51 mg

Ingredient	Weight	Measure	Issue
BEEF,POT ROAST,RAW	40 lbs		
WATER	15-2/3 lbs	1 gal 3-1/2 qts	
VINEGAR,DISTILLED	9-3/8 lbs	1 gal 1/2 qts	
SUGAR,BROWN,PACKED	1-1/4 lbs	1 qts	
SALT	3 oz	1/4 cup 1 tbsp	
MUSTARD,DRY	4 oz	1/2 cup 2 tbsp	
CLOVES,GROUND	7/8 oz	1/4 cup 1/3 tbsp	
PEPPER,BLACK,GROUND	1/3 oz	1 tbsp	
GARLIC POWDER	1/3 oz	1 tbsp	
BAY LEAF,WHOLE,DRIED	3/8 oz	12 each	
ONIONS,FRESH,CHOPPED	4 lbs	2 qts 3-3/8 cup	4-1/2 lbs
CARROTS,FRESH,SLICED	3-1/8 lbs	2 qts 3-1/8 cup	3-3/4 lbs
CELERY,FRESH,CHOPPED	2-3/8 lbs	2 qts 1 cup	3-1/4 lbs
RESERVED LIQUID	27-1/8 lbs	3 gal 1 qts	
COOKIES,GINGERSNAPS,CRUSHED	2 lbs		

Method

1. Place roasts in a steam-jacketed kettle or stock pot.
2. Combine water, vinegar, sugar, salt, mustard flour, cloves, pepper, garlic, bay leaves, onions, carrots and celery.
3. Pour mixture over beef; bring to a boil; cover. Simmer 3-1/2 to 4 hours or until tender; turn roasts every hour. CCP: Internal temperature must reach 145 F. or higher for 15 seconds.
4. Remove beef and bay leaves. Reserve 3-1/4 gallons marinade. Let beef stand 20 minutes; slice 1/8-inch thick. Arrange 50 portions in each steam table pan.
5. Bring marinade to a boil; add cookie crumbs, simmer until crumbs are dissolved, stirring constantly. CCP: Temperature must reach 165 F. or higher for 15 seconds.
6. Pour 6-1/2 quarts gravy over beef in each pan. CCP: Hold for service at 140 F. or higher.

Notes

1. In Step 5, a mixture of 1 pound 8 ounces or 1-1/2 quarts flour and 1 quart water may be used for thickening instead of cookie crumbs.

MEAT. FISH. AND POULTRY No.L 067 00

GLAZED HAM LOAF

Yield 100   Portion 5 Ounces

Calories	Carbohydrates	Protein	Fat	Cholesterol	Sodium	Calcium
297 cal	8 g	28 g	16 g	114 mg	1126 mg	47 mg

Ingredient	Weight	Measure	Issue
MILK,NONFAT,DRY	3 oz	1-1/4 cup	
WATER	3-1/2 lbs	1 qts 2-5/8 cup	
BREAD,WHITE,SLICED	1-1/2 lbs	1 gal 1 qts	
ONIONS,FRESH,CHOPPED	1-1/3 lbs	3-3/4 cup	1-1/2 lbs
EGGS,WHOLE,FROZEN	1-3/4 lbs	3-1/4 cup	
PORK,HAM,CURED,GROUND	18 lbs	3 gal 3/8 qts	
PORK,GROUND,RAW	12 lbs		
PEPPER,BLACK,GROUND	1/4 oz	1 tbsp	
SUGAR,BROWN,PACKED	10-7/8 oz	2-1/8 cup	
MUSTARD,DRY	3-1/8 oz	1/2 cup	
VINEGAR,DISTILLED	12-1/2 oz	1-1/2 cup	

Method

1. Reconstitute milk in mixer bowl.
2. Add bread; mix to moisten; let stand 5 minutes; mix until smooth.
3. Add onions, eggs, ham, pork and pepper. Mix at medium speed until well blended.
4. Shape into 8-4 pound 14 ounce loaves; place 4 loaves, crosswise, in each roasting pan.
5. Combine brown sugar, mustard flour and vinegar. Blend well. Spoon 6 tablespoons mixture over each loaf.
6. Bake 1-1/2 hours at 350 F.; baste each loaf with brown sugar mixture at least twice during a cooking period. CCP: Internal temperature must reach 145 F. or higher for 15 seconds.
7. Remove excess liquid. Cool slightly. Cut 13 slices per loaf. CCP: Hold for service at 140 F. or higher.

MEAT, FISH, AND POULTRY No.L 068 00

SCALLOPED HAM AND NOODLES

Yield 100   **Portion** 1 Cup

Calories	Carbohydrates	Protein	Fat	Cholesterol	Sodium	Calcium
238 cal	18 g	22 g	8 g	61 mg	1208 mg	71 mg

Ingredient	Weight	Measure	Issue
NOODLES,EGG	3 lbs	2 gal 1 qts	
WATER,BOILING	33-1/2 lbs	4 gal	
HAM,CANNED,CHUNKS	20 lbs		
MILK,NONFAT,DRY	14-3/8 oz	1 qts 2 cup	
WATER,WARM	15-2/3 lbs	1 gal 3-1/2 qts	
WATER,COLD	3-1/8 lbs	1 qts 2 cup	
FLOUR,WHEAT,GENERAL PURPOSE	1-1/8 lbs	1 qts	
ONIONS,FRESH,CHOPPED	1-5/8 lbs	1 qts 5/8 cup	1-3/4 lbs
PEPPERS,GREEN,FRESH,CHOPPED	1 lbs	3 cup	1-1/4 lbs
BUTTER,MELTED	3 oz	1/4 cup 2-1/3 tbsp	
BREADCRUMBS,DRY,GROUND,FINE	7-5/8 oz	2 cup	
PAPRIKA,GROUND	1/8 oz	1/4 tsp	

Method

1. Cook noodles in boiling salted water 8 to 10 miuntes until tender. Drain.
2. Drain ham chunks, cut into bite-sized pieces. Reserve 2 qt ham juices.
3. Reconstitute milk in warm water; add reserved ham juices and mustard powder. Heat. DO NOT BOIL.
4. Blend flour and cold water stirring to make a slurry. Add slurry to hot milk and ham juices stirring constantly. Bring to a boil. Cover; reduce heat; simmer 10 minutes or until thickened, stirring frequently.
5. Stir onions, peppers, ham and noodles into thickened sauce. Cover; reduce heat; simmer 5 minutes.
6. Pour 1-2/3 gallons of ham and noodle mixture evenly into each ungreased pan.
7. Combine bread crumbs, paprika and butter or margarine. Sprinkle crumb mixture evenly over ham and noodles in each pan.
8. Sprinkle crumb mixture over each pan.
9. Using a convection oven, bake 20 minutes or until lightly browned at 325 F. on high fan, open vent. CCP: Internal temperature must reach 165 F. or higher for 15 seconds. Hold for service at 140 F. or higher.

Notes

1. In Step 1, 6 pounds of macaroni may be used.

MEAT, FISH, AND POULTRY No.L 069 00

BAKED HAM

Yield 100   **Portion**   3 Ounces

Calories	Carbohydrates	Protein	Fat	Cholesterol	Sodium	Calcium
172 cal	0 g	22 g	9 g	57 mg	1446 mg	8 mg

Ingredient   Weight   Measure   Issue
HAM,COOKED,BONELESS   25 lbs

Method
1. Split casing; peel from ham; place hams in pans.
2. Insert meat thermometer into center of ham. DO NOT ADD WATER; DO NOT COVER.
3. Using a convection oven, bake 2 hours uncovered at 300 F. on high fan, closed vent. CCP: Internal temperature must reach 145 F. or higher for 15 seconds.
4. Let stand 20 minutes before slicing.

MEAT, FISH, AND POULTRY No.L 069 01

GRILLED HAM STEAK

Yield 100 Portion 3 Ounces

Calories	Carbohydrates	Protein	Fat	Cholesterol	Sodium	Calcium
174 cal	0 g	22 g	9 g	57 mg	1446 mg	8 mg

Ingredient	**Weight**	**Measure**	**Issue**
HAM,COOKED,BONELESS	25 lbs		
SHORTENING	7/8 oz	2 tbsp	

Method

1 Slice ham into 4 ounce steaks.
2 Cut edge of each steak in several places to prevent curling. Grill ham on lightly greased 350 F. griddle about 1-1/2 minutes on each side or until browned. Remove fat from griddle as it accumulates. CCP: Internal temperature must reach 145 F. or higher for 15 seconds. Hold for service at 140 F. or higher.

MEAT. FISH. AND POULTRY No.L 070 00

BARBECUED HAM STEAK

Yield 100   Portion 3 Ounces

Calories	Carbohydrates	Protein	Fat	Cholesterol	Sodium	Calcium
224 cal	13 g	22 g	9 g	57 mg	1567 mg	19 mg

Ingredient	Weight	Measure	Issue
HAM,COOKED,BONELESS	25 lbs		
SHORTENING	7/8 oz	2 tbsp	
SUGAR,GRANULATED	2 lbs	1 qts 1/2 cup	
CHILI POWDER,DARK,GROUND	1-5/8 oz	1/4 cup 2-1/3 tbsp	
CLOVES,GROUND	1/2 oz	2 tbsp	
ALLSPICE,GROUND	3/8 oz	2 tbsp	
PEPPER,RED,GROUND	1/4 oz	1 tbsp	
MUSTARD,PREPARED	13-1/4 oz	1-1/2 cup	
TOMATO PASTE,CANNED	2 lbs	3-1/2 cup	
VINEGAR,DISTILLED	3-1/8 lbs	1 qts 2 cup	
ONIONS,FRESH,CHOPPED	8-1/2 oz	1-1/2 cup	9-1/2 oz
CELERY,FRESH,CHOPPED	6-1/3 oz	1-1/2 cup	8-2/3 oz

Method

1 Split casing; peel from hams. Cut hams into steaks weighing about 4 ounces each.
2 Grill 1-1/2 minutes on each side on a lightly greased 350 F. griddle. Place steaks, overlapping, in pans. CCP: Hold for service at 140 F. or higher.
3 Combine sugar, chili powder, cloves, allspice, red pepper, mustard, tomato paste, vinegar, onions and celery. Bring to a boil; reduce heat; simmer 15 minutes.
4 Pour 2 quarts sauce over steaks in each pan. Cover.
5 Bake at 300 F. for 15 minutes or until thoroughly heated. CCP: Internal temperature must reach 145 F. or higher for 15 seconds. Serve sauce with steaks. CCP: Hold for service at 140 F. or higher.

MEAT, FISH, AND POULTRY No. L 070 01

BARBECUED HAM STEAK CANNED HAM

Yield 100   **Portion** 3 Ounces

Calories	Carbohydrates	Protein	Fat	Cholesterol	Sodium	Calcium
224 cal	13 g	22 g	9 g	57 mg	1567 mg	19 mg

Ingredient	**Weight**	**Measure**	**Issue**
HAM,CANNED,CHUNKS	25 lbs		
SHORTENING	7/8 oz	2 tbsp	
SUGAR,GRANULATED	2 lbs	1 qts 1/2 cup	
CHILI POWDER,DARK,GROUND	1-5/8 oz	1/4 cup 2-1/3 tbsp	
CLOVES,GROUND	1/2 oz	2 tbsp	
ALLSPICE,GROUND	3/8 oz	2 tbsp	
PEPPER,RED,GROUND	1/4 oz	1 tbsp	
MUSTARD,PREPARED	13-1/4 oz	1-1/2 cup	
TOMATO PASTE,CANNED	2 lbs	3-1/2 cup	
VINEGAR,DISTILLED	3-1/8 lbs	1 qts 2 cup	
ONIONS,FRESH,CHOPPED	8-1/2 oz	1-1/2 cup	9-1/2 oz
CELERY,FRESH,CHOPPED	6-1/3 oz	1-1/2 cup	8-2/3 oz

Method

1. Cut ham into 3 ounce steaks.
2. Grill 1-1/2 minutes on each side on a lightly greased 350 F. griddle. Place steaks, overlapping, in pans.
3. Combine sugar, chili powder, cloves, allspice, red pepper, mustard, tomato paste, vinegar, onions and celery. Bring to a boil; reduce heat; simmer 15 minutes.
4. Pour 2 quarts sauce over steaks in each pan. Cover.
5. Bake at 300 F. for 15 minutes or until thoroughly heated. CCP: Internal temperature must reach 145 F. or higher for 15 seconds. Serve sauce with steaks. CCP: Hold for service at 140 F. or higher.

MEAT, FISH, AND POULTRY No.L 071 00

BAKED CANNED HAM

Yield 100   Portion   3-1/2 Ounces

Calories	Carbohydrates	Protein	Fat	Cholesterol	Sodium	Calcium
184 cal	3 g	22 g	9 g	57 mg	1447 mg	11 mg

Ingredient   **Weight**   **Measure**   **Issue**

HAM,CANNED,COOKED   25 lbs
SUGAR,BROWN,PACKED   10-7/8 oz   2-1/8 cup
VINEGAR,DISTILLED   4-1/8 oz   1/2 cup
CLOVES,GROUND   1/4 oz   1 tbsp

Method

1. Remove wrapping. Place hams in roasting pans.
2. Score ham with knife 1/8-inch deep in diamond shape pattern, allowing 1 inch between scores. Insert meat thermometer in center of ham.
3. Using a convection oven, bake uncovered on high fan, closed vent, at 300 F. for 1 hour.
4. Combine sugar, vinegar and ground cloves. Spread mixture evenly over hams in pan.
5. Bake uncovered 30 to 40 minutes. CCP: Internal temperature must reach 145 F. or higher for 15 seconds.
6. Let stand 20 minutes before slicing. Slice about 1/8-inch thick. CCP: Hold for service at 140 F. or higher.

MEAT, FISH, AND POULTRY   No.L 071 01

BAKED HAM STEAK (CANNED HAM)

Yield 100					Portion 3-1/2 Ounces

Calories	Carbohydrates	Protein	Fat	Cholesterol	Sodium	Calcium
184 cal	3 g	22 g	9 g	57 mg	1447 mg	11 mg

Ingredient	Weight	Measure	Issue
HAM,CANNED,COOKED	25 lbs		
SUGAR,BROWN,PACKED	10-7/8 oz	2-1/8 cup	
VINEGAR,DISTILLED	4-1/8 oz	1/2 cup	
CLOVES,GROUND	1/4 oz	1 tbsp	

Method

1 Slice ham into 4 ounce steaks about 1/4-inch thick.
2 Overlap steaks in roasting pans.
3 Combine sugar, vinegar, and ground cloves. Spread mixture evenly over steaks in each pan.
4 Using a convection oven, bake at 325 F. for 30 minutes on high fan, closed vent. CCP: Internal temperature must reach 145 F. or higher for 15 seconds. Hold for service at 140 F. or higher.

MEAT. FISH. AND POULTRY No.L 071 02

GRILLED HAM STEAK (CANNED HAM)

Yield 100   Portion 3 Ounces

Calories	Carbohydrates	Protein	Fat	Cholesterol	Sodium	Calcium
178 cal	3 g	17 g	10 g	55 mg	1269 mg	7 mg

Ingredient	**Weight**	**Measure**	**Issue**
HAM,COOKED,SLICED	25 lbs	4 gal 1-7/8 qts	
SHORTENING	7/8 oz	2 tbsp	

Method

1. Slice ham into 100 slices.
2. Grill ham on lightly greased 350 F. griddle 1-1/2 minutes on each side or until browned. CCP: Internal temperature must reach 145 F. or higher for 15 seconds. Hold for service at 140 F. or higher.

MEAT, FISH, AND POULTRY No.L 071 03

GRILLED HAM SLICE (CANNED HAM)

Yield 100　　　　　　　　　　　　　　　　　　　**Portion**　2-1/2 Ounces

Calories	Carbohydrates	Protein	Fat	Cholesterol	Sodium	Calcium
140 cal	0 g	17 g	7 g	45 mg	1157 mg	6 mg

Ingredient	**Weight**	**Measure**	**Issue**
HAM,CANNED,COOKED	20 lbs		
SHORTENING	7/8 oz	2 tbsp	

Method

1. Slice ham into about 3-1/4 ounce slices. Grill ham on a lightly greased 350 F. griddle about 1 minute on each side. CCP: Internal temperature must reach 145 F. or higher for 15 seconds. Hold for service at 140 F. or higher.

MEAT, FISH, AND POULTRY No.L 072 00

BAKED HAM, MACARONI, AND TOMATOES (CANNED HAM)

Yield 100   Portion 1 Cup

Calories	Carbohydrates	Protein	Fat	Cholesterol	Sodium	Calcium
258 cal	25 g	16 g	11 g	36 mg	947 mg	130 mg

Ingredient	Weight	Measure	Issue
MACARONI NOODLES,ELBOW,DRY	4-1/2 lbs	1 gal 7/8 qts	
SALT	1 oz	1 tbsp	
WATER,BOILING	25-1/8 lbs	3 gal	
ONIONS,FRESH,CHOPPED	4-1/4 lbs	3 qts	4-3/4 lbs
PEPPERS,GREEN,FRESH,CHOPPED	2 lbs	1 qts 2 cup	2-3/8 lbs
MUSHROOMS,CANNED,DRAINED	2 lbs	1 qts 2 cup	
GARLIC POWDER	1/4 oz	3/8 tsp	
TOMATOES,CANNED,CRUSHED,INCL LIQUIDS	19-7/8 lbs	2 gal 1 qts	
SUGAR,GRANULATED	2-1/4 oz	1/4 cup 1-1/3 tbsp	
OREGANO,CRUSHED	1/4 oz	1 tbsp	
OIL,SALAD	7-2/3 oz	1 cup	
HAM,CANNED,COOKED,DICED	10 lbs		
CHEESE,AMERICAN,SHREDDED	3 lbs	3 qts	

Method

1 Add macaroni slowly to boiling, salted water; stir occasionally, until water returns to a boil. Boil 10 to 12 minutes. Drain. Use in Step 5.
2 Sautee onions, peppers, mushrooms and garlic in salad oil or shortening until tender.
3 Combine sauteed vegetables, tomatoes, sugar, and oregano. Bring to a boil; reduce heat; simmer 10 minutes.
4 Combine ham, vegetables and macaroni.
5 Place about 1-1/2 gallons in each steam table pan.
6 Bake 25 minutes at 350 F. CCP: Internal temperature must reach 165 F. or higher for 15 seconds.
7 Sprinkle 3 cups cheese evenly over mixture in each pan. Bake 10 minutes or until cheese is lightly browned. CCP: Hold for service at 140 F. or higher.

MEAT, FISH, AND POULTRY No. L 072 01

BAKED LUNCHEON MEAT, MACARONI, AND CHEESE

Yield 100   Portion 1 Cup

Calories	Carbohydrates	Protein	Fat	Cholesterol	Sodium	Calcium
350 cal	26 g	13 g	22 g	38 mg	955 mg	131 mg

Ingredient	**Weight**	**Measure**	**Issue**
MACARONI NOODLES,ELBOW,DRY	4-1/2 lbs	1 gal 7/8 qts	
SALT	1 oz	1 tbsp	
WATER,BOILING	25-1/8 lbs	3 gal	
ONIONS,FRESH,CHOPPED	4-1/4 lbs	3 qts	4-3/4 lbs
PEPPERS,GREEN,FRESH,CHOPPED	2 lbs	1 qts 2 cup	2-3/8 lbs
MUSHROOMS,CANNED,DRAINED	2 lbs	1 qts 2 cup	
OIL,SALAD	7-2/3 oz	1 cup	
GARLIC POWDER	1/4 oz	3/8 tsp	
TOMATOES,CANNED,CRUSHED,INCL LIQUIDS	19-7/8 lbs	2 gal 1 qts	
SUGAR,GRANULATED	2-1/4 oz	1/4 cup 1-1/3 tbsp	
OREGANO,CRUSHED	1/4 oz	1 tbsp	
LUNCHEON MEAT,CANNED	10 lbs		
CHEESE,AMERICAN,SHREDDED	3 lbs	3 qts	

Method

1. Add macaroni slowly to boiling salted water; stir occasionally, until water returns to a boil. Boil 10 to 15 minutes. Drain.
2. Saute onions, peppers, mushrooms and garlic in salad oil or shortening until tender.
3. Combine sauteed vegetables, tomatoes, sugar, and oregano. Bring to a boil; reduce heat; simmer 10 minutes.
4. Combine luncheon meat, vegetables and macaroni.
5. Place about 1-1/2 gallons in each steam table pan.
6. Bake 25 minutes at 350 F. CCP: Internal temperature must reach 165 F. or higher for 15 seconds.
7. Sprinkle 3 cups cheese evenly over mixture in each pan. Bake 10 minutes or until cheese is lightly browned. CCP: Hold for service at 140 F. or higher.

MEAT, FISH, AND POULTRY No.L 072 02

BAKED HAM, MACARONI AND TOMATOES (CANNED CHUNKS)

Yield 100   Portion 1 Cup

Calories	Carbohydrates	Protein	Fat	Cholesterol	Sodium	Calcium
327 cal	25 g	25 g	14 g	58 mg	1525 mg	133 mg

Ingredient	Weight	Measure	Issue
MACARONI NOODLES,ELBOW,DRY	4-1/2 lbs	1 gal 7/8 qts	
SALT	1 oz	1 tbsp	
WATER,BOILING	25-1/8 lbs	3 gal	
ONIONS,FRESH,CHOPPED	4-1/4 lbs	3 qts	4-2/3 lbs
PEPPERS,GREEN,FRESH,CHOPPED	2 lbs	1 qts 2 cup	2-3/8 lbs
MUSHROOMS,CANNED,DRAINED	2 lbs	1 qts 2 cup	
OIL,SALAD	7-2/3 oz	1 cup	
GARLIC POWDER	1/4 oz	3/8 tsp	
TOMATOES,CANNED,CRUSHED,INCL LIQUIDS	19-7/8 lbs	2 gal 1 qts	
SUGAR,GRANULATED	2-1/4 oz	1/4 cup 1-1/3 tbsp	
OREGANO,CRUSHED	1/4 oz	1 tbsp	
HAM,CANNED,COOKED,DICED	20 lbs		
CHEESE,AMERICAN,SHREDDED	3 lbs	3 qts	

Method

1. Add macaroni slowly to boiling salted water; stir occasionally, until water returns to a boil. Boil 10 to 15 minutes. Drain. Use in Step 5.
2. Saute onions, peppers, mushrooms and garlic in salad oil or shortening until tender.
3. Combine vegetables, tomatoes, sugar, and oregano. Bring to a boil; reduce heat; simmer 10 minutes.
4. Combine drained, canned ham chunks, vegetables and macaroni.
5. Place about 1-1/2 gallons in each steam table pan.
6. Bake 25 minutes at 350 F. CCP: Internal temperature must reach 165 F. or higher for 15 seconds.
7. Sprinkle 3 cups cheese evenly over mixture in each pan. Bake 10 minutes or until cheese is lightly browned. CCP: Hold for service at 140 F. or higher.

MEAT. FISH. AND POULTRY No.L 073 00

SCALLOPED HAM AND POTATOES (CANNED HAM)

Yield 100   Portion   1 Cup

Calories	Carbohydrates	Protein	Fat	Cholesterol	Sodium	Calcium
335 cal	12 g	23 g	21 g	86 mg	1363 mg	153 mg

Ingredient | Weight | Measure | Issue

Ingredient	Weight	Measure	Issue
HAM,CANNED,CHUNKS	20 lbs		
POTATO,WHITE,DEHYDRATED,DICED	3-1/2 lbs		
WATER	23 lbs	2 gal 3 qts	
MILK,NONFAT,DRY	1-1/3 lbs	2 qts 5/8 cup	
WATER,WARM	5-1/4 lbs	2 qts 2 cup	
RESERVED STOCK	2-5/8 lbs	1 qts 1 cup	
BUTTER,MELTED	3 lbs	1 qts 2 cup	
FLOUR,WHEAT,GENERAL PURPOSE	1-2/3 lbs	1 qts 2 cup	
ONIONS,FRESH,CHOPPED	1 lbs	2-7/8 cup	1-1/8 lbs
CHEESE,CHEDDAR,SHREDDED	2 lbs	2 qts	

Method

1. Drain ham chunks. Reserve 1-1/4 quart of liquid for use in Step 3. Cut ham into bite-sized pieces; use in Step 6.
2. Add potatoes to water. Cover. Bring quickly to a boil; simmer 15 minutes. Drain; set aside for use in Step 6.
3. Reconstitute milk. Add reserved stock; heat to just below boiling. DO NOT BOIL.
4. Blend butter or margarine and flour together until smooth. Add to hot milk, stirring constantly.
5. Add onions; simmer sauce 5 minutes or until thickened.
6. Combine ham, potatoes, and sauce. Place 6-1/4 quarts ham-potato mixture in each pan.
7. Bake 25 minutes at 350 F. CCP: Internal temperature must reach 165 F. or higher for 15 seconds.
8. Sprinkle 2 cups cheese evenly over mixture in each steam table pan.
9. Bake an additional 10 minutes or until cheese is lightly browned. CCP: Hold for service at 140 F. or higher.

MEAT, FISH, AND POULTRY   No. L 074 00

CHILIES RELLENOS

Yield 100   Portion   1 Serving

Calories	Carbohydrates	Protein	Fat	Cholesterol	Sodium	Calcium
278 cal	27 g	7 g	17 g	15 mg	582 mg	148 mg

Ingredient   Weight   Measure   Issue

CHILIES RELLENOS, FROZEN, 4 OZ   25 lbs

Method

1. Put Chilies Rellenos in basket in single layer to prevent overcooking and bursting of filling.
2. Fry 5 minutes at 350 F. or until golden brown. CCP: Internal temperature must reach 165 F. or higher for 15 seconds.
3. Drain well in basket or on absorbent paper. CCP: Hold for service at 140 F. or higher.

MEAT, FISH, AND POULTRY No.L 075 00

BROCCOLI, CHEESE, AND RICE

Yield 100   Portion 1 Cup

Calories	Carbohydrates	Protein	Fat	Cholesterol	Sodium	Calcium
317 cal	28 g	15 g	17 g	35 mg	718 mg	346 mg

Ingredient | Weight | Measure | Issue

Ingredient	Weight	Measure	Issue
RICE,LONG GRAIN	3-5/8 lbs	2 qts 3/4 cup	
WATER	9-3/8 lbs	1 gal 1/2 qts	
ONIONS,FRESH,CHOPPED	2-2/3 lbs	1 qts 3-1/2 cup	3 lbs
SALT	7/8 oz	1 tbsp	
MILK,NONFAT,DRY	7-3/4 oz	3-1/4 cup	
WATER,WARM	8-1/3 lbs	1 gal	
SOUP,CONDENSED,CREAM OF MUSHROOM	9-1/2 lbs	1 gal 1/3 qts	
GARLIC POWDER	7/8 oz	3 tbsp	
PEPPER,BLACK,GROUND	1/2 oz	2 tbsp	
OREGANO,CRUSHED	1/2 oz	3 tbsp	
BROCCOLI,FROZEN,SPEARS,THAWED,1/2""	26-7/8 lbs	4 gal 3-1/2 qts	
CHEESE,AMERICAN	8 lbs	2 gal <1/16th qts	
MARGARINE,MELTED	8 oz	1 cup	
BREADCRUMBS,DRY,GROUND,FINE	1 lbs	1 qts	

Method

1 Combine rice, water, onions, and salt in steam-jacketed kettle or stock pot; bring to a boil. Stir occasionally.
2 Cover tightly; reduce heat; simmer 20 to 25 minutes. DO NOT STIR.
3 Reconstitute milk. Blend in soup, garlic powder, pepper, and oregano. Combine with rice mixture, stirring well. Bring to a boil stirring constantly.
4 Add broccoli; bring to a boil, stirring constantly; simmer 5 minutes or until broccoli is almost tender.
5 Reduce heat; add cheese, stirring constantly until cheese is melted.
6 Pour 5-1/2 quarts mixture into each steam table pan.
7 Combine butter or margarine and bread crumbs. Mix well. Sprinkle 4-1/2 ounces or 1-1/2 cups crumbs evenly over each pan.
8 Using a convection oven, bake on high fan, closed vent 15 to 20 minutes at 350 F. or until sauce is bubbly and crumbs are lightly browned. DO NOT OVERBAKE. CCP: Internal temperature must reach 145 F. or higher for 15 seconds. Hold for service at 140 F. or higher.

Notes

1 In Step 1, 4 pounds of parboiled brown rice may be used per 100 portions. Cook 30-35 minutes or until most of the water is absorbed.

MEAT. FISH. AND POULTRY No.L 076 00

BEEF MANICOTTI (CANNELLONI)

Yield 100　　　　　　　　　　　　　　　　　　　**Portion** 2 Shells

Calories	Carbohydrates	Protein	Fat	Cholesterol	Sodium	Calcium
588 cal	63 g	27 g	26 g	63 mg	1037 mg	212 mg

Ingredient	Weight	Measure	Issue
SAUCE,PIZZA,CANNED	36-1/8 lbs	3 gal 3 qts	
MANICOTTI,BEEF,W/O SAUCE,FROZEN	46-7/8 lbs		

Method

1. Spread 1-1/2 cups pizza sauce in thin layer over bottom of each steam table pan.
2. Place 20 frozen manicotti in each pan.
3. Pour 4-1/2 cups pizza sauce over each pan to cover manicotti.
4. Cover pans.
5. Using a convection oven, bake 30 minutes on high fan, closed vent at 350 F. Remove cover; bake 5 minutes longer. CCP: Internal temperature must reach 165 F. or higher for 15 seconds. Hold for service at 140 F. or higher.

MEAT, FISH, AND POULTRY No. L 076 01

CHEESE MANICOTTI

Yield 100   **Portion** 2 Shells

Calories	Carbohydrates	Protein	Fat	Cholesterol	Sodium	Calcium
307 cal	34 g	16 g	12 g	37 mg	1132 mg	344 mg

Ingredient   Weight   Measure   Issue

Ingredient	Weight	Measure	Issue
SAUCE,PIZZA,CANNED	36-1/8 lbs	3 gal 3 qts	
MANICOTTI,CHEESE,W/O SAUCE,FROZEN	46-7/8 lbs		

Method

1. Spread 1-1/2 cups pizza sauce in thin layer over bottom of each steam table pan.
2. Place 20 frozen manicotti in each pan.
3. Pour 4-1/2 cups pizza sauce over each pan to cover manicotti.
4. Cover pans.
5. Using a convection oven, bake 30 minutes on high fan, closed vent at 350 F. Remove cover; bake 5 minutes longer. CCP: Internal temperature must reach 165 F. or higher for 15 seconds. Hold for service at 140 F. or higher.

MEAT. FISH. AND POULTRY   No.L 077 00

SAVORY ROAST LAMB

Yield 100   Portion   4 Ounces

Calories	Carbohydrates	Protein	Fat	Cholesterol	Sodium	Calcium
333 cal	1 g	37 g	19 g	122 mg	92 mg	29 mg

Ingredient   **Weight**   **Measure**   **Issue**

LAMB,LEG,BONELESS   39 lbs
GARLIC POWDER   1/4 oz   1/3 tsp
PEPPER,BLACK,GROUND   1/2 oz   2 tbsp
PAPRIKA,GROUND   3/4 oz   3 tbsp
OREGANO,CRUSHED   1 oz   1/4 cup 2-1/3 tbsp
VINEGAR,DISTILLED   1-1/3 lbs   2-1/2 cup
OIL,SALAD   1-1/4 lbs   2-1/2 cup

Method

1 Place roasts fat side up in pans. Combine garlic, oregano, paprika, pepper, vinegar and salad oil. Refrigerate 2 hours; turn occasionally. DO NOT ADD WATER; DO NOT COVER.
2 Roast 3 to 4 hours at 325 F. Insert meat thermometer after 2 hours of cooking; continue to roast until the thermometer registers the desired degree of doneness. CCP: Internal temperature must reach 145 F. or higher for 15 seconds.
3 Let roasts stand 20 minutes. Remove string or netting before slicing. CCP: Hold for service at 140 F. or higher.

MEAT, FISH, AND POULTRY No.L 078 00

CHICKEN ADOBO (8 PC)

Yield 100   Portion 7 Ounces

Calories	Carbohydrates	Protein	Fat	Cholesterol	Sodium	Calcium
321 cal	14 g	41 g	11 g	119 mg	1091 mg	36 mg

Ingredient	Weight	Measure	Issue
CHICKEN, 8 PC CUT, SKIN REMOVED	82 lbs		
VINEGAR,DISTILLED	5-1/4 lbs	2 qts 2 cup	
WATER	12-1/2 lbs	1 gal 2 qts	
SOY SAUCE	2-7/8 lbs	1 qts 1/2 cup	
GINGER,GROUND	2 oz	1/2 cup 2-2/3 tbsp	
PEPPER,BLACK,GROUND	7/8 oz	1/4 cup 1/3 tbsp	
GARLIC POWDER	5/8 oz	2 tbsp	
COOKING SPRAY,NONSTICK	2-1/8 oz	1/4 cup 2/3 tbsp	
CHICKEN BROTH		3 qts 3 cup	
PEPPERS,GREEN,FRESH,CHOPPED	4-3/8 lbs	3 qts 1-1/2 cup	5-3/8 lbs
ONIONS,FRESH,CHOPPED	3-1/2 lbs	2 qts 1-7/8 cup	3-7/8 lbs
SUGAR,GRANULATED	10-5/8 oz	1-1/2 cup	
WATER,COLD	3-2/3 lbs	1 qts 3 cup	
CORNSTARCH	1-1/4 lbs	1 qts 3/8 cup	

Method

1. Wash chicken thoroughly under cold running water. Drain well. Remove excess fat. Place approximately 22 pounds chicken in each roasting pan; cover.
2. Combine vinegar, water, soy sauce, ginger, pepper, and garlic powder; stir to blend.
3. Ladle 3 quarts marinade over chicken in each pan. CCP: Cover; marinate under refrigeration at 41 F. or lower for 45 minutes, turning once.
4. Drain chicken. CCP: Reserve marinade under refrigeration at 41 F. or lower for use in Step 7.
5. Place chicken, meat side up, on lightly sprayed sheet pans. Lightly spray chicken with cooking spray.
6. Using a convection oven, bake for 40 minutes on 325 F. on high fan, closed vent. CCP: Internal temperature must reach 165 F. or higher for 15 seconds. Transfer chicken to steam table pans. CCP: Hold at 140 F. or higher for use in Step 9.
7. Combine marinade, broth, peppers, onions and sugar in a steam-jacketed kettle or stockpot. Bring to a boil. Cover, reduce heat; simmer 8-10 minutes until tender.
8. Blend cornstarch and cold water together to make a smooth slurry. Add slurry to broth and vegetable mixture. Bring to a boil. Cover, reduce heat; simmer 3 minutes or until thickened, stirring frequently to prevent sticking. CCP: Temperature must reach 165 F. or higher for 15 seconds.
9. Pour 2-3/4 quarts sauce evenly over chicken in each pan. CCP: Hold for service at 140 F. or higher.

MEAT, FISH, AND POULTRY No. L 079 00

SWEET AND SOUR PORK CHOPS

Yield 100   Portion   3 Ounces

Calories	Carbohydrates	Protein	Fat	Cholesterol	Sodium	Calcium
355 cal	18 g	29 g	18 g	81 mg	126 mg	12 mg

Ingredient	Weight	Measure	Issue
PINEAPPLE,CANNED,CHUNKS,JUICE PACK,INCL LIQUIDS	6-5/8 lbs	3 qts	
VINEGAR,DISTILLED	1-2/3 lbs	3-1/4 cup	
SOY SAUCE	5-1/8 oz	1/2 cup	
RESERVED LIQUID	4-1/8 lbs	2 qts	
SUGAR,GRANULATED	2-1/4 lbs	1 qts 1 cup	
GINGER,GROUND	1/4 oz	1 tbsp	
PEPPERS,GREEN,FRESH,JULIENNE	1-1/4 lbs	3-3/4 cup	1-1/2 lbs
CORNSTARCH	7-7/8 oz	1-3/4 cup	
WATER	1-5/8 lbs	3 cup	
PORK CHOP,BONELESS,5 OZ	31-1/4 lbs		
COOKING SPRAY,NONSTICK	2 oz	1/4 cup 1/3 tbsp	

Method

1. Drain pineapple; reserve juice for Step 2; reserve pineapple for use in Step 3.
2. Combine vinegar, soy sauce, reserved pineapple juice and water, sugar and ginger. Bring to a boil; reduce heat; simmer 5 minutes.
3. Add pineapple and peppers to sauce. Bring to a boil; reduce heat; simmer 5 minutes or until peppers are almost tender, stirring constantly.
4. Dissolve cornstarch in water; stir until smooth. Add to sauce, stirring constantly. Cook until thick and clear. Keep hot for Step 6.
5. Brown chops 11 minutes on each side on 375 F. griddle. CCP: Internal temperature must reach 145 F. or higher for 15 seconds.
6. CCP: Hold for service at 140 F. or higher. Serve with 1/4 cup sauce.

MEAT, FISH, AND POULTRY   No.L 079 01

SWEET AND SOUR CHICKEN (8 PC)

Yield 100   Portion   8 Ounces

Calories	Carbohydrates	Protein	Fat	Cholesterol	Sodium	Calcium
351 cal	23 g	39 g	10 g	119 mg	198 mg	26 mg

Ingredient

Ingredient	Weight	Measure	Issue
CHICKEN, 8 PC CUT, SKIN REMOVED	82 lbs		
COOKING SPRAY,NONSTICK	2-1/8 oz	1/4 cup 2/3 tbsp	
PINEAPPLE,CANNED,CHUNKS,JUICE PACK,INCL LIQUIDS	5-1/4 lbs	2 qts 1-1/2 cup	
RESERVED LIQUID	4-1/8 lbs	2 qts	
SUGAR,GRANULATED	2-1/4 lbs	1 qts 1 cup	
VINEGAR,DISTILLED	1-2/3 lbs	3-1/4 cup	
SOY SAUCE	5-1/8 oz	1/2 cup	
GINGER,GROUND	3/8 oz	2 tbsp	
PEPPERS,GREEN,FRESH,JULIENNE	2 lbs	1 qts 2 cup	2-3/8 lbs
CORNSTARCH	2 lbs	1 qts 3 cup	
WATER,COLD	1-5/8 lbs	3 cup	

Method

1. Wash chicken thoroughly under cold running water. Drain well.
2. Lightly spray chicken with cooking spray. Place chicken on lightly sprayed sheet pan.
3. Using a convection oven, bake 40 minutes at 325 F. on high fan, closed vent. CCP: Internal temperature must reach 165 F. or higher for 15 seconds. Hold at 140 F. or higher for use in Step 8.
4. Drain pineapple. Reserve juice.
5. Combine reserved pineapple juice, sugar, vinegar, soy sauce and ginger in steam-jacketed kettle or stockpot. Bring to a boil. Cover; reduce heat; simmer 5 minutes.
6. Add pineapple and peppers to sauce. Bring to a boil; reduce heat; simmer 5 minutes or until peppers are almost tender, stirring constantly.
7. Blend cornstarch and cold water together to make a smooth slurry. Add slurry to hot sauce, stirring constantly. Cover; reduce heat; simmer 3 to 5 minutes or until thickened, stirring frequently to prevent sticking.
8. Transfer chicken to steam table pans. Pour sauce evenly over chicken in each pan. CCP: Hold for service at 140 F. or higher.

MEAT, FISH, AND POULTRY No.L 079 02

SWEET AND SOUR CHICKEN (COOKED DICED)

Yield 100   **Portion** 8 Ounces

Calories	Carbohydrates	Protein	Fat	Cholesterol	Sodium	Calcium
420 cal	68 g	25 g	6 g	73 mg	398 mg	35 mg

Ingredient	**Weight**	**Measure**	**Issue**
PINEAPPLE,CANNED,CHUNKS,JUICE PACK,INCL LIQUIDS	20-7/8 lbs	2 gal 1-1/2 qts
RESERVED LIQUID	16-3/4 lbs	2 gal
SUGAR,GRANULATED	8-7/8 lbs	1 gal 1 qts
VINEGAR,DISTILLED	6-3/4 lbs	3 qts 1 cup
SOY SAUCE	1-1/4 lbs	2 cup
GINGER,GROUND	1-1/2 oz	1/2 cup
PEPPERS,GREEN,FRESH,JULIENNE	7-7/8 lbs	1 gal 2 qts	9-5/8 lbs
CORNSTARCH	2 lbs	1 qts 3 cup
WATER,COLD	6-1/4 lbs	3 qts
CHICKEN,COOKED,DICED	18 lbs	

Method

1. Drain pineapple.
2. Combine reserved pineapple juice, sugar, vinegar, soy sauce and ginger in steam-jacketed kettle or stockpot. Bring to a boil. Cover; reduce heat; simmer 5 minutes.
3. Add pineapple and peppers to sauce. Bring to a boil; reduce heat; simmer 5 minutes or until peppers are almost tender, stirring constantly.
4. Blend cornstarch and cold water together to make a smooth slurry. Add slurry to hot sauce, stirring constantly. Cover; reduce heat; simmer 3 to 5 minutes or until thickened, stirring frequently to prevent sticking.
5. Stir chicken gently into thickened sauce. Cover; reduce heat; simmer 2 minutes. CCP: Internal temperature must reach 165 F. or higher for 15 seconds.
6. Pour sweet and sour chicken into ungreased steam table pans. CCP: Hold for service at 140 F. or higher.

MEAT, FISH, AND POULTRY No.L 080 00

PORK CHOP SUEY

Yield 100   Portion 1-1/4 Cups

Calories	Carbohydrates	Protein	Fat	Cholesterol	Sodium	Calcium
384 cal	23 g	32 g	18 g	98 mg	859 mg	58 mg

Ingredient	Weight	Measure	Issue
PORK CUBES,RAW	32 lbs		
PEPPER,BLACK,GROUND	1/2 oz	2 tbsp	
GINGER,GROUND	<1/16th oz	1/8 tsp	
WATER	10-1/2 lbs	1 gal 1 qts	
SOY SAUCE	2-1/2 lbs	1 qts	
MOLASSES	5-3/4 oz	1/2 cup	
ONIONS,FRESH,SLICED	12-1/8 lbs	2 gal 4 qts	13-1/2 lbs
CELERY,FRESH,SLICED	8-1/2 lbs	2 gal <1/16th qts	11-5/8 lbs
CABBAGE,GREEN,FRESH,WEDGED	4-1/3 lbs	1 gal 3 qts	5-3/8 lbs
CORNSTARCH	1 lbs	3-3/4 cup	
WATER	4-1/8 lbs	2 qts	
BEAN SPROUTS,CANNED,DRAINED	3-1/8 lbs	1 gal 1-3/4 qts	
NOODLES,CHOW MEIN,CANNED	3-5/8 lbs	2 gal 1 qts	

Method

1. Brown pork in steam-jacketed kettle or stock pot.
2. Sprinkle pork with pepper and ginger.
3. Add water, soy sauce, and molasses; bring to a boil; cover; simmer 1 hour or until tender. CCP: Internal temperature must reach 145 F. or higher for 15 seconds.
4. Add onions, celery and cabbage to pork mixture; mix well; cover; bring to a boil; reduce heat; simmer 10 minutes.
5. Combine cornstarch and water. Stir slowly into hot mixture, stirring constantly. Cook 3 to 5 minutes or until thickened.
6. Add bean sprouts; mix well; bring to a simmer. CCP: Hold for service at 140 F. or higher.
7. Serve with 1/3 cup chow mein noodles and steamed rice.

MEAT, FISH, AND POULTRY No.L 080 01

SHRIMP CHOP SUEY

Yield 100   Portion 1-1/4 Cups

Calories	Carbohydrates	Protein	Fat	Cholesterol	Sodium	Calcium
221 cal	23 g	19 g	6 g	140 mg	1640 mg	79 mg

Ingredient	Weight	Measure	Issue
SHRIMP,FROZEN,RAW,PEELED,DEVEINED	20 lbs		
WATER,BOILING	25-1/8 lbs	3 gal	
RESERVED LIQUID	20-7/8 lbs	2 gal 2 qts	
ONIONS,FRESH,CHOPPED	10-1/8 lbs	1 gal 3-1/8 qts	11-1/4 lbs
CELERY,FRESH,CHOPPED	9-1/2 lbs	2 gal 1 qts	13 lbs
CABBAGE,GREEN,FRESH,SHREDDED	3-1/3 lbs	1 gal 1-3/8 qts	4-1/8 lbs
SOY SAUCE	3-3/4 lbs	1 qts 2 cup	
MOLASSES	5-3/4 oz	1/2 cup	
SALT	3-3/8 oz	1/4 cup 1-2/3 tbsp	
GARLIC POWDER	1/4 oz	1/3 tsp	
PEPPER,BLACK,GROUND	1/8 oz	1/3 tsp	
GINGER,GROUND	<1/16th oz	1/8 tsp	
CORNSTARCH	1-1/8 lbs	1 qts	
WATER	4-1/8 lbs	2 qts	
NOODLES,CHOW MEIN,CANNED	3-5/8 lbs	2 gal 1 qts	
BEAN SPROUTS,CANNED,DRAINED	6-3/4 lbs	3 qts 1/4 cup	

Method

1 Place shrimp in boiling water; cover; return to boil. Reduce heat; simmer 3 to 5 minutes. Immediately remove shrimp from cooking liquid and rinse in cold water or ice bath for 2 minutes. Drain shrimp. Reserve shrimp cooking liquid for use in Step 2. CCP: Refrigerate shrimp at 41 F. for use in Step 4.
2 Combine cooking liquid, onions, celery, cabbage, soy sauce, molasses, salt, garlic powder, pepper, and ginger. Bring to a boil; reduce heat; simmer, covered, for 10 minutes, stirring occasionally until vegetables are tender-crisp. CCP: Internal temperature must reach 165 F. or higher for 15 seconds.
3 Combine cornstarch and water. Stir to make a smooth slurry. Add slurry to hot mixture, stirring constantly. Bring to a boil. Cook gently 3 to 5 minutes, stirring occasionally.
4 Add shrimp and bean sprouts; mix well. Simmer 2 to 3 minutes. CCP: Hold for service at 140 F. or higher. Serve over chow mein noodles.

MEAT, FISH, AND POULTRY No. L 081 00

ROAST PORK

Yield 100 Portion 3-1/2 Ounces

Calories	Carbohydrates	Protein	Fat	Cholesterol	Sodium	Calcium
247 cal	0 g	27 g	14 g	81 mg	59 mg	20 mg

Ingredient | Weight | Measure | Issue

Ingredient	Weight	Measure	Issue
PORK,LOIN,BONELESS,RAW	31-1/4 lbs		
PEPPER,BLACK,GROUND	7/8 oz	1/4 cup 1/3 tbsp	

Method

1. Place roasts, fat side up in pans without crowding. Sprinkle with pepper.
2. Insert meat thermometer in center of the thickest part of a roast. DO NOT ADD WATER. DO NOT COVER.
3. Using a convection oven, roast 1 1/2 hours to 2 hours at 325 F. on high fan, closed vent, depending on size of roasts. CCP: Internal temperature must reach 145 F. or higher for 15 seconds.
4. Let stand 20 minutes. Remove strings or netting before slicing. CCP: Hold for service at 140 F. or higher.

MEAT, FISH, AND POULTRY No.L 081 01

ROAST PORK TENDERLOIN

Yield 100   Portion 4 Ounces

Calories	Carbohydrates	Protein	Fat	Cholesterol	Sodium	Calcium
186 cal	0 g	30 g	6 g	93 mg	65 mg	6 mg

Ingredient	Weight	Measure	Issue
PORK, TENDERLOIN	31-1/4 lbs		
PEPPER,BLACK,GROUND	7/8 oz	1/4 cup 1/3 tbsp	

Method

1. Tie roasts. Place roasts fat side up in roasting pans. Sprinkle with pepper.
2. Insert meat thermometer in center of the thickest part of a roast. DO NOT ADD WATER. DO NOT COVER.
3. Using a convection oven, roast 45 - 60 minutes at 325 F. on high fan, closed vent. CCP: Internal temperature must reach 145 F. or higher for 15 seconds. Hold for service at 140 F. or higher.

MEAT. FISH. AND POULTRY No.L 081 02

BARBECUED PORK LOIN

Yield 100   **Portion** 4 Ounces

Calories	Carbohydrates	Protein	Fat	Cholesterol	Sodium	Calcium
303 cal	14 g	28 g	15 g	81 mg	628 mg	39 mg

Ingredient   Weight   Measure   Issue

PORK,LOIN,BONELESS,RAW   31-1/4 lbs
PEPPER,BLACK,GROUND   7/8 oz   1/4 cup 1/3 tbsp
BARBECUE SAUCE   1 gal 2-1/4 qts

Method

1. Place roasts, fat side up in pans without crowding. Sprinkle with pepper.
2. Insert meat thermometer in center of the thickest part of a roast. DO NOT ADD WATER. DO NOT COVER.
3. Roast 2 to 4 hours at 325 F., depending on size of roasts. CCP: Internal temperature must reach 145 F. or higher for 15 seconds.
4. Let stand 20 minutes. Remove strings or netting before slicing.
5. Prepare Barbecue Sauce, Recipe No. O 002 00, or use prepared Barbecue Sauce.
6. Serve 1/4 cup hot Barbecue Sauce over pork. CCP: Hold for service at 140 F. or higher.

MEAT, FISH, AND POULTRY No. L 082 00

SWEET AND SOUR PORK

Yield 100   Portion 1 Cup

Calories	Carbohydrates	Protein	Fat	Cholesterol	Sodium	Calcium
348 cal	28 g	29 g	13 g	98 mg	329 mg	26 mg

Ingredient	Weight	Measure	Issue
BEAN SPROUTS,CANNED,INCL LIQUIDS	13 lbs	2 gal 3-7/8 qts	
PINEAPPLE,CANNED,CHUNKS,JUICE PACK,INCL LIQUIDS	6-3/4 lbs	3 qts 1/4 cup	
RESERVED LIQUID	7-5/8 lbs	3 qts 2-1/2 cup	
SUGAR,GRANULATED	3-1/2 lbs	2 qts	
VINEGAR,DISTILLED	3-1/8 lbs	1 qts 2 cup	
SOY SAUCE	10-1/8 oz	1 cup	
GINGER,GROUND	2/3 oz	1/4 cup	
GARLIC POWDER	1/8 oz	1/8 tsp	
PORK CUBES,RAW	32 lbs		
CORNSTARCH	14-2/3 oz	3-1/4 cup	
WATER	4-1/8 lbs	2 qts	
PEPPERS,GREEN,FRESH,MEDIUM,SLICED,THIN	4 lbs	3 qts	4-3/4 lbs

Method

1. Drain bean sprouts and pineapple; reserve juices for use in Step 2.
2. Combine reserved juices, sugar, vinegar, soy sauce, ginger and garlic. Blend well. Set aside for use in Step 4.
3. Cook pork in steam-jacketed kettle or stock pot about 10 minutes or until tender. DO NOT OVERCOOK. CCP: Internal temperature must reach 145 F. or higher for 15 seconds.
4. Add soy sauce mixture to pork. Bring to a boil; reduce heat; simmer 3 minutes.
5. Dissolve cornstarch in water; stir until smooth. Add to pork mixture. Bring to boil; reduce heat; simmer 5 minutes or until thickened, stirring constantly.
6. Add green peppers, pineapple and bean sprouts. Bring to a boil; reduce heat; cook 5 minutes. CCP: Hold for service at 140 F. or higher. Serve with steamed rice.

MEAT, FISH, AND POULTRY No.L 083 00

CREOLE PORK CHOPS

Yield 100   Portion   3 Ounces

Calories	Carbohydrates	Protein	Fat	Cholesterol	Sodium	Calcium
326 cal	9 g	31 g	18 g	81 mg	282 mg	33 mg

Ingredient	Weight	Measure	Issue
PORK CHOP,BONELESS,5 OZ	31-1/4 lbs		
COOKING SPRAY,NONSTICK	2 oz	1/4 cup 1/3 tbsp	
TOMATOES,CANNED,CRUSHED,INCL LIQUIDS	13-1/4 lbs	1 gal 2 qts	
TOMATO PASTE,CANNED	1 lbs	1-3/4 cup	
ONIONS,FRESH,CHOPPED	1-5/8 lbs	1 qts 5/8 cup	1-3/4 lbs
PEPPERS,GREEN,FRESH,CHOPPED	1-1/2 lbs	1 qts 1/2 cup	1-7/8 lbs
CELERY,FRESH,CHOPPED	1-1/4 lbs	1 qts 3/4 cup	1-3/4 lbs
FLOUR,WHEAT,GENERAL PURPOSE	8-7/8 oz	2 cup	
WORCESTERSHIRE SAUCE	2-1/8 oz	1/4 cup 1/3 tbsp	
SUGAR,GRANULATED	1-3/4 oz	1/4 cup 1/3 tbsp	
SALT	1 oz	1 tbsp	
PEPPER,BLACK,GROUND	1/4 oz	1 tbsp	
PEPPER,RED,GROUND	1/8 oz	1/3 tsp	

Method

1. Lightly spray griddle with non-stick cooking spray. Brown chops on griddle for 5 minutes on each side.
2. Place an equal number of chops in each pan.
3. Combine tomatoes, tomato paste, onions, peppers, celery, flour, Worcestershire sauce, sugar, salt and peppers.
4. Bring to a boil; stir well; reduce heat; cover; simmer 5 minutes or until thickened.
5. Pour about 3 quart sauce over chops in each pan. Cover.
6. Using a convection oven, bake 1-1/4 hours on high fan, closed vent or until tender in 325 F. oven. CCP: Internal temperature must reach 145 F. or higher for 15 seconds. Hold for service at 140 F. or higher.

MEAT, FISH, AND POULTRY No.L 083 01

BARBECUED PORK CHOPS

Yield 100   Portion 3 Ounces

Calories	Carbohydrates	Protein	Fat	Cholesterol	Sodium	Calcium
324 cal	6 g	30 g	19 g	81 mg	452 mg	15 mg

Ingredient	Weight	Measure	Issue
PORK CHOP,BONELESS,5 OZ	31-1/4 lbs		
COOKING SPRAY,NONSTICK	2 oz	1/4 cup 1/3 tbsp	
SAUCE,BARBECUE	11 lbs	1 gal 1 qts	

Method

1 Lightly spray griddle with non-stick cooking spray. Brown chops at 375 F. griddle for 5 minutes on each side.
2 Place an equal number of chops in each steam table pan.
3 Heat prepared barbecue sauce.
4 Pour 6-1/2 cups sauce over chops in each pan. Cover pans.
5 Using a convection oven, bake 1-1/4 hours or until tender in 325 F. oven on high fan, closed vent. CCP: Internal temperature must reach 145 F. or higher for 15 seconds. Hold for service at 140 F. or higher.

MEAT. FISH. AND POULTRY No.L 084 00

BAKED STUFFED PORK CHOPS

Yield 100   Portion   3 Ounces

Calories	Carbohydrates	Protein	Fat	Cholesterol	Sodium	Calcium
342 cal	10 g	31 g	19 g	91 mg	209 mg	29 mg

Ingredient	Weight	Measure	Issue
PORK CHOP,BONELESS,5 OZ	31-1/4 lbs		
COOKING SPRAY,NONSTICK	2 oz	1/4 cup 1/3 tbsp	
BREAD,WHITE,SLICED	3-2/3 lbs	3 gal	
ONIONS,FRESH,CHOPPED	1 lbs	3 cup	1-1/8 lbs
SALT	5/8 oz	1 tbsp	
PEPPER,BLACK,GROUND	1/8 oz	1/3 tsp	
SEASONING,POULTRY	1/4 oz	2 tbsp	
EGGS,WHOLE,FROZEN	8-5/8 oz	1 cup	
WATER	5-1/4 lbs	2 qts 2 cup	
PEPPERS,GREEN,FRESH,RINGS	4 lbs	3 qts 1/8 cup	4-7/8 lbs
WATER	2-5/8 lbs	1 qts 1 cup	

Method

1. Lightly spray griddle with non-stick cooking spray. Brown chops on griddle for 5 minutes on each side.
2. Place 20 chops in each steam table pan.
3. Combine bread, onions, salt, pepper, poultry seasoning, eggs, and water; mix lightly but thoroughly.
4. Place 1 pepper ring on each chop; top with 1/4 cup bread mixture.
5. Pour 1 cup water in each pan.
6. Using a convection oven, bake 40-45 minutes on high fan, closed vent or until tender in 325 F. oven. CCP: Internal temperature must reach 145 F. or higher for 15 seconds. Hold for service at 140 F. or higher.

MEAT, FISH, AND POULTRY No.L 084 01

PORK CHOPS WITH APPLE RINGS

Yield 100   Portion 3 Ounces

Calories	Carbohydrates	Protein	Fat	Cholesterol	Sodium	Calcium
315 cal	7 g	29 g	18 g	81 mg	45 mg	8 mg

Ingredient	Weight	Measure	Issue
PORK CHOP,BONELESS,5 OZ	31-1/4 lbs		
COOKING SPRAY,NONSTICK	2 oz	1/4 cup 1/3 tbsp	
APPLES,COOKING,FRESH,UNPEELED	7 lbs	25 each	8-1/4 lbs
SUGAR,GRANULATED	8-7/8 oz	1-1/4 cup	
WATER	2-5/8 lbs	1 qts 1 cup	

Method

1. Lightly spray griddle with non-stick cooking spray. Brown chops on griddle for 6 minutes on each side.
2. Place 20 chops in each steam table pan.
3. Core apples; slice crosswise into rings 1/2-inch thick, 4 rings per apple.
4. Place 1 ring on each chop; sprinkle about 1/4 cup sugar over apples in each pan.
5. Pour 2-1/2 cups water in each pan.
6. Using a convection oven, bake 18 to 20 minutes at 325 F. oven or until apples are tender. CCP: Internal temperature must reach 145 F. or higher for 15 seconds. Hold for service at 140 F. or higher.

MEAT, FISH, AND POULTRY No.L 085 00

BRAISED PORK CHOPS

Yield 100　　　　　　　　　　　　　　　　　　　Portion 3 Ounces

Calories	Carbohydrates	Protein	Fat	Cholesterol	Sodium	Calcium
286 cal	0 g	29 g	18 g	81 mg	184 mg	6 mg

Ingredient　　　　　　　　　　　　　　Weight　　　　　Measure　　　　　Issue

Ingredient	Weight	Measure
PORK CHOP,BONELESS,5 OZ	31-1/4 lbs	
COOKING SPRAY,NONSTICK	2 oz	1/4 cup 1/3 tbsp
SALT	1-1/4 oz	2 tbsp
PEPPER,BLACK,GROUND	1/8 oz	1/3 tsp
WATER	3-1/8 lbs	1 qts 2 cup

Method

1. Lightly spray griddle with non-stick cooking spray. Brown chops on griddle for 5 minutes on each side.
2. Sprinkle mixture of salt and pepper evenly over pork chops.
3. Place an equal quantity of chops in each steam table pan.
4. Pour 3 cups water in each pan. Cover.
5. Using a convection oven, bake in 325 F. oven for 1-1/4 hours or until done on high fan, closed vent. CCP: Internal temperature must reach 145 F. or higher for 15 seconds. Hold for service at 140 F. or higher.

MEAT, FISH, AND POULTRY No.L 085 01

GRILLED PORK CHOPS

Yield 100   Portion 3 Ounces

Calories	Carbohydrates	Protein	Fat	Cholesterol	Sodium	Calcium
286 cal	0 g	29 g	18 g	81 mg	44 mg	6 mg

Ingredient   **Weight**   **Measure**   **Issue**

PORK CHOP,BONELESS,5 OZ   31-1/4 lbs
COOKING SPRAY,NONSTICK   2 oz   1/4 cup 1/3 tbsp

Method

1 Lightly spray griddle with non-stick cooking spray. Grill chops on griddle 11 minutes on each side or until browned and thoroughly cooked. CCP: Internal temperature must reach 145 F. or higher for 15 seconds. Hold for service at 140 F. or higher.

MEAT, FISH, AND POULTRY No.L 085 02

PORK CHOPS WITH MUSHROOM GRAVY

Yield 100   Portion 3 Ounces

Calories	Carbohydrates	Protein	Fat	Cholesterol	Sodium	Calcium
344 cal	6 g	30 g	21 g	81 mg	343 mg	18 mg

Ingredient	Weight	Measure	Issue
PORK CHOP,BONELESS,5 OZ	31-1/4 lbs		
COOKING SPRAY,NONSTICK	2 oz	1/4 cup 1/3 tbsp	
PEPPER,BLACK,GROUND	1/8 oz	1/3 tsp	
SOUP,CONDENSED,CREAM OF MUSHROOM	9-1/2 lbs	1 gal 1/4 qts	
FLOUR,WHEAT,GENERAL PURPOSE	13-1/4 oz	3 cup	
WATER	3-2/3 lbs	1 qts 3 cup	

Method

1 Lightly spray griddle with non-stick cooking spray. Brown chops on griddle for 5 minutes on each side.
2 Place an equal quantity of chops in each steam table pan.
3 Combine pepper, soup, and flour. Mix well. Add water; bring to a boil, stirring constantly.
4 Pour 2 quarts gravy over chops in each pan. Cover.
5 Using a convection oven, bake in 325 F. oven for 1-1/4 hours on high fan, closed vent or until done. CCP: Internal temperature must reach 145 F. or higher for 15 seconds. Hold for service at 140 F. or higher.

MEAT, FISH, AND POULTRY No. L 086 01

CREOLE PORK STEAKS (FROZEN BREADED PORK STEAKS)

Yield 100 **Portion** 4 Ounces

Calories	Carbohydrates	Protein	Fat	Cholesterol	Sodium	Calcium
409 cal	29 g	25 g	22 g	114 mg	484 mg	81 mg

Ingredient	Weight	Measure	Issue
CREOLE SAUCE		2 gal	
PORK,STEAK,BREADED,FROZEN	35 lbs		

Method

1. Prepare 1 recipe Creole Sauce, Recipe No. O 005 00 per 100 portions.
2. Deep fat fry at 350 F. pork steak 7 to 8 minutes or until done. CCP: Internal temperature must reach 145 F. or higher for 15 seconds.
3. Drain well in basket or on absorbent paper. Place on sheet pans. CCP: Hold for service at 140 F. or higher.
4. Serve 1/3 cup of sauce with each steak.

Notes

1. For oven method, bake in a 325 F. convection oven for 20 minutes on high fan, closed vent. CCP: Internal temperature must reach 145 F. or higher for 15 seconds.

MEAT, FISH, AND POULTRY   No. L 086 02

BREADED PORK STEAKS (FROZEN)

Yield 100   **Portion** 6-1/2 Ounces

Calories	Carbohydrates	Protein	Fat	Cholesterol	Sodium	Calcium
509 cal	31 g	33 g	27 g	163 mg	401 mg	77 mg

Ingredient   Weight   Measure   Issue

PORK,STEAK,BREADED,FROZEN   35 lbs

Method

1. Deep fat fry at 350 F. pork steak 7 to 8 minutes or until done.   CCP: Internal temperature must reach 145 F. or higher for 15 seconds.
2. Drain in basket or on absorbent paper. Place on pans. CCP: Hold for service at 140 F. or higher.

Notes

1. For oven method, bake at 325 F. in a convection oven, for 20 minutes on high fan, and closed vent. CCP: Internal temperature must reach 145 F. or higher for 15 seconds.

MEAT, FISH, AND POULTRY No.L 086 03

PORK SCHNITZEL (FROZEN BREADED PORK STEAKS)

Yield 100   Portion 4 Ounces

Calories	Carbohydrates	Protein	Fat	Cholesterol	Sodium	Calcium
371 cal	23 g	24 g	20 g	114 mg	281 mg	63 mg

Ingredient	Weight	Measure	Issue
PORK,STEAK,BREADED,FROZEN	35 lbs		
LEMONS,FRESH	3-1/2 lbs	9 each	

Method

1. Deep fat fry at 350 F. pork steak 7 to 8 minutes or until done CCP: Internal temperature must reach 145 F. or higher for 15 seconds.
2. Drain well in basket or on absorbent paper. Place on sheet pans. CCP: Hold for service at 140 F. or higher.
3. Remove ends of lemons. Cut 11 to 12 slices per lemon. Serve 1 lemon slice with each steak.

Notes

1. For oven method, bake in 325 F. convection oven for 20 minutes on high fan, closed vent. CCP: Internal temperature must reach 145 F. or higher for 15 seconds.

MEAT, FISH, AND POULTRY No. L 087 00

PORK CHOPS MEXICANA

Yield 100 Portion 5 Ounces

Calories	Carbohydrates	Protein	Fat	Cholesterol	Sodium	Calcium
319 cal	7 g	30 g	19 g	81 mg	376 mg	22 mg

Ingredient	Weight	Measure	Issue
PORK CHOP,BONELESS,5 OZ	31-1/4 lbs		
COOKING SPRAY,NONSTICK	2 oz	1/4 cup 1/3 tbsp	
WATER	4-2/3 lbs	2 qts 1 cup	
CATSUP,TOMATO,CANNED	2-7/8 lbs	1 qts 1-3/8 cup	
SOY SAUCE	10-1/8 oz	1 cup	
VINEGAR,DISTILLED	1 lbs	2 cup	
ONIONS,FRESH,CHOPPED	1-3/4 lbs	1 qts 1 cup	2 lbs
PEPPERS,GREEN,FRESH,CHOPPED	14-1/2 oz	2-3/4 cup	1-1/8 lbs
CHILI POWDER,DARK,GROUND	6-1/3 oz	1-1/2 cup	
PAPRIKA,GROUND	1-1/2 oz	1/4 cup 2-1/3 tbsp	
GARLIC POWDER	3/8 oz	1 tbsp	
SUGAR,GRANULATED	7/8 oz	2 tbsp	
MUSTARD,DRY	3-1/8 oz	1/2 cup	

Method

1 Lightly spray griddle with non-stick cooking spray. Brown chops 5 minutes on both sides on 375 F. griddle.
2 Place an equal number of chops in each steam table pan.
3 Combine water, catsup, soy sauce, vinegar, onions, peppers, chili powder, paprika, garlic, sugar, and mustard flour; mix thoroughly. Bring to boil; reduce heat; cover; simmer 5 minutes.
4 Pour 2 quarts mixture over chops in each pan.
5 Bake in 375 F. oven for 1-3/4 to 2 hours or until tender. CCP: Internal temperature must reach 145 F. or higher for 15 seconds.
6 Skim excess fat from sauce; serve sauce over chops. CCP: Hold for service at 140 F. or higher.

MEAT, FISH, AND POULTRY No. L 088 00

GRILLED POLISH SAUSAGE

Yield 100   Portion 3 Ounces

Calories	Carbohydrates	Protein	Fat	Cholesterol	Sodium	Calcium
202 cal	1 g	9 g	18 g	43 mg	544 mg	7 mg

Ingredient	Weight	Measure	Issue
SAUSAGE,POLISH,PORK,RAW	18-3/4 lbs		

Method

1. Cut sausage into 3 ounce pieces or cut diagonally in 1/2-inch thick slices.
2. Grill until thoroughly cooked and browned. Turn frequently to ensure even browning. CCP: Internal temperature must reach 145 F. or higher for 15 seconds. Hold for service at 140 F. or higher.

Notes

1. Sausage may be simmered. Pierce each sausage. Cover with water in a steam jacketed kettle or stock pot. Cover; bring to a boil; reduce heat; simmer 10 minutes. Drain, leaving enough water to cover bottom of container.

MEAT, FISH, AND POULTRY No.L 088 01

BAKED ITALIAN SAUSAGE (HOT OR SWEET)

Yield 100   **Portion** 1 Each

Calories	Carbohydrates	Protein	Fat	Cholesterol	Sodium	Calcium
201 cal	1 g	12 g	16 g	48 mg	573 mg	15 mg

Ingredient   Weight   Measure   Issue

SAUSAGE,ITALIAN,SWEET,RAW   18-3/4 lbs
WATER   1 lbs   2 cup

Method

1. Place Italian sausage links in single layers on sheet pans. Pierce each sausage.
2. Pour 1 cup hot water over sausages in each pan. Cover; bake in 400 F. oven 20 minutes.
3. Remove cover; bake 15 minutes or until browned. CCP: Internal temperature must reach 145 F. or higher for 15 seconds. Hold for service at 140 F. or higher.

MEAT, FISH, AND POULTRY   No.L 088 02

GRILLED FRANKFURTERS

Yield 100 **Portion** 2 Each

Calories	Carbohydrates	Protein	Fat	Cholesterol	Sodium	Calcium
290 cal	2 g	10 g	26 g	45 mg	1016 mg	10 mg

Ingredient **Weight** **Measure** **Issue**

FRANKFURTERS   20 lbs

Method

1. Pierce each frankfurter before grilling.
2. Grill until thoroughly cooked and browned. Turn frequently to ensure even browning. CCP: Internal temperature must reach 145 F. or higher for 15 seconds. Hold for service at 140 F. or higher.

Notes

1. Frankfurters may be simmered. Pierce each frankfurter. Cover with water in a steam jacketed kettle or stock pot. Cover; bring to a boil; 10 minutes. Drain, leaving enough water to cover bottom of container.

MEAT, FISH, AND POULTRY No.L 088 03

GRILLED BRATWURST

Yield 100　　　　　　　　　　　　　　　　　**Portion** 1 Each

Calories	Carbohydrates	Protein	Fat	Cholesterol	Sodium	Calcium
256 cal	2 g	12 g	22 g	51 mg	474 mg	37 mg

Ingredient	**Weight**	**Measure**	**Issue**
BRATWURST	18-3/4 lbs		

Method

1. Pierce each bratwurst before grilling.
2. Grill until thoroughly cooked and browned. Turn frequently to ensure even browning. CCP: Internal temperature must reach 145 F. or higher for 15 seconds. Hold for service at 140 F. or higher.

MEAT, FISH, AND POULTRY No.L 088 05

SIMMERED KNOCKWURST

Yield 100 **Portion** 1 Each

Calories	Carbohydrates	Protein	Fat	Cholesterol	Sodium	Calcium
279 cal	2 g	11 g	25 g	53 mg	916 mg	10 mg

Ingredient   **Weight**   **Measure**   **Issue**

KNOCKWURST,3 OZ   20 lbs

Method

1. Pierce each knockwurst; cover with water in steam-jacketed kettle or stock pot. Cover. Bring to a boil; reduce heat; simmer 10 minutes.
2. Drain, leaving enough water to cover bottom of container. Keep hot until served. CCP: Internal temperature must reach 145 F. or higher for 15 seconds. Hold for service at 140 F. or higher.

MEAT, FISH, AND POULTRY No.L 089 00

GRILLED SAUSAGE PATTIES

Yield 100   **Portion** 2 Patties

Calories	Carbohydrates	Protein	Fat	Cholesterol	Sodium	Calcium
197 cal	0 g	10 g	17 g	44 mg	690 mg	17 mg

Ingredient	**Weight**	**Measure**	**Issue**
SAUSAGE,PORK,RAW	25 lbs		

Method

1. Slice sausage into 2 ounce patties.
2. Grill 12 minutes or until patties are browned and well done. Turn frequently to ensure even browning. CCP: Internal temperature must reach 145 F. or higher for 15 seconds. Hold for service at 140 F. or higher.

Notes

1. Patties may be baked in a 350 F. oven for 25 minutes or until well done. CCP: Internal temperature must reach 145 F. or higher for 15 seconds. Hold for service at 140 F. or higher.

MEAT, FISH, AND POULTRY   No. L 089 02

GRILLED SAUSAGE PATTIES (PREFORMED)

Yield 100   Portion   1 Patty

Calories	Carbohydrates	Protein	Fat	Cholesterol	Sodium	Calcium
147 cal	0 g	8 g	12 g	33 mg	517 mg	13 mg

Ingredient	**Weight**	**Measure**	**Issue**
SAUSAGE PATTY,PORK,RAW,3 OZ	18-3/4 lbs		

Method

1. Use frozen preformed pork sausage patties.
2. Grill 7 minutes or until well done. Turn frequently to ensure even browning. CCP: Internal temperature must reach 145 F. or higher for 15 seconds. Hold for service at 140 F. or higher.

Notes

1. Patties may be baked at 325 F. in convection oven, for 7 minutes on low fan. CCP: Internal temperature must reach 145 F. or higher for 15 seconds. Hold for service at 140 F. or higher.

MEAT, FISH, AND POULTRY No.L 091 00

GRILLED SAUSAGE LINKS (COOKED PORK AND BEEF)

Yield 100 **Portion** 2 Pieces

Calories	Carbohydrates	Protein	Fat	Cholesterol	Sodium	Calcium
176 cal	1 g	6 g	16 g	38 mg	461 mg	5 mg

Ingredient **Weight** **Measure** **Issue**

SAUSAGE,PORK AND BEEF,SMOKED   12 lbs

Method

1. Heat sausage on griddle about 5 minutes.
2. Turn frequently to ensure even browning. CCP: Internal temperature must reach 145 F. or higher for 15 seconds. Hold for service at 140 F. or higher.

Notes

1. Sausages may be baked in a 400 F. oven for 10 minutes or until heated. CCP: Internal temperature must reach 145 F. or higher for 15 seconds. Hold for service at 140 F. or higher.

MEAT, FISH, AND POULTRY No. L 092 00

BARBECUED SPARERIBS

Yield 100   Portion 7 Ounces

Calories	Carbohydrates	Protein	Fat	Cholesterol	Sodium	Calcium
594 cal	16 g	40 g	41 g	161 mg	1022 mg	81 mg

Ingredient	Weight	Measure	Issue
PORK,SPARERIBS,FROZEN,RAW	75 lbs		
WATER	33-1/2 lbs	4 gal	
SAUCE,CHILI	2-1/8 lbs	3-3/4 cup	
CATSUP	11-1/8 lbs	1 gal 1-1/4 qts	
WORCESTERSHIRE SAUCE	14-7/8 oz	1-3/4 cup	
MUSTARD,PREPARED	6-5/8 oz	3/4 cup	
VINEGAR,DISTILLED	1-1/3 lbs	2-1/2 cup	
SALT	1-7/8 oz	3 tbsp	
PEPPER,BLACK,GROUND	2/3 oz	3 tbsp	
PEPPER,RED,GROUND	1/4 oz	1 tbsp	

Method

1 Cut ribs into serving size portions 10 to 12 ounces raw weight total or 2 to 4 ribs. Place ribs in steam-jacketed kettle or stock pot.
2 Cover with water; bring to a boil; reduce heat; simmer 45 minutes or until tender. Drain ribs.
3 Combine chili sauce, catsup, Worcestershire sauce, mustard, vinegar, salt, black and red pepper; bring to a boil. Reduce heat; simmer 5 minutes.
4 Dip ribs in sauce to coat well. Overlap ribs in rows fat side up, in pans.
5 Pour remaining sauce evenly over ribs in each steam table pan; cover pans.
6 Bake 1 hour in 325 F. oven, uncover pans; bake 30 minutes longer. CCP: Internal temperature must reach 145 F. or higher for 15 seconds.
7 Skim off excess fat before serving. CCP: Hold for service at 140 F. or higher.

MEAT, FISH, AND POULTRY No.L 093 00

BRAISED SPARERIBS

Yield 100   Portion   7 Ounces

Calories	Carbohydrates	Protein	Fat	Cholesterol	Sodium	Calcium
536 cal	2 g	39 g	40 g	161 mg	451 mg	68 mg

Ingredient	Weight	Measure	Issue
PORK,SPARERIBS,FROZEN,RAW	75 lbs		
ONIONS,FRESH,CHOPPED	5-1/4 lbs	3 qts 2-7/8 cup	5-7/8 lbs
SALT	3 oz	1/4 cup 1 tbsp	
PEPPER,BLACK,GROUND	1/2 oz	2 tbsp	
WATER	6-1/4 lbs	3 qts	

Method

1 Cut ribs into 10 to 12 ounce portions, about 2 to 4 ribs. Overlap ribs in rows, fat side up, in pans. Using a convection oven, bake at 375 F. for 20 minutes on high fan, open vent or until golden brown.
2 Drain or skim off excess fat.
3 Sprinkle onions, salt and pepper over ribs. Add water to cover bottom of each pan. Cover.
4 Using convection oven, bake at 300 F. for 2 hours on low fan, closed vent until tender. CCP: Internal temperature must reach 145 F. or higher for 15 seconds. Hold for service at 140 F. or higher.

MEAT, FISH, AND POULTRY No. L 093 01

SPARERIBS AND SAUERKRAUT

Yield 100 **Portion** 7 Ounces

Calories	Carbohydrates	Protein	Fat	Cholesterol	Sodium	Calcium
548 cal	5 g	40 g	40 g	161 mg	865 mg	96 mg

Ingredient	Weight	Measure	Issue
PORK,SPARERIBS,FROZEN,RAW	75 lbs		
PEPPER,BLACK,GROUND	1/4 oz	1 tbsp	
SAUERKRAUT,SHREDDED,CANNED,DRAINED	24-3/4 lbs	4 gal 3-3/4 qts	

Method

1 Cut ribs into 10 to 12 ounce portions or 2 to 4 ribs. Overlap ribs in rows, fat side up, in pans. Bake at 400 F. for 30 minutes or until golden brown in roasting pans.
2 Drain or skim off excess fat.
3 Place sauerkraut and pepper over ribs in each pan. Cover.
4 Using a convection oven, bake at 325 F. 2 hours on low fan closed vent or until tender. CCP: Internal temperature must reach 145 F. or higher for 15 seconds. Hold for service at 140 F. or higher.

MEAT, FISH, AND POULTRY No.L 094 00

SWEET AND SOUR SPARERIBS

Yield 100   Portion 7 Ounces

Calories	Carbohydrates	Protein	Fat	Cholesterol	Sodium	Calcium
607 cal	21 g	39 g	40 g	161 mg	294 mg	84 mg

Ingredient	Weight	Measure	Issue
PORK,SPARERIBS,FROZEN,RAW	75 lbs		
WATER	33-1/2 lbs	4 gal	
CORNSTARCH	6 oz	1-3/8 cup	
WATER	3-1/8 lbs	1 qts 2 cup	
SUGAR,BROWN,PACKED	1-7/8 lbs	1 qts 2 cup	
GINGER,GROUND	1/2 oz	2-2/3 tbsp	
SOY SAUCE	10-1/8 oz	1 cup	
VINEGAR,DISTILLED	4-1/8 lbs	2 qts	
PEPPER,BLACK,GROUND	1/8 oz	1/4 tsp	
GARLIC POWDER	1/8 oz	1/4 tsp	
PINEAPPLE,CANNED,CRUSHED,JUICE PACK,INCL LIQUIDS	13-1/8 lbs	1 gal 2 qts	

Method

1 Cut ribs into serving size portions, 2 to 4 ribs, 10 to 12 ounces raw weight total. Place ribs in steam-jacketed kettle or stock pot.
2 Cover with water; bring to a boil; reduce heat; simmer 45 minutes or until tender.
3 Drain.
4 Dissolve cornstarch in water. Add sugar, ginger, soy sauce, vinegar, pepper, and garlic powder. Cook at medium heat until sauce thickens. Stir frequently.
5 Combine pineapple with sauce. Bring to a boil.
6 Overlap ribs in rows, fat side up, in pans. Pour sauce evenly over ribs in each pan.
7 Using convection oven, bake at 325 F., covered for 1 hour on high fan, closed vent; uncover; bake for 15 minutes longer. CCP: Internal temperature must reach 145 F. or higher for 15 seconds.
8 Skim off excess fat before serving. CCP: Hold for service at 140 F. or higher.

MEAT, FISH, AND POULTRY No. L 095 00

CANTONESE SPARERIBS

Yield 100   Portion 8 Ounces

Calories	Carbohydrates	Protein	Fat	Cholesterol	Sodium	Calcium
585 cal	13 g	41 g	40 g	161 mg	1529 mg	72 mg

Ingredient

Ingredient	Weight	Measure	Issue
PORK,SPARERIBS,FROZEN,RAW	75 lbs		
WATER	33-1/2 lbs	4 gal	
SOY SAUCE	5-1/8 lbs	2 qts	
SUGAR,GRANULATED	2 lbs	1 qts 1/2 cup	
CATSUP	2-1/8 lbs	1 qts	

Method

1. Cut ribs into 10 to 12 ounce pieces, about 2 to 4 ribs. Place in stock pot or steam-jacketed kettle.
2. Cover with cold water; bring to a boil; cook 30 minutes. Drain.
3. Place ribs in stainless steel pan. Combine soy sauce, sugar, and catsup. Pour marinade over ribs; marinate at least 1 hour. CCP: Marinate under refrigeration at 41 F. or lower.
4. Remove ribs from marinade; place an equal quantity of ribs in each steam table pan.
5. Bake at 400 F. for 1-1/2 to 2 hours, basting ribs frequently with marinade. CCP: Internal temperature must reach 145 F. or higher for 15 seconds. Hold for service at 140 F. or higher.

MEAT, FISH, AND POULTRY No. L 096 00

ROAST FRESH HAM

Yield 100 Portion 4 Ounces

Calories	Carbohydrates	Protein	Fat	Cholesterol	Sodium	Calcium
247 cal	0 g	31 g	12 g	82 mg	2082 mg	11 mg

Ingredient Weight Measure Issue

PORK,HAM,FRESH,BONELESS,RAW 45 lbs 6 gal 7/8 qts

Method

1. Place hams in pans.
2. Insert meat thermometer into thickest part of ham. DO NOT ADD WATER. DO NOT COVER.
3. Using a convection oven, bake at 300 F. for 4 hours. CCP: Internal temperature must reach 145 F. or higher for 15 seconds.
4. Let stand 20 minutes. Remove string or netting before slicing. CCP: Hold for service at 140 F. or higher.

MEAT, FISH, AND POULTRY No.L 097 00

SHRIMP JAMBALAYA

Yield 100   Portion   1-1/2 Cups

Calories	Carbohydrates	Protein	Fat	Cholesterol	Sodium	Calcium
352 cal	49 g	25 g	6 g	100 mg	1792 mg	132 mg

Ingredient	Weight	Measure	Issue
SHRIMP,RAW,PEELED,DEVEINED	10 lbs		
COOKING SPRAY,NONSTICK	1-1/2 oz	3 tbsp	
ONIONS,FRESH,CHOPPED	12 lbs	2 gal 1/2 qts	13-1/3 lbs
CELERY,FRESH,CHOPPED	1-3/4 lbs	1 qts 2-5/8 cup	2-3/8 lbs
PEPPERS,GREEN,FRESH,CHOPPED	2 lbs	1 qts 2-1/8 cup	2-1/2 lbs
GARLIC POWDER	5/8 oz	2 tbsp	
TOMATOES,CANNED,CRUSHED,INCL LIQUIDS	26-1/2 lbs	3 gal	
TOMATO PASTE,CANNED	1-1/2 lbs	2-1/2 cup	
SALT	1-7/8 oz	3 tbsp	
BASIL,SWEET,WHOLE,CRUSHED	3-1/8 oz	1-1/4 cup	
MARJORAM,SWEET,GROUND	1/3 oz	1/4 cup 1-2/3 tbsp	
THYME,FRESH	2/3 oz	1/2 cup	
OREGANO,CRUSHED	7/8 oz	1/4 cup 1-2/3 tbsp	
PEPPER,RED,GROUND	1/4 oz	1 tbsp	
BAY LEAF,FRESH	1/4 oz	8 each	
CHICKEN BROTH		2 gal	
RICE,LONG GRAIN	8-1/2 lbs	1 gal 1-1/4 qts	
HAM,COOKED,BONELESS	13 lbs		

Method

1. Thoroughly rinse and drain shrimp. CCP: Refrigerate at 41 F. or lower for use in Step 5.
2. Lightly spray steam jacketed kettle or stockpot with nonstick cooking spray. Stir-cook onions, peppers, and celery in a steam-jacketed kettle or stock pot 8 to 10 minutes or until tender, stirring constantly.
3. Add tomatoes, chicken broth, tomato paste, basil, salt, marjoram, thyme, oregano, garlic powder, red pepper, and bay leaves to cooked vegetables. Stir to blend. Bring to a boil. Reduce heat; simmer 10 minutes.
4. Add ham and rice to sauce mixture. Stir to blend. Bring to a boil. Cover; reduce heat; simmer 30 minutes or until rice is tender, stirring occasionally. CCP: Internal temperature must reach 145 F. or higher for 15 seconds.
5. Add shrimp to sauce and rice mixture. Stir to blend. Bring to a boil. Cover; reduce heat; simmer 6 to 8 minutes or until shrimp is just done. Do not overcook the shrimp. CCP: Internal temperature of the shrimp must reach 145 F. or higher for 15 seconds. Remove the bay leaves. CCP: Hold for service at 140 F. or higher.

MEAT. FISH. AND POULTRY No.L 099 00

PORK ADOBO

Yield 100 **Portion** 5 Ounces

Calories	Carbohydrates	Protein	Fat	Cholesterol	Sodium	Calcium
262 cal	6 g	28 g	13 g	98 mg	325 mg	16 mg

Ingredient	Weight	Measure	Issue
PORK CUBES,RAW	32 lbs		
SOY SAUCE	1 lbs	1-1/2 cup	
VINEGAR,DISTILLED	2-1/8 lbs	1 qts	
GARLIC POWDER	1/8 oz	1/8 tsp	
GINGER,GROUND	3/4 oz	1/4 cup 1/3 tbsp	
BAY LEAF,FRESH	1/8 oz	4 each	
PEPPER,BLACK,GROUND	1/2 oz	2 tbsp	
CORNSTARCH	11-1/4 oz	2-1/2 cup	
WATER,COLD	2-1/8 lbs	1 qts	
ONIONS,FRESH,SLICED	3 lbs	2 qts 3-7/8 cup	3-1/3 lbs
PEPPERS,GREEN,FRESH,MEDIUM,SLICED,THIN	4 lbs	3 qts 1/8 cup	4-7/8 lbs

Method

1. Place pork in steam jacketed kettle or stock pot.
2. Combine soy sauce, vinegar, garlic, ginger, bay leaves, and pepper. Pour over pork; mix well. Cover; bring to a boil; reduce heat; simmer 30 minutes. Skim off excess fat. Remove bay leaves.
3. Dissolve cornstarch in water; stir into pork mixture. Bring to a boil, reduce heat; cook 5 minutes or until thickened.
4. Add onions and peppers; cook until tender, about 20 minutes. CCP: Internal temperature of pork must reach 145 F. or higher for 15 seconds. Hold for service at 140 F. or higher.

MEAT, FISH, AND POULTRY No.L 100 00

SIMMERED PORK HOCKS (HAM HOCKS)

Yield 100 Portion 7 Ounces

Calories	Carbohydrates	Protein	Fat	Cholesterol	Sodium	Calcium
461 cal	1 g	33 g	35 g	94 mg	2171 mg	27 mg

Ingredient	Weight	Measure	Issue
PORK,HOCKS,(CURED & SMOKED),FROZEN	64 lbs		
WATER,BOILING	66-7/8 lbs	8 gal	
SALT	3-3/4 oz	1/4 cup 2-1/3 tbsp	
BAY LEAF,FRESH	1/3 oz	9 each	
GARLIC POWDER	1/4 oz	1/3 tsp	
PEPPER,BLACK,GROUND	2/3 oz	3 tbsp	
ONIONS,FRESH,QUARTERED	3 lbs	2 qts 3-7/8 cup	3-1/3 lbs

Method

1. Place frozen pork hocks in steam-jacketed kettle or stock pot. Add water, salt, bay leaves, garlic, pepper, and onions. Cover; bring to a boil; reduce heat; simmer 2-1/2 hours or until tender. CCP: Internal temperature must reach 145 F. or higher for 15 seconds.
2. Place pork hocks in serving pans. Add enough cooking liquid to half cover pork hocks. Remove bay leaves before serving. CCP: Hold for service at 140 F. or higher.

MEAT, FISH, AND POULTRY No.L 101 00

ITALIAN STYLE VEAL STEAKS

Yield 100　　　　　　　　　　　　　　　　　　**Portion**　3 Ounces

Calories	Carbohydrates	Protein	Fat	Cholesterol	Sodium	Calcium
190 cal	9 g	13 g	12 g	39 mg	471 mg	46 mg

Ingredient	Weight	Measure	Issue
COOKING SPRAY,NONSTICK	2 oz	1/4 cup 1/3 tbsp	
VEAL,PATTY,UNBREADED	25 lbs		
COOKING SPRAY,NONSTICK	2 oz	1/4 cup 1/3 tbsp	
ONIONS,FRESH,CHOPPED	2 lbs	1 qts 1-5/8 cup	2-1/4 lbs
PEPPERS,GREEN,FRESH,CHOPPED	1-1/2 lbs	1 qts 1/2 cup	1-7/8 lbs
BEEF BROTH		2 qts	
TOMATOES,CANNED,DICED,DRAINED	13-1/4 lbs	1 gal 2 qts	
PARSLEY,FRESH,BUNCH,CHOPPED	4 oz	1-7/8 cup	4-1/4 oz
SUGAR,GRANULATED	3-1/2 oz	1/2 cup	
SALT	1 oz	1 tbsp	
OREGANO,CRUSHED	1/3 oz	2 tbsp	
BASIL,SWEET,WHOLE,CRUSHED	5/8 oz	1/4 cup 1/3 tbsp	
GARLIC POWDER	5/8 oz	2 tbsp	

Method

1. Lightly spray griddle with cooking spray. Grill veal steaks 8 minutes.
2. Evenly shingle 25 veal steaks into each ungreased steam table pan.
3. Stir-cook peppers and onions in a lightly sprayed steam jacketed kettle or stock pot about 8 to 10 minutes, stirring constantly.
4. Add tomatoes, broth, parsley, sugar, salt, sweet basil, oregano and garlic powder to cooked vegetables; stir to blend. Mix well; bring to a boil. Reduce heat. Simmer 5 minutes.
5. Pour 2-1/4 quart sauce over steaks in each pan.
6. Using a convection oven, bake at 325 F. 20 minutes on high fan, closed vent. CCP: Internal temperature must reach 145 F. or higher for 15 seconds. Hold for service at 140 F. or higher.

MEAT. FISH. AND POULTRY No.L 102 00

VEAL PAPRIKA STEAK

Yield 100 Portion 3 Ounces

Calories	Carbohydrates	Protein	Fat	Cholesterol	Sodium	Calcium
229 cal	9 g	14 g	15 g	47 mg	659 mg	46 mg

Ingredient	Weight	Measure	Issue
VEAL,PATTY,UNBREADED	25 lbs		
COOKING SPRAY,NONSTICK	3/4 oz	1 tbsp	
COOKING SPRAY,NONSTICK	2 oz	1/4 cup 1/3 tbsp	
ONIONS,FRESH,SLICED	4 lbs	3 qts 3-3/4 cup	4-1/2 lbs
BEEF BROTH		1 gal 1 qts	
PAPRIKA,GROUND	1-1/3 oz	1/4 cup 1-2/3 tbsp	
GARLIC POWDER	1/4 oz	1/3 tsp	
SALT	1-1/4 oz	2 tbsp	
WATER	2-1/8 lbs	1 qts	
FLOUR,WHEAT,GENERAL PURPOSE	1-1/8 lbs	1 qts	
MUSHROOMS,CANNED,DRAINED	2-3/4 lbs	2 qts	
SOUR CREAM	4 lbs	2 qts	
PAPRIKA	3/4 oz	3 tbsp	

Method

1. Lightly spray griddle with cooking spray. Grill veal steaks for 8 minutes.
2. Shingle 25 veal steaks into each ungreased steam table pan.
3. Stir-cook onions in a lightly sprayed steam jacketed kettle or stock pot 8 to 10 minutes; stirring constantly.
4. Add broth, paprika, salt and garlic powder to cooked onions; stir to blend. Bring to a boil, reduce heat to a simmer.
5. Blend flour and water together; stir to make a slurry. Add slurry to broth and onions, stirring constantly. Bring to a boil; reduce heat; simmer 5 minutes or until thickened stirring constantly.
6. Stir chopped mushrooms into thickened gravy; heat to a simmer. Remove from heat.
7. Blend sour cream with 1 qt gravy. Combine remaining gravy. Mix well.
8. Pour 2-1/2 qt of mushroom/onion gravy over steaks in each pan. Sprinkle 2-1/4 tsp paprika over steaks in each pan.
9. Cover; using a convection oven, bake at 325 F. 20 minutes or until thoroughly heated on high fan, closed vent. CCP: Internal temperature must reach 145 F. or higher for 15 seconds. Hold for service at 140 F. or higher.

MEAT, FISH, AND POULTRY No.L 103 00

VEAL PARMESAN

Yield 100   **Portion** 6-1/2 Ounces

Calories	Carbohydrates	Protein	Fat	Cholesterol	Sodium	Calcium
416 cal	17 g	28 g	26 g	109 mg	747 mg	146 mg

Ingredient	Weight	Measure	Issue
TOMATO SAUCE		1 gal 2-1/2 qts	
VEAL,STEAKS,BREADED,FROZEN	37-1/2 lbs		
CHEESE,MOZZARELLA,SLICED	3-1/8 lbs	3 qts 1/2 cup	
CHEESE,PARMESAN,GRATED	7 oz	2 cup	

Method

1. Prepare 1 recipe Tomato Sauce, Recipe No. O 015 00 per 100 portions. Keep hot.
2. Place steaks on sheet pans. Using convection oven, bake at 400 F. for 10 minutes on high fan, closed vent. Turn steaks. Bake 6-8 minutes or until thoroughly heated and browned on high fan, closed vent. CCP: Internal temperature must reach 145 F. or higher.
3. Cut mozzarella cheese slices in half. Place 1/2 slice cheese on each steak.
4. Pour 1-1/4 quarts sauce over steaks in each pan.
5. Sprinkle about 6 tablespoons parmesan cheese over steaks in each pan.
6. Using convection oven, bake at 325 F. 6-8 minutes or until cheese is melted. Hold for service at 140 F. or higher.

MEAT. FISH. AND POULTRY No.L 103 01

VEAL STEAK

Yield 100 **Portion** 6-1/2 Ounces

Calories	Carbohydrates	Protein	Fat	Cholesterol	Sodium	Calcium
358 cal	10 g	24 g	24 g	96 mg	383 mg	36 mg

Ingredient **Weight** **Measure** **Issue**

VEAL,STEAKS,BREADED,FROZEN   37-1/2 lbs

Method

1. Deep fry veal steaks at 350 F. about 5 minutes or until golden brown.
2. CCP: Internal temperature must reach 145 F. or higher for 15 seconds. Hold for service at 140 F. or higher.

MEAT, FISH, AND POULTRY No.L 104 00

JAEGERSCHNITZEL

Yield 100　　　　　　　　　　　　　　　　　**Portion**　4-1/2 Ounces

Calories	Carbohydrates	Protein	Fat	Cholesterol	Sodium	Calcium
408 cal	17 g	25 g	26 g	99 mg	850 mg	45 mg

Ingredient	Weight	Measure	Issue
BUTTER,MELTED	4 oz	1/2 cup	
PEPPERS,GREEN,FRESH,CHOPPED	1-1/4 lbs	3-3/4 cup	1-1/2 lbs
MUSHROOMS,CANNED,DRAINED	3-1/2 lbs	2 qts 2-1/8 cup	
PIMIENTO,CANNED,DRAINED,CHOPPED	7 oz	1 cup	
GARLIC POWDER	1/8 oz	1/4 tsp	
PEPPER,BLACK,GROUND	1/4 oz	3/8 tsp	
PARSLEY,DEHYDRATED,FLAKED	1/2 oz	1/2 cup 2-2/3 tbsp	
BROWN GRAVY		1 gal 2-1/4 qts	
TOMATO PASTE,CANNED	11-1/2 oz	1-1/4 cup	
VEAL,STEAKS,BREADED,FROZEN	37-1/2 lbs		

Method

1 Saute peppers, mushrooms, pimientos and garlic in butter or margarine 3 minutes. Add pepper and parsley. Cook 2 minutes.
2 Prepare 1 recipe Brown Gravy per 100 portions, Recipe No. O 016 00. Add tomato paste; mix well. Bring to boil, stirring constantly.
3 Add gravy mixture to mushroom mixture. Stir. CCP: Hold at 140 F. or higher for use in Step 5.
4 Place veal steaks on sheet pans. Using a convection oven, bake for 10 minutes at 400 F. high fan, closed vent. Turn steaks; bake 6-8 minutes or until thoroughly heated and browned on high fan, closed vent. CCP: Internal temperature must reach 155 F. or higher for 15 seconds.
5 Serve each steak with 1/4 cup hot mushroom sauce. CCP: Hold for service at 140 F. or higher.

MEAT, FISH, AND POULTRY No.L 105 00

VEAL CUBES PARMESAN

Yield 100   Portion   5-1/2 Ounces

Calories	Carbohydrates	Protein	Fat	Cholesterol	Sodium	Calcium
276 cal	6 g	29 g	15 g	114 mg	568 mg	89 mg

Ingredient	Weight	Measure	Issue
VEAL,ROAST,BONELESS,THAWED,DICED	30 lbs		
ONIONS,FRESH,CHOPPED	2-1/8 lbs	1 qts 2 cup	2-1/3 lbs
SALT	1-7/8 oz	3 tbsp	
SUGAR,GRANULATED	1-3/4 oz	1/4 cup 1/3 tbsp	
PEPPER,RED,GROUND	<1/16th oz	1/8 tsp	
GARLIC POWDER	1/8 oz	1/8 tsp	
OREGANO,CRUSHED	1/8 oz	1 tbsp	
BASIL,SWEET,WHOLE,CRUSHED	1/8 oz	1 tbsp	
TOMATO PASTE,CANNED	5 lbs	2 qts 3/4 cup	
WATER	18-3/4 lbs	2 gal 1 qts	
CHEESE,PARMESAN,GRATED	14-1/8 oz	1 qts	

Method

1. Brown veal in steam-jacketed kettle. Drain or skim off excess fat.
2. Add onions; saute until tender.
3. Mix salt, sugar, red pepper, garlic, oregano, basil, tomato paste, and water. Add to veal; bring to a boil. Reduce heat; cover; simmer 1 hour 15 minutes or until veal is tender. CCP: Internal temperature must reach 145 F. or higher for 15 seconds.
4. Place 4-1/4 quarts of veal mixture in each steam table pan.
5. Sprinkle 1 cup cheese over mixture in each pan. CCP: Hold for service at 140 F. or higher.

MEAT, FISH, AND POULTRY No. L 106 00

ROAST VEAL

Yield 100   Portion 4 Ounces

Calories	Carbohydrates	Protein	Fat	Cholesterol	Sodium	Calcium
296 cal	0 g	34 g	17 g	140 mg	127 mg	26 mg

Ingredient   **Weight**   **Measure**   **Issue**

VEAL,ROAST,BONELESS,RAW   38 lbs
PEPPER,BLACK,GROUND   1/2 oz   2 tbsp

Method

1. Place roasts fat side up in pans without crowding. Sprinkle roasts with pepper.
2. Insert meat thermometer into roasts. DO NOT ADD WATER; DO NOT COVER.
3. Using a convection oven, bake at 325 F. 3-1/2 hours on high fan, closed vent. CCP: Internal temperature must reach 145 F. or higher for 15 seconds.
4. Let stand 20 minutes. Remove netting before slicing. CCP: Hold for service at 140 F. or higher.

MEAT, FISH, AND POULTRY No.L 106 01

ROAST VEAL WITH HERBS

Yield 100   Portion 4 Ounces

Calories	Carbohydrates	Protein	Fat	Cholesterol	Sodium	Calcium
296 cal	0 g	34 g	17 g	140 mg	127 mg	29 mg

Ingredient	Weight	Measure	Issue
VEAL,ROAST,BONELESS,RAW	38 lbs		
PEPPER,BLACK,GROUND	1/3 oz	1 tbsp	
THYME,GROUND	1/4 oz	1 tbsp	
GARLIC POWDER	1/8 oz	1/4 tsp	
TARRAGON,GROUND	1/8 oz	1 tbsp	
DILL WEED,DRIED	1/8 oz	1 tbsp	

Method

1 Place roasts fat side up in pans. Rub roasts with pepper, ground thyme, garlic powder, ground tarragon and dill weed.
2 Insert meat thermometer into roasts. DO NOT ADD WATER; DO NOT COVER.
3 Using a convection oven, roast at 325 F. 3-1/2 hours on high fan, closed vent. CCP: Internal temperature must reach 145 F. or higher for 15 seconds.
4 Let stand 20 minutes. Remove netting before slicing. CCP: Hold for service at 140 F. or higher.

MEAT, FISH, AND POULTRY No.L 107 00

BRAISED LIVER WITH ONIONS

Yield 100   Portion 4 Ounces

Calories	Carbohydrates	Protein	Fat	Cholesterol	Sodium	Calcium
268 cal	15 g	22 g	13 g	326 mg	480 mg	17 mg

Ingredient

Ingredient	Weight	Measure	Issue
COOKING SPRAY,NONSTICK	2 oz	1/4 cup 1/3 tbsp	
BEEF,LIVER,RAW,SLICED,4 OZ	25 lbs		
FLOUR,WHEAT,GENERAL PURPOSE	2-1/2 lbs	2 qts 1 cup	
SALT	3-3/4 oz	1/4 cup 2-1/3 tbsp	
PEPPER,BLACK,GROUND	1/2 oz	2 tbsp	
PAPRIKA,GROUND	1 oz	1/4 cup 1/3 tbsp	
SHORTENING	1-3/4 lbs	1 qts	
ONIONS,FRESH,SLICED	8 lbs	1 gal 3-7/8 qts	8-7/8 lbs
WATER	8-1/3 lbs	1 gal	

Method

1. Lightly spray griddle with non-stick cooking spray. Dredge liver in mixture of flour, salt, pepper, and paprika; shake off excess. Brown on a 375 F. griddle.
2. Overlap about 50 slices in each pan.
3. Saute onions in shortening or salad oil until tender; spread an equal quantity over liver in each pan.
4. Pour hot water over liver and onions in each roasting pan; cover.
5. Bake 30 minutes in 350 F. oven or until liver is fork-tender. CCP: Internal temperature must reach 145 F. or higher for 15 seconds. Hold for service at 140 F. or higher.

MEAT, FISH, AND POULTRY No.L 107 01

GRILLED LIVER

Yield 100 **Portion** 4 Ounces

Calories	Carbohydrates	Protein	Fat	Cholesterol	Sodium	Calcium
182 cal	12 g	22 g	5 g	326 mg	478 mg	9 mg

Ingredient	Weight	Measure	Issue
COOKING SPRAY,NONSTICK	2 oz	1/4 cup 1/3 tbsp	
BEEF,LIVER,RAW,SLICED,4 OZ	25 lbs		
FLOUR,WHEAT,GENERAL PURPOSE	2-1/2 lbs	2 qts 1 cup	
SALT	3-3/4 oz	1/4 cup 2-1/3 tbsp	
PEPPER,BLACK,GROUND	1/2 oz	2 tbsp	
PAPRIKA,GROUND	1 oz	1/4 cup 1/3 tbsp	

Method

1. Lightly spray griddle with non-stick cooking spray. Dredge liver in mixture of flour, salt, pepper, and paprika; shake off excess. Brown evenly on both sides on a 375 F. griddle. CCP: Internal temperature must reach 145 F. or higher for 15 seconds. Hold at 140 F. or higher for service.

MEAT, FISH, AND POULTRY No.L 108 00

BREADED LIVER

Yield 100   Portion   4-1/2 Ounces

Calories	Carbohydrates	Protein	Fat	Cholesterol	Sodium	Calcium
337 cal	20 g	24 g	18 g	357 mg	552 mg	29 mg

Ingredient / Weight / Measure / Issue

Ingredient	Weight	Measure	Issue
MILK,NONFAT,DRY	7/8 oz	1/4 cup 2-1/3 tbsp	
WATER,WARM	1 lbs	1-7/8 cup	
EGGS,WHOLE,FROZEN	1-1/2 lbs	2-7/8 cup	
BEEF,LIVER,RAW,SLICED,4 OZ	25 lbs		
BREADCRUMBS,DRY,GROUND,FINE	2-5/8 lbs	2 qts 3 cup	
FLOUR,WHEAT,GENERAL PURPOSE	3 lbs	2 qts 3 cup	
SALT	3-3/4 oz	1/4 cup 2-1/3 tbsp	
PEPPER,BLACK,GROUND	1/4 oz	1 tbsp	
SHORTENING	2-3/4 lbs	1 qts 2 cup	

Method

1. Reconstitute milk; add eggs.
2. Dip liver in milk and egg mixture. Drain.
3. Dredge liver in mixture of crumbs, flour, salt and pepper; shake off excess.
4. Brown slices on lightly greased griddle about 5 minutes per side at 375 F. CCP: Internal temperature must reach 145 F. or higher for 15 seconds.
5. CCP: Hold for service at 140 F. or higher.

MEAT, FISH, AND POULTRY No. L 108 01

BREADED LIVER WITH ONION AND MUSHROOM GRAVY

Yield 100   Portion   4-1/2 Ounces

Calories	Carbohydrates	Protein	Fat	Cholesterol	Sodium	Calcium
414 cal	25 g	25 g	23 g	357 mg	917 mg	33 mg

Ingredient	Weight	Measure	Issue
MILK,NONFAT,DRY	7/8 oz	1/4 cup 2-1/3 tbsp	
WATER,WARM	1 lbs	1-7/8 cup	
EGGS,WHOLE,FROZEN	1-1/2 lbs	2-7/8 cup	
BEEF,LIVER,RAW,SLICED,4 OZ	25 lbs		
BREADCRUMBS,DRY,GROUND,FINE	2-5/8 lbs	2 qts 3 cup	
FLOUR,WHEAT,GENERAL PURPOSE	3 lbs	2 qts 3 cup	
SALT	3-3/4 oz	1/4 cup 2-1/3 tbsp	
PEPPER,BLACK,GROUND	1/4 oz	1 tbsp	
SHORTENING	2-3/4 lbs	1 qts 2 cup	
ONION AND MUSHROOM GRAVY		1 gal 2 qts	

Method

1 Reconstitute milk; add eggs.
2 Dip liver in milk and egg mixture. Drain.
3 Dredge liver in mixture of crumbs, flour, salt and pepper; shake off excess.
4 Brown slices on lightly greased 375 F. griddle about 5 minutes per side. CCP: Internal temperature must reach 145 F. or higher for 15 seconds.
5 Serve with 1 recipe Onion and Mushroom Gravy, Recipe No. O 016 09, per 100 portions. Each portion is 1 slice of liver plus 1/4 cup of gravy.

MEAT, FISH, AND POULTRY No.L 109 00

OVEN FRIED CHICKEN FILLETS (3 OZ)

Yield 100　　　　　　　　　　　　　　　　　　　　**Portion** 2 Fillets

Calories	Carbohydrates	Protein	Fat	Cholesterol	Sodium	Calcium
551 cal	21 g	28 g	39 g	73 mg	1165 mg	49 mg

Ingredient	Weight	Measure	Issue
CHICKEN FILLET,BREADED,PRECOOKED,FROZEN,3 OZ	37-1/2 lbs		

Method

1. Place fillets on pans. Using a convection oven, bake 12 to 14 minutes or until thoroughly heated in a 375 F. oven on high fan, closed vent. CCP: Internal temperature must reach 165 F. or higher for 15 seconds. Hold for service at 140 F. or higher.

MEAT, FISH, AND POULTRY No.L 109 01

FRIED CHICKEN FILLETS (3 OZ)

Yield 100   **Portion** 2 Fillets

Calories	Carbohydrates	Protein	Fat	Cholesterol	Sodium	Calcium
611 cal	21 g	28 g	46 g	73 mg	1165 mg	49 mg

Ingredient   Weight   Measure   Issue

CHICKEN FILLET,BREADED,PRECOOKED,FROZEN,3 OZ   37-1/2 lbs

Method

1. Fry fillets in 350 F. deep fat fryer for 4 minutes or until thoroughly heated. CCP: Internal temperature must reach 165 F. or higher for 15 seconds.
2. Drain in basket or on absorbent paper. CCP: Hold for service at 140 F. or higher.

MEAT. FISH. AND POULTRY No.L 109 02

OVEN FRIED CHICKEN FILLET (5 OZ)

Yield 100　　　　　　　　　　　　　　　　　　　　　**Portion**　4-1/2 Ounces

Calories	Carbohydrates	Protein	Fat	Cholesterol	Sodium	Calcium
470 cal	18 g	24 g	33 g	62 mg	994 mg	42 mg

Ingredient　　　　　　　　　　　　　　　　　　Weight　　　　Measure　　　　Issue

CHICKEN FILLET,BREADED,PRECOOKED,FROZEN,5 OZ　　32 lbs

Method

1. Place fillets on sheet pans. Using a convection oven, bake 12 to 14 minutes at 375 F. on high fan, closed vent or until thoroughly heated. CCP: Internal temperature must reach 165 F. or higher for 15 seconds. Hold for service at 140 F. or higher.

MEAT, FISH, AND POULTRY No.L 109 03

FRIED CHICKEN FILLETS (5 OZ)

Yield 100   Portion   4 Ounces

Calories	Carbohydrates	Protein	Fat	Cholesterol	Sodium	Calcium
510 cal	18 g	24 g	38 g	62 mg	994 mg	42 mg

Ingredient   **Weight**   **Measure**   **Issue**

CHICKEN FILLET,BREADED,PRECOOKED,FROZEN,5 OZ   32 lbs

Method

1 Fry fillets in 350 F. deep fat fryer 5 minutes or until thoroughly heated. CCP: Internal temperature must reach 165 F. or higher for 15 seconds.
2 Drain in basket or on absorbent paper. CCP: Hold for service at 140 F. or higher.

MEAT, FISH, AND POULTRY No.L 109 04

OVEN FRIED CHICKEN FILLET NUGGETS

Yield 100 Portion 10 Each

Calories	Carbohydrates	Protein	Fat	Cholesterol	Sodium	Calcium
481 cal	20 g	24 g	34 g	57 mg	1020 mg	53 mg

Ingredient	Weight	Measure	Issue
CHICKEN NUGGET,BREADED,PRECOOKED,IQF	32-1/4 lbs		

Method

1. Place nuggets on sheet pans.
2. Using a convection oven, bake at 375 F. for 13 to 15 minutes or until thoroughly heated on high fan, closed vent. CCP: Internal temperature must reach 165 F. or higher for 15 seconds. Hold for service at 140 F. or higher.

MEAT, FISH, AND POULTRY No.L 109 05

FRIED CHICKEN FILLET NUGGETS

Yield 100   Portion 10 Each

Calories	Carbohydrates	Protein	Fat	Cholesterol	Sodium	Calcium
521 cal	20 g	24 g	39 g	57 mg	1020 mg	53 mg

Ingredient   **Weight**   **Measure**   **Issue**

CHICKEN NUGGET,BREADED,PRECOOKED,IQF   32-1/4 lbs

Method

1. Fry nuggets at 350 F. in deep fat fryer for 2-1/2 to 3 minutes or until thoroughly heated. CCP: Internal temperature must reach 165 F. or higher for 15 seconds.
2. Drain in basket or on absorbent paper. CCP: Hold for service at 140 F. or higher.

MEAT, FISH, AND POULTRY No.L 110 00

CORNED BEEF HASH

Yield 100   **Portion** 3-1/2 Ounces

Calories	Carbohydrates	Protein	Fat	Cholesterol	Sodium	Calcium
208 cal	11 g	12 g	13 g	58 mg	730 mg	13 mg

Ingredient	Weight	Measure	Issue
BEEF,CORNED,RAW	15 lbs		
ONIONS,FRESH,CHOPPED	2-1/2 lbs	1 qts 3-1/8 cup	2-3/4 lbs
PEPPERS,GREEN,FRESH,CHOPPED	1-1/2 lbs	1 qts 1/2 cup	1-7/8 lbs
SHORTENING		1/2 cup	
POTATOES,WHITE,FRESH	10 lbs	1 gal 3-1/4 qts	
WATER,BOILING	14-5/8 lbs	1 gal 3 qts	
SALT	1/2 oz	3/8 tsp	
RESERVED STOCK	1-5/8 lbs	3 cup	
PEPPER,BLACK,GROUND	1/8 oz	1/4 tsp	
COOKING SPRAY,NONSTICK	2 oz	1/4 cup 1/3 tbsp	

Method

1. Place whole pieces of corned beef in steam-jacketed kettle or stock pot; cover with water. Bring to a boil. Cover; reduce heat; simmer 2-1/2 hours. Remove scum as it rises to surface. Remove; reserve stock for use in Step 5. CCP: Hold stock at 140 F. or higher.
2. Let corned beef stand 12 to 20 minutes; chop finely.
3. Saute onions and peppers in shortening or salad oil about 10 minutes or until tender. Stir frequently.
4. Place potatoes in boiling salted water. Return to a boil. Reduce heat; cook 10 minutes or until tender, drain.
5. Combine beef, vegetables, potatoes, stock and pepper; mix thoroughly.
6. Lightly spray each pan with non-stick cooking spray. Place about 1-1/2 gallons corned beef mixture into each lightly sprayed steam table pan.
7. Using a convection oven, bake 25 minutes in 325 F. oven or until lightly browned high fan, open vent. CCP: Internal temperature must reach 145 F. or higher for 15 minutes.

Notes

1. In Steps 1 and 2, 9 pounds 15 ounces precooked corned beef, may be used per 100 portions. Follow Steps 3 and 4. In Step 5, use 3 cups water for reserved stock. Follow Steps 6 and 7.

MEAT. FISH. AND POULTRY No.L 110 01

CORNED BEEF HASH (CANNED)

Yield 100   Portion 4 Ounces

Calories	Carbohydrates	Protein	Fat	Cholesterol	Sodium	Calcium
258 cal	6 g	12 g	20 g	50 mg	443 mg	0 mg

Ingredient   **Weight**   **Measure**   **Issue**

CORNED BEEF HASH   27 lbs

Method

1 Prepare according to instructions on container. CCP: Hold for service at 140 F. or higher for 15 seconds.

MEAT, FISH, AND POULTRY No.L 111 00

NEW ENGLAND BOILED DINNER

Yield 100   **Portion** 1 Serving

Calories	Carbohydrates	Protein	Fat	Cholesterol	Sodium	Calcium
626 cal	46 g	36 g	33 g	168 mg	2008 mg	130 mg

Ingredient   Weight   Measure   Issue

BEEF,CORNED,RAW   43-1/2 lbs
WATER   33-1/2 lbs   4 gal
CABBAGE,GREEN,FRESH,WEDGED   30 lbs   12 gal 5/8 qts   37-1/2 lbs
CARROTS,FRESH,2"" STRIPS   10 lbs   2 gal 2-1/3 qts   12-1/4 lbs
RUTABAGAS,FRESH,CHOPPED   10 lbs   2 gal 1/8 qts   11-3/4 lbs
POTATOES,FRESH,PEELED,CUBED   30-1/4 lbs   5 gal 2 qts   37-1/3 lbs
ONIONS,FRESH,QUARTERED   5 lbs   1 gal 7/8 qts   5-1/2 lbs

Method

1 Place whole pieces of corned beef in steam-jacketed kettle or stock pot; cover with water.
2 Bring to a boil. Cover; reduce heat; simmer 2-1/2 hours. Remove scum as it rises to surface.
3 Remove corned beef from liquid. Reserve liquid for use in Step 7.
4 Place corned beef in roasting pans.
5 Bake at 325 F. 1 hour or until tender. CCP: Internal temperature must reach 145 F. or higher for 15 seconds.
6 Let stand 15 to 20 minutes before slicing. Slice corned beef across grain into 3/16-inch slices.
7 Bring reserved liquid to a boil. CCP: Internal temperature must reach 165 F. or higher for 15 seconds. Add cabbage; return to a boil; cook 12 to 15 minutes or until tender. Remove from liquid. Cover; keep warm.
8 Add carrots and rutabagas to reserved liquid; return to a boil; continue to cook 5 minutes.
9 Add potatoes; return to a boil; cook 10 minutes.
10 Add onions; return to a boil; continue to cook 15 minutes or until vegetables are tender. CCP: Hold for service at 140 F. or higher. Each portion: 1 wedge cabbage, 1-1/3 cup other vegetables topped with 3 to 4 thin slices of corned beef.

MEAT, FISH, AND POULTRY No. L 111 01

NEW ENGLAND BOILED DINNER (PRECOOKED FROZEN BEEF)

Yield 100   Portion 1 Serving

Calories	Carbohydrates	Protein	Fat	Cholesterol	Sodium	Calcium
545 cal	47 g	30 g	27 g	135 mg	2522 mg	134 mg

Ingredient	Weight	Measure	Issue
BEEF,CORNED,COOKED	30 lbs		
HAM BROTH (FROM MIX)		8 gal	
CABBAGE,GREEN,FRESH,WEDGED	30 lbs	12 gal 5/8 qts	37-1/2 lbs
CARROTS,FRESH,2" STRIPS	10 lbs	2 gal	12-1/4 lbs
RUTABAGAS,FRESH,CHOPPED	10 lbs	2 gal	11-3/4 lbs
RESERVED LIQUID	62-2/3 lbs	7 gal 2 qts	
POTATOES,FRESH,PEELED,CUBED	30-1/4 lbs	5 gal 2 qts	37-1/3 lbs
ONIONS,FRESH,QUARTERED	5 lbs	3 qts 3-3/4 cup	5-1/2 lbs

Method

1. Place precooked corned beef on sheet pans.
2. Using a convection oven, bake 30 to 35 minutes at 300 F. on high fan, closed vent. CCP: Internal temperature must reach 165 F. or higher for 15 seconds.
3. Let stand 15 to 20 minutes before slicing. Slice corned beef across grain into 3/16-inch slices.
4. Prepare stock according to recipe to make reserved liquid. Bring reserved liquid to a boil. CCP: Internal temperature must reach 165 F. or higher for 15 seconds. Add cabbage; return to a boil; cook 12 to 15 minutes or until tender. Remove from liquid. Cover; keep warm.
5. Add carrots and rutabagas to reserved liquid; return to a boil; continue to cook 5 minutes.
6. Add potatoes; return to a boil; cook 10 minutes.
7. Add onions; return to a boil; continue to cook 15 minutes or until vegetables are tender. CCP: Hold for service at 140 F. or higher. Each portion: 1 wedge cabbage, 1-1/3 cup other vegetables topped with 3 to 4 thin slices of corned beef.

Notes

1. Due to the grain of brisket being varied within a cut, turn piece of meat while carving to ensure cutting across grain to prevent shredding.

MEAT, FISH, AND POULTRY No.L 112 00

SIMMERED CORNED BEEF

Yield 100   Portion   4 Ounces

Calories	Carbohydrates	Protein	Fat	Cholesterol	Sodium	Calcium
431 cal	1 g	31 g	33 g	168 mg	1952 mg	18 mg

Ingredient   **Weight**   **Measure**   **Issue**

BEEF,CORNED,RAW   43-1/2 lbs
WATER   41-3/4 lbs   5 gal

Method

1. Place whole pieces of corned beef in steam-jacketed kettle or stock pot; cover with water.
2. Bring to a boil. Cover; reduce heat; simmer 2-1/2 hours. Remove scum as it rises to surface.
3. Remove corned beef from liquid.
4. Place corned beef in roasting pans.
5. Bake 1 hour or until tender in 325 F. oven. CCP: Internal temperature must reach 145 F. or higher for 15 seconds.
6. Let stand 15 to 20 minutes before slicing. CCP: Hold for service at 140 F. or higher.

MEAT, FISH, AND POULTRY No.L 112 01

APPLE GLAZED CORNED BEEF

Yield 100   Portion 4 Ounces

Calories	Carbohydrates	Protein	Fat	Cholesterol	Sodium	Calcium
467 cal	10 g	32 g	33 g	168 mg	2038 mg	27 mg

Ingredient	Weight	Measure	Issue
BEEF,CORNED,RAW	43-1/2 lbs		
WATER	41-3/4 lbs	5 gal	
JUICE,APPLE,CANNED	9-1/2 lbs	1 gal 1/3 qts	
SOY SAUCE	5-1/8 oz	1/2 cup	
WORCESTERSHIRE SAUCE	6-1/3 oz	3/4 cup	
VINEGAR,DISTILLED	1 lbs	2 cup	
MUSTARD,DRY	2 oz	1/4 cup 1-1/3 tbsp	
SUGAR,BROWN,PACKED	10-7/8 oz	2-1/8 cup	

Method

1. Place whole pieces of corned beef in steam-jacketed kettle or stock pot; cover with water.
2. Bring to a boil. Cover; reduce heat; simmer 2-1/2 hours. Remove scum as it rises to surface.
3. Remove corned beef from liquid.
4. Combine canned apple juice, soy sauce, Worcestershire sauce, vinegar, mustard, and packed brown sugar; blend well; pour over meat in roasting pans.
5. Bake 1 hour or until tender. CCP: Internal temperature must reach 145 F. or higher for 15 seconds. Baste every 15 minutes.
6. Let stand 15 to 20 minutes before slicing. CCP: Hold for service at 140 F. or higher.

MEAT, FISH, AND POULTRY No.L 112 02

BAKED CORNED BEEF (PRECOOKED FROZEN)

Yield 100 **Portion** 4 Ounces

Calories	Carbohydrates	Protein	Fat	Cholesterol	Sodium	Calcium
342 cal	1 g	25 g	26 g	133 mg	1543 mg	11 mg

Ingredient	Weight	Measure	Issue
BEEF,CORNED,COOKED	30 lbs		

Method

1. Place thawed precooked corned beef on sheet pans. Using a convection oven, bake at 300 F. for 30 to 35 minutes. CCP: Internal temperature must reach 145 F. or higher for 15 seconds.
2. Let stand 15 to 20 minutes before slicing. CCP: Hold for service at 140 F. or higher.

MEAT, FISH, AND POULTRY No. L 113 00

BAKED FRANKFURTERS WITH SAUERKRAUT

Yield 100 **Portion** 2 Each

Calories	Carbohydrates	Protein	Fat	Cholesterol	Sodium	Calcium
312 cal	7 g	11 g	27 g	45 mg	1765 mg	44 mg

Ingredient | **Weight** | **Measure** | **Issue**

SAUERKRAUT,SHREDDED,CANNED,INCL LIQUIDS   25 lbs   3 gal
FRANKFURTERS   20 lbs

Method

1. Heat sauerkraut to a simmer. Drain excess liquid.
2. Place 3 quarts sauerkraut in each steam table pan. Arrange 50 frankfurters on top of sauerkraut in each pan.
3. Using a convection oven, bake 20 to 25 minutes at 300 F. on low fan, open vent. CCP: Internal temperature must reach 145 F. or higher for 15 seconds. Hold for service at 140 F. or higher.

MEAT, FISH, AND POULTRY No.L 113 01

BAKED KNOCKWURST WITH SAUERKRAUT

Yield 100   **Portion** 1 Each

Calories	Carbohydrates	Protein	Fat	Cholesterol	Sodium	Calcium
301 cal	6 g	12 g	25 g	53 mg	1665 mg	44 mg

Ingredient | Weight | Measure | Issue

SAUERKRAUT,SHREDDED,CANNED,INCL LIQUIDS — 25 lbs — 3 gal
KNOCKWURST,3 OZ — 20 lbs

Method

1. Heat sauerkraut to a simmer. Drain excess liquid.
2. Place 3 quarts sauerkraut in each pan. Arrange knockwurst on top of sauerkraut in each pan.
3. Using a convection oven, bake 20 to 25 minutes at 300 F. on low fan, open vent. CCP: Internal temperature must reach 155 F. or higher for 15 seconds. Hold for service at 140 F. or higher.

MEAT. FISH. AND POULTRY No.L 114 00

TERIYAKI CHICKEN (8 PC)

Yield 100   Portion 8 Ounces

Calories	Carbohydrates	Protein	Fat	Cholesterol	Sodium	Calcium
296 cal	6 g	42 g	10 g	119 mg	1726 mg	34 mg

Ingredient	Weight	Measure	Issue
CHICKEN, 8 PC CUT, SKIN REMOVED	82 lbs		
WATER	11 lbs	1 gal 1-1/4 qts	
SOY SAUCE	6-1/3 lbs	2 qts 2 cup	
JUICE,PINEAPPLE,CANNED,UNSWEETENED	5 lbs	2 qts 1 cup	
GINGER,GROUND	4-5/8 oz	1-1/2 cup	
PEPPER,BLACK,GROUND	1-1/3 oz	1/4 cup 2-1/3 tbsp	
GARLIC POWDER	1-1/4 oz	1/4 cup 1/3 tbsp	
COOKING SPRAY,NONSTICK	2-1/8 oz	1/4 cup 2/3 tbsp	

Method

1. Wash chicken thoroughly under cold running water. Drain well. Remove excess fat. Place approximately 22 pounds in each roasting pan.
2. Combine water, soy sauce, pineapple juice, ginger, pepper, and garlic powder; mix well. Pour 3-1/2 qt marinade over chicken in each pan; cover. CCP: Marinate under refrigeration at 41 F. or lower for 45 minutes, turning once.
3. Drain chicken. CCP: Reserve marinade under refrigeration at 41 F. or lower for use in Step 5.
4. Lightly spray chicken with cooking spray. Place chicken, meat side up, on lightly sprayed sheet pans.
5. Using a convection oven, bake 20 minutes at 325 F. on high fan, closed vent. Baste chicken with 1 cup reserved marinade per pan. Discard remaining marinade. Bake an additional 20 minutes for a total of 40 minutes. CCP: Internal temperature must reach 165 F. or higher for 15 seconds. Hold for service at 140 F. or higher.

Notes

1. In Step 2, 2 gallons of prepared teriyaki sauce may be used per 100 portions.

MEAT, FISH, AND POULTRY No.L 114 01

TERIYAKI CHICKEN (THIGHS)

Yield 100   Portion 4 Ounces

Calories	Carbohydrates	Protein	Fat	Cholesterol	Sodium	Calcium
308 cal	1 g	38 g	16 g	135 mg	487 mg	20 mg

Ingredient

Ingredient	Weight	Measure	Issue
CHICKEN,THIGHS,BNLS/SKNLS,RAW	31-1/4 lbs		
WATER	2-1/3 lbs	1 qts 1/2 cup	
SOY SAUCE	1-3/8 lbs	2-1/4 cup	
JUICE,PINEAPPLE,CANNED,UNSWEETENED	1-1/8 lbs	2 cup	
GINGER,GROUND	1 oz	1/4 cup 1-2/3 tbsp	
PEPPER,BLACK,GROUND	1/3 oz	1 tbsp	
GARLIC POWDER	1/4 oz	3/8 tsp	
COOKING SPRAY,NONSTICK	1-1/2 oz	3 tbsp	

Method

1. Wash chicken thoroughly under cold running water. Drain well. Remove excess fat. Place chicken in roasting pans, cover.
2. Combine water, soy sauce, pineapple juice, ginger, pepper, and garlic powder; mix well.
3. Pour teriyaki sauce over chicken in each pan; cover. CCP: Marinate under refrigeration at 41 F. or lower for 45 minutes.
4. Place chicken thighs on each lightly sprayed sheet pan. Lightly spray chicken with cooking spray. Discard remaining teriyaki sauce.
5. Using a convection oven, bake 12 to 14 minutes at 325 F. on high fan, closed vent. CCP: Internal temperature must reach 165 F. or higher for 15 seconds.
6. Transfer chicken to steam table pans. Hold for service at 140 F. or higher.

Notes

1. In Step 2, 2 gallons of prepared teriyaki sauce may be used per 100 portions.

MEAT, FISH, AND POULTRY No.L 115 00

SPICY BAKED FISH

Yield 100 Portion 4-1/2 Ounces

Calories	Carbohydrates	Protein	Fat	Cholesterol	Sodium	Calcium
191 cal	8 g	27 g	5 g	72 mg	585 mg	32 mg

Ingredient	Weight	Measure	Issue
FISH,FLOUNDER/SOLE FILLET,RAW	30 lbs		
COOKING SPRAY,NONSTICK	2 oz	1/4 cup 1/3 tbsp	
ONIONS,FRESH,CHOPPED	2 lbs	1 qts 1-5/8 cup	2-1/4 lbs
OIL,SALAD	7-2/3 oz	1 cup	
SAUCE,BARBECUE	9-7/8 lbs	1 gal 1/2 qts	
MUSHROOMS,CANNED,STEMS & PIECES,CHOPPED,DRAINED	5-1/2 lbs	1 gal	
JUICE,LEMON	4-1/3 oz	1/2 cup	

Method

1. Separate fillets or steaks; cut into 4-1/2 ounce portions, if necessary. Lightly spray pans with non-stick cooking spray. Arrange single layers of fish on sheet pans.
2. Saute onions in shortening or salad oil in stock pot or steam-jacketed kettle until tender.
3. Add barbecue sauce, mushrooms, and lemon juice to sauteed onions. Bring sauce to a boil; reduce heat; simmer 10 minutes.
4. Pour 7-1/2 cups sauce evenly over fish in each pan. Cover.
5. Bake 10 minutes; uncover; bake 10 minutes or until done in 375 F. oven. CCP: Internal temperature must reach 145 F. or higher for 15 seconds. Hold for service at 140 F. or higher.
6. Serve fish with 1/4 cup sauce.

MEAT, FISH, AND POULTRY No.L 116 00

MACARONI TUNA SALAD

Yield 100   Portion   3/4 Cup

Calories	Carbohydrates	Protein	Fat	Cholesterol	Sodium	Calcium
201 cal	13 g	14 g	10 g	66 mg	367 mg	21 mg

Ingredient	Weight	Measure	Issue
WATER	14-5/8 lbs	1 gal 3 qts	
SALT	1/2 oz	3/8 tsp	
OIL,SALAD	1/3 oz	1/3 tsp	
MACARONI NOODLES,ELBOW,DRY	2-1/3 lbs	2 qts 2 cup	
FISH,TUNA,CANNED,WATER PACK,INCL LIQUIDS	9-1/2 lbs	1 gal 3 qts	
CELERY,FRESH,CHOPPED	3-1/8 lbs	2 qts 3-3/4 cup	4-1/4 lbs
ONIONS,FRESH,CHOPPED	1 lbs	2-7/8 cup	1-1/8 lbs
PIMIENTO,CANNED,DRAINED,CHOPPED	3-3/8 oz	1/2 cup	
PEPPER,BLACK,GROUND	1/4 oz	1 tbsp	
JUICE,LEMON	8-5/8 oz	1 cup	
PICKLE RELISH,SWEET	1-1/8 lbs	2 cup	
SALAD DRESSING,MAYONNAISE TYPE	3-5/8 lbs	1 qts 3-3/8 cup	
EGG,HARD COOKED,CHOPPED	2-1/2 lbs	2 qts 1/4 cup	
PARSLEY,FRESH,BUNCH,CHOPPED	1/2 oz	1/4 cup	1/2 oz
PAPRIKA,GROUND	1/4 oz	1 tbsp	

Method

1. Add salt and salad oil to water; heat to a rolling boil.
2. Add macaroni slowly while stirring constantly until water boils again. Cook about 8 to 10 minutes or until tender; stir occasionally. DO NOT OVERCOOK.
3. Drain. Rinse with cold water; drain thoroughly. Use immediately in recipe preparation or place in shallow containers and cover.
4. Combine tuna, macaroni, celery, onions, and pimientos. Mix lightly but thoroughly.
5. Combine salad dressing, pickle relish, lemon juice and pepper. Stir to blend thoroughly.
6. Add chopped eggs and salad dressing mixture to tuna mixture. Mix lightly.
7. Garnish with parsley and paprika; cover. CCP: Refrigerate product at 41 F. or lower until ready to serve.

MEAT. FISH. AND POULTRY No.L 116 01

CHICKEN ROTINI SALAD (CANNED CHICKEN)

Yield 100   Portion   3/4 Cup

Calories	Carbohydrates	Protein	Fat	Cholesterol	Sodium	Calcium
281 cal	16 g	17 g	16 g	94 mg	692 mg	28 mg

Ingredient	Weight	Measure	Issue
WATER	20-7/8 lbs	2 gal 2 qts	
SALT	5/8 oz	1 tbsp	
OIL,SALAD	1/2 oz	1 tbsp	
MACARONI NOODLES,ROTINI,DRY	3-1/8 lbs	3 qts 1-1/2 cup	
CHICKEN,BONED,CANNED,PIECES	15-1/2 lbs	1 gal 2-1/8 qts	
CELERY,FRESH,CHOPPED	4 lbs	3 qts 3-1/8 cup	5-1/2 lbs
ONIONS,FRESH,CHOPPED	1 lbs	2-7/8 cup	1-1/8 lbs
PIMIENTO,CANNED,DRAINED,CHOPPED	3-3/8 oz	1/2 cup	
SALAD DRESSING,MAYONNAISE TYPE	4-1/8 lbs	2 qts 3/8 cup	
PICKLE RELISH,SWEET	1-1/8 lbs	2 cup	
JUICE,LEMON	8-5/8 oz	1 cup	
SALT	1 oz	1 tbsp	
PEPPER,BLACK,GROUND	1/4 oz	1 tbsp	
EGG,HARD COOKED,CHOPPED	2-1/2 lbs	2 qts 1/4 cup	
PARSLEY,FRESH,BUNCH,CHOPPED	1/2 oz	1/4 cup	1/2 oz
PAPRIKA,GROUND	1/4 oz	1 tbsp	

Method

1. Add salt and salad oil to water; heat to a rolling boil.
2. Add rotini slowly while stirring constantly until water boils again. Cook about 10 to 12 minutes or until tender; stir occasionally. DO NOT OVER COOK.
3. Drain. Rinse with cold water; drain thoroughly.
4. Cut chicken into 1/2-inch pieces.
5. Combine chicken, rotini, celery, onions and pimientos. Mix lightly but thoroughly.
6. Combine salad dressing, pickle relish, lemon juice, salt and pepper. Stir to blend thoroughly.
7. Add chopped eggs and salad dressing mixture to chicken mixture. Mix lightly.
8. Garnish with parsley and paprika; cover. CCP: Refrigerate product at 41 F. or lower until ready to serve.

MEAT, FISH, AND POULTRY No.L 116 02

CHICKEN ROTINI SALAD (COOKED DICED)

Yield 100   Portion 3/4 Cup

Calories	Carbohydrates	Protein	Fat	Cholesterol	Sodium	Calcium
268 cal	16 g	19 g	14 g	102 mg	403 mg	26 mg

Ingredient	Weight	Measure	Issue
WATER	20-7/8 lbs	2 gal 2 qts	
SALT	5/8 oz	1 tbsp	
OIL,SALAD	1/2 oz	1 tbsp	
MACARONI NOODLES,ROTINI,DRY	3-1/8 lbs	3 qts 1-1/2 cup	
CHICKEN,COOKED,DICED	12 lbs		
CELERY,FRESH,CHOPPED	3-1/8 lbs	2 qts 3-3/4 cup	4-1/4 lbs
ONIONS,FRESH,CHOPPED	1 lbs	2-7/8 cup	1-1/8 lbs
PIMIENTO,CANNED,DRAINED,CHOPPED	3-3/8 oz	1/2 cup	
SALAD DRESSING,MAYONNAISE TYPE	3-5/8 lbs	1 qts 3-3/8 cup	
PICKLE RELISH,SWEET	1-1/8 lbs	2 cup	
JUICE,LEMON	8-5/8 oz	1 cup	
SALT	1 oz	1 tbsp	
PEPPER,BLACK,GROUND	1/4 oz	1 tbsp	
EGG,HARD COOKED,CHOPPED	2-1/2 lbs	2 qts 1/4 cup	
PARSLEY,FRESH,BUNCH,CHOPPED	1/2 oz	1/4 cup	1/2 oz
PAPRIKA,GROUND	1/4 oz	1 tbsp	

Method

1 Add salt and salad oil to water; heat to a rolling boil.
2 Add rotini slowly while stirring constantly until water boils again. Cook about 10 to 12 minutes or until tender. Stir occasionally. DO NOT OVERCOOK.
3 Drain. Rinse with cold water; drain thoroughly.
4 Combine chicken, rotini, celery, onions and pimientos. Mix lightly but thoroughly.
5 Combine salad dressing, pickle relish, lemon juice, salt and pepper. Stir to blend thoroughly.
6 Add chopped eggs and salad dressing mixture to chicken mixture. Mix lightly.
7 Garnish with parsley and paprika; cover. CCP: Refrigerate product at 41 F. or lower until ready to serve.

MEAT, FISH, AND POULTRY No.L 117 01

GRILLED LUNCHEON MEAT

Yield 100   **Portion** 3 Ounces

Calories	Carbohydrates	Protein	Fat	Cholesterol	Sodium	Calcium
407 cal	2 g	14 g	38 g	60 mg	1408 mg	10 mg

Ingredient	Weight	Measure	Issue
LUNCHEON MEAT, CANNED	24 lbs		
SHORTENING	9 oz	1-1/4 cup	

Method

1. Cut luncheon meat into 1-3/4 ounce slices.
2. Grill meat on a lightly greased 350 F. griddle 1 minute per side or until lightly browned. CCP: Internal temperature must reach 145 F. or higher for 15 seconds. Hold for service at 140 F. or higher.

Notes

1. Luncheon meat may be oven cooked. Using a convection oven, bake at 325 F. 5 minutes on low fan, open vent.

MEAT, FISH, AND POULTRY No.L 119 00

BAKED FISH

Yield 100   Portion 4 Ounces

Calories	Carbohydrates	Protein	Fat	Cholesterol	Sodium	Calcium
162 cal	0 g	26 g	6 g	72 mg	364 mg	22 mg

Ingredient	Weight	Measure	Issue
COOKING SPRAY,NONSTICK	2 oz	1/4 cup 1/3 tbsp	
FISH,FLOUNDER/SOLE FILLET,RAW	30 lbs		
JUICE,LEMON	12-7/8 oz	1-1/2 cup	
MARGARINE,MELTED	1 lbs	2 cup	
SALT	1-7/8 oz	3 tbsp	
PAPRIKA,GROUND	1/2 oz	2 tbsp	
PARSLEY,FRESH,BUNCH,CHOPPED	1 oz	1/4 cup	1 oz

Method

1. Separate fillets or steaks; cut into 4-1/2 ounce portions, if necessary. Lightly spray pans with non-stick cooking spray. Arrange single layers of fish on pans.
2. Combine lemon juice, butter or margarine, salt and paprika. Mix well. Drizzle 3/4 cup mixture over fish in each pan.
3. Using a convection oven, bake 7 minutes at 325 F. on high fan, closed vent or until lightly browned. CCP: Internal temperature must reach 145 F. or higher for 15 seconds. Hold for service at 140 F. or higher. Garnish with parsley before serving.

MEAT. FISH. AND POULTRY   No.L 119 01

BAKED FISH WITH GARLIC BUTTER

Yield 100   Portion   4 Ounces

Calories	Carbohydrates	Protein	Fat	Cholesterol	Sodium	Calcium
178 cal	0 g	26 g	8 g	72 mg	246 mg	22 mg

Ingredient	Weight	Measure	Issue
FISH,FLOUNDER/SOLE FILLET,RAW	30 lbs		
COOKING SPRAY,NONSTICK	2 oz	1/4 cup 1/3 tbsp	
JUICE,LEMON	4-1/3 oz	1/2 cup	
MARGARINE,MELTED	1-1/2 lbs	3 cup	
SALT	5/8 oz	1 tbsp	
GARLIC POWDER	7/8 oz	3 tbsp	
PARSLEY,FRESH,BUNCH,CHOPPED	1 oz	1/4 cup	1 oz

Method

1. Separate fillets or steaks; cut into 4-1/2 ounce portions, if necessary. Lightly spray pans with non-stick cooking spray. Arrange single layers of fish on sheet pans.
2. Combine lemon juice, butter or margarine, salt and garlic powder. Mix well. Drizzle 3/4 cup mixture over fish in each pan.
3. Using a convection oven, bake at 325 F. for 7 minutes or until lightly browned on high fan, closed vent. CCP: Internal temperature must reach 145 F. or higher for 15 seconds. Hold for service at 140 F. or higher. Garnish with parsley before serving.

MEAT, FISH, AND POULTRY No.L 119 02

ONION-LEMON BAKED FISH

Yield 100   **Portion**   4 Ounces

Calories	Carbohydrates	Protein	Fat	Cholesterol	Sodium	Calcium
166 cal	1 g	26 g	6 g	72 mg	365 mg	24 mg

Ingredient	Weight	Measure	Issue
FISH,FLOUNDER/SOLE FILLET,RAW	30 lbs		
COOKING SPRAY,NONSTICK	2 oz	1/4 cup 1/3 tbsp	
JUICE,LEMON	12-7/8 oz	1-1/2 cup	
MARGARINE,MELTED	12 oz	1-1/2 cup	
SALT	1-7/8 oz	3 tbsp	
PAPRIKA,GROUND	1/2 oz	2 tbsp	
ONIONS,FRESH,CHOPPED	2-1/8 lbs	1 qts 2 cup	2-1/3 lbs
MARGARINE,MELTED	4 oz	1/2 cup	
PARSLEY,FRESH,BUNCH,CHOPPED	1 oz	1/4 cup	1 oz

Method

1 Separate fillets or steaks; cut into 4-1/2 ounce portions, if necessary. Lightly spray pans with non-stick cooking spray. Arrange single layers of fish on sheet pans.
2 Combine lemon juice, butter or margarine, salt, and paprika. Drizzle 3/4 cup mixture over fish in each pan.
3 Saute finely chopped onions in butter or margarine until tender. Distribute 1 cup sauteed onions over top of fish in each pan.
4 Using a convection oven, bake 7 minutes or until lightly browned in 325 F. oven on high fan, closed vent. CCP: Internal temperature must reach 145 F. or higher for 15 seconds. Hold for service at 140 F. or higher. Garnish with parsley before serving.

MEAT, FISH, AND POULTRY   No.L 119 03

LEMON BAKED FISH

Yield 100   **Portion**   4 Ounces

Calories	Carbohydrates	Protein	Fat	Cholesterol	Sodium	Calcium
154 cal	0 g	26 g	5 g	72 mg	354 mg	21 mg

Ingredient	**Weight**	**Measure**	**Issue**
FISH,FLOUNDER/SOLE FILLET,RAW	30 lbs		
COOKING SPRAY,NONSTICK	2 oz	1/4 cup 1/3 tbsp	
JUICE,LEMON	1-1/8 lbs	2 cup	
MARGARINE,MELTED	12 oz	1-1/2 cup	
SALT	1-7/8 oz	3 tbsp	
PAPRIKA,GROUND	1/2 oz	2 tbsp	
PARSLEY,FRESH,BUNCH,CHOPPED	1 oz	1/4 cup	1 oz

Method

1. Separate fillets or steaks; cut into 4-1/2 ounce portions, if necessary. Lightly spray pans with non-stick cooking spray. Arrange single layers of fish on sheet pans.
2. Combine lemon juice, butter or margarine, salt and paprika. Mix well. Drizzle 3/4 cup mixture over fish in each pan.
3. Using a convection oven, bake 7 minutes at 325 F. on high fan, closed vent, or until lightly browned. CCP: Internal temperature must reach 145 F. or higher for 15 seconds. Hold for service at 140 F. or higher. Garnish with parsley before serving.

MEAT, FISH, AND POULTRY No. L 119 04

HERBED BAKED FISH

Yield 100   **Portion** 4 Ounces

Calories	Carbohydrates	Protein	Fat	Cholesterol	Sodium	Calcium
162 cal	0 g	26 g	6 g	72 mg	364 mg	22 mg

Ingredient	Weight	Measure	Issue
FISH,FLOUNDER/SOLE FILLET,RAW	30 lbs		
COOKING SPRAY,NONSTICK	2 oz	1/4 cup 1/3 tbsp	
JUICE,LEMON	12-7/8 oz	1-1/2 cup	
MARGARINE,MELTED	1 lbs	2 cup	
SALT	1-7/8 oz	3 tbsp	
BASIL,DRIED,CRUSHED	<1/16th oz	1/8 tsp	
THYME,GROUND	<1/16th oz	1/8 tsp	
TARRAGON,GROUND	<1/16th oz	1/8 tsp	
MARJORAM,SWEET,GROUND	<1/16th oz	1/8 tsp	
DILL WEED,DRIED	<1/16th oz	1/8 tsp	

Method

1. Separate fillets or steaks; cut into 4-1/2 ounce portions, if necessary. Lightly spray pans with non-stick cooking spray. Arrange single layers of fish on sheet pans.
2. Combine lemon juice, butter or margarine, salt, ground basil, ground thyme, ground tarragon, ground marjoram and whole dill weed. Mix well. Drizzle 3/4 cup mixture over fish in each pan.
3. Using a convection oven, bake 7 minutes at 325 F. on high fan, closed vent or until lightly browned. CCP: Internal temperature must reach 145 F. or higher for 15 seconds. Hold for service at 140 F. or higher.

MEAT. FISH. AND POULTRY No.L 119 05

MUSTARD-DILL BAKED FISH

Yield 100   Portion   4 Ounces

Calories	Carbohydrates	Protein	Fat	Cholesterol	Sodium	Calcium
168 cal	2 g	26 g	6 g	72 mg	183 mg	24 mg

Ingredient	Weight	Measure	Issue
COOKING SPRAY,NONSTICK	2 oz	1/4 cup 1/3 tbsp	
FISH,FLOUNDER/SOLE FILLET,RAW	30 lbs		
JUICE,LEMON	1-1/8 lbs	2 cup	
MARGARINE,MELTED	1 lbs	2 cup	
MUSTARD,PREPARED	8-7/8 oz	1 cup	
SUGAR,GRANULATED	3-1/2 oz	1/2 cup	
DILL WEED,DRIED	1/4 oz	2 tbsp	
GARLIC POWDER	1/8 oz	1/4 tsp	

Method

1. Separate fillets or steaks; cut into 4-1/2 ounce portions, if necessary. Lightly spray pans with non-stick cooking spray. Arrange single layers of fish on sheet pans.
2. Combine lemon juice, melted butter or margarine, prepared mustard, granulated sugar, whole dillweed and garlic powder. Stir to blend ingredients well. Drizzle about 1-1/3 cups sauce mixture over fish in each pan.
3. Using a convection oven, bake at 325 F. 7 minutes on high fan, closed vent or until lightly browned. CCP: Internal temperature must reach 145 F. or higher for 15 seconds. Hold for service at 140 F. or higher.

MEAT, FISH, AND POULTRY No.L 119 06

FISH AMANDINE

Yield 100   **Portion** 4 Ounces

Calories	Carbohydrates	Protein	Fat	Cholesterol	Sodium	Calcium
181 cal	1 g	26 g	7 g	72 mg	364 mg	29 mg

Ingredient	Weight	Measure	Issue
ALMONDS,SLIVERED	11-3/8 oz	3 cup	
FISH,FLOUNDER/SOLE FILLET,RAW	30 lbs		
COOKING SPRAY,NONSTICK	2 oz	1/4 cup 1/3 tbsp	
JUICE,LEMON	12-7/8 oz	1-1/2 cup	
MARGARINE,MELTED	1 lbs	2 cup	
SALT	1-7/8 oz	3 tbsp	
PAPRIKA,GROUND	1/2 oz	2 tbsp	

Method

1. Spread shelled slivered almonds on a sheet pan in a thin layer. Using a convection oven, bake at 300 F. 12 to 15 minutes on high fan, open vent, stirring occasionally until almonds are lightly browned. Remove from oven.
2. Separate fillets or steaks; cut into 4-1/2 ounce portions, if necessary. Lightly spray pans with non-stick cooking spray. Arrange single layers of fish on pans.
3. Combine lemon juice, butter or margarine, salt, and paprika. Mix well. Drizzle 3/4 cup mixture over fish in each pan.
4. Using a convection oven, bake 7 minutes at 325 F. on high fan, closed vent or until lightly browned. CCP: Internal temperature must reach 145 F. or higher for 15 seconds. Hold for service at 140 F. or higher.
5. Sprinkle 3/4 cup toasted almonds over fish in each pan.

MEAT, FISH, AND POULTRY No.L 119 07

CAJUN BAKED FISH

Yield 100   Portion 4 Ounces

Calories	Carbohydrates	Protein	Fat	Cholesterol	Sodium	Calcium
163 cal	1 g	26 g	6 g	72 mg	364 mg	24 mg

Ingredient	Weight	Measure	Issue
FISH,FLOUNDER/SOLE FILLET,RAW	30 lbs		
COOKING SPRAY,NONSTICK	2 oz	1/4 cup 1/3 tbsp	
JUICE,LEMON	12-7/8 oz	1-1/2 cup	
MARGARINE,MELTED	1 lbs	2 cup	
SALT	1-7/8 oz	3 tbsp	
ONION POWDER	1/2 oz	2 tbsp	
OREGANO,CRUSHED	1/3 oz	2 tbsp	
PAPRIKA,GROUND	1/2 oz	2 tbsp	
GARLIC POWDER	1/3 oz	1 tbsp	
PEPPER,RED,GROUND	1/4 oz	1 tbsp	
PEPPER,BLACK,GROUND	1/4 oz	1 tbsp	

Method

1. Separate fillets or steaks; cut into 4-1/2 ounce portions, if necessary. Lightly spray pans with non-stick cooking spray. Arrange single layers of fish on sheet pans.
2. Combine lemon juice, butter or margarine, salt, onion powder, crushed oregano, ground paprika, garlic powder, red pepper and black pepper. Mix well. Drizzle 3/4 cup mixture over fish in each pan.
3. Using a convection oven, bake 7 minutes at 325 F. on high fan, closed vent or until lightly browned. CCP: Internal temperature must reach 145 F. or higher for 15 seconds. Hold for service at 140 F. or higher.

MEAT, FISH, AND POULTRY No.L 120 00

BAKED STUFFED FISH

Yield 100   Portion   4-1/2 Ounces

Calories	Carbohydrates	Protein	Fat	Cholesterol	Sodium	Calcium
275 cal	22 g	28 g	7 g	85 mg	380 mg	32 mg

Ingredient	Weight	Measure	Issue
CELERY,FRESH,CHOPPED	1 lbs	3-3/4 cup	1-3/8 lbs
ONIONS,FRESH,CHOPPED	1-5/8 lbs	1 qts 5/8 cup	1-3/4 lbs
BUTTER,MELTED	12 oz	1-1/2 cup	
CRACKER CRUMBS	5-7/8 lbs	1 gal 1-3/4 qts	
PEPPER,BLACK,GROUND	1/4 oz	3/8 tsp	
THYME,GROUND	1/3 oz	2 tbsp	
WATER	2-1/8 lbs	1 qts	
FISH,FLOUNDER/SOLE FILLET,RAW	30 lbs		
COOKING SPRAY,NONSTICK	2 oz	1/4 cup 1/3 tbsp	
JUICE,LEMON	6-1/2 oz	3/4 cup	
BUTTER,MELTED	8 oz	1 cup	
SALT	1-7/8 oz	3 tbsp	
PAPRIKA,GROUND	1/2 oz	2 tbsp	

Method

1 Saute celery and onions in butter or margarine until tender.
2 Combine cracker crumbs, pepper, and thyme; add to vegetables.
3 Add water to vegetable-crumb mixture; toss mixture but do not pack.
4 Lightly spray each sheet pan with non-stick cooking spray. Separate fillets; cut into 2-1/4 ounce pieces. Place 50 pieces on each pan.
5 Place 1/4 cup vegetable crumb mixture on each piece. Cover with second fish piece.
6 Combine lemon juice and butter or margarine; pour over fish in each pan.
7 Sprinkle salt and paprika over fish.
8 Bake about 25 minutes in 375 F. oven or until lightly browned. CCP: Internal temperature must reach 145 F. or higher for 15 seconds. Hold for service at 140 F. or higher.

MEAT, FISH, AND POULTRY No.L 121 00

SHRIMP SCAMPI

Yield 100　　　　　　　　　　　　　　　　　　　**Portion** 5-1/2 Ounces

Calories	Carbohydrates	Protein	Fat	Cholesterol	Sodium	Calcium
199 cal	6 g	24 g	9 g	210 mg	583 mg	61 mg

Ingredient	Weight	Measure	Issue
SHRIMP,RAW,PEELED,DEVEINED	30 lbs		
TOMATOES,CANNED,DICED,DRAINED	3-3/4 lbs	1 qts 2-3/4 cup	
MARGARINE,MELTED	2 lbs	1 qts	
JUICE,LEMON	1 lbs	1-7/8 cup	
GARLIC POWDER	9-1/2 oz	2 cup	
SALT	1-7/8 oz	3 tbsp	
PARSLEY,DEHYDRATED,FLAKED	5/8 oz	3/4 cup 2 tbsp	
PEPPER,BLACK,GROUND	3/8 oz	1 tbsp	
BREADCRUMBS	1 lbs	1 qts	

Method

1. Rinse shrimp; drain. Place 7-1/2 pounds shrimp in each steam table pan.
2. Add 2 cups tomatoes to each pan.
3. Combine margarine or butter, lemon juice, garlic, salt, parsley, and pepper. Blend well; mixture will separate.
4. While stirring, ladle about 14 ounces scampi sauce over shrimp and tomatoes in each pan. Toss lightly but thoroughly.
5. Cover, using a convection oven, bake 15 minutes at 350 F. on high fan, closed vent; uncover, stir; bake 5 minutes or until shrimp are done. DO NOT OVERCOOK. CCP: Internal temperature must reach 145 F. or higher for 15 seconds. Remove from oven.
6. Evenly sprinkle 2 cups breadcrumbs over top of each pan. Stir to blend crumbs with liquid to thicken sauce. Serve with steamed rice or pasta.

MEAT, FISH, AND POULTRY No.L 122 00

PAN FRIED FISH

Yield 100   **Portion** 4 Ounces

Calories	Carbohydrates	Protein	Fat	Cholesterol	Sodium	Calcium
239 cal	8 g	27 g	10 g	72 mg	309 mg	28 mg

Ingredient	Weight	Measure	Issue
FISH,FLOUNDER/SOLE FILLET,RAW	30 lbs		
BREADCRUMBS	1-3/8 lbs	1 qts 2 cup	
FLOUR,WHEAT,GENERAL PURPOSE	1-1/2 lbs	1 qts 1-1/2 cup	
SALT	1-1/2 oz	2-1/3 tbsp	
PEPPER,BLACK,GROUND	1/3 oz	1 tbsp	
SHORTENING,VEGETABLE,MELTED	1-3/4 lbs	1 qts	

Method

1. Separate fillets or steaks; cut into 4-1/2 ounce portions, if necessary.
2. Dredge fish in mixture of crumbs, flour, salt, and pepper; shake off excess.
3. Fry fish in hot shallow fat, 1/8-inch deep. Brown 2 to 4 minutes on each side; turn carefully. CCP: Internal temperature must reach 145 F. or higher for 15 seconds.
4. Drain well on absorbent paper. CCP: Hold for service at 140 F. or higher.

MEAT, FISH, AND POULTRY No.L 122 01

TEMPURA FISH

Yield 100 **Portion** 4 Ounces

Calories	Carbohydrates	Protein	Fat	Cholesterol	Sodium	Calcium
255 cal	11 g	28 g	10 g	94 mg	382 mg	55 mg

Ingredient   **Weight**   **Measure**   **Issue**

FISH,FLOUNDER/SOLE FILLET,RAW   30 lbs
TEMPURA BATTER   1 gal

Method

1 Separate fillets or steaks; cut into 4-1/2 ounce portions, if necessary.
2 Prepare 1 recipe Tempura Batter per 100 portions, Recipe No. D 038 00.
3 Dip fish into batter. Drain. Fry in 365 F. deep fat fryer for 2 to 4 minutes or until golden brown. CCP: Internal temperature must reach 145 F. or higher for 15 seconds. Frying time for fish will vary with type and thickness of fish.
4 Drain well on absorbent paper. CCP: Hold for service at 140 F. or higher.

MEAT. FISH. AND POULTRY No.L 122 02

DEEP FAT FRIED FISH

Yield 100 **Portion** 4 Ounces

Calories	Carbohydrates	Protein	Fat	Cholesterol	Sodium	Calcium
247 cal	8 g	27 g	11 g	72 mg	309 mg	28 mg

Ingredient

Ingredient	Weight	Measure	Issue
FISH,FLOUNDER/SOLE FILLET,RAW	30 lbs		
BREADCRUMBS	1-3/8 lbs	1 qts 2 cup	
FLOUR,WHEAT,GENERAL PURPOSE	1-1/2 lbs	1 qts 1-1/2 cup	
SALT	1-1/2 oz	2-1/3 tbsp	
PEPPER,BLACK,GROUND	1/3 oz	1 tbsp	

Method

1. Separate fillets; cut into 4-1/2 ounce portions, if necessary.
2. Dredge fish in crumbs, flour, salt, and pepper mixture; shake off excess.
3. Fry in 365 F. deep fat fryer for 2 to 4 minutes or until golden brown. CCP: Internal temperature must reach 145 F. or higher for 15 seconds. Frying time for fish will vary with type and thickness of fish.
4. Drain well in basket or on absorbent paper. CCP: Hold for service at 140 F. or higher.

MEAT. FISH. AND POULTRY No.L 123 00

OVEN FRIED FISH

Yield 100   **Portion** 4 Ounces

Calories	Carbohydrates	Protein	Fat	Cholesterol	Sodium	Calcium
236 cal	9 g	27 g	9 g	73 mg	373 mg	50 mg

Ingredient	**Weight**	**Measure**	**Issue**
FISH,FLOUNDER/SOLE FILLET,RAW	30 lbs		
MILK,NONFAT,DRY	3-1/4 oz	1-3/8 cup	
WATER,WARM	3-7/8 lbs	1 qts 3-1/2 cup	
BREADCRUMBS	3-3/4 lbs	1 gal	
SALT	1-1/2 oz	2-1/3 tbsp	
PEPPER,BLACK,GROUND	1/3 oz	1 tbsp	
COOKING SPRAY,NONSTICK	2 oz	1/4 cup 1/3 tbsp	
OIL,SALAD	1-1/2 lbs	3 cup	

Method

1. Separate fish fillets; cut into 4-1/2 ounce portions, if necessary.
2. Reconstitute milk. Dip fillets into milk mixture. Drain.
3. Dredge fillets in crumb mixture; shake off excess.
4. Lightly spray sheet pans with non-stick cooking spray. Place fillets in a single layer on each sprayed pan.
5. Sprinkle 3/4 cup salad oil, shortening or margarine over fillets in each pan.
6. Using a convection oven, bake 10 to 15 minutes at 350 F. on high fan, open vent or until lightly browned. CCP: Internal temperature must reach 145 F. or higher for 15 seconds. Hold for service at 140 F. or higher.

MEAT, FISH, AND POULTRY   No. L 124 00

BAKED FISH PORTIONS

Yield 100   Portion   3-1/2 Ounces

Calories	Carbohydrates	Protein	Fat	Cholesterol	Sodium	Calcium
278 cal	24 g	16 g	12 g	114 mg	594 mg	20 mg

Ingredient   **Weight**   **Measure**   **Issue**

FISH,PORTIONS,BREADED,FRZ   25 lbs

Method

1. Place fish on ungreased sheet pans. Portions should not touch. DO NOT THAW FISH before baking.
2. Using a convection oven, bake 20 to 22 minutes at 400 F. or until browned on high fan, open vent. CCP: Internal temperature must reach 145 F. or higher for 15 seconds. Hold for service at 140 F. or higher.

MEAT. FISH. AND POULTRY No.L 124 01

BAKED FISH PORTIONS (BATTER DIPPED)

Yield 100 **Portion** 6 Ounces

Calories	Carbohydrates	Protein	Fat	Cholesterol	Sodium	Calcium
416 cal	36 g	24 g	19 g	171 mg	891 mg	31 mg

Ingredient   **Weight**   **Measure**   **Issue**

FISH,BATTER DIPPED,FROZEN   37-1/2 lbs

Method

1. Place fish on sheet pans. Portions should not touch. DO NOT THAW FISH before baking.
2. Using a convection oven, bake at 400 F. 20 to 22 minutes or until browned on high fan, open vent. CCP: Internal temperature must reach 145 F. or higher for 15 seconds. Hold for service at 140 F. or higher.

MEAT, FISH, AND POULTRY No.L 124 02

FRENCH FRIED FISH PORTIONS

Yield 100   **Portion**   3-1/2 Ounces

Calories	Carbohydrates	Protein	Fat	Cholesterol	Sodium	Calcium
318 cal	24 g	16 g	17 g	114 mg	594 mg	20 mg

Ingredient   Weight   Measure   Issue

FISH,PORTIONS,BREADED,FRZ   25 lbs

Method

1. Fry fish portions in 350 F. deep fat fryer for 4 to 4-1/2 minutes or until lightly browned. DO NOT thaw fish portions before frying. CCP: Internal temperature must reach 145 F. or higher for 15 seconds.
2. Drain well in basket or on absorbent paper. CCP: Hold for service at 140 F. or higher.

MEAT, FISH, AND POULTRY No. L 124 03

FRENCH FRIED FISH PORTIONS (BATTER DIP)

Yield 100 **Portion** 6 Ounces

Calories	Carbohydrates	Protein	Fat	Cholesterol	Sodium	Calcium
503 cal	40 g	27 g	25 g	191 mg	990 mg	34 mg

Ingredient   Weight   Measure   Issue

FISH, BATTER DIPPED, FROZEN   37-1/2 lbs

Method

1. Fry breaded fish portions in 350 F. deep fat fryer or 4 to 4-1/2 minutes or until lightly browned. CCP: Internal temperature must reach 145 F. or higher for 15 seconds.
2. Drain well in basket or absorbent paper. CCP: Hold for service at 140 F. or higher.

MEAT, FISH, AND POULTRY   No.L 124 04

FISH AND CHIPS

Yield 100 **Portion** 6 Ounces

Calories	Carbohydrates	Protein	Fat	Cholesterol	Sodium	Calcium
701 cal	70 g	27 g	35 g	171 mg	902 mg	46 mg

Ingredient   Weight   Measure   Issue

FISH,BATTER DIPPED,FROZEN   37-1/2 lbs
FRENCH FRIED POTATOES (3-1/2 OUNCE)   25-3/4 kg   100 unit

Method

1 Place fish on ungreased pans. Bake for 35 minutes in 425 F. oven. CCP: Internal temperature must reach 145 F. or higher for 15 seconds.

2 Prepare Recipe Nos. Q 045 01 or Q 045 05, French Fried Potatoes. Each portion is 6 ounces of fish and 1 cup French Fries.

MEAT, FISH, AND POULTRY No.L 124 05

BAKED FISH NUGGETS

Yield 100 **Portion** 4 Ounces

Calories	Carbohydrates	Protein	Fat	Cholesterol	Sodium	Calcium
262 cal	18 g	14 g	14 g	54 mg	452 mg	123 mg

Ingredient **Weight** **Measure** **Issue**

FISH NUGGETS,BREADED,FROZEN   34 lbs

Method

1 Place about 5 pounds 10 ounces nuggets on each sheet pan. Bake in 450 F. oven 16 to 18 minutes or in 425 F. convection oven on high fan, closed vent 14 to 16 minutes or until lightly browned. CCP: Internal temperature must reach 145 F. or higher for 15 seconds. Hold for service at 140 F. or higher.

MEAT, FISH, AND POULTRY No.L 124 06

FRENCH FRIED FISH NUGGETS

Yield 100 Portion 4-1/2 Ounces

Calories	Carbohydrates	Protein	Fat	Cholesterol	Sodium	Calcium
302 cal	18 g	14 g	19 g	54 mg	452 mg	123 mg

Ingredient **Weight** **Measure** **Issue**

FISH NUGGETS,BREADED,FROZEN 34 lbs

Method

1. Fry fish nuggets in 350 F. deep fat fryer 4 minutes or until lightly browned. CCP: Internal temperature must reach 145 F. or higher for 15 seconds.
2. Drain well in basket or on absorbent paper. CCP: Hold for service at 140 F. or higher.

MEAT, FISH, AND POULTRY No. L 125 00

CHIPPER FISH

Yield 100   Portion 4-1/2 Ounces

Calories	Carbohydrates	Protein	Fat	Cholesterol	Sodium	Calcium
251 cal	10 g	29 g	10 g	85 mg	406 mg	113 mg

Ingredient   Weight   Measure   Issue

Ingredient	Weight	Measure	Issue
FISH,FLOUNDER/SOLE FILLET,RAW	30 lbs		
SALAD DRESSING,FRENCH,PREPARED,L/C	4-5/8 lbs	2 qts	
POTATO CHIPS	2 lbs		
CHEESE,CHEDDAR,SHREDDED	2-3/4 lbs	2 qts 3 cup	

Method

1. Separate fillets; cut into 4-1/2 ounce portions, if necessary. Dip fillets in French dressing; place in single layers on sheet pans.
2. Crush chips. Combine chips and cheese. Sprinkle about 1 quart mixture over fish in each pan.
3. Using a convection oven, bake 7 minutes at 350 F. on high fan, closed vent, or until done. CCP: Internal temperature must reach 145 F. or higher for 15 seconds. Hold for service at 140 F. or higher.

MEAT, FISH, AND POULTRY No.L 126 00

FRIED OYSTERS

Yield 100   **Portion** 6 Each

Calories	Carbohydrates	Protein	Fat	Cholesterol	Sodium	Calcium
407 cal	54 g	16 g	13 g	89 mg	457 mg	30 mg

Ingredient	Weight	Measure	Issue
OYSTERS,FROZEN	14 lbs		
FLOUR,WHEAT,GENERAL PURPOSE	4-3/8 lbs	1 gal	
SALT	3 oz	1/4 cup 1 tbsp	
PEPPER,BLACK,GROUND	1/8 oz	1/8 tsp	
MILK,NONFAT,DRY	1-1/4 oz	1/2 cup	
WATER,WARM	1-1/2 lbs	2-3/4 cup	
EGGS,WHOLE,FROZEN	2 lbs	3-3/4 cup	
CRACKER CRUMBS	9-1/8 lbs	2 gal 1 qts	

Method

1 Dredge oysters in mixture of flour, salt, and pepper; shake off excess.
2 Reconstitute milk; add eggs.
3 Dip floured oysters in milk and egg mixture; drain.
4 Dredge oysters in cracker crumbs until well coated; shake off excess.
5 Fry about 5 minutes or until lightly browned in 375 F. deep fat. CCP: Internal temperature must reach 145 F. or higher for 15 seconds.
6 Drain well in basket or on absorbent paper. CCP: Hold for service at 140 F. or higher.

Notes

1 In Step 4, 5 pounds or 1 gallon of dry bread crumbs or 13-3/4 cups of cornmeal may be used for cracker crumbs.

MEAT. FISH. AND POULTRY No.L 126 01

FRIED OYSTERS (BREADED, FROZEN)

Yield 100   **Portion** 6 Each

Calories	Carbohydrates	Protein	Fat	Cholesterol	Sodium	Calcium
219 cal	10 g	8 g	16 g	73 mg	378 mg	56 mg

Ingredient   **Weight**   **Measure**   **Issue**

OYSTERS,BREADED,IQF   25 lbs

Method

1. Fry oysters for 3 to 5 minutes or until lightly browned in 375 F. deep fat. CCP: Internal temperature must reach 145 F. or higher for 15 seconds.
2. Drain well in basket or on absorbent paper. CCP: Hold for service at 140 F. or higher.

MEAT, FISH, AND POULTRY No.L 127 00

BOILED LOBSTER, WHOLE

Yield 100   Portion 16 Ounces

Calories	Carbohydrates	Protein	Fat	Cholesterol	Sodium	Calcium
76 cal	1 g	16 g	0 g	56 mg	930 mg	56 mg

Ingredient	**Weight**	**Measure**	**Issue**
WATER,BOILING	58-1/2 lbs	7 gal	
SALT	5-3/4 oz	1/2 cup 1 tbsp	
LOBSTER,WHOLE,FROZEN	100 lbs		
BAY LEAF,WHOLE,DRIED	1/8 oz	3 lf	
PARSLEY,FRESH,BUNCH,CHOPPED	8 oz	3-3/4 cup	8-3/8 oz

Method

1. Plunge the first batch, about 25 lobsters, individually into steam-jacketed kettle of fast boiling water. Water should cover lobsters. Add salt and bay leaves to water, if desired.
2. Cover kettle. Bring water to a boil; reduce heat; simmer 15 minutes or until lobsters turn a brilliant red. CCP: Internal temperature must reach 145 F. or higher for 15 seconds. Remove lobsters.
3. Follow Steps 1 and 2 for remaining batches. Replenish water as needed to ensure lobsters are covered.
4. TO PREPARE FOR SERVING: Place lobster on back. Using a sharp knife, make quick incision at the mouth; draw knife quickly down entire length of body and tail. Be careful not to break the stomach or lady, a small sac just back of the head.
5. Spread the body flat. Remove, with your hand, the black colored intestinal vein which runs from the head to the tail; throw away. Remove and discard the lobster's stomach or lady and the spongy tissue. Leave the green liver and the red coral roe, if any.
6. Crack claws with a mallet.

Notes

1. Cook lobsters in batches of 25.
2. If using fresh, live lobsters, be sure each lobster is alive. When picked up, if the tail is stretched out flat, it should snap back.
3. Garnish with parsley.
4. Lobsters may be steamed. Steam lobster for 6 to 8 minutes in a 5 pound PSI steamer or for 4 to 6 minutes in a 15 pound PSI steamer. CCP: Internal temperature must reach 145 F. or higher for 15 seconds.

MEAT. FISH. AND POULTRY No.L 127 01

BOILED LOBSTER TAIL, FROZEN

Yield 100   Portion 8 Ounces

Calories	Carbohydrates	Protein	Fat	Cholesterol	Sodium	Calcium
174 cal	2 g	36 g	1 g	127 mg	957 mg	114 mg

Ingredient	Weight	Measure	Issue
LOBSTER,FROZEN,SPINY,TAIL	50 lbs		
WATER,BOILING	33-1/2 lbs	4 gal	
SALT	2-1/2 oz	1/4 cup 1/3 tbsp	
PARSLEY,FRESH,BUNCH,CHOPPED	8 oz	3-3/4 cup	8-3/8 oz

Method

1 Drop frozen tails into boiling salt water to cover, allow 1-1/3 tablespoons salt per gallon of water.
2 Return water to a boil; simmer 15 minutes or until tails turn a brilliant red or bright orange. CCP: Internal temperature must reach 145 F. or higher for 15 seconds. Drain.
3 Slit underside of tail lengthwise; remove membrane.
4 Garnish with parsley. NOTES: Lobster tails may be steamed. Steam in a 5 pound PSI steamer for 12 to 15 minutes or in a 15 pound PSI steamer for 10 to 12 minutes. CCP: Internal temperature must reach 145 F. or higher for 15 seconds.

MEAT, FISH, AND POULTRY No.L 127 03

BOILED CRAB LEGS, ALASKAN KING, FROZEN

Yield 100   Portion 5 Ounces

Calories	Carbohydrates	Protein	Fat	Cholesterol	Sodium	Calcium
173 cal	0 g	34 g	3 g	94 mg	1906 mg	113 mg

Ingredient	Weight	Measure	Issue
CRAB LEGS,ALASKAN KING	50 lbs		
WATER,BOILING	58-1/2 lbs	7 gal	
BAY LEAF,WHOLE,DRIED	1/4 oz	6 lf	
JUICE,LEMON	4-1/3 oz	1/2 cup	
PARSLEY,FRESH,BUNCH,CHOPPED	8 oz	3-3/4 cup	8-3/8 oz

Method

1 Drop legs in boiling water in steam-jacketed kettle or larger stock pot. If desired, add 6 bay leaves and lemon juice.
2 Bring water to boil; reduce heat. Cover. Simmer 10 minutes. CCP: Internal temperature must reach 145 F. or higher for 15 seconds. Remove legs. Garnish with parsley. NOTES: Crab legs may be steamed. Steam in a 5 pound PSI for 6 to 8 minutes or in a 15 pound PSI for 4 to 5 minutes. CCP: Internal temperature must reach 145 F. or higher for 15 seconds.

MEAT, FISH, AND POULTRY No.L 127 04

BOILED SHRIMP, FROZEN

Yield 100   Portion 7 Shrimp

Calories	Carbohydrates	Protein	Fat	Cholesterol	Sodium	Calcium
49 cal	0 g	10 g	0 g	97 mg	112 mg	19 mg

Ingredient	Weight	Measure	Issue
SHRIMP,FROZEN,RAW,UNPEELED	25 lbs		

Method

1. Place shellfish in perforated pans. Place perforated pans inside solid pans.
2. Boil until done, approximately 6 minutes. CCP: Internal temperature must reach 145 F. or higher for 15 seconds.
3. Do not over cook. Serve shrimp immediately. Over cooking will cause shellfish to be tough, rubbery, and dry.

Notes

1. Shrimp may be steamed. Steam in a 5 pound PSI steamer for 14 to 16 minutes or in a 15 pound PSI steamer for 10 to 12 minutes. CCP: Internal temperature must reach 145 F. or higher for 15 seconds.

MEAT, FISH, AND POULTRY No.L 128 00

SALMON CAKES

Yield 100   Portion 2 Cakes

Calories	Carbohydrates	Protein	Fat	Cholesterol	Sodium	Calcium
212 cal	7 g	21 g	11 g	78 mg	319 mg	240 mg

Ingredient	Weight	Measure	Issue
SALMON,CANNED,PINK	19 lbs	2 gal 3-2/3 qts	
POTATO,WHITE,INSTANT,GRANULES	6-3/4 oz	1 qts	
MILK,NONFAT,DRY	1-3/4 oz	3/4 cup	
SALT	1-1/4 oz	2 tbsp	
BUTTER	8 oz	1 cup	
EGGS,WHOLE,FROZEN	2 lbs	3-3/4 cup	
ONIONS,FRESH,CHOPPED	1-3/8 lbs	3-7/8 cup	1-1/2 lbs
PEPPER,BLACK,GROUND	1/3 oz	1 tbsp	
PARSLEY,DEHYDRATED,FLAKED	1/8 oz	1/4 cup 1/3 tbsp	
PAPRIKA,GROUND	1/4 oz	1 tbsp	
MARGARINE,MELTED	12 oz	1-1/2 cup	
BREADCRUMBS,DRY,GROUND,FINE	1-7/8 lbs	2 qts	
COOKING SPRAY,NONSTICK	1 oz	2 tbsp	

Method

1. Drain salmon; reserve liquid for use in Step 3. Remove and discard skin and bones from salmon. Flake salmon; cover.
2. Combine potatoes, milk, and salt; cover.
3. Blend salmon liquid and butter or margarine. Mix well. Take liquid mixture and rapidly add water to equal 2-1/2 qts per 100 portions to potato mixture. Whip until smooth.
4. Combine salmon, potato mixture, eggs, onions, pepper and parsley flakes. Mix thoroughly. Scoop and shape salmon into 3 inch diameter cakes by 1-1/2 inch thick, weighing about 2-1/2 ounces each.
5. Combine crumbs, paprika and margarine or butter; cover.
6. Lightly spray each sheet pan with non-stick cooking spray. Lightly coat each cake with crumb mixture. Brush off excess crumbs to ensure a thin coating. Place 34 cakes on each lightly sprayed sheet pan; cover.
7. Using a convection oven, bake 16-18 minutes or until lightly browned on high fan, open vent. CCP: Internal temperature must reach 145 F. or higher for 15 seconds. Serve immediately or hold for service at 140 F. or higher.

Notes

1. In Step 7, cakes may be cooked on a preheated 350 F. griddle. Lightly spray griddle with cooking spray. Grill salmon cakes 9 minutes; turn; grill second side 6 minutes. CCP: Internal temperature must reach 145 F. or higher for 15 seconds. Hold for service at 140 F. or higher.

MEAT, FISH, AND POULTRY No.L 129 00

SALMON LOAF

Yield 100   Portion 4-1/2 Ounces

Calories	Carbohydrates	Protein	Fat	Cholesterol	Sodium	Calcium
191 cal	9 g	21 g	7 g	73 mg	176 mg	244 mg

Ingredient	Weight	Measure	Issue
SALMON,CANNED,PINK	19 lbs	2 gal 3-2/3 qts	
CELERY,FRESH,CHOPPED	1-1/3 lbs	1 qts 1 cup	1-7/8 lbs
COOKING SPRAY,NONSTICK	2 oz	1/4 cup 1/3 tbsp	
ONIONS,FRESH,CHOPPED	1-1/3 lbs	3-3/4 cup	1-1/2 lbs
RESERVED LIQUID	5-1/4 lbs	2 qts 2 cup	
BREADCRUMBS	3-3/4 lbs	1 gal	
EGGS,WHOLE,FROZEN	2 lbs	3-3/4 cup	
PARSLEY,DEHYDRATED,FLAKED	1/8 oz	1/4 cup 1/3 tbsp	
PEPPER,BLACK,GROUND	1/8 oz	1/4 tsp	
COOKING SPRAY,NONSTICK	3/4 oz	1 tbsp	

Method

1 Drain salmon; reserve 2-1/2 qt of salmon liquid for use in Step 3. Remove and discard skin and bones from salmon. Flake salmon; cover salmon and salmon liquid.
2 Stir-cook celery and onions in a lightly sprayed steam-jacketed kettle or stock pot about 8 to 10 minutes, stirring constantly.
3 Combine salmon, salmon liquid and cooked vegetables with bread crumbs, eggs, pepper and parsley. Mix lightly but thoroughly. DO NOT OVERMIX.
4 Lightly spray each sheet pan with non-stick cooking spray. Firmly and evenly pack 8 lb 2 oz salmon mixture into each sheet pan. Divide into 2 equal loaves (about 7 inches wide) across the pan. Space evenly; smooth top and sides; cover.
5 Using a convection oven, bake 35 to 40 minutes at 325 F. or until lightly browned on high fan, closed vent. CCP: Internal temperature must reach 145 F. or higher for 15 seconds.
6 Let stand 10 minutes before slicing. Cut 13 slices per loaf. CCP: Hold for service at 140 F. or higher.

MEAT, FISH, AND POULTRY No.L 130 00

SCALLOPED SALMON AND PEAS

Yield 100 **Portion** 6-1/2 Ounces

Calories	Carbohydrates	Protein	Fat	Cholesterol	Sodium	Calcium
273 cal	15 g	23 g	13 g	55 mg	465 mg	288 mg

Ingredient	Weight	Measure	Issue
SALMON,CANNED,PINK	19-3/8 lbs	2 gal 3-7/8 qts	
COOKING SPRAY,NONSTICK	2 oz	1/4 cup 1/3 tbsp	
MILK,NONFAT,DRY	13-3/4 oz	1 qts 1-3/4 cup	
WATER,WARM	15-2/3 lbs	1 gal 3-1/2 qts	
FLOUR,WHEAT,GENERAL PURPOSE	1-1/8 lbs	1 qts	
SALT	1-1/2 oz	2-1/3 tbsp	
BUTTER,MELTED	1 lbs	2 cup	
ONIONS,FRESH,CHOPPED	1 lbs	2-7/8 cup	1-1/8 lbs
PAPRIKA,GROUND	1/4 oz	1 tbsp	
PEAS,GREEN,CANNED,DRAINED	9 lbs	1 gal 2 qts	
BREADCRUMBS	1-3/8 lbs	1 qts 2 cup	
BUTTER,MELTED	1 lbs	2 cup	

Method

1 Lightly spray each steam table pan with non-stick spray. Place 2-1/2 quarts salmon in each steam table pan.
2 Reconstitute milk; heat to just below boiling. DO NOT BOIL.
3 Blend flour, salt, and butter or margarine together; stir until smooth.
4 Add flour, salt, and butter or margarine mixture to milk stirring constantly. Cook 5 to 10 minutes or until thickened. Stir as necessary.
5 Add onions and paprika to sauce; cook 5 minutes.
6 Place 1-1/2 quarts peas over salmon. Mix carefully. Pour sauce over mixture; stir until lightly mixed.
7 Combine bread crumbs and melted butter or margarine. Sprinkle 3 cups buttered crumbs over each pan.
8 Using a convection oven, bake at 325 F. for 20 minutes on low fan, open vent or until browned. CCP: Internal temperature must reach 145 F. or higher for 15 seconds. Hold for service at 140 F. or higher.

Notes

1 In Step 1, 18 lbs canned tuna may be used instead of salmon.

MEAT, FISH, AND POULTRY No.L 131 00

CHOPSTICK TUNA

Yield 100   Portion 1 Cup

Calories	Carbohydrates	Protein	Fat	Cholesterol	Sodium	Calcium
387 cal	26 g	25 g	21 g	20 mg	666 mg	61 mg

Ingredient	Weight	Measure	Issue
FISH,TUNA,CANNED,WATER PACK,DRAINED	14-1/2 lbs	2 gal 2-2/3 qts	
CELERY,FRESH,SLICED	10-1/2 lbs	2 gal 1-7/8 qts	14-3/8 lbs
ONIONS,FRESH,SLICED	3 lbs	2 qts 3-7/8 cup	3-1/3 lbs
NUTS,UNSALTED,CHOPPED,COARSELY	4 lbs	3 qts 1/2 cup	
NOODLES,CHOW MEIN,CANNED	3-1/8 lbs	2 gal	
SOUP,CONDENSED,CREAM OF MUSHROOM	8-3/4 lbs	3 qts 3-3/4 cup	
NOODLES,CHOW MEIN,CANNED	3-1/8 lbs	2 gal	

Method

1 Drain tuna; discard juice. Flake tuna; combine with celery, onions, nuts, and chow mein noodles.
2 Combine soup with tuna mixture.
3 Pour an equal quantity of tuna-soup mixture into each steam table pan.
4 Sprinkle about 2 quart noodles over mixture in each pan.
5 Bake 20 to 25 minutes at 375 F. or until heated thoroughly. CCP: Internal temperature must reach 145 F. or higher for 15 seconds. Hold for service at 140 F. or higher.

MEAT, FISH, AND POULTRY No.L 132 00

TUNA SALAD

Yield 100   Portion 3/4 Cup

Calories	Carbohydrates	Protein	Fat	Cholesterol	Sodium	Calcium
240 cal	10 g	22 g	12 g	110 mg	526 mg	47 mg

Ingredient	Weight	Measure	Issue
FISH,TUNA,CANNED,WATER PACK,INCL LIQUIDS	16-1/2 lbs	3 gal 1/8 qts	
CELERY,FRESH,CHOPPED	8 lbs	1 gal 3-5/8 qts	11 lbs
ONIONS,FRESH,CHOPPED	1-3/8 lbs	1 qts	1-5/8 lbs
PICKLE RELISH,SWEET,DRAINED	2-2/3 lbs	1 qts 1 cup	
SALAD DRESSING,MAYONNAISE TYPE	4-1/4 lbs	2 qts 1/2 cup	
PEPPER,BLACK,GROUND	1/3 oz	1 tbsp	
JUICE,LEMON	1-1/4 lbs	2-3/8 cup	
EGG,HARD COOKED,CHOPPED	4-1/4 lbs	38 Eggs	
LETTUCE,LEAF,FRESH,HEAD	4 lbs		6-1/4 lbs

Method

1. Combine tuna, celery and onions. Mix lightly but thoroughly.
2. Combine salad dressing, pickle relish, lemon juice and pepper. Stir to blend thoroughly.
3. Add chopped eggs and salad dressing mixture to tuna mixture. Mix lightly.
4. Place 1 lettuce leaf on each serving dish. Top with 3/4 cup tuna salad; cover. CCP: Refrigerate product at 41 F. or lower until ready to serve.

MEAT. FISH. AND POULTRY   No.L 132 01

SALMON SALAD (CANNED SALMON)

Yield 100   Portion   3/4 Cup

Calories	Carbohydrates	Protein	Fat	Cholesterol	Sodium	Calcium
264 cal	10 g	20 g	16 g	118 mg	332 mg	234 mg

Ingredient	Weight	Measure	Issue
SALMON,CANNED,PINK	17-1/4 lbs	2 gal 2-5/8 qts	
ONIONS,FRESH,CHOPPED	1-1/3 lbs	3-3/4 cup	1-1/2 lbs
CELERY,FRESH,CHOPPED	8 lbs	1 gal 3-5/8 qts	11 lbs
PICKLE RELISH,SWEET,DRAINED	2-2/3 lbs	1 qts 1 cup	
JUICE,LEMON	1-1/4 lbs	2-3/8 cup	
PEPPER,BLACK,GROUND	1/3 oz	1 tbsp	
SALAD DRESSING,MAYONNAISE TYPE	4-1/4 lbs	2 qts 1/2 cup	
EGG,HARD COOKED,CHOPPED	4-1/4 lbs	38 Eggs	
LETTUCE,LEAF,FRESH,HEAD	4 lbs		6-1/4 lbs

Method

1. Remove and discard skin and bones from salmon. Flake salmon. Coarsely chop salmon into 1 inch pieces. Cover.
2. Combine salmon, onions and celery. Mix lightly but thoroughly.
3. Combine salad dressing, pickle relish, lemon juice and pepper. Stir to blend thoroughly.
4. Add chopped eggs and salad dressing mixture to salmon mixture. Mix lightly.
5. Place 1 lettuce leaf on each serving dish. Top with 3/4 cup salmon salad; cover. CCP: Refrigerate product at 41 F. or lower until ready to serve.

MEAT. FISH. AND POULTRY No.L 133 00

BAKED TUNA AND NOODLES

Yield 100   **Portion** 1 Cup

Calories	Carbohydrates	Protein	Fat	Cholesterol	Sodium	Calcium
305 cal	25 g	24 g	12 g	45 mg	606 mg	99 mg

Ingredient	Weight	Measure	Issue
FISH,TUNA,CANNED,WATER PACK,DRAINED	15-1/2 lbs	2 gal 3-3/8 qts	
NOODLES,EGG	4-1/2 lbs	3 gal 1-1/2 qts	
WATER,BOILING	18-3/4 lbs	2 gal 1 qts	
SALT	7/8 oz	1 tbsp	
FLOUR,WHEAT,GENERAL PURPOSE	1-3/8 lbs	1 qts 1 cup	
SALT	1-2/3 oz	2-2/3 tbsp	
SHORTENING,VEGETABLE,MELTED	1-3/4 lbs	1 qts	
MILK,NONFAT,DRY	1-1/4 lbs	2 qts	
WATER,WARM	20-7/8 lbs	2 gal 2 qts	
CELERY,FRESH,SLICED	4-3/8 lbs	1 gal 1/8 qts	6 lbs
ONIONS,FRESH,CHOPPED	11-1/4 oz	2 cup	12-1/2 oz
PIMIENTO,CANNED,DRAINED,CHOPPED	11-1/4 oz	1-5/8 cup	
COOKING SPRAY,NONSTICK	2 oz	1/4 cup 1/3 tbsp	
BREADCRUMBS	11-3/8 oz	3 cup	
BUTTER,MELTED	6 oz	3/4 cup	
PAPRIKA,GROUND	3/4 oz	3 tbsp	

Method

1. Drain tuna; flake.
2. Cook noodles in boiling salted water 8 minutes or until tender. Drain. Set aside for use in Step 7.
3. Blend flour, salt, and shortening or salad oil together using a wire whip; stir until smooth.
4. Reconstitute milk; heat to just below boiling. DO NOT BOIL.
5. Add milk to roux, stirring constantly. Simmer 10 to 15 minutes or until thickened. Stir as necessary.
6. Add celery and onions to sauce; bring to a boil, stirring constantly.
7. Combine tuna, noodles and pimientos with sauce. Mix well.
8. Lightly spray non-stick cooking spray in steam table pans. Pour about 6-1/2 quarts mixture into each steam table pan.
9. Combine crumbs, butter or margarine and paprika. Sprinkle about 1 cup over mixture in each pan.
   Using a convection oven, bake at 325 F. 35 minutes on high fan, closed vent or until lightly browned and bubbly. CCP: Internal temperature must reach 145 F. or higher for 15 seconds. Hold for service at 140 F. or higher.

MEAT. FISH. AND POULTRY   No.L 133 01
BAKED TUNA AND NOODLES (CREAM OF MUSHROOM SOUP)

Yield 100   Portion   1 Cup

Calories	Carbohydrates	Protein	Fat	Cholesterol	Sodium	Calcium
285 cal	25 g	23 g	10 g	45 mg	1010 mg	75 mg

Ingredient	Weight	Measure	Issue
FISH,TUNA,CANNED,WATER PACK,DRAINED	15-1/2 lbs	2 gal 3-3/8 qts	
NOODLES,EGG	4-1/2 lbs	3 gal 1-1/2 qts	
WATER,BOILING	18-3/4 lbs	2 gal 1 qts	
SALT	1 oz	1 tbsp	
SOUP,CONDENSED,CREAM OF MUSHROOM	18-3/4 lbs	2 gal 1/2 qts	
MILK,NONFAT,DRY	6-5/8 oz	2-3/4 cup	
WATER,WARM	7-1/3 lbs	3 qts 2 cup	
CELERY,FRESH,SLICED	4-3/8 lbs	1 gal 1/8 qts	6 lbs
ONIONS,FRESH,CHOPPED	11-1/4 oz	2 cup	12-1/2 oz
PIMIENTO,CANNED,DRAINED,CHOPPED	11-1/4 oz	1-5/8 cup	
COOKING SPRAY,NONSTICK	2 oz	1/4 cup 1/3 tbsp	
BREADCRUMBS	11-3/8 oz	3 cup	
BUTTER,MELTED	6 oz	3/4 cup	
PAPRIKA,GROUND	3/4 oz	3 tbsp	

Method

1. Drain tuna; flake.
2. Cook noodles in boiling salted water 8 minutes or until tender. Drain. Set aside for use in Step 4.
3. Use canned condensed cream of mushroom soup. Reconstitute nonfat dry milk with warm water. Add milk, celery and onions to soup. Blend; cover; heat to a simmer.
4. Combine tuna, noodles and pimientos with sauce. Mix well.
5. Lightly spray each steam table pan with non-stick cooking spray. Pour about 6-1/2 quarts mixture into each steam table pan.
6. Combine crumbs, butter or margarine and paprika. Sprinkle about 1 cup over mixture in each pan.
7. Using a convection oven, bake at 325 F. 35 minutes on high fan, closed vent or until browned and bubbly. CCP: Internal temperature must reach 145 F. or higher for 15 seconds. Hold for service at 140 F. or higher.

MEAT, FISH, AND POULTRY No.L 134 00

FRIED SCALLOPS

Yield 100   Portion   5 Ounces

Calories	Carbohydrates	Protein	Fat	Cholesterol	Sodium	Calcium
269 cal	24 g	18 g	11 g	51 mg	769 mg	44 mg

Ingredient

Ingredient	Weight	Measure	Issue
SCALLOPS,SEA,RAW	30 lbs	2 gal 1 qts	
FLOUR,WHEAT,GENERAL PURPOSE	4-3/8 lbs	1 gal	
SALT	5-1/8 oz	1/2 cup	
PEPPER,BLACK,GROUND	1/4 oz	1 tbsp	
PAPRIKA,GROUND	1/2 oz	2 tbsp	
MILK,NONFAT,DRY	7/8 oz	1/4 cup 2-1/3 tbsp	
WATER,WARM	1 lbs	1-7/8 cup	
EGGS,WHOLE,FROZEN	1-1/4 lbs	2-1/4 cup	
BREADCRUMBS	2-7/8 lbs	3 qts	

Method

1. Wash scallops thoroughly; cut large ones in half. Drain well.
2. Dredge scallops in mixture of flour, salt, pepper and paprika; shake off excess.
3. Reconstitute milk; add eggs.
4. Dip floured scallops in milk and egg mixture. Drain.
5. Dredge scallops in crumbs until well coated.
6. Fry 3 minutes or until golden brown in 350 F. deep fat. CCP: Internal temperature must reach 145 F. or higher for 15 seconds.
7. Drain well in basket or on absorbent paper. CCP: Hold for service at 140 F. or higher.

MEAT, FISH, AND POULTRY No. L 135 00

CREOLE SCALLOPS

Yield 100　　　　　　　　　　　　　　　　　　Portion 1 Cup

Calories	Carbohydrates	Protein	Fat	Cholesterol	Sodium	Calcium
166 cal	18 g	18 g	3 g	31 mg	571 mg	80 mg

Ingredient	Weight	Measure	Issue
CREOLE SAUCE		4 gal 1 qts	
SCALLOPS,SEA,RAW	34 lbs	2 gal 2-1/4 qts	
WATER	29-1/4 lbs	3 gal 2 qts	

Method

1 Prepare 2 recipes Creole Sauce, Recipe No. O 005 00 per 100 portions for use in Step 4.
2 Wash scallops thoroughly; cut large ones in half. Drain well.
3 Cook scallops in steam-jacketed kettle or stock pot 3 to 4 minutes. DO NOT OVERCOOK. Drain well.
4 Add cooked scallops to sauce; bring to simmer; cook 2 to 3 minutes. CCP: Internal temperature must reach 145 F. or higher.
5 CCP: Hold for service at 140 F. or higher.

MEAT, FISH, AND POULTRY No.L 135 01

CREOLE FISH

Yield 100 **Portion** 3-1/2 Ounces

Calories	Carbohydrates	Protein	Fat	Cholesterol	Sodium	Calcium
360 cal	32 g	17 g	18 g	114 mg	803 mg	48 mg

Ingredient	Weight	Measure	Issue
CREOLE SAUCE		2 gal 1/2 qts	
FISH,PORTIONS,BREADED,FRZ	25 lbs		

Method

1 Prepare 1 recipe Creole Sauce per 100 portions, Recipe No. O 005 00 for use in Step 3.
2 Fry fish in 350 F. deep fat for 3 minutes or until lightly browned. CCP: Internal temperature must reach 145 F. or higher for 15 seconds.
3 Drain well on absorbent paper. Ladle 2 ounces or 1/4 cup sauce over each fish portion just before serving. CCP: Hold for service at 140 F. or higher.

MEAT. FISH. AND POULTRY No.L 135 02

CREOLE FISH FILLETS

Yield 100   **Portion** 4-1/2 Ounces

Calories	Carbohydrates	Protein	Fat	Cholesterol	Sodium	Calcium
167 cal	8 g	27 g	3 g	72 mg	321 mg	47 mg

Ingredient | **Weight** | **Measure** | **Issue**

CREOLE SAUCE   2 gal 1/2 qts
FISH,FLOUNDER/SOLE FILLET,RAW   30 lbs

Method

1 Prepare 1 recipe Creole Sauce per 100 portions, Recipe No. O 005 00 for use in Step 3.
2 Separate fillets, cut into 4-1/2 ounces. Arrange in single layers in steam table pans.
3 Ladle 2 ounces or 1/4 cup hot sauce over each portion.
4 Using a convection oven, bake for 15 minutes at 325 F. on high fan, closed vent or until thoroughly heated. CCP: Internal temperature must reach 145 F. or higher for 15 seconds. Hold for service at 140 F. or higher.

MEAT, FISH, AND POULTRY No.L 136 00

CREOLE SHRIMP

Yield 100   **Portion** 8 Ounces

Calories	Carbohydrates	Protein	Fat	Cholesterol	Sodium	Calcium
319 cal	50 g	21 g	4 g	140 mg	870 mg	121 mg

Ingredient   Weight   Measure   Issue

Ingredient	Weight	Measure	Issue
SHRIMP,FROZEN,RAW,PEELED,DEVEINED	20 lbs		
WATER,BOILING	25-1/8 lbs	3 gal	
CREOLE SAUCE		5 gal	
RICE,LONG GRAIN	8-1/2 lbs	1 gal 1-1/4 qts	
WATER,COLD	23 lbs	2 gal 3 qts	
SALT	1-7/8 oz	3 tbsp	
OIL,SALAD	1-1/2 oz	3 tbsp	

Method

1. Place shrimp in boiling water; cover; return to boil. Reduce heat; simmer 5 minutes; drain. CCP: Refrigerate at 41 F. or lower for use in Step 5.
2. Prepare 2-1/2 recipes Creole Sauce, Recipe No. O 005 00 per 100 portions.
3. Combine rice, water, salt and salad oil; bring to a boil. Stir occasionally.
4. Cover tightly; simmer 20 to 25 minutes. DO NOT STIR.
5. Add shrimp to sauce; simmer until shrimp are heated through. DO NOT OVERCOOK. Stir occasionally. CCP: Internal temperature must reach 165 F. or higher for 15 seconds.
6. Serve over rice. CCP: Hold for service at 140 F. or higher.

MEAT. FISH. AND POULTRY No.L 137 00

FRENCH FRIED SHRIMP

Yield 100   Portion 4 Each

Calories	Carbohydrates	Protein	Fat	Cholesterol	Sodium	Calcium
272 cal	22 g	19 g	12 g	179 mg	569 mg	51 mg

Ingredient	Weight	Measure	Issue
SHRIMP,FROZEN,RAW,PEELED,DEVEINED	35 lbs		
FLOUR,WHEAT,GENERAL PURPOSE	4-3/8 lbs	1 gal	
SALT	3 oz	1/4 cup 1 tbsp	
PEPPER,BLACK,GROUND	1/4 oz	1 tbsp	
PAPRIKA,GROUND	1/3 oz	1 tbsp	
EGGS,WHOLE,FROZEN	2 lbs	3-3/4 cup	
WATER	2-1/8 lbs	1 qts	
BREADCRUMBS	2-7/8 lbs	3 qts	

Method

1. Wash shrimp; drain well.
2. Dredge shrimp in mixture of flour, salt, pepper, and paprika; shake off excess.
3. Combine beaten eggs and water. Dip shrimp in egg and water mixture; drain well.
4. Dredge shrimp in crumbs until well coated; shake off excess.
5. Deep fry 2 minutes or until golden brown. CCP: Internal temperature must reach 145 F. or higher for 15 seconds.
6. Drain well in basket or an absorbent paper. CCP: Hold for service at 140 F. or higher.

MEAT, FISH, AND POULTRY No. L 137 01

TEMPURA SHRIMP

Yield 100 **Portion** 4 Shrimp

Calories	Carbohydrates	Protein	Fat	Cholesterol	Sodium	Calcium
238 cal	16 g	18 g	11 g	173 mg	567 mg	81 mg

Ingredient **Weight** **Measure** **Issue**

SHRIMP,FROZEN,RAW,PEELED,DEVEINED   20 lbs
TEMPURA BATTER   1 gal 2 qts

Method

1. Wash shrimp; drain well.
2. Prepare Tempura Batter, Recipe No. D 052 00. Dip shrimp into batter; deep fat fry at 350 F. for 2-1/2 minutes or until golden brown. CCP: Internal temperature must reach 145 F. or higher for 15 seconds.
3. Drain well in basket or an absorbent paper. CCP: Hold for service at 140 F. or higher.

MEAT, FISH, AND POULTRY No.L 137 02

FRENCH FRIED SHRIMP (BREADED, FROZEN)

Yield 100   Portion 4 Each

Calories	Carbohydrates	Protein	Fat	Cholesterol	Sodium	Calcium
332 cal	14 g	26 g	19 g	214 mg	415 mg	81 mg

Ingredient   **Weight**   **Measure**   **Issue**

SHRIMP,BREADED,FROZEN   38 lbs

Method

1. Use shrimp, breaded, frozen. Do not allow shrimp to thaw before cooking.
2. Fry at 350 F. for 3 to 4 minutes or until golden brown. CCP: Internal temperature must reach 145 F. or higher for 15 seconds.
3. Drain well in basket or on absorbent paper. CCP: Hold for service at 140 F. or higher.

MEAT, FISH, AND POULTRY No.L 138 00

SHRIMP CURRY

Yield 100   Portion 3/4 Cup

Calories	Carbohydrates	Protein	Fat	Cholesterol	Sodium	Calcium
191 cal	12 g	16 g	9 g	140 mg	480 mg	45 mg

Ingredient

Ingredient	Weight	Measure	Issue
SHRIMP,FROZEN,RAW,PEELED,DEVEINED	20 lbs		
WATER,BOILING	25-1/8 lbs	3 gal	
ONIONS,FRESH,CHOPPED	3 lbs	2 qts 1/2 cup	3-1/3 lbs
PEPPERS,GREEN,FRESH,CHOPPED	2 lbs	1 qts 2-1/8 cup	2-1/2 lbs
OIL,SALAD	3-7/8 oz	1/2 cup	
FLOUR,WHEAT,GENERAL PURPOSE	1-2/3 lbs	1 qts 2 cup	
OIL,SALAD	1-1/2 lbs	3 cup	
WATER,WARM	20-7/8 lbs	2 gal 2 qts	
APPLES,FRESH,MEDIUM,PEELED,CORED,CHOPPED	4-3/8 lbs	0 gal 4 qts	5-5/8 lbs
CELERY,FRESH,CHOPPED	1-5/8 lbs	1 qts 2-1/8 cup	2-1/4 lbs
CURRY POWDER	1-1/2 oz	1/4 cup 3 tbsp	
GINGER,GROUND	3/8 oz	2 tbsp	
PEPPER,RED,GROUND	1/8 oz	1/3 tsp	
GARLIC POWDER	3/4 oz	2-2/3 tbsp	
HORSERADISH,PREPARED	1-5/8 oz	3 tbsp	
SALT	2-1/2 oz	1/4 cup 1/3 tbsp	
MUSHROOMS,CANNED,STEMS & PIECES,CHOPPED,DRAINED	1-1/4 lbs	3-3/4 cup	
JUICE,LEMON	6-1/2 oz	3/4 cup	

Method

1. Place shrimp in boiling water; cover, return to a boil. Reduce heat; simmer 3 to 5 minutes; drain. DO NOT OVERCOOK. CCP: Hold at 41 F. or lower for use in Step 5.
2. Saute onions and peppers in salad oil or shortening 10 minutes or until tender.
3. Add flour to salad oil or shortening; blend thoroughly.
4. Cook until well browned, stirring frequently.
5. Gradually add water to flour mixture; cook until thick and smooth, stirring constantly.
6. Add sauteed vegetables.
7. Add apples, celery, curry powder, ginger, red pepper, garlic, horseradish and salt; simmer 20 minutes.
8. Add shrimp, mushrooms and lemon juice; simmer 2 to 3 minutes, stirring constantly. CCP: Internal temperature must reach 165 F. or higher for 15 seconds. Hold for service at 140 F. or higher.

MEAT, FISH, AND POULTRY No. L 139 00

SHRIMP SALAD

Yield 100   **Portion** 1/2 Cup

Calories	Carbohydrates	Protein	Fat	Cholesterol	Sodium	Calcium
124 cal	3 g	16 g	5 g	143 mg	408 mg	55 mg

Ingredient	Weight	Measure	Issue
SHRIMP,FROZEN,RAW,PEELED,DEVEINED	20 lbs		
WATER,BOILING	25-1/8 lbs	3 gal	
CELERY,FRESH,CHOPPED	6-1/3 lbs	1 gal 2 qts	8-2/3 lbs
JUICE,LEMON	8-5/8 oz	1 cup	
SALT	1-1/2 oz	2-1/3 tbsp	
PEPPER,BLACK,GROUND	1/8 oz	1/3 tsp	
SALAD DRESSING,MAYONNAISE TYPE	2 lbs	1 qts	
LETTUCE,LEAF,FRESH,HEAD	4 lbs		6-1/4 lbs

Method

1 Place shrimp in boiling water; cover; return to boil; reduce heat; simmer 3 to 5 minutes; drain. CCP: Internal temperature must reach 145 F. or higher for 15 seconds. Chill.
2 Cut shrimp into halves or quarters.
3 Combine shrimp, celery, lemon juice, salt, and pepper.
4 Cover; refrigerate to chill thoroughly. CCP: Refrigerate at 41 F. or lower.
5 Just before serving, add salad dressing; toss lightly. CCP: Hold for service at 41 F. or lower.
6 Optional: Place 1 lettuce leaf on each serving dish; add salad, cover; refrigerate until ready to serve.

MEAT, FISH, AND POULTRY No.L 140 00

SEAFOOD NEWBURG

Yield 100					Portion 6 Ounces

Calories	Carbohydrates	Protein	Fat	Cholesterol	Sodium	Calcium
210 cal	7 g	24 g	9 g	128 mg	401 mg	91 mg

Ingredient | Weight | Measure | Issue

Ingredient	Weight	Measure	Issue
FISH,FLOUNDER/SOLE FILLET,RAW	14 lbs		
SCALLOPS,SEA,RAW	8 lbs	2 qts 1-5/8 cup	
SHRIMP,RAW,PEELED,DEVEINED	8 lbs		
WATER,BOILING	33-1/2 lbs	4 gal	
MILK,NONFAT,DRY	1 lbs	1 qts 2-5/8 cup	
RESERVED LIQUID	17-3/4 lbs	2 gal 1/2 qts	
BUTTER,MELTED	2 lbs	1 qts	
FLOUR,WHEAT,GENERAL PURPOSE	1-1/8 lbs	1 qts	
SALT	1-1/4 oz	2 tbsp	
PAPRIKA,GROUND	1 oz	1/4 cup 1/3 tbsp	
NUTMEG,GROUND	1/8 oz	1/3 tsp	
EGG YOLK,BEATEN	8-3/4 oz	15 egylk	

Method

1. Add fish, scallops and shrimp to boiling water in steam-jacketed kettle or stock pot. Return to a boil. Reduce heat; simmer 5 minutes. CCP: Internal temperature must reach 145 F. or higher for 15 seconds.
2. Drain. Reserve liquid for use in Step 3. Place 6 pounds 12 ounces fish, scallops and shrimp in each steam table pan. CCP: Hold at 140 F. or higher for use in Step 7.
3. Reconstitute milk with reserved liquid. Heat to just below boiling. DO NOT BOIL.
4. Blend butter or margarine and flour to make roux; stir until smooth. Add milk to roux stirring constantly.
5. Add salt, paprika and nutmeg. Simmer 10 to 15 minutes or until thickened. Stir as necessary.
6. Add about 1 quart sauce to egg yolks while constantly stirring. Pour egg mixture slowly back into remaining sauce. Stir to blend well.
7. Pour 3-1/4 quarts sauce over seafood in each pan. Stir gently. CCP: Hold for service at 140 F. or higher.

MEAT, FISH, AND POULTRY No. L 141 00

CRAB CAKES

Yield 100   Portion 5 Ounces

Calories	Carbohydrates	Protein	Fat	Cholesterol	Sodium	Calcium
404 cal	48 g	20 g	14 g	128 mg	1474 mg	154 mg

Ingredient	Weight	Measure	Issue
CRAB MEAT,COOKED	15 lbs		
BREADCRUMBS	18-1/8 lbs	4 gal 3 qts	
MUSTARD,PREPARED	2-1/4 oz	1/4 cup 1/3 tbsp	
SALAD DRESSING,MAYONNAISE TYPE	3 oz	1/4 cup 2-1/3 tbsp	
BUTTER,MELTED	2 lbs	1 qts	
EGGS,WHOLE,FROZEN	2 lbs	3-3/4 cup	
SALT	2-1/2 oz	1/4 cup 1/3 tbsp	
PEPPER,BLACK,GROUND	1/3 oz	1 tbsp	
MILK,NONFAT,DRY	1-3/4 oz	3/4 cup	
WATER	2 lbs	3-3/4 cup	
EGGS,WHOLE,FROZEN	2 lbs	3-3/4 cup	
BREADCRUMBS	2-7/8 lbs	3 qts	

Method

1. Remove any shell or cartilage from crab meat.
2. Add bread crumbs, mustard, salad dressing, butter or margarine, eggs, salt, and pepper; mix lightly.
3. For each cake, measure 1/4 cup of mixture. Form into cakes 1/2 to 3/4-inch thick, about 2 ounce each. CCP: Refrigerate at 41 F. or lower.
4. Reconstitute milk; add eggs; mix well.
5. Dip chilled crab cakes in milk and egg mixture, then in bread crumbs; shake off excess.
6. Fry at 350 F. for 2 to 3 minutes or until golden brown. CCP: Internal temperature must reach 145 F. or higher for 15 seconds. Drain well in basket or on absorbent paper. CCP: Hold for service at 140 F. or higher.

MEAT, FISH, AND POULTRY No.L 142 00

HONEY GLAZED ROCK CORNISH HENS

Yield 100   Portion 6 Ounces

Calories	Carbohydrates	Protein	Fat	Cholesterol	Sodium	Calcium
342 cal	12 g	25 g	21 g	147 mg	92 mg	22 mg

Ingredient   **Weight**   **Measure**   **Issue**

CORNISH HEN,ROCK,RAW,WHOLE   78-1/8 lbs
COOKING SPRAY,NONSTICK   2 oz   1/4 cup 1/3 tbsp
SUGAR,BROWN,PACKED   1-1/2 lbs   1 qts 1/2 cup
HONEY   1-1/2 lbs   2 cup
JUICE,ORANGE   1-1/8 lbs   2 cup

Method

1. Remove necks and giblets. Wash hens, inside and out, thoroughly under cold running water. Drain well; pat dry.
2. Using sharp boning knife or cleaver, split hens in half lengthwise.
3. Lightly spray sheet pans with non-stick cooking spray. Place each half skin side up, on sheet pans.
4. Using a convection oven, bake at 325 F. for 30 minutes on high fan, closed vent.
5. Heat brown sugar, honey and orange juice until sugar is melted to make a glaze.
6. Remove hens from oven; brush tops with glaze.
7. Return to convection oven; bake 20 minutes or until golden brown or done. CCP: Internal temperature must reach 165 F. or higher for 15 seconds.
8. Brush remaining glaze over hens in each pan before serving. CCP: Hold for service at 140 F. or higher.

MEAT, FISH, AND POULTRY   No.L 142 01

ROCK CORNISH HENS WITH SYRUP GLAZE

Yield 100   Portion   6 Ounces

Calories	Carbohydrates	Protein	Fat	Cholesterol	Sodium	Calcium
349 cal	14 g	25 g	21 g	147 mg	104 mg	16 mg

Ingredient | Weight | Measure | Issue

CORNISH HEN,ROCK,RAW,WHOLE   78-1/8 lbs

COOKING SPRAY,NONSTICK   2 oz   1/4 cup 1/3 tbsp

SYRUP,PANCAKE & WAFFLE   4-1/8 lbs   1 qts 2 cup

Method

1. Remove necks and giblets. Wash hens, inside and out, thoroughly under cold running water. Drain well; pat dry.
2. Using sharp boning knife or cleaver, split hens in half lengthwise.
3. Lightly spray sheet pans with non-stick cooking spray. Place each half on sheet pans.
4. Using a convection oven, bake at 325 F. 30 minutes on high fan, closed vent.
5. Remove hens from oven; brush tops with maple syrup, or use Recipe No. D 050 00, Maple Syrup.
6. Return to oven; bake 20 minutes or until done in 325 F. oven. CCP: Internal temperature must reach 165 F. or higher for 15 seconds.
7. Brush remaining warm syrup over hens in each pan before serving. CCP: Hold for service at 140 F. or higher.

MEAT, FISH, AND POULTRY No.L 142 02

HERBED CORNISH HENS

Yield 100   Portion   6 Ounces

Calories	Carbohydrates	Protein	Fat	Cholesterol	Sodium	Calcium
295 cal	1 g	25 g	20 g	147 mg	369 mg	26 mg

Ingredient	Weight	Measure	Issue
CORNISH HEN,ROCK,RAW,WHOLE	78-1/8 lbs		
JUICE,LEMON	1-1/8 lbs	2 cup	
SALT	2-1/2 oz	1/4 cup 1/3 tbsp	
GARLIC POWDER	3/4 oz	2-1/3 tbsp	
PEPPER,BLACK,GROUND	1/2 oz	2 tbsp	
PAPRIKA,GROUND	1/2 oz	2 tbsp	
ONION POWDER	1/2 oz	2 tbsp	
CELERY SEED	1/2 oz	2 tbsp	
SEASONING,POULTRY	1/4 oz	2 tbsp	
THYME,GROUND	1/3 oz	2 tbsp	
BASIL,SWEET,WHOLE,CRUSHED	1/2 oz	3 tbsp	

Method

1 Remove necks and giblets. Wash hens, inside and out, thoroughly under cold running water. Drain well; pat dry.
2 Using sharp boning knife or cleaver, split hens in half lengthwise.
3 Place each half skin side up, on ungreased pans; brush hens with lemon juice.
4 Combine salt, garlic, black pepper, ground paprika, onion powder, celery seed, ground poultry seasoning, ground thyme, and crushed sweet basil; mix well. Sprinkle 3 tbsp mixture evenly over hens in each pan.
5 Using a convection oven, bake 1 to 1-1/4 hours at 325 F. on high fan, closed vent for 40 minutes or until done. CCP: Internal temperature must reach 165 F. or higher for 15 seconds. Hold for service at 140 F. or higher.

MEAT, FISH, AND POULTRY   No.L 143 00

BAKED CHICKEN (8 PC)

Yield 100   Portion   8 Ounces

Calories	Carbohydrates	Protein	Fat	Cholesterol	Sodium	Calcium
260 cal	0 g	39 g	10 g	119 mg	394 mg	21 mg

Ingredient   **Weight**   **Measure**   **Issue**

CHICKEN, 8 PC CUT, SKIN REMOVED   82 lbs
COOKING SPRAY,NONSTICK   2-1/8 oz   1/4 cup 2/3 tbsp
SALT   2-1/2 oz   1/4 cup 1/3 tbsp
PEPPER,BLACK,GROUND   7/8 oz   1/4 cup 1/3 tbsp

Method

1. Wash chicken thoroughly under cold running water. Drain well. Remove excess fat.
2. Lightly spray sheet pans with non-stick cooking spray. Place chicken meat side up on each sheet pan.
3. Combine salt and pepper; mix well.
4. Sprinkle 1 tbsp seasoning mixture evenly over chicken in each pan. Lightly spray chicken with cooking spray.
5. Using a convection oven, bake 40 minutes at 325 F. on high fan, closed vent. CCP: Internal temperature must reach 165 F. or higher for 15 seconds.
6. Transfer chicken to steam table pans. CCP: Hold for service at 140 F. or higher.

MEAT, FISH, AND POULTRY No. L 143 01

MEXICAN BAKED CHICKEN (8 PC)

Yield 100   Portion 2 Pieces

Calories	Carbohydrates	Protein	Fat	Cholesterol	Sodium	Calcium
267 cal	1 g	39 g	11 g	119 mg	403 mg	35 mg

Ingredient	Weight	Measure	Issue
CHICKEN, 8 PC CUT, SKIN REMOVED	82 lbs		
COOKING SPRAY, NONSTICK	2-1/8 oz	1/4 cup 2/3 tbsp	
CHILI POWDER, DARK, GROUND	2-2/3 oz	1/2 cup 2 tbsp	
SALT	2-1/2 oz	1/4 cup 1/3 tbsp	
CUMIN, GROUND	2-1/8 oz	1/2 cup 2 tbsp	
GARLIC POWDER	1-3/4 oz	1/4 cup 2-1/3 tbsp	
OREGANO, CRUSHED	1-5/8 oz	1/2 cup 2 tbsp	

Method

1. Wash chicken thoroughly under cold running water. Drain well. Remove excess fat.
2. Lightly spray sheet pans with non-stick cooking spray. Place chicken meat side up on each sheet pan.
3. Combine chili powder, salt, ground cumin, garlic powder and crushed oregano; mix well.
4. Sprinkle 4-2/3 tbsp seasoning mixture evenly over chicken in each pan. Lightly spray chicken with cooking spray.
5. Using a convection oven, bake 40 minutes at 325 F. on high fan, closed vent. CCP: Internal temperature must reach 165 F. or higher for 15 seconds.
6. Transfer chicken to steam table pans. CCP: Hold for service at 140 F. or higher.

MEAT, FISH, AND POULTRY No.L 143 02

HERBED BAKED CHICKEN (8 PC)

Yield 100   Portion   2 Pieces

Calories	Carbohydrates	Protein	Fat	Cholesterol	Sodium	Calcium
262 cal	1 g	39 g	10 g	119 mg	395 mg	34 mg

Ingredient	Weight	Measure	Issue
CHICKEN, 8 PC CUT, SKIN REMOVED	82 lbs		
COOKING SPRAY,NONSTICK	2-1/8 oz	1/4 cup 2/3 tbsp	
SALT	2-1/2 oz	1/4 cup 1/3 tbsp	
PEPPER,BLACK,GROUND	7/8 oz	1/4 cup 1/3 tbsp	
OREGANO,CRUSHED	1-7/8 oz	3/4 cup	
MARJORAM,SWEET,GROUND	1/3 oz	1/4 cup 2-1/3 tbsp	
ROSEMARY,GROUND	1/2 oz	1/4 cup 1/3 tbsp	

Method

1. Wash chicken thoroughly under cold running water. Drain well. Remove excess fat.
2. Lightly spray sheet pans with non-stick cooking spray. Place chicken meat side up on each sheet pan.
3. Combine salt, pepper, crushed oregano, ground marjoram and ground rosemary; mix well.
4. Sprinkle 2-2/3 tbsp seasoning mixture evenly over chicken in each pan. Lightly spray chicken with cooking spray.
5. Using a convection oven, bake 40 minutes at 325 F. on high fan, closed vent. CCP: Internal temperature must reach 165 F. or higher for 15 seconds.
6. Transfer chicken to steam table pans. CCP: Hold for service at 140 F. or higher.

MEAT, FISH, AND POULTRY No.L 143 03

BAKED CHICKEN (BREAST BONELESS)

Yield 100   Portion 5 Ounces

Calories	Carbohydrates	Protein	Fat	Cholesterol	Sodium	Calcium
179 cal	0 g	32 g	4 g	88 mg	286 mg	16 mg

Ingredient

Ingredient	Weight	Measure	Issue
CHICKEN,BREAST,BNLS/SKNLS,5 OZ	31-1/4 lbs		
COOKING SPRAY,NONSTICK	2 oz	1/4 cup 1/3 tbsp	
SALT	1-7/8 oz	3 tbsp	
PEPPER,BLACK,GROUND	2/3 oz	3 tbsp	
COOKING SPRAY,NONSTICK	1-1/2 oz	3 tbsp	

Method

1. Wash chicken thoroughly under cold running water. Drain well. Remove excess fat.
2. Lightly spray sheet pans with non-stick cooking spray. Place chicken breasts on each sheet pan.
3. Combine salt and pepper; mix well.
4. Sprinkle 1-1/2 tbsp seasoning mixture evenly over chicken in each pan. Lightly spray chicken with cooking spray.
5. Using a convection oven, bake 10 to 12 minutes at 325 F. on high fan, closed vent. CCP: Internal temperature must reach 165 F. or higher for 15 seconds.
6. Transfer chicken to steam table pans. CCP: Hold for service at 140 F. or higher.

MEAT, FISH, AND POULTRY No.L 143 04

MEXICAN BAKED CHICKEN (BREAST BONELESS)

Yield 100   Portion 5 Ounces

Calories	Carbohydrates	Protein	Fat	Cholesterol	Sodium	Calcium
184 cal	1 g	32 g	5 g	88 mg	293 mg	28 mg

Ingredient	Weight	Measure	Issue
CHICKEN,BREAST,BNLS/SKNLS,5 OZ	31-1/4 lbs		
COOKING SPRAY,NONSTICK	1-1/2 oz	3 tbsp	
CHILI POWDER,DARK,GROUND	2-1/8 oz	1/2 cup	
SALT	1-7/8 oz	3 tbsp	
CUMIN,GROUND	1-2/3 oz	1/2 cup	
GARLIC POWDER	1-1/4 oz	1/4 cup 1/3 tbsp	
OREGANO,CRUSHED	1-1/4 oz	1/2 cup	
COOKING SPRAY,NONSTICK	2 oz	1/4 cup 1/3 tbsp	

Method

1. Wash chicken thoroughly under cold running water. Drain well. Remove excess fat.
2. Lightly spray sheet pans with non-stick cooking spray. Place chicken breasts on each sheet pan.
3. Combine chili powder, salt, ground cumin, garlic powder and crushed oregano; mix well.
4. Sprinkle 5-2/3 tbsp seasoning mixture evenly over chicken in each pan. Lightly spray chicken with cooking spray.
5. Using a convection oven, bake for 10-12 minutes at 325 F. on high fan, closed vent. CCP: Internal temperature must reach 165 F. or higher for 15 seconds.
6. Transfer chicken to steam table pans. CCP: Hold for service at 140 F. or higher.

MEAT, FISH, AND POULTRY No.L 143 05

HERBED BAKED CHICKEN (BREAST BONELESS)

Yield 100　　　　　　　　　　　　　　　　　　　**Portion** 5 Ounces

Calories	Carbohydrates	Protein	Fat	Cholesterol	Sodium	Calcium
181 cal	0 g	32 g	4 g	88 mg	286 mg	25 mg

Ingredient　　　　　　　　　　　　　　　Weight　　　　Measure　　　　Issue

Ingredient	Weight	Measure
CHICKEN,BREAST,BNLS/SKNLS,5 OZ	31-1/4 lbs	
COOKING SPRAY,NONSTICK	1-1/2 oz	3 tbsp
SALT	1-7/8 oz	3 tbsp
PEPPER,BLACK,GROUND	7/8 oz	1/4 cup 1/3 tbsp
OREGANO,CRUSHED	1-1/4 oz	1/2 cup
MARJORAM,SWEET,GROUND	1/4 oz	1/4 cup 1/3 tbsp
ROSEMARY,GROUND	1/2 oz	1/4 cup 1/3 tbsp
COOKING SPRAY,NONSTICK	2 oz	1/4 cup 1/3 tbsp

Method

1. Wash chicken thoroughly under cold running water. Drain well. Remove excess fat.
2. Lightly spray sheet pans with non-stick cooking spray. Place chicken breasts on each sheet pan.
3. Combine salt, pepper, crushed oregano, ground marjoram and ground rosemary; mix well.
4. Sprinkle 3 tbsp seasoning mixture evenly over chicken in each pan. Lightly spray chicken with cooking spray.
5. Using a convection oven, bake for 10-12 minutes at 325 F. on high fan, closed vent. CCP: Internal temperature must reach 165 F. or higher for 15 seconds.
6. Transfer chicken to steam table pans. CCP: Hold for service at 140 F. or higher.

MEAT. FISH. AND POULTRY No.L 144 00

BAKED TURKEY AND NOODLES

Yield 100   Portion 1 Cup

Calories	Carbohydrates	Protein	Fat	Cholesterol	Sodium	Calcium
277 cal	22 g	23 g	10 g	74 mg	985 mg	91 mg

Ingredient	Weight	Measure	Issue
TURKEY,BNLS,WHITE AND DARK MEAT	23 lbs		
WATER,BOILING	31-1/3 lbs	3 gal 3 qts	
ONIONS,FRESH,CHOPPED	4-1/4 lbs	3 qts	4-3/4 lbs
BAY LEAF,WHOLE,DRIED	1/3 oz	9 lf	
WATER,BOILING	25-1/8 lbs	3 gal	
NOODLES,EGG	2-1/4 lbs	1 gal 2-3/4 qts	
SALT	7/8 oz	1 tbsp	
FLOUR,WHEAT,GENERAL PURPOSE	2-1/4 lbs	2 qts	
WATER,COLD	4-1/8 lbs	2 qts	
MILK,NONFAT,DRY	5-3/8 oz	2-1/4 cup	
SALT	1-7/8 oz	3 tbsp	
PEPPER,BLACK,GROUND	1/4 oz	1 tbsp	
GARLIC POWDER	1/4 oz	1/3 tsp	
BASIL,SWEET,WHOLE,CRUSHED	1/3 oz	2 tbsp	
BREADCRUMBS	1 lbs	1 qts	
BUTTER,MELTED	6 oz	3/4 cup	
CHEESE,CHEDDAR,SHREDDED	8 oz	2 cup	

Method

1. Cut turkey into 3/4 to 1-inch cubes.
2. Place turkey in stock pot or steam-jacketed kettle; add water, onion, and bay leaves. Bring to a boil. Cover; reduce heat; simmer 35 to 40 minutes.
3. Remove bay leaves and discard. Drain turkey and onions. Reserve 2-1/2 gal stock for use in Step 7. CCP: Hold reserved stock at 140 F. or higher for use in Step 5. Hold turkey at 140 F. or higher for use in Step 7.
4. Cook noodles in boiling salted water 8 to 10 minutes or until tender. Drain. Use immediately in recipe preparation or rinse with cold water, drain thoroughly; place in shallow containers, cover and refrigerate.
5. Blend flour and cold water together to make a smooth slurry. Add slurry to stock stirring constantly. Bring to a boil. Cover; reduce heat; simmer 10 minutes or until thickened, stirring frequently to prevent sticking.
6. Reconstitute milk. Add salt, pepper, garlic powder and basil; stir milk mixture into thickened stock. Bring to a boil. Cover; reduce heat; simmer 2 minutes.
7. Stir turkey, onions, and noodles gently into thickened sauce. Heat to a simmer.
8. Pour turkey and noodle mixture into ungreased steam table pans.
9. Combine crumbs, margarine (or butter) and cheese. Sprinkle 2 cups crumb mixture evenly over turkey and noodles in each pan.
10. Using a convection oven, bake 25 minutes at 325 F. on high fan, closed vent or until lightly browned and thoroughly heated. CCP: Internal temperature must reach 165 F. or higher for 15 seconds. Hold for service at 140 F. or higher.

MEAT, FISH, AND POULTRY No.L 144 01

BAKED CHICKEN AND NOODLES (CANNED CHICKEN)

Yield 100   Portion 1 Cup

Calories	Carbohydrates	Protein	Fat	Cholesterol	Sodium	Calcium
250 cal	20 g	21 g	9 g	62 mg	1401 mg	72 mg

Ingredient	Weight	Measure	Issue
WATER,BOILING	25-1/8 lbs	3 gal	
NOODLES,EGG	2-1/4 lbs	1 gal 2-3/4 qts	
SALT	7/8 oz	1 tbsp	
CHICKEN,BONED,CANNED,PIECES	18 lbs	1 gal 3-1/8 qts	
CHICKEN BROTH		2 gal 2 qts	
ONIONS,FRESH,QUARTERED	3 lbs	2 qts 3-7/8 cup	3-1/3 lbs
FLOUR,WHEAT,GENERAL PURPOSE	2-1/4 lbs	2 qts	
WATER,COLD	4-1/8 lbs	2 qts	
WATER,WARM	6 lbs	2 qts 3-1/2 cup	
MILK,NONFAT,DRY	5-3/8 oz	2-1/4 cup	
SALT	1-7/8 oz	3 tbsp	
PEPPER,BLACK,GROUND	1/4 oz	1 tbsp	
GARLIC POWDER	1/4 oz	1/3 tsp	
BASIL,SWEET,WHOLE,CRUSHED	1/3 oz	2 tbsp	
BREADCRUMBS	1 lbs	1 qts	
BUTTER,MELTED	6 oz	3/4 cup	
CHEESE,CHEDDAR,SHREDDED	8 oz	2 cup	

Method

1. Cook noodles in boiling salted water 8 to 10 minutes or until tender. Drain. Use immediately in recipe preparation or rinse with cold water, drain thoroughly, place in shallow containers, cover, and refrigerate.
2. Cut chicken into 1 inch pieces.
3. Add onions to broth and bring to a boil. Cover; reduce heat; simmer 8 to 10 minutes until tender.
4. Blend flour and cold water; stir to make a smooth slurry. Add slurry to broth and onion mixture stirring constantly. Bring to a boil. Cover; reduce heat; simmer 10 minutes or until thickened, stirring frequently to prevent sticking.
5. Reconstitute milk in warm water. Add salt, pepper, garlic powder and basil; stir milk mixture into thickened broth. Bring to a boil. Cover; reduce heat; simmer 2 minutes.
6. Stir chicken and noodles gently into thickened sauce. Heat to a simmer.
7. Pour chicken and noodle mixture into ungreased steam table pans.
8. Combine crumbs, margarine (or butter) and cheese. Sprinkle crumb mixture evenly over chicken and noodles in each pan.
9. Using a convection oven, bake 25 minutes at 325 F. on high fan, open vent. CCP: Internal temperature must reach 165 F. or higher for 15 seconds. Hold for service at 140 F. or higher.

MEAT, FISH, AND POULTRY No.L 144 03

BAKED CHICKEN AND NOODLES (COOKED DICED)

Yield 100   Portion 1 Cup

Calories	Carbohydrates	Protein	Fat	Cholesterol	Sodium	Calcium
284 cal	20 g	28 g	9 g	89 mg	1101 mg	74 mg

Ingredient	Weight	Measure	Issue
WATER,BOILING	25-1/8 lbs	3 gal	
NOODLES,EGG	2-1/4 lbs	1 gal 2-3/4 qts	
SALT	7/8 oz	1 tbsp	
CHICKEN BROTH		2 gal 2 qts	
ONIONS,FRESH,QUARTERED	3 lbs	3 qts	3-3/8 lbs
FLOUR,WHEAT,GENERAL PURPOSE	2-1/4 lbs	2 qts	
WATER,COLD	4-1/8 lbs	2 qts	
WATER,WARM	6 lbs	2 qts 3-1/2 cup	
MILK,NONFAT,DRY	5-3/8 oz	2-1/4 cup	
SALT	1-7/8 oz	3 tbsp	
PEPPER,BLACK,GROUND	1/4 oz	1 tbsp	
GARLIC POWDER	1/4 oz	1/3 tsp	
BASIL,SWEET,WHOLE,CRUSHED	1/3 oz	2 tbsp	
CHICKEN,COOKED,DICED	18 lbs		
BREADCRUMBS	1 lbs	1 qts	
BUTTER,MELTED	6 oz	3/4 cup	
CHEESE,CHEDDAR,SHREDDED	8 oz	2 cup	

Method

1. Cook noodles in boiling salted water 8 to 10 minutes until tender. Drain. Use immediately in recipe preparation or rinse with cold water, drain thoroughly, place in shallow containers, cover, and refrigerate.
2. Add onions to broth and bring to a boil. Cover; reduce heat; simmer 8 to 10 minutes until tender.
3. Blend flour and cold water together to make a smooth slurry. Add slurry to broth and onion mixture stirring constantly. Bring to a boil. Cover; reduce heat; simmer 10 minutes or until thickened, stirring frequently to prevent sticking.
4. Reconstitute milk in warm water. Add salt, pepper, garlic powder and basil; stir milk mixture into thickened broth. Bring to a boil. Cover; reduce heat; simmer 2 minutes.
5. Stir chicken and noodles gently into thickened sauce. Heat to a simmer.
6. Pour chicken and noodle mixture into ungreased steam table pans.
7. Combine crumbs, margarine or butter and cheese. Sprinkle crumb mixture evenly over chicken and noodles in each pan.
8. Using a convection oven, bake 25 minutes at 325 F. on high fan, open vent or until lightly browned. CCP: Internal temperature must reach 165 F. or higher for 15 seconds. Hold for service at 140 F. or higher.

MEAT, FISH, AND POULTRY No.L 145 00

CHICKEN VEGA (8 PC)

Yield 100   Portion 9 Ounces

Calories	Carbohydrates	Protein	Fat	Cholesterol	Sodium	Calcium
479 cal	45 g	45 g	12 g	121 mg	1249 mg	131 mg

Ingredient	Weight	Measure	Issue
CHICKEN, 8 PC CUT, SKIN REMOVED	82 lbs		
COOKING SPRAY,NONSTICK	2-1/8 oz	1/4 cup 2/3 tbsp	
WATER,WARM	20-7/8 lbs	2 gal 2 qts	
MILK,NONFAT,DRY	1-1/4 lbs	2 qts	
SOUP,DEHYDRATED,ONION	12 oz	2-5/8 cup	
WATER,COLD	3-2/3 lbs	1 qts 3 cup	
FLOUR,WHEAT,GENERAL PURPOSE	1-1/8 lbs	1 qts	
CHICKEN BROTH		3 gal	
RICE,LONG GRAIN	9-5/8 lbs	1 gal 1-7/8 qts	

Method

1. Wash chicken thoroughly under cold running water. Drain well. Remove excess fat. Place chicken, meat side up, on each lightly sprayed sheet pan. Lightly spray chicken with cooking spray.
2. Using a convection oven, bake chicken 40 minutes at 350 F. on high fan, closed vent. Hold at 140 F. or higher for use in Step 6.
3. Reconstitute milk in warm water. Heat milk to a simmer. Do not boil. Add dehydrated onion soup; mix well.
4. Blend flour and cold water together; stir to make a smooth slurry. Add slurry to hot seasoned milk mixture stirring constantly. Bring to a boil. Cover; reduce heat; simmer 5 minutes or until thickened; stirring frequently to prevent sticking.
5. Place 2-1/3 uncooked rice evenly in each of 10 steam table pans. Pour hot chicken broth over rice in each pan; stir well.
6. Place 20 pieces of pre-baked chicken evenly over rice mixture in each pan.
7. Pour 1-1/4 quarts of sauce evenly over chicken in each pan.
8. Cover; using a convection oven, bake 35 minutes at 350 F. or until rice is tender, on high fan, closed vent. CCP: Internal temperature must reach 165 F. or higher for 15 seconds. Hold for service at 140 F. or higher.

MEAT. FISH. AND POULTRY No.L 146 00

BARBECUED CHICKEN (8 PC)

Yield 100   Portion   8 Ounces

Calories	Carbohydrates	Protein	Fat	Cholesterol	Sodium	Calcium
324 cal	16 g	40 g	11 g	120 mg	981 mg	35 mg

Ingredient	Weight	Measure	Issue
CHICKEN, 8 PC CUT, SKIN REMOVED	82 lbs		
COOKING SPRAY,NONSTICK	2-1/8 oz	1/4 cup 2/3 tbsp	
CATSUP	10-5/8 lbs	1 gal 1 qts	
SAUCE,CHILI	2-1/8 lbs	3-3/4 cup	
VINEGAR,DISTILLED	1-1/3 lbs	2-1/2 cup	
WORCESTERSHIRE SAUCE	14-7/8 oz	1-3/4 cup	
MUSTARD,PREPARED	6-5/8 oz	3/4 cup	
PEPPER,BLACK,GROUND	2/3 oz	3 tbsp	
SALT	1-7/8 oz	3 tbsp	
PEPPER,RED,GROUND	1/4 oz	1 tbsp	

Method

1. Wash chicken thoroughly under cold running water. Drain well. Remove excess fat.
2. Combine catsup, chili sauce, vinegar, Worcestershire sauce, mustard, salt, black pepper, and red pepper in a steam-jacketed kettle or stockpot. Bring to a boil. Cover, reduce heat; simmer 5 minutes.
3. Place chicken, meat side up, on lightly sprayed sheet pans. Lightly spray chicken with cooking spray. Using a convection oven, bake 20 minutes at 325 F. on high fan, closed vent.
4. Dip chicken in barbecue sauce to coat well; place chicken, meat side up, on sheet pans. Bake an additional 20 minutes, for a total of 40 minutes. CCP: Internal temperature must reach 165 F. or higher for 15 seconds. Transfer chicken to steam table pans.
5. Bring remaining barbecue sauce to a boil.
6. Pour barbecue sauce evenly over chicken in each pan. CCP: Hold for service at 140 F. or higher.

Notes

1. In Step 2, 2 gallons of prepared BBQ sauce may be used per 100 portions.

MEAT, FISH, AND POULTRY No.L 146 01

BARBECUED CHICKEN (BREAST BONELESS)

Yield 100 **Portion** 5 Ounces

Calories	Carbohydrates	Protein	Fat	Cholesterol	Sodium	Calcium
237 cal	16 g	33 g	4 g	89 mg	942 mg	31 mg

Ingredient	**Weight**	**Measure**	**Issue**
CHICKEN,BREAST,BNLS/SKNLS,5 OZ	31-1/4 lbs		
CATSUP	10-5/8 lbs	1 gal 1 qts	
SAUCE,CHILI	2-1/8 lbs	3-3/4 cup	
VINEGAR,DISTILLED	1-1/3 lbs	2-1/2 cup	
WORCESTERSHIRE SAUCE	14-7/8 oz	1-3/4 cup	
MUSTARD,PREPARED	6-5/8 oz	3/4 cup	
PEPPER,BLACK,GROUND	2/3 oz	3 tbsp	
SALT	1-7/8 oz	3 tbsp	
PEPPER,RED,GROUND	1/4 oz	1 tbsp	
COOKING SPRAY,NONSTICK	3/4 oz	1 tbsp	

Method

1 Wash chicken thoroughly under cold running water. Drain well. Remove excess fat. Place chicken in roasting pans.
2 Combine catsup, chili sauce, vinegar, Worcestershire sauce, mustard, salt, black pepper, and red pepper; mix well.
3 Pour 1 gallon barbecue sauce evenly over chicken in each pan; cover.
4 Place chicken breasts on lightly sprayed sheet pans.
5 Using a convection oven, bake 12 to 14 minutes at 325 F. on high fan, open vent. CCP: Internal temperature must reach 165 F. or higher for 15 seconds. Transfer chicken to steam table pans.
6 Bring remaining barbecue sauce to a boil.
7 Pour 3-1/2 cups of barbecue sauce over chicken in each pan. CCP: Hold for service at 140 F. or higher.

Notes

1 In Step 2, 2 gallons of prepared BBQ sauce may be used per 100 portions.

MEAT, FISH, AND POULTRY No.L 147 00

CHICKEN A LA KING (COOKED DICED)

Yield 100   Portion 1 Cup

Calories	Carbohydrates	Protein	Fat	Cholesterol	Sodium	Calcium
246 cal	17 g	27 g	7 g	74 mg	1235 mg	62 mg

Ingredient	Weight	Measure	Issue
CHICKEN BROTH		2 gal 3 qts	
CELERY,FRESH,CHOPPED	4 lbs	3 qts 3-1/8 cup	5-1/2 lbs
ONIONS,FRESH,CHOPPED	1 lbs	2-7/8 cup	1-1/8 lbs
SALT	1-7/8 oz	3 tbsp	
PEPPER,WHITE,GROUND	1/3 oz	1 tbsp	
WATER,WARM	7-7/8 lbs	3 qts 3 cup	
MILK,NONFAT,DRY	7-1/4 oz	3 cup	
CHICKEN BROTH		3 qts	
FLOUR,WHEAT,GENERAL PURPOSE	3-7/8 lbs	3 qts 2 cup	
CHICKEN,COOKED,DICED	18 lbs		
PEPPERS,GREEN,FRESH,CHOPPED	1 lbs	3 cup	1-1/4 lbs
PIMIENTO,CANNED,DRAINED,CHOPPED	8-1/2 oz	1-1/4 cup	

Method

1. Place broth, celery, onions, salt and pepper in a steam jacketed kettle or stockpot; bring to a boil. Cover; reduce heat; simmer 8-10 minutes until tender.
2. Reconstitute milk in warm water. Stir milk into cooked vegetables and broth.
3. Blend flour and broth together to make a smooth slurry. Add slurry to vegetables, broth, and milk mixture, stirring constantly. Bring to a boil. Cover; reduce heat; simmer 10 minutes or until thickened, stirring frequently to prevent sticking.
4. Stir chicken, peppers, and pimientos gently into thickened sauce. Cover, reduce heat; simmer 2 minutes. CCP: Internal temperature must reach 165 F. or higher for 15 seconds.
5. Pour 2-1/2 gallons of chicken a la king into ungreased steam table pans. CCP: Hold for service at 140 F. or higher.

MEAT. FISH. AND POULTRY No.L 147 01

CHICKEN A LA KING (CANNED CHICKEN)

Yield 100   Portion 1 Cup

Calories	Carbohydrates	Protein	Fat	Cholesterol	Sodium	Calcium
212 cal	17 g	20 g	7 g	47 mg	1325 mg	60 mg

Ingredient	**Weight**	**Measure**	**Issue**
CHICKEN,BONED,CANNED,PIECES	18 lbs	1 gal 3-1/8 qts	
CHICKEN BROTH		2 gal 3 qts	
CELERY,FRESH,CHOPPED	4 lbs	3 qts 3-1/8 cup	5-1/2 lbs
ONIONS,FRESH,CHOPPED	1 lbs	2-7/8 cup	1-1/8 lbs
PEPPER,WHITE,GROUND	1/3 oz	1 tbsp	
WATER,WARM	7-7/8 lbs	3 qts 3 cup	
MILK,NONFAT,DRY	7-1/4 oz	3 cup	
CHICKEN BROTH		3 qts	
FLOUR,WHEAT,GENERAL PURPOSE	3-7/8 lbs	3 qts 2 cup	
PEPPERS,GREEN,FRESH,CHOPPED	1 lbs	3 cup	1-1/4 lbs
PIMIENTO,CANNED,DRAINED,CHOPPED	8-1/2 oz	1-1/4 cup	

Method

1 Cut chicken into 1 inch pieces.
2 Place broth, celery, onions, and pepper in a steam jacketed kettle or stockpot; bring to a boil. Cover; reduce heat; simmer 8-10 minutes until tender.
3 Reconstitute milk in warm water. Stir milk into cooked vegetables and broth.
4 Blend flour and second broth together; stir to make a smooth slurry. Add slurry to vegetables, broth, and milk mixture stirring constantly. Bring to a boil. Cover; reduce heat; simmer 10 minutes or until thickened, stirring frequently to prevent sticking.
5 Stir chicken, peppers, and pimientos gently into thickened sauce. Cover; reduce heat; simmer 2 minutes. CCP: Internal temperature must reach 165 F. or higher for 15 seconds.
6 Pour 2-1/2 gallons of chicken a la king into ungreased steam table pans. CCP: Hold for service at 140 F. or higher.

MEAT. FISH. AND POULTRY No.L 147 02

TURKEY A LA KING

Yield 100   Portion   1 Cup

Calories	Carbohydrates	Protein	Fat	Cholesterol	Sodium	Calcium
213 cal	19 g	18 g	7 g	46 mg	2278 mg	81 mg

Ingredient	Weight	Measure	Issue
TURKEY,BNLS,WHITE AND DARK MEAT,DICED	18 lbs		
WATER	50-1/8 lbs	6 gal	
SALT	5-3/4 oz	1/2 cup 1 tbsp	
BAY LEAF,WHOLE,DRIED	1/3 oz	9 each	
CHICKEN BROTH		2 gal 3 qts	
CELERY,FRESH,CHOPPED	4 lbs	3 qts 3-1/8 cup	5-1/2 lbs
ONIONS,FRESH,CHOPPED	1 lbs	2-7/8 cup	1-1/8 lbs
SALT	1-7/8 oz	3 tbsp	
PEPPER,BLACK,GROUND	1/3 oz	1 tbsp	
WATER,WARM	7-7/8 lbs	3 qts 3 cup	
MILK,NONFAT,DRY	7-1/4 oz	3 cup	
CHICKEN BROTH		3 qts	
FLOUR,WHEAT,GENERAL PURPOSE	3-7/8 lbs	3 qts 2 cup	
PEPPERS,GREEN,FRESH,CHOPPED	1 lbs	3 cup	1-1/4 lbs
PIMIENTO,CANNED,DRAINED,CHOPPED	8-1/2 oz	1-1/4 cup	

Method

1. Place turkey in stock pot or steam jacketed kettle; add water, salt and bay leaves. Bring to a boil; reduce heat; simmer turkey in 6 gallons water 3 to 4 hours or until tender. CCP: Internal temperature must reach 165 F. or higher for 15 seconds. Drain. Cool. Dice cooked turkey. CCP: Refrigerate at 41 F. or lower for use in Step 5.
2. Place broth, celery, onions, salt, and pepper in a steam jacketed kettle or stockpot; bring to a boil. Cover; reduce heat; simmer 8-10 minutes until tender.
3. Reconstitute milk in warm water. Stir milk into cooked vegetables and broth.
4. Blend flour and broth together; stir to make a smooth slurry. Add slurry to vegetables, broth, and milk mixture, stirring constantly. Bring to a boil. Cover; reduce heat; simmer 10 minutes or until thickened, stirring frequently to prevent sticking.
5. Stir turkey, peppers, and pimientos gently into thickened sauce. Cover, reduce heat; simmer 2 minutes. CCP: Internal temperature must reach 165 F. or higher for 15 seconds.
6. Pour 2-1/2 gallons of turkey a la king into ungreased steam table pans. CCP: Hold for service at 140 F. or higher.

MEAT, FISH, AND POULTRY No.L 148 00

CHICKEN CACCIATORE (8 PC)

Yield 100   **Portion** 8 Ounces

Calories	Carbohydrates	Protein	Fat	Cholesterol	Sodium	Calcium
348 cal	21 g	42 g	11 g	119 mg	764 mg	87 mg

Ingredient	Weight	Measure	Issue
TOMATOES,CANNED,CRUSHED,INCL LIQUIDS	26-1/2 lbs	3 gal	
TOMATO PASTE,CANNED	7-3/4 lbs	3 qts 1-1/2 cup	
ONIONS,FRESH,1/4"" STRIPS	3-1/3 lbs	3 qts 1-1/8 cup	3-2/3 lbs
WATER	3-1/8 lbs	1 qts 2 cup	
PEPPERS,GREEN,FRESH,JULIENNE	4-1/4 lbs	3 qts 7/8 cup	5-1/8 lbs
SUGAR,GRANULATED	7 oz	1 cup	
SALT	1-7/8 oz	3 tbsp	
GARLIC POWDER	1-1/8 oz	1/4 cup	
PEPPER,BLACK,GROUND	1/2 oz	2 tbsp	
OREGANO,CRUSHED	1/2 oz	3 tbsp	
THYME,GROUND	1/2 oz	3 tbsp	
BASIL,SWEET,WHOLE,CRUSHED	3/8 oz	2-2/3 tbsp	
BAY LEAF,WHOLE,DRIED	1/4 oz	6 lf	
CHICKEN, 8 PC CUT, SKIN REMOVED	82 lbs		
COOKING SPRAY,NONSTICK	2-1/8 oz	1/4 cup 2/3 tbsp	

Method

1. Combine tomatoes, tomato paste, onions, water, sweet peppers, sugar, salt, garlic, pepper, oregano, thyme, basil and bay leaves in steam-jacketed kettle or stock pot. Bring to boil; cover; reduce heat; simmer 1 hour. Remove bay leaves.
2. Wash chicken thoroughly under cold running water. Drain well. Remove excess fat.
3. Place chicken, meat side up, on lightly sprayed sheet pans. Lightly spray chicken with cooking spray.
4. Using a convection oven, bake 20 minutes at 325 F. on high fan, closed vent.
5. Transfer chicken to steam table pans. Pour 3-1/4 quart sauce evenly over chicken in each pan.
6. Cover; using a convection oven, bake 30 to 35 minutes at 325 F. on high fan, closed vent. CCP: Internal temperature must reach 165 F. or higher for 15 seconds.
7. CCP: Hold for service at 140 F. or higher. Serve with 1/2 cup sauce.

MEAT, FISH, AND POULTRY No. L 148 01

CHICKEN CACCIATORE (COOKED DICED)

Yield 100   Portion   1-1/4 Cups

Calories	Carbohydrates	Protein	Fat	Cholesterol	Sodium	Calcium
245 cal	21 g	27 g	7 g	73 mg	721 mg	81 mg

Ingredient	Weight	Measure	Issue
TOMATOES,CANNED,CRUSHED,INCL LIQUIDS	26-1/2 lbs	3 gal	
WATER	20-7/8 lbs	2 gal 2 qts	
TOMATO PASTE,CANNED	7-3/4 lbs	3 qts 1-1/2 cup	
ONIONS,FRESH,1/4"" STRIPS	4 lbs	1 gal	4-1/2 lbs
PEPPERS,GREEN,FRESH,JULIENNE	4-1/4 lbs	3 qts 1 cup	5-1/4 lbs
SUGAR,GRANULATED	7 oz	1 cup	
SALT	1-7/8 oz	3 tbsp	
GARLIC POWDER	1-1/8 oz	1/4 cup	
PEPPER,BLACK,GROUND	1/2 oz	2 tbsp	
OREGANO,CRUSHED	1/2 oz	3 tbsp	
THYME,GROUND	1/2 oz	3 tbsp	
BASIL,SWEET,WHOLE,CRUSHED	3/8 oz	2-2/3 tbsp	
BAY LEAF,WHOLE,DRIED	1/4 oz	6 lf	
CHICKEN,COOKED,DICED	18 lbs		

Method

1 Combine tomatoes, water, tomato paste, onions, sweet peppers, sugar, salt, garlic, pepper, oregano, thyme, basil and bay leaves in steam-jacketed kettle or stock pot. Bring to boil; cover; reduce heat; simmer 1 hour. Remove bay leaves.
2 Stir chicken gently into cacciatore sauce. Cover; reduce heat; simmer 2 minutes. CCP: Internal temperature must reach 165 F. or higher for 15 seconds.
3 Pour 2-1/2 gal chicken cacciatore mixture into ungreased steam table pans. CCP: Hold for service at 140 F. or higher.

MEAT, FISH, AND POULTRY No.L 149 00

BAKED CHICKEN AND GRAVY (8 PC)

Yield 100　　　　　　　　　　　　　　　　Portion 2 Pieces

Calories	Carbohydrates	Protein	Fat	Cholesterol	Sodium	Calcium
300 cal	7 g	40 g	11 g	120 mg	1025 mg	44 mg

Ingredient	Weight	Measure	Issue
CHICKEN, 8 PC CUT, SKIN REMOVED	82 lbs		
COOKING SPRAY,NONSTICK	2-1/8 oz	1/4 cup 2/3 tbsp	
SALT	1-7/8 oz	3 tbsp	
PEPPER,BLACK,GROUND	1/2 oz	2 tbsp	
FLOUR,WHEAT,GENERAL PURPOSE	1-2/3 lbs	1 qts 2 cup	
CHICKEN BROTH		2 gal 1 qts	
WATER	3-7/8 lbs	1 qts 3-1/2 cup	
MILK,NONFAT,DRY	3-5/8 oz	1-1/2 cup	
CHICKEN BROTH		1 qts 2 cup	
PAPRIKA,GROUND	1/4 oz	1 tbsp	

Method

1. Wash chicken thoroughly under cold running water. Drain well. Remove excess fat.
2. Place chicken, meat side up, on lightly sprayed sheet pans. Sprinkle pieces of chicken with mixture of salt and pepper. Lightly spray chicken with cooking spray.
3. Using a convection oven, bake 20 minutes on high fan, closed vent at 325 F. Transfer chicken to roasting pans. CCP: Hold at 140 F. or higher for use in Step 8.
4. Lightly brown flour in a roasting pan on top of a gas range for 10 to 12 minutes; in a 350 F. tilting fry pan for 16 to 18 minutes; or in a roasting pan using a convection oven, at 350 F on low fan, open vent for 25 to 27 minutes. Use a wire whip to stir and distribute flour for even browning.
5. Heat chicken broth in a steam-jacketed kettle or stockpot.
6. Reconstitute milk in warm water; stir milk into hot broth.
7. Blend flour and second broth together; stir to make a smooth slurry. Add slurry to broth and milk mixture. Bring to a boil. Cover; reduce heat; simmer 10 minutes or until thickened, stirring frequently to prevent sticking.
8. Pour 3-1/2 qt gravy evenly over chicken in each pan. Sprinkle one teaspoon of paprika over each pan.
9. Cover. Using a convection oven, bake at 325 F. 30 to 35 minutes on high fan, closed vent. CCP: Internal temperature must reach 165 F. or higher for 15 seconds. Hold for service at 140 F. or higher.

MEAT. FISH. AND POULTRY No.L 149 01

BAKED CHICKEN WITH MUSHROOM GRAVY (8 PC)

Yield 100   Portion   2 Pieces

Calories	Carbohydrates	Protein	Fat	Cholesterol	Sodium	Calcium
308 cal	9 g	41 g	11 g	120 mg	1065 mg	45 mg

Ingredient	Weight	Measure	Issue
CHICKEN, 8 PC CUT, SKIN REMOVED	82 lbs		
COOKING SPRAY,NONSTICK	2-1/8 oz	1/4 cup 2/3 tbsp	
SALT	1-7/8 oz	3 tbsp	
PEPPER,BLACK,GROUND	1/2 oz	2 tbsp	
MUSHROOMS,CANNED,STEMS & PIECES,CHOPPED,DRAINED	2 lbs	1 qts 2 cup	
WATER,WARM	3-7/8 lbs	1 qts 3-1/2 cup	
MILK,NONFAT,DRY	3-5/8 oz	1-1/2 cup	
FLOUR,WHEAT,GENERAL PURPOSE	2 lbs	1 qts 2 cup	
CHICKEN BROTH		2 gal 1 qts	
CHICKEN BROTH		1 qts 2 cup	
PAPRIKA,GROUND	1/4 oz	1 tbsp	

Method

1. Wash chicken thoroughly under cold running water. Drain well. Remove excess fat.
2. Place chicken, meat side up, on lightly sprayed sheet pans. Sprinkle chicken with mixture of salt and pepper. Lightly spray chicken with cooking spray.
3. Using a convection oven, bake 20 minutes at 325 F. on high fan, closed vent. Transfer chicken to roasting pans. CCP: Hold at 140 F. or higher for use in Step 9.
4. Drain mushrooms and reserve liquid for use in Step 5. Chop mushrooms.
5. Combine mushroom liquid and enough warm water to equal 7-1/2 cups. Reconstitute milk with mushroom liquid and warm water mixture.
6. Lightly brown flour in a roasting pan on top of a gas range for 10 to 12 minutes; a 350 F. convection oven on low fan, open vent for 25 to 27 minutes or in a 350 F. tilting fry pan for 16 to 18 minutes. Use a wire whip to stir and distribute flour for even browning. Cool; set aside for use in Step 8.
7. Heat chicken broth to a simmer in a steam-jacketed kettle or stockpot; stir milk into hot broth.
8. Blend flour and second chicken broth together to make a smooth slurry. Add slurry to broth and milk mixture. Bring to a boil. Cover; reduce heat; simmer 10 minutes or until thickened, stirring frequently to prevent sticking. Stir chopped mushrooms gently into gravy, heat to a simmer.
9. Pour 3-1/2 qt gravy evenly over chicken in each pan. Sprinkle one teaspoon of paprika over each pan.
10. Cover. Using a convection oven, bake at 325 F. 30 to 35 minutes on high fan, closed vent. CCP: Internal temperature must reach 165 F. or higher for 15 seconds. Hold for service at 140 F. or higher.

MEAT. FISH. AND POULTRY No.L 149 02

BAKED CHICKEN WITH MUSHROOM GRAVY (8 PC CND SOUP)

Yield 100   Portion   2 Pieces

Calories	Carbohydrates	Protein	Fat	Cholesterol	Sodium	Calcium
332 cal	5 g	40 g	16 g	120 mg	603 mg	40 mg

Ingredient

Ingredient	Weight	Measure	Issue
CHICKEN, 8 PC CUT, SKIN REMOVED	82 lbs		
COOKING SPRAY,NONSTICK	2-1/8 oz	1/4 cup 2/3 tbsp	
PEPPER,BLACK,GROUND	1/2 oz	2 tbsp	
SOUP,CONDENSED,CREAM OF MUSHROOM	15-1/2 lbs	1 gal 3 qts	
WATER	9-3/8 lbs	1 gal 1/2 qts	
PAPRIKA,GROUND	1/4 oz	1 tbsp	

Method

1. Wash chicken thoroughly under cold running water. Drain well. Remove excess fat.
2. Place chicken, meat side up, on lightly sprayed sheet pans. Sprinkle chicken with pepper. Lightly spray chicken with cooking spray.
3. Using a convection oven, bake at 325 F. for 20 minutes on high fan, closed vent. Transfer chicken to roasting pans.
4. Combine mushroom soup and water. Bring to a boil.
5. Pour 3-1/2 qt gravy evenly over chicken in each pan. Sprinkle one teaspoon of paprika over each pan.
6. Cover. Using a convection oven, bake 30 to 35 minutes at 325 F. on high fan, closed vent. CCP: Internal temperature must reach 165 F. or higher for 15 seconds. Hold for service at 140 F. or higher.

MEAT. FISH. AND POULTRY   No.L 150 00
TURKEY POT PIE

Yield 100   Portion   1 Cup

Calories	Carbohydrates	Protein	Fat	Cholesterol	Sodium	Calcium
368 cal	40 g	28 g	10 g	65 mg	964 mg	128 mg

Ingredient	Weight	Measure	Issue
TURKEY,BNLS,WHITE AND DARK MEAT	26 lbs		
COOKING SPRAY,NONSTICK	2 oz	1/4 cup 1/3 tbsp	
ONIONS,FRESH,CHOPPED	2 lbs	1 qts 1-5/8 cup	2-1/4 lbs
SALT	5/8 oz	1 tbsp	
PEPPER,BLACK,GROUND	1/4 oz	1 tbsp	
THYME,GROUND	1/8 oz	1 tbsp	
BAY LEAF,WHOLE,DRIED	1/3 oz	9 lf	
WATER,WARM	4-7/8 lbs	2 qts 1-3/8 cup	
POTATOES,FRESH,PEELED,CUBED	8 lbs	1 gal 1-7/8 qts	9-7/8 lbs
CARROTS,FRESH,CHOPPED	8 lbs	1 gal 3-1/8 qts	9-3/4 lbs
CELERY,FRESH,CHOPPED	2 lbs	1 qts 3-1/2 cup	2-3/4 lbs
WATER,COLD	4-1/8 lbs	2 qts	
FLOUR,WHEAT,GENERAL PURPOSE	2-1/3 lbs	2 qts 1/2 cup	
PEAS,GREEN,FROZEN	5-3/4 lbs	1 gal 1/2 qts	
FLOUR,WHEAT,GENERAL PURPOSE	3-1/3 lbs	3 qts	
SUGAR,GRANULATED	2-1/3 oz	1/4 cup 1-2/3 tbsp	
BAKING POWDER	2-3/8 oz	1/4 cup 1-1/3 tbsp	
SALT	5/8 oz	1 tbsp	
WATER,WARM	4-7/8 lbs	2 qts 1-3/8 cup	
MILK,NONFAT,DRY	4-3/4 oz	2 cup	
EGG WHITES	2-1/8 lbs	1 qts	
MARGARINE,MELTED	4 oz	1/2 cup	

Method

1. Cut turkey into 3/4 to 1-inch cubes.
2. Place turkey, onions, salt, pepper, thyme, and bay leaves in a steam-jacketed kettle or stockpot. Cook 15 minutes, stirring occasionally, until onions are lightly browned and turkey is partially cooked and slightly tender.
3. Add water, potatoes, carrots and celery. Bring to a boil. Cover; reduce heat; simmer 15 minutes or until potatoes are almost tender. Remove bay leaves.
4. Blend cold water and flour together; stir to make a smooth slurry. Add slurry to turkey mixture stirring constantly. Bring to a boil. Cover; reduce heat; simmer 8-10 minutes or until thickened, stirring frequently to prevent sticking.
5. Add peas; stir; bring to a simmer.
6. Pour 1-1/3 gallons of turkey mixture into each ungreased steam table pan.
7. For batter topping, sift together flour, sugar, baking powder and salt into mixer bowl.
8. Reconstitute milk in warm water. Combine milk, egg whites and margarine or butter. Add to dry ingredients; mix at low speed until dry ingredients are moistened, about 30 seconds. Do not overmix.
9. Pour 3-1/4 cups of batter evenly over top of turkey mixture in each pan.
10. Using a convection oven, bake 20 to 25 minutes at 400 F. or until lightly browned on low fan, open vent. CCP: Internal temperature must reach 165 F. or higher for 15 seconds.
11. Cut 3 x 6. CCP: Hold for service at 140 F. or higher.

Notes

1. In Step 1, 18 lb (3 1/2 gal) cooked, diced turkey may be substituted. In Step 3, use 16 lb (2 gal) chicken broth in place of water.
2. In Step 3, 8 lbs 8 ounces drained canned sliced carrots (13 lb 2 oz, 2-No.10 cn A.P.) or 8 lbs frozen carrots may be used per 100 portions. Add carrots to sauce in Step 5.
3. In Step 9, batter will be very thin. DO NOT add additional flour. CCP: If prepared in advance, refrigerate at 41 F. or lower until use.
4. Baking Powder Biscuits may be used for topping. Omit Steps 7 through 10. Prepare Recipe No. D 001 00 or D 001 01; place baked biscuits over top of hot turkey mixture in each pan. Bake 10 to 15 minutes or until biscuits are lightly browned.

MEAT, FISH, AND POULTRY No. L 150 01

CHICKEN POT PIE (CANNED CHICKEN)

Yield 100   Portion   1 Cup

Calories	Carbohydrates	Protein	Fat	Cholesterol	Sodium	Calcium
317 cal	38 g	23 g	8 g	46 mg	1101 mg	107 mg

Ingredient	Weight	Measure	Issue
CHICKEN,BONED,CANNED,PIECES	18 lbs	1 gal 3-1/8 qts	
COOKING SPRAY,NONSTICK	1/8 oz	1/8 tsp	
ONIONS,FRESH,CHOPPED	2 lbs	1 qts 1-5/8 cup	2-1/4 lbs
PEPPER,BLACK,GROUND	1/4 oz	1 tbsp	
THYME,GROUND	1/8 oz	1 tbsp	
BAY LEAF,WHOLE,DRIED	1/3 oz	9 lf	
CHICKEN BROTH		2 gal	
POTATOES,FRESH,PEELED,CUBED	8 lbs	1 gal 1-7/8 qts	9-7/8 lbs
CARROTS,FRESH,CHOPPED	8 lbs	1 gal 3-1/8 qts	9-3/4 lbs
CELERY,FRESH,CHOPPED	2 lbs	1 qts 3-1/2 cup	2-3/4 lbs
WATER,COLD	4-1/8 lbs	2 qts	
FLOUR,WHEAT,GENERAL PURPOSE	2-1/3 lbs	2 qts 1/2 cup	
PEAS,GREEN,FROZEN	5-3/4 lbs	1 gal 1/2 qts	
FLOUR,WHEAT,GENERAL PURPOSE	3-1/3 lbs	3 qts	
SUGAR,GRANULATED	2-1/3 oz	1/4 cup 1-2/3 tbsp	
BAKING POWDER	2-3/8 oz	1/4 cup 1-1/3 tbsp	
SALT	5/8 oz	1 tbsp	
WATER,WARM	4-7/8 lbs	2 qts 1-3/8 cup	
MILK,NONFAT,DRY	4-3/4 oz	2 cup	
EGG WHITES	2-1/8 lbs	1 qts	
MARGARINE,MELTED	4 oz	1/2 cup	

Method

1. Cut chicken into 1 inch pieces; cover.
2. Lightly spray steam-jacketed kettle or stockpot with non-stick spray. Add onions, pepper, thyme, and bay leaves. Stir-cook 5 minutes until onions are tender.
3. Add broth, potatoes, carrots and celery. Bring to a boil. Cover; reduce heat; simmer 15 minutes or until potatoes are almost tender. Remove bay leaves.
4. Blend flour and cold water together; stir to make a slurry. Add slurry vegetable mixture stirring constantly. Bring to boil. Cover; reduce heat; simmer 8 to 10 minutes or until thickened, stirring frequently to prevent sticking.
5. Fold in chicken and peas. Bring to a boil. Cover; reduce heat; simmer 5 to 10 minutes.
6. Pour 1-1/3 gallons of mixture into each ungreased pan.
7. For batter topping, sift together flour, sugar, baking powder, and salt into mixer bowl.
8. Reconstitute milk in warm water. Combine milk, egg whites and margarine or butter. Add to dry ingredients; mix at low speed until dry ingredients are moistened, about 30 seconds. Do not overmix.
9. Pour 3-1/4 cups of batter evenly over top of chicken mixture in each pan.
10. Using a convection oven, bake 20 to 25 minutes at 400 F. or until lightly browned on low fan, open vent. CCP: Internal temperature must reach 165 F. or higher for 15 seconds.
11. Cut 3 x 6. CCP: Hold for service at 140 F. or higher.

Notes

1. In Step 3, 8 lbs 8 oz drained sliced carrots (13 lb 2 oz, 2-No. 10 cn A.P.) or 8 lbs frozen carrots may be used per 100 portions. Add carrots to sauce in Step 5.
2. In Step 9, batter will be very thin. Do not add additional flour. If prepared in advance, refrigerate at 41 F. or lower until ready to use.
3. Baking powder biscuits may be used for topping. Omit Steps 7 through 9. Prepare Recipe No. D 001 00 or D 001 01; place 18 biscuits over top of hot mixture in each pan. Bake 10 to 15 minutes or until biscuits are lightly browned.

MEAT, FISH, AND POULTRY No.L 150 03

CHICKEN POT PIE (COOKED DICED)

Yield 100   Portion 1 Cup

Calories	Carbohydrates	Protein	Fat	Cholesterol	Sodium	Calcium
351 cal	38 g	31 g	8 g	73 mg	802 mg	109 mg

Ingredient	Weight	Measure	Issue
COOKING SPRAY,NONSTICK	1/8 oz	1/8 tsp	
ONIONS,FRESH,CHOPPED	2 lbs	1 qts 1-5/8 cup	2-1/4 lbs
PEPPER,BLACK,GROUND	1/4 oz	1 tbsp	
THYME,GROUND	1/8 oz	1 tbsp	
BAY LEAF,WHOLE,DRIED	1/3 oz	9 lf	
CHICKEN BROTH		2 gal	
POTATOES,FRESH,PEELED,CUBED	8 lbs	1 gal 1-7/8 qts	9-7/8 lbs
CARROTS,FRESH,CHOPPED	8 lbs	1 gal 3-1/8 qts	9-3/4 lbs
CELERY,FRESH,CHOPPED	2 lbs	1 qts 3-1/2 cup	2-3/4 lbs
WATER,COLD	4-1/8 lbs	2 qts	
FLOUR,WHEAT,GENERAL PURPOSE	2-1/3 lbs	2 qts 1/2 cup	
CHICKEN,COOKED,DICED	18 lbs		
PEAS,GREEN,FROZEN	5-3/4 lbs	1 gal 1/2 qts	
FLOUR,WHEAT,GENERAL PURPOSE	3-1/3 lbs	3 qts	
SUGAR,GRANULATED	2-1/3 oz	1/4 cup 1-2/3 tbsp	
BAKING POWDER	2-3/8 oz	1/4 cup 1-1/3 tbsp	
SALT	5/8 oz	1 tbsp	
WATER,WARM	4-7/8 lbs	2 qts 1-3/8 cup	
MILK,NONFAT,DRY	4-3/4 oz	2 cup	
EGG WHITES	2-1/8 lbs	1 qts	
MARGARINE,MELTED	4 oz	1/2 cup	

Method

1 Lightly spray steam-jacketed kettle or stockpot with non-stick spray. Add onions, pepper, thyme, and bay leaves. Stir-cook 5 minutes until onions are tender.
2 Add broth, potatoes, carrots and celery. Bring to a boil. Cover; reduce heat; simmer 15 minutes or until potatoes are almost tender. Remove bay leaves.
3 Blend flour and cold water together; stir to make a slurry. Add slurry vegetable mixture stirring constantly. Bring to boil. Cover; reduce heat; simmer 8 to 10 minutes or until thickened, stirring frequently to prevent sticking.
4 Fold in chicken and peas. Bring to a boil. Cover; reduce heat; simmer 5 to 10 minutes.
5 Pour 1-1/3 gallons of mixture into each ungreased pan.
6 For batter topping, sift together flour, sugar, baking powder, and salt into mixer bowl.
7 Reconstitute milk in warm water. Combine milk, egg whites and margarine or butter. Add to dry ingredients; mix at low speed until dry ingredients are moistened, about 30 seconds. Do not overmix.
8 Pour 3-1/4 cups of batter evenly over top of chicken mixture in each pan.
9 Using a convection oven, bake 20 to 25 minutes at 400 F. or until lightly browned on low fan, open vent. CCP: Internal temperature must reach 165 F. or higher for 15 seconds.
10 Cut 3 x 6. CCP: Hold for service at 140 F. or higher.

Notes

1 In Step 3, 8 lbs 8 oz drained sliced carrots (13 lb 2 oz, 2-No. 10 cn A.P.) or 8 lbs frozen carrots may be used per 100 portions. Add carrots to sauce in Step 5.
2 In Step 9, batter will be very thin. Do not add additional flour. If prepared in advance, refrigerate at 41 F. or lower until ready to use.
3 Baking powder biscuits may be used for topping. Omit Steps 7 through 9. Prepare Recipe No. D 001 00 or D 001 01; place 18 biscuits over top of hot mixture in each pan. Bake 10 to15 minutes or until biscuits are lightly browned.

MEAT, FISH, AND POULTRY No.L 151 00

CHICKEN SALAD (COOKED DICED)

Yield 100 **Portion** 3/4 Cup

Calories	Carbohydrates	Protein	Fat	Cholesterol	Sodium	Calcium
229 cal	4 g	24 g	12 g	77 mg	371 mg	43 mg

Ingredient

Ingredient	Weight	Measure	Issue
CHICKEN,COOKED,DICED	18 lbs		
CELERY,FRESH,CHOPPED	9-1/2 lbs	2 gal 1 qts	13 lbs
SALAD DRESSING,MAYONNAISE TYPE	2-3/4 lbs	1 qts 1-1/2 cup	
ONIONS,FRESH,CHOPPED	15 oz	2-5/8 cup	1 lbs
JUICE,LEMON	8-5/8 oz	1 cup	
SALT	1-2/3 oz	2-2/3 tbsp	
PEPPER,BLACK,GROUND	1/4 oz	1 tbsp	
LETTUCE,LEAF,FRESH,CHOPPED	4 lbs	2 gal 1/8 qts	6-1/4 lbs

Method

1 Combine chicken, celery, salad dressing, onions, lemon juice, salt, and pepper. Mix lightly but thoroughly.
2 Place 1 lettuce leaf on serving dish. Top with 3/4 cup salad. CCP: Refrigerate product at 41 F. or lower until served.

MEAT. FISH. AND POULTRY No.L 151 01

CHICKEN SALAD (CANNED CHICKEN)

Yield 100   Portion 3/4 Cup

Calories	Carbohydrates	Protein	Fat	Cholesterol	Sodium	Calcium
249 cal	5 g	21 g	16 g	64 mg	621 mg	48 mg

Ingredient	Weight	Measure	Issue
CHICKEN,BONED,CANNED,PIECES	23-1/4 lbs	2 gal 1-1/8 qts	
CELERY,FRESH,CHOPPED	11-5/8 lbs	2 gal 3 qts	15-7/8 lbs
SALAD DRESSING,MAYONNAISE TYPE	3-1/2 lbs	1 qts 3 cup	
ONIONS,FRESH,CHOPPED	1 lbs	2-7/8 cup	1-1/8 lbs
JUICE,LEMON	8-5/8 oz	1 cup	
PEPPER,BLACK,GROUND	1/3 oz	1 tbsp	
LETTUCE,LEAF,FRESH,HEAD	4 lbs		6-1/4 lbs

Method

1 Drain. Cut chicken into 1/2 inch pieces.
2 Combine chicken, celery, salad dressing, onions, lemon juice, and pepper. Mix lightly but thoroughly.
3 Place lettuce leaf on serving dish. Top with 3/4 cup salad. CCP: Refrigerate product at 41 F. or lower until served.

MEAT. FISH. AND POULTRY No.L 151 02

TURKEY SALAD (BONELESS, FROZEN)

Yield 100 **Portion** 3/4 Cup

Calories	Carbohydrates	Protein	Fat	Cholesterol	Sodium	Calcium
193 cal	6 g	15 g	12 g	49 mg	770 mg	52 mg

Ingredient	Weight	Measure	Issue
TURKEY,BNLS,WHITE AND DARK MEAT,DICED	18 lbs		
CELERY,FRESH,CHOPPED	7-1/8 lbs	1 gal 2-3/4 qts	9-3/4 lbs
SALAD DRESSING,MAYONNAISE TYPE	2-3/4 lbs	1 qts 1-1/2 cup	
ONIONS,FRESH,CHOPPED	6-1/3 oz	1-1/8 cup	7 oz
JUICE,LEMON	8-5/8 oz	1 cup	
SALT	1-2/3 oz	2-2/3 tbsp	
PEPPER,BLACK,GROUND	1/4 oz	1 tbsp	
LETTUCE,LEAF,FRESH,CHOPPED	4 lbs	2 gal 1/8 qts	6-1/4 lbs

Method

1. Combine turkey, celery, salad dressing, onions, lemon juice, salt and pepper. Mix lightly but thoroughly.
2. Place lettuce leaf on serving dish. Top with 3/4 cup salad. CCP: Refrigerate product at 41 F. or lower until served.

MEAT. FISH. AND POULTRY No.L 152 00

CHICKEN TETRAZZINI (CANNED CHICKEN)

Yield 100   Portion 1 Cup

Calories	Carbohydrates	Protein	Fat	Cholesterol	Sodium	Calcium
271 cal	28 g	21 g	7 g	43 mg	1090 mg	108 mg

Ingredient	Weight	Measure	Issue
WATER,BOILING	25-1/8 lbs	3 gal	
SALT	1 oz	1 tbsp	
OIL,SALAD	1/2 oz	1 tbsp	
SPAGHETTI NOODLES,DRY	5 lbs	1 gal 1-3/8 qts	
ONIONS,FRESH,CHOPPED	8-1/2 oz	1-1/2 cup	9-1/2 oz
PEPPERS,GREEN,FRESH,CHOPPED	7-7/8 oz	1-1/2 cup	9-5/8 oz
COOKING SPRAY,NONSTICK	1/8 oz	1/8 tsp	
CHICKEN BROTH		1 gal 3 qts	
FLOUR,WHEAT,BREAD	2-3/8 lbs	2 qts	
WATER,COLD	4-1/8 lbs	2 qts	
WATER,WARM	7-7/8 lbs	3 qts 3 cup	
MILK,NONFAT,DRY	7-1/4 oz	3 cup	
NUTMEG,GROUND	1/2 oz	2 tbsp	
PEPPER,BLACK,GROUND	1/4 oz	1 tbsp	
CHICKEN,BONED,CANNED,PIECES	15-1/2 lbs	1 gal 2-1/8 qts	
MUSHROOMS,CANNED,STEMS & PIECES,INCL LIQUIDS	5-1/8 lbs	3 qts 3 cup	
PIMIENTO,CANNED,DRAINED,CHOPPED	11-7/8 oz	1-3/4 cup	
CHEESE,PARMESAN,GRATED	14-1/8 oz	1 qts	

Method

1. Add salt and salad oil to water; heat to a rolling boil.
2. Add spaghetti slowly while stirring constantly until water boils again. Cook 10 to 12 minutes or until tender; stir occasionally. DO NOT OVERCOOK.
3. Drain. Rinse with cold water; drain thoroughly.
4. Stir-cook onions and peppers in a lightly sprayed steam-jacketed kettle or stockpot 3 minutes or until tender, stirring constantly.
5. Add chicken broth to cooked vegetables; stir to blend. Bring to a boil; reduce heat to a simmer.
6. Blend flour and water together to make a smooth slurry. Add slurry to broth and onions, stirring constantly. Bring to a boil. Cover; reduce heat; simmer 10 minutes or until thickened, stirring frequently to prevent sticking.
7. Reconstitute milk in warm water. Add nutmeg and pepper; stir milk mixture into thickened broth. Bring to a boil. Cover; reduce heat; simmer 2 minutes.
8. Cut chicken into 1 inch pieces. Stir chicken, spaghetti, mushrooms and pimientos gently into thickened sauce. Heat to a simmer.
9. Pour chicken and spaghetti mixture into ungreased steam-table pans. Sprinkle parmesan cheese evenly over chicken and spaghetti mixture in each pan.
10. Using a convection oven, bake at 325 F. for 15 minutes or until lightly browned on high fan, open vent. CCP: Internal temperature must reach 165 F. or higher for 15 seconds. Hold for service at 140 F. or higher.

MEAT, FISH, AND POULTRY No.L 152 01

TUNA TETRAZZINI (CANNED TUNA)

Yield 100   Portion 1 Cup

Calories	Carbohydrates	Protein	Fat	Cholesterol	Sodium	Calcium
223 cal	28 g	21 g	3 g	20 mg	950 mg	105 mg

Ingredient	Weight	Measure	Issue
WATER,BOILING	25-1/8 lbs	3 gal	
SALT	1 oz	1 tbsp	
OIL,SALAD	1/2 oz	1 tbsp	
SPAGHETTI NOODLES,DRY	5 lbs	1 gal 1-3/8 qts	
ONIONS,FRESH,CHOPPED	8 oz	1-3/8 cup	8-7/8 oz
PEPPERS,GREEN,FRESH,CHOPPED	8 oz	1-1/2 cup	9-3/4 oz
COOKING SPRAY,NONSTICK	1/8 oz	1/8 tsp	
CHICKEN BROTH		1 gal 3 qts	
WATER,COLD	4-1/8 lbs	2 qts	
FLOUR,WHEAT,GENERAL PURPOSE	2-1/4 lbs	2 qts	
MILK,NONFAT,DRY	7-1/4 oz	3 cup	
WATER,WARM	7-7/8 lbs	3 qts 3 cup	
PEPPER,BLACK,GROUND	1/4 oz	1 tbsp	
FISH,TUNA,CANNED,WATER PACK,DRAINED	11-5/8 lbs	2 gal 5/8 qts	
MUSHROOMS,CANNED,STEMS & PIECES,INCL LIQUIDS	5-1/8 lbs	3 qts 3 cup	
PIMIENTO,CANNED,DRAINED,CHOPPED	11-7/8 oz	1-3/4 cup	
CHEESE,PARMESAN,GRATED	14-1/8 oz	1 qts	

Method

1 Add salt and salad oil to water; heat to a rolling boil.
2 Add spaghetti slowly while stirring constantly until water boils again. Cook about 10 to 12 minutes or until tender; stir occasionally. DO NOT OVERCOOK.
3 Drain. Rinse with cold water; drain thoroughly. Use immediately in recipe preparation or place in shallow containers and cover.
4 Stir-cook onions and peppers in a lightly sprayed steam-jacketed kettle or stockpot 3 minutes or until tender, stirring constantly.
5 Add chicken broth to cooked vegetables; stir to blend. Bring to a boil; reduce heat to a simmer.
6 Blend flour and water together to make a smooth slurry. Add slurry to broth and onions, stirring constantly. Bring to a boil. Cover, reduce heat; simmer 10 minutes or until thickened, stirring frequently to prevent sticking.
7 Reconstitute milk in warm water. Add pepper; stir milk mixture into thickened broth. Bring to a boil. Cover; reduce heat; simmer 2 minutes.
8 Stir tuna, spaghetti, mushrooms, and pimientos gently into thickened sauce. Heat to a simmer.
9 Pour 1-1/2 gal tuna and spaghetti mixture into each ungreased pan. Sprinkle 1 cup parmesan cheese over tuna and spaghetti mixture in each pan.
10 Using a convection oven, bake at 325 F. for 15 minutes or until lightly browned on high fan, open vent. CCP: Internal temperature must register 145 F. or higher for 15 seconds. Hold for service at 140 F. or higher.

MEAT, FISH, AND POULTRY No.L 152 02

CHICKEN TETRAZZINI (COOKED DICED)

Yield 100 Portion 1 Cup

Calories	Carbohydrates	Protein	Fat	Cholesterol	Sodium	Calcium
270 cal	28 g	23 g	6 g	52 mg	1028 mg	108 mg

Ingredient	Weight	Measure	Issue
WATER,BOILING	25-1/8 lbs	3 gal	
SALT	1 oz	1 tbsp	
OIL,SALAD	1/2 oz	1 tbsp	
SPAGHETTI NOODLES,DRY	5 lbs	1 gal 1-3/8 qts	
ONIONS,FRESH,CHOPPED	8-1/2 oz	1-1/2 cup	9-3/8 oz
PEPPERS,GREEN,FRESH,CHOPPED	7-7/8 oz	1-1/2 cup	9-5/8 oz
COOKING SPRAY,NONSTICK	1/8 oz	1/8 tsp	
CHICKEN BROTH		1 gal 3 qts	
WATER,COLD	4-1/8 lbs	2 qts	
FLOUR,WHEAT,BREAD	2-3/8 lbs	2 qts	
WATER,WARM	7-7/8 lbs	3 qts 3 cup	
MILK,NONFAT,DRY	7-1/4 oz	3 cup	
SALT	1-7/8 oz	3 tbsp	
NUTMEG,GROUND	1/2 oz	2 tbsp	
PEPPER,BLACK,GROUND	1/4 oz	1 tbsp	
CHICKEN,COOKED,DICED	12 lbs		
MUSHROOMS,CANNED,STEMS & PIECES,INCL LIQUIDS	5-1/8 lbs	3 qts 3 cup	
PIMIENTO,CANNED,DRAINED,CHOPPED	11-7/8 oz	1-3/4 cup	
CHEESE,PARMESAN,GRATED	14-1/8 oz	1 qts	

Method

1. Add salt and salad oil to water; heat to a rolling boil.
2. Add spaghetti slowly while stirring constantly until water boils again. Cook 10 to 12 minutes or until tender; stir occasionally. DO NOT OVERCOOK.
3. Drain. Rinse with cold water; drain thoroughly. Reserve for use in Step 8.
4. Stir-cook onions and peppers in a lightly sprayed steam-jacketed kettle or stockpot 3 minutes or until tender, stirring constantly.
5. Add chicken broth to cooked vegetables; stir to blend. Bring to a boil; reduce heat to a simmer.
6. Blend flour and water together to make a smooth slurry. Add slurry to broth and onions, stirring constantly. Bring to a boil. Cover; reduce heat; simmer 10 minutes or until thickened, stirring frequently to prevent sticking.
7. Reconstitute milk in warm water. Add salt, nutmeg and pepper; stir milk mixture into thickened broth. Bring to a boil. Cover; reduce heat; simmer 2 minutes.
8. Stir chicken, spaghetti, mushrooms and pimientos gently into thickened sauce. Heat to a simmer.
9. Pour 1-1/2 gallons of chicken and spaghetti mixture into ungreased steam table pans. Sprinkle 1 cup parmesan cheese evenly over chicken and spaghetti mixture in each pan.
10. Using a convection oven, bake at 325 F. for 15 minutes or until lightly browned on high fan, open vent. CCP: Internal temperature must reach 165 F. or higher for 15 seconds. Hold for service at 140 F. or higher.

MEAT, FISH, AND POULTRY No.L 153 00

CHINESE FIVE-SPICE CHICKEN (8 PC)

Yield 100   **Portion** 2 Pieces

Calories	Carbohydrates	Protein	Fat	Cholesterol	Sodium	Calcium
273 cal	2 g	40 g	10 g	119 mg	760 mg	31 mg

Ingredient	Weight	Measure	Issue
CHICKEN, 8 PC CUT, SKIN REMOVED	82 lbs		
SOY SAUCE	2-1/2 lbs	1 qts	
ONIONS,FRESH,CHOPPED	1-7/8 lbs	1 qts 1-3/8 cup	2-1/8 lbs
GINGER,GROUND	2 oz	1/2 cup 2-2/3 tbsp	
CINNAMON,GROUND	1 oz	1/4 cup 1/3 tbsp	
GARLIC POWDER	5/8 oz	2 tbsp	
FENNEL,GROUND	1/2 oz	2-1/3 tbsp	
CLOVES,GROUND	1/4 oz	1 tbsp	
PEPPER,BLACK,GROUND	1/4 oz	1 tbsp	
COOKING SPRAY,NONSTICK	2-1/8 oz	1/4 cup 2/3 tbsp	

Method

1. Wash chicken thoroughly under cold running water. Drain well; remove excess fat. Place chicken in roasting pans.
2. Combine soy sauce, onions, ginger; cinnamon, garlic powder, fennel, cloves, and pepper; mix well. Pour 3-1/2 cups marinade over chicken in each pan; cover. CCP: Marinate under refrigeration at 41 F. or lower for 45 minutes, turning once.
3. Drain chicken. CCP: Reserve marinade under refrigeration at 41 F. or lower for use in Step 5.
4. Place chicken, meat side up, on lightly sprayed sheet pans. Lightly spray chicken with cooking spray.
5. Using a convection oven, bake 20 minutes at 325 F. on high fan, closed vent. Baste chicken with reserved marinade. Discard remaining marinade. Bake an additional 20 minutes, for a total of 40 minutes. CCP: Internal temperature must reach 165 F. or higher for 15 seconds. Hold for service at 140 F. or higher.

MEAT, FISH, AND POULTRY No. L 154 00

CREOLE CHICKEN (8 PC)

Yield 100   Portion 2 Pieces

Calories	Carbohydrates	Protein	Fat	Cholesterol	Sodium	Calcium
308 cal	11 g	41 g	11 g	119 mg	420 mg	62 mg

Ingredient	Weight	Measure	Issue
ONIONS, FRESH, CHOPPED	2-1/4 lbs	1 qts 2-3/8 cup	2-1/2 lbs
PEPPERS, GREEN, FRESH, CHOPPED	2-1/4 lbs	1 qts 2-7/8 cup	2-3/4 lbs
CELERY, FRESH, CHOPPED	2-1/4 lbs	2 qts 1/2 cup	3-1/8 lbs
COOKING SPRAY, NONSTICK	1/4 oz	1/4 tsp	
TOMATOES, CANNED, CRUSHED, INCL LIQUIDS	22 lbs	2 gal 2 qts	
SUGAR, GRANULATED	3-1/2 oz	1/2 cup	
WORCESTERSHIRE SAUCE	1-5/8 oz	3 tbsp	
SALT	1-1/2 oz	2-1/3 tbsp	
PEPPER, BLACK, GROUND	3/8 oz	1 tbsp	
FLOUR, WHEAT, GENERAL PURPOSE	3-1/3 oz	3/4 cup	
WATER, COLD	12-1/2 oz	1-1/2 cup	
CHICKEN, 8 PC CUT, SKIN REMOVED	82 lbs		
COOKING SPRAY, NONSTICK	2-1/8 oz	1/4 cup 2/3 tbsp	

Method

1. Stir-cook onions, peppers, and celery in a lightly sprayed steam-jacketed kettle or stockpot 10 minutes or until tender, stirring constantly.
2. Add tomatoes, sugar, Worcestershire sauce, salt and pepper to vegetables. Bring to a boil. Cover; reduce heat; simmer 10 minutes.
3. Blend flour and cold water to make a smooth slurry. Add slurry to vegetable and tomato mixture. Bring to a boil. Cover; reduce heat; simmer 5 minutes or until thickened, stirring frequently to prevent sticking.
4. Wash chicken thoroughly under cold running water. Drain well. Remove excess fat. Place chicken, meat side up, on lightly sprayed sheet pans. Lightly spray chicken with cooking spray.
5. Using a convection oven, bake 20 minutes at 325 F. on high fan, closed vent.
6. Transfer chicken to steam table pans. Pour 2 quarts sauce evenly over chicken in each pan.
7. Cover; using a convection oven, bake 30 to 35 minutes at 350 F. on high fan, closed vent. CCP: Internal temperature must reach 165 F. or higher for 15 seconds.
8. CCP: Hold for service at 140 F. or higher.
9. Serve over cooked rice Recipe No. E 005 00.

MEAT, FISH, AND POULTRY No.L 154 01

CREOLE CHICKEN (COOKED DICED)

Yield 100   Portion 1-1/4 Cups

Calories	Carbohydrates	Protein	Fat	Cholesterol	Sodium	Calcium
204 cal	11 g	26 g	6 g	73 mg	375 mg	54 mg

Ingredient	Weight	Measure	Issue
ONIONS,FRESH,CHOPPED	2-1/4 lbs	1 qts 2-3/8 cup	2-1/2 lbs
PEPPERS,GREEN,FRESH,CHOPPED	2-1/4 lbs	1 qts 2-7/8 cup	2-3/4 lbs
CELERY,FRESH,CHOPPED	2-1/4 lbs	2 qts 1/2 cup	3-1/8 lbs
COOKING SPRAY,NONSTICK	1/4 oz	1/4 tsp	
TOMATOES,CANNED,CRUSHED,INCL LIQUIDS	22 lbs	2 gal 2 qts	
SUGAR,GRANULATED	3-1/2 oz	1/2 cup	
WORCESTERSHIRE SAUCE	1-5/8 oz	3 tbsp	
SALT	1-1/2 oz	2-1/3 tbsp	
PEPPER,BLACK,GROUND	3/8 oz	1 tbsp	
FLOUR,WHEAT,GENERAL PURPOSE	3-1/3 oz	3/4 cup	
WATER,COLD	12-1/2 oz	1-1/2 cup	
CHICKEN,COOKED,DICED	18 lbs		

Method

1. Stir-cook onions, peppers, and celery in a lightly sprayed steam-jacketed kettle or stockpot 10 minutes or until tender, stirring constantly.
2. Add tomatoes, sugar, Worcestershire sauce, salt and pepper to vegetables. Bring to a boil. Cover; reduce heat; simmer 10 minutes.
3. Blend flour and cold water to make a smooth slurry. Add slurry to vegetable and tomato mixture. Bring to a boil. Cover; reduce heat; simmer 5 minutes or until thickened, stirring frequently to prevent sticking.
4. Stir chicken gently into thickened creole sauce. Cover, reduce heat; simmer 2 minutes. CCP: Internal temperature must reach 165 F. or higher for 15 seconds.
5. Pour 2-1/2 gallon creole chicken into ungreased steam table pans. CCP: Hold for service at 140 F. or higher.
6. Serve over cooked rice Recipe No. E 005 00.

MEAT, FISH, AND POULTRY No.L 155 00

FRIED CHICKEN (8 PC)

Yield 100   Portion 8 Ounces

Calories	Carbohydrates	Protein	Fat	Cholesterol	Sodium	Calcium
496 cal	14 g	46 g	27 g	144 mg	553 mg	29 mg

Ingredient	Weight	Measure	Issue
CHICKEN, 8 PIECE CUT	82 lbs		
FLOUR,WHEAT,GENERAL PURPOSE	3-7/8 lbs	3 qts 2 cup	
SALT	3-3/4 oz	1/4 cup 2-1/3 tbsp	
PEPPER,BLACK,GROUND	7/8 oz	1/4 cup 1/3 tbsp	
PAPRIKA,GROUND	1/2 oz	2 tbsp	

Method

1 Wash chicken thoroughly under cold running water. Drain well
2 Dredge chicken pieces in mixture of flour, salt, pepper and paprika; shake off excess.
3 Fry until golden brown or until done in 325 F. deep fat. CCP: Internal temperature must reach 165 F. or higher for 15 seconds.
4 Drain well in basket or on absorbent paper. CCP: Hold for service at 140 F. or higher.

Notes

1 Approximate frying time for cut-up 8 piece chicken is: Wings, 5 to 7 minutes; Legs, 10 to 13 minutes; Thighs, 10 to 15 minutes; Breasts, 10 to 15 minutes.

MEAT. FISH. AND POULTRY No.L 155 01

SOUTHERN FRIED CHICKEN (8 PC)

Yield 100   Portion   2 Pieces

Calories	Carbohydrates	Protein	Fat	Cholesterol	Sodium	Calcium
496 cal	14 g	46 g	27 g	144 mg	553 mg	29 mg

Ingredient	Weight	Measure	Issue
CHICKEN, 8 PIECE CUT	82 lbs		
FLOUR,WHEAT,GENERAL PURPOSE	3-7/8 lbs	3 qts 2 cup	
SALT	3-3/4 oz	1/4 cup 2-1/3 tbsp	
PEPPER,BLACK,GROUND	7/8 oz	1/4 cup 1/3 tbsp	
PAPRIKA,GROUND	1/2 oz	2 tbsp	

Method

1 Wash chicken thoroughly under cold running water. Drain well.
2 Dredge chicken pieces in mixture of flour, salt, pepper and paprika; shake off excess.
3 Brown chicken in batches in 325 F. deep fat. For each type of piece, fry according to minimum times in Note 1.
4 Place chicken on sheet pans. Using a convection oven, bake uncovered at 350 F. for 15 minutes or until done on high fan, open vent. CCP: Internal temperature must reach 165 F. or higher for 15 seconds.
5 Drain well on absorbent paper. CCP: Hold for service at 140 F. or higher.

Notes

1 Approximate frying time for cut-up 8 piece chicken is: Wings, 5 to 7 minutes; Legs, 10 to 13 minutes; Thighs, 10 to 15 minutes; Breasts, 10 to 15 minutes.

MEAT, FISH, AND POULTRY No. L 155 02

FRIED CHICKEN (PRECKD BRDED, FZN FOR DEEP FAT FRY)

Yield 100 Portion 2 Pieces

Calories	Carbohydrates	Protein	Fat	Cholesterol	Sodium	Calcium
664 cal	18 g	46 g	44 g	207 mg	569 mg	42 mg

Ingredient **Weight** **Measure** **Issue**

CHICKEN,BREADED,PRECOOKED,FRYER 65 lbs

Method

1. Fry chicken in 350 F. deep fat 5 to 6 minutes or until browned and heated thoroughly. CCP: Temperature must reach 165 F. or higher for 15 seconds.
2. Drain well on absorbent paper. CCP: Hold for service at 140 F. or higher.

MEAT, FISH, AND POULTRY No.L 156 00

OVEN BAKED CHICKEN (8 PC)

Yield 100 **Portion** 2 Pieces

Calories	Carbohydrates	Protein	Fat	Cholesterol	Sodium	Calcium
341 cal	15 g	41 g	12 g	119 mg	487 mg	54 mg

Ingredient	Weight	Measure	Issue
CHICKEN, 8 PC CUT, SKIN REMOVED	82 lbs		
BREADCRUMBS	6-2/3 lbs	1 gal 3 qts	
SALT	1-7/8 oz	3 tbsp	
PEPPER, BLACK, GROUND	7/8 oz	1/4 cup 1/3 tbsp	
PAPRIKA, GROUND	1/2 oz	2 tbsp	
COOKING SPRAY, NONSTICK	2-1/8 oz	1/4 cup 2/3 tbsp	

Method

1. Wash chicken thoroughly under cold running water. Drain well. Remove excess fat.
2. Combine breadcrumbs, salt, pepper and paprika; mix well.
3. Dredge chicken pieces in breadcrumb mixture; shake off excess.
4. Place chicken, meat side up, on lightly sprayed sheet pans. Lightly spray chicken with cooking spray.
5. Using a convection oven, bake at 325 F. for 40 minutes on high fan, open vent. CCP: Internal temperature must reach 165 F. or higher for 15 seconds.
6. Transfer chicken to steam table pans. CCP: Hold for service at 140 F. or higher.

Notes

1. In Step 2, 7 lb corn flake crumbs may be substituted for breadcrumbs.

MEAT, FISH, AND POULTRY No.L 156 01

FRIED CHICKEN (PRECKED, BREAD CHIX, FRZ FOR OVEN)

Yield 100   Portion 2 Pieces

Calories	Carbohydrates	Protein	Fat	Cholesterol	Sodium	Calcium
583 cal	18 g	46 g	35 g	207 mg	569 mg	42 mg

Ingredient   **Weight**   **Measure**   **Issue**

CHICKEN,BREADED,PRECOOKED,FRYER   65 lbs

Method

1. Using a convection oven, bake at 350 F. for 25 to 30 minutes on high fan, closed vent. CCP: Internal temperature must reach 165 F. or higher for 15 seconds.

MEAT, FISH, AND POULTRY No.L 157 00

PINEAPPLE CHICKEN (8 PC)

Yield 100　　　　　　　　　　　　　　　　　　Portion 2 Pieces

Calories	Carbohydrates	Protein	Fat	Cholesterol	Sodium	Calcium
338 cal	20 g	40 g	10 g	119 mg	317 mg	38 mg

Ingredient	Weight	Measure	Issue
CHICKEN, 8 PC CUT, SKIN REMOVED	82 lbs		
COOKING SPRAY,NONSTICK	2-1/8 oz	1/4 cup 2/3 tbsp	
PINEAPPLE,CANNED,CRUSHED	19-3/4 lbs	2 gal 1 qts	
JUICE,PINEAPPLE,CANNED,UNSWEETENED	6-1/3 lbs	2 qts 3-1/2 cup	
SOY SAUCE	12-2/3 oz	1-1/4 cup	
SUGAR,GRANULATED	6-1/8 oz	3/4 cup 2 tbsp	

Method

1. Wash chicken pieces thoroughly under cold running water. Drain well. Remove excess fat.
2. Place chicken, meat side up, on lightly sprayed sheet pans. Lightly spray chicken with cooking spray.
3. Using a convection oven, bake 40 minutes at 325 F. on high fan, closed vent. CCP: Internal temperature must reach 165 F. or higher for 15 seconds. Hold at 140 F. or higher for use in Step 5.
4. Combine pineapple, pineapple juice, soy sauce, and sugar. Bring to a boil. Cover, reduce heat; simmer for 5 minutes.
5. Transfer chicken to steam table pans. Pour 2 quart sauce evenly over chicken in each pan.
6. CCP: Hold for service at 140 F. or higher. Serve with 1/4 cup sauce.

MEAT, FISH, AND POULTRY No.L 158 00

SAVORY BAKED CHICKEN (8 PC)

Yield 100   **Portion** 2 Pieces

Calories	Carbohydrates	Protein	Fat	Cholesterol	Sodium	Calcium
269 cal	2 g	40 g	10 g	119 mg	658 mg	25 mg

Ingredient	Weight	Measure	Issue
CHICKEN, 8 PC CUT, SKIN REMOVED	82 lbs		
SOY SAUCE	1-7/8 lbs	3 cup	
WORCESTERSHIRE SAUCE	1-5/8 lbs	3 cup	
CHICKEN BROTH		3 cup	
GARLIC POWDER	1 oz	3-1/3 tbsp	
PEPPER,BLACK,GROUND	2/3 oz	3 tbsp	
COOKING SPRAY,NONSTICK	2-1/8 oz	1/4 cup 2/3 tbsp	
PARSLEY,FRESH,BUNCH,CHOPPED	1 oz	1/4 cup	1 oz

Method

1. Wash chicken thoroughly under cold running water. Drain well. Remove excess fat. Place chicken in roasting pans.
2. Combine soy sauce, Worcestershire sauce, chicken broth, garlic powder and pepper; mix well. Pour marinade over chicken in each pan; cover. CCP: Marinate under refrigeration at 41 F. or lower for 45 minutes, turning once.
3. Drain chicken. CCP: Reserve marinade under refrigeration at 41 F. or lower for use in Step 5.
4. Place chicken, meat side up, on lightly sprayed sheet pans. Lightly spray chicken with cooking spray.
5. Using a convection oven, bake 20 minutes at 325 F. on high fan, closed vent. Baste chicken with reserved marinade. Discard remaining marinade. Bake an additional 20 minutes, for a total of 40 minutes. CCP: Internal temperature must reach 165 F. or higher for 15 seconds.
6. Sprinkle with parsley. CCP: Hold for service at 140 F. or higher.

MEAT, FISH, AND POULTRY No.L 158 01

SAVORY BAKED CHICKEN (THIGHS)

Yield 100   Portion   4 Ounces

Calories	Carbohydrates	Protein	Fat	Cholesterol	Sodium	Calcium
310 cal	2 g	38 g	16 g	135 mg	667 mg	22 mg

Ingredient	Weight	Measure	Issue
CHICKEN,THIGHS,BNLS/SKNLS,RAW	31-1/4 lbs		
SOY SAUCE	1-7/8 lbs	3 cup	
WORCESTERSHIRE SAUCE	1-5/8 lbs	3 cup	
CHICKEN BROTH		3 cup	
GARLIC POWDER	1 oz	3-1/3 tbsp	
PEPPER,BLACK,GROUND	2/3 oz	3 tbsp	
COOKING SPRAY,NONSTICK	1-1/2 oz	3 tbsp	
PARSLEY,FRESH,BUNCH,CHOPPED	1 oz	1/2 cup	1-1/8 oz

Method

1 Wash chicken thoroughly under cold running water. Drain well. Remove excess fat. Place chicken in roasting pans.
2 Combine soy sauce, Worcestershire sauce, chicken broth, garlic powder and pepper; mix well.
3 Pour marinade over chicken in each pan; cover. CCP: Marinate under refrigeration at 41 F. or lower for 45 minutes.
4 Place chicken thighs on lightly sprayed sheet pans. Lightly spray chicken with cooking spray. Discard remaining marinade.
5 Using a convection oven, bake 12-14 minutes at 325 F. on high fan, closed vent. CCP: Internal temperature must reach 165 F. or higher for 15 seconds.
6 Transfer chicken to steam table pans. Sprinkle with parsley. CCP: Hold for service at 140 F. or higher.

MEAT, FISH, AND POULTRY   No.L 159 00

SZECHWAN CHICKEN (8 PC)

Yield 100   Portion   2 Pieces

Calories	Carbohydrates	Protein	Fat	Cholesterol	Sodium	Calcium
311 cal	12 g	40 g	10 g	119 mg	751 mg	26 mg

Ingredient	Weight	Measure	Issue
CHICKEN, 8 PC CUT, SKIN REMOVED	82 lbs		
WATER	4-2/3 lbs	2 qts 1 cup	
CHICKEN BROTH		1 qts	
VINEGAR,DISTILLED	2-1/8 lbs	1 qts	
SOY SAUCE	1-7/8 lbs	3 cup	
CATSUP	1-5/8 lbs	3 cup	
SUGAR,GRANULATED	1-3/4 lbs	1 qts	
PEPPER,RED,CRUSHED	2/3 oz	1/2 cup	
COOKING SPRAY,NONSTICK	2 oz	1/4 cup 1/3 tbsp	
WATER,COLD	1-1/3 lbs	2-1/2 cup	
CORNSTARCH	5-1/8 oz	1-1/8 cup	

Method

1. Wash chicken thoroughly under cold water. Drain well. Remove excess fat. Place chicken in roasting pans.
2. Combine water, chicken broth, vinegar, soy sauce, catsup, sugar and red pepper in a steam-jacketed kettle or stockpot. Bring to a boil. Cover; reduce heat; simmer 5 minutes.
3. Pour 8-1/2 cups marinade over chicken in each pan; cover. CCP: Marinate under refrigeration at 41 F. or lower for 45 minutes, turning once.
4. Drain chicken. CCP: Reserve marinade under refrigeration at 41 F. or lower for use in Step 7.
5. Place chicken, meat side up, on lightly sprayed sheet pans.
6. Using a convection oven, bake 40 minutes at 325 F. on high fan, closed vent. CCP: Internal temperature must reach 165 F. or higher for 15 seconds. Transfer chicken to steam table pans.
7. Bring remaining marinade to a boil.
8. Blend cornstarch and cold water together to make a smooth slurry. Add slurry to marinade; bring to a boil. Cover; reduce heat; simmer 3 minutes or until thickened, stirring frequently to prevent sticking. CCP: Temperature must register 165 F. or higher for 15 seconds.
9. Pour 5-3/4 cups sauce evenly over chicken in each pan. CCP: Hold for service at 140 F. or higher.

MEAT, FISH, AND POULTRY No.L 159 01

SZECHWAN CHICKEN (BREAST BONELESS)

Yield 100   Portion 5 Ounces

Calories	Carbohydrates	Protein	Fat	Cholesterol	Sodium	Calcium
226 cal	12 g	33 g	4 g	88 mg	712 mg	21 mg

Ingredient	Weight	Measure	Issue
CHICKEN,BREAST,BNLS/SKNLS,5 OZ	31-1/4 lbs		
WATER	4-2/3 lbs	2 qts 1 cup	
CHICKEN BROTH		1 qts	
VINEGAR,DISTILLED	2-1/8 lbs	1 qts	
SOY SAUCE	1-7/8 lbs	3 cup	
CATSUP	1-5/8 lbs	3 cup	
SUGAR,GRANULATED	1-3/4 lbs	1 qts	
PEPPER,RED,CRUSHED	2/3 oz	1/2 cup	
COOKING SPRAY,NONSTICK	1-1/2 oz	3 tbsp	
WATER,COLD	1-1/3 lbs	2-1/2 cup	
CORNSTARCH	5-1/8 oz	1-1/8 cup	

Method

1. Wash chicken thoroughly under cold water. Drain well. Remove excess fat. Place chicken in roasting pans.
2. Combine water, chicken broth, vinegar, soy sauce, catsup, sugar, and red pepper in a steam-jacketed kettle or stockpot. Bring to a boil. Cover; reduce heat; simmer 5 minutes.
3. Pour marinade over chicken in each pan; cover. CCP: Marinate under refrigeration at 41 F. or lower for 45 minutes.
4. Drain chicken. CCP: Reserve marinade under refrigeration at 41 F. or lower for use in Step 7.
5. Place chicken breasts on each lightly sprayed sheet pan. Lightly spray chicken with cooking spray.
6. Using a convection oven, bake 12 to 14 minutes at 325 F. on high fan, closed vent. CCP: Internal temperature must reach 165 F. or higher for 15 seconds. Transfer chicken to steam table pans.
7. Bring remaining marinade to a boil.
8. Blend cornstarch and cold water together to make a smooth slurry. Add slurry to marinade; bring to a boil. Cover; reduce heat; simmer 3 minutes or until thickened, stirring frequently to prevent sticking. CCP: Temperature must register 165 F. or higher for 15 seconds.
9. Pour 6 cups sauce evenly over chicken in each pan. CCP: Hold for service at 140 F. or higher.

MEAT. FISH. AND POULTRY No.L 160 00

CHICKEN CHOW MEIN (COOKED DICED)

Yield 100   Portion   1 Cup

Calories	Carbohydrates	Protein	Fat	Cholesterol	Sodium	Calcium
230 cal	14 g	28 g	7 g	73 mg	1762 mg	68 mg

Ingredient	Weight	Measure	Issue
CHICKEN BROTH		2 gal 2 qts	
ONIONS,FRESH,SLICED	12-1/8 lbs	2 gal 4 qts	13-1/2 lbs
CELERY,FRESH,SLICED	9-1/2 lbs	2 gal 1 qts	13 lbs
CABBAGE,GREEN,FRESH,CHOPPED	4-1/8 lbs	1 gal 2-2/3 qts	5-1/8 lbs
SOY SAUCE	3-3/4 lbs	1 qts 2 cup	
MOLASSES	5-3/4 oz	1/2 cup	
GINGER,GROUND	3/8 oz	2 tbsp	
GARLIC POWDER	1/4 oz	1/3 tsp	
PEPPER,BLACK,GROUND	1/8 oz	1/3 tsp	
CORNSTARCH	14-2/3 oz	3-1/4 cup	
WATER,COLD	4-1/8 lbs	2 qts	
CHICKEN,COOKED,DICED	18 lbs		
BEAN SPROUTS,CANNED,DRAINED	3-1/3 lbs	1 gal 2 qts	

Method

1. Combine chicken broth, onions, celery, cabbage, soy sauce, molasses, ginger, garlic powder and pepper in a steam jacketed kettle or stockpot. Bring to a boil. Cover; reduce heat; simmer 8 to 10 minutes until vegetables are tender.
2. Blend cornstarch and cold water together to make a smooth slurry. Add slurry to hot broth and vegetable mixture, stirring constantly. Cover; reduce heat; simmer 3 to 5 minutes or until thickened, stirring frequently to prevent sticking.
3. Stir chicken and bean sprouts gently into thickened sauce. Cover; reduce heat; simmer 2 minutes. CCP: Internal temperature must reach 165 F. or higher for 15 seconds.
4. Pour 2-1/2 gal chicken chow mein into ungreased steam table pans. CCP: Hold for service at 140 F. or higher. Serve over steamed rice. Optional: Top each serving with 1/3 cup chow mein noodles.

MEAT, FISH, AND POULTRY No.L 160 01

CHICKEN CHOW MEIN (CANNED CHICKEN)

Yield 100 **Portion** 1 Cup

Calories	Carbohydrates	Protein	Fat	Cholesterol	Sodium	Calcium
235 cal	15 g	24 g	8 g	59 mg	2169 mg	69 mg

Ingredient	Weight	Measure	Issue
CHICKEN BROTH		2 gal 2 qts	
ONIONS,FRESH,SLICED	12-1/8 lbs	2 gal 4 qts	13-1/2 lbs
CELERY,FRESH,SLICED	9-1/2 lbs	2 gal 1 qts	13 lbs
CABBAGE,GREEN,FRESH,CHOPPED	4-1/8 lbs	1 gal 2-2/3 qts	5-1/8 lbs
SOY SAUCE	3-3/4 lbs	1 qts 2 cup	
MOLASSES	5-3/4 oz	1/2 cup	
GINGER,GROUND	3/8 oz	2 tbsp	
GARLIC POWDER	1/4 oz	1/3 tsp	
PEPPER,BLACK,GROUND	1/8 oz	1/3 tsp	
CORNSTARCH	1-1/8 lbs	1 qts	
WATER,COLD	4-1/8 lbs	2 qts	
CHICKEN,BONED,CANNED,PIECES	23-1/4 lbs	2 gal 1-1/8 qts	
BEAN SPROUTS,CANNED,DRAINED	3-1/3 lbs	1 gal 2 qts	

Method

1. Combine chicken broth, onions, celery, soy sauce, molasses, ginger, garlic powder and pepper in a steam jacketed kettle or stockpot. Bring to a boil. Cover; reduce heat; simmer 8 to 10 minutes until vegetables are tender.
2. Blend cornstarch and cold water together to make a smooth slurry. Add slurry to hot broth and vegetable mixture, stirring constantly. Cover; reduce heat; simmer 3 to 5 minutes or until thickened, stirring frequently to prevent sticking.
3. Cut chicken into 1-inch pieces.
4. Stir chicken and bean sprouts gently into thickened sauce. Cover; reduce heat; simmer 2 minutes. CCP: Temperature must reach 165 F. or higher for 15 seconds.
5. Pour 2-1/2 gal chicken chow mein into ungreased steam table pans. CCP: Hold for service at 140 F. or higher. Serve over steamed rice. Optional: Top each serving with 1/3 cup chow mein noodles.

MEAT. FISH. AND POULTRY No.L 161 00

ROAST TURKEY

Yield 100	Portion 4 Ounces

Calories	Carbohydrates	Protein	Fat	Cholesterol	Sodium	Calcium
172 cal	0 g	27 g	7 g	69 mg	901 mg	23 mg

Ingredient
Ingredient	Weight	Measure	Issue
TURKEY,WHOLE,READY-TO-COOK,RAW	65 lbs		
SALT	7-5/8 oz	3/4 cup	
SHORTENING,VEGETABLE,MELTED	7-1/4 oz	1 cup	

Method

1. Remove bands from legs; open turkey cavity. Cut off wing tips.
2. Wash turkey thoroughly inside and out, under cold running water. Drain well.
3. Rub cavity with salt.
4. Tuck legs and tail into cavity. Place in roasting pans, breast side up. Turkeys should not touch each other.
5. Rub skin with salad oil or melted shortening. DO NOT ADD WATER.
6. Insert meat thermometer in center of inside thigh muscle of smallest bird.
7. Roast uncovered. CCP: Internal temperature OF ALL TURKEYS must reach 165 F. or higher for 15 seconds.
8. Baste frequently with drippings. CCP: Hold for service at 140 F. or higher.

MEAT, FISH, AND POULTRY No.L 162 00

ROAST TURKEY (BONELESS TURKEY)

Yield 100   **Portion**   3-1/2 Ounces

Calories	Carbohydrates	Protein	Fat	Cholesterol	Sodium	Calcium
257 cal	4 g	31 g	12 g	95 mg	1010 mg	55 mg

Ingredient   Weight   Measure   Issue

TURKEY,BNLS,WHITE AND DARK MEAT   38 lbs

Method

1. Place turkeys in roasting pans.
2. Using a convection oven, roast 2-1/2 to 3-1/2 hours in 325 F. oven, on high fan, closed vent. Baste occasionally with drippings, uncovered. CCP: Internal temperature must reach 165 F. or higher for 15 seconds. Hold for service at 140 F. or higher.

Notes

1. When roasted, remove from oven; let stand at least 15 to 20 minutes to absorb juices and for ease in slicing.

MEAT, FISH, AND POULTRY No.L 162 01

ROAST TURKEY WITH BARBECUE SAUCE

Yield 100   Portion   3-1/2 Ounces

Calories	Carbohydrates	Protein	Fat	Cholesterol	Sodium	Calcium
309 cal	13 g	32 g	13 g	95 mg	1581 mg	68 mg

Ingredient	Weight	Measure	Issue
TURKEY,BNLS,WHITE AND DARK MEAT	38 lbs		
SAUCE,BARBECUE	15-3/8 lbs	1 gal 3 qts	

Method

1 Place turkey in pans.
2 Using a convection oven, roast 2-1/2 to 3-1/2 hours in 325 F. oven, on high fan, closed vent. Baste occasionally with drippings. CCP: Internal temperature must reach 165 F. or higher for 15 seconds. Hold for service at 140 F. or higher.
3 Use prepared Barbecue Sauce or Barbecue Sauce, Recipe No. O 002 00. Bring sauce to a boil; reduce heat; cover; simmer about 5 minutes or until heated thoroughly.
4 Slice turkey about 1/4-inch thick. CCP: Hold for service at 140 F. or higher. Serve 1/4 cup sauce over turkey slices.

Notes

1 When roasted, remove from oven; let stand at least 15 to 20 minutes to absorb juices and for ease in slicing.

MEAT, FISH, AND POULTRY No.L 163 00

TURKEY NUGGETS

Yield 100   Portion   3-1/2 Ounces

Calories	Carbohydrates	Protein	Fat	Cholesterol	Sodium	Calcium
284 cal	23 g	25 g	9 g	65 mg	1631 mg	68 mg

Ingredient	Weight	Measure	Issue
TURKEY,BNLS,WHITE AND DARK MEAT	26 lbs		
FLOUR,WHEAT,GENERAL PURPOSE	3-1/3 lbs	3 qts	
SALT	5-1/8 oz	1/2 cup	
GARLIC POWDER	1-5/8 oz	1/4 cup 1-2/3 tbsp	
SEASONING,POULTRY	1/3 oz	2-2/3 tbsp	
PEPPER,BLACK,GROUND	1/3 oz	1 tbsp	
PAPRIKA,GROUND	1/3 oz	1 tbsp	
MILK,NONFAT,DRY	1-3/4 oz	3/4 cup	
WATER,WARM	2 lbs	3-3/4 cup	
EGG WHITES	1-5/8 lbs	3 cup	
SALT	2-1/2 oz	1/4 cup 1/3 tbsp	
PARSLEY,DEHYDRATED,FLAKED	1/8 oz	1/4 cup 1/3 tbsp	
BREADCRUMBS	3-5/8 lbs	3 qts 3 cup	
COOKING SPRAY,NONSTICK	1 oz	2 tbsp	

Method

1. Cut turkey into 1-1/2 to 2-inch strips.
2. Dredge turkey in mixture of flour, salt, garlic powder, poultry seasoning, pepper and paprika.
3. Reconstitute milk; add egg whites; mix well.
4. Dip floured turkey in milk and egg white mixture. Drain well.
5. Blend second salt, parsley and breadcrumbs to create breadcrumb mixture. Roll turkey in bread crumb mixture until well coated; shake off excess.
6. Lightly spray sheet pans with non-stick cooking spray. Place turkey nuggets onto sprayed sheet pans.
7. Spray turkey nuggets with cooking spray to ensure even browning.
8. Using a convection oven, bake 10 to 12 minutes at 375 F. on high fan, closed vent. CCP: Internal temperature must reach 165 F. or higher for 15 seconds. Hold at 140 F. or higher for service. Serve with sweet and sour sauce, barbecue sauce or mustard sauce.

MEAT. FISH. AND POULTRY No.L 164 00

ROAST DUCK

Yield 100   **Portion** 7 Ounces

Calories	Carbohydrates	Protein	Fat	Cholesterol	Sodium	Calcium
657 cal	0 g	37 g	55 g	164 mg	115 mg	22 mg

Ingredient	Weight	Measure	Issue
DUCK,WHOLE,READY TO COOK	100 lbs		
PEPPER,BLACK,GROUND	1/8 oz	1/3 tsp	

Method

1. Wash duck thoroughly, inside and out, under cold running water. Drain well.
2. Rub cavity of duck with pepper.
3. Place duck, breast side up on sheet pans without crowding. Prick skin of duck.
4. Roast 2 hours or until duck is done in 325 F. oven. CCP: Internal temperature must reach 165 F. or higher for 15 seconds.
5. Pour off fat frequently during roasting period. CCP: Hold for service at 140 F. or higher.

MEAT, FISH, AND POULTRY No.L 164 01

HAWAIIAN BAKED DUCK

Yield 100   Portion 7 Ounces

Calories	Carbohydrates	Protein	Fat	Cholesterol	Sodium	Calcium
677 cal	5 g	37 g	55 g	164 mg	116 mg	27 mg

Ingredient	Weight	Measure	Issue
DUCK,WHOLE,READY TO COOK	100 lbs		
PEPPER,BLACK,GROUND	1/8 oz	1/3 tsp	
GINGER,GROUND	1/4 oz	1 tbsp	
JUICE,ORANGE	4-3/8 lbs	2 qts	
JUICE,PINEAPPLE,CANNED,UNSWEETENED	4-3/8 lbs	2 qts	

Method

1. Wash duck thoroughly, inside and out, under cold running water. Drain well.
2. Rub cavity of duck with a mixture of pepper and ginger.
3. Place duck, breast side up, in pans without crowding. Prick skin of duck.
4. Combine orange juice with canned pineapple juice.
5. Roast 2 hours, basting frequently with juice mixture, until duck is done in 325 F. oven. CCP: Internal temperature must reach 165 F. or higher for 15 seconds.
6. Pour off fat frequently during roasting period. CCP: Hold for service at 140 F. or higher.

MEAT. FISH. AND POULTRY No.L 164 02

ROAST DUCK WITH APPLE JELLY GLAZE

Yield 100   Portion   7 Ounces

Calories	Carbohydrates	Protein	Fat	Cholesterol	Sodium	Calcium
690 cal	8 g	37 g	56 g	165 mg	145 mg	23 mg

Ingredient	Weight	Measure	Issue
DUCK,WHOLE,READY TO COOK	100 lbs		
PEPPER,BLACK,GROUND	1/8 oz	1/3 tsp	
BUTTER	2 oz	1/4 cup 1/3 tbsp	
JELLY,APPLE	2 lbs	3 cup	
JUICE,APPLE,CANNED	13-1/8 oz	1-1/2 cup	
JUICE,LEMON	2-1/8 oz	1/4 cup 1/3 tbsp	
JUICE,ORANGE	4-3/8 oz	1/2 cup	
CATSUP	6-1/3 oz	3/4 cup	
VINEGAR,DISTILLED	1 oz	2 tbsp	

Method

1. Wash duck thoroughly, inside and out, under cold running water. Drain well.
2. Rub cavity of duck with pepper.
3. Place duck, breast side up, in pans without crowding. Prick skin of duck.
4. Roast 1-1/2 hours at 325 F. Pour off fat frequently during roasting period.
5. Melt butter or margarine. Add apple jelly and canned apple juice. Stir to break up jelly; continue stirring until jelly is melted. Remove from heat.
6. Add lemon juice, orange juice, tomato catsup and vinegar. Stir until well blended. Increase oven temperature to 375 F.
7. Brush skin evenly with 1/2 of the glaze; roast 15 minutes. Repeat with remaining glaze; roast an additional 15 minutes or until tender. CCP: Internal temperature must reach 165 F. or higher for 15 seconds. Hold for service at 140 F. or higher.

MEAT, FISH, AND POULTRY   No.L 164 03

HONEY GLAZED DUCK

Yield 100   **Portion** 7 Ounces

Calories	Carbohydrates	Protein	Fat	Cholesterol	Sodium	Calcium
701 cal	11 g	37 g	55 g	164 mg	370 mg	23 mg

Ingredient   **Weight**   **Measure**   **Issue**
DUCK,WHOLE,READY TO COOK   100 lbs
PEPPER,BLACK,GROUND   1/8 oz   1/3 tsp
HONEY   3 lbs   1 qts
SOY SAUCE   10-1/8 oz   1 cup
GINGER,GROUND   1/4 oz   1 tbsp
SALT   7/8 oz   1 tbsp
PEPPER,BLACK,GROUND   1/8 oz   1/3 tsp

Method

1. Wash duck thoroughly, inside and out, under cold running water. Drain well.
2. Rub cavity of duck with pepper.
3. Place duck, breast side up, in pans without crowding. Prick skin of duck.
4. Roast 1-1/2 hours at 325 F.
5. Combine honey, soy sauce, ground ginger, salt and black pepper. Stir until well blended.
6. Increase oven temperature to 375 F. Brush skin of ducks evenly with 1/2 of glaze. Roast 15 minutes. Repeat with remaining glaze. Roast an additional 15 minutes or until tender. CCP: Internal temperature must reach 165 F. or higher for 15 seconds. Hold for service at 140 F. or higher.

MEAT, FISH, AND POULTRY No.L 165 00

PIZZA

Yield 100   Portion 1 Slice

Calories	Carbohydrates	Protein	Fat	Cholesterol	Sodium	Calcium
226 cal	28 g	9 g	9 g	16 mg	449 mg	146 mg

Ingredient	Weight	Measure	Issue
PIZZA SAUCE		1 gal	
YEAST,ACTIVE,DRY	2-3/8 oz	1/4 cup 2 tbsp	
WATER,WARM	9-3/8 oz	1-1/8 cup	
WATER,COLD	3-1/8 lbs	1 qts 2 cup	
FLOUR,WHEAT,BREAD	6-5/8 lbs	1 gal 1-1/2 qts	
SALT	1 oz	1 tbsp	
SUGAR,GRANULATED	2-1/3 oz	1/4 cup 1-2/3 tbsp	
OIL,SALAD	7-2/3 oz	1 cup	
OIL,SALAD	1-7/8 oz	1/4 cup 1/3 tbsp	
OIL,SALAD	1-7/8 oz	1/4 cup 1/3 tbsp	
CHEESE,MOZZARELLA,SHREDDED	4 lbs	1 gal	
CHEESE,PARMESAN,GRATED	7 oz	2 cup	

Method

1. Prepare 1 recipe Pizza Sauce, Recipe No. O 012 00 or use prepared pizza sauce..
2. Sprinkle yeast over water. DO NOT USE TEMPERATURES ABOVE 110 F. Mix well. Let stand 5 minutes, stir.
3. Place water, flour, salt, sugar and salad oil or melted shortening in mixer bowl in order listed. Add yeast solution.
4. Using a dough hook, mix at low speed about 8 minutes until dough is smooth and elastic. Dough temperature should be 86 F. to 88 F.
5. Divide dough; shape into four 2 pound 7 ounce balls. Cover; let rise in warm place 1-1/2 to 2 hours or until double in bulk.
6. Coat bottom and sides of each pan with 1 tablespoon salad oil or melted shortening.
7. Place dough balls on lightly floured working surface. Roll out each ball to 1/8-inch thickness. Transfer dough to 18x26 sheet pans pushing dough slightly up edges of pan. Using 1 tablespoon oil per pan, lightly brush dough. Gently prick dough to prevent bubbling.
8. Using a convection oven, bake at 450 F. 7 minutes on high fan, closed vent or until slightly brown.
9. Spread 1 quart sauce evenly over dough in each pan.
10. Sprinkle 1 quart shredded cheese over each pan.
11. Sprinkle 1/2 cup grated cheese over mixture in each pan.
12. Using a convection oven, bake 8 minutes at 450 F. on high fan, closed vent or until crust is browned and cheese starts to turn golden.
13. Cut 5 by 5. CCP: Hold for service at 140 F. or higher.

MEAT, FISH, AND POULTRY No.L 165 01

PIZZA (THICK CRUST)

Yield 100   Portion 4-1/2 Ounces

Calories	Carbohydrates	Protein	Fat	Cholesterol	Sodium	Calcium
362 cal	50 g	14 g	12 g	19 mg	607 mg	175 mg

Ingredient	Weight	Measure	Issue
PIZZA SAUCE		1 gal	
YEAST,ACTIVE,DRY	4-2/3 oz	1/2 cup 3 tbsp	
WATER,WARM	1-1/8 lbs	2-1/4 cup	
WATER,COLD	6-1/4 lbs	3 qts	
FLOUR,WHEAT,BREAD	13 lbs	2 gal 2-3/4 qts	
SALT	2-1/3 oz	1/4 cup	
SUGAR,GRANULATED	5-1/4 oz	3/4 cup	
OIL,SALAD	1-1/8 lbs	2-1/4 cup	
CHEESE,MOZZARELLA,SHREDDED	5 lbs	1 gal 1 qts	
CHEESE,PARMESAN,GRATED	7 oz	2 cup	

Method

1. Prepare 1 recipe Pizza Sauce, Recipe No. O 012 00 or use prepared pizza sauce.
2. Sprinkle yeast over water. DO NOT USE TEMPERATURES ABOVE 110 F. Mix well. Let stand 5 minutes, stir.
3. Place water, flour, salt, sugar and salad oil in mixer bowl in order listed. Add yeast solution.
4. Using a dough hook, mix at low speed about 10 minutes until dough is smooth and elastic. Dough temperature should be 86 F. to 88 F.
5. Divide dough; shape into four 4 pound 10 ounce balls. Cover; let rise in warm place 1-1/2 to 2 hours or until double in bulk.
6. Coat bottom and sides of each pan with 1 tablespoon salad oil or melted shortening.
7. Place dough balls on lightly floured working surface. Roll out each ball to 1/4-inch thickness. Transfer dough to sheet pans pushing dough slightly up edges of pan. Using 1 tablespoon oil per pan, lightly brush dough. Gently prick dough to prevent bubbling.
8. Using a convection oven, bake at 450 F. 7 minutes on high fan, closed vent or until slightly brown.
9. Spread 1 quart sauce evenly over dough in each pan.
10. Sprinkle 1-1/4 quart shredded cheese over each pan.
11. Sprinkle 1/2 cup grated cheese over mixture in each pan.
12. Using a convection oven, bake at 450 F. about 8 minutes or until crust is browned and cheese starts to turn golden.
13. Cut 5 by 5. CCP: Hold for service at 140 F. or higher.

MEAT, FISH, AND POULTRY No.L 165 02

MUSHROOM, GREEN PEPPER AND ONION PIZZA

Yield 100   Portion   4 Ounces

Calories	Carbohydrates	Protein	Fat	Cholesterol	Sodium	Calcium
235 cal	31 g	10 g	9 g	16 mg	476 mg	150 mg

Ingredient	Weight	Measure	Issue
PIZZA SAUCE		1 gal	
YEAST,ACTIVE,DRY	2-3/8 oz	1/4 cup 2 tbsp	
WATER,WARM	9-3/8 oz	1-1/8 cup	
WATER,COLD	3-1/8 lbs	1 qts 2 cup	
FLOUR,WHEAT,BREAD	6-5/8 lbs	1 gal 1-1/2 qts	
SALT	1 oz	1 tbsp	
SUGAR,GRANULATED	2-1/3 oz	1/4 cup 1-2/3 tbsp	
OIL,SALAD	7-2/3 oz	1 cup	
OIL,SALAD	1-7/8 oz	1/4 cup 1/3 tbsp	
OIL,SALAD	1-7/8 oz	1/4 cup 1/3 tbsp	
CHEESE,MOZZARELLA,SHREDDED	4 lbs	1 gal	
MUSHROOMS,CANNED,SLICED,DRAINED	1-3/8 lbs	1 qts	
PEPPERS,GREEN,FRESH,MEDIUM,SLICED,THIN	4 lbs	3 qts 1/8 cup	4-7/8 lbs
ONIONS,FRESH,SLICED	1-3/4 lbs	1 qts 2-7/8 cup	2 lbs
CHEESE,PARMESAN,GRATED	7 oz	2 cup	

Method

1 Prepare 1 recipe Pizza Sauce, Recipe No. O 012 00 or use prepared pizza sauce.
2 Sprinkle yeast over water. DO NOT USE TEMPERATURES ABOVE 110 F. Mix well. Let stand 5 minutes, stir.
3 Place water, flour, salt, sugar and salad oil in mixer bowl in order listed. Add yeast solution.
4 Using a dough hook, mix at low speed about 8 minutes until dough is smooth and elastic. Dough temperature should be 86 F. to 88 F.
5 Divide dough; shape into 2 pound 7 ounce balls. Cover; let rise in warm place 1-1/2 to 2 hours or until double in bulk.
6 Coat bottom and sides of each pan with 1 tablespoon salad oil or melted shortening.
7 Place dough balls on lightly floured working surface. Roll out each ball to 1/8-inch thickness. Transfer dough to sheet pans pushing dough slightly up edges of pan. Using 1 tablespoon oil per pan, lightly brush dough. Gently prick dough to prevent bubbling.
8 Using a convection oven, bake at 450 F. 7 minutes or until slightly brown on high fan, closed vent.
9 Spread 1 quart sauce evenly over dough in each pan.
10 Sprinkle 1 quart shredded cheese over each pan.
11 Drain mushrooms; slice peppers and onions. Evenly distribute 1 cup mushrooms, 3 cups green peppers, and 1-3/4 cups onion over cheese in each pan.
12 Sprinkle 1/2 cup grated cheese over mixture in each pan.
13 Using a convection oven, bake at 450 F. about 8 minutes or until crust is browned and cheese starts to turn golden on high fan, closed vent.
14 Cut 5 by 5. CCP: Hold for service at 140 F. or higher.

MEAT, FISH, AND POULTRY No.L 165 03

HAMBURGER PIZZA

Yield 100   Portion 4 Ounces

Calories	Carbohydrates	Protein	Fat	Cholesterol	Sodium	Calcium
300 cal	29 g	17 g	13 g	44 mg	466 mg	150 mg

Ingredient	Weight	Measure	Issue
PIZZA SAUCE		1 gal	
YEAST,ACTIVE,DRY	2-3/8 oz	1/4 cup 2 tbsp	
WATER,WARM	9-3/8 oz	1-1/8 cup	
WATER,COLD	3-1/8 lbs	1 qts 2 cup	
FLOUR,WHEAT,BREAD	6-5/8 lbs	1 gal 1-1/2 qts	
SALT	1 oz	1 tbsp	
SUGAR,GRANULATED	2-1/3 oz	1/4 cup 1-2/3 tbsp	
OIL,SALAD	7-2/3 oz	1 cup	
OIL,SALAD	1-7/8 oz	1/4 cup 1/3 tbsp	
OIL,SALAD	1-7/8 oz	1/4 cup 1/3 tbsp	
CHEESE,MOZZARELLA,SHREDDED	4 lbs	1 gal	
BEEF,GROUND,BULK,RAW,90% LEAN	8 lbs		
ONIONS,FRESH,CHOPPED	12-2/3 oz	2-1/4 cup	14-1/8 oz
PEPPER,BLACK,GROUND	1/8 oz	1/8 tsp	
OREGANO,CRUSHED	1/8 oz	1 tbsp	
CHEESE,PARMESAN,GRATED	7 oz	2 cup	

Method

1. Prepare 1 recipe Pizza Sauce, Recipe No. O 012 00 or use prepared pizza sauce.
2. Sprinkle yeast over water. DO NOT USE TEMPERATURES ABOVE 110 F. Mix well. Let stand 5 minutes, stir.
3. Place water, flour, salt, sugar and salad oil in mixer bowl in order listed. Add yeast solution.
4. Using a dough hook, mix at low speed about 8 minutes until dough is smooth and elastic. Dough temperature should be 86 F. to 88 F.
5. Divide dough; shape into four 2 pound 7 ounce balls. Cover; let rise in warm place 1-1/2 to 2 hours or until double in bulk.
6. Coat bottom and sides of each pan with 1 tablespoon salad oil or melted shortening.
7. Place dough balls on lightly floured working surface. Roll out each ball to 1/8-inch thickness. Transfer dough to sheet pans pushing dough slightly up edges of pan. Using 1 tablespoon oil per pan, lightly brush dough. Gently prick dough to prevent bubbling.
8. Using a convection oven, bake at 450 F. 7 minutes or until slightly brown on high fan, closed vent.
9. Spread 1 quart sauce evenly over dough in each pan.
10. Sprinkle 1 quart shredded cheese over each pan.
11. Saute thawed ground beef with onions. Drain or skim off excess fat; add black pepper, crushed oregano. Blend well. CCF: Internal temperature must reach 155 F. or higher for 15 seconds. Sprinkle 1 quart of meat mixture in each pan.
12. Sprinkle 1/2 cup grated cheese over mixture in each pan.
13. Using a convection oven, bake at 450 F. about 8 minutes or until crust is browned and cheese starts to turn golden on high fan, closed vent. CCP: Hold for service at 140 F. or higher.
14. Cut 5 by 5.

MEAT, FISH, AND POULTRY No. L 165 04

PEPPERONI, GREEN PEPPER, AND MUSHROOM PIZZA

Yield 100   Portion 3-1/2 Ounces

Calories	Carbohydrates	Protein	Fat	Cholesterol	Sodium	Calcium
255 cal	30 g	10 g	11 g	19 mg	568 mg	149 mg

Ingredient	Weight	Measure	Issue
PIZZA SAUCE		1 gal	
YEAST,ACTIVE,DRY	2-3/8 oz	1/4 cup 2 tbsp	
WATER,WARM	9-3/8 oz	1-1/8 cup	
WATER,COLD	3-1/8 lbs	1 qts 2 cup	
FLOUR,WHEAT,BREAD	6-5/8 lbs	1 gal 1-1/2 qts	
SALT	1 oz	1 tbsp	
SUGAR,GRANULATED	2-1/3 oz	1/4 cup 1-2/3 tbsp	
OIL,SALAD	7-2/3 oz	1 cup	
OIL,SALAD	1-7/8 oz	1/4 cup 1/3 tbsp	
OIL,SALAD	1-7/8 oz	1/4 cup 1/3 tbsp	
CHEESE,MOZZARELLA,SHREDDED	4 lbs	1 gal	
MUSHROOMS,CANNED,SLICED,DRAINED	1-3/8 lbs	1 qts	
PEPPERS,GREEN,FRESH,MEDIUM,SLICED,THIN	4 lbs	3 qts 1/8 cup	4-7/8 lbs
PEPPERONI	1 lbs		
CHEESE,PARMESAN,GRATED	7 oz	2 cup	

Method

1. Prepare 1 recipe Pizza Sauce, Recipe No. O 012 00 or use prepared pizza sauce.
2. Sprinkle yeast over water. DO NOT USE TEMPERATURES ABOVE 110 F. Mix well. Let stand 5 minutes, stir.
3. Place water, flour, salt, sugar and salad oil in mixer bowl in order listed. Add yeast solution.
4. Using a dough hook, mix at low speed about 8 minutes until dough is smooth and elastic. Dough temperature should be 86 F. to 88 F.
5. Divide dough; shape into four 2 pound 7 ounce balls. Cover; let rise in warm place 1-1/2 to 2 hours or until double in bulk.
6. Coat bottom and sides of each pan with 1 tablespoon salad oil or melted shortening.
7. Place dough balls on lightly floured working surface. Roll out each ball to 1/8-inch thickness. Transfer dough to sheet pans pushing dough slightly up edges of pan. Using 1 tablespoon oil per pan, lightly brush dough. Gently prick dough to prevent bubbling.
8. Using a convection oven bake at 450 F. 7 minutes or until slightly brown on high fan, closed vent.
9. Spread 1 quart sauce evenly over dough in each pan.
10. Sprinkle 1 quart shredded cheese over each pan.
11. Drain mushrooms, slice peppers, slice pepperoni. Evenly distribute 1 cup mushrooms, 3 cups green peppers and 4 ounces pepperoni over cheese in each pan.
12. Sprinkle 1/2 cup grated cheese over mixture in each pan.
13. Using a convection oven, bake at 450 F. about 8 minutes or until crust is browned and cheese starts to turn golden on high fan, closed vent. CCP: Hold for service at 140 F. or higher.
14. Cut 5 by 5.

MEAT, FISH, AND POULTRY No.L 165 05

PEPPERONI PIZZA

Yield 100   Portion 1 Slice

Calories	Carbohydrates	Protein	Fat	Cholesterol	Sodium	Calcium
248 cal	29 g	10 g	11 g	19 mg	541 mg	147 mg

Ingredient

Ingredient	Weight	Measure	Issue
PIZZA SAUCE		1 gal	
YEAST,ACTIVE,DRY	2-3/8 oz	1/4 cup 2 tbsp	
WATER,WARM	9-3/8 oz	1-1/8 cup	
WATER,COLD	3-1/8 lbs	1 qts 2 cup	
FLOUR,WHEAT,BREAD	6-5/8 lbs	1 gal 1-1/2 qts	
SALT	1 oz	1 tbsp	
SUGAR,GRANULATED	2-1/3 oz	1/4 cup 1-2/3 tbsp	
OIL,SALAD	7-2/3 oz	1 cup	
OIL,SALAD	1-7/8 oz	1/4 cup 1/3 tbsp	
OIL,SALAD	1-7/8 oz	1/4 cup 1/3 tbsp	
CHEESE,MOZZARELLA,SHREDDED	4 lbs	1 gal	
PEPPERONI	1 lbs		
CHEESE,PARMESAN,GRATED	7 oz	2 cup	

Method

1. Prepare 1 recipe Pizza Sauce, Recipe No. O 012 00 or use prepared pizza sauce.
2. Sprinkle yeast over water. DO NOT USE TEMPERATURES ABOVE 110 F. Mix well. Let stand 5 minutes, stir.
3. Place water, flour, salt, sugar and salad oil in mixer bowl in order listed. Add yeast solution.
4. Using a dough hook, mix at low speed about 8 minutes until dough is smooth and elastic. Dough temperature should be 86 F. to 88 F.
5. Divide dough; shape into four 2 pound 7 ounce balls. Cover; let rise in warm place 1-1/2 to 2 hours or until double in bulk.
6. Coat bottom and sides of each pan with 1 tablespoon salad oil or melted shortening.
7. Place dough balls on lightly floured working surface. Roll out each ball to 1/8-inch thickness. Transfer dough to pans pushing dough slightly up edges of pan. Using 1 tablespoon oil per pan, lightly brush dough. Gently prick dough to prevent bubbling.
8. Using a convection oven, bake at 450 F. 7 minutes or until slightly brown on high fan, closed vent.
9. Spread 1 quart sauce evenly over dough in each pan.
10. Sprinkle 1 quart shredded cheese over each pan.
11. Thinly slice pepperoni; evenly distribute 4 ounces over cheese in each pan.
12. Sprinkle 1/2 cup grated cheese over mixture in each pan.
13. Using a convection oven, bake at 450 F. 8 minutes or until crust is browned and cheese starts to turn golden on high fan, closed vent. CCP: Hold for service at 140 F. or higher.
14. Cut 5 by 5.

MEAT, FISH, AND POULTRY No.L 165 06

PIZZA (ROLL MIX)

Yield 100   Portion 4 Ounces

Calories	Carbohydrates	Protein	Fat	Cholesterol	Sodium	Calcium
211 cal	27 g	9 g	8 g	16 mg	519 mg	161 mg

Ingredient	Weight	Measure	Issue
PIZZA SAUCE		1 gal	
ROLL,MIX	6-3/4 lbs		
YEAST,ACTIVE,DRY	2-1/4 oz	1/4 cup 1-2/3 tbsp	
WATER	3-3/4 lbs	1 qts 3-1/8 cup	
OIL,SALAD	1-7/8 oz	1/4 cup 1/3 tbsp	
OIL,SALAD	1-7/8 oz	1/4 cup 1/3 tbsp	
CHEESE,MOZZARELLA,SHREDDED	4 lbs	1 gal	
CHEESE,PARMESAN,GRATED	7 oz	2 cup	

Method

1. Prepare 1 recipe Pizza Sauce, Recipe No. O 012 00 or use prepared pizza sauce.
2. Combine roll mix, yeast, and water. Follow directions on containers.
3. Shape into four 2 pound 10 ounce balls.
4. Coat bottom and sides of each pan with 1 tablespoon salad oil or melted shortening.
5. Place dough balls on lightly floured working surface. Roll out each ball to 1/8-inch thickness. Transfer dough to pans pushing dough slightly up edges of pan. Using 1 tablespoon oil per pan, lightly brush dough. Gently prick dough to prevent bubbling.
6. Using a convection oven, bake at 450 F. 7 minutes or until slightly brown on high fan, closed vent.
7. Spread 1 quart sauce evenly over dough in each pan.
8. Sprinkle 1 quart shredded cheese over each pan.
9. Sprinkle 1/2 cup grated cheese over mixture in each pan.
10. Using a convection oven, bake at 450 F. 10 minutes or until crust is browned and cheese starts to turn golden on high fan, closed vent. CCP: Hold for service at 140 F. or higher.
11. Cut 5 by 5.

MEAT. FISH. AND POULTRY No.L 165 07

PORK OR ITALIAN SAUSAGE PIZZA

Yield 100   Portion 1 Slice

Calories	Carbohydrates	Protein	Fat	Cholesterol	Sodium	Calcium
265 cal	29 g	11 g	12 g	25 mg	545 mg	160 mg

Ingredient	Weight	Measure	Issue
PIZZA SAUCE		1 gal	
YEAST,ACTIVE,DRY	2-3/8 oz	1/4 cup 2 tbsp	
WATER,WARM	9-3/8 oz	1-1/8 cup	
WATER,COLD	3-1/8 lbs	1 qts 2 cup	
FLOUR,WHEAT,BREAD	6-5/8 lbs	1 gal 1-1/2 qts	
SALT	1 oz	1 tbsp	
SUGAR,GRANULATED	2-1/3 oz	1/4 cup 1-2/3 tbsp	
OIL,SALAD	7-2/3 oz	1 cup	
OIL,SALAD	1-7/8 oz	1/4 cup 1/3 tbsp	
OIL,SALAD	1-7/8 oz	1/4 cup 1/3 tbsp	
CHEESE,MOZZARELLA	4-1/2 lbs	3 qts 3-1/2 cup	
SAUSAGE,POLISH,PORK,RAW	3 lbs		
CHEESE,PARMESAN,GRATED	7 oz	2 cup	

Method

1. Prepare 1 recipe Pizza Sauce, Recipe No. O 012 00 or use prepared pizza sauce.
2. Sprinkle yeast over water. DO NOT USE TEMPERATURES ABOVE 110 F. Mix well. Let stand 5 minutes, stir.
3. Place water, flour, salt, sugar and salad oil in mixer bowl in order listed. Add yeast solution.
4. Using a dough hook, mix at low speed about 8 minutes until dough is smooth and elastic. Dough temperature should be 86 F. to 88 F.
5. Divide dough; shape into four 2 pound 7 ounce balls. Cover; let rise in warm place 1-1/2 to 2 hours or until double in bulk.
6. Coat bottom and sides of each pan with 1 tablespoon salad oil or melted shortening.
7. Place dough balls on lightly floured working surface. Roll out each ball to 1/8-inch thickness. Transfer dough to pans pushing dough slightly up edges of pan. Using 1 tablespoon oil per pan, lightly brush dough. Gently prick dough to prevent bubbling.
8. Using a convection oven, bake at 450 F. 7 minutes or until slightly brown on high fan, closed vent.
9. Spread 1 quart sauce evenly over dough in each pan.
10. Sprinkle 1 quart shredded cheese over each pan.
11. Saute pork or sausage until light brown; drain or skim off excess fat. CCP: Internal temperature must reach 155 F. or higher for 15 seconds. Evenly distribute 1-1/2 cups sausage over cheese in each pan.
12. Sprinkle 1/2 cup grated cheese over mixture in each pan.
13. Using a convection oven, bake at 450 F. about 8 minutes or until crust is browned and cheese starts to turn golden on high fan, closed vent. CCP: Hold for service at 140 F. or higher.
14. Cut 5 by 5.

MEAT, FISH, AND POULTRY No.L 165 08

FRENCH BREAD PIZZA

Yield 100   Portion 4 Ounces

Calories	Carbohydrates	Protein	Fat	Cholesterol	Sodium	Calcium
323 cal	46 g	14 g	9 g	24 mg	827 mg	245 mg

Ingredient	Weight	Measure	Issue
PIZZA SAUCE		1 gal	
BREAD,FRENCH	17 lbs		
CHEESE,MOZZARELLA,SHREDDED	6-1/2 lbs	1 gal 2-1/2 qts	
CHEESE,PARMESAN,GRATED	4 oz	1-1/8 cup	

Method

1 Prepare 1 recipe Pizza Sauce, Recipe No. O 012 00 or use prepared pizza sauce.
2 Cut each loaf of bread lengthwise and divide each half into 3 pieces. Place 12 pieces on each pan.
3 Spread 2-1/3 tablespoons sauce over each piece.
4 Evenly distribute 1 ounce or 1/4 cup shredded cheese over each piece.
5 Sprinkle 1/2 teaspoon grated cheese over mixture on each piece.
6 Using a convection oven, bake at 400 F. for 6 minutes or until cheese starts to turn golden on high fan, closed vent. CCP: Hold for service at 140 F. or higher.

MEAT, FISH, AND POULTRY   No. L 165 09

SAUSAGE, GREEN PEPPER, AND ONION PIZZA

Yield 100   Portion   1 Slice

Calories	Carbohydrates	Protein	Fat	Cholesterol	Sodium	Calcium
266 cal	30 g	11 g	11 g	23 mg	541 mg	152 mg

Ingredient | Weight | Measure | Issue

Ingredient	Weight	Measure	Issue
PIZZA SAUCE		1 gal	
YEAST,ACTIVE,DRY	2-3/8 oz	1/4 cup 2 tbsp	
WATER,WARM	9-3/8 oz	1-1/8 cup	
WATER,COLD	3-1/8 lbs	1 qts 2 cup	
FLOUR,WHEAT,BREAD	6-5/8 lbs	1 gal 1-1/2 qts	
SALT	1 oz	1 tbsp	
SUGAR,GRANULATED	2-1/3 oz	1/4 cup 1-2/3 tbsp	
OIL,SALAD	7-2/3 oz	1 cup	
OIL,SALAD	1-7/8 oz	1/4 cup 1/3 tbsp	
OIL,SALAD	1-7/8 oz	1/4 cup 1/3 tbsp	
CHEESE,MOZZARELLA,SHREDDED	4 lbs	1 gal	
SAUSAGE,ITALIAN,HOT	3 lbs		
PEPPERS,GREEN,FRESH,MEDIUM,SLICED,THIN	4 lbs	3 qts 1/8 cup	4-7/8 lbs
ONIONS,FRESH,SLICED	1-3/4 lbs	1 qts 2-7/8 cup	2 lbs
CHEESE,PARMESAN,GRATED	7 oz	2 cup	

Method

1. Prepare 1 recipe Pizza Sauce, Recipe No. O 012 00 or use prepared pizza sauce.
2. Sprinkle yeast over water. DO NOT USE TEMPERATURES ABOVE 110 F. Mix well. Let stand 5 minutes, stir.
3. Place water, flour, salt, sugar and salad oil in mixer bowl in order listed. Add yeast solution.
4. Using a dough hook, mix at low speed about 8 minutes until dough is smooth and elastic. Dough temperature should be 86 F. to 88 F.
5. Divide dough; shape into four 2 pound 7 ounce balls. Cover; let rise in warm place 1-1/2 to 2 hours or until double in bulk.
6. Coat bottom and sides of each pan with 1 tablespoon salad oil or melted shortening.
7. Place dough balls on lightly floured working surface. Roll out each ball to 1/8-inch thickness. Transfer dough to pans pushing dough slightly up edges of pan. Using 1 tablespoon oil per pan, lightly brush dough. Gently prick dough to prevent bubbling.
8. Using a convection oven, bake at 450 F. 7 minutes or until slightly brown on high fan, closed vent.
9. Spread 1 quart sauce evenly over dough in each pan.
10. Sprinkle 1 quart shredded cheese over each pan.
11. Saute pork or Italian sausage until light brown. CCP: Internal temperature must reach 155 F. or higher for 15 seconds. Drain or skim off excess fat. Evenly distribute 1-1/2 cups sausage, 3 cups green peppers, and 1-3/4 cups onions over cheese in each pan.
12. Sprinkle 1/2 cup grated cheese over mixture in each pan.
13. Using a convection oven, bake at 450 F. 8 minutes or until crust is browned and cheese starts to turn golden on high fan, closed vent. CCP: Hold for service at 140 F. or higher.
14. Cut 5 by 5.

MEAT. FISH. AND POULTRY No.L 165 10

PIZZA (POURABLE PIZZA CRUST)

Yield 100   Portion 1 Slice

Calories	Carbohydrates	Protein	Fat	Cholesterol	Sodium	Calcium
260 cal	39 g	11 g	7 g	16 mg	419 mg	185 mg

Ingredient	Weight	Measure	Issue
PIZZA SAUCE		1 gal	
YEAST,ACTIVE,DRY	3-3/8 oz	1/2 cup	
WATER,WARM	8-1/3 lbs	1 gal	
FLOUR,WHEAT,BREAD	8-1/2 lbs	1 gal 3 qts	
MILK,NONFAT,DRY	10-3/8 oz	1 qts 3/8 cup	
SUGAR,GRANULATED	8-7/8 oz	1-1/4 cup	
SALT	5/8 oz	1 tbsp	
OIL,SALAD	1-7/8 oz	1/4 cup 1/3 tbsp	
OIL,SALAD	1-7/8 oz	1/4 cup 1/3 tbsp	
CORN MEAL	4-7/8 oz	1 cup	
CHEESE,MOZZARELLA,SHREDDED	4 lbs	1 gal	
CHEESE,PARMESAN,GRATED	7 oz	2 cup	

Method

1. Prepare 1 recipe Pizza Sauce, Recipe No. O 012 00 or use prepared pizza sauce.
2. Sprinkle yeast over water. DO NOT USE TEMPERATURES ABOVE 110 F. Mix well. Let stand 5 minutes, stir.
3. Sift together flour, nonfat dry milk, sugar, and salt. Add yeast solution and salad oil or melted shortening.
4. Using wire whip, blend at medium speed 10 minutes. Batter will be lumpy.
5. Coat bottom and sides of each pan with 1 tablespoon salad oil or melted shortening.
6. Sprinkle 1/4 cup cornmeal evenly into each pan. Pour 1-3/4 quart pizza dough batter into each pan. Spread evenly. Let stand 20 minutes.
7. Using a convection oven, bake at 450 F. 12 minutes or until slightly brown on high fan, open vent.
8. Spread 1 quart sauce evenly over dough in each pan.
9. Sprinkle 1 quart shredded cheese over each pan.
10. Sprinkle 1/2 cup grated cheese over mixture in each pan.
11. Using a convection oven, bake at 450 F. 8 minutes or until crust is browned and cheese starts to turn golden on high fan, closed vent. CCP: Hold for service at 140 F. or higher.
12. Cut 5 by 5.

MEAT, FISH, AND POULTRY No.L 166 00

PIZZA (12 INCH FROZEN CRUST)

Yield 100   **Portion** 4 Ounces

Calories	Carbohydrates	Protein	Fat	Cholesterol	Sodium	Calcium
259 cal	35 g	11 g	8 g	20 mg	456 mg	203 mg

Ingredient

Ingredient	Weight	Measure	Issue
PIZZA CRUST,12"",FROZEN	12-1/2 lbs		
COOKING SPRAY,NONSTICK	2 oz	1/4 cup 1/3 tbsp	
SAUCE,PIZZA,CANNED	10-7/8 lbs	1 gal 1/2 qts	
CHEESE,MOZZARELLA	4-2/3 lbs	1 gal	
CHEESE,PARMESAN,GRATED	7 oz	2 cup	

Method

1. Place 2 crusts on each greased sheet pan.
2. Pour 3/4 cup sauce over each crust.
3. Sprinkle about 2/3 cup cheese over each pizza.
4. Sprinkle about 1-1/4 tablespoon grated cheese over mixture in each pan.
5. Bake at 450 F. about 20 minutes or until crust is browned and crisp.
6. Cut each pizza into 4 wedges. CCP: Hold for service at 140 F. or higher.

MEAT, FISH, AND POULTRY No.L 167 00

CHUCK WAGON STEW (BEANS WITH BEEF)

Yield 100　　　　　　　　　　　　　　　　　　　**Portion**　1-1/4 Cups

Calories	Carbohydrates	Protein	Fat	Cholesterol	Sodium	Calcium
416 cal	54 g	28 g	12 g	70 mg	1138 mg	141 mg

Ingredient	Weight	Measure	Issue
BEEF,GROUND,BULK,RAW,90% LEAN	15 lbs		
ONIONS,FRESH,CHOPPED	6 lbs	1 gal 1/4 qts	6-2/3 lbs
PEPPERS,GREEN,FRESH,CHOPPED	3 lbs	2 qts 1-1/8 cup	3-2/3 lbs
CATSUP	1-7/8 lbs	3-1/2 cup	
BEANS,BAKED,W/PORK,CANNED	53-1/2 lbs	6 gal	

Method

1. Cook beef with onions and peppers until it loses its pink color, stirring to break apart, in steam-jacketed kettle or stock pot. Drain or skim off excess fat.
2. Add catsup and beans to beef, onion and pepper mixture. Stir well.
3. Simmer for 20 minutes. CCP: Internal temperature must reach 155 F. or higher for 15 seconds. Hold for service at 140 F. or higher.

MEAT, FISH, AND POULTRY No.L 168 00

BAKED SCALLOPS

Yield 100   Portion   4-1/2 Ounces

Calories	Carbohydrates	Protein	Fat	Cholesterol	Sodium	Calcium
124 cal	6 g	14 g	5 g	37 mg	257 mg	32 mg

Ingredient   Weight   Measure   Issue

Ingredient	Weight	Measure
SCALLOPS,SEA,RAW	30 lbs	2 gal 1 qts
JUICE,LEMON	11-1/2 oz	1-3/8 cup
BREADCRUMBS	1-3/8 lbs	1 qts 2 cup
SALT	1/2 oz	3/8 tsp
PEPPER,BLACK,GROUND	1/8 oz	1/3 tsp
PAPRIKA,GROUND	1/8 oz	1/8 tsp
BASIL,SWEET,WHOLE,CRUSHED	1/4 oz	1 tbsp
GARLIC POWDER	3/4 oz	2-2/3 tbsp
BUTTER,MELTED	1 lbs	2 cup
PARSLEY,DEHYDRATED,FLAKED	3/8 oz	1/2 cup

Method

1 Wash scallops thoroughly; cut large ones in half. Drain well.
2 Marinate scallops in lemon juice 5 to 10 minutes.
3 Mix bread crumbs, salt, pepper, paprika, basil and garlic.
4 Drain scallops. Dredge scallops in seasoned bread crumbs. Place an equal quantity of scallops in each steam table pan.
5 Drizzle 1/2 cup melted butter or margarine over top of scallops in each pan. Using a convection oven, bake at 350 F. 20 minutes on high fan, closed vent. CCP: Internal temperature must reach 145 F. or higher for 15 seconds.
6 Remove from oven; sprinkle each pan with 2 tablespoons parsley. CCP: Hold at 140 F. or higher for service.

MEAT, FISH, AND POULTRY No.L 169 00

BAKED WHOLE TROUT

Yield 100 **Portion** 10 Ounces

Calories	Carbohydrates	Protein	Fat	Cholesterol	Sodium	Calcium
269 cal	6 g	23 g	17 g	87 mg	326 mg	94 mg

Ingredient	Weight	Measure	Issue
FISH,RAINBOW TROUT,WHOLE,RAW	63 lbs		
COOKING SPRAY,NONSTICK	2 oz	1/4 cup 1/3 tbsp	
BUTTER,MELTED	2-1/2 lbs	1 qts 1 cup	
DILL WEED,DRIED	1/4 oz	2 tbsp	
PEPPER,BLACK,GROUND	1/8 oz	1/3 tsp	
JUICE,LEMON	2-1/8 lbs	1 qts	
BREADCRUMBS	2-1/8 lbs	2 qts 1 cup	
SALT	1-1/4 oz	2 tbsp	

Method

1. Place single layer of fish on pans sprayed with non-stick cooking spray.
2. Combine butter or margarine, dill weed, and pepper; add lemon juice. Use 1 cup lemon-butter mixture for each pan of fish. Lightly brush inside and top of each fish.
3. Combine bread crumbs and salt. Use 1 cup bread crumbs per pan; evenly sprinkle on inside and outside of fish.
4. Bake 15 minutes in 375 F. convection oven. CCP: Internal temperature must reach 145 F. or higher for 15 seconds. Hold for service at 140 F. or higher.

Notes

1. Since trout does not hold well in serving line for long periods of time, prepare by progressive cooking methods in small batches.

MEAT, FISH, AND POULTRY   No.L 169 01

BAKED TROUT FILLETS

Yield 100   **Portion** 5 Ounces

Calories	Carbohydrates	Protein	Fat	Cholesterol	Sodium	Calcium
298 cal	4 g	32 g	16 g	105 mg	247 mg	122 mg

Ingredient	Weight	Measure	Issue
FISH,RAINBOW TROUT,FILLET,RAW,5 OZ	32 lbs		
COOKING SPRAY,NONSTICK	2 oz	1/4 cup 1/3 tbsp	
BUTTER,MELTED	1-2/3 lbs	3-3/8 cup	
PEPPER,BLACK,GROUND	1/8 oz	1/4 tsp	
JUICE,LEMON	1-3/8 lbs	2-5/8 cup	
BREADCRUMBS	1-3/8 lbs	1 qts 2 cup	
SALT	7/8 oz	1 tbsp	

Method

1. Place single layer of fish on pans sprayed with non-stick cooking spray in rows, skin side down.
2. Combine butter or margarine and pepper; add lemon juice. Use 1 cup lemon butter mixture for each pan of fish. Evenly brush inside and top of each fish.
3. Combine breadcrumbs and salt. Use 1 cup bread crumbs per pan; evenly sprinkle over top of each fish.
4. Bake 9 minutes in 375 F. oven. CCP: Internal temperature must reach 145 F. or higher for 15 seconds.

Notes

1. Since trout does not hold well in serving line for long periods of time, prepare by progressive cooking methods in small batches.

MEAT, FISH, AND POULTRY No.L 170 00

CHILI (WITHOUT BEANS)

Yield 100   Portion 1 Cup

Calories	Carbohydrates	Protein	Fat	Cholesterol	Sodium	Calcium
346 cal	16 g	34 g	17 g	106 mg	677 mg	70 mg

Ingredient	Weight	Measure	Issue
BEEF,GROUND,BULK,RAW,90% LEAN	30 lbs		
TOMATOES,CANNED,CRUSHED,INCL LIQUIDS	19-7/8 lbs	2 gal 1 qts	
TOMATO PASTE,CANNED	7-1/8 lbs	3 qts 1/4 cup	
ONIONS,FRESH,CHOPPED	3-1/8 lbs	2 qts 7/8 cup	3-1/2 lbs
CHILI POWDER,DARK,GROUND	9-7/8 oz	2-3/8 cup	
CUMIN,GROUND	2-1/4 oz	1/2 cup 2-2/3 tbsp	
PAPRIKA,GROUND	2 oz	1/2 cup	
SALT	1-7/8 oz	3 tbsp	
PEPPER,RED,GROUND	2/3 oz	1/4 cup	
GARLIC POWDER	1/3 oz	1 tbsp	
WATER	14-5/8 lbs	1 gal 3 qts	

Method

1. Cook beef until it loses its pink color, stirring to break apart. Drain or skim off excess fat.
2. Add tomatoes, tomato paste, onions, chili powder, cumin, paprika, salt, pepper, garlic and water; stir. Bring to a simmer; cook 1 hour, stirring occasionally. DO NOT BOIL. CCP: Internal temperature must reach 155 F. or higher for 15 seconds. Hold for service at 140 F. or higher.

MEAT. FISH. AND POULTRY No.L 171 00

CHEESE PITA PIZZA

Yield 100   **Portion** 2-1/2 Ounces

Calories	Carbohydrates	Protein	Fat	Cholesterol	Sodium	Calcium
234 cal	37 g	10 g	5 g	15 mg	450 mg	163 mg

Ingredient	Weight	Measure	Issue
BREAD,PITA,WHITE,5-INCH	13-1/4 lbs	100 each	
SAUCE,PIZZA,CANNED	7-1/4 lbs	3 qts	
CHEESE,MOZZARELLA,SHREDDED	4 lbs	1 gal	

Method

1. Place 15 pitas on each sheet pan.
2. Spread 2 tablespoons pizza sauce evenly on each pita.
3. Sprinkle 1/4 cup cheese over sauce on each pizza.
4. Using a convection oven, bake at 450 F. 5 minutes on high fan, closed vent or until cheese starts to turn golden.

MEAT. FISH. AND POULTRY No.L 171 01

MUSHROOM, ONION, AND GREEN PEPPER PITA PIZZA

Yield 100   Portion 4 Ounces

Calories	Carbohydrates	Protein	Fat	Cholesterol	Sodium	Calcium
242 cal	38 g	10 g	5 g	15 mg	490 mg	167 mg

Ingredient	Weight	Measure	Issue
BREAD,PITA,WHITE,5-INCH	13-1/4 lbs	100 each	
SAUCE,PIZZA,CANNED	7-1/4 lbs	3 qts	
CHEESE,MOZZARELLA,SHREDDED	4 lbs	1 gal	
MUSHROOMS,CANNED,SLICED,DRAINED	2 lbs	1 qts 2 cup	
ONIONS,FRESH,CHOPPED	2-1/8 lbs	1 qts 2 cup	2-1/3 lbs
PEPPERS,GREEN,FRESH,CHOPPED	2 lbs	1 qts 2-1/8 cup	2-1/2 lbs

Method

1. Place 15 pitas on each sheet pan.
2. Spread 2 tablespoons pizza sauce evenly on each pita.
3. Sprinkle about 1 ounce or 1/4 cup cheese, 1 tablespoon mushrooms, 1 tablespoon onions and 1 tablespoon peppers over sauce on each pita.
4. Using a convection oven bake at 450 F. 5 minutes on high fan, closed vent or until cheese starts to turn golden.

MEAT. FISH. AND POULTRY No.L 172 00

BEEF STEW (CANNED BEEF CHUNKS)

Yield 100   Portion   1-1/4 Cups

Calories	Carbohydrates	Protein	Fat	Cholesterol	Sodium	Calcium
387 cal	19 g	38 g	17 g	104 mg	152 mg	39 mg

Ingredient

Ingredient	Weight	Measure	Issue
BEEF,CANNED,CHUNKS,W/NATURAL JUICE,DRAINED	29 lbs	6 gal 2-1/2 qts	
PEPPER,BLACK,GROUND	1/2 oz	2 tbsp	
GARLIC POWDER	1/2 oz	1 tbsp	
WATER	16-3/4 lbs	2 gal	
TOMATOES,CANNED,CRUSHED,INCL LIQUIDS	6-5/8 lbs	3 qts	
THYME,GROUND	1/8 oz	1 tbsp	
BAY LEAF,WHOLE,DRIED	1/8 oz	4 lf	
CARROTS,FRESH,SLICED	3-3/8 lbs	2 qts 4 cup	4-1/8 lbs
CELERY,FRESH,CHOPPED	4-1/4 lbs	1 gal	5-7/8 lbs
ONIONS,FRESH,QUARTERED	2-1/2 lbs	2 qts 1-7/8 cup	2-3/4 lbs
POTATOES,FRESH,CHOPPED	10-1/3 lbs	1 gal 3-1/2 qts	12-3/4 lbs
FLOUR,WHEAT,GENERAL PURPOSE	1-1/4 lbs	1 qts 1/2 cup	
WATER,COLD	3-1/8 lbs	1 qts 2 cup	

Method

1 Place beef, pepper and garlic in steam-jacketed kettle or stock pot.
2 Add water, tomatoes, thyme and bay leaves. Bring to a boil; reduce heat.
3 Add carrots to beef mixture. Cover; simmer 15 minutes.
4 Add celery, onions and potatoes to beef mixture. Stir to mix. Cover; simmer 20 minutes or until vegetables are tender.
5 Thicken gravy, if desired. Combine flour and water. Add to stew while stirring; cook 5 minutes or until thickened. CCP: Internal temperature must reach 165 F. or higher for 15 seconds. Hold for service at 140 F. or higher.

MEAT. FISH. AND POULTRY No.L 173 00

CHEESE TORTELLINI MARINARA

Yield 100   Portion 1 Cup

Calories	Carbohydrates	Protein	Fat	Cholesterol	Sodium	Calcium
273 cal	46 g	13 g	5 g	26 mg	1004 mg	205 mg

Ingredient	Weight	Measure	Issue
MARINARA SAUCE		3 gal 2-1/4 qts	
WATER,BOILING	58-1/2 lbs	7 gal	
SALT	1-1/4 oz	2 tbsp	
TORTELLINI,FROZEN,CHEESE	14 lbs		

Method

1. Prepare 3/4 recipe Marinara Sauce, Recipe No. O 004 00 or use prepared marinara sauce.
2. Add tortellini slowly to boiling salted water, stirring constantly until water boils again. Cook according to package instructions; DO NOT OVERCOOK. Drain thoroughly.
3. Add tortellini to sauce. Stir gently but thoroughly.
4. Simmer 5 minutes or until thoroughly heated. CCP: Hold for service at 140 F. or higher.

MEAT, FISH, AND POULTRY No.L 173 01

SPINACH TORTELLINI MARINARA (FROZEN)

Yield 100　　　　　　　　　　　　　　　　　　　Portion 1 Cup

Calories	Carbohydrates	Protein	Fat	Cholesterol	Sodium	Calcium
341 cal	33 g	19 g	15 g	111 mg	640 mg	371 mg

Ingredient	Weight	Measure	Issue
MARINARA SAUCE		3 gal 2-1/4 qts	
WATER,BOILING	58-1/2 lbs	7 gal	
SALT	1-1/4 oz	2 tbsp	
TORTELLINI,FROZEN,SPINACH	14 lbs		

Method

1. Prepare 3/4 recipe Marinara Sauce, Recipe No. O 004 00 or use prepared marinara sauce.
2. Add spinach filled tortellini slowly to boiling salted water, stirring constantly until water boils again. Cook according to package instructions. DO NOT OVERCOOK. Drain thoroughly.
3. Add tortellini to sauce. Stir gently but thoroughly.
4. Simmer 5 minutes or until thoroughly heated. CCP: Hold for service at 140 F. or higher.

MEAT. FISH. AND POULTRY No.L 173 02

CHEESE TORTELLINI MARINARA (DEHYDRATED)

Yield 100   **Portion** 1 Cup

Calories	Carbohydrates	Protein	Fat	Cholesterol	Sodium	Calcium
272 cal	46 g	13 g	5 g	26 mg	997 mg	205 mg

Ingredient	Weight	Measure	Issue
MARINARA SAUCE		3 gal 2-1/4 qts	
WATER, BOILING	58-1/2 lbs	7 gal	
SALT	1-1/4 oz	2 tbsp	
TORTELLINI, CHEESE, DRY	9-1/4 lbs		

Method

1. Prepare 3/4 recipe Marinara Sauce, Recipe No. O 004 00 or use prepared marinara sauce.
2. Add tortellini slowly to boiling salted water, stirring constantly until water boils again. Cook according to package instructions. DO NOT OVERCOOK. Drain thoroughly.
3. Add tortellini to sauce. Stir gently but thoroughly.
4. Simmer 5 minutes or until thoroughly heated. CCP: Hold for service at 140 F. or higher.

MEAT. FISH. AND POULTRY No.L 174 00

RICE FRITTATA

Yield 100 **Portion** 11 Ounces

Calories	Carbohydrates	Protein	Fat	Cholesterol	Sodium	Calcium
391 cal	25 g	24 g	22 g	220 mg	805 mg	430 mg

Ingredient	Weight	Measure	Issue
RICE,LONG GRAIN	3-3/4 lbs	2 qts 1-3/8 cup	
WATER,BOILING	10-1/2 lbs	1 gal 1 qts	
SALT	3/4 oz	1 tbsp	
OIL,SALAD	3/4 oz	1 tbsp	
TOMATOES,FRESH,CHOPPED	15-7/8 lbs	2 gal 2 qts	16-1/4 lbs
CHEESE,CHEDDAR,SHREDDED	11 lbs	2 gal 3 qts	
MUSHROOMS,CANNED,DRAINED	7 lbs	1 gal 1-1/8 qts	
PEPPERS,GREEN,FRESH,CHOPPED	4-7/8 lbs	3 qts 2-7/8 cup	6 lbs
COOKING SPRAY,NONSTICK	2 oz	1/4 cup 1/3 tbsp	
ONIONS,FRESH,CHOPPED	5-1/4 lbs	3 qts 2-7/8 cup	5-7/8 lbs
MILK,NONFAT,DRY	5-5/8 oz	2-3/8 cup	
WATER,WARM	5-3/4 lbs	2 qts 3 cup	
EGG WHITES	8-1/2 lbs	1 gal	
EGGS,WHOLE,FROZEN	8-5/8 lbs	1 gal	
PEPPER,BLACK,GROUND	7/8 oz	1/4 cup 1/3 tbsp	
SALT	1-1/4 oz	2 tbsp	

Method

1. Combine rice, water, salt and salad oil. Bring to a boil. Stir occasionally. Cover tightly; simmer 20 to 25 minutes.
2. Combine tomatoes, cheese, mushrooms, peppers, onions and rice. Mix well. Place 5-1/2 quarts mixture in each lightly sprayed steam table pan.
3. Reconstitute milk.
4. Thaw egg products. Combine milk, eggs, pepper, and salt. Mix well.
5. Pour 1-1/2 quart egg mixture over rice mixture in each pan. Stir to distribute evenly.
6. Using a convection oven, bake at 325 F. for 45 minutes or until eggs are completely set on low fan, closed vent. CCP: Internal temperature must reach 145 F. or higher for 15 seconds. Hold for service at 140 F. or higher. Cut 3 by 5.

MEAT, FISH, AND POULTRY No.L 175 00

POTATO FRITTATA

Yield 100   Portion 12 Ounces

Calories	Carbohydrates	Protein	Fat	Cholesterol	Sodium	Calcium
243 cal	28 g	18 g	7 g	213 mg	568 mg	175 mg

Ingredient	Weight	Measure	Issue
POTATOES,FRESH,PEELED,CUBED	16-1/2 lbs	3 gal	20-3/8 lbs
WATER,BOILING	16-3/4 lbs	2 gal	
SALT	1 oz	1 tbsp	
BROCCOLI,FRESH,CHOPPED	16-1/2 lbs	5 gal 1-1/4 qts	27 lbs
WATER,BOILING	20-7/8 lbs	2 gal 2 qts	
SALT	1/2 oz	3/8 tsp	
TOMATOES,FRESH,SLICED	19 lbs	2 gal 4 qts	19-3/8 lbs
ONIONS,FRESH,CHOPPED	6-1/3 lbs	1 gal 1/2 qts	7 lbs
CHEESE,PARMESAN,GRATED	14-1/8 oz	1 qts	
COOKING SPRAY,NONSTICK	2 oz	1/4 cup 1/3 tbsp	
MILK,NONFAT,DRY	7-1/4 oz	3 cup	
WATER,WARM	7-1/3 lbs	3 qts 2 cup	
EGG WHITES	10-2/3 lbs	1 gal 1 qts	
EGGS,WHOLE,FROZEN	10-3/4 lbs	1 gal 1 qts	
PARSLEY,FRESH,BUNCH,CHOPPED	5-1/4 oz	2-1/2 cup	5-1/2 oz
SALT	1-1/4 oz	2 tbsp	
PEPPER,BLACK,GROUND	7/8 oz	1/4 cup 1/3 tbsp	
GARLIC POWDER	1-1/8 oz	1/4 cup	
PAPRIKA,GROUND	5/8 oz	2-1/3 tbsp	
BASIL,SWEET,WHOLE,CRUSHED	5/8 oz	1/4 cup 1/3 tbsp	

Method

1. Add potatoes to boiling salted water, bring to a boil; reduce heat. Cover; simmer 8 minutes or until tender. Drain well.
2. Add broccoli to boiling salted water. Return to a boil; reduce heat. Simmer 5 minutes until tender; cool.
3. Combine potatoes, broccoli, tomatoes, onions, and parmesan cheese. Toss lightly. Place 1-1/2 gallon mixture in each lightly sprayed steam table pan.
4. Reconstitute milk.
5. Thaw eggs. Combine milk, eggs, parsley, salt, pepper, garlic, paprika, and basil. Mix well.
6. Pour 2 quarts egg mixture over potato mixture in each pan. Stir to distribute evenly.
7. Using a convection oven, bake at 325 F. for 40-45 minutes or until eggs are set on low fan, closed vent. CCP: Internal temperature must reach 145 F. or higher for 15 seconds. Hold for service at 140 F. or higher. Cut 3 by 5.

MEAT, FISH, AND POULTRY No.L 176 00
VEGETABLE STUFFED PEPPERS

Yield 100   Portion   2 Halves

Calories	Carbohydrates	Protein	Fat	Cholesterol	Sodium	Calcium
368 cal	50 g	13 g	15 g	21 mg	1163 mg	218 mg

Ingredient	Weight	Measure	Issue
TOMATO SAUCE		3 gal 2 qts	
PEPPERS,GREEN,FRESH	32 lbs	6 gal 3/8 qts	39 lbs
WATER,BOILING	83-5/8 lbs	10 gal	
CORN,CANNED,WHOLE KERNEL,DRAINED	8-3/4 lbs	1 gal 2 qts	
BEANS,KIDNEY,DARK RED,CANNED,DRAINED	8-1/4 lbs	1 gal 1-1/4 qts	
STEAMED RICE		1 gal 2 qts	
TOMATOES,FRESH,SLICED	2 lbs	1 qts 1 cup	2 lbs
ONIONS,FRESH,CHOPPED	2-7/8 lbs	2 qts 1/8 cup	3-1/4 lbs
GARLIC POWDER	1/3 oz	1 tbsp	
PARSLEY,FRESH,BUNCH,CHOPPED	4-1/4 oz	2 cup	4-1/2 oz
SALT	1-1/4 oz	2 tbsp	
PEPPER,BLACK,GROUND	5/8 oz	2-2/3 tbsp	
PEPPER,RED,GROUND	1/8 oz	1/3 tsp	
CUMIN,GROUND	7/8 oz	1/4 cup 1/3 tbsp	
CHILI POWDER,DARK,GROUND	3-1/8 oz	3/4 cup	
CHEESE,CHEDDAR,SHREDDED	4-1/2 lbs	1 gal 1/2 qts	
OIL,SALAD	7-2/3 oz	1 cup	
WATER	3-2/3 lbs	1 qts 3 cup	

Method

1. Prepare 2 recipes Tomato Sauce, Recipe No. O 015 00 or use prepared tomato sauce. CCP: Hold at 140 F. or higher.
2. Cut each pepper in half lengthwise; remove core.
3. Place peppers in boiling water. Return to a boil; cook 1 minute. Drain well.
4. Drain corn and beans. Rinse beans. Drain well.
5. Combine corn, beans, cooked rice, tomatoes, onions, garlic, parsley, salt, peppers, cumin, chili powder, oil, and cheese. Mix lightly.
6. Fill each pepper with 1/2 cup vegetable-rice mixture. Place peppers in pans.
7. Pour 1/2 cup water around peppers in each steam table pan.
8. Pour 3-1/3 cups tomato sauce over peppers each pan. Cover.
9. Using a convection oven bake at 325 F. for 40 minutes or until thoroughly heated on high fan, closed vent. CCP: Internal temperature must reach 145 F. or higher for 15 seconds. Hold for service at 140 F. or higher.

Notes

1. In Step 4, 12 pounds canned pinto or black beans may be used per 100 portions. Drain beans.
2. In Step 4, 9-1/8 pounds frozen corn may be used, per 100 servings.
3. In Step 6, 9 pounds cooked brown rice may be used, per 100 servings. Cook according to Recipe No. E 005 05.

MEAT, FISH, AND POULTRY   No.L 177 00

BOMBAY CHICKEN (8 PC)

Yield 100 **Portion** 1 Piece

Calories	Carbohydrates	Protein	Fat	Cholesterol	Sodium	Calcium
368 cal	31 g	39 g	10 g	119 mg	119 mg	27 mg

Ingredient   Weight   Measure   Issue

Ingredient	Weight	Measure	Issue
HONEY	7-7/8 lbs	2 qts 2-1/2 cup	
JUICE, LIME	3-3/8 lbs	1 qts 2-3/4 cup	
CURRY POWDER	2 oz	1/2 cup 1 tbsp	
CHICKEN, 8 PC CUT, SKIN REMOVED	82 lbs		

Method

1. Combine honey, lime juice and curry; mix well. Bring to a boil. Cover; reduce heat; simmer 10 minutes.
2. Wash chicken thoroughly under cold running water. Drain well. Remove excess fat. Place chicken breasts on lightly sprayed sheet pans.
3. Using a convection oven, bake at 350 F. for 40 minutes or until done on high fan, closed vent. Transfer chicken to steam table pans. Hold at 140 F. or higher for use in Step 4.
4. Pour 1-1/4 qt sauce evenly over chicken in each pan. Bake at 350 F. 10 to 15 minutes. CCP: Internal temperature must reach 165 F. or higher for 15 seconds. Hold for service at 140 F. or higher.

MEAT, FISH, AND POULTRY   No.L 177 01

BOMBAY CHICKEN (BREAST BONELESS)

Yield 100   Portion 5 Ounces

Calories	Carbohydrates	Protein	Fat	Cholesterol	Sodium	Calcium
287 cal	31 g	32 g	4 g	88 mg	81 mg	22 mg

Ingredient

Ingredient	Weight	Measure	Issue
CHICKEN,BREAST,BNLS/SKNLS,5 OZ	31-1/4 lbs		
COOKING SPRAY,NONSTICK	1-1/2 oz	3 tbsp	
HONEY	7-7/8 lbs	2 qts 2-1/2 cup	
JUICE,LIME	3-3/8 lbs	1 qts 2-3/4 cup	
CURRY POWDER	2 oz	1/2 cup 1 tbsp	

Method

1. Wash chicken thoroughly under cold running water. Drain well. Remove excess fat.
2. Place chicken breasts on lightly sprayed sheet pans. Lightly spray chicken with cooking spray.
3. Using a convection oven, bake at 325 F. 10 to 12 minutes on high fan, closed vent. Transfer chicken to steam table pans. CCP: Internal temperature must reach 165 F. or higher for 15 seconds. Hold at 140 F. or higher for use in Step 5.
4. Combine honey, lime juice and curry; mix well. Bring to a boil. Cover; reduce heat; simmer 10 minutes.
5. Pour 1-1/4 qt sauce evenly over chicken in each pan. CCP: Hold for service at 140 F. or higher.

MEAT, FISH, AND POULTRY No. L 178 00

TROPICAL CHICKEN SALAD (COOKED DICED)

Yield 100 **Portion** 1 Cup

Calories	Carbohydrates	Protein	Fat	Cholesterol	Sodium	Calcium
419 cal	24 g	26 g	25 g	82 mg	254 mg	43 mg

Ingredient	Weight	Measure	Issue
PINEAPPLE,CANNED,CHUNKS,JUICE PACK,INCL LIQUIDS	9-7/8 lbs	1 gal 1/2 qts	
SALAD DRESSING,MAYONNAISE TYPE	5-7/8 lbs	3 qts	
RESERVED LIQUID	8-1/3 oz	1 cup	
CURRY POWDER	1-3/4 oz	1/2 cup	
CHICKEN,COOKED,DICED	18 lbs		
APPLES,FRESH,MEDIUM,PEELED,CORED,CHOPPED	12-3/8 lbs	2 gal 3-1/4 qts	15-7/8 lbs
COCONUT,PREPARED,SWEETENED FLAKES	1-1/2 lbs	1 qts 3-1/2 cup	
ALMONDS,SLIVERED	1-1/4 lbs	1 qts 1 cup	
LETTUCE,ICEBERG,FRESH	4 lbs		4-1/3 lbs

Method

1 Drain pineapple. Reserve juice for use in Step 2 and pineapple for use in Step 3.
2 Blend salad dressing, reserved pineapple juice, curry powder and salt together; cover.
3 Combine chicken, apples, pineapple, coconut, and almonds. Mix lightly.
4 Add salad dressing mixture to chicken mixture. Mix lightly but thoroughly; cover. CCP: Refrigerate at 41 F. or lower.
5 Place 1 lettuce leaf on each serving dish; place 1 cup chicken mixture on top of lettuce; cover. CCP: Refrigerate at 41 F. or lower until ready to serve.

Notes

1 In Step 2, 6-1/2 pound (3 quarts) low fat plain yogurt may be used for salad dressing per 100 servings.

MEAT, FISH, AND POULTRY No. L 178 01

TROPICAL CHICKEN SALAD (CANNED CHICKEN)

Yield 100 Portion 1 Cup

Calories	Carbohydrates	Protein	Fat	Cholesterol	Sodium	Calcium
434 cal	26 g	23 g	27 g	69 mg	786 mg	45 mg

Ingredient	Weight	Measure	Issue
PINEAPPLE,CANNED,CHUNKS,JUICE PACK,INCL LIQUIDS	10-3/8 lbs	1 gal 3/4 qts	
SALAD DRESSING,MAYONNAISE TYPE	6-1/8 lbs	3 qts 1/2 cup	
RESERVED LIQUID	8-1/3 oz	1 cup	
CURRY POWDER	1-7/8 oz	1/2 cup 1/3 tbsp	
SALT	1 oz	1 tbsp	
CHICKEN,BONED,CANNED,PIECES	23-1/4 lbs	2 gal 1-1/8 qts	
APPLES,FRESH,MEDIUM,PEELED,CORED,CHOPPED	13-1/4 lbs	3 gal <1/16th qts	17 lbs
COCONUT,PREPARED,SWEETENED FLAKES	1-5/8 lbs	2 qts	
ALMONDS,SLIVERED	1-1/4 lbs	1 qts 1-1/4 cup	
LETTUCE,ICEBERG,FRESH	4 lbs		4-1/3 lbs

Method

1 Drain pineapple. Reserve juice for use in Step 2 and pineapple for use in Step 4.
2 Blend salad dressing, reserved pineapple juice, curry powder and salt together; cover.
3 Combine chicken, apples, pineapple, coconut, and almonds. Mix lightly.
4 Add salad dressing mixture to chicken mixture. Mix lightly but thoroughly; cover. CCP: Refrigerate at 41 F. or lower.
5 Place 1 lettuce leaf on each serving dish; place 1 cup chicken salad mixture on top of lettuce; cover. CCP: Refrigerate at 41 F. or lower until ready to serve.

Notes

1 In Step 2, 6-1/2 pounds (3 quarts) low fat plain yogurt may be used for salad dressing per 100 servings.

MEAT, FISH, AND POULTRY No.L 179 00

HONEY GINGER CHICKEN (BREAST BONELESS)

Yield 100   Portion 5 Ounces

Calories	Carbohydrates	Protein	Fat	Cholesterol	Sodium	Calcium
217 cal	11 g	33 g	4 g	88 mg	481 mg	21 mg

Ingredient | Weight | Measure | Issue

Ingredient	Weight	Measure	Issue
CHICKEN,BREAST,BNLS/SKNLS,5 OZ	31-1/4 lbs		
COOKING SPRAY,NONSTICK	3/4 oz	1 tbsp	
HONEY	2-1/4 lbs	3 cup	
SOY SAUCE	1-5/8 lbs	2-1/2 cup	
JUICE,LEMON	1-1/3 lbs	2-1/2 cup	
GARLIC POWDER	2-3/8 oz	1/2 cup	
ONION POWDER	1-7/8 oz	1/2 cup	
GINGER,GROUND	1-1/2 oz	1/2 cup	
WATER,COLD	8-1/3 oz	1 cup	
CORNSTARCH	2-1/4 oz	1/2 cup	

Method

1. Wash chicken thoroughly under cold running water. Drain well. Remove excess fat.
2. Arrange chicken breasts shingle-style in lightly sprayed steam table pans.
3. Combine honey, soy sauce, lemon juice, garlic powder, onion powder, and ground ginger; mix well. Pour sauce over chicken in each pan.
4. Using a convection oven, bake at 325 F. for 12 to 14 minutes on high fan, closed vent. CCP: Internal temperature must reach 165 F. or higher for 15 seconds.
5. Transfer chicken to steam table pans. Drain sauce. Reserve sauce.
6. Bring reserved sauce to a boil. Blend cornstarch and cold water together to make a smooth slurry. Add slurry to hot sauce, stirring constantly. Cover; reduce heat; simmer 3 to 5 minutes or until thickened, stirring frequently to prevent sticking. CCP: Temperature must reach 165 F. or higher for 15 seconds.
7. Pour 1 qt sauce evenly over chicken in each pan. CCP: Hold for service at 140 F. or higher.

MEAT, FISH, AND POULTRY No.L 180 00

TURKEY SAUSAGE PATTIES

Yield 100   Portion 2 Ounces

Calories	Carbohydrates	Protein	Fat	Cholesterol	Sodium	Calcium
106 cal	3 g	13 g	5 g	42 mg	244 mg	30 mg

Ingredient	Weight	Measure	Issue
TURKEY,GROUND,90% LEAN,RAW	15 lbs		
BREADCRUMBS	1 lbs	1 qts	
SALT	1-1/2 oz	2-1/3 tbsp	
PEPPER,BLACK,GROUND	7/8 oz	1/4 cup	
GARLIC POWDER	3/4 oz	2-2/3 tbsp	
BASIL,SWEET,WHOLE,CRUSHED	2/3 oz	1/4 cup 2/3 tbsp	
SEASONING,POULTRY	2-1/8 oz	1 cup	

Method

1 Place turkey in mixer bowl.
2 Combine breadcrumbs, poultry seasoning, salt, pepper, garlic powder, and basil. Add to turkey.
3 Mix on low speed 3 to 4 minutes or until thoroughly blended.
4 Shape into 2-1/2 ounce balls. Place 20 balls on each sheet pan. Flatten each ball into a 4-inch patty.
5 Using a convection oven, bake at 325 F. for 9 minutes on high fan, closed vent. CCP: Internal temperature must reach 165 F. or higher for 15 seconds. Hold for service at 140 F. or higher.

Notes

1 Grill patties on 350 F. ungreased griddle for 3 minutes on each side. CCP: Internal temperature must reach 165 F. or higher for 15 seconds. Hold for service at 140 F. or higher.

MEAT, FISH, AND POULTRY No.L 181 00

CHICKEN IN ORANGE SAUCE (BREAST BONELESS)

Yield 100   Portion 5 Ounces

Calories	Carbohydrates	Protein	Fat	Cholesterol	Sodium	Calcium
208 cal	8 g	33 g	4 g	88 mg	264 mg	22 mg

Ingredient	Weight	Measure	Issue
CHICKEN,BREAST,BNLS/SKNLS,5 OZ	31-1/4 lbs		
COOKING SPRAY,NONSTICK	1-1/4 oz	2-2/3 tbsp	
ONIONS,FRESH,CHOPPED	1-1/4 lbs	3-1/2 cup	1-3/8 lbs
JUICE,ORANGE	8-3/4 lbs	1 gal	
FLOUR,WHEAT,GENERAL PURPOSE	6-5/8 oz	1-1/2 cup	
SUGAR,GRANULATED	7 oz	1 cup	
SALT	1-2/3 oz	2-2/3 tbsp	
PAPRIKA,GROUND	2/3 oz	2-2/3 tbsp	
PEPPER,BLACK,GROUND	1/3 oz	1 tbsp	
ROSEMARY,GROUND	1/8 oz	1 tbsp	
GARLIC POWDER	1/8 oz	1/8 tsp	

Method

1 Wash chicken thoroughly under cold running water. Drain well. Remove excess fat.
2 Place chicken breasts in lightly sprayed steam table pans. DO NOT OVERLAP.
3 Stir-cook onions in a lightly sprayed steam-jacketed kettle or stockpot 3 minutes or until tender, stirring constantly.
4 Combine orange juice, flour, sugar, salt, paprika, pepper, rosemary, and garlic powder; mix well. Add onions; stir to blend.
5 Ladle 2-3/4 cups orange mixture over chicken in each pan.
6 Using a convection oven, bake 12 to 14 minutes at 325 F. on high fan, closed vent. CCP: Internal temperature must reach 165 F. or higher for 15 seconds. Hold for service at 140 F. or higher.

MEAT, FISH, AND POULTRY No.L 182 00

FIESTA CHICKEN (FAJITA STRIPS)

Yield 100   Portion 6 Ounces

Calories	Carbohydrates	Protein	Fat	Cholesterol	Sodium	Calcium
150 cal	14 g	18 g	2 g	44 mg	488 mg	38 mg

Ingredient	Weight	Measure	Issue
COOKING SPRAY,NONSTICK	1 oz	2 tbsp	
ONIONS,FRESH,CHOPPED	5-5/8 lbs	0 gal 4 qts	6-1/4 lbs
PEPPERS,GREEN,FRESH,CHOPPED	2-5/8 lbs	2 qts	3-1/4 lbs
WATER	10-1/2 lbs	1 gal 1 qts	
TOMATOES,CANNED,DICED,INCL LIQUIDS	4-5/8 lbs	2 qts	
TOMATO PASTE,CANNED	3-7/8 lbs	1 qts 2-5/8 cup	
JUICE,ORANGE	8-3/4 lbs	1 gal	
PIMIENTO,CANNED,DRAINED,CHOPPED	1 lbs	2-1/4 cup	
SUGAR,GRANULATED	3-1/2 oz	1/2 cup	
SALT	2-1/3 oz	1/4 cup	
CHILI POWDER,DARK,GROUND	2-1/8 oz	1/2 cup	
GARLIC POWDER	3/4 oz	2-2/3 tbsp	
CUMIN,GROUND	3/8 oz	2 tbsp	
OREGANO,CRUSHED	1 oz	1/4 cup 2-1/3 tbsp	
PEPPER,RED,GROUND	1/4 oz	1 tbsp	
PEPPER,BLACK,GROUND	1/4 oz	1 tbsp	
CHICKEN,FAJITA STRIPS	15-5/8 lbs		
WATER	8-1/3 oz	1 cup	
FLOUR,WHOLE WHEAT	4-1/4 oz	1 cup	

Method

1 Stir-cook onions and peppers in a lightly sprayed steam-jacketed kettle or stockpot 3 minutes, stirring constantly.
2 Add water, tomatoes, tomato paste, orange juice, pimentos, sugar, salt, chili power, garlic powder, cumin, oregano, red pepper, and black pepper. Bring to a boil. Cover; reduce heat; simmer 20 to 25 minutes.
3 Stir chicken gently into vegetable and tomato mixture. Cover; reduce heat; simmer 10 minutes.
4 Blend flour and cold water together to make a smooth slurry. Add slurry to chicken mixture stirring constantly. Bring to a boil. Cover; reduce heat; simmer 5 minutes or until thickened, stirring frequently to prevent sticking. CCP: Internal temperature must reach 165 F. or higher for 15 seconds.
5 Pour 2 gal chicken and tomato mixture into ungreased steam table pans. CCP: Hold for service at 140 F. or higher.
6 Serve over 1 recipe Tossed Green Rice, Recipe No. E 005 02.

MEAT, FISH, AND POULTRY No. L 183 00

BUFFALO CHICKEN (8 PC)

Yield 100   Portion   8 Ounces

Calories	Carbohydrates	Protein	Fat	Cholesterol	Sodium	Calcium
421 cal	7 g	45 g	23 g	144 mg	363 mg	32 mg

Ingredient	Weight	Measure	Issue
CHICKEN, 8 PIECE CUT	82 lbs		
COOKING SPRAY, NONSTICK	1 oz	2 tbsp	
CATSUP	4-1/4 lbs	2 qts	
VINEGAR, DISTILLED	4-1/8 lbs	2 qts	
PEPPER, RED, GROUND	6 oz	2 cup	

Method

1. Wash chicken thoroughly under cold running water; drain well. Remove excess fat.
2. Place chicken, skin side up, on lightly sprayed sheet pans. Using a convection oven, bake at 350 F. for 20 minutes on high fan, closed vent.
3. Combine catsup, vinegar, and red pepper; mix well.
4. Dip chicken in buffalo sauce to coat well; place chicken, skin side up, on sheet pans. Discard remaining buffalo sauce. Bake an additional 20 minutes, for a total of 40 minutes. CCP: Internal temperature must reach 165 F. or higher for 15 seconds.
5. Transfer chicken to steam table pans. CCP: Hold for service at 140 F. or higher.

Notes

1. In Step 3, 2 gallons of prepared buffalo sauce can be used per 100 portions.

MEAT, FISH, AND POULTRY No. L 184 00

GRILLED TURKEY PATTIES (GROUND TURKEY)

Yield 100   **Portion**   4-1/2 Ounces

Calories	Carbohydrates	Protein	Fat	Cholesterol	Sodium	Calcium
204 cal	9 g	23 g	9 g	72 mg	313 mg	45 mg

Ingredient

Ingredient	Weight	Measure	Issue
TURKEY,GROUND,90% LEAN,RAW	25-1/2 lbs		
BREADCRUMBS	3-1/8 lbs	3 qts 1 cup	
ONIONS,FRESH,CHOPPED	2-7/8 lbs	2 qts 1/8 cup	3-1/4 lbs
PARSLEY,FRESH,BUNCH,CHOPPED	3-1/2 oz	1-5/8 cup	3-2/3 oz
SALT	1-1/4 oz	2 tbsp	
GARLIC POWDER	1-1/4 oz	1/4 cup 1/3 tbsp	
PEPPER,WHITE,GROUND	1/2 oz	2 tbsp	
MUSTARD,DRY	3/4 oz	2 tbsp	
COOKING SPRAY,NONSTICK	2 oz	1/4 cup 1/3 tbsp	

Method

1. Combine turkey, breadcrumbs, onions, parsley, salt, garlic, pepper and mustard; mix thoroughly.
2. Shape mixture into oval patties 1/2-inch thick weighing approximately 5 ounces each.
3. Grill patties on lightly greased griddle 8 minutes on each side. CCP: Internal temperature must reach 165 F. or higher for 15 seconds. Hold for service at 140 F. or higher.

Notes

1. In Step 3, turkey patties may be baked in a convection oven at 325 F. for 20 to 25 minutes on high fan, open vent.

MEAT, FISH, AND POULTRY No.L 185 00

CARIBBEAN CATFISH

Yield 100   Portion   4 Ounces

Calories	Carbohydrates	Protein	Fat	Cholesterol	Sodium	Calcium
229 cal	9 g	21 g	12 g	68 mg	183 mg	34 mg

Ingredient	Weight	Measure	Issue
PEPPERS,GREEN,FRESH,CHOPPED	2 lbs	1 qts 2-1/8 cup	2-1/2 lbs
MARGARINE	6 oz	3/4 cup	
ONIONS,FRESH,CHOPPED	2-1/8 lbs	1 qts 2 cup	2-1/3 lbs
BREAD,WHITE,CUBED	2-5/8 lbs	2 gal 1/2 qts	
OREGANO,CRUSHED	1/3 oz	2 tbsp	
PEPPER,RED,GROUND	1/4 oz	1 tbsp	
GARLIC POWDER	7/8 oz	3 tbsp	
CILANTRO,DRY	1-1/3 oz	1-1/4 cup	
JUICE,LIME	12 oz	1-1/2 cup	
FISH,CATFISH,FILLET	30 lbs		
JUICE,LIME	12 oz	1-1/2 cup	
MARGARINE,MELTED	6 oz	3/4 cup	
GARLIC POWDER	7/8 oz	3 tbsp	
PEPPER,RED,GROUND	1/4 oz	1 tbsp	
LIMES,FRESH	4-3/4 oz	2 each	

Method

1. Saute green peppers and onions in margarine or butter, 10 minutes or until onions are transparent.
2. Combine onion mixture with bread cubes, lime juice, cilantro, garlic powder, red pepper and oregano; mix well.
3. Place 1-1/4 ounces filling in center of each catfish fillet, skin side up. Roll up and place seam side down in staggered row, 4 by 5, in steam table pans.
4. Combine margarine or butter with lime juice. Add garlic powder and red pepper. Brush evenly over fish in each pan. If desried sprinkle 1 tablespoon grated lime rind over fish in each pan.
5. Using a convection oven, bake at 350 F. for 25 to 30 minutes or until done on high fan, closed vent. CCP: Internal temperature must reach 165 F. or higher for 15 seconds. Hold for service at 140 F. or higher.

MEAT, FISH, AND POULTRY No.L 185 01

CARIBBEAN FLOUNDER

Yield 100   Portion 4 Ounces

Calories	Carbohydrates	Protein	Fat	Cholesterol	Sodium	Calcium
192 cal	9 g	27 g	5 g	72 mg	210 mg	44 mg

Ingredient	Weight	Measure	Issue
PEPPERS,GREEN,FRESH,CHOPPED	2 lbs	1 qts 2-1/8 cup	2-1/2 lbs
MARGARINE	6 oz	3/4 cup	
ONIONS,FRESH,CHOPPED	2-1/8 lbs	1 qts 2 cup	2-1/3 lbs
BREAD,WHITE,CUBED	2-5/8 lbs	2 gal 1/2 qts	
OREGANO,CRUSHED	1/3 oz	2 tbsp	
PEPPER,RED,GROUND	1/4 oz	1 tbsp	
GARLIC POWDER	7/8 oz	3 tbsp	
CILANTRO,DRY	1-1/3 oz	1-1/4 cup	
JUICE,LIME	12 oz	1-1/2 cup	
FISH,FLOUNDER/SOLE FILLET,RAW	30 lbs		
JUICE,LIME	12 oz	1-1/2 cup	
MARGARINE,MELTED	6 oz	3/4 cup	
GARLIC POWDER	7/8 oz	3 tbsp	
PEPPER,RED,GROUND	1/4 oz	1 tbsp	
LIMES,FRESH	4-3/4 oz	2 each	

Method

1. Saute green peppers and onions in margarine or butter, 10 minutes or until onions are transparent.
2. Combine onion mixture with bread cubes, lime juice, cilantro, garlic powder, red pepper and oregano; mix well.
3. Place 1-1/4 ounces filling in center of each flounder fillet, skin side up. Roll up and place seam side down in staggered row, 4 by 5, in steam table pans.
4. Combine margarine or butter with lime juice. Add garlic powder and red pepper. Brush evenly over fish in each pan. If desired, sprinkle 1 tablespoon grated lime rind over fish in each pan.
5. Using a convection oven, bake at 350 F. for 25 to 30 minutes or until done on high fan, closed vent. CCP: Internal temperature must reach 165 F. or higher for 15 seconds. Hold for service at 140 F. or higher.

MEAT, FISH, AND POULTRY   No. L 186 00

BAKED YOGURT CHICKEN (BREAST BONELESS)

Yield 100   Portion   5 Ounces

Calories	Carbohydrates	Protein	Fat	Cholesterol	Sodium	Calcium
257 cal	15 g	35 g	5 g	89 mg	230 mg	81 mg

Ingredient	Weight	Measure	Issue
CHICKEN,BREAST,BNLS/SKNLS,5 OZ	31-1/4 lbs		
BREADCRUMBS	5-3/4 lbs	1 gal 2 qts	
GARLIC POWDER	5/8 oz	2 tbsp	
ONION POWDER	1/2 oz	2 tbsp	
PAPRIKA,GROUND	1/2 oz	2 tbsp	
PEPPER,RED,GROUND	1/4 oz	1 tbsp	
GINGER,GROUND	1/8 oz	1/4 tsp	
YOGURT,PLAIN,LOWFAT	4-1/3 lbs	2 qts	
GARLIC POWDER	5/8 oz	2 tbsp	
ONION POWDER	1/2 oz	2 tbsp	
PEPPER,RED,GROUND	1/4 oz	1 tbsp	
GINGER,GROUND	1/8 oz	1/4 tsp	
COOKING SPRAY,NONSTICK	1 oz	2 tbsp	

Method

1 Wash chicken thoroughly under cold running water. Drain well. Remove excess fat.
2 Combine bread crumbs, garlic powder, onion powder, paprika, red pepper and ginger; mix well. Set aside for use in Step 4.
3 Combine yogurt, garlic powder, onion powder, red pepper and ginger in shallow pan; mix well.
4 Dip chicken in yogurt mixture, then in crumb mixture; shake off excess.
5 Place chicken breasts 1 inch apart on each lightly sprayed sheet pan.
6 Using a convection oven, bake 12 to 14 minutes at 325 F. on high fan, closed vent. CCP: Internal temperature must reach 165 F. or higher for 15 seconds.
7 Transfer and shingle chicken in steam table pans with bottom side up. CCP: Hold for service at 140 F. or higher.

MEAT, FISH, AND POULTRY No.L 187 00

HOT AND SPICY CHICKEN (8 PC)

Yield 100   Portion   8 Ounces

Calories	Carbohydrates	Protein	Fat	Cholesterol	Sodium	Calcium
430 cal	35 g	45 g	11 g	120 mg	627 mg	85 mg

Ingredient | Weight | Measure | Issue

Ingredient	Weight	Measure	Issue
CHICKEN, 8 PC CUT, SKIN REMOVED	82 lbs		
WATER, WARM	7-1/3 lbs	3 qts 2 cup	
MILK, NONFAT, DRY	13 oz	1 qts 1-3/8 cup	
FLOUR, WHEAT, GENERAL PURPOSE	8-7/8 lbs	2 gal	
SALT	4-1/2 oz	1/4 cup 3-1/3 tbsp	
GARLIC POWDER	4-1/8 oz	3/4 cup 2 tbsp	
ONION POWDER	3-1/4 oz	3/4 cup 2 tbsp	
PEPPER, BLACK, GROUND	2-3/8 oz	1/2 cup 2-2/3 tbsp	
THYME, FRESH	2/3 oz	1/2 cup	
PAPRIKA	1-1/4 oz	1/4 cup 1-1/3 tbsp	
PEPPER, RED, GROUND	1-1/8 oz	1/4 cup 2-1/3 tbsp	
MARJORAM, SWEET, GROUND	1/2 oz	1/2 cup	
PEPPER, WHITE, GROUND	1 oz	1/4 cup 1/3 tbsp	
COOKING SPRAY, NONSTICK	2-1/8 oz	1/4 cup 2/3 tbsp	

Method

1. Wash chicken thoroughly under cold running water. Drain well. Remove excess fat.
2. Reconstitute milk in warm water.
3. Combine flour, salt, garlic powder, onion powder, black pepper, thyme, paprika, red pepper, marjoram, and white pepper; mix thoroughly.
4. Dip chicken in milk; drain; Dredge chicken in flour mixture. Shake off excess.
5. Place chicken, meat side up, on lightly sprayed sheet pans. Lightly spray chicken with cooking spray.
6. Using a convection oven, bake 20 minutes at 325 F. on high fan, open vent. Turn chicken pieces over. Bake an additional 20 minutes, for a total of 40 minutes. CCP: Internal temperature must reach 165 F. or higher for 15 seconds.
7. Transfer chicken to steam table pans. CCP: Hold for service at 140 F. or higher.

MEAT, FISH, AND POULTRY No.L 188 00

TURKEY FINGERS

Yield 100   Portion   3-1/2 Ounces

Calories	Carbohydrates	Protein	Fat	Cholesterol	Sodium	Calcium
253 cal	18 g	24 g	9 g	65 mg	692 mg	43 mg

Ingredient	Weight	Measure	Issue
TURKEY,BNLS,WHITE AND DARK MEAT	26 lbs		
GARLIC POWDER	1-5/8 oz	1/4 cup 1-2/3 tbsp	
SEASONING,POULTRY	1/3 oz	2-2/3 tbsp	
PAPRIKA,GROUND	1/3 oz	1 tbsp	
PEPPER,BLACK,GROUND	1/3 oz	1 tbsp	
FLOUR,WHEAT,GENERAL PURPOSE	4-3/8 lbs	1 gal	
COOKING SPRAY,NONSTICK	1 oz	2 tbsp	

Method

1. Cut turkey into 1/2-inch thick slices. Cut slices into 1/4-inch strips, 2 or 3 inches long.
2. Combine flour, garlic powder, poultry seasoning, paprika and pepper; mix thoroughly.
3. Dredge turkey strips in seasoned flour. Shake off excess. Spray grill with cooking spray.
4. Grill turkey strips about 12 to 15 minutes or until done on a well greased griddle, turning frequently. CCP: Internal temperature must reach 165 F. or higher for 15 seconds.
5. CCP: Hold for service at 140 F. or higher. Serve with a sauce such as Sweet and Sour Sauce, Recipe No. O 008 00, Barbecue Sauce, Recipe No. O 002 00, Mustard Sauce, Recipe No. O 006 00, Honey Mustard Sauce, Recipe No. O 029 00, Horseradish Dijon Sauce, Recipe No. O 028 00, or Tropical Fruit Salsa, Recipe No. O 030 00.

MEAT, FISH, AND POULTRY No.L 189 00

ITALIAN BROCCOLI PASTA

Yield 100   Portion   11 Ounces

Calories	Carbohydrates	Protein	Fat	Cholesterol	Sodium	Calcium
464 cal	90 g	21 g	4 g	5 mg	1382 mg	278 mg

Ingredient

Ingredient	Weight	Measure	Issue
COOKING SPRAY,NONSTICK	2 oz	1/4 cup 1/3 tbsp	
ONIONS,FRESH,CHOPPED	8-1/2 lbs	1 gal 2 qts	9-1/2 lbs
TOMATOES,CANNED,DICED,DRAINED	46-1/4 lbs	5 gal 1 qts	
TOMATO PASTE,CANNED	2-7/8 lbs	1 qts 1 cup	
SUGAR,GRANULATED	10-5/8 oz	1-1/2 cup	
SALT	5-1/8 oz	1/2 cup	
GARLIC POWDER	2-3/8 oz	1/2 cup	
PEPPER,BLACK,GROUND	1-1/4 oz	1/4 cup 1-2/3 tbsp	
BASIL,DRIED,CRUSHED	1-1/4 oz	1/2 cup	
OREGANO,CRUSHED	1-1/4 oz	1/2 cup	
THYME,GROUND	1/3 oz	2 tbsp	
BROCCOLI,FROZEN,SPEARS,THAWED,1/2""	31 lbs	5 gal 2-1/2 qts	
WATER	83-5/8 lbs	10 gal	
SALT	2-1/2 oz	1/4 cup 1/3 tbsp	
MACARONI NOODLES,ROTINI,DRY	16-2/3 lbs	4 gal 2 qts	
ONIONS,GREEN,FRESH,CHOPPED	1-1/2 lbs	1 qts 2-3/4 cup	1-2/3 lbs
CHEESE,PARMESAN,GRATED	1-1/3 lbs	1 qts 2 cup	

Method

1 Spray steam jacketed kettle with non-stick spray. Add onions. Stir well. Cover; cook 10 minutes or until onions are tender, stirring constantly.
2 Add tomatoes, tomato paste, sugar, salt, garlic powder, pepper, basil, oregano, and thyme to onions. Bring to a boil; reduce heat; simmer 25 to 30 minutes; stirring occasionally.
3 Add broccoli; stir well; return to a simmer; simmer 3 to 5 minutes or until thoroughly heated. Do not overcook. CCP: Hold for service at 140 F. or higher.
4 Add salt to water; heat to a rolling boil. Slowly add rotini while stirring constantly until water boils again. Cook 10 to 12 minutes or until tender; stir occasionally. Drain. Rinse with warm water; drain thoroughly.
5 Each portion: Ladle 1-1/4 cups of tomato-broccoli sauce over 1-1/4 cups rotini. Sprinkle 1 tablespoon green onion and 1 tablespoon parmesan cheese over top of each portion.

MEAT, FISH, AND POULTRY No.L 190 00

CRANBERRY GLAZED CHICKEN (BREAST BONELESS)

Yield 100   Portion 5 Ounces

Calories	Carbohydrates	Protein	Fat	Cholesterol	Sodium	Calcium
254 cal	21 g	32 g	4 g	88 mg	88 mg	20 mg

Ingredient	Weight	Measure	Issue
CHICKEN,BREAST,BNLS/SKNLS,5 OZ	31-1/4 lbs		
COOKING SPRAY,NONSTICK	1-1/2 oz	3 tbsp	
CRANBERRY SAUCE,JELLIED	7-1/3 lbs	3 qts	
HONEY	2 lbs	2-5/8 cup	
JUICE,LIME	1-5/8 lbs	3-1/4 cup	
WATER,COLD	1-3/4 lbs	3-3/8 cup	
CINNAMON,GROUND	1/2 oz	2 tbsp	
GINGER,GROUND	1/8 oz	3/8 tsp	
CLOVES,GROUND	<1/16th oz	1/8 tsp	

Method

1. Wash chicken thoroughly under cold running water. Drain well. Remove excess fat.
2. Place chicken breasts on lightly sprayed sheet pans. Lightly spray chicken with cooking spray.
3. Using a convection oven, bake at 325 F. 10 to 12 minutes on high fan, closed vent. CCP: Internal temperature must reach 165 F. or higher for 15 seconds. Transfer chicken to steam table pans. Hold at 140 F. or higher for use in Step 5.
4. Break up cranberry sauce with wire whip. Add honey, lime juice, water, cinnamon, ginger and cloves; blend well. Bring to a boil. Cover; reduce heat; simmer 10 minutes.
5. Pour 1-1/4 qt sauce evenly over chicken in each pan. CCP: Hold for service at 140 F. or higher.

MEAT. FISH. AND POULTRY No.L 191 00

CHICKEN & ITALIAN VEGETABLE PASTA (FAJITA STRIPS)

Yield 100   Portion 1-1/4 Cups

Calories	Carbohydrates	Protein	Fat	Cholesterol	Sodium	Calcium
329 cal	48 g	24 g	5 g	41 mg	665 mg	187 mg

Ingredient	Weight	Measure	Issue
COOKING SPRAY,NONSTICK	1 oz	2 tbsp	
CHICKEN,FAJITA STRIPS	12-1/2 lbs		
ONIONS,FRESH,CHOPPED	2-1/8 lbs	1 qts 2 cup	2-1/3 lbs
TOMATOES,CANNED,DICED,DRAINED	26-1/2 lbs	3 gal	
CATSUP	2-3/8 lbs	1 qts 1/2 cup	
GARLIC POWDER	1-1/2 oz	1/4 cup 1-1/3 tbsp	
BASIL,DRIED,CRUSHED	1-7/8 oz	3/4 cup	
OREGANO,CRUSHED	2-1/2 oz	1 cup	
SALT	1 oz	1 tbsp	
PEPPER,BLACK,GROUND	2/3 oz	3 tbsp	
WATER,COLD	1-5/8 lbs	3 cup	
FLOUR,WHEAT,GENERAL PURPOSE	9-7/8 oz	2-1/4 cup	
WATER	41-3/4 lbs	5 gal	
SALT	1 oz	1 tbsp	
MACARONI NOODLES,ROTINI,DRY	8-1/3 lbs	2 gal 1 qts	
VEGETABLES,MIXED,FROZEN,ITALIAN	9 lbs	2 gal	
CHEESE,PARMESAN,GRATED	10-5/8 oz	3 cup	
CHEESE,MOZZARELLA,PART SKIM,SHREDDED	1-1/2 lbs	1 qts 2 cup	

Method

1. Stir-cook chicken and onions in a lightly sprayed steam-jacketed kettle or stockpot about 10 minutes, or until chicken is partially cooked and slightly tender.
2. Add tomatoes, catsup, garlic powder, basil, oregano, salt, and pepper. Bring to a boil. Reduce heat; simmer uncovered 5 minutes.
3. Blend flour and cold water together to make a smooth slurry. Add slurry to chicken mixture stirring constantly. Bring to a boil. Cover, reduce heat; simmer 8 to 10 minutes or until thickened, stirring frequently to prevent sticking.
4. Add salt to water; heat to a rolling boil. Slowly add pasta while stirring constantly until water boils again. Cook about 10 to 12 minutes or until tender; stir occasionally. Drain. Rinse with cold water; drain thoroughly.
5. Place 1/2 gal cooked pasta into each steam table pan. Add 1-1/4 qt Italian mixed vegetables and 1/2 cup parmesan cheese evenly to each pan; stir to combine. Pour 3-1/4 chicken and tomato mixture evenly over pasta in each pan; stir to combine.
6. Sprinkle 1 cup shredded mozzarella evenly over pasta mixture in each pan. Using a convection oven, bake 15 to 20 minutes at 350 F. on high fan, open vent. CCP: Internal temperature must reach 165 F. or higher for 15 seconds. Hold for service at 140 F. or higher.

MEAT. FISH. AND POULTRY No.L 192 00

HONEY LEMON CHICKEN BREAST (BREAST BONELESS)

Yield 100   Portion   5 Ounces

Calories	Carbohydrates	Protein	Fat	Cholesterol	Sodium	Calcium
219 cal	13 g	32 g	4 g	88 mg	88 mg	19 mg

Ingredient	Weight	Measure	Issue
CHICKEN,BREAST,BNLS/SKNLS,5 OZ	31-1/4 lbs		
HONEY	3 lbs	1 qts	
JUICE,LEMON	2-1/8 lbs	1 qts	
MUSTARD,DIJON	2-7/8 oz	1/4 cup 1-2/3 tbsp	
LEMON RIND,GRATED	1-1/8 oz	1/4 cup 1-2/3 tbsp	
CURRY POWDER	3/8 oz	1 tbsp	
GINGER,GROUND	1/3 oz	1 tbsp	
COOKING SPRAY,NONSTICK	3/4 oz	1 tbsp	
WATER,COLD	1 lbs	2 cup	
CORNSTARCH	2-1/4 oz	1/2 cup	

Method

1. Wash chicken thoroughly under cold running water. Drain well. Remove excess fat. Place chicken in roasting pans.
2. Combine honey, lemon juice, dijon mustard, lemon rind, curry and ginger; mix well.
3. Pour marinade over chicken in each roasting pan; cover. CCP: Marinate under refrigeration 41 F. or lower for 45 minutes.
4. Place chicken breasts on each lightly sprayed sheet pan. Lightly spray chicken with cooking spray. CCP: Refrigerate remaining marinade at 41 F. or lower for use in Step 7.
5. Using a convection oven, bake 12 to 14 minutes at 325 F. on high fan, closed vent. CCP: Internal temperature must reach 165 F. or higher for 15 seconds.
6. Transfer chicken to steam table pans. CCP: Hold at 140 F. or higher for use in Step 8. Drain chicken drippings.
7. Bring chicken drippings and reserved marinade to a boil. Blend cornstarch and cold water together to make a smooth slurry. Add slurry to hot sauce, stirring constantly. Cover; reduce heat; simmer 3 to 5 minutes or until thickened, stirring frequently to prevent sticking. CCP: Temperature must reach 165 F. or higher for 15 seconds.
8. Pour 3-1/4 cups sauce evenly over chicken in each pan. CCP: Hold for service at 140 F. or higher.

MEAT, FISH, AND POULTRY No.L 193 00

CAJUN ROAST BEEF

Yield 100   Portion   4 Ounces

Calories	Carbohydrates	Protein	Fat	Cholesterol	Sodium	Calcium
280 cal	1 g	39 g	12 g	112 mg	226 mg	20 mg

Ingredient	Weight	Measure	Issue
SALT	1-1/4 oz	2 tbsp	
GARLIC POWDER	1-1/8 oz	1/4 cup	
PEPPER,RED,GROUND	2/3 oz	1/4 cup	
PEPPER,WHITE,GROUND	2/3 oz	2-2/3 tbsp	
PEPPER,BLACK,GROUND	5/8 oz	2-2/3 tbsp	
ONION POWDER	5/8 oz	2-2/3 tbsp	
THYME,GROUND	3/8 oz	2-2/3 tbsp	
BASIL,DRIED,CRUSHED	3/8 oz	2-2/3 tbsp	
OREGANO,CRUSHED	3/8 oz	2-2/3 tbsp	
BEEF,OVEN ROAST,TEMPERED	40 lbs		

Method

1. Combine salt, garlic powder, red pepper, white pepper, black pepper, onion powder, thyme, basil, and oregano. Mix until well blended.
2. Trim excess fat from the roasts. Place in pan without crowding.
3. Sprinkle cajun spice mixture evenly over entire surface of the roast. Arrange in pan fat side up. Be sure entire surface of roast is covered with spice mixture.
4. Insert meat thermometer in the center of the thickest part of the main muscle.
5. Using a convection oven, roast 1 hour 45 minutes - 2-1/2 hours, depending on size of roast, at 300 F. on high fan, closed vent. CCP: Internal temperature must reach 145 F. or higher for 15 seconds. Let stand 20 minutes before slicing.
6. Cut 8 slices per pound. CCP: Hold at 140 F. or higher for service.

Notes

1. Arrange roasts in pans according to size. Allow 18 minutes per pound for rare, 20 minutes per pound for medium.

MEAT, FISH, AND POULTRY No.L 193 01

CAJUN ROAST TENDERLOIN OF BEEF

Yield 100   **Portion** 4 Ounces

Calories	Carbohydrates	Protein	Fat	Cholesterol	Sodium	Calcium
292 cal	1 g	34 g	16 g	104 mg	216 mg	18 mg

Ingredient	Weight	Measure	Issue
SALT	1-1/4 oz	2 tbsp	
GARLIC POWDER	1-1/8 oz	1/4 cup	
PEPPER,RED,CRUSHED	1/3 oz	1/4 cup	
PEPPER,WHITE,GROUND	2/3 oz	2-2/3 tbsp	
PEPPER,BLACK,GROUND	5/8 oz	2-2/3 tbsp	
ONION POWDER	5/8 oz	2-2/3 tbsp	
THYME,GROUND	3/8 oz	2-2/3 tbsp	
BASIL,DRIED,CRUSHED	3/8 oz	2-2/3 tbsp	
OREGANO,CRUSHED	3/8 oz	2-2/3 tbsp	
BEEF,TENDERLOIN,RAW	36 lbs		

Method

1. Combine salt, garlic powder, red pepper, white pepper, black pepper, onion powder, thyme, basil and oregano. Mix until well blended.
2. Trim excess fat and silverskin membrane from the roasts. Place in pans without crowding.
3. Sprinkle cajun spice mixture evenly over entire roast. Fold thin end under to make roast an even thickness throughout. Be sure entire surface of tenderloin is covered with spice mixture.
4. Insert meat thermometer in the thickest end of roast.
5. Using a convection oven, roast at 375 F. for 45 minutes, depending on size of roast. CCP: Internal temperature must reach 145 F. or higher for 15 seconds. Let stand in a warm place 15 minutes before slicing.
6. Cut 8 slice per pound. CCP: Hold at 140 F. or higher for service.

Notes

1. Arrange tenderloins in pans according to size. Allow 9 to 10 minutes per pound.

MEAT, FISH, AND POULTRY No.L 194 00

TROPICAL BAKED PORK CHOPS

Yield 100   Portion 3 Ounces

Calories	Carbohydrates	Protein	Fat	Cholesterol	Sodium	Calcium
374 cal	13 g	40 g	17 g	99 mg	172 mg	25 mg

Ingredient	Weight	Measure	Issue
FRUIT COCKTAIL,CANNED,JUICE PACK,INCL LIQUIDS	4 lbs	1 qts 3-5/8 cup	
RESERVED LIQUID	2-1/2 lbs	1 qts 7/8 cup	
JUICE,PINEAPPLE,CANNED,UNSWEETENED	3-7/8 lbs	1 qts 3-1/8 cup	
JUICE,LIME	2 lbs	1 qts	
ONIONS,FRESH,CHOPPED	3 lbs	2 qts 1/2 cup	3-1/3 lbs
SUGAR,BROWN,PACKED	10-1/4 oz	2 cup	
GARLIC POWDER	1-5/8 oz	1/4 cup 1-2/3 tbsp	
SALT	7/8 oz	1 tbsp	
GINGER,GROUND	7/8 oz	1/4 cup 2/3 tbsp	
ALLSPICE,GROUND	1/4 oz	1 tbsp	
CINNAMON,GROUND	1/4 oz	3/8 tsp	
NUTMEG,GROUND	1/8 oz	1/3 tsp	
PEPPER,RED,GROUND	1/8 oz	1/4 tsp	
COOKING SPRAY,NONSTICK	2 oz	1/4 cup 1/3 tbsp	
PORK,LOIN CHOPS,5 OZ	31-1/4 lbs		
CORNSTARCH	7-7/8 oz	1-3/4 cup	
JUICE,PINEAPPLE,CANNED,UNSWEETENED	11 oz	1-1/4 cup	
CILANTRO,DRY	1/8 oz	2-2/3 tbsp	
ONIONS,GREEN,FRESH,SLICED	14-3/8 oz	1 qts 1/8 cup	1 lbs
LIMES,FRESH	12-1/2 oz		

Method

1. Drain fruit; reserve juice for use in Step 2. Coarsely chop fruit.
2. Combine reserved juice with pineapple juice to make 3 quarts. Add lime juice, onions, brown sugar, garlic powder, salt, ginger, allspice, cinnamon, nutmeg, and red pepper. Stir well to blend.
3. Lightly spray griddle with non-stick cooking spray. Grill pork chops 5 minutes on each side or until browned.
4. Shingle 50 chops in each steam table pan.
5. Pour 3 qts juice mixture over chops in each pan; cover.
6. Using a convection oven, bake at 325 F. for 50 minutes on high fan, closed vent or until tender. CCP: Internal temperature must reach 145 F. or higher for 15 seconds. Remove chops to serving pans. CCP: Hold at 140 F. or higher for use in Step 9.
7. Dissolve cornstarch in pineapple juice.
8. Pour drippings from pork chops into steam-jacketed kettle or stock pot. Skim off fat. Bring to boil; slowly add cornstarch mixture, stirring constantly. Bring to a boil; cook 5 minutes or until slightly thickened and clear. Add tropical fruit and cilantro; simmer 1 minute.
9. Pour 8 cups sauce over chops in each pan.
10. Cut lime slices in half. Serve each chop with 1/3 cup sauce, 2 teaspoons sliced green onions and 1/2 slice of lime. CCP: Hold at 140 F. or higher for service.

MEAT, FISH, AND POULTRY No.L 195 00

TERIYAKI BEEF STRIPS

Yield 100   Portion 5 Ounces

Calories	Carbohydrates	Protein	Fat	Cholesterol	Sodium	Calcium
213 cal	8 g	26 g	8 g	70 mg	699 mg	20 mg

Ingredient	Weight	Measure	Issue
WATER	2-1/8 lbs	1 qts	
JUICE,PINEAPPLE,CANNED,UNSWEETENED	3-1/8 lbs	1 qts 1-3/4 cup	
SOY SAUCE	2-1/2 lbs	1 qts	
GINGER,GROUND	1-1/2 oz	1/2 cup	
PEPPER,BLACK,GROUND	1/2 oz	2 tbsp	
GARLIC POWDER	3/8 oz	1 tbsp	
BEEF,OVEN ROAST,TEMPERED	25 lbs		
WATER	8-1/3 oz	1 cup	
CORNSTARCH	2-1/4 oz	1/2 cup	
ONIONS,FRESH,SLICED	6-1/8 lbs	1 gal 2 qts	6-3/4 lbs
PEPPERS,GREEN,FRESH,MEDIUM,SLICED,THIN	4-3/4 lbs	3 qts 2-1/2 cup	5-3/4 lbs
COOKING SPRAY,NONSTICK	2 oz	1/4 cup 1/3 tbsp	

Method

1 Combine water, pineapple juice, soy sauce, ginger, pepper, and garlic; mix well. Divide teriyaki sauce in half.
2 Cut beef into 1/4-inch thin slices; cut slices into 1/2-inch strips, 3 to 4 inches long.
3 Pour 1-3/4 quarts teriyaki sauce over beef strips. CCP: Cover; marinate under refrigeration at 41 F. or lower. Drain well.
4 Bring reserved teriyaki sauce to a boil. Combine cornstarch and water; add to teriyaki sauce. Simmer 5 minutes or until thickened. CCP: Hold at 140 F. or higher for use in Step 7.
5 Saute onions and peppers about 2 minutes or until almost transparent.
6 Combine beef strips with sauteed onion and peppers. Brown 1 to 2 minutes on lightly sprayed 400 F. griddle to desired degree of doneness, turning frequently. CCP: Internal temperature must reach 145 F. or higher for 15 seconds.
7 Pour thickened teriyaki sauce over beef mixture. CCP: Hold at 140 F. or higher for service.

Notes

1 In Step 6, brown strips in batches. Use 6 cups onions and pepper mixture for 6 pounds 12 ounces of beef strips per 100 portions.

MEAT, FISH, AND POULTRY No.L 195 01

TERIYAKI BEEF STRIPS (FAJITA STRIPS)

Yield 100 Portion 5 Ounces

Calories	Carbohydrates	Protein	Fat	Cholesterol	Sodium	Calcium
269 cal	8 g	34 g	11 g	96 mg	718 mg	22 mg

Ingredient	Weight	Measure	Issue
WATER	2-1/8 lbs	1 qts	
JUICE,PINEAPPLE,CANNED,UNSWEETENED	3-1/8 lbs	1 qts 1-3/4 cup	
SOY SAUCE	2-1/2 lbs	1 qts	
GINGER,GROUND	1-1/2 oz	1/2 cup	
PEPPER,BLACK,GROUND	1/2 oz	2 tbsp	
GARLIC POWDER	3/8 oz	1 tbsp	
BEEF,FAJITA STRIPS	33-3/4 lbs		
WATER	8-1/3 oz	1 cup	
CORNSTARCH	2-1/4 oz	1/2 cup	
ONIONS,FRESH,SLICED	6-1/8 lbs	1 gal 2 qts	6-3/4 lbs
PEPPERS,GREEN,FRESH,MEDIUM,SLICED,THIN	4-3/4 lbs	3 qts 2-1/2 cup	5-3/4 lbs
COOKING SPRAY,NONSTICK	2 oz	1/4 cup 1/3 tbsp	

Method

1 Combine water, pineapple juice, soy sauce, ginger, pepper, and garlic; mix well. Divide teriyaki sauce in half.
2 Cut beef into 1/4-inch thin slices; cut slices into 1/2-inch strips, 3 to 4 inches long.
3 Pour 1-3/4 quarts teriyaki sauce over beef strips. CCP: Cover; marinate under refrigeration at 41 F. or lower. Drain well.
4 Bring reserved teriyaki sauce to a boil. Combine cornstarch and water; add to teriyaki sauce. Simmer 5 minutes or until thickened. CCP: Hold at 140 F. or higher for use in Step 7.
5 Saute onions and peppers about 2 minutes or until almost transparent.
6 Combine beef strips with sauteed onions and peppers. Brown 1 to 2 minutes on lightly sprayed 400 F. griddle to desired degree of doneness, turning frequently. CCP: Internal temperature must reach 155 F. or higher for 15 seconds.
7 Pour thickened teriyaki sauce over beef mixture. CCP: Hold at 140 F. or higher for service.

Notes

1 In Step 6, brown strips in batches. Use 6 cups onions and pepper mixture for 6 pounds 12 ounces of beef strips per 100 portions.

MEAT, FISH, AND POULTRY No. L 196 00

SOUTHWESTERN SWEET POTATOES, BLACK BEANS, AND CORN

Yield 100   Portion 1-1/4 Cups

Calories	Carbohydrates	Protein	Fat	Cholesterol	Sodium	Calcium
356 cal	74 g	15 g	2 g	0 mg	197 mg	102 mg

Ingredient	Weight	Measure	Issue
COOKING SPRAY,NONSTICK	2 oz	1/4 cup 1/3 tbsp	
ONIONS,FRESH,CHOPPED	8-1/2 lbs	1 gal 2 qts	9-1/2 lbs
SWEET POTATOES,FROZEN,THAWED,CUBED	25 lbs	4 gal 1/8 qts	
WATER	6-1/4 lbs	3 qts	
PEPPERS,JALAPENOS,CANNED,DRAINED,CHOPPED	8-3/8 oz	1-3/4 cup	
CUMIN,GROUND	4-1/4 oz	1-1/4 cup	
GARLIC POWDER	3-1/2 oz	3/4 cup	
SALT	1-1/4 oz	2 tbsp	
PEPPER,BLACK,GROUND	2/3 oz	3 tbsp	
BEANS,BLACK,CANNED,DRAINED	26 lbs	2 gal 3-1/2 qts	
CORN,FROZEN,WHOLE KERNEL	13-3/4 lbs	2 gal 1-1/2 qts	
CILANTRO,DRY	1 oz	3/4 cup 2 tbsp	
LIMES,FRESH	2-1/3 lbs	15-1/2 each	

Method

1 Lightly spray steam jacketed kettle or tilting fry pan with non-stick spray.
2 Add onions; stir; cover; cook 5 minutes or until tender, stirring occasionally.
3 Add potatoes, water, jalapenos, cumin, garlic powder, salt and pepper. Stir; cover; cook 7 to 10 minutes or until potatoes are almost tender.
4 Add beans and corn; stir; cook 15 minutes or until thoroughly heated, stirring occasionally. CCP: Internal temperature must reach 145 F. or higher for 15 seconds.
5 Add cilantro; stir. Transfer to serving pans.
6 Serve each portion with lime wedge. CCP: Hold for service at 140 F. or higher.

Notes

1 In Step 4, 10 pounds dry black beans and 4 gallons water may be used per 100 portions. Follow Steps 1 through 5 of Recipe No. Q 003 00 , Boston Baked Beans.
2 In Step 3, 25 pounds fresh sweet potatoes may be used per 100 portions. Cook 10 to 15 minutes or until tender.

MEAT, FISH, AND POULTRY No.L 196 01

SOUTHWESTERN SWEET POTATOES,BLACK BEAN,CORN(CND)

Yield 100   Portion   1-1/4 Cups

Calories	Carbohydrates	Protein	Fat	Cholesterol	Sodium	Calcium
351 cal	74 g	15 g	2 g	0 mg	243 mg	82 mg

Ingredient	Weight	Measure	Issue
SWEET POTATOES,CANNED,W/SYRUP	28-1/8 lbs	3 gal 2 qts	
BEANS,BLACK,CANNED,DRAINED	26 lbs	2 gal 3-1/2 qts	
CORN,FROZEN,WHOLE KERNEL	13-3/4 lbs	2 gal 1-1/2 qts	
ONIONS,FRESH,CHOPPED	8-1/2 lbs	1 gal 2 qts	9-3/8 lbs
WATER	4-1/8 lbs	2 qts	
PEPPERS,JALAPENOS,CANNED,DRAINED,CHOPPED	8-3/8 oz	1-3/4 cup	
CUMIN,GROUND	4-1/4 oz	1-1/4 cup	
GARLIC POWDER	3-1/2 oz	3/4 cup	
SALT	1-1/4 oz	2 tbsp	
CILANTRO,DRY	1 oz	3/4 cup 2 tbsp	
PEPPER,BLACK,GROUND	2/3 oz	3 tbsp	
LIMES,FRESH	2-1/3 lbs	15-1/2 each	

Method

1. Cut potatoes into 3/4-inch pieces. Set aside for use in Step 4.
2. Combine beans, corn, onions, water, jalapeno peppers, cumin, garlic powder, salt, dry cilantro, and black pepper.
3. Place 4-1/4 quarts mixture in each pan.
4. Add 9-3/4 cups sweet potatoes to each pan. Gently fold potatoes into bean and corn mixture to evenly distribute ingredients.
5. Cover; using a convection oven bake at 350 F. for 1 hour or until thoroughly heated. CCP: Internal temperature must reach 145 F. or higher for 15 seconds.
6. CCP: Hold for service at 140 F. or higher.
7. Serve each portion with lime wedges.

Notes

1. In Step 2, 10 pounds dry black beans and 4 gallons water may be used per 100 portions. Follow Steps 1 through 5 of Recipe No. Q 003 00, Boston Baked Beans.
2. In Step 1, 25 pounds fresh sweet potatoes may be used per 100 portions. Cook 10 to 15 minutes or until tender.

MEAT, FISH, AND POULTRY No.L 197 00

DIJON BAKED PORK CHOPS

Yield 100   **Portion** 3 Ounces

Calories	Carbohydrates	Protein	Fat	Cholesterol	Sodium	Calcium
315 cal	11 g	33 g	15 g	78 mg	422 mg	43 mg

Ingredient	Weight	Measure	Issue
MUSTARD,DIJON	4-3/4 lbs	2 qts 1 cup	
JUICE,APPLE,CANNED	1 lbs	1-7/8 cup	
SALAD DRESSING,MAYONNAISE TYPE,FAT FREE	1-1/8 lbs	2 cup	
ONIONS,FRESH,CHOPPED	11-1/4 oz	2 cup	12-1/2 oz
BREADCRUMBS	3-1/3 lbs	3 qts 2 cup	
PARSLEY,DEHYDRATED,FLAKED	3/4 oz	1 cup	
COOKING SPRAY,NONSTICK	2 oz	1/4 cup 1/3 tbsp	
PORK,LOIN CHOPS,5 OZ	31-1/4 lbs		

Method

1. Combine mustard, apple juice, salad dressing, and onions in mixer bowl. Beat at low speed 30 seconds. Beat at medium speed 1 minute or until well blended.
2. Combine crumbs and parsley.
3. Lightly spray each sheet pan with nonstick cooking spray.
4. Dip chops in mustard mixture. Dredge in bread crumb mixture.
5. Place 20 chops on each sheet pan. Using a convection oven bake at 350 F. for 30 minutes or until chops are tender and well done. CCP: Internal temperature must reach 145 F. or higher for 15 seconds. Hold for service at 140 F. or higher.

MEAT. FISH. AND POULTRY No.L 198 00

GREEK LEMON TURKEY PASTA

Yield 100   Portion 1-1/3 Cups

Calories	Carbohydrates	Protein	Fat	Cholesterol	Sodium	Calcium
354 cal	50 g	23 g	7 g	40 mg	1585 mg	127 mg

Ingredient	Weight	Measure	Issue
SOY SAUCE	10-1/8 oz	1 cup	
JUICE,LEMON	8-5/8 oz	1 cup	
GARLIC CLOVES,FRESH,MINCED	1-3/4 oz	1/4 cup 2-1/3 tbsp	2 oz
PEPPER,BLACK,GROUND	1-5/8 oz	1/4 cup 3-1/3 tbsp	
LEMON RIND,GRATED	1 oz	1/4 cup 1 tbsp	
TURKEY,BNLS,WHITE AND DARK MEAT	16 lbs		
WATER	66-7/8 lbs	8 gal	
SALT	2-1/2 oz	1/4 cup 1/3 tbsp	
MACARONI NOODLES,ROTINI,DRY	12 lbs	3 gal 1 qts	
WATER	1-1/3 lbs	2-1/2 cup	
CORNSTARCH	9 oz	2 cup	
CHICKEN BROTH		2 gal 2 qts	
COOKING SPRAY,NONSTICK	1/2 oz	1 tbsp	
SPINACH,FROZEN	10-7/8 lbs	1 gal 2-1/2 qts	
ONIONS,FRESH,SLICED	2-5/8 lbs	2 qts 2-3/8 cup	2-7/8 lbs
JUICE,LEMON	1-1/3 lbs	2-1/2 cup	

Method

1. Combine soy sauce, lemon juice, garlic, pepper, and lemon rind. Mix well.
2. Slice tempered turkey into 1/2-inch slices; cut slices into 1/2-inch strips; 2 to 3-inches in length. Add marinade. Toss to coat turkey evenly. CCP: Cover; marinate under refrigeration at 41 F. or lower for use in Step 6.
3. Bring water to a boil; add salt. Slowly add rotini while stirring until water boils again. Cook 10 to 12 minutes or until almost tender, stir occasionally. Drain. Rinse in cold water. Drain thoroughly.
4. Dissolve cornstarch in water.
5. Prepare chicken stock according to package directions.
6. Spray steam jacketed kettle or tilt fry pan with nonstick spray. Add turkey and marinade. Stir-cook until turkey is no longer pink. CCP: Internal temperature must reach 165 F. or higher for 15 seconds.
7. Add spinach and stock; bring to a boil stirring. Reduce heat; slowly add cornstarch mixture, constantly about 5 minutes or until slightly thickened.
8. Add green onions, lemon juice, and rotini, stirring until ingredients are well distributed. CCP: Internal temperature must reach 165 F. or higher for 15 seconds. Transfer to serving pans. CCP: Hold for service at 140 F. or higher.

MEAT. FISH. AND POULTRY No.L 200 00

GRILLED TURKEY SAUSAGE LINKS

Yield 100 **Portion** 2 Ounces

Calories	Carbohydrates	Protein	Fat	Cholesterol	Sodium	Calcium
82 cal	0 g	9 g	4 g	30 mg	176 mg	14 mg

Ingredient   Weight   Measure   Issue

SAUSAGE LINK,TURKEY,RAW   12-1/2 lbs

Method

1. Grill 12 minutes. CCP: Internal temperature must reach 165 F. or higher for 15 seconds.
2. Turn frequently to ensure even browning.
3. CCP: Hold for service at 140 F. or higher.

Notes

1. In Step 1, turkey sausages can be baked in a convection oven, at 350 F. for 10 minutes on high fan, closed vent.

MEAT, FISH, AND POULTRY No.L 201 00

TAMALE PIE (TURKEY)

Yield 100   Portion   9-1/2 Ounces

Calories	Carbohydrates	Protein	Fat	Cholesterol	Sodium	Calcium
305 cal	36 g	21 g	9 g	48 mg	856 mg	122 mg

Ingredient	Weight	Measure	Issue
CORN MEAL	6-1/2 lbs	1 gal 1-1/3 qts	
CHILI POWDER,DARK,GROUND	4-1/4 oz	1 cup	
SALT	3 oz	1/4 cup 1 tbsp	
WATER,BOILING	25-1/8 lbs	3 gal	
TURKEY,GROUND,90% LEAN,RAW	16 lbs		
ONIONS,FRESH,CHOPPED	2-1/8 lbs	1 qts 2 cup	2-1/3 lbs
PEPPERS,GREEN,FRESH,CHOPPED	14-1/2 oz	2-3/4 cup	1-1/8 lbs
TOMATOES,CANNED,DICED,DRAINED	13-1/4 lbs	1 gal 2 qts	
CORN,CANNED,WHOLE KERNEL,DRAINED	5-3/4 lbs	1 gal	
OLIVES,RIPE,PITTED,SLICED,INCL LIQUIDS	3-1/4 lbs	2 qts 3 cup	
CHILI POWDER,DARK,GROUND	4-1/4 oz	1 cup	
CUMIN,GROUND	1-1/4 oz	1/4 cup 2-1/3 tbsp	
GARLIC POWDER	1 oz	3-1/3 tbsp	
SALT	7/8 oz	1 tbsp	
PEPPER,RED,GROUND	1/4 oz	1 tbsp	
COOKING SPRAY,NONSTICK	2 oz	1/4 cup 1/3 tbsp	
CHEESE,CHEDDAR,LOWFAT,SHREDDED	3 lbs	3 qts	

Method

1 Mix cornmeal, chili powder and salt together; gradually stir into water. Bring to a boil.
2 Reduce heat; simmer 25 minutes, stirring frequently until a stiff paste is formed. Set aside for use in Step 5.
3 Cook turkey with onions and peppers until turkey loses its pink color, stirring to break apart. Drain or skim off excess fat.
4 Add tomatoes, corn, olives, chili powder, cumin, salt, garlic powder, and red pepper to turkey mixture; simmer 15 minutes, stirring frequently. CCP: Hold at 140 F. or higher for use in Step 6.
5 Spread 2-1/3 cups cornmeal paste over bottom and sides of each lightly sprayed steam table pan to form a thin crust.
6 Pour 2 quarts meat mixture over crust in each pan.
7 Spread 4-2/3 cups cornmeal paste evenly over meat mixture in each pan.
8 Using a convection oven, bake at 325 F. 50 to 60 minutes, on low fan, open vent. CCP: Internal temperature must reach 165 F. or higher for 15 seconds. Remove from oven.
9 Sprinkle 1-1/2 cups cheese evenly over each pan.
10 Let stand 10 to 15 minutes to allow filling to firm and cheese to melt.
11 Cut 3 by 4. CCP: Hold for service at 140 F. or higher.

Notes

1 In Step 4, 7 pounds 5 ounces canned, ripe, whole pitted olives, drained and chopped may be used per 100 portions.

MEAT, FISH, AND POULTRY No.L 202 00

ORIENTAL TUNA PATTIES

Yield 100   Portion   4-1/2 Ounces

Calories	Carbohydrates	Protein	Fat	Cholesterol	Sodium	Calcium
225 cal	20 g	24 g	5 g	105 mg	674 mg	66 mg

Ingredient	Weight	Measure	Issue
WATER	3-1/8 lbs	1 qts 2 cup	
SOY SAUCE	1 lbs	1-1/2 cup	
JUICE,LIME	12 oz	1-1/2 cup	
SUGAR,BROWN,PACKED	3-7/8 oz	3/4 cup	
GARLIC POWDER	5/8 oz	2 tbsp	
GINGER,GROUND	3/8 oz	2 tbsp	
PEPPER,RED,GROUND	<1/16th oz	1/8 tsp	
WATER	1 lbs	2 cup	
CORNSTARCH	4-1/2 oz	1 cup	
ONIONS,GREEN,FRESH,SLICED	8 oz	2-1/4 cup	8-7/8 oz
PEANUTS,SHELLED	7-3/4 oz	1-1/2 cup	
FISH,TUNA,CANNED,WATER PACK,INCL LIQUIDS	15-3/8 lbs	2 gal 3-1/3 qts	
BREADCRUMBS	6-1/4 lbs	1 gal 2-1/2 qts	
EGGS,WHOLE,FROZEN	4-1/4 lbs	2 qts	
ONIONS,FRESH,CHOPPED	3 lbs	2 qts 1/2 cup	3-1/3 lbs
CELERY,FRESH,CHOPPED	2-2/3 lbs	2 qts 2-1/8 cup	3-2/3 lbs
HORSERADISH,PREPARED	7-3/8 oz	3/4 cup 2 tbsp	
GARLIC CLOVES,FRESH,MINCED	3-1/4 oz	1/2 cup 3 tbsp	3-3/4 oz
COOKING SPRAY,NONSTICK	2 oz	1/4 cup 1/3 tbsp	

Method

1 Combine water, soy sauce, lime juice, brown sugar, garlic powder, ginger and red pepper. Bring to a boil. Reduce heat.
2 Combine water and cornstarch. Blend until smooth. Add to sauce mixture while stirring. Simmer 3 minutes or until thickened.
3 Add green onions and peanuts. Stir well.
4 Drain tuna; place drained tuna in a mixer bowl. Flake tuna on low speed about 30 seconds or until tuna chunks begin to flake.
5 Add bread crumbs, eggs, onions, celery, horseradish and garlic. Mix 2 minutes at low speed or until ingredients are combined. Do not overmix.
6 Shape into 100 4-3/4 ounce balls; place 20 balls on each sheet pan. Cover with parchment paper; flatten into patties by pressing down with another sheet pan to a thickness of 1/2-inch. CCP: Refrigerate at 41 F. or lower until ready to grill.
7 Grill patties on lightly sprayed 350 F. griddle 4 to 5 minutes per side or until golden brown. CCP: Internal temperature must reach 145 F. or higher for 15 seconds. Hold for service at 140 F. or higher.
8 Serve with 2 tablespoons Oriental Sauce. (Recipe O 026 00).

Notes

1 In Step 7, the patties may be baked in 350 F. convection oven for 20 minutes. CCP: Internal temperature must be heated to 145 F. or higher for 15 seconds on high fan, closed vent.

MEAT, FISH, AND POULTRY No.L 203 00

VEGETABLE CURRY WITH RICE

Yield 100   Portion   2-1/2 Cups

Calories	Carbohydrates	Protein	Fat	Cholesterol	Sodium	Calcium
458 cal	98 g	12 g	3 g	0 mg	688 mg	109 mg

Ingredient	Weight	Measure	Issue
VEGETABLE BROTH		2 gal 1 qts	
TOMATOES,CANNED,DICED,DRAINED	4-1/8 lbs	1 qts 3-1/2 cup	
POTATOES,FRESH,CHOPPED	12 lbs	2 gal 3/4 qts	14-7/8 lbs
SQUASH,BUTTERNUT,FRESH,CUBED	6-1/8 lbs	1 gal 1 qts	7-1/3 lbs
BEANS,GARBANZO,CANNED,DRAINED	11-5/8 lbs	1 gal 1-1/2 qts	
APPLESAUCE,CANNED,SWEETENED	7 lbs	3 qts 1/2 cup	
CAULIFLOWER,FROZEN	5 lbs		
CARROTS,FROZEN,SLICED	5-1/8 lbs	1 gal 1/2 qts	
RAISINS	1-1/8 lbs	3-1/2 cup	
CURRY POWDER	10-1/4 oz	2-7/8 cup	
GARLIC POWDER	13-5/8 oz	2-7/8 cup	
SALT	7/8 oz	1 tbsp	
CUMIN,GROUND	1/3 oz	1 tbsp	
ONIONS,FRESH,SLICED	4 lbs	1 gal	4-1/2 lbs
PEPPERS,GREEN,FRESH,MEDIUM,SLICED,THIN	10-1/2 lbs	2 gal	12-7/8 lbs
SQUASH,ZUCCHINI,FRESH,JULIENNE	4 lbs	1 gal	4-1/4 lbs
PEAS,GREEN,FROZEN	4-1/2 lbs	3 qts 2 cup	
RICE,BROWN,LONG GRAIN,DRY	12-1/2 lbs	1 gal 3-2/3 qts	
WATER	33-1/2 lbs	4 gal	
SALT	2-1/2 oz	1/4 cup 1/3 tbsp	

Method

1. Drain the diced tomatoes, save the juice, set tomatoes aside. Prepare the stock according to package directions using the reserved juice from the tomatoes and water.
2. In a steam-jacketed kettle or stock pot, add vegetable broth, potatoes and squash, bring to a boil. Cover, cook 8 minutes or until potatoes are tender.
3. Add chick peas, applesauce, cauliflower, carrots, tomatoes, raisins, curry powder, garlic powder, salt, cumin and onions. Stir well. Bring to a boil; reduce heat; simmer 7 minutes. Stir occasionally.
4. Add peppers and zucchini; return to a simmer. Simmer 9 to 11 minutes or until all vegetables are tender. Stir occasionally.
5. Add peas; stir; simmer 3 minutes or until peas are heated through. CCP: Hold for service at 140 F. or higher.
6. Combine rice, water, and salt. Bring to a boil; stir; cover tightly; simmer 25 minutes or until most of water is absorbed. Do not
7. Remove from heat; transfer to shallow pans. Cover. CCP: Hold for service at 140 F. or higher.
8. Serve 1-1/2 cups vegetable curry over 1 cup rice.

Notes

1. In Step 2, 7 pounds frozen butternut squash cubes may be used per 100 portions. Add with pepper-onion blend in Step 4.
2. In Step 3, 5 pounds fresh cauliflower florets may be used per 100 portions. Add in Step 2.
3. In Step 3, 5 pound fresh, peeled 1/4-inch sliced carrots may be used per 100 portions. Add in Step 2.
4. In Step 4, 5 pounds frozen sliced zucchini or summer squash may be used per 100 portions.

MEAT, FISH, AND POULTRY No. L 204 00

TURKEY PEACH PASTA SALAD (ENTREE)

Yield 100   Portion 1-1/2 Cups

Calories	Carbohydrates	Protein	Fat	Cholesterol	Sodium	Calcium
313 cal	41 g	22 g	7 g	51 mg	749 mg	111 mg

Ingredient	Weight	Measure	Issue
TURKEY,BNLS,WHITE AND DARK MEAT	20 lbs		
SALT	7/8 oz	1 tbsp	
OIL,SALAD	5/8 oz	1 tbsp	
WATER,BOILING	25-1/8 lbs	3 gal	
MACARONI NOODLES,ROTINI,DRY	4 lbs	1 gal 1/3 qts	
PEACHES,CANNED,HALVES,LIGHT SYRUP	40-1/2 lbs	4 gal 2-1/2 qts	
YOGURT,PLAIN,LOWFAT	4-1/3 lbs	2 qts	
SALAD DRESSING,MAYONNAISE TYPE,FAT FREE	9 oz	1 cup	
MUSTARD,DIJON	2-1/8 oz	1/4 cup 1/3 tbsp	
SALT	3/8 oz	1/3 tsp	
THYME LEAVES,DRIED,GROUND	1/3 oz	2 tbsp	
CELERY,FRESH,CHOPPED	5 lbs	1 gal 3/4 qts	6-7/8 lbs
CARROTS,FRESH,SHREDDED	2 lbs	2 qts 1/4 cup	2-1/2 lbs
ONIONS,GREEN,FRESH,CHOPPED	1 lbs	1 qts 1/2 cup	1-1/8 lbs
CILANTRO,DRY	3/4 oz	1/2 cup 2-2/3 tbsp	
LETTUCE,LEAF,FRESH,HEAD	4 lbs		6-1/4 lbs

Method

1. Cut turkey into 3/8 inch thick slices. Cut slices into 3/8 strips, 2 inches long. Cover. CCP: Refrigerate at 41 F. or lower for use in Step 7.
2. Add salt and salad oil to water; heat to a rolling boil. Add rotini slowly while stirring constantly until water boils again. Cook about 10 to 12 minutes or until tender; stir occasionally. Do not overcook.
3. Drain rotini. Rinse with cold water; drain thoroughly. Place in shallow containers; refrigerate and cover.
4. Drain peaches; reserve 1 cup peach juice for use in Step 5. Coarsely chop slices into 1-inch pieces. Set aside for use in Step 7.
5. Combine yogurt, peach juice, salad dressing, mustard, salt and thyme in mixer bowl. Blend at medium speed until smooth about 2 minutes.
6. Combine turkey, rotini, peaches, celery, carrots, onions and cilantro.
7. Add yogurt dressing to turkey peach mixture. Mix thoroughly but lightly to coat all ingredients with dressing.
8. Place lettuce leaf on each serving dish. Top with 1-1/2 cups of turkey peach pasta salad; cover. CCP: Refrigerate product at 41 F. or lower until ready to serve.

Notes

1. In Step 7, 3 oz (1-1/2 cup) trimmed, chopped, fresh cilantro (4 oz A.P.) may be used.

MEAT, FISH, AND POULTRY No.L 205 00

ITALIAN RICE AND BEEF

Yield 100   Portion 9 Ounces

Calories	Carbohydrates	Protein	Fat	Cholesterol	Sodium	Calcium
342 cal	20 g	29 g	15 g	92 mg	368 mg	132 mg

Ingredient	Weight	Measure	Issue
BEEF,GROUND,BULK,RAW,90% LEAN	24 lbs		
TOMATOES,CANNED,DICED,DRAINED		3 gal	
ONIONS,FRESH,CHOPPED	5 lbs	3 qts 2-1/8 cup	5-1/2 lbs
PEPPERS,GREEN,FRESH,CHOPPED	4 lbs	3 qts 1/8 cup	4-7/8 lbs
RICE,LONG GRAIN	3-7/8 lbs	2 qts 1-1/2 cup	
BEEF BROTH		1 qts 2 cup	
SUGAR,GRANULATED	5-1/4 oz	3/4 cup	
GARLIC POWDER	2-3/8 oz	1/2 cup	
SALT	1-1/4 oz	2 tbsp	
BASIL,SWEET,WHOLE,CRUSHED	1-1/4 oz	1/2 cup	
OREGANO,CRUSHED	1-1/4 oz	1/2 cup	
PEPPER,BLACK,GROUND	3/8 oz	1 tbsp	
CHEESE,MOZZARELLA,PART SKIM,SHREDDED	2 lbs	2 qts	
CHEESE,PARMESAN,GRATED	7 oz	2 cup	

Method

1. Cook beef in a steam jacketed kettle or stock pot until it loses its pink color, stirring to break apart. Drain fat.
2. Add tomatoes, onions, peppers, rice, beef stock, sugar, garlic powder, salt, basil, oregano and pepper. Stir to blend. Bring to a boil. Cover tightly; reduce heat; simmer 20 to 25 minutes or until rice is tender. Do not stir.
3. Place 1-2/3 gal cooked beef mixture into each ungreased steam table pan.
4. Sprinkle 1/2 cup parmesan cheese evenly over beef mixture in each pan.
5. Using a convection oven, bake at 325 F. 15 to 20 minutes on high fan, open vent. CCP: Internal temperature must reach 155 F. or higher for 15 seconds.
6. Distribute 2 cups shredded mozzarella cheese evenly over parmesan cheese in each pan. Bake an additional 3 minutes to melt cheese. CCP: Hold for service at 140 F. or higher.

MEAT, FISH, AND POULTRY No.L 206 00

BAYOU CHICKEN (BREAST BONELESS)

Yield 100 **Portion** 5 Ounces

Calories	Carbohydrates	Protein	Fat	Cholesterol	Sodium	Calcium
186 cal	3 g	32 g	4 g	88 mg	435 mg	27 mg

Ingredient | Weight | Measure | Issue

Ingredient	Weight	Measure	Issue
CHICKEN,BREAST,BNLS/SKNLS,5 OZ	31-1/4 lbs		
JUICE,LEMON	2-2/3 lbs	1 qts 1 cup	
SEASONING,CAJUN	10-2/3 oz	2 cup	
HOT SAUCE	1-1/2 lbs	3 cup	
COOKING SPRAY,NONSTICK	2 oz	1/4 cup 1/3 tbsp	

Method

1. Wash chicken breasts thoroughly under cold running water. Drain well. Remove excess fat.
2. Combine lemon juice, hot pepper sauce and cajun seasoning. Stir until well blended.
3. Pour 1 quart marinade over chicken breasts in each pan; cover. CCP: Marinate under refrigeration at 41 F. or lower for 45 minutes.
4. Lightly spray each pan with non-stick cooking spray. Place 25 chicken breasts on each sheet pan. Lightly spray breasts with non-stick cooking spray. Pour 1 quart marinade over chicken breasts in each pan; cover. Discard any remaining marinade.
5. Using a convection oven, bake 12 to 14 minutes on high fan, closed vent. CCP: Internal temperature must reach 165 F. or higher for 15 seconds.
6. Transfer chicken to steam table pans. CCP: Hold for service at 140 F. or higher.

MEAT, FISH, AND POULTRY No.L 207 00

SOUTHWESTERN SHRIMP LINGUINE

Yield 100   **Portion** 10 Ounces

Calories	Carbohydrates	Protein	Fat	Cholesterol	Sodium	Calcium
290 cal	36 g	24 g	5 g	168 mg	656 mg	178 mg

Ingredient

Ingredient	Weight	Measure	Issue
WATER	66-7/8 lbs	8 gal	
SALT	1-7/8 oz	3 tbsp	
OIL,SALAD	1 oz	2 tbsp	
PASTA,LINGUINE	12 lbs	8 gal 3-7/8 qts	
WATER	25-1/8 lbs	3 gal	
SHRIMP,RAW,PEELED,DEVEINED	20 lbs		
COOKING SPRAY,NONSTICK	2 oz	1/4 cup 1/3 tbsp	
PEPPERS,RED,FRESH	6 lbs	1 gal 5/8 qts	7-1/3 lbs
ONIONS,FRESH,CHOPPED	6 lbs	1 gal 1/4 qts	6-2/3 lbs
RESERVED LIQUID	4-1/8 lbs	2 qts	
JUICE,LIME	1-1/2 lbs	3 cup	
PEPPERS,JALAPENOS,CANNED,DRAINED,CHOPPED	4-3/4 oz	1 cup	
GARLIC POWDER	4-3/4 oz	1 cup	
PARSLEY,DEHYDRATED,FLAKED	2-7/8 oz	1 qts	
CILANTRO,DRY	1-5/8 oz	1-1/2 cup	
PEPPER,BLACK,GROUND	1/4 oz	1 tbsp	
CHEESE,PARMESAN,GRATED	1-1/3 lbs	1 qts 2 cup	

Method

1. Add salt and salad oil to water; heat to a rolling boil.
2. Add pasta slowly while stirring constantly until water boils again. Cook 10 to 12 minutes or until tender; stirring occasionally. DO NOT OVERCOOK. Drain. Rinse with cold water. Drain thoroughly.
3. Place shrimp in boiling water; cover; return to a boil. Reduce heat; simmer 3 to 5 minutes. DO NOT OVERCOOK.
4. Immediately remove shrimp from cooking liquid and rinse with cold water or ice bath for 2 minutes. Drain shrimp. Reserve 2 quarts shrimp cooking liquid for use in Step 6. Refrigerate shrimp at 41 F. or lower for use in Step 7.
5. Stir-cook sweet red peppers and onions in a lightly sprayed steam jacketed kettle or stock pot for 8 to 10 minutes or until tender, stirring constantly.
6. Add reserved shrimp liquid, lime juice, jalapeno peppers, garlic powder, parsley, cilantro and black pepper to cooked sweet peppers and onions. Stir well to blend. Bring to a boil; reduce heat to a simmer.
7. Add linguine and shrimp to the hot broth and vegetable mixture. Heat to a simmer while gently tossing for 1 minute to coat the linguine and shrimp with the sauce. CCP: Temperature must reach 145 F. or higher for 15 seconds.
8. Pour 2-1/4 gal shrimp-linguine mixture into each ungreased steam table pan. Sprinkle 1-1/2 cups parmesan cheese over shrimp linguine mixture in each pan. CCP: Hold for service at 140 F. or higher.

MEAT, FISH, AND POULTRY   No. L 208 00

PASTA TOSCANO

Yield 100   Portion   9 Ounces

Calories	Carbohydrates	Protein	Fat	Cholesterol	Sodium	Calcium
447 cal	33 g	24 g	24 g	85 mg	1341 mg	188 mg

Ingredient	Weight	Measure	Issue
WATER	54-1/3 lbs	6 gal 2 qts	
SALT	1-1/2 oz	2-1/3 tbsp	
OIL,SALAD	1 oz	2 tbsp	
PASTA,PENNE	10 lbs	7 gal 1-7/8 qts	
SAUSAGE,ITALIAN,HOT	22 lbs		
COOKING SPRAY,NONSTICK	2 oz	1/4 cup 1/3 tbsp	
PEPPERS,GREEN,FRESH,CHOPPED	6 lbs	1 gal 5/8 qts	7-1/3 lbs
ONIONS,FRESH,CHOPPED	6 lbs	1 gal 1/4 qts	6-2/3 lbs
PEPPERS,RED,FRESH,SLICED	6 lbs	1 gal 3-3/8 qts	7-1/3 lbs
CHICKEN BROTH		1 gal	
ONION POWDER	7/8 oz	1/4 cup 1/3 tbsp	
PEPPER,BLACK,GROUND	1/2 oz	2 tbsp	
FENNEL,GROUND	1/4 oz	1 tbsp	
CHEESE,MOZZARELLA,PART SKIM,SHREDDED	2 lbs	2 qts	
CHEESE,PARMESAN,GRATED	14-1/8 oz	1 qts	
PARSLEY,DEHYDRATED,FLAKED	3/4 oz	1 cup	

Method

1. Add salt and salad oil to water. Heat to a rolling boil.
2. Add pasta slowly while stirring constantly until water boils again. Cook 10 to 12 minutes or according to package instructions. Drain well. Hold for use in Step 7.
3. Place sausage in single layer on 2 ungreased sheet pans. Using a convection oven, bake at 325 F. 10 to 12 minutes on high fan, closed vent. Remove from oven and let sausage stand for 3 minutes. Cut sausage diagonally into 1/2 inch slices.
4. Lightly spray griddle or tilt frying pan with non-stick cooking spray. Grill peppers, onions and sausage 6 to 8 minutes while tossing intermittently; lightly spray with cooking spray if needed.
5. Combine chicken broth, onion powder, pepper, and fennel. Stir well to blend. Heat to a simmer.
6. Combine pasta and cooked sausage/pepper-onion mixture. Pour seasoned broth over pasta/sausage mixture. Add mozzarella and parmesan cheeses and parsley. Toss lightly to evenly distribute all ingredients.
7. Place 2-1/3 gal cooked pasta/sausage mixture into each ungreased steam table pan; cover.
8. Using a convection oven, bake at 325 F. 20 to 25 minutes on high fan, open vent. CCP: Internal temperature must reach 145 F. or higher for 15 seconds. Hold for service at 140 F. or higher.

MEAT, FISH, AND POULTRY No.L 209 00
SEAFOOD STEW

Yield 100   **Portion** 1-1/2 Cups

Calories	Carbohydrates	Protein	Fat	Cholesterol	Sodium	Calcium
247 cal	30 g	27 g	2 g	101 mg	807 mg	67 mg

Ingredient	Weight	Measure	Issue
COOKING SPRAY,NONSTICK	1-1/2 oz	3 tbsp	
PEPPERS,GREEN,FRESH,CHOPPED	6-1/8 lbs	1 gal 2/3 qts	7-1/2 lbs
PEPPERS,RED,FRESH,CHOPPED	6-1/8 lbs	1 gal 2/3 qts	
ONIONS,FRESH,CHOPPED	3-3/4 lbs	2 qts 1-1/2 cup	
CHILI POWDER,DARK,GROUND	3-1/8 oz	3/4 cup	
SEASONING, OLD BAY	3-1/4 oz		
GARLIC POWDER	1-1/4 oz	1/4 cup 1/3 tbsp	
JUICE,ORANGE	8-3/4 lbs	1 gal	
STOCK,CHICKEN	16-7/8 lbs	2 gal	
TOMATOES,CANNED,DICED,INCL LIQUIDS	20-3/4 lbs	2 gal 1 qts	
RICE,LONG GRAIN & WILD	4-1/4 lbs	3 qts	
FISH,COD FILLETS,FROZEN,SKINLESS	16 lbs		
SHRIMP,FROZEN,RAW,PEELED,DEVEINED	10 lbs		

Method

1. Stir-cook peppers and onions in a lightly sprayed steam jacketed kettle or stockpot 8 to 10 minutes or until tender, stirring constantly. Add the chili powder, Old Bay seasoning and garlic powder. Stir-cook for 1 minute. Add orange juice to mixture; stir; cover.
2. Add chicken broth, tomatoes, and rice to cooked vegetable mixture. Bring to a boil. Cover; reduce heat; simmer 25 minutes or until rice is tender. CCP: Temperature must register 165 F. or higher for 15 seconds.
3. Add fish and simmer gently 4 minutes. Add shrimp and simmer gently 2 to 3 minutes. DO NOT OVERCOOK. CCP: Temperature must register 145 F. or higher for 15 seconds.
4. Pour 2-1/3 gal into each steam table pan. CCP: Hold for service at 140 F. or higher.

MEAT. FISH. AND POULTRY   No.L 210 00

SANTE FE GLAZED CHICKEN (BREAST BONELESS)

Yield 100   Portion   4 Ounces

Calories	Carbohydrates	Protein	Fat	Cholesterol	Sodium	Calcium
262 cal	21 g	33 g	5 g	88 mg	288 mg	27 mg

Ingredient	Weight	Measure	Issue
CHICKEN,BREAST,BNLS/SKNLS,5 OZ	31-1/4 lbs		
COOKING SPRAY,NONSTICK	2 oz	1/4 cup 1/3 tbsp	
ONIONS,FRESH,CHOPPED	2-1/4 lbs	1 qts 2-3/8 cup	2-1/2 lbs
WATER	6-1/4 lbs	3 qts	
JUICE,ORANGE	8-3/4 lbs	1 gal	
HOT SAUCE	3 oz	1/4 cup 2-1/3 tbsp	
SALT	1-2/3 oz	2-2/3 tbsp	
GARLIC POWDER	7/8 oz	3 tbsp	
MARJORAM,SWEET,GROUND	1/8 oz	3 tbsp	
ROSEMARY,GROUND	1/3 oz	3 tbsp	
CILANTRO,DRY	1/4 oz	1/4 cup 2/3 tbsp	
THYME,GROUND	1/4 oz	1 tbsp	
COOKING SPRAY,NONSTICK	2 oz	1/4 cup 1/3 tbsp	
HONEY	4 lbs	1 qts 1-3/8 cup	
CORNSTARCH	4-1/2 oz	1 cup	
WATER	2-1/8 lbs	1 qts	

Method

1. Wash chicken thoroughly under cold running water. Drain well.
2. Lightly spray steam jacketed kettle with non-stick cooking spray. Stir-cook onions in a steam jacketed kettle or stock pot 3 minutes stirring constantly.
3. Combine onions, water, orange juice, hot pepper sauce, salt, garlic powder, marjoram, rosemary, cilantro and thyme.
4. Pour 2-1/4 qt marinade over chicken in each roasting pan; cover. CCP: Marinate under refrigeration at 41 F. or lower for 45 minutes.
5. Remove chicken from marinade. Reserve marinade for use in Step 8. CCP: Refrigerate marinade at 41 F. or lower.
6. Place 25 chicken breasts on each lightly sprayed sheet pan. Lightly spray chicken with cooking spray.
7. Using a convection oven, bake 12 to 14 minutes on high fan, closed vent. CCP: Internal temperature must reach 165 F. or higher for 15 seconds.
8. Transfer chicken to steam table pans. Hold at 140 F. or higher for use in Step 9.
9. Bring reserved marinade and honey to a boil. Blend cornstarch and cold water together, stir to make a smooth slurry. Add slurry to hot sauce, stirring constantly. Cover; reduce heat; simmer 3 to 5 minutes or until thickened, stirring frequently to prevent sticking. CCP: Temperature must reach 165 F. or higher for 15 seconds.
10. Pour 1-1/4 quart sauce evenly over chicken in each pan. CCP: Hold for service at 140 F. or higher.

MEAT, FISH, AND POULTRY No.L 212 00

WHITE BEAN CHICKEN CHILI (COOKED DICED)

Yield 100   Portion 1-1/2 Cups

Calories	Carbohydrates	Protein	Fat	Cholesterol	Sodium	Calcium
350 cal	35 g	35 g	8 g	73 mg	972 mg	91 mg

Ingredient	Weight	Measure	Issue
BEANS, CANNELLINI,CANNED,DRAINED	33-1/8 lbs	3 gal 2-2/3 qts	
COOKING SPRAY,NONSTICK	1-1/2 oz	3 tbsp	
ONIONS,FRESH,CHOPPED	6 lbs	1 gal 1/4 qts	6-2/3 lbs
PEPPERS,GREEN,FRESH,CHOPPED	4-7/8 lbs	3 qts 2-7/8 cup	6 lbs
PEPPERS,RED,FRESH,CHOPPED	4-7/8 lbs	3 qts 2-7/8 cup	
SEASONING, SANTE FE	4-5/8 oz	1-3/8 cup	
STOCK,CHICKEN	19 lbs	2 gal 1 qts	
CHICKEN,COOKED,DICED	18 lbs		
TOMATOES,CANNED,DICED,INCL LIQUIDS	10-1/3 lbs	1 gal 1/2 qts	
OREGANO,CRUSHED	1-1/4 oz	1/2 cup	
FLOUR,WHEAT,GENERAL PURPOSE	1 lbs	3-5/8 cup	
WATER,COLD	2-1/8 lbs	1 qts	

Method

1. Rinse cannellini beans in cold water, drain well. Set aside for use in Step 3.
2. Stir-cook onions and peppers in a lightly sprayed steam jacketed kettle or stockpot for 8 to 10 minutes or until tender, stirring constantly. Add the Sante Fe Style seasoning. Stir-cook for 1 minute to release the volatile oils.
3. Add the cannellini beans, chicken broth, chicken, tomatoes and oregano to cooked onion and pepper mixture. Bring to a boil. Cover, reduce heat; simmer 15 minutes.
4. Blend flour and cold water together; stir to make a smooth slurry. Add slurry to white bean chicken chili stirring constantly. Bring to a boil. Cover; reduce heat; simmer 10 minutes or until thickened, stirring frequently to prevent sticking. CCP: Internal temperature must reach 165 F. or higher for 15 seconds.
5. Pour 3 gallons white bean chicken chili into each ungreased pan. CCP: Hold for service at 140 F. or higher.

MEAT, FISH, AND POULTRY No.L 213 00

CHICKEN BRIYANI (COOKED DICED)

Yield 100   Portion   12 Ounces

Calories	Carbohydrates	Protein	Fat	Cholesterol	Sodium	Calcium
379 cal	42 g	29 g	10 g	77 mg	882 mg	112 mg

Ingredient	Weight	Measure	Issue
BUTTER,MELTED	6 oz	1/2 cup	
OIL,SALAD	5-3/4 oz	3/4 cup	
ONIONS,FRESH,CHOPPED	7 lbs	1 gal 1 qts	7-3/4 lbs
RICE,LONG GRAIN	9 lbs	1 gal 1-1/2 qts	
CUMIN,GROUND	1-1/8 oz	1/4 cup 1-2/3 tbsp	
ALLSPICE,GROUND	3-1/2 oz	1 cup	
CHILI POWDER,DARK,GROUND	3/4 oz	3 tbsp	
GARLIC POWDER	3/8 oz	1 tbsp	
PEPPER,RED,GROUND	1/4 oz	1 tbsp	
CHICKEN,COOKED,DICED	18 lbs		
SPINACH,CHOPPED,FROZEN	4 lbs	2 qts 3-5/8 cup	
TOMATOES,CANNED,DICED,DRAINED	16-7/8 lbs	2 gal	
CHICKEN BROTH		2 gal 2 qts	

Method

1. Melt butter or margarine. Add salad oil and onions. Stir well. Saute until onions for 5 minutes or until they are tender.
2. Add rice. Cook rice 10 minutes or until lightly browned, stirring constantly. Add cumin, all spice, chili powder, garlic powder and red pepper.
3. Place 2-1/2 quart seasoned onion and rice mixture into ungreased steam table pans. Add 9 cups tomatoes, 1 gallon chicken and 1 quart spinach to each steam table pan. Stir to combine.
4. Pour 2-1/2 quart hot broth over rice, tomato, chicken and spinach mixture in each pan; stir well.
5. Cover, using a convection oven, bake at 350 F. for 55 to 60 minutes on high fan, closed vent. CCP: Internal temperature must register 145 F. or higher for 15 seconds. Hold for service at 140 F. or higher.

MEAT, FISH, AND POULTRY No.L 216 00

CHEDDAR CHICKEN AND BROCCOLI (COOKED DICED)

Yield 100   Portion 10 Ounces

Calories	Carbohydrates	Protein	Fat	Cholesterol	Sodium	Calcium
302 cal	21 g	32 g	10 g	83 mg	1035 mg	134 mg

Ingredient	Weight	Measure	Issue
ONIONS,FRESH,CHOPPED	4-1/2 lbs	3 qts 3/4 cup	5 lbs
CELERY,FRESH,CHOPPED	5-1/2 lbs	1 gal 1-1/4 qts	7-1/2 lbs
COOKING SPRAY,NONSTICK	2 oz	1/4 cup 1/3 tbsp	
CHICKEN BROTH		2 gal 2 qts	
WATER	8-1/3 lbs	1 gal	
RICE,LONG GRAIN & WILD	4-1/4 lbs	3 qts	
SALT	1-7/8 oz	3 tbsp	
GARLIC POWDER	7/8 oz	3 tbsp	
PEPPER,BLACK,GROUND	1/2 oz	2 tbsp	
CHICKEN,COOKED,DICED	18 lbs		
BROCCOLI,FROZEN,CHOPPED	12 lbs	2 gal	
CHEESE,CHEDDAR,SHREDDED	2 lbs	2 qts	

Method

1. Stir-cook onions and celery in a lightly sprayed steam jacketed kettle or stock pot for 8 to 10 minutes or until tender stirring constantly.
2. Add broth, water, rice, salt, garlic powder and pepper to cooked onions and celery; bring to a boil. Cover tightly; reduce heat; simmer 20 minutes. There will be excess cooking liquid in cooked rice mixture.
3. Stir chicken and broccoli into cooked rice mixture. Cover; simmer an additional 15 minutes. CCP: Internal temperature must reach 145 F. or higher for 15 seconds.
4. Pour 2 gallon rice, chicken and broccoli mixture into each ungreased steam table pan. Distribute 2 cups shredded cheddar cheese evenly over rice, chicken and broccoli mixture into each pan.
5. Bake for 3 minutes to melt cheese. CCP: Hold for service at 140 F. or higher.

MEAT, FISH, AND POULTRY   No.L 217 00

ASIAN BARBECUE TURKEY

Yield 100   **Portion**   4 Ounces

Calories	Carbohydrates	Protein	Fat	Cholesterol	Sodium	Calcium
184 cal	6 g	34 g	2 g	89 mg	823 mg	22 mg

Ingredient	Weight	Measure	Issue
TURKEY,BREAST,FILLET	31-1/4 lbs		
SAUCE,BARBECUE	5-1/2 lbs	2 qts 2 cup	
SOY SAUCE	2-1/4 lbs	3-1/2 cup	
JUICE,ORANGE	2-1/4 lbs	1 qts	
GARLIC POWDER	3-1/2 oz	3/4 cup	
GINGER,GROUND	1 oz	1/4 cup 1-2/3 tbsp	
COOKING SPRAY,NONSTICK	2 oz	1/4 cup 1/3 tbsp	
ONIONS,GREEN,FRESH,CHOPPED	7 oz	2 cup	7-7/8 oz

Method

1. Wash turkey thoroughly under cold running water. Drain well. Place approximately 16 pounds of turkey breast fillets in each roasting pan; cover.
2. Combine barbecue sauce, soy sauce, orange juice, garlic powder and ginger; mix well.
3. Pour 2 qt seasoned barbecue sauce over turkey in each pan; cover. CCP: Marinate under refrigeration at 41 F. or lower for 45 minutes.
4. Place 25 turkey breast fillets on each lightly sprayed sheet pan.
5. Using a convection oven, bake 12 to 14 minutes on high fan, open vent. CCP: Internal temperature must reach 165 F. or higher for 15 seconds. Transfer turkey to steam table pans.
6. Garnish each pan with 1/2 cup chopped green onions.

MEAT, FISH, AND POULTRY No.L 219 00

LEMON N' HERB TURKEY FILLETS

Yield 100   Portion 4 Ounces

Calories	Carbohydrates	Protein	Fat	Cholesterol	Sodium	Calcium
280 cal	12 g	36 g	9 g	134 mg	255 mg	58 mg

Ingredient | Weight | Measure | Issue

Ingredient	Weight	Measure	Issue
TURKEY,BREAST,FILLET	31-1/4 lbs		
BREADCRUMBS	4-1/2 lbs	1 gal 3/4 qts	
MARGARINE	1-3/4 lbs	3-1/2 cup	
SEASONING,LEMON N' HERB	3-5/8 oz	3/4 cup	
PEPPER,BLACK,GROUND	1-3/4 oz	1/2 cup	
BASIL,DRIED,CRUSHED	5/8 oz	1/4 cup 1/3 tbsp	
OREGANO,CRUSHED	5/8 oz	1/4 cup 1/3 tbsp	
EGGS,WHOLE,FROZEN	2-1/4 lbs	1 qts 1/4 cup	
COOKING SPRAY,NONSTICK	1 oz	2 tbsp	

Method

1. Wash turkey fillets thoroughly under cold running water. Drain well.
2. Combine breadcrumbs, lemon n' herb seasoning, pepper, basil, oregano and margarine. Mix well.
3. Dip turkey fillets in eggs, then in crumb mixture. Shake off excess.
4. Lightly spray each sheet pan with non-stick cooking spray. Place 17 fillets on each sheet pan.
5. Using a convection oven, bake 18 to 20 minutes at 325 F. on high fan, open vent. CCP: Internal temperature must reach 165 F. or higher for 15 seconds.
6. Transfer and shingle turkey fillets in steam table pans. CCP: Hold for service at 140 F. or higher.

MEAT, FISH, AND POULTRY   No.L 221 00

TURKEY DIVAN

Yield 100   **Portion**   3 Ounces

Calories	Carbohydrates	Protein	Fat	Cholesterol	Sodium	Calcium
241 cal	10 g	38 g	5 g	61 mg	2357 mg	140 mg

Ingredient	Weight	Measure	Issue
TURKEY BREAST,BNLS,PRECKD	30 lbs		
WATER	6-1/4 lbs	3 qts	
BROCCOLI,FROZEN,SPEARS	24 lbs	4 gal 1-1/2 qts	
MILK,NONFAT,DRY	7-1/4 oz	3 cup	
SOUP,CONDENSED,CREAM OF CHICKEN	8-7/8 lbs	1 gal	
PEPPER,BLACK,GROUND	2/3 oz	3 tbsp	
CHEESE,CHEDDAR,LOWFAT,SHREDDED	2 lbs	2 qts	
PAPRIKA,GROUND	1/3 oz	1 tbsp	

Method

1. Place turkey in roasting pans.
2. Using a convection oven, bake at 300 F. with fan on, for 25 minutes. Baste occasionally with drippings. CCP: Internal temperature must reach 165 F. or higher for 15 seconds.
3. Let roasts stand 15 to 20 minutes to absorb juices and for ease in slicing. Slice turkey breast into 3 ounce slices.
4. Cook broccoli spears 5 to 8 minutes in steamer at 5 lb P.S.I. until tender -crisp. DO NOT OVERCOOK! Remove from steamer. Arrange into 4 oz portions and cover.
5. Reconstitute the milk with warm water.
6. Add milk to condensed soup in a steam jacketed kettle or stock pot. Stir to blend. Bring to a boil; reduce heat; simmer 2 minutes.
7. Ladle 2 cups sauce into bottom of each steam table pan. Spread evenly.
8. Arrange 12 - 4 ounce portions of broccoli spears evenly over sauce in each pan. Sprinkle 1 tsp pepper evenly over broccoli spears in each pan.
9. Fold 3 ounces turkey slices over each portion of broccoli spears. Serving will be easier if edges of turkey are folded under broccoli portions.
10. Pour 4-1/2 cups soup mixture evenly over broccoli/turkey in each pan.
11. Top each portion with 1/3 tablespoon cheese. Sprinkle 1/2 teaspoon paprika evenly over cheese in each pan. Cover.
12. Using a convection oven, bake 20-25 minutes on high fan, closed vent or until sauce is bubbly. CCP: Internal temperature must reach 165 F. or higher for 15 seconds. Hold for service at 140 F. or higher.

MEAT, FISH, AND POULTRY No.L 222 00

SPICY ITALIAN PORK CHOPS

Yield 100   Portion   4 Ounces

Calories	Carbohydrates	Protein	Fat	Cholesterol	Sodium	Calcium
471 cal	9 g	30 g	35 g	81 mg	605 mg	20 mg

Ingredient   Weight   Measure   Issue

PORK CHOP,BONELESS,5 OZ   31-1/4 lbs
SAUCE,BARBECUE   7-3/4 lbs   3 qts 2 cup
SALAD DRESSING,ITALIAN   7-1/4 lbs   3 qts 2 cup
CHILI POWDER,DARK,GROUND   5-5/8 oz   1-3/8 cup
COOKING SPRAY,NONSTICK   2 oz   1/4 cup 1/3 tbsp

Method

1. Place pork chops in 2 roasting pans.
2. Combine barbecue sauce, Italian dressing and chili powder. Mix well.
3. Pour 3-1/2 quarts of barbecue sauce mixture over pork chops in each roasting pan; cover. CCP: Marinate under refrigeration at 41 F. lower for 45 minutes.
4. Drain pork chops. Bring reserved marinade to a boil for one minute.
5. Lightly spray griddle with non-stick cooking spray. Grill pork chops on griddle for 4 minutes on each side.
6. Transfer pork chops to steam table pans. Pour 6-1/2 cups hot barbecue sauce over pork chops in each pan; cover.
7. Using a convection oven, bake 20 to 25 minutes at 325 F. on high fan, open vent. CCP: Internal temperature must reach 145 F. or higher for 15 seconds. Hold for service at 140 F. or higher.

MEAT, FISH, AND POULTRY No.L 223 00

LIME CHICKEN SOFT TACOS (FAJITA STRIPS)

Yield 100   Portion 7 Ounces

Calories	Carbohydrates	Protein	Fat	Cholesterol	Sodium	Calcium
318 cal	36 g	31 g	6 g	66 mg	854 mg	154 mg

Ingredient	Weight	Measure	Issue
JUICE,LIME	1-1/4 lbs	2-1/2 cup	
SUGAR,GRANULATED	3-1/2 oz	1/2 cup	
SALT	2-1/3 oz	1/4 cup	
GARLIC POWDER	2 oz	1/4 cup 3 tbsp	
ONION POWDER	7/8 oz	1/4 cup	
PEPPER,BLACK,GROUND	5/8 oz	2-2/3 tbsp	
OREGANO,CRUSHED	1-1/4 oz	1/2 cup	
CHICKEN,FAJITA STRIPS	23 lbs		
TOMATOES,FRESH	12-1/2 lbs		12-3/4 lbs
ONIONS,GREEN,FRESH,CHOPPED	1-3/8 lbs	1 qts 2-1/4 cup	1-1/2 lbs
TORTILLAS,WHEAT,10 INCH	12-3/8 lbs	100 each	
COOKING SPRAY,NONSTICK	1-1/2 oz	3 tbsp	
CHEESE,MONTEREY JACK,REDUCED FAT,SHREDDED	1-1/2 lbs	1 qts 2-1/4 cup	
SALSA		3 qts 1 cup	

Method

1. Combine lime juice, sugar, salt garlic powder, onion powder, pepper and oregano. Stir well to blend.
2. Pour mixture over chicken strips. Mix thoroughly to evenly distribute seasonings around all surfaces of chicken. Cover. Marinate under refrigeration at 41 F. or lower for 45 minutes for use in Step 5.
3. Dice tomatoes. Combine tomatoes and green onions.
4. Wrap tortillas in foil; place in warm oven (150 F.) or in a warmer for 15 minutes or until pliable.
5. Lightly spray griddle with cooking spray. Grill chicken strips 5-7 minutes or until lightly browned while tossing intermittently; lightly spray chicken with cooking spray as needed. CCP: Internal temperature must register 165 F. or higher for 15 seconds.
6. Place 6 to 7 cooked fajita strips (2 oz), 1 tablespoon Monterey Jack cheese and 4 tablespoons tomato/green onion mixture into each tortilla. If desired, top each tortilla with salsa.
7. Roll tortilla; wrap in foil. CCP: Serve immediately or hold for service at 140 F. or higher.

MEAT, FISH, AND POULTRY No.L 224 00

SAUSAGE, BEANS AND GREENS

Yield 100   Portion   1-1/2 Cups

Calories	Carbohydrates	Protein	Fat	Cholesterol	Sodium	Calcium
299 cal	27 g	30 g	8 g	72 mg	1341 mg	102 mg

Ingredient	Weight	Measure	Issue
BEANS, CANNELLINI,CANNED,DRAINED	15 lbs	1 gal 2-5/8 qts	
SAUSAGE LINK,TURKEY,RAW	5-1/2 lbs		
ONIONS,FRESH,CHOPPED	3 lbs	2 qts 1/2 cup	3-1/3 lbs
THYME,GROUND	3/4 oz	1/4 cup 1-1/3 tbsp	
SAGE,GROUND	1/2 oz	1/4 cup 3-1/3 tbsp	
PEPPER,RED,GROUND	1/8 oz	1/3 tsp	
GARLIC POWDER	1/3 oz	1 tbsp	
ONION POWDER	1/8 oz	1/8 tsp	
CHICKEN BROTH		3 gal	
WATER	18-3/4 lbs	2 gal 1 qts	
POTATOES,FROZEN,DICED	13 lbs		
CHICKEN,COOKED,PULLED,WHITE/DARK	14-1/2 lbs		
KALE,FRESH,CHOPPED	6 lbs	2 gal 2-1/8 qts	8-1/2 lbs
CARROTS,FROZEN,SLICED	3 lbs	2 qts 2-5/8 cup	
VINEGAR,CIDER	4-1/4 oz	1/2 cup	
SUGAR,GRANULATED		1/2 cup	
SALT	1-1/4 oz	2 tbsp	
PEPPER,BLACK,GROUND	1/3 oz	1 tbsp	

Method

1. Rinse cannellini beans in cold water; drain well.
2. Cut turkey sausage links into 1/2 inch slices. Stir-cook in a steam jacketed kettle or stockpot until itt losses its pink color.
3. Add onions to turkey sausage and continue to cook 4-5 minutes or until transparent; stirring constantly. Add the thyme, sage, red pepper, garlic powder and onion powder. Stir-cook for 1 minute.
4. Add the chicken broth, water, cannellini beans, potatoes, chicken, kale, carrots, vinegar, sugar, salt and black pepper to cooked sausage and onion mixture. Bring to a boil. Cover; reduce heat; simmer 20-25 minutes until potatoes are tender. CCP: Temperature must reach 165 F. or higher for 15 seconds.
5. Pour 2-1/3 gal into each pan. CCP: Hold for service at 140 F. or higher.

MEAT. FISH. AND POULTRY No.L 225 00

ORANGE & ROSEMARY HONEY GLAZED PORK CHOPS

Yield 100   Portion 1 Chop

Calories	Carbohydrates	Protein	Fat	Cholesterol	Sodium	Calcium
458 cal	28 g	40 g	20 g	99 mg	171 mg	18 mg

Ingredient	Weight	Measure	Issue
PORK,LOIN CHOPS,5 OZ	31-1/4 lbs		
OIL,OLIVE	11-3/8 oz	1-1/2 cup	
JUICE,ORANGE	6-5/8 lbs	3 qts	
JUICE,LEMON	2-1/8 lbs	1 qts	
GARLIC POWDER	1-1/4 oz	1/4 cup 2/3 tbsp	
ROSEMARY,GROUND	1-1/3 oz	1/2 cup 3-1/3 tbsp	
SALT	7/8 oz	1 tbsp	
PEPPER,BLACK,GROUND	1/3 oz	1 tbsp	
COOKING SPRAY,NONSTICK	2 oz	1/4 cup 1/3 tbsp	
HONEY	6 lbs	2 qts	
CORNSTARCH	6-3/4 oz	1-1/2 cup	
WATER,COLD	3-1/8 lbs	1 qts 2 cup	

Method

1. Place 15 lb 10 oz pork chops in each roasting pan; cover.
2. Combine olive oil, orange juice, lemon juice, garlic powder, rosemary, salt and pepper; mix well.
3. Pour 2 quarts marinade over pork chops in each roasting pan; cover. CCP: Marinate under refrigeration at 41 F. or lower for 45 minutes.
4. Drain pork chops. Reserve marinade. CCP: Refrigerate remaining marinade at 41 F. or lower for use in Step 7.
5. Spray griddle with cooking spray. Grill pork chops 2 minutes in each side.
6. Transfer pork chops to 4-12x20x2-1/2 inch steam table pans.
7. Bring reserved marinade to a boil. Add honey and stir to blend. Blend cornstarch and cold water together; stir to make a smooth slurry. Add slurry to hot sauce, stirring constantly. Cover; reduce heat; simmer 3 to 5 minutes or until thickened, stirring frequently to prevent sticking.
8. Pour 7-1/2 cups thickened sauce over pork chops in each pan; cover.
9. Using a convection oven, bake 20 to 25 minutes on high fan, open vent. CCP: Internal temperature must register 155 F. or higher for 15 seconds. Hold for service at 140 F. or higher.

MEAT, FISH, AND POULTRY   No.L 500 00

RUSSIAN TURKEY STEW

Yield 100   Portion   1 Cup

Calories	Carbohydrates	Protein	Fat	Cholesterol	Sodium	Calcium
327 cal	39 g	24 g	8 g	79 mg	711 mg	185 mg

Ingredient

Ingredient	Weight	Measure	Issue
TURKEY,BNLS,WHITE AND DARK MEAT	18 lbs		
COOKING SPRAY,NONSTICK	2 oz	1/4 cup 1/3 tbsp	
ONIONS,FRESH,CHOPPED	5-1/4 lbs	3 qts 3 cup	5-7/8 lbs
PEPPERS,GREEN,FRESH,CHOPPED	2 lbs	1 qts 2 cup	2-3/8 lbs
GARLIC POWDER	1-1/2 oz	1/4 cup 1-1/3 tbsp	
MUSHROOMS,FRESH,WHOLE,SLICED	14-7/8 oz	1 qts 2 cup	1 lbs
FLOUR,WHEAT,GENERAL PURPOSE	4-3/8 oz	1 cup	
SALT	1-1/4 oz	2 tbsp	
TOMATOES,CANNED,DICED,DRAINED	4-3/8 lbs	2 qts	
PAPRIKA,GROUND	2-7/8 oz	3/4 cup	
SEASONING,POULTRY	1 oz	1/2 cup	
PEPPER,BLACK,GROUND	1/2 oz	2 tbsp	
MILK,EVAPORATED,SKIM,CANNED	8-7/8 lbs	1 gal	
NOODLES,EGG,DRY	8 lbs	5 gal 3-7/8 qts	
WATER	66-7/8 lbs	8 gal	
PARSLEY,DEHYDRATED,FLAKED	1-1/2 oz	2 cup	
ONIONS,GREEN,FRESH,SLICED	3-1/2 oz	1 cup	3-7/8 oz

Method

1. Cut thawed turkey into 1 inch cubes. Saute turkey in a lightly greased steam-jacketed kettle. Cook until slightly browned, remove from pan. CCP: Internal temperature must reach 165 F. or higher for 15 seconds. CCP: Hold at 140 F. or higher for use in Step 4.
2. Add onions, bell peppers, and garlic to kettle. Saute until onions are translucent. Add mushrooms and saute 5 more minutes. Sprinkle flour over vegetables, stir, saute for one minute.
3. Add tomatoes, paprika, poultry seasoning, salt, and pepper. Saute 5 minutes.
4. Return turkey back into the pan, add milk and stir well. CCP: Hold for service at 140 F. or higher.
5. Bring a kettle of water to boil. Add noodles and cook until soft. Drain pasta and toss in parsley and green onions. Serve turkey sauce over pasta.

MEAT. FISH. AND POULTRY No.L 501 00

PASTA PRIMAVERA

Yield 100   **Portion** 1 Cup

Calories	Carbohydrates	Protein	Fat	Cholesterol	Sodium	Calcium
288 cal	54 g	12 g	4 g	2 mg	691 mg	169 mg

Ingredient	Weight	Measure	Issue
SALT	1-1/2 oz	2-1/3 tbsp	
OIL,SALAD	1/3 oz	1/3 tsp	
WATER	54-1/3 lbs	6 gal 2 qts	
MACARONI NOODLES,ROTINI,DRY	10 lbs	2 gal 2-7/8 qts	
OIL,SALAD	5-3/4 oz	3/4 cup	
FLOUR,WHEAT,GENERAL PURPOSE	14-2/3 oz	3-3/8 cup	
TOMATOES,CANNED,DICED,DRAINED	19-7/8 lbs	2 gal 1 qts	
MILK,EVAPORATED,SKIM,CANNED	5 lbs	2 qts 1 cup	
CHEESE,PARMESAN,GRATED	7 oz	2 cup	
SALT	1-2/3 oz	2-2/3 tbsp	
GARLIC POWDER	5/8 oz	2 tbsp	
PEPPER,BLACK,GROUND	1/2 oz	2 tbsp	
OREGANO,CRUSHED	1/8 oz	1 tbsp	
BASIL,SWEET,WHOLE,CRUSHED	5/8 oz	1/4 cup 1/3 tbsp	
ONIONS,FRESH,CHOPPED	3-1/2 lbs	2 qts 1-7/8 cup	3-7/8 lbs
MUSHROOMS,CANNED,DRAINED	6-7/8 lbs	1 gal 1 qts	
CARROTS,FRESH	2-3/4 lbs		3-1/3 lbs
SQUASH,ZUCCHINI,FRESH,SLICED	2-1/2 lbs	2 qts 2 cup	2-5/8 lbs
SQUASH,FRESH,SUMMER,SLICED	2-1/2 lbs	2 qts 2 cup	2-5/8 lbs
PEPPERS,GREEN,FRESH,CHOPPED	2 lbs	1 qts 2-1/8 cup	2-1/2 lbs
PEPPERS,RED,FRESH,CHOPPED	2 lbs	1 qts 2-1/8 cup	
CELERY,FRESH,SLICED	2 lbs	1 qts 3-1/2 cup	2-3/4 lbs
BROCCOLI,FRESH,FLORETS	2 lbs	2 qts 2-3/8 cup	3-1/4 lbs
PARSLEY,DEHYDRATED,FLAKED	3/4 oz	1 cup	

Method

1. Add salt and salad oil to water; heat to rolling boil.
2. Add rotini to a steam-jacketed kettle and cook for 10 to 12 minutes, or until tender. Drain. Rinse with cold water; drain thoroughly. Use immediately in recipe preparation or place in shallow containers and cover.
3. Blend salad oil and flour together to form a roux; using a wire whip, stir until smooth. Cook roux for 3 minutes stirring constantly.
4. Drain tomatoes. Reserve 1 gal tomato liquid. Set aside for use in Step 7.
5. Reconstitute milk in water. Gradually add milk and tomato liquid roux while stirring constantly. Bring to a boil. Cover; reduce heat; simmer 5 minutes or until thickened, stirring frequently to prevent sticking.
6. Add parmesan cheese, salt, garlic powder, pepper, oregano and basil to thickened sauce. Stir to blend well.
7. Add tomatoes, onions, mushrooms, carrots, zucchini, yellow squash, green peppers, red peppers, celery, broccoli and parsley to thickened sauce. Stir, bring to a boil. Cover; reduce heat; simmer 7 to 10 minutes until tender.
8. Add rotini to thickened sauce and vegetable mixture. Heat to a simmer while stirring for 1 minute to coat the rotini with the vegetable sauce. CCP: Temperature must reach 165 F. or higher for 15 seconds.
9. Pour 3 gallons vegetable rontini mixture into each ungreased pan. CCP: Hold for service at 140 F. or higher.

MEAT, FISH, AND POULTRY No.L 502 00

FISH FLORENTINE

Yield 100   **Portion** 4 Ounces

Calories	Carbohydrates	Protein	Fat	Cholesterol	Sodium	Calcium
194 cal	5 g	21 g	10 g	68 mg	448 mg	50 mg

Ingredient

Ingredient	Weight	Measure	Issue
ONIONS,FRESH,SLICED	5 lbs	1 gal 1/2 qts	
COOKING SPRAY,NONSTICK	2 oz	1/4 cup 1/3 tbsp	
SUGAR,GRANULATED	7/8 oz	2 tbsp	
PEPPER,BLACK,GROUND	1/8 oz	1/3 tsp	
SOY SAUCE	10-1/8 oz	1 cup	
SPINACH,FROZEN	5 lbs	3 qts	
EGG WHITES	5-2/3 oz	1/2 cup 2-2/3 tbsp	
JUICE,LEMON	1-5/8 oz	3 tbsp	
NUTMEG,GROUND	1/4 oz	1 tbsp	
FISH,CATFISH,FILLET	30 lbs		
OIL,SALAD	1 oz	2 tbsp	
PEPPER,RED,GROUND	1/8 oz	1/4 tsp	
PAPRIKA,GROUND	1/2 oz	2 tbsp	
SALT	3/8 oz	1/3 tsp	
CHICKEN BROTH		2 qts	
CORNSTARCH	4-1/2 oz	1 cup	
WATER	1 lbs	2 cup	

Method

1. Stir-cook onions with sugar and pepper in a lightly sprayed steam-jacketed kettle or stockpot 8 to 10 minutes, or until tender. Add soy sauce; stir and remove from heat.
2. Drain spinach. Press out excess liquid. Blend spinach, beaten egg whites, lemon juice, nutmeg.
3. Combine spinach with onion mixture. Divide evenly among 5 steam table pans. Roll fish filets and place on top of spinach mixture, placing 20 rolls per pan. Set aside for use in Step 6. CCP: Refrigerate at 40 F. or lower.
4. Heat oil with paprika and red pepper in steam-jacketed kettle or stockpot. Add chicken broth and salt; stir to blend well. Bring to a boil. Reduce heat to a simmer.
5. Blend cornstarch and cold water; stir to make smooth slurry. Add slurry to hot liquid, stirring constantly. Reduce heat; simmer 3 to 5 minutes or until thickened, stirring frequently to prevent sticking. CCP: Temperature must register 165 F. or higher for 15 seconds.
6. Pour 1-3/4 cups red pepper sauce/glaze evenly over fish to coat top and sides. Cover. Using a convection oven, bake at 325 F. 25-30 minutes on high fan, closed vent. CCP: Internal temperature must reach 145 F. or higher for 15 seconds. DO NOT OVERCOOK.
7. Remove from oven. Serve one fish roll with spinach/onion mixture. CCP: Hold at 140 F. or higher for service.

Notes

1. Any white flesh fish fillet can be substituted for catfish.

MEAT, FISH, AND POULTRY No. L 503 00

JAMAICAN RUM CHICKEN (BREAST BONELESS)

Yield 100   Portion 5 Ounces

Calories	Carbohydrates	Protein	Fat	Cholesterol	Sodium	Calcium
212 cal	9 g	33 g	4 g	88 mg	344 mg	36 mg

Ingredient	Weight	Measure	Issue
CHICKEN,BREAST,BNLS/SKNLS,5 OZ	31-1/4 lbs		
VINEGAR,RED WINE	2-1/8 lbs	1 qts	
SALT	1-1/4 oz	2 tbsp	
PEPPER,BLACK,GROUND	2/3 oz	3 tbsp	
COOKING SPRAY,NONSTICK	1-1/2 oz	3 tbsp	
TOMATOES,CANNED,DICED,DRAINED	4-1/8 lbs	1 qts 3-1/2 cup	
CHICKEN BROTH		1 qts 2 cup	
SUGAR,BROWN,LIGHT	1-1/4 lbs	3-3/4 cup	
ONIONS,FRESH,CHOPPED	4-1/4 oz	3/4 cup	4-2/3 oz
FLAVORING,RUM	2-1/2 oz	1/4 cup 1-2/3 tbsp	
PARSLEY,DEHYDRATED,FLAKED	1-1/8 oz	1-1/2 cup	
GARLIC POWDER	7/8 oz	3 tbsp	
NUTMEG,GROUND	1/3 oz	1 tbsp	
CUMIN,GROUND	1/4 oz	1 tbsp	
SAGE,GROUND	1/8 oz	1 tbsp	
CORNSTARCH	3-3/8 oz	3/4 cup	
WATER,COLD	12-1/2 oz	1-1/2 cup	

Method

1. Wash chicken thoroughly under cold running water. Drain well. Remove excess fat. Place chicken in roasting pans.
2. Combine vinegar, salt, and pepper; stir to blend.
3. Ladle marinade over chicken in each roasting pan; cover. CCP: Marinate under refrigeration at 41 F. or lower for 45 minutes.
4. Place chicken breasts on each lightly sprayed sheet pan. Lightly spray chicken with cooking spray. Discard remaining marinade.
5. Using a convection oven, bake 12 to 14 minutes at 325 F. on high fan, closed vent. CCP: Internal temperature must reach 165 F. or higher for 15 seconds.
6. Transfer chicken to steam table pans. Hold at 140 F. or higher for use in Step 9.
7. Combine tomatoes, broth, sugar, onions, rum flavoring, parsley, garlic, nutmeg, cumin and sage. Bring to a boil. Cover; reduce heat; simmer 15 minutes.
8. Blend cornstarch and cold water together to make a smooth slurry. Add slurry to hot sauce, stirring constantly. Cover; reduce heat; simmer 3 to 5 minutes or until thickened, stirring frequently to prevent sticking.
9. Pour 1 quart sauce evenly over chicken in each pan. CCP: Hold for service at 140 F. or higher.

MEAT, FISH, AND POULTRY No.L 504 00

BAKED FISH SCANDIA

Yield 100　　　　　　　　　　　　　　　　　**Portion** 4 Ounces

Calories	Carbohydrates	Protein	Fat	Cholesterol	Sodium	Calcium
179 cal	5 g	28 g	4 g	79 mg	192 mg	102 mg

Ingredient	Weight	Measure	Issue
BREADCRUMBS	1-1/4 lbs	1 qts 1-3/8 cup	
JUICE,LEMON	11-1/2 oz	1-3/8 cup	
ONION POWDER	1/2 oz	2 tbsp	
PEPPER,WHITE,GROUND	1/8 oz	1/8 tsp	
HOT SAUCE	1/8 oz	1/8 tsp	
PARSLEY,DEHYDRATED,FLAKED	1/2 oz	1/2 cup 2-2/3 tbsp	
COOKING SPRAY,NONSTICK	2 oz	1/4 cup 1/3 tbsp	
FISH,FLOUNDER/SOLE FILLET,RAW	30 lbs		
YOGURT,PLAIN,LOWFAT	4-1/3 lbs	2 qts	
CHEESE,CHEDDAR	1-1/8 lbs	1 qts	

Method

1. In a bowl, mix bread crumbs, lemon juice, onion powder, white pepper, hot pepper sauce and parsley.
2. Spray each steam table pan with non-stick cooking spray.
3. Place 25 fish portions into each steam table pan.
4. Cover each portion with 1 tablespoon of low fat yogurt.
5. Sprinkle 3 ounces cheese per pan, on top of yogurt.
6. Sprinkle 1 tablespoon of crumb mixture onto each portion.
7. Using a convection oven, bake 25 minutes at 350 F. or until fish flakes easily with a fork. CCP: Internal temperature must reach 145 F. or higher for 15 seconds. Hold for service at 140 F. or higher.

MEAT, FISH, AND POULTRY No.L 506 00

THAI BEEF SALAD

Yield 100   Portion   1-1/2 Cups

Calories	Carbohydrates	Protein	Fat	Cholesterol	Sodium	Calcium
297 cal	25 g	29 g	9 g	74 mg	631 mg	51 mg

Ingredient	Weight	Measure	Issue
BEEF,FAJITA STRIPS	25 lbs		
GARLIC POWDER	1-1/4 oz	1/4 cup 1/3 tbsp	
SOY SAUCE	10-1/8 oz	1 cup	
GINGER,GROUND	3/4 oz	1/4 cup 1/3 tbsp	
PEPPER,BLACK,GROUND	1/4 oz	1 tbsp	
CABBAGE,GREEN,FRESH,SHREDDED	8-5/8 lbs	3 gal 2 qts	10-3/4 lbs
LETTUCE,ICEBERG,FRESH	3-7/8 lbs		4-1/8 lbs
PEPPERS,GREEN,FRESH,MEDIUM,SLICED,THIN	4 lbs	3 qts 1/8 cup	4-7/8 lbs
CARROTS,FRESH,SHREDDED	2 lbs	2 qts 1/4 cup	2-1/2 lbs
ONIONS,FRESH,SLICED	1-1/2 lbs	1 qts 1-7/8 cup	1-2/3 lbs
EGG ROLL WRAPPERS	7 lbs	100 each	
BEEF BROTH		1 qts	
SOY SAUCE	10-1/8 oz	1 cup	
GINGER,GROUND	3/4 oz	1/4 cup 1/3 tbsp	
PEPPER,RED,CRUSHED	1/8 oz	1 tbsp	
OIL,SALAD	1-7/8 oz	1/4 cup 1/3 tbsp	

Method

1. Combine beef strips with garlic, soy sauce, ginger, black pepper and red pepper. CCP: Marinate under refrigeration at or below 41 F. for at least 30 minutes.
2. Combine cabbage, lettuce, bell pepper, carrots, and onion. Lay in bottom of serving pans.
3. Slice egg roll wrappers into thin strips and bake in 325 F. convection oven until crisp and golden.
4. Combine beef broth, soy sauce, ginger, vegetable oil.
5. Heat grill until hot, sear beef until brown. CCP: Internal temperature must reach 145 F. or higher for 15 seconds.
6. CCP: Hold beef at 140 F. or higher for service. Lay warm beef strips over salad, and pour sauce over. Place toasted egg roll wrappers around the edges.

MEAT, FISH, AND POULTRY No. L 507 00

VEGETARIAN BURRITO

Yield 100　　　　　　　　　　　　　　　　　Portion 2 Burritos

Calories	Carbohydrates	Protein	Fat	Cholesterol	Sodium	Calcium
521 cal	61 g	25 g	20 g	40 mg	1060 mg	416 mg

Ingredient	Weight	Measure	Issue
COOKING SPRAY,NONSTICK	2 oz	1/4 cup 1/3 tbsp	
ONIONS,FRESH,CHOPPED	1-3/8 lbs	1 qts	1-5/8 lbs
GARLIC POWDER	5/8 oz	2 tbsp	
CHILI POWDER,DARK,GROUND	3/4 oz	3 tbsp	
CUMIN,GROUND	7/8 oz	1/4 cup 1/3 tbsp	
BEANS,PINTO,CANNED,DRAINED	8-1/2 lbs	1 gal	
BEANS,KIDNEY,DARK RED,CANNED,DRAINED	6-1/4 lbs	1 gal	
BEANS,BLACK,CANNED,DRAINED	9 lbs	1 gal	
TORTILLAS,WHEAT,6 INCH	14-1/8 lbs	200 each	
LETTUCE,ICEBERG,FRESH,SHREDDED	4 lbs	2 gal 1/4 qts	4-1/3 lbs
TOMATOES,FRESH,CHOPPED	10 lbs	1 gal 2-1/4 qts	10-1/4 lbs
CHEESE,MONTEREY JACK	10 lbs	2 gal 5/8 qts	
SAUCE,SALSA	10-3/4 lbs	1 gal 1 qts	

Method

1. Spray a steam-jacketed kettle with vegetable spray. Saute onion until translucent. Add garlic, chili powder, and cumin. Cook until aroma is released.
2. Add drained beans and simmer for 20 minutes.
3. To assemble burritos, place a scoop of bean mixture down the center of the tortilla, add 2 tablespoons lettuce, 1 tablespoons tomato and 3 tablespoons grated Monterey jack cheese. To roll, turn the lower lip of the burrito up, and roll laterally.
4. Wrap each burrito in foil deli papers and place in a pan for service. CCP: Hold for service at 41 F. or lower.
5. Serve with 3 tablespoons salsa on the side.

MEAT, FISH, AND POULTRY No. L 508 00

VEGETABLE LASAGNA

Yield 100   Portion   8 Ounces

Calories	Carbohydrates	Protein	Fat	Cholesterol	Sodium	Calcium
292 cal	38 g	22 g	7 g	16 mg	1036 mg	242 mg

Ingredient	Weight	Measure	Issue
NOODLES,LASAGNA,UNCOOKED	5-3/4 lbs	1 gal 2-1/4 qts	
OIL, CANOLA	3-7/8 oz	1/2 cup	
SQUASH,FRESH,SUMMER,SLICED	1-1/4 lbs	1 qts 1 cup	1-1/3 lbs
MUSHROOMS,CANNED,SLICED,WHITE	11 oz	2 cup	
ONIONS,FRESH,CHOPPED	1 lbs	3 cup	1-1/8 lbs
FLOUR,WHEAT,GENERAL PURPOSE	4-3/8 oz	1 cup	
BROCCOLI,FROZEN,CUT	6-7/8 lbs	1 gal 1 qts	
SAUCE,TOMATO,CANNED	16-1/8 lbs	1 gal 3-1/2 qts	
TOMATO PASTE,CANNED	4 lbs	1 qts 3 cup	
OREGANO,CRUSHED	1-7/8 oz	3/4 cup	
GARLIC POWDER	1/2 oz	1 tbsp	
CHEESE,COTTAGE,LOWFAT	16 lbs	2 gal	
PARSLEY,DEHYDRATED,FLAKED	3/8 oz	1/2 cup	
GARLIC POWDER	1/3 oz	1 tbsp	
BREADCRUMBS	1 lbs	1 qts	
CHEESE,PARMESAN,GRATED	3-1/2 oz	1 cup	
CHEESE,MOZZARELLA,PART SKIM,SHREDDED	3-3/4 lbs	3 qts 3 cup	

Method

1. Cook lasagna noodles in a steam-jacketed kettle for 10 to 12 minutes, until tender. Drain. Hold in cold water.
2. In a small kettle, heat vegetable oil. Add the zucchini, drained mushrooms, and onions. Saute for 3 minutes or until zucchini is slightly tender. Stir in flour, cook 3 minutes. Remove from heat and set aside.
3. Place broccoli in a steam table pan and steam for 6 minutes, or until tender. Drain well and set aside.
4. In a steam kettle, heat the tomato sauce and tomato paste. Add oregano and garlic powder. Simmer, uncovered for 30 minutes.
5. Add the sauteed vegetables and steamed broccoli to the tomato sauce. Stir to combine. Simmer for 10 minutes.
6. In a large bowl, combine the cottage cheese, parsley, garlic powder, and bread crumbs. Mix well.
7. Combine parmesan cheese and mozzarella cheese.
8. Spread 1 cup vegetable sauce on the bottom of each steam table pan to prevent sticking.
9. Assembly: First layer: 7-1/2 lasagna noodles; 1 quart of cottage cheese mixture; 1 quart and 1 cup of vegetable sauce; 2-1/4 cups parmesan-mozzarella cheese mixture; Second layer: repeat first layer; Third layer: 7-1/2 lasagna noodles; 2-1/2 cups vegetable sauce.
10. Sprinkle 1/2 cup parmesan cheese over each pan of lasagna. Cover with wrap or foil. Using a convection oven, bake at 350 F. for 40 minutes until bubbling. CCP: Internal temperature must reach 145 F. or higher for 15 seconds.
11. Remove from oven and allow to set for 15 minutes before serving. Cut each pan 5 by 5 (25 portions per pan). CCP: Hold for service at 140 F. or higher.

MEAT, FISH, AND POULTRY No.L 510 00

TUNA PLATE TRIO

Yield 100 Portion 1 Plate

Calories	Carbohydrates	Protein	Fat	Cholesterol	Sodium	Calcium
367 cal	24 g	27 g	19 g	106 mg	767 mg	89 mg

Ingredient	Weight	Measure	Issue
FISH,TUNA,CANNED,WATER PACK,INCL LIQUIDS	20 lbs	3 gal 2-3/4 qts	
EGG,HARD COOKED,CHOPPED	3-1/2 lbs	2 qts 3-5/8 cup	
CELERY,FRESH,CHOPPED	11-5/8 lbs	2 gal 3 qts	15-7/8 lbs
PIMIENTO,CANNED,DRAINED,CHOPPED	2-3/8 oz	1/4 cup 2 tbsp	
PICKLE RELISH,SWEET	2-2/3 lbs	1 qts 1 cup	
SALAD DRESSING,MAYONNAISE TYPE	5 lbs	2 qts 2 cup	
MILK,NONFAT,DRY	1-3/4 oz	3/4 cup	
WATER,WARM	14-5/8 oz	1-3/4 cup	
SALAD DRESSING,MAYONNAISE TYPE	2 lbs	1 qts	
PEPPER,BLACK,GROUND	1/8 oz	1/3 tsp	
MUSTARD,PREPARED	1-1/8 oz	2 tbsp	
SALT	5/8 oz	1 tbsp	
SUGAR,GRANULATED	12-1/3 oz	1-3/4 cup	
VINEGAR,DISTILLED	8-1/3 oz	1 cup	
CABBAGE,GREEN,FRESH,SHREDDED	12 lbs	4 gal 3-1/2 qts	15 lbs
LETTUCE,ICEBERG,FRESH	6-1/4 lbs		6-3/4 lbs
CARROTS,FRESH	8 lbs		9-3/4 lbs
TOMATOES,FRESH	8 lbs		8-1/8 lbs

Method

1 Tuna salad: Drain tuna and flake. Discard liquid.
2 Add eggs, celery, chopped pimientos, and pickles to tuna. Toss lightly until well blended. CCP: Refrigerate at 41 F. or lower.
3 Add salad dressing to tuna mixture. Toss lightly.
4 Coleslaw: Reconstitute milk, add salad dressing, pepper, mustard, salt, and sugar; mix well. Add vinegar gradually; blend well. Pour dressing over cabbage; toss lightly until well mixed. CCP: Cover and refrigerate product at 41 F. or lower until ready for service.
5 To prepare salad plate: Line plate with lettuce, portion 1/2 cup coleslaw and 3/4 cup tuna salad on top of the lettuce. Arrange 2 to 3 carrot sticks and tomato wedges on top of the lettuce. CCP: Hold for service at 41 F. or lower.

MEAT, FISH, AND POULTRY No.L 512 00

GRILLED TURKEY SAUSAGE PATTY (PRE-MADE)

Yield 100   Portion 1 Patty

Calories	Carbohydrates	Protein	Fat	Cholesterol	Sodium	Calcium
122 cal	0 g	14 g	6 g	45 mg	264 mg	22 mg

Ingredient	Weight	Measure	Issue
SAUSAGE PATTY,TURKEY,RAW,2 OZ	18-3/4 lbs		

Method

1. Preheat grill to 350 F.
2. Grill 12 minutes or until patties are browned and well done. Turn frequently to ensure even browning. CCP: Internal temperature must reach 165 F. or higher for 15 seconds. Hold for service at 140 F. or higher.

MEAT, FISH, AND POULTRY No.L 515 00

OVEN FRIED TURKEY BACON

Yield 100   Portion 2 Slices

Calories	Carbohydrates	Protein	Fat	Cholesterol	Sodium	Calcium
57 cal	0 g	4 g	4 g	21 mg	308 mg	9 mg

Ingredient   Weight   Measure   Issue
BACON,TURKEY,RAW   12 lbs

Method
1 Arrange slices in rows down the length of each sheet pan, with fat edges slightly overlapping lean edges.
2 Bake 25 minutes at 375 F. Drain excess fat. Bake additional 5 to 10 minutes or until bacon is slightly crisp. Do not overcook.
3 Drain thoroughly. Place on absorbent paper or in perforated steam table pan. CCP: Hold for service at 140 F. or higher.

Notes
1 In Step 2, if convection oven is used, bake at 325 F. for 25 minutes on high fan, open vent. Drain fat. Bake additional 5 to 10 minutes.

MEAT, FISH, AND POULTRY No. L 523 00

MAMBO PORK ROAST

Yield 100 **Portion** 4 Ounces

Calories	Carbohydrates	Protein	Fat	Cholesterol	Sodium	Calcium
358 cal	26 g	29 g	15 g	81 mg	204 mg	43 mg

Ingredient	Weight	Measure	Issue
PEACHES,CANNED,QUARTERS,INCL LIQUIDS	6-1/2 lbs	3 qts	
PINEAPPLE,CANNED,CRUSHED,JUICE PACK,INCL LIQUIDS	5-1/2 lbs	2 qts 2 cup	
VINEGAR,RED WINE	1-1/4 lbs	2-1/4 cup	
CINNAMON,GROUND	1/4 oz	1 tbsp	
PARSLEY,DEHYDRATED,FLAKED	3/8 oz	1/2 cup	
ALLSPICE,GROUND	1/8 oz	1/8 tsp	
PORK,LOIN,BONELESS,RAW	31-1/4 lbs		
SALT	1-1/4 oz	2 tbsp	
PEPPER,BLACK,GROUND	1/2 oz	2 tbsp	
WATER	2-1/8 lbs	1 qts	
RICE,LONG GRAIN	4-7/8 lbs	3 qts	
WATER,COOL	20-7/8 lbs	2 gal 2 qts	

Method

1. Combine peaches, pineapple, juice from fruits, vinegar, cinnamon, parsley, and allspice in a large bowl. Reserve for use in Step 2.
2. Rub the pork roasts with salt and pepper. Place roasts with at least 3 inches space dividing each roast. Divide the fruit sauce among roasting pans. Cover, roast in 350 F. oven for 2-1/2 hours, basting with fruit sauce every 45 minutes. If pan dries out, add 2 inches of water, recover, and continue cooking. CCP: Internal temperature must reach 145 F. or higher for 15 seconds.
3. Rinse the rice in cool water. Place the rice and water in a steamer, cover and steam for 15 to 20 minutes or until tender.
4. Slice pork roasts in 1-ounce slices. Place 2 slices of pork with 1/4 cup of the fruit sauce over 1/2 cup rice.

MEAT, FISH, AND POULTRY No.L 524 00

WHITE FISH WITH MUSHROOMS

Yield 100   Portion   4 Ounces

Calories	Carbohydrates	Protein	Fat	Cholesterol	Sodium	Calcium
142 cal	3 g	27 g	2 g	72 mg	409 mg	27 mg

Ingredient	Weight	Measure	Issue
FISH,FLOUNDER/SOLE FILLET,RAW	30 lbs		
SALT	5/8 oz	1 tbsp	
PEPPER,WHITE,GROUND	1/2 oz	2 tbsp	
OIL, CANOLA	1-1/2 oz	3 tbsp	
ONIONS,FRESH,CHOPPED	1 lbs	3 cup	1-1/8 lbs
GARLIC POWDER	7/8 oz	3 tbsp	
MUSHROOMS,CANNED,SLICED,INCL LIQUIDS	7-5/8 lbs	1 gal 1-1/2 qts	
SOY SAUCE	5-1/8 oz	1/2 cup	
JUICE,LEMON	1-1/8 lbs	2 cup	
ONIONS,GREEN,FRESH,SLICED	10-5/8 oz	3 cup	11-3/4 oz

Method

1. Season fish with salt and pepper. Drizzle with oil.
2. In a tilt griddle, saute onions, add garlic and mushrooms. Cook for 5 minutes.
3. Add soy sauce and 1 cup of lemon juice. Continue to cook into a dark mushroom broth for 3 minutes.
4. Broil or bake fish in 400 F. conventional oven on sheet pans for 20 minutes. CCP: Internal temperature must reach 145 F. or higher for 15 seconds.
5. Place mushrooms on the bottom of the 2-inch steam table pans. Layer fish on top, garnish with raw scallions and remaining lemon juice. CCP: Hold for service at 140 F. or higher.

Notes

1. Boneless, skinless chicken may be substituted for fish.

www.ingramcontent.com/pod-product-compliance
Lightning Source LLC
Chambersburg PA
CBHW081341070526
44578CB00005B/687